Silicon-Based Polymer Science

ADVANCES IN CHEMISTRY SERIES **224**

Silicon-Based Polymer Science
A Comprehensive Resource

John M. Zeigler, EDITOR
Sandia National Laboratories

F. W. Gordon Fearon, EDITOR
Dow Corning Japan, Ltd.

Developed from the International Topical Workshop
"Advances in Silicon-Based Polymer Science"
sponsored by the Division of Polymer Chemistry, Inc.,
of the American Chemical Society,
Makaha, Oahu, Hawaii,
November 21–24, 1987

American Chemical Society, Washington, DC 1990

Library of Congress Cataloging-in-Publication Data

Silicon-based polymer science: a comprehensive resource.
John M. Zeigler, F. W. Gordon Fearon, editors.
p. cm.—(Advances in chemistry series, 0065–2393; 224)
Developed from a symposium sponsored by the Division of Polymer Chemistry, Inc., of the American Chemical Society.

ISBN 0–8412–1546–4

1. Silicon polymers—Congresses.

I. Zeigler, John M. (Martin), 1951– . II. Fearon, F. W. Gordon. III. American Chemical Society. Division of Polymer Chemistry, Inc. IV. Series.

QD1.A355 no. 224
[QD383.S54]
540 s—dc20
[547.7] 89–17843
CIP

The paper used in this publication meets the minimum requirements of American National Standard for Information Sciences—Permanence of Paper for Printed Library Materials, ANSI Z39.48–1984. ∞

PRINTED IN THE UNITED STATES OF AMERICA

Second printing 1991

Advances in Chemistry Series

M. Joan Comstock, *Series Editor*

FOREWORD

The ADVANCES IN CHEMISTRY SERIES was founded in 1949 by the American Chemical Society as an outlet for symposia and collections of data in special areas of topical interest that could not be accommodated in the Society's journals. It provides a medium for symposia that would otherwise be fragmented because their papers would be distributed among several journals or not published at all. Papers are reviewed critically according to ACS editorial standards and receive the careful attention and processing characteristic of ACS publications. Volumes in the ADVANCES IN CHEMISTRY SERIES maintain the integrity of the symposia on which they are based; however, verbatim reproductions of previously published papers are not accepted. Papers may include reports of research as well as reviews, because symposia may embrace both types of presentation.

ABOUT THE EDITORS

JOHN M. ZEIGLER is supervisor of the Physical Chemistry and Mechanical Properties of Polymers Division of Sandia National Laboratories in Albuquerque, NM. He received his B.A. in chemistry (summa cum laude) in 1974 from Wabash College and his Ph.D. in organic chemistry from the University of Illinois in 1979. After two years with the American Cyanamid Company in Stamford, CT, working on carbocation chemistry in liquid HF, he joined Sandia in 1981. He assumed his present position in 1985. His research interests center on the design, synthesis, properties, and applications of polymeric inorganic and organic electronic materials. His current research involves polysilanes, conductive polymers, and polymers in microelectronics processing.

He has coauthored more than 40 papers and holds seven U.S. and five foreign patents in these areas. He has been a plenary lecturer at a number of international meetings and received an IR-100 award in 1985 for the discovery of self-developing polysilane photoresists. In addition to serving as coorganizer of the International Topical Workshop "Advances in Silicon-Based Polymer Science" in 1987, he has organized sessions dealing with σ-conjugated polymers (polysilanes) at several international meetings and has been a member of organizing committees for meetings dealing with microelectronics processing.

He is a member of Phi Beta Kappa, the American Chemical Society, the American Physical Society, the American Association for the Advancement of Science, SPIE (Society of Photo-Instrumentation Engineers), Sigma Xi, and Alpha Chi Sigma.

F. W. GORDON FEARON is technical director and vice president for Dow Corning in Japan. He received his B.S. in chemistry from the University of Leeds and his Ph.D. in chemistry from the University of Wales for research in the silicon field. He carried out postdoctoral research in silicon chemistry at Iowa State University. He joined Dow Corning Corporation in 1968, and since then he has held a wide variety of research, development, and business management positions, with particular emphasis on expanding the scope of useful silicon science and technology.

CONTENTS

INDEXES

PREFACE

THE WORLD IS BUILT FROM SILICON-BASED polymers. Silicate materials in many shapes or forms account for more than 90% of the land mass. Technologies associated with these materials have been developed through recorded history, but the science underlying these materials is, relatively, still in its infancy. The first systematic study of silicon chemistry was carried out in the first two decades of this century. The first useful synthetic organosilicon polymers, the polysiloxanes, were developed in the late 1930s, and a new industry based on these synthetic polymers was born in the early 1940s. This industry has grown rapidly worldwide to the point that it is now a multi-billion-dollar endeavor.

Scientists in this field recognize that silicon polymers that have been developed commercially constitute merely "the tip of an iceberg" or, possibly, a silicon tetrahedron. The bulk of the silicon industry today is, in fact, based on one polymer system, poly(dimethylsiloxane) and a large family of materials obtained by minor modifications of this polymer. Imagine the prospects for other polymer systems based on silicon!

History will show that during the last decade of the 20th century new families of silicon-based materials were harnessed to provide the growth that, in the first decades of the 21st century, firmly established synthetic silicon polymers as unique materials complementary to organic polymer systems. Is this a rash prediction? Maybe.

In the 1987 *Chemical Abstracts*, the section on silicon polymers contains more than 3000 entries and lists more than 1000 authors. The topics span polysilalkylenes to polysilazanes. To pull together a representative group of workers from this broad and, as yet, ununified field was an exciting challenge. The International Topical Workshop "Advances in Silicon-Based Polymer Science," held in Makaha, HI, and on which this volume was based, was designed to summarize the recent progress in the understanding of silicon polymers, seek unifying principles, and provide insight into the possible directions for future scientific and technological advances in the field. The Workshop drew together current research in the areas of synthesis, properties, chemistry, and applications of silicon-based polymers. Plenary presentations provided critical overviews and perspectives of rapidly advancing fields. Poster sessions provided a forum for the presentation of important new research results.

The challenge of silicon-based polymers is all the more intriguing because of the rapid growth in the past decade in the number of structural types available and our increasing understanding of the special and highly flexible role that silicon plays in these materials. The familiar polysiloxanes are renowned for their stability and insulating qualities. The study of polysilanes (polysilylenes) is driven by useful and highly adjustable photo-instability and capabilities for charge and energy conduction. Millennium-old silicate ceramics have now been supplemented by the new ceramic materials derived from pyrolysis of formable preceramic polymers based on silicon. Progress has been made in converting natural silicates directly to organosilicon polymers without the intermediacy of the direct process for the formation of silicon monomers. The presence of silicon as a pendant group on polyacetylene backbones gives novel polymers that are unequaled in their capacity for selective gas permeation. As in the past, the remarkable level of recent progress has been founded on the realization that silicon lends unique characteristics to polymers. Silicon is most emphatically not simply a "high-atomic-number carbon" when incorporated in polymers.

Various themes are evident in this volume; silicon-based polymer science is presented from the fundamentals of synthesis and structure to the often highly complex technological applications. In the first chapter, Barton and Boudjouk state: "Our collective experience requires that we warn the reader that the forthcoming material may well prove to be addictive!" We hope they are correct.

We thank all the authors and participants for their contributions. We particularly thank the sponsors of the Workshop for providing an unusually high level of support, which permitted us to assist many who would otherwise have not been able to attend.

Rumor has it that another group will be gathering in the same location in 1991 to mine once more the golden sands of Makaha for polymeric treasures. If our prediction for the future of silicon-based materials is true, much remains to be discovered.

JOHN M. ZEIGLER
Sandia National Laboratories
Albuquerque, NM 87185

F. W. GORDON FEARON
Dow Corning Japan, Ltd.
Tokyo, Japan

July 14, 1989

ACKNOWLEDGMENTS

WE GRATEFULLY ACKNOWLEDGE the support provided to the International Topical Workshop "Advances in Silicon-Based Polymer Science" by the following organizations:

Allied Corporation
Allied–Signal
Corning Glass Works
Dow Corning Corporation
Dynamit Nobel–Petrarch Systems
General Electric Company
General Electric Company, Engineered Materials Group, Silicone Products Division
James River Corporation
Nippon Carbon Company, Ltd.
Shin–Etsu Chemical Company
Sandia National Laboratories
Wacker Silicones Corporation
Toray Silicone Company, Ltd.
Toshiba Silicones Corporation, Ltd.
Union Carbide Corporation
United States Navy, Office of Naval Research
Wacker–Chemie, GmbH

JOHN M. ZEIGLER
Sandia National Laboratories
Albuquerque, NM 87185

F. W. GORDON FEARON
Dow Corning Japan, Ltd.
Tokyo, Japan

July 14, 1989

Introduction: My Favorite Element

Eugene G. Rochow

Myerlee Manor, No. 107, 1499 Brandy Wine Circle, Fort Myers, FL 33919

MY FAVORITE ELEMENT IS SILICON, and I am very happy to make this contribution to this volume, because it brings together so many loose threads from past years. Some threads that I had thought had been abandoned in the past now suddenly come to light. Now I find not only siloxane polymers as a topic of importance but also ceramic chemistry in all its broad aspects, polysilanes, and hyperpure silicon made from intermediates for silicones. Polysilanes made from those same intermediates are now being used as coatings on more hyperpure silicon that came from the same direct process. Not only have we come full circle, but we have circles within circles. This development is astonishing and gives me a happy and contented feeling about my favorite element.

Silicon: An Element with Many Possibilities

For some years I made a living as a ceramic chemist. I did not claim to be one, but at least I got a job in that area in 1935. I always believed that organosilicon chemistry should overlap with ceramic chemistry in various ways. Very few people have that same point of view. When I left General Electric (GE) and went to Harvard University, I did some experiments with molten silicates; one of these was the measurement of electrode potentials in molten silicates, the way pH is measured in aqueous solutions. I thought that this field was very interesting, and the American Ceramic Society gave me an award for the most significant paper in ceramic chemistry for that year. Do you suppose I could get any student interested in pursuing the field? No!

Many things can be done with silicon, and an overlap between disciplines exists. Many of my attempts to bridge the gap were failures. I tried to make the analogues of alloys by mixing molten silicates with melted thermally stable silicones, but the silicates always melted at about 100 °C above the decomposition point of the silicones. I gave up that line of research, and now, the same thing is being accomplished, not by the metallurgical technique of mixing liquids and allowing them to solidify, but by a sort of powder metallurgy technique going through sols and gels. When a silicate gel and an organosilicon gel are brought together and dehydrated under the appropriate conditions, a heterogeneous ceramic alloy with new and interesting properties is formed.

My old friend methylphenylsilicone, which I patented in 1941, has reappeared in a new and more elegant form as a block copolymer prepared in a fancier way, with the accompanying improvement in properties. I also investigated polyphenylsilanes as possible competitors of silicones, but the polyphenylsilanes were less stable than the siloxanes, and so I abandoned them. (All the emphasis at GE was on high-temperature insulation, and I liked and needed the job!) Now the big leap to polyphenylsilenes has been made, and the inherent instability of polyphenylsilenes is exploited by using them as photosensitive coatings on silicon chips. Thus, my negative finding has been turned upside down to good advantage.

For decades I avoided mixtures or copolymers of organic resins with silicones, even abhorred them, because I did not want the pristine properties of my silicone brain children to be degraded or ruined by adulteration. Now, all manner of siloxane–organic copolymers (such as the polyimide block copolymers) with disciplined structures and admirable properties have been developed.

I neglected a lot of possibilities involving my favorite element. Some of the neglect was for reasons that seemed compelling at the time. A vast unexplored area of silicon–nitrogen compounds beckoned, and for several years, a number of young co-workers and I fussed with methylsilazanes and ways of polymerizing them. We made some progress and considered it a great day when the hydrolyzability of the average SiN compound was licked. We recrystallized one of our silazanes from 95% alcohol and recovered it intact! But the search for stable methylsilazanes seemed to stretch on endlessly into the horizon, and of course, most young people live for the present and cannot be bothered by the distant future. Eventually the work petered out; we had to adopt Ulrich Wannagat's view that the field of organosilicon–nitrogen compounds (which he called the "field of sin") was really an unending swamp, a morass that could swallow up any number of man-years of work. The field was unproductive.

Now see what has happened! Many researchers have reverted to good, solid inorganic chemistry and have used the methylsilazanes and other organosilicon compounds as precursors of silicon nitride and silicon carbide, not in the massive lump form so familiar in ceramic chemistry, but as thin coherent fibers that retain fantastic strength even at red heat and are useful in ultramodern composites. A blessed bridge has sprung up between two enormous areas of silicon chemistry, whose adherents had little or no intercommunication whatever before this! Beautiful, ubiquitous, friendly silicon gets another chance to show its true worth. The second-most abundant element on this beautiful earth shows how versatile and how largely unexplored it is!

Like a good bridge player, I look at all the wonderful things happening today involving my favorite element silicon, and I begin to think hard about what could possibly ruin the play of this magnificent hand. What could possibly stem the great advances reported in this volume? The answer became very obvious: a loss of support could cripple or kill the advances now being made with silicon. If you do not believe me, just look at what happened to the American space program.

Communicating with the Public

The great advances in silicon-based polymer chemistry are mostly mysterious and incomprehensible to the general public, and we ourselves must take the trouble to describe those advances in the popular, common language if we want to keep the research momentum going. If we do not write for the public, it will lose interest and withdraw its commercial and governmental support.

Scientific and technical language is absolutely essential for our professional meetings and in our journals.

But the general public does not always understand chemical language and may even harbor an active hostility toward technical jargon. Since the beginning of the human race, people have always feared that which they do not understand. The fear is expressed in many forms, ranging all the way from ridicule to mob violence, with all stages in between. You have all witnessed this or experienced it.

During the height of antiscience feeling in the early 1970s, I found my own children hooting with laughter as they tried to read the list of ingredients on a package of cookies and made their way slowly through the hydrogenated soybean oil to the diglycerides and lecithin. Why were they deriding these ingredients? "Well," they said, (1) "They sound so funny!" (2) "They are chemicals and so cannot be good for you," and (3) "They are probably synthetic and, therefore, downright poisonous."

Let us deal with these points in order. First, "the names sound so funny." Well, foreign words always sound different the first time we hear them, and I caution my own children never to poke fun at foreign words or at the people who utter them, for to do so is only to reveal one's ignorance and mark one as a fool. However, the reaction of ridicule for unfamiliar-sounding names and expressions will go on.

The second objection is the entrenched view that anything with a chemical name is not good for you. This objection is raised by the great scientifically illiterate masses. The objection extends to all economic and social levels and even among professionals. I once heard a Harvard professor of English get up and say in the presence of the president and all the deans that he had never taken a course in science and was proud of it. So widespread is the antiscience feeling and the entrenched opposition to scientific terminology. This opposition arises from fear of the unknown.

Let us take a moment to examine why the ignorance of even elementary science is so widespread. In the 25 years that I taught Chemistry I at Harvard, I never found a student who was incapable of understanding the subject. Everyone who graduates from Harvard is required to know how to swim, and some of the students have a hard time. Of course, those with physical handicaps are excused. Eventually, though, the faculty had to make some other exceptions. They found that some individuals simply sank and could not be taught to swim. But no one in my experience has ever been unable to learn some chemistry. Regardless of background, preparation, or the lack of it, all college students can learn chemistry, or enough of it to understand the world around them and to be intelligent voters. Furthermore, I have dealt with enough classes of high-school and grade-school students to know that they, too, can grasp logical principles and apply them, if the principles are delivered in their own language.

Why then do so few college students pursue degrees in science? Mostly, it is because they are turned away by parents, advisors, and even teachers, who have blocked out science from their own lives earlier and then advise the youngsters to do the same. They say science is too hard. The attitude is if you have trouble with your backhand in tennis use a forehand shot all the time. This attitude does not work in today's world. So why are the parents, teachers, and advisors so ignorant in science? Because they, too, were advised to shy away from it by their parents and teachers, and so on. It is a self-perpetuating delusion, and we must all work to overcome it.

The third objection is that anything with a chemical name is probably synthetic and, therefore, assuredly poisonous. To many thousands of college students and to anyone else who would listen, I have explained that giving another name to a substance, a material, or a food does not change its character or its properties. Water is water whether we name it in English, German, Sanskrit, or the precise terminology of the International Union of Pure and Applied Chemistry. And yes, water is a poison if taken in sufficient amount. Taken in gross excess, too much water disturbs the ionic balance in the blood, leading to convulsions, coma, and even death. In excess, every drug is a poison; conversely, every poison is a potential drug in the treatment of one disease or another. Everything can be poisonous, whether we call it by its chemical name or not; it is all a matter of quantity or proportion. We must have due regard for proportion.

Then we come to that other epithet, "synthetic". If we think about it for a moment, is not everything we eat synthetic? Food is synthesized in many stages, starting with the conversion of carbon dioxide and water to simple sugars through the agency of sunlight in green plants. Every plant product is synthetic, and because every animal in the world depends on the plant life, every animal is a synthetic product. All of us are synthetic, too.

Originally, we were synthesized by our mothers. Now, we ourselves synthesize our body parts from the amino acids, carbohydrates, and minerals obtained from ingested plant and animal foods that contain what we need for the purpose.

Some may challenge me on the matter of mineral constituents of the human body. How about the calcium for bones and teeth? How about the sodium chloride we eat? How about the potassium, the phosphorus, and the iron in all of us? All these elements were synthesized, too. They were synthesized in stars like our sun, long ago. Everything has been synthesized from the primordial hydrogen by nuclear reactions far off and long ago. When I explained this concept to freshmen in Chemistry I in connection with teaching them some elementary nuclear chemistry, one shiny-eyed Radcliffe girl came up afterward and said, "Am I really made of stardust?" I said, "Yes, and it actually shows!"

Cows produce synthesized milk, and bees produce synthetic honey. Because milk and honey are synthetic, are they bad for us? No, we do not consider them so. Then, why does this dichotomy exist? Why are the synthetic products of plants and animals considered good and pure and wholesome and natural, whereas the synthetic products of mankind are considered, ipso facto, vile and poisonous? Why does this colossal put-down of mankind exist? Why are plants and bees and beef cattle natural and noble and virtuous and benevolent, whereas only people are evil and malevolent?

Nothing is inherently wrong with what man manufactures; its goodness or badness depends merely on what use we make of it. There is nothing bad about vitamin C just because it is manufactured; it is still ascorbic acid (or one of its salts) and is just as effective as any other kind. So let us get rid of this myth about "synthetic". At every opportunity, we should explain patiently why it is nonsensical to equate "synthetic" with "toxic".

Lest you think I have wandered too far afield, let me summarize my thesis this way. First, every practicing chemist is an ambassador for science in general and for chemistry in particular. We must explain things to the nonchemical public and correct misconceptions gently and patiently at every turn, or else nothing will ever change. Second, the message about bland, benevolent, and ubiquitous silicon must get across to the man and woman in the street, for it is the public who will buy the products that come out of the research discussed in this volume and who will instruct congressmen about appropriations for research. Third, I urge researchers to write a simple, straightforward account, in two or three sentences in nontechnical language, about what they have done, why they have done it, and why it is important. When they have written this summary, I ask them to try it out on their family and friends, so that they understand and can spread the good word about silicon, too.

I expect many workers to plead that they are too busy making rapid advances in a very active field to take time to enlighten the general public.

They may say that they can best serve science and silicon chemistry by pressing on with their demanding laboratory work. Perhaps they are right, but this view goes counter to my own experience.

In 1945, a publisher said to me, "This is a tremendously exciting new field you are working in. Write a little book about it! Even if it is only 60 pages long, we'll publish it!" The result was *An Introduction to the Chemistry of the Silicones*, which (in its second edition) was translated into six languages and influenced many a book and many a researcher thereafter. I still find phrases and features from it appearing in new publications, probably through subconscious recollection on the part of the young author.

That little book was written at home, at night; it did not interfere with laboratory time. It did lead to endless requests for American Chemical Society (ACS) local-section talks, which involved dozens of long train trips and many weeks away from home. Those ACS tours took a lot of time away from experiments, and I found them nothing short of exhausting. Was it worth the effort to preach the gospel of organosilicon chemistry by pen and in the lecture room? The answer is a resounding yes!

I'll wager that many practitioners of silicon chemistry would not be where they are if their mentor in graduate school (or his mentor before him) had not been influenced by that effort. And the process is still going on. I was asked two years ago to write "a slim volume on silicon and silicones for the general reader," that is, for the public rather than the committed chemist. The result was a book (my 14th book) with that very title. If it sells only a dozen copies but influences just one young person to take up organosilicon chemistry, I shall have considered the effort worthwhile.

RECEIVED May 27, 1988. ACCEPTED revised manuscript October 25, 1988.

FUNDAMENTALS

1

Organosilicon Chemistry

A Brief Overview

Thomas J. Barton[1] and Philip Boudjouk[2]

[1]Chemistry Department and Ames Laboratory (Department of Energy), Iowa State University, Ames, IA 50011
[2]Chemistry Department, North Dakota State University, Fargo, ND 58105

The chapter gives a brief overview of organosilicon chemistry and serves as an introduction to the rest of the material in the volume.

ORGANOSILICON CHEMISTRY is a rich stratum under the long-plowed fields of organic chemistry and contains largely untapped veins that are interlinked to a wide variety of chemical, physical, and engineering disciplines. Our collective experience requires that we warn the reader that the forthcoming material may well prove to be addictive! The reader is also warned that simple extrapolation from organic chemistry to organosilicon chemistry is a dangerous predictive tool. As expected from their proximity in the periodic chart, carbon and silicon often display close similarities in structures and reactions, but it is the numerous differences that are most interesting and important.

In this chapter, we will be skimming the rough topology of organosilicon chemistry, and thus, we will give only the peaks of the highest mountains and completely miss the valleys and often exclude the smaller hills. Our purpose is not to be exhaustive but simply to prepare the reader for some of the material that lies ahead in this volume, and we hope to whet your interest in what we believe is a most fascinating and important subject.

0065–2393/90/0224–0003$12.00/0

Nomenclature

Organosilicon compounds can always be named by the oxa–aza convention as organic compounds containing silicon atoms substituted for carbons in the longest continuous chain. For example,

$$(CH_3)_3Si\text{–}CH_2\text{–}Si(CH_3)_2\text{–}CH_2\text{–}Si(CH_3)_3$$

2,2,4,4,6,6-hexamethyl-2,4,6-trisilaheptane

However, this method is rarely used, because at least for relatively uncomplicated molecules, the compounds are easier to name as derivatives of silane, SiH_4. For example,

Me_2SiCl_2	dichlorodimethylsilane
$Me_3SiSiMe_3$	hexamethyldisilane

If the silicon group is to be named as a substituent, the radicals are named as follows:

$H_3Si\text{–}$	silyl
$H_2Si{<}$	silylene
$H_3Si\text{–}SiH_2\text{–}$	disilanyl
$H_2SiO\text{–}$	siloxy
$H_3SiNH\text{–}$	silylamino
$H_3SiOSiH_2\text{–}$	disiloxanyl
$H_3SiOSiH_2O\text{–}$	disiloxanoxy

For example:

$$Et_3Si\text{-}CH_2\text{-}\underset{\substack{|\\ OSiMe_3}}{CH}\text{-}CH_2CH_3$$

1-triethylsilyl-2-trimethylsiloxybutane

Cyclosilanes can be named as illustrated by the following examples:

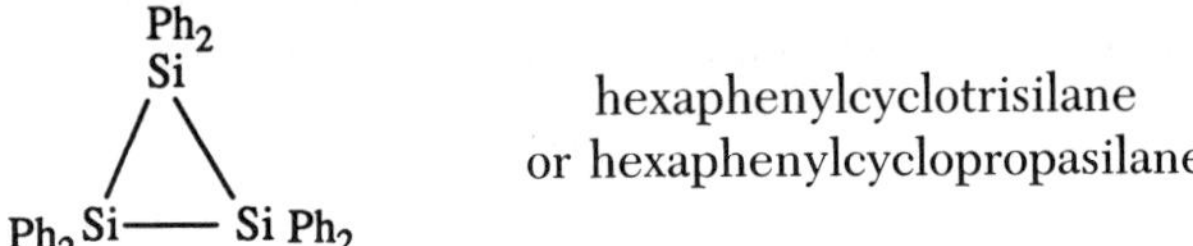

hexaphenylcyclotrisilane
or hexaphenylcyclopropasilane

$$\begin{array}{ccc} & Me_2 & \\ & Si & \\ Me_2Si & & SiMe_2 \\ | & & | \\ Me_2Si & & SiMe_2 \\ & Si & \\ & Me_2 & \end{array}$$

dodecamethylcyclohexasilane

Hydroxy derivatives are named analogously to alcohols by using the -ol ending:

Me_3SiOH	trimethylsilanol
$Me_2Si(OH)_2$	dimethylsilane diol
$Me_3Si–O–SiMe_2OH$	pentamethyldisiloxanol
Ph_3SiONa	sodium triphenylsilanolate

Substituents other than hydroxyl are named exactly as they are in standard organic nomenclature. An exception is hydrogen, because H bonded to Si is a functional atom. Therefore, it is often useful to refer to *organosilicon hydrides*. The term *organosilane* does not indicate the presence of an Si–H bond.

$CH_3CH_2SiH_2SiH_3$	ethyldisilane
Me_3SiNH_2	aminotrimethylsilane

Silazanes [$H_3Si(NHSiH_2)_nNHSiH_3$] are called disilazane, trisilazane, etc., depending on the number of silicons. For example,

$$\begin{array}{c} Me_3Si\text{-}NH_2\text{-}Si\text{-}NH_2\text{-}SiMe_3 \\ Me_2 \end{array}$$

1,1,1,3,3,5,5,5-octamethyltrisilazane

Siloxanes are named in a similar fashion:

$H_3Si(OSiH_2)_nOSiH_3$	$n = 0$, disiloxane
	$n = 4$, hexasiloxane

The naming of cyclosiloxanes is very simple, as shown by the following example:

methoxypentamethylcyclotrisiloxane

A special shorthand method, the "General Electric siloxane notation", is commonly used for methylsiloxanes. In this notation, the groups are abbreviated as follows:

Me_3SiO-	$-OSi(Me)_2O-$	$-O-Si(Me)(O)-O-$	$-O-Si(O)(O)-O-$
M	D	T	Q

For example:

$Me_3Si-O-SiMe_3$	MM
$Me_3Si(OSiMe_2)_8OSiMe_3$	MD_8M
$(Me_3SiO)_3Si-O-SiMe_2-O-SiMe(OSiMe_3)_2$	M_3QDTM_2
$(Me_2SiO)_4$	D_4

The Nature(s) of Silicon Bonding

Most organosilanes are tetrahedral about silicon, which is consistent with the use of sp^3-hybridized orbitals on silicon.

SiR_4 Si $3s^2\ 3p^2\ 3d^0$

However, in contrast to carbon, numerous compounds are known in which Si is penta- or hexacoordinated.

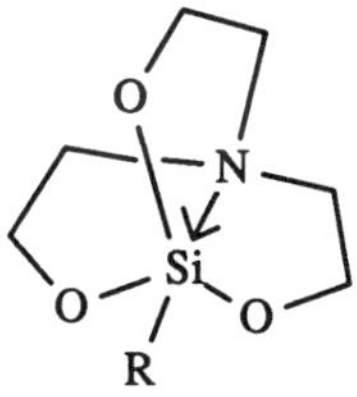

(a silatrane)

$K_2[RSiF_5]$

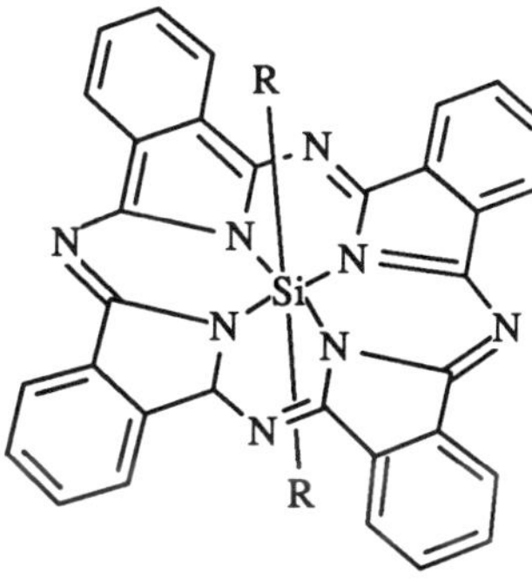

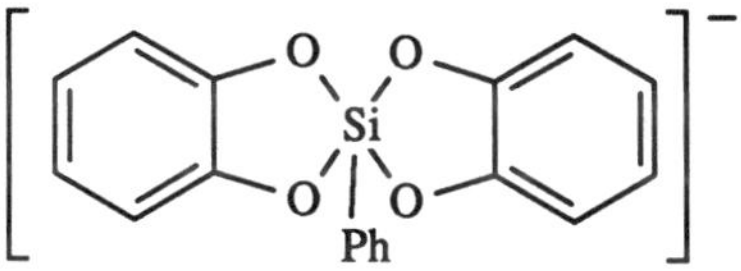

The formation of higher coordinated silicon is often explained in terms of *d*-orbital involvement ($3d_{z^2}$ in trigonal bipyramidal complexes or $3d_{z^2}$ and $d_{x^2-y^2}$ in octahedral complexes). However, it has long been recognized that it is equally possible to describe the bonding without 3*d* orbitals by using three-center molecular orbitals. No compelling data are available for either bonding possibility, but it is of interest to note that if *d* orbitals are to be used, some contraction must occur before they can be effectively hybridized with the *s* and *p* orbitals. Indeed, in keeping with this requirement, higher coordinated Si compounds are most easily formed from the halides, and the ease of formation and complex stability both increase with increasing electronegativity (I < Br < Cl < F). Thus,

$$SiCl_4 + 2\,NC_5H_5 \rightarrow SiCl_4 \cdot 2NC_5H_5$$

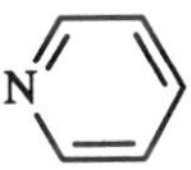

but

$$MeSiCl_3 + NC_5H_5 \rightarrow \text{no reaction}$$

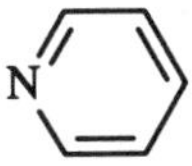

(*p–d*)π Bonding. Experimental Si–X bond lengths are always shorter than the sum of the respective covalent radii. The Si–C bond length is brought very close to the experimental value if polar effects from the electronegativity differences are included, but Si–X bonds for more electronegative Xs are still predicted to be too long, as shown by the following bond length data:

Si–O in $(H_3Si)_2O$	1.634 Å (experimental)
	1.77 Å (calculated)
Si–F in SiF_4	1.553 Å (experimental)
	1.71 Å (calculated)

The most frequently used rationalization for these differences is $(p–d)\pi$ bonding, that is, the donation of lone-pair electrons from X into the vacant $3d$ orbitals of silicon. Thus, the electronegative atom X would both contract the d orbitals and transfer electrons to create additional bonding. This concept has been used to rationalize a number of unique structural features such as the planarity about nitrogen in $(H_3Si)_3N$.

Si–X ↔ Si=X

H_3Si 120° N—SiH_3 H_3Si

Acceptance of $(p–d)\pi$ bonding in silicon compounds has not been universal and is now being challenged frequently by theoretical studies. The question seems to be down to $(p–d)\pi$ bonding versus polar effects, either of which can explain most of the data. Whereas ab initio calculations have better reproduced experimental bond lengths and dipole moments by inclusion of d functions in the Si basis set, this better reproduction may simply be the result of compensation for an inadequate $s–p$ basis set.

Of particular importance to this volume is the Si–O bond. In addition to the previously mentioned difference in calculated and experimental bond lengths, replacement of the carbons in dimethyl ether with silicons also produces dramatic changes in bond angles, and this effect has also been attributed to $p \rightarrow d$ back bonding.

H_3C—O—CH_3 111.5° H_3Si—O—CH_3 120.6° H_3Si—O—SiH_3 144.1°

Once again, however, ab initio calculations reproduced both bond angles and lengths without the use of d orbitals. It was concluded that, evidently, the Si–O bond is much more polar than estimated from electronegativities. Thus, both angles and lengths can be explained by coulombic repulsions.

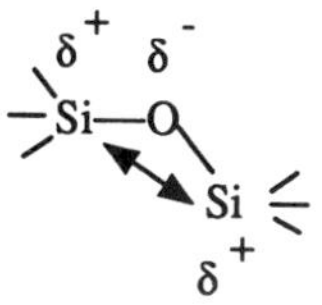

Most recently, an ^{17}O NMR study of quadrupole coupling constants in silicates has been interpreted as strong evidence for $(p{-}d)\pi$ bonding between silicon and oxygen. This question will be subject to considerable scrutiny for some time to come.

Another important system for which $(p{-}d)\pi$ bonding has been assumed, but calculations deny, is the α-silyl anion, $R_3Si{-}CH_2^-$, which is unusually stable and easy to form. Current thinking is that α-carbanion stabilization is due to the high polarizability of Si and the presence of low-lying σ* orbitals.

$$(CH_3)_4Si \xrightarrow[\text{THF/TMEDA}]{n\text{-BuLi}} (CH_3)_3Si{-}CH_2Li$$

$$(CH_3)_4C \xrightarrow[\text{THF/TMEDA}]{n\text{-BuLi}} \text{no reaction}$$

In the previous reactions, THF is tetrahydrofuran, and TMEDA is tetramethylenediamine.

σ–π Conjugation. An extremely important aspect of silicon bonding is *hyperconjugation* or σ–π conjugation. Hyperconjugative interaction of the Si–C bond with various π systems is well documented in allylic silanes by a variety of spectroscopic methods. A particularly telling demonstration is the dramatic lowering of the ionization potential (IP) of **1** (in which hyperconjugation is possible) and **2** (in which the β Si–C bond is locked in the π nodal plane).

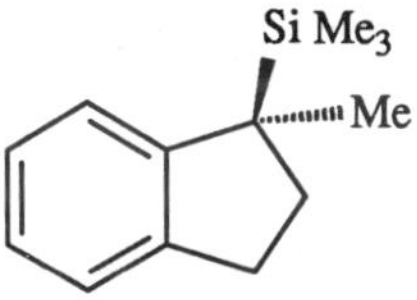

1, IP = 8.13 eV

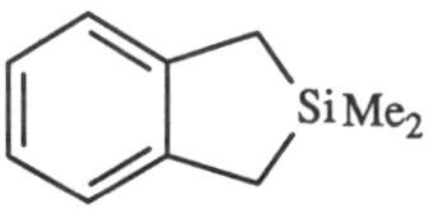

2, IP = 8.42 eV

Of equal importance and probably of more pertinence to the subject of this volume is the ability of the Si–Si bond to hyperconjugatively interact with π systems. In the case of $Ar{-}SiR_3$, hyperconjugative electron release by Si–R is possibly masked by the stronger electron acceptance by Si through $(p{-}d)\pi$ bonding. However, for the $Ar{-}SiR_2SiR_3$ system, much greater electron donation is expected because of the higher polarizability of the Si–Si bond relative to that of Si–C, as clearly seen in the following examples:

$Me_3Si\text{–}CH{=}CH_2$ λ_{max} 202 nm

$Me_3Si\text{–}SiMe_2\text{–}CH{=}CH_3$ λ_{max} 223 nm

Si Me_2 Si Me_2

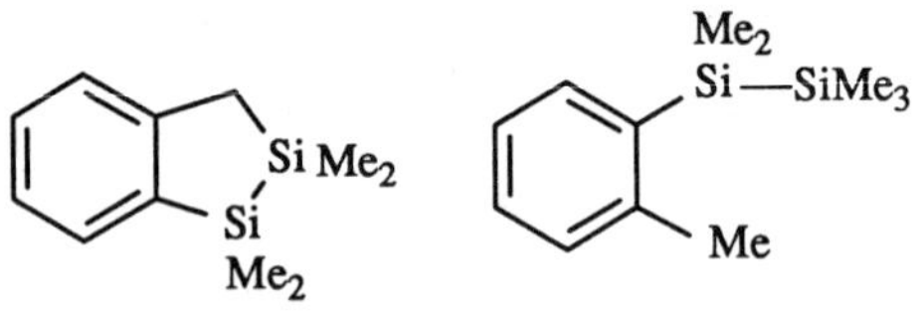

λ_{max} <200 nm

Me_2 Si—$SiMe_3$ Me

λ_{max} 233 nm

One of the most important manifestations of Si–C hyperconjugation is the well-established β stabilization of carbon-centered radicals and carbonium ions by silyl groups.

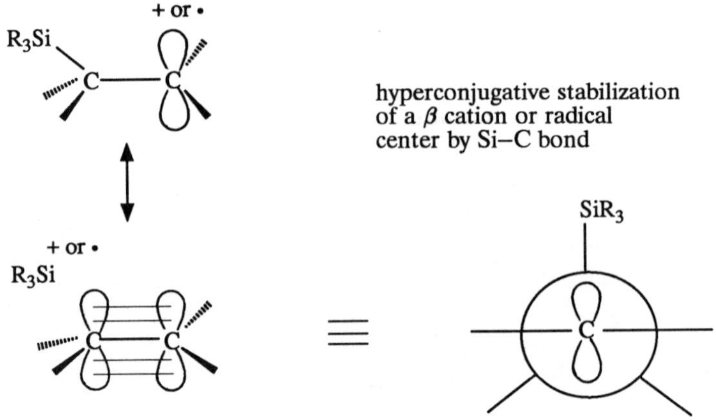

+ or •

R_3Si

C C

hyperconjugative stabilization of a β cation or radical center by Si–C bond

+ or •

R_3Si

SiR_3

C

This stabilization is extremely important in controlling the chemistry and reactivity of a wide variety of organosilicon reactions. Relative to carbon, an α-silyl group retards solvolysis, and a β-silyl group strongly accelerates solvolysis. Therefore, tremendous control over carbocation chemistry is possible simply by the judicious placement of silyl groups. An excellent example of this use of silyl groups is the extensive use in organic synthesis of allylsilanes through addition of an electrophile to the terminal carbon (anti-Markovnikov addition!) to produce a β-silyl cation, which then eliminates silicon to form a double bond at the other end.

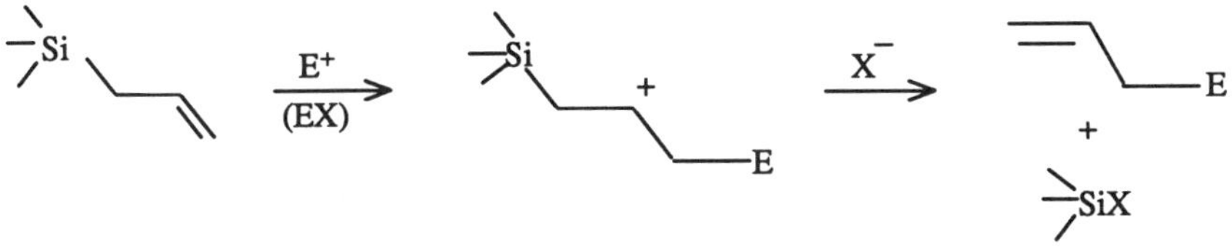

For example,

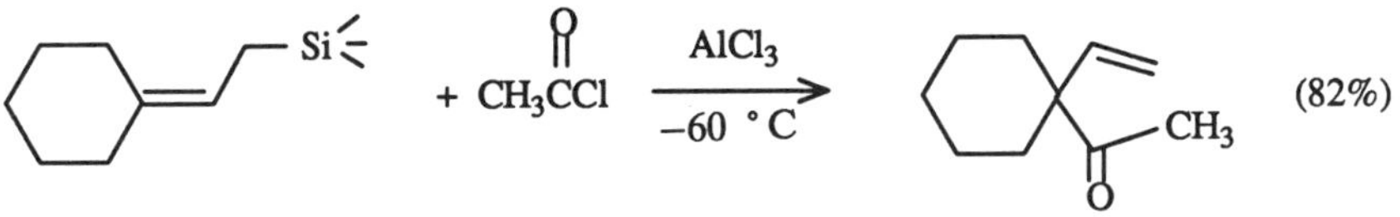

(82%)

Over the past few years, the debate over the origin of the β-silicon effect on carbocations has narrowed to one of the relative magnitudes of inductive and hyperconjugative factors. Theory and experiment are finally in agreement that hyperconjugation is by far the dominant factor—29 kcal/mol calculated to be from β-stabilization (!) versus 9 kcal/mol from induction and polarization. The realization of these effects is dramatically revealed in the S_N1 solvolyses of the conformationally locked cyclohexyl trifluoroacetates (OTFA) (**3**–**5**). The relative solvolysis rates at 25 °C for compounds **3**–**5** are 1, 4 × 10^4, and 2.4 × 10^{12}, respectively. Compound **4** cannot attain the necessary anti-coplanar relationship of the Si–C and C–O bonds, which is present in **5** and required for full hyperconjugative interaction with the cation formed as the C–O bond suffers heterolysis.

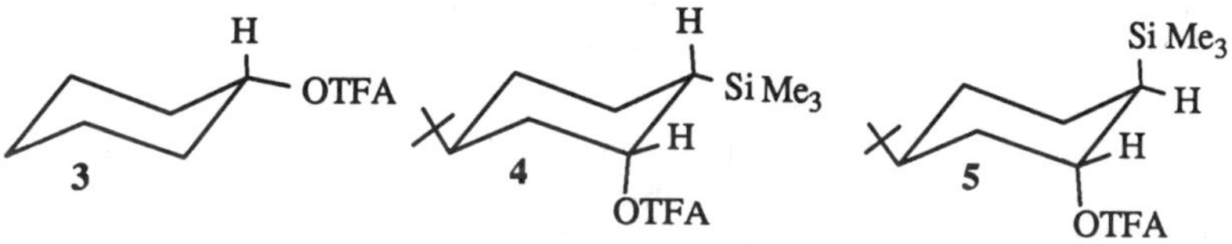

Both α- and β-silyl radicals are stabilized relative to the all-carbon systems. Although these stabilizations are sufficiently large to control a great deal of chemistry, their magnitudes (probably ~3 kcal/mol) are far less than for the analogous cations. The relative reactivities for H· abstraction by the *tert*-butoxy radical are as follows:

$$Et_3C\text{–}\overset{4.2}{C}H_2\text{–}\overset{1.0}{C}H_3$$

$$Et_3Si\text{–}\overset{22}{C}H_2\text{–}\overset{5.5}{C}H_3$$

The energies of various silicon bonds are given in Table I.

The experimental Si–H bond strengths are most interesting in that they are remarkably insensitive to substitution on Si, with the dramatic exception

Table I. Silicon Bond Energies

Bond	*Compound*	*Bond Energy (kcal/mol)*
Si–H	$H_3Si–H$	90.3
	$Me_3Si–H$	90.3
	$Cl_3Si–H$	91.3
	$PhSiH_2–H$	88.2
	$Me_3SiSiH_2–H$[a]	85.3
	$(Me_3Si)_3Si–H$[a]	79.0
Si–C	$H_3Si–CH_3$	88.2
	$Me_3Si–CH_3$	89.4 (74.8)[b]
Si–Si	$H_3Si–SiH_3$	74
	$Me_3Si–SiMe_3$	80.5 (63.0)[b]
Si–X	$Me_3Si–Cl$	113
	$Me_3Si–Br$	96
	$Me_3Si–I$	77
	$Cl_3Si–Cl$	111
	$F_3Si–F$	160
	$Me_3Si–OH$	128
	$Me_3Si–NHMe$	100
	$Me_3Si–SBu$	99

[a]Data were obtained by a photoacoustic technique.
[b]Values in parentheses were derived from ΔH_f (enthalpy of fusion) data obtained by mass spectroscopy and probably have some inherent error.

of silyl substitution. At this time it is unclear why successive silyl substitution progressively weakens the Si–H bond.

Many of the various silicon bonds have strengthened considerably over the past 10–20 years! It seems that they are now leveling off, but different techniques continue to produce different numbers.

The Si–Si bond is quite different from the C–C bond and actually resembles more the C=C bond in its chemistry and properties. For example, disilanes readily undergo electrophilic cleavage by the same reagents that add to olefins by cleavage of the π bond.

$$Me_3Si–SiMe_3 \xrightarrow{Br_2} 2Me_3SiBr$$

$$Me_3Si–SiMe_3 \xrightarrow{PhCO_3H} Me_3Si–O–SiMe_3$$

The Si–Si bond provided the first example of a σ-bond donor in the formation of charge-transfer complexes with TCNE (tetracyanoethylene). Conversely, the Si–Si bond can act as an electron acceptor to form disilanyl radical anions.

$$Me_3SiSiMe_3 + (CN)_2C=C(CN)_2 \longrightarrow [Me_3Si–SiMe_3]^{+\cdot}\ [(NC)_2C=C(CN)_2]^{\dot{-}}$$

$$(\lambda_{CT} = 417\ nm)$$

$$Me_3SiSiMe_3 \xrightarrow{K} [Me_3SiSiMe_3]^{\dot{-}}\ K^+$$

Perhaps most striking is the demonstration that the system $(\text{-Si-})_n$ is electronically "conjugated". An intense UV absorption results from a $\sigma \rightarrow \sigma^*$ (or $\sigma \rightarrow 3d\pi$) transition, which shifts position bathochromically with increased silicon catenation (Table II). These examples show that the λ_{max} quickly approaches a limiting value with increasing chain length. Thus, alkylsilane polymers absorb at 305–320 nm.

Table II. Bathochromic Shift in Silicon Catenation

Silane	λ_{max} *(nm)*
$Me_3Si\text{-}SiMe_3$	ca. 200
$Me(Me_2Si)_3Me$	215
$Me(Me_2Si)_4Me$	234
$Me(Me_2Si)_6Me$	260
$Me(Me_2Si)_{12}Me$	285
$Me(Me_2Si)_{18}Me$	291
$Me(Me_2Si)_{24}Me$	293

Making and Breaking of Bonds to Silicon

The Rochow Process. Rochow found that alkyl and aryl halides react directly with silicon when their vapors contacted silicon at elevated temperatures to produce complex mixtures of organosilicon halides. The reaction is promoted by a wide variety of metals from both the main group and the transition series, but the most efficient catalyst is copper. The most studied reaction of this type is the reaction between methyl chloride and silicon to give dimethyldichlorosilane and methyltrichlorosilane. Dimethyldichlorosilane is major feedstock silane for methylsilicon polymers.

$$MeCl + Si \xrightarrow[Cu]{>300\ ^\circ C} \begin{cases} Me_2SiCl_2 \text{ (bp 70 °C; 30–800\%)} \\ MeSiCl_3 \text{ (bp 66 °C; 10–40\%)} \\ Me_3SiCl \text{ (bp 57.7 °C)} \\ SiCl_4 \text{ (bp 57.6 °C)} \\ MeSiHCl_2 \text{ (bp 40.7 °C)} \\ SiCl_3H \text{ (bp 31.8 °C)} \end{cases}$$

$$Me_2SiCl_2 \xrightarrow{H_2O} \text{-}Me_2SiO(SiMe_2)_nOSiMe_2\text{-}$$

Since the "direct reaction" produces all substitution possibilities of methylchlorosilanes, it is, indeed, remarkable that selectivity for the most desired Me_2SiCl_2 can be >90%! The mechanism of the direct reaction has not been fully elucidated, but evidence points to the formation of a Si–Cu intermetallic compound that more readily polarizes the C–Cl bond than either silicon or copper alone to generate highly reactive silicon subchlorides and methyl radicals. The following scheme is a reasonable pathway to tetrasubstituted silanes:

$$MeCl + (Si\text{–}Cu) \rightarrow [MeCuCl] \rightarrow Me^{\bullet} + CuCl$$

$$CuCl + Si \rightarrow [SiCl] + Cu$$

$$[Si]Cl + Me^{\bullet} \rightarrow MeSiCl \rightarrow Me_xSiCl_{4-x}\ (x = 1\text{–}3)$$

Hydrosilation. The second most important method for preparing organosilanes on a large scale is the addition of hydrosilanes across a carbon–carbon double bond. The reaction is quite general and applies to a wide variety of substituted alkenes, dienes, and alkynes.

$$R_3Si\text{—}H + \text{>C=C<} \xrightarrow{\text{catalyst}} R_3Si\text{–}\overset{|}{C}\text{–}\overset{|}{C}\text{–}H$$

These reactions, which are catalyzed by a broad spectrum of agents including pure and supported metals, metal salts, bases, ultraviolet light, and free radical initiators, can give high yields of product at less than 100 °C and often at room temperature. Typically, homogeneous catalysts are used, the most efficient of which is chloroplatinic acid, H_2PtCl_6, known as Speier's catalyst. As an example, a 10^{-6} M concentration of this catalyst relative to silane can produce quantitative hydrosilation of terminal alkenes within minutes. Unfortunately, the catalyst is not recoverable, and on an industrial scale, this loss adds significantly to the cost of production (~$0.12/lb or $0.26/kg). Often, hydrosilation of dienes is better accomplished with palladium [e.g., $Pd(PPh_3)_4$] catalysis.

$$R_3Si\text{–}H + \text{CH}_2\text{=CH–}R' \xrightarrow{H_2PtCl_6} \underset{\text{major}}{R_3Si\text{–CH}_2\text{–CH}(H)\text{–}R'} + \underset{\text{minor}}{H\text{–CH}_2\text{–CH}(SiR_3)\text{–}R'}$$

$$R_3Si\text{–}H + \text{CH}_2\text{=CH–CH=CH}_2 \xrightarrow[100\ ^{\circ}C]{PPh_3 + Pd} R_3Si\text{–CH}_2\text{–CH}(H)\text{–CH=CH}_2 + \underset{94\%}{R_3Si\text{–CH}_2\text{–CH=CH–CH}_2\text{–}H}$$

Normally, the less hindered addition product dominates, although some control of the isomer distribution is possible by careful selection of the silane, olefin, and/or catalyst. The hydrosilation of carbonyl compounds is also well known:

$$R_3Si\text{–}H + O{=}CR'_2 \xrightarrow{Rh(PPh_3)_3Cl} R_3Si\text{–}O\text{-}CHR'_2$$

R_3Si–H + [α,β-unsaturated ketone, O] → [silyl enol ether, $OSiR_3$]

The mechanisms of all metal-catalyzed hydrosilations are thought to be very similar. The pathway probably involves an adduct composed of the silane, the alkene, and the metal. Transfer of the silicon to the carbon is believed to occur after the π-bonded olefin rearranges to a σ complex. Whereas the mechanism displayed in the following scheme involves olefin insertion into Pt–H, equally possible is insertion into Pt–Si followed by reductive elimination of the alkyl silane.

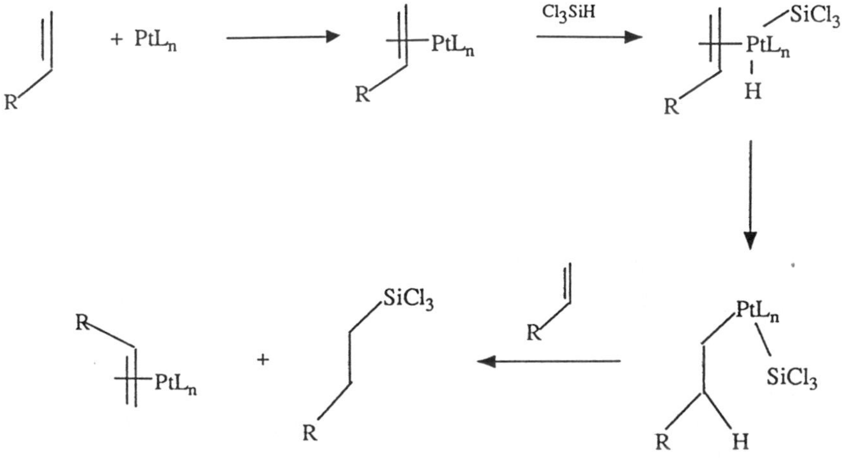

R + PtL_n → R, PtL_n —Cl_3SiH→ R, $SiCl_3$, PtL_n, H

↓

R, PtL_n + R, $SiCl_3$ ← R (olefin) ← R, H, PtL_n, $SiCl_3$

Hydrosilation can be promoted by peroxides that initiate the reaction by hydrogen abstraction from the silane and propagate in a fashion similar to that of classical organic free radical additions:

initiation

$$(RCOO)_2 \rightarrow R^{\cdot} + 2CO_2$$

$$R^{\cdot} + Cl_3Si\text{–}H \rightarrow Cl_3Si^{\cdot} + H\text{–}R$$

propagation

$$Cl_3Si^{\cdot} + CH_2{=}CH_2 \rightarrow Cl_3SiCH_2CH_2^{\cdot}$$

$$Cl_3SiCH_2CH_2^{\cdot} + HSiCl_3 \rightarrow Cl_3SiCH_2CH_3 + {}^{\cdot}SiCl_3$$

If the olefin is perhalogenated, for example, tetrafluoroethylene, then the β radical, $Cl_3SiCF_2CF_2^{\cdot}$, can easily polymerize.

Reactions with alkynes may be controlled to add 1 or 2 moles of silane. *n*-Butylacetylene gives a mixture of monoaddition products in which the dominant isomer is usually the β adduct:

$$R_3Si\text{-}H + C_4H_9C{\equiv}C\text{–}H \longrightarrow \underset{\alpha}{(R_3Si)(H)C{=}C(H)(C_4H_9)} + \underset{\beta}{(R_3Si)(C_4H_9)C{=}CH_2}$$

Monoaddition to phenylacetylene in the presence of platinum on carbon or chloroplatinic acid generally gives the *trans* product, whereas under free radical conditions, *cis* products are obtained.

The vast majority of hydrosilations have been carried out with Cl_3SiH, which is usually quite reactive, whereas R_3SiH is often very unreactive. However, activation of platinum catalysts with oxygen dramatically enhances additions of alkylsilanes (*1*).

Redistribution Reactions. A redistribution reaction is one in which there is no net change in the number and type of chemical bonds:

$$MX_4 \leftrightharpoons MX_3Y \leftrightharpoons MX_2Y_2 \leftrightharpoons MXY_3 \leftrightharpoons MY_4$$

For group IV metals, the ease of redistribution parallels the size of the central atom. Thus for systems in which M = Sn and X and Y are alkyl, aryl, hydrogen, or electronegative groups such as halogens or alkoxy groups, equilibration can often be reached under very mild conditions, that is, <200 °C in the absence of a catalyst. The redistribution of groups on silicon, however, is a feasible process only in the presence of a catalyst, normally a Lewis acid such as aluminum chloride, and at elevated temperatures.

Because of the commercial importance of halosilanes, particularly the dihalo and trihalo derivatives, in the production of silicone polymers, the reaction in which M = Si has received considerable attention. For example, the direct process produces Me_3SiCl and $MeSiCl_3$ as part of a complex mixture including the valuable Me_2SiCl_2, which is difficult to separate because of similar boiling points. Fortunately, the equilibration of alkyl groups on silicon is not statistical, a fact that permits a very useful application of the redistribution reaction:

$$Me_3SiCl + MeSiCl_3 \leftrightharpoons Me_2SiCl_2$$

Dimethyldichlorosilane can be obtained in 77% yield by using $AlCl_3$ as the catalyst at 300 °C and 1000 atm (101,000 kPa). Remarkably, the reaction temperature can be reduced to 150 °C if a small quantity of CH_3HSiCl_2 is added.

The Si–H bond is particularly labile under Lewis acid conditions, and when stoichiometric quantities of hydrosilanes are used, the reaction can be hazardous. For example, the redistribution of the very stable phenylsilane generates silane, which reacts explosively with air.

$$PhSiH_3 \xrightarrow{AlCl_3} Ph_4Si + SiH_4$$

Whereas numerous reactivity studies have been reported for a wide variety of silicon compounds and catalysts, caution is advised when comparing rate data and accepting mechanistic proposals. The uncertainty arises from the lack of uniformity of reaction conditions, especially regarding the presence of water and oxygen, both of which have significant effects on rate and product distribution.

Nevertheless, some general trends have emerged. The ease of migration follows the order H > aryl > alkyl, and the migration of each of these groups becomes less facile if an electron-withdrawing group is present on the silicon. The activity of catalysts follows the order $Al_2Br_6 > Al_2Cl_6 > Al_2I_6 > Ga_2Br_6 > Ga_2Cl_6$, BCl_3, Fe_2Cl_6. Protonic acids such as sulfuric and sulfonic acids are usually more reactive than Lewis acids.

Cyclosilanes merit special comment because of their tendency to yield polymeric products. Three- and four-membered rings containing silicon polymerize readily under very mild conditions. Silacyclopentanes, on the other hand, give linear polymeric materials [MW (molecular weight) = 1000–2500] when heated with aluminum halides.

$$\text{(cyclo-}C_4H_8\text{)}SiMe_2 \xrightarrow[AlCl_3]{20\text{–}80\ ^\circ C} -\!\!\left[CH_2CH_2CH_2CH_2SiR_2\right]_n\!\!- \xrightarrow{>80\ ^\circ C} \text{cross-linked polymers}$$

Prolonged heating of the product mixture leads to extensive cross-linking. Silacycloheptanes also polymerize under these conditions. The silacyclohexanes, however, resist ring opening and polymerization and react with Lewis acids to give primarily molecular-disproportionation products. Complex organosilanes that are not accessible by other routes can be prepared in high yield via redistribution reactions:

Me Me
Si
Me_2Si $SiMe_2$
$AlCl_3$
Me
Si
Me-Si
Si-Me
Si
Me
+ Me_4Si

$$\text{Me}_3\text{Si-substituted cyclohexane (Me}_3\text{Si, Me}_3\text{Si, SiMe}_3) \xrightarrow{AlCl_3} \text{1-methyl-1-silaadamantane (Me–Si)} + 2Me_4Si$$

Interesting polymers containing phenyl rings as part of the backbone have been prepared from 2-silaindane derivatives:

$$C_6H_4(CH_2)_2SiMe_2 \longrightarrow \left(CH_2-C_6H_4-CH_2SiMe_2 \right)_n$$

Extensive rate studies on aluminum-halide-catalyzed redistributions of organosilanes support a mechanism in which polarization of alkyl silicon bonds and an associative step are essential:

$$R_3Si{-}R'\cdots\cdots AlBr_3 + R_3Si{-}R' \longrightarrow R_2R'{-}Si\,(R)\,\overset{\delta^+}{SiR_3}\,(R'\cdots \overset{\delta^-}{AlBr_3}) \longrightarrow R_4Si + R_2SiR'_2$$

Unlike Lewis-acid-catalyzed rearrangements of organic compounds, ionic intermediates are not important in the mechanism. No evidence has been found to support the existence of silylenium ions, R_3Si^+, and careful studies have ruled out carbocations as well.

Base-catalyzed redistributions are also efficient and have been used successfully in the synthesis of fluorosilanes. The high volatility of fluorosilanes allows the easy separation of products.

$$C_6H_5SiF_3 \xrightarrow[B:]{R_3SiCl} R_3SiF\uparrow + C_6H_5SiCl_3$$

$$C_6H_5SiF_3 \xrightarrow[B:]{SiCl_4} SiF_4\uparrow + C_6H_5SiCl_3$$

Organometallic Coupling Reactions. The most popular laboratory method of generating silicon–carbon bonds is the reaction of an organometallic compound with a functionalized silane. For example, a Grignard reagent added to a chlorosilane in a polar solvent will give high yields of the coupled product:

$$SiCl_4 + RMgX \rightarrow R\text{–}SiCl_3$$

These reactions usually follow the stoichiometry permitting stepwise substitution:

$$MeSiCl_3 + 1\text{-}C_{10}H_7MgBr \rightarrow 1\text{-}C_{10}H_7SiMeCl_2$$

$$1\text{-}C_{10}H_7SiMeCl_2 + PhMgBr \rightarrow (1\text{-}C_{10}H_7)PhSiMeCl$$

Other organometallic reagents, such as organolithium, organosodium, and organozinc compounds, will also function in this capacity. Organolithium reagents are often preferred because of their greater reactivity. Additionally, the inherent reactivity of silicon allows the use of easily accessible leaving groups. Thus OR, OC(O)R, and SR, in which R is an alkyl or aryl group, can be displaced readily by an organometallic reagent.

$$\left.\begin{array}{l} R_3Si\text{-}OR' \\ R_3Si\text{–}OC(O)R' \\ R_3Si\text{–}SR \end{array}\right] \xrightarrow{\overset{\delta^-\quad\delta^+}{R''\text{–}M}} R_3Si\text{–}R''$$

In some organosilanes, particularly the strained cyclosilanes, even the hydride is an efficient leaving group:

$$\text{(silacyclobutane)}\ Si(R)(H) \xrightarrow{R'\text{–}M} \text{(silacyclobutane)}\ Si(R)(R')$$

Silicon–carbon bonds can be prepared by reductive silylation with an active metal, a chlorosilane, and an organic substrate with electronegative substituents. Although the mechanisms have not been elucidated, organometallic intermediates are probably essential to the transformation:

$$Cl_3CHO + Me_3SiCl \xrightarrow{Mg} (Me_3Si)(Cl)C{=}C(H)(OSiMe_3)$$

$$\text{1,2-(OMe)}_2\text{C}_6\text{H}_4 \xrightarrow[\text{Me}_3\text{SiCl}]{\text{Li}} \text{1,2-(OMe)}_2\text{-3,3,6,6-(SiMe}_3)_4\text{-cyclohexa-1,4-diene}$$

Silylmetallic compounds can be prepared in synthetically useful quantities and used to generate new silicon–carbon bonds:

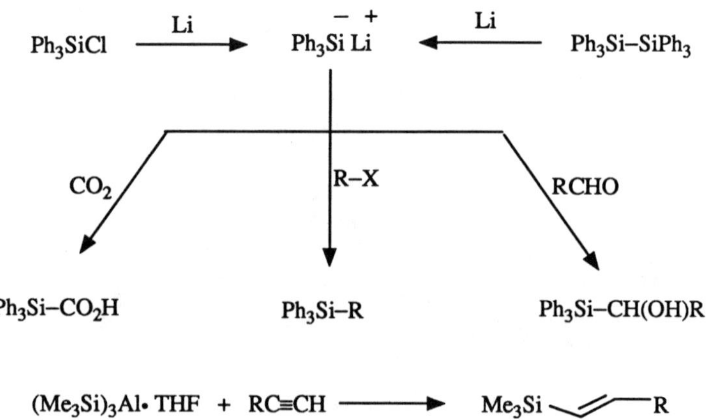

Reactions of Organosilanes

Nucleophilic Substitution. Silicon readily expands its valence shell, a property allowing organosilicon compounds to undergo nucleophilic substitution more easily than their carbon analogues. Chlorosilanes are the most common substrates for displacement reactions producing high yields of substitution products, even with weak nucleophiles under mild conditions:

$$R_3Si\text{–}SR' \xleftarrow{R'SH} R_3SiCl \xrightarrow{R'_2NH} R_3Si\text{–}NR'_2$$

$$R_3Si\text{–}OR' \xleftarrow{R'OH} R_3SiCl \xrightarrow{H_2O} R_3Si\text{–}OH$$

$$R_3Si\text{–}OC(O)CH_3 \xleftarrow{CH_3CO_2^-} R_3SiCl \xrightarrow{LiAlH_4} R_3Si\text{–}H$$

Double displacements are commonly used for ring synthesis in which X is normally a halogen and Y is one of the common nucleophilic groups or carbon.

$$R_2SiX_2 + \begin{matrix}-Y \\ -Y\end{matrix}\Big) \xrightarrow{-2X^-} R_2Si\begin{matrix}Y \\ Y\end{matrix}\Big)$$

Substitution at silicon centers is invariably bimolecular, there being no well-documented examples of unimolecular, that is, S_N1-type, displacement reactions that are commonplace in carbon chemistry. The ease with which silicon expands its valence shell significantly lowers the energy of the transition states that require the attacking nucleophile to form a bond to the silicon atom. The size of the silicon atom permits the nucleophile to approach from different directions. Thus, not only does silicon undergo facile backside attack to give the inversion product via a trigonal bipyramid geometry, but silicon will also permit "flank" attack, which leads to a different trigonal bipyramid and the retention product:

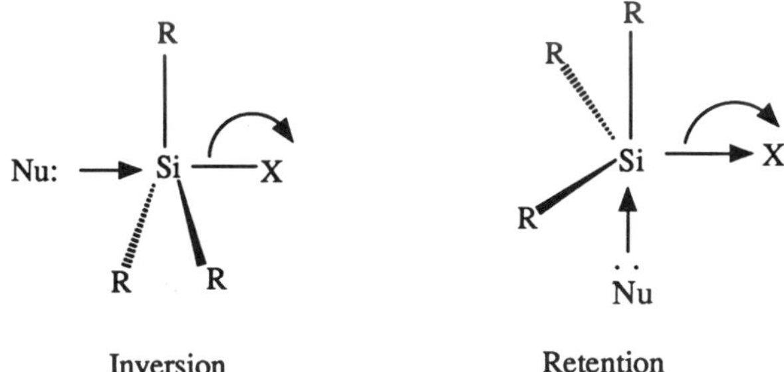

A comparison of the relative reactivities of norbornane and silanorbornane illustrates the point. 1-Halonorbornane derivatives are very resistant to nucleophilic substitution reactions under strenuously applied S_N1 and S_N2 conditions. The low reactivities of these compounds result from the cage structure that prohibits deformation to the planar geometry required for the carbocation intermediates of unimolecular reactions and inhibits the backside attack required for bimolecular substitutions at carbon.

1-Chloro-1-silanorbornane, on the other hand, is very reactive, at least 10^6 times more so than the carbon analogue and many times more reactive than most chlorosilanes. Inspection of the geometry of the bridgehead silicon shows that the bond angles about the silicon are very close to those of a trigonal bipyramid. Thus, because only four groups are attached to the silicon atom, the attacking nucleophile can approach and start to form a bond to the silicon at the vacant apical position. With the ground-state geometry very close to that of the transition state, the activation energy for substitution is lower than that for silanes with normal tetrahedral geometry.

Various theoretical studies (2, 3) reveal that nucleophilic substitution on silicon always proceeds through a five-coordinate silicon adduct intermediate. Thus the S_N2 mechanism on Si never resembles that of carbon.

Because of the polar nature of the Si–C bond, it is cleaved more readily than the C–C bond by ionic reagents. Cleavage of the Si–C bond can be achieved by either nucleophilic attack on Si or electrophilic attack on C. A general rule is as follows: If a particular C–H bond is broken by an ionic reagent, that reagent will more easily cleave the corresponding C–Si bond.

Nucleophilic cleavage of alkyl groups from Si is difficult and requires forcing conditions. However when a stabilized carbanion is formed, this cleavage can be quite facile. For example:

$$Me_3Si{-}\overset{O}{\overset{\|}{C}}R' \xrightarrow[RO^-]{ROH} Me_3SiOR + R'CHO$$

$$Me_3Si{-}CH_2CH{=}CH_2 \xrightarrow{F^-} Me_3SiF + {}^-CH_2CH{=}CH_2$$

$$Ph_3Si{-}C{\equiv}CPh \xrightarrow[H_2O]{^-OH} Ph_3SiOH + HC{\equiv}CPh$$

Electrophilic Substitution. The silicon–hydrogen bond is far more polarized than the carbon–hydrogen bond. The electronegativities of silicon and hydrogen result in a negatively polarized hydrogen, that is, a hydride. This condition explains the occasional use of hydride as the leaving group in nucleophilic substitution at silicon. The increased electron density on hydrogen when attached to silicon also makes it more polarizable, however, and thus susceptible to electrophilic substitution:

$$R_3\overset{\delta+}{Si}{-}\overset{\delta-}{H} + X_2 \rightarrow R_3Si{-}X + HX$$

The relative rates of electrophilic substitution for halogens are 130:8:1 for Br–Cl, Br_2, and Cl_2. The halogenation reactions are nearly quantitative at or below room temperature. As a result, silicon hydrides are useful for storing silanes for prolonged periods and eliminating loss through adventitious hydrolysis. These substitutions usually proceed with retention of configuration presumably through an intermediate, like

$$\left[R_3Si \overset{+}{\cdots} \begin{matrix} H \\ \vdots \\ Br \end{matrix} \right] \cdots\cdots Cl^-$$

A similar intermediate is proposed to explain the retentive oxidation of Si–H bonds with perbenzoic acid. In this reaction, Br^+ is replaced with OH^+ and Cl^- is replaced with $C_6H_5CO_2^-$. Cleavage of alkyl groups from silicon can be accomplished with strongly electrophilic reagents, such as hydrogen halides in the presence of aluminum halides:

$$Me_4Si + HCl + AlCl_3 \rightarrow Me_3SiCl + CH_4$$

Aryl groups, on the other hand, cleave from silicon more easily. Bromine, for example, will perform the task quantitatively at room temperature:

$$Ph_4Si + Br_2 \rightarrow Ph_3SiBr + PhBr$$

The Si–C bond is also broken in various elimination reactions. Thermally induced α elimination of R_3SiX can occur either from saturated or unsaturated carbon to produce the corresponding carbene.

$$H_3C\text{–}CH(SiMe_3)(OMe) \xrightarrow{\Delta} [H_3C\text{–}\ddot{C}H] + Me_3SiOMe$$

$$H_2C{=}C(SiMe_3)(OMe) \xrightarrow{\Delta} [H_2C{=}C\colon] + Me_3SiOMe$$

β-Chlorosilanes are often thermally unstable toward elimination. For example, $Et_3SiCH_2CH_2Cl$ cannot be distilled at atmospheric pressure.

$$R_3Si\text{–}\overset{|}{\underset{|}{C}}\text{–}\overset{|}{\underset{|}{C}}\text{–}Cl \xrightarrow{\Delta \text{ or } ^-OH} R_3Si\text{–}Cl + {>}C{=}C{<}$$

γ-Halosubstituted silanes can also undergo thermal elimination, although AlX_3 catalysis is often used.

$$Me_3SiCH_2CH_2CH_2Br \xrightarrow{AlBr_3} Me_3SiBr + \Delta\ (92\%)$$

Reactive Intermediates

Reactive intermediates are species that are so kinetically unstable that they cannot be isolated or observed under normal conditions. In organic chemistry, the radicals ($R_3C^{\cdot}$), carbanions (R_3C^-), carbocations (R_3C^+), and carbenes ($R_2C{:}$) are the more-common intermediates. The area, and the resulting chemistry, is far richer in organosilicon chemistry, because many of the bonding situations that produce quite stable organic compounds are highly reactive when Si replaces C.

Silenes ($R_2Si{=}CR_2$). Although a few stable molecules containing the Si=C unit have been prepared recently as crystals from which structural parameters could be obtained, the vast majority of the chemistry of silenes has been investigated by indirect means with transient molecules. Silenes can be generated from a wide variety of precursors through gas-phase thermolysis,

Me_2Si (silacyclobutane) $\xrightarrow[N_2]{600\ ^\circ C}$ $Me_2Si{=}CH_2$ + $H_2C{=}CH_2$ ⟶ Me_2Si $SiMe_2$

(Si-Me, allyl) $\xrightarrow[\text{vacuum}]{>400\ ^\circ C}$ − (propene) (retroene) ⟶ Si-Me $\xrightarrow{F_3C{-}C{\equiv}C{-}CF_3}$ (Si, Me, CF_3, CF_3)

Me_2Si (R, R) $\xrightarrow[\text{vacuum (retro 2 + 4)}]{>400\ ^\circ C}$ $Me_2Si{=}CH_2$ + (R, R) ; $Me_2Si{=}CH_2$ → dimer

($SiMe_3$, $Si(Me_2){-}OMe$) $\xrightarrow[10^{-4}\ \text{torr}\ (\beta\ \text{elimination})]{740\ ^\circ C}$ (=$SiMe_2$) + Me_3SiOMe ; → dimer

via thermal rearrangements,

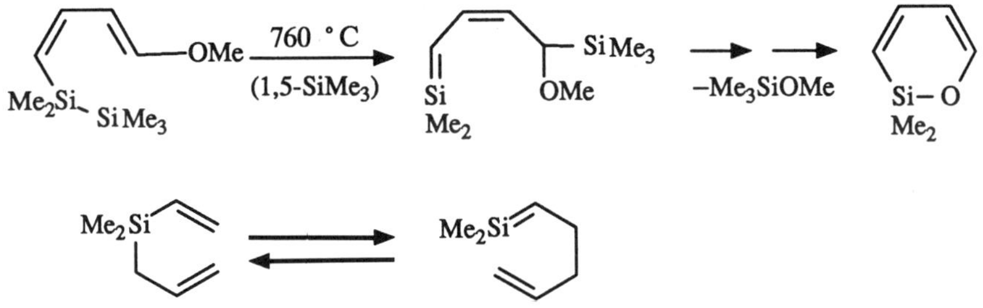

760 °C (1,5-$SiMe_3$) Me_2Si $SiMe_3$ OMe Si Me_2 $SiMe_3$ OMe $-Me_3SiOMe$ Si–O Me_2

Me_2Si Me_2Si

via photochemical isomerizations,

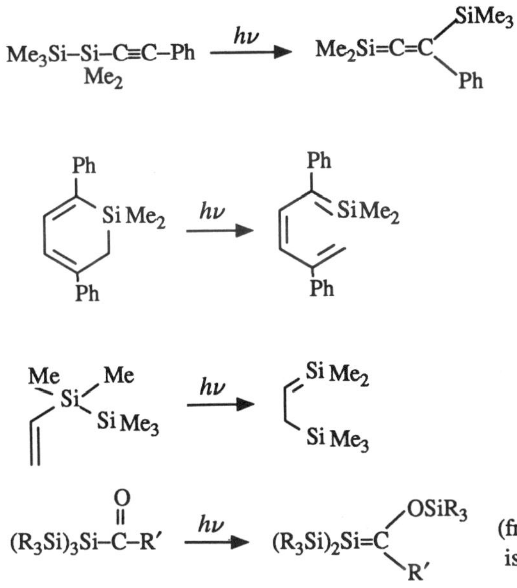

Me_3Si–Si–C≡C–Ph Me_2 $h\nu$ Me_2Si=C=C $SiMe_3$ Ph

Ph $SiMe_2$ Ph $h\nu$ Ph $SiMe_2$ Ph

Me Me Si $SiMe_3$ $h\nu$ $SiMe_2$ $SiMe_3$

O $(R_3Si)_3Si$–C–R′ $h\nu$ $(R_3Si)_2Si$=C $OSiR_3$ R′ (from this photoisomerization stable, isolable silenes have been produced)

via eliminations from silyl halides or esters,

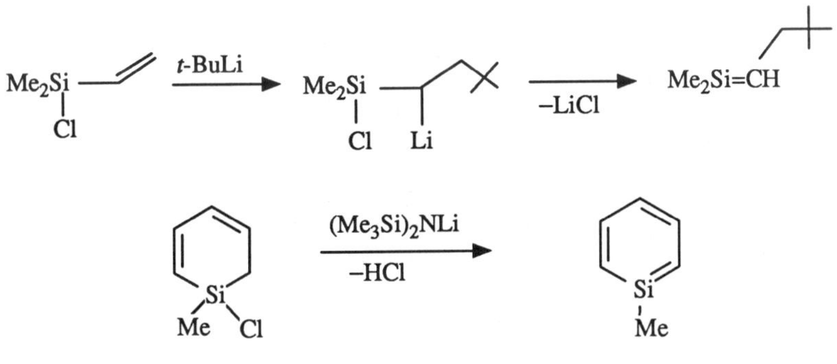

Me_2Si Cl t-BuLi Me_2Si Cl Li $-LiCl$ Me_2Si=CH

Si Me Cl $(Me_3Si)_2NLi$ $-HCl$ Si Me

$$Me_2Si(F)\text{—}C(Li)(SiMe_3)(SiMeBu^t_2) \xrightarrow{-LiF} Me_2Si{=}C(SiMe_3)(SiMeBu^t_2) \quad \text{(isolable)}$$

and via rearrangement of α-silyl carbenes.

$$Me_3Si\text{–}C(=N_2)\text{–}SiMe_3 \xrightarrow{\Delta \text{ or } h\nu} Me_3Si\text{–}\ddot{C}\text{–}SiMe_3 \longrightarrow Me_2Si{=}C(Me)(SiMe_3)$$

The chemistry of silenes can generally be predicted from known olefin chemistry and from a consideration of the polar nature of the Si = C unit.

Reactions of $\overset{\delta+}{Si}{=}\overset{\delta-}{C}$:

- $R_2C{=}O$ → $R_2C\text{–}O$, $\text{–}C\text{–}Si\text{–}$ (ring)
- diene → silacyclohexene
- ROD → $Si(OR)\text{—}C(D)$
- R_3SiOR' → $R_3SiCSiOR'$
- THF → $O^+\text{–}Si\text{—}C^-$ (oxonium)
- O_2 → $Si\text{—}C$, $O\text{–}O$ (ring)
- $Si{=}C$ → 1,3-disilacyclobutane
- RN_3 → $R\text{–}N$, Si, $N{=}N$ (triazoline ring)

The reactions of silenes should not be viewed merely as laboratory curiosities. The synthetic routes to silenes are often quite efficient, and the reactions usually proceed with excellent yield. Thus, silenes represent important building blocks in organosilicon chemistry.

Silenes have been the object of numerous theoretical studies. It now seems generally agreed that the silicon–carbon π bond strength is 35–36 kcal/mol or roughly half that of the carbon–carbon π bond. It should be noted that whereas the inclusion of *d* orbitals does improve the computed geometry, the π bond strength is unchanged.

Perhaps the most interesting reaction of silenes is one that finds little counterpart in carbon chemistry. Appropriately substituted silenes can isomerize to silylenes by a 1,2-shift of H (or $SiMe_3$).

$$\mathrm{H(R)Si{=}CH_2} \rightleftharpoons \mathrm{R\ddot{S}i{-}CH_3}$$

Although the rearrangement is almost isothermal, the barrier is a considerable 41 kcal/mol.

Although the resonance energy of silabenzene is calculated to be about three-fourths of that of benzene, silabenzenes are extremely reactive and, thus, have been examined only in frozen matrices and in the gas phase. By far the simplest route to silabenzenes is pyrolysis of the 1,4-dihydro derivatives.

$$\text{1,4-dihydro-1,4-dimethyl-1,4-disila derivative (Si–H, Si–Me)} \xrightarrow[\text{vacuum}]{800\ ^\circ\text{C}} H_2 + \text{1,4-dimethyl-1,4-disilabenzene derivative}$$

Silylenes (R_2Si:). Silylenes, which are considerably more stable than the analogous carbenes, have a singlet ground state. With few exceptions, their reactions are analogous to those of carbenes. Silylenes are generated by α eliminations,

$$Me_3Si{-}SiMe_2(OMe) \xrightarrow{\Delta} Me_3SiOMe + {:}SiMe_2$$

(migrating group can also be H, X or vinyl)

by thermal extrusions,

$$\text{7,7-dimethyl-7-silanorbornadiene derivative} \xrightarrow{\Delta \text{ or } h\nu} \text{benzene} + Me_2Si{:}$$

$$\text{(Me}_2\text{C)}_2\text{SiMe}_2 \xrightarrow[14\ h]{108\ ^\circ C} Me_2C{=}CMe_2 + Me_2Si{:}$$

by photochemical extrusions,

$$(Me_2Si)_6 \xrightarrow{h\nu} (Me_2Si)_5 + Me_2Si{:}$$

and by a variety of isomerizations.

$$Me_3Si(Me)Si{=}CH_2 \xrightarrow{\Delta} Me\ddot{S}i{-}CH_2SiMe_3$$

$$H_2Si(CH_2)_3 \xrightarrow{\Delta} H{-}\ddot{S}i{-}CH_2CH_2CH_3$$

$$\xrightarrow{\Delta}$$

$$\xrightarrow{} \quad (X = H \text{ or } SiMe_3)$$

The majority of silylene reactions are insertions that immediately reestablish a tetracoordinated silicon. The most frequently observed insertions are into Si–O,

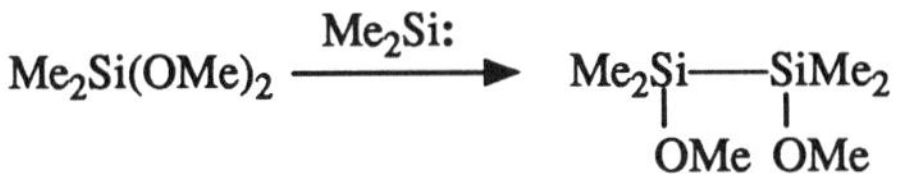

$$Me_2Si(OMe)_2 \xrightarrow{Me_2Si:} Me_2Si(OMe){-}SiMe_2(OMe)$$

into Si–H,

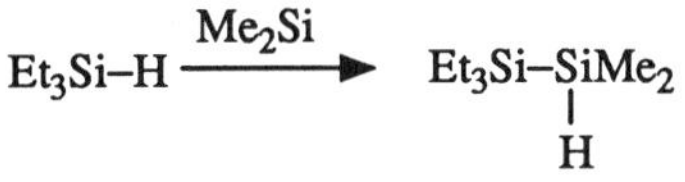

$$Et_3Si{-}H \xrightarrow{Me_2Si} Et_3Si{-}SiMe_2H$$

into O–H,

$$ROH \xrightarrow{Me_2Si:} ROSiMe_2H$$

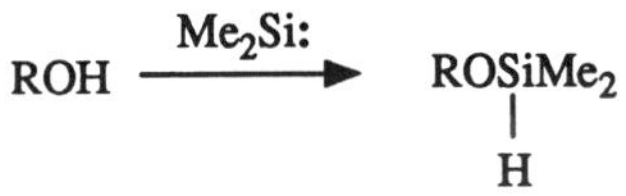

into C–H (actually quite rare),

$$H_3C{-}SiMe_2{-}CH_2{-}\ddot{Si}{-}Me \longrightarrow \text{cyclo-}[H_2C{-}SiMe_2{-}CH_2{-}Si(Me)(H)]$$

into C=C,

$$Me_2C{=}CMe_2 \xrightarrow{Me_2Si:} \text{cyclo-}[Me_2C{-}CMe_2{-}SiMe_2]$$

$$CH_2{=}CH{-}CH_2{-}\ddot{Si}{-}Me \longrightarrow \text{1-H-1-Me-silacyclobutene}$$

$$\text{(pentadienyl)}\ddot{Si}{-}Me \longrightarrow \text{1-H-1-Me-silacyclohexadiene}$$

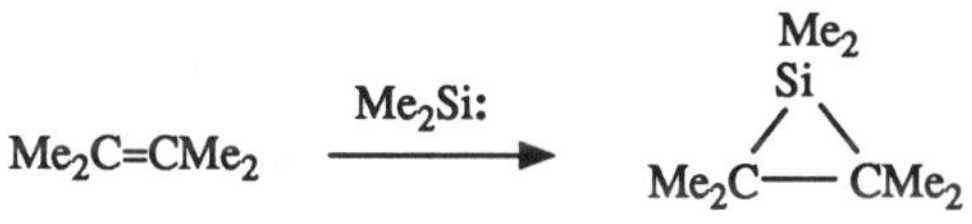

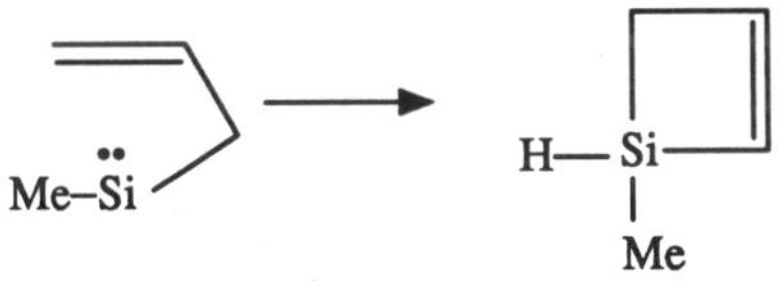

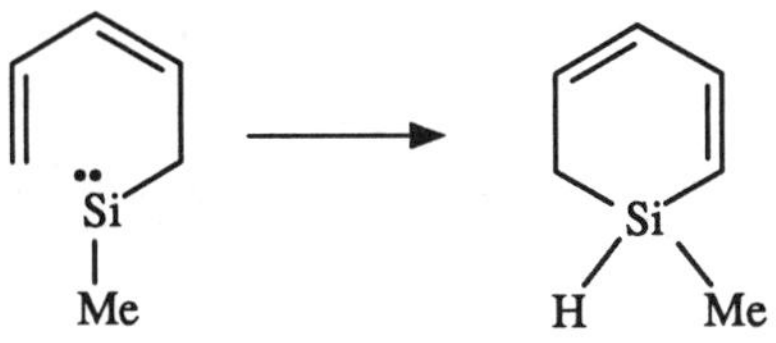

into C=C–C=C,

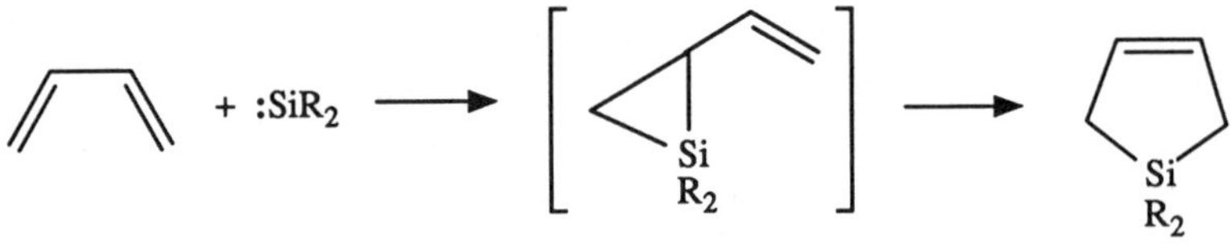

$$\text{C=C–C=C} + :SiR_2 \longrightarrow [\text{vinylsilirane}] \longrightarrow \text{silacyclopent-3-ene } (SiR_2)$$

into C≡C,

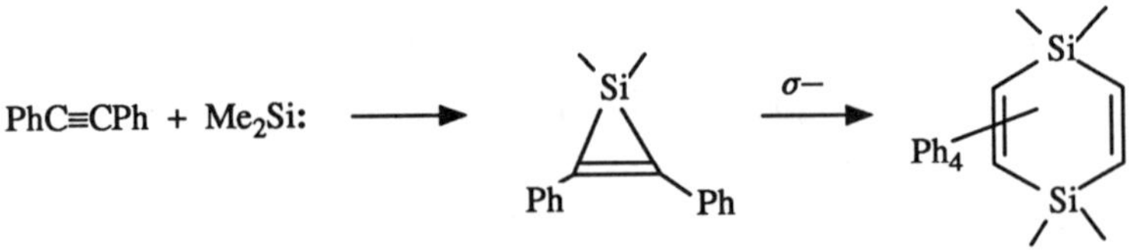

$$PhC{\equiv}CPh + Me_2Si: \longrightarrow \text{silacyclopropene} \xrightarrow{\sigma-} Ph_4\text{-1,4-disilacyclohexadiene}$$

into C=O,

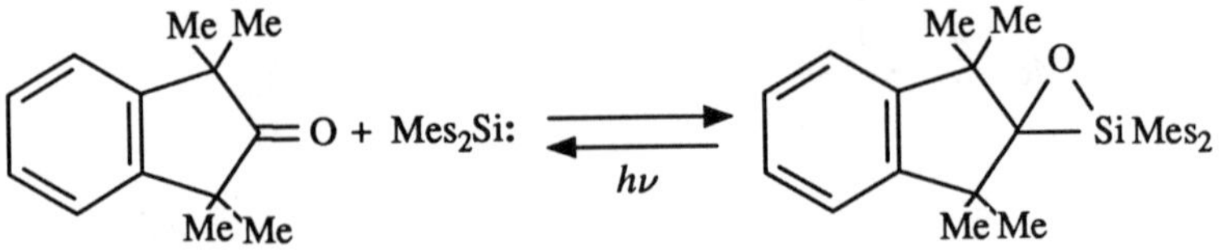

$$\text{ketone} = O + Mes_2Si: \underset{h\nu}{\rightleftharpoons} \text{oxasilirane } SiMes_2$$

and into C–C.

$$\text{cyclopropyl–}\ddot{S}i\text{–Me} \longrightarrow \text{silacyclobutene} \longrightarrow \text{vinyl(Me)Si=CH}_2 \longrightarrow \text{1,3-disilacyclobutane}$$

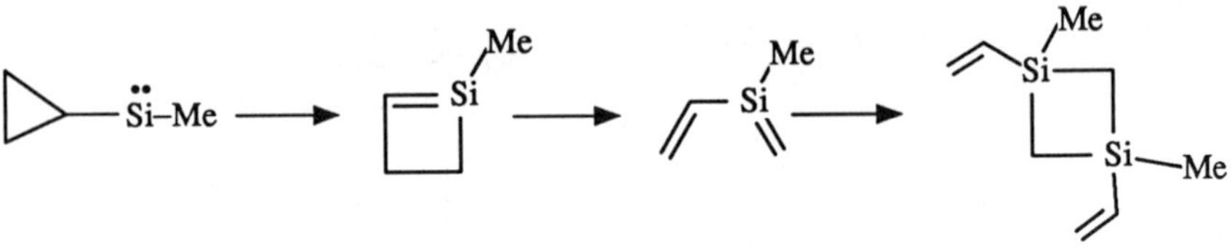

Disilenes ($R_2Si=SiR_2$). Most of the excitement in disilene chemistry in recent years stems from the discovery that relatively stable, isolable disilenes can be prepared. Disilenes with small substituents are unstable even at very low temperatures, but those with bulky substituents are stable in solution and, in extreme cases, can be isolated as crystalline solids. Examples of the preparation of "stable" disilenes are

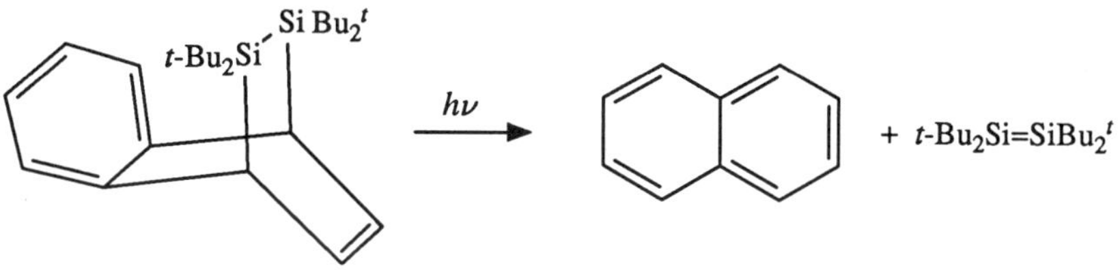

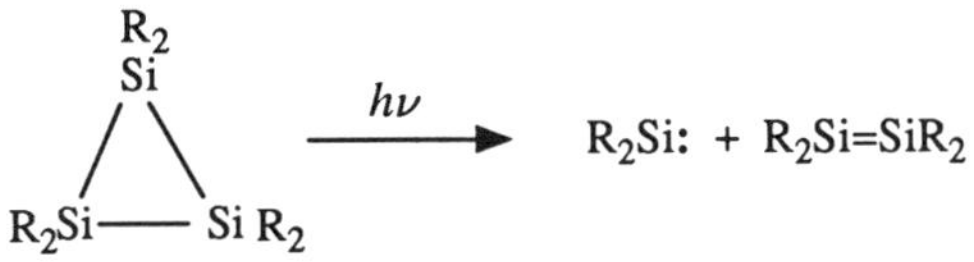

$t\text{-}Bu_2Si\text{-}SiBu_2^t$ $\xrightarrow{h\nu}$ naphthalene + $t\text{-}Bu_2Si{=}SiBu_2^t$

$(R_2Si)_3$ $\xrightarrow{h\nu}$ $R_2Si{:}$ + $R_2Si{=}SiR_2$

(R = neopentyl)

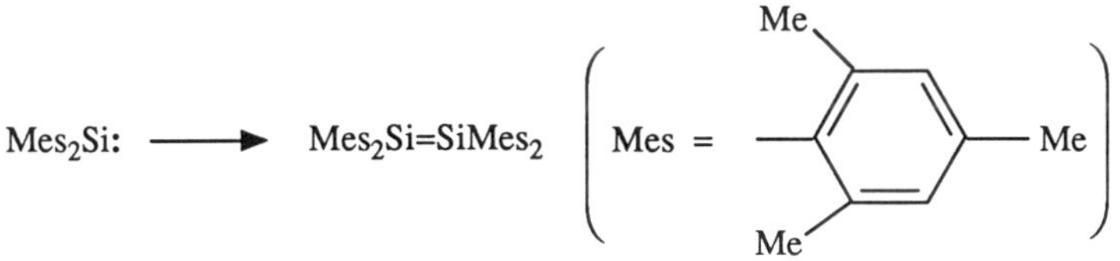

$Mes_2Si{:}$ $\longrightarrow$ $Mes_2Si{=}SiMes_2$ (Mes = 2,4,6-trimethylphenyl: Me, Me, Me)

Disilenes can undergo a variety of intramolecular reactions:

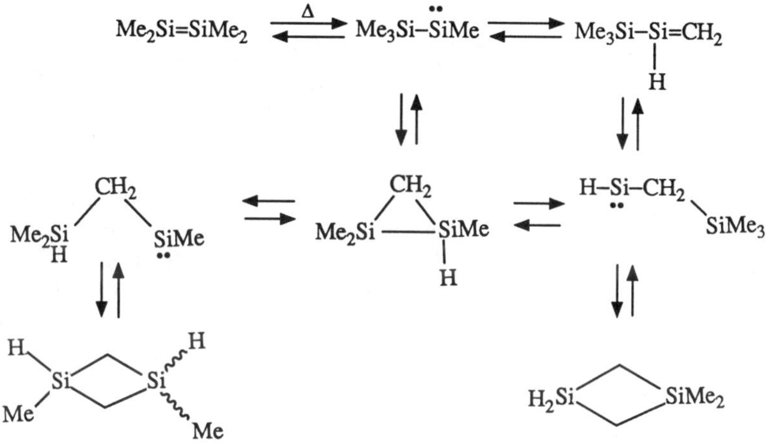

$Me_2Si{=}SiMe_2 \overset{\Delta}{\rightleftharpoons} Me_3Si\text{–}\ddot{S}iMe \rightleftharpoons Me_3Si\text{–}Si(H){=}CH_2$

CH_2 / $Me_2Si(H)$ / $\ddot{S}iMe$ $\rightleftharpoons$ CH_2 / Me_2Si——$SiMe(H)$ $\rightleftharpoons$ $H\text{–}\ddot{S}i\text{–}CH_2\text{–}SiMe_3$

H, Si, Si, H, Me, Me ; H_2Si, $SiMe_2$

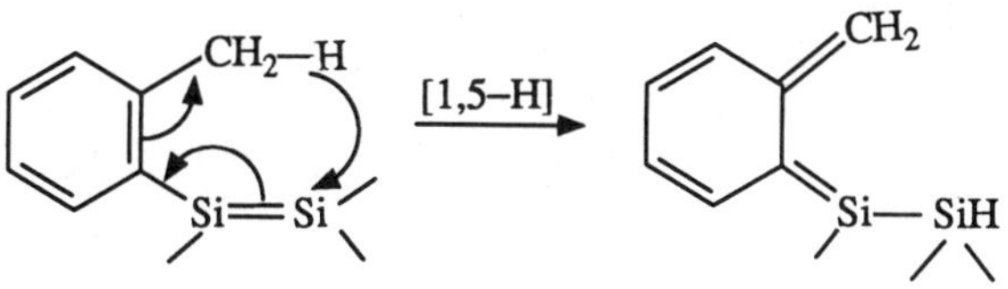

$CH_2\text{–}H$, Si=Si $\xrightarrow{[1,5\text{–}H]}$ CH_2, Si— SiH

The bimolecular processes of disilenes are rather predictable,

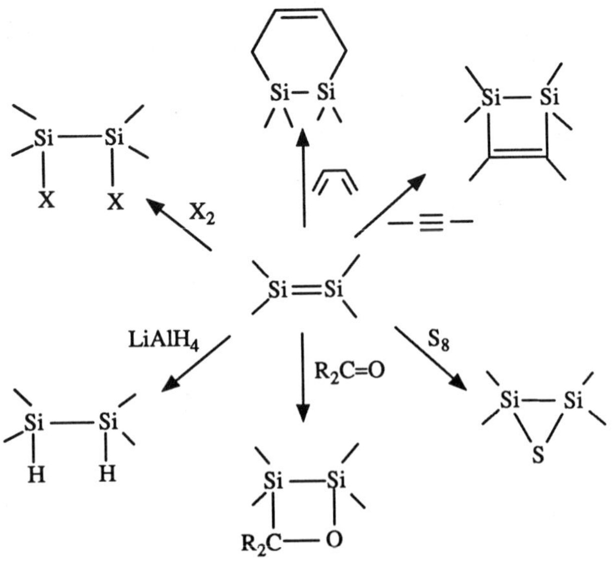

with at least one notable exception:

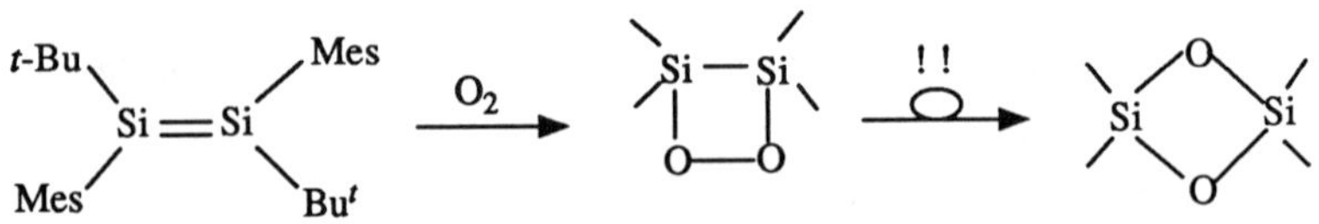

Because H–$\ddot{Si}$–SiH_3 is nearly isoenergetic with $H_2Si{=}SiH_2$, it is not obvious which structure would be favored for hexasilabenzene, Si_6H_6. Whereas proponents of aromaticity may take heart from the theoretical calculations that show the hexasilabenzene structure to be 34 kcal/mol more stable than the tris(silylene) form, they will be reminded of how little aromaticity has to do with the magic of six π electrons by the revelation that the hexasilaprismane is calculated to be 14 kcal/mol more stable yet.

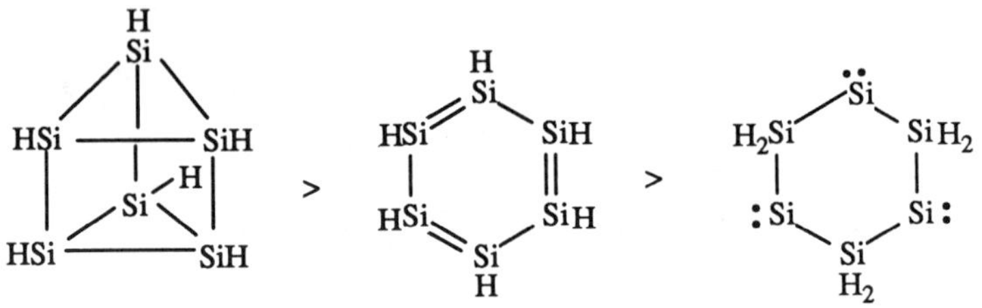

most stable > > least stable

Silanones ($R_2Si=O$). Despite the fact that the past few years have yielded several reports of spectroscopic observation of several silanones in frozen matrices, all we know of silanone chemistry to date comes from the rationalization of reaction product structures in terms of silanone intermediates. The sheer weight of these data, coupled with theoretical calculations, makes it difficult not to believe in the existence of free silanones. However, the reader should be aware that often there exist alternative mechanisms that do not involve the postulated silanone intermediates.

Silanones have apparently been produced by a wide variety of reactions, a few of which are shown in the following schemes:

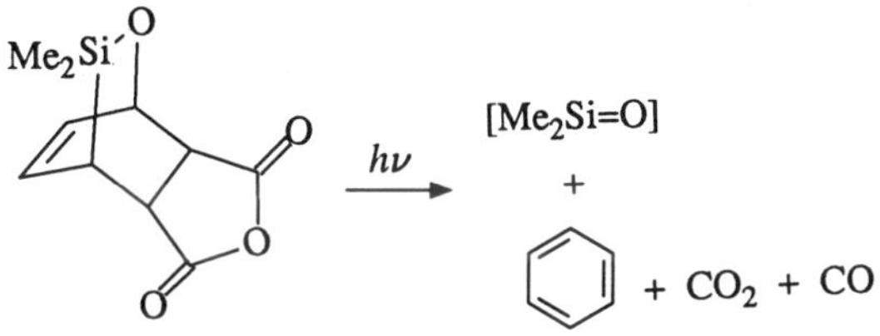

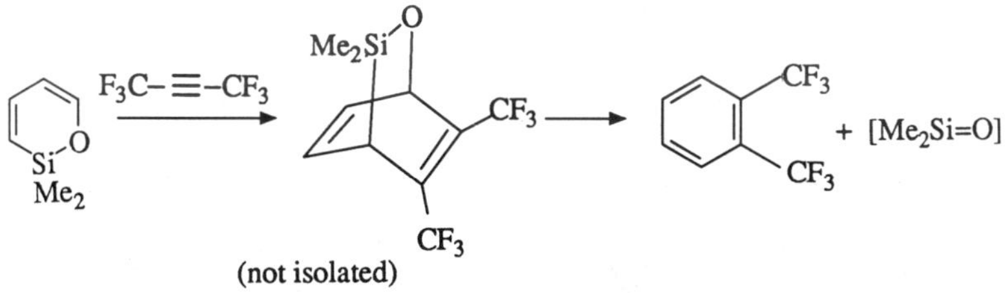

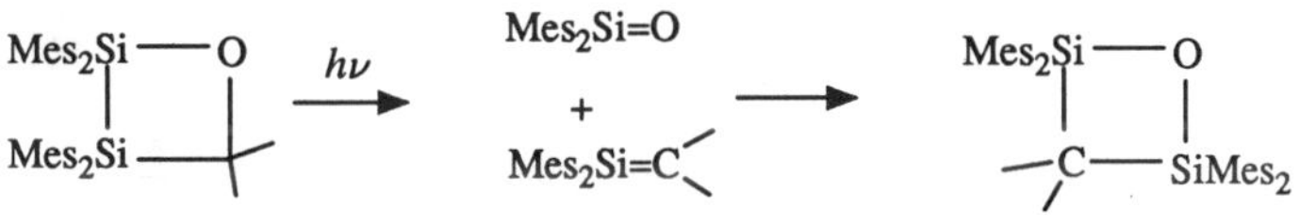

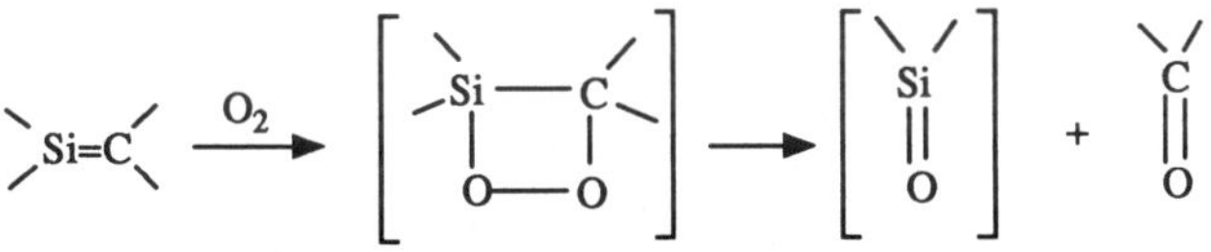

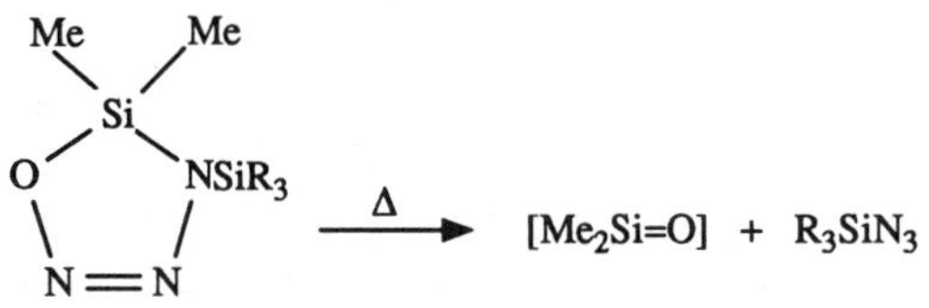

$\xrightarrow{h\nu}$ $[Me_2Si=O]$ + benzene + CO_2 + CO

$\xrightarrow{F_3C-\equiv-CF_3}$ (not isolated) $\rightarrow$ + $[Me_2Si=O]$

$Mes_2Si=O$ + $Mes_2Si=C<$

$>Si=C< \xrightarrow{O_2}$

$\xrightarrow{\Delta}$ $[Me_2Si=O]$ + R_3SiN_3

$[R_2Si{=}O]$ + $>C{=}C<$ may well not be a unimolecular process

$Me_2Si{:}$ $\xrightarrow[\text{or } R_3N=O]{Me_2S=O}$ $[Me_2Si{=}O]$ + Me_2S (or R_3N)

$\xrightarrow{\Delta}$ + $[Me_2Si{=}O]$

Very few reactions of silanones have been proposed. Left to its own devices, $H_2Si{=}O$ will dimerize with no calculated barrier and an enthalpic gain of ~106 kcal/mol. The dimer is also subject to facile insertion by silanone, and thus the observation of D_3 and D_4 products is often cited as evidence for silanone intermediacy.

$Me_2Si{=}O$ + $O{=}SiMe_2$ $\xrightarrow{E_a \simeq 0}$ (Me₂SiO)₂ $\xrightarrow{Me_2Si=O}$ (Me₂SiO)₃

Silanones are readily trapped by the Si–O, O–H, and Si–Cl bonds and apparently by not much else. For example:

$[Me_2Si{=}O]$ $\xrightarrow{Et_3SiH}$ $Et_3Si{-}OSiHMe_2$

$[Me_2Si{=}O]$ $\xrightarrow{ROH}$ $Me_2Si(OR)OH$

$[Me_2Si{=}O]$ $\xrightarrow{Me_3SiCl}$ $Me_3Si{-}O{-}SiMe_2Cl$

$[Me_2Si{=}O]$ $\xrightarrow{Me_2Si(OMe)_2}$ $MeOMe_2Si{-}O{-}SiMe_2OMe$

Silicon-Centered Radicals ($R_3Si^{\bullet}$). Silyl radical chemistry is not nearly as developed as its carbon counterpart. In striking contrast to its famous carbon analogue, $Ph_3SiSiPh_3$ does not undergo thermolysis to $Ph_3Si^\bullet$ [although Mes_6Si_2 (Mes is mesityl) ruptures the Si–Si bond homolytically]. Numerous reports of the process $R_3SiH \rightarrow R_3Si^\bullet + H^\bullet$ are most likely to involve chain abstraction, because the Si–H bond is now recognized to be stronger than the Si–C bond. Very few clean routes to $R_3Si^\bullet$ are available:

$$(R_3Si)_2Hg \xrightarrow{h\nu} [R_3Si^\bullet] \xleftarrow{h\nu} R_3Si\text{–}C(=O)\text{–}Me$$

$$R_3Si\text{–}N{=}N\text{–}SiR_3 \xrightarrow{h\nu} [R_3Si^\bullet] \xleftarrow{Hg/h\nu} R_3SiH$$

The most common route, and usually the most convenient, is radical abstraction of H from R_3SiH. For example:

$$Cl_3SiH \xrightarrow[\Delta \text{ or } h\nu]{ROOR} Cl_3Si^\bullet$$

A comparison of the properties of $^\bullet CH_3$ and $^\bullet SiH_3$ is given in Table III. Unlike carbon-centered radicals, chiral $R_3Si^\bullet$ reacts with substantial retention of configuration. From the following example,

$$H\text{—}Si(Me)(Np)\text{—}Ph \xrightarrow[\Delta]{(PhCO_2)_2} R_3Si^\bullet \xrightarrow{CCl_4} R_3SiCl^*$$

$\alpha = +33.7°$ (starting silane); $\alpha = -5.3°$ (product)

(∴ >90% retention of configuration)

(Np is naphthyl.)

it may be concluded that the rate of inversion of silyl radicals is slow compared with the rate of Cl abstraction from CCl_4. Other reactions include

$$[R_3Si^\bullet] \xrightarrow{\text{benzene}} C_6H_5\text{—}SiR_3$$

$$[R_3Si^\bullet] \xrightarrow{R'_2C{=}O} R_3Si\text{–}O\text{–}\dot{C}R'_2$$

$$[R_3Si^\bullet] \xrightarrow{-\beta\text{-H}} R_2Si{=}C{<}$$

$$[R_3Si^\bullet] \xrightarrow{>C{=}C<} R_3Si\text{–}\overset{|}{\underset{|}{C}}\text{–}\dot{C}{<}$$

Table III. Properties of C and Si Radicals

Radical	*% s-Character of Unpaired Electron*	*Bond Angle*	*Structure*
$\cdot CH_3$	0	120°	planar
$\cdot SiH_3$	21	111°	pyramidal

Silyl Anions ($R_3Si:^-$). For years aryl substitution was thought to be necessary for the formation of silyl alkali compounds. This is not the case, and trialkylsilyl alkali metal compounds are now readily available. The most general and convenient method of generation is disilane cleavage. For example:

$$Me_3Si{-}SiMe_3 \xrightarrow[HMPA]{MeLi} Me_4Si + Me_3SiLi$$

In the previous reaction, HMPA is hexamethylphosphoramide.

The reactions of R_3Si^- are usually similar to those of R_3C^-.

$$R_3SiLi \xrightarrow{CO_2} R_3SiCO_2H$$

$$R_3SiLi \xrightarrow{Ph_2C{=}CH_2} Ph_2CHCH_2SiR_3$$

$$R_3SiLi \xrightarrow{Me_2C{=}O} Me_2C(OH){-}SiR_3$$

$$R_3SiLi \xrightarrow{EtCl} R_3SiEt$$

$$R_3SiLi \xrightarrow{\text{cyclohexenone}} \text{3-}(R_3Si)\text{cyclohexanone}$$

Trialkylsilyl alkali metal compounds are readily oxidized by a variety of aromatic compounds, ketones, amides, and anhydrides. Thus electron transfer is likely to be the first step of the majority of R_3Si^- reactions.

$$Me_6Si_2 \xrightarrow[HMPA]{NaOMe} Me_3SiNa \xrightarrow{\text{naphthalene}} Me_3Si^{\bullet} + [\text{naphthalene}]^{\bullet -}$$

Of considerable synthetic potential are alkylpentafluorosilicates:

$$R{-}SiCl_3 \xrightarrow{KF} [R{-}SiF_5]^{=}\ K_2^{+}$$

$$PhCH{=}CHBr \xleftarrow{NBS} PhCH{=}CHSiF_5^{=} \xrightarrow{Cu(SCN)_2} PhCH{=}CHSCN$$

(NBS is *N*-bromosuccinimide.)

Silylenium Ions (R_3Si^+). Because silicon is considerably more electropositive than carbon, R_3Si^+ can be expected to be significantly more stable than R_3C^+. Indeed R_3Si^+ is always encountered in the mass spectra of organosilicon compounds, and theoretical calculations support the stability of R_3Si^+.

$$H_3Si\text{–}\overset{+}{C}H_2 \rightarrow H_2\overset{+}{Si}\text{–}CH_3 \text{ (thus, } H_3Si\text{–}CH_2^+ \text{ is incapable of existence!)}$$

$$E_a \text{ (activation energy)} \simeq 0;\ \Delta E = 49 \text{ kcal/mol (from calculations)}$$

Only in 1982 was apparently solid evidence presented for the direct observation of a silylenium ion:

$$(n\text{-PrS})_3SiH + Ph_3C^+ClO_4^- \rightarrow (n\text{-Pr})_3Si^+ClO_4^- \text{ (1982)}$$

$$Ph_3SiH + PhC^+ClO_4^- \rightarrow Ph_3Si^+ClO_4^- \text{ (1986)}$$

At the time of this writing, this area remains controversial. On the one hand, we have ^{13}C and ^{1}H NMR spectra, along with conductance studies, in support of the presence of free silylenium ions in solution, whereas ^{29}Si and ^{35}Cl NMR data are used to support the claim that Ph_3SiClO_4 is simply a covalent ester in solution, as it has long been known to be in the solid state.

Molecular Rearrangements

Organosilicon compounds are, in general, quite prone to intramolecular rearrangements. The majority of these rearrangements are actually intramolecular nucleophilic substitutions on silicon and can be viewed as

$$R_3Si \quad \ddot{X} \longrightarrow R_3\overset{-}{Si}\text{–}\overset{+}{X} \longrightarrow R_3Si\text{–}X$$

Thus, most rearrangements proceed with retention on silicon. We will, in this section, simply try to provide at least one example each of the most common organosilicon rearrangements.

1,2 Rearrangements. ***Dyatropic (Two Concurrent Group Migrations).***

$$R_2C(SiMe_3)\text{–}O\text{–}Si(CD_3)_3 \underset{E_a\,=\,31\text{ kcal}}{\overset{140\text{–}170\ ^\circ C}{\rightleftharpoons}} R_2C(Si(CD_3)_3)\text{–}O\text{–}SiMe_3$$

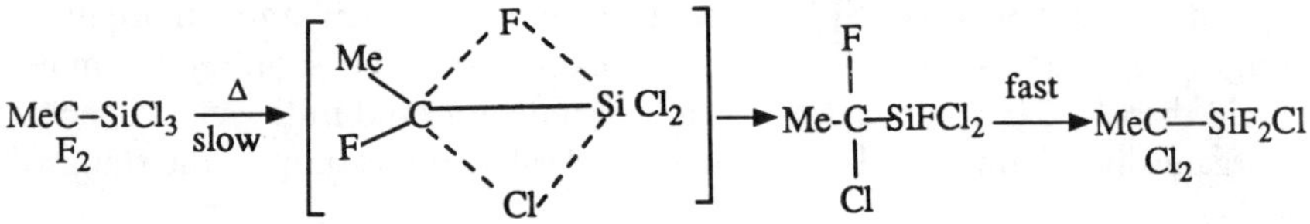

$$MeCF_2{-}SiCl_3 \xrightarrow[\text{slow}]{\Delta} [\text{Me, F–C}\cdots\text{F, Cl}\cdots\text{SiCl}_2] \rightarrow Me{-}CFCl{-}SiFCl_2 \xrightarrow{\text{fast}} MeCCl_2{-}SiF_2Cl$$

Anionic.

$$R_3Si{-}CR'_2(OH) \xrightarrow{\text{Na, K, RLi, }R_3N\text{, etc.}} HCR_2{-}OSiR'_3$$

($\Delta S^{\ddagger}$ = −35–40 eu; therefore, concerted)

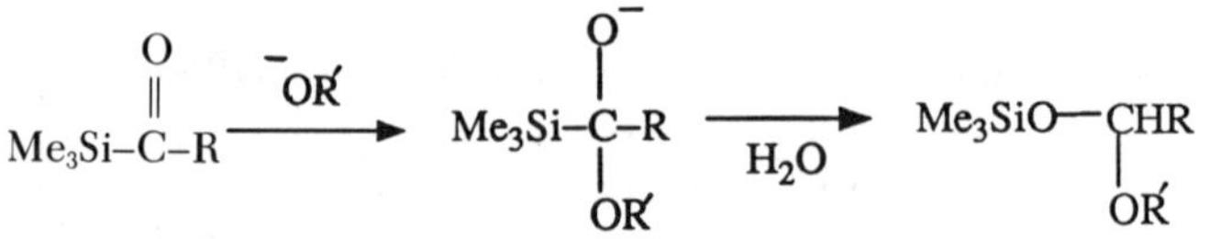

$$Me_3Si{-}C(=O){-}R \xrightarrow{^{-}OR'} Me_3Si{-}C(O^-)(OR'){-}R \xrightarrow{H_2O} Me_3SiO{-}CHR(OR')$$

Cationic.

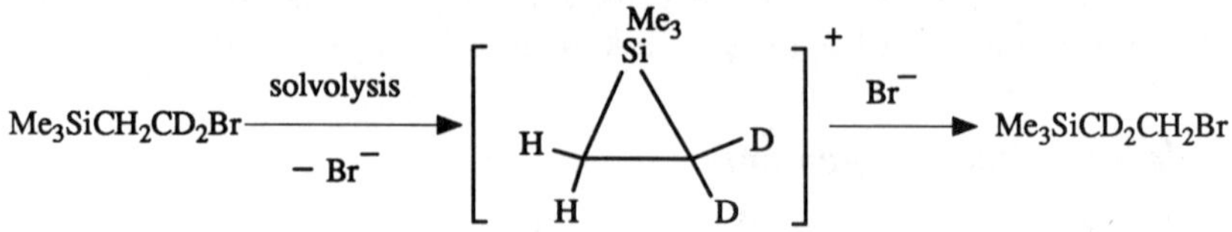

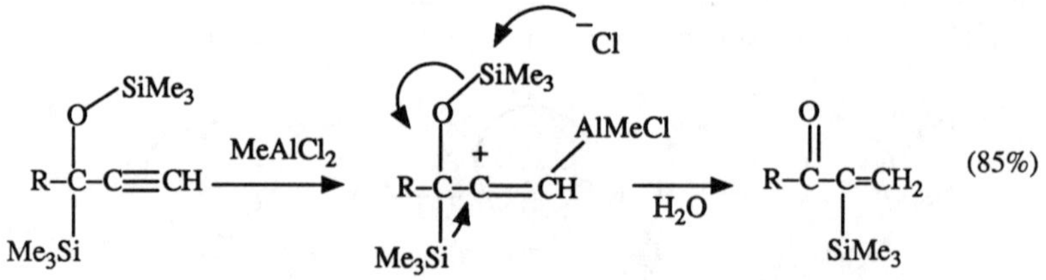

$$Me_3SiCH_2CD_2Br \xrightarrow[-\,Br^-]{\text{solvolysis}} [\text{bridged } Me_3Si^+ \text{ (H, H, D, D)}]^+ \xrightarrow{Br^-} Me_3SiCD_2CH_2Br$$

$$R{-}C(OSiMe_3)(Me_3Si){-}C{\equiv}CH \xrightarrow{MeAlCl_2} R{-}C(OSiMe_3)(Me_3Si){-}C^+{=}CH(AlMeCl) \;\; (^-Cl) \xrightarrow{H_2O} R{-}C(=O){-}C(SiMe_3){=}CH_2 \quad (85\%)$$

Involving Carbenes.

$$\text{2-(SiMe}_3\text{)-4,5-dihydrofuran} \xrightarrow{\Delta} \text{carbene (H, O–SiMe}_3\text{)} \rightarrow Me_3SiO(CH_2)_2C{\equiv}C{-}H \quad (49\%)$$

$$R_3Si{-}C(=O)R' \underset{\text{dark}}{\overset{h\nu}{\rightleftharpoons}} :C(OSiR_3){-}R' \xrightarrow{R''OH} R_3SiO{-}CHR'(OR'')$$

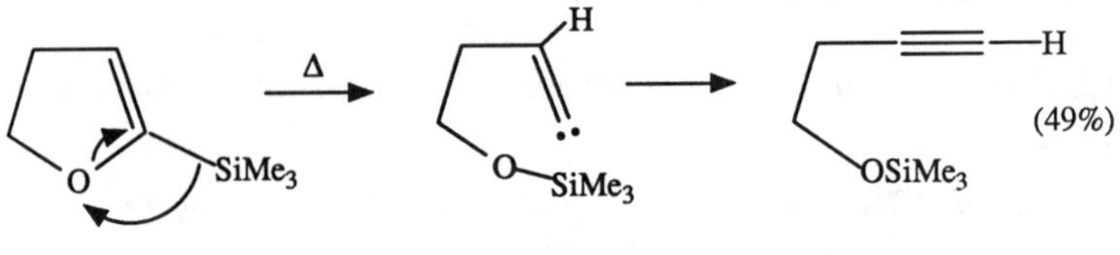

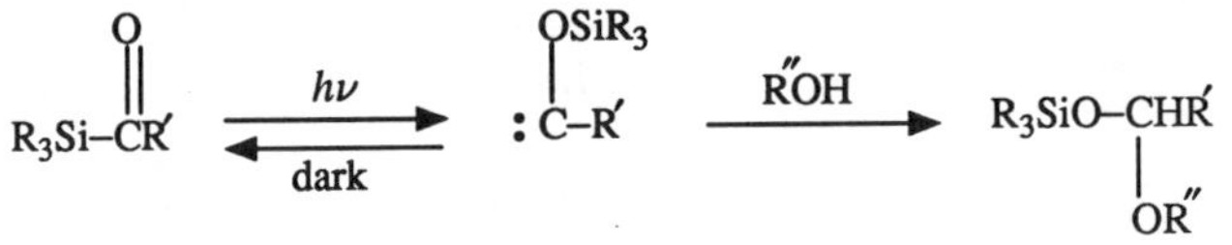

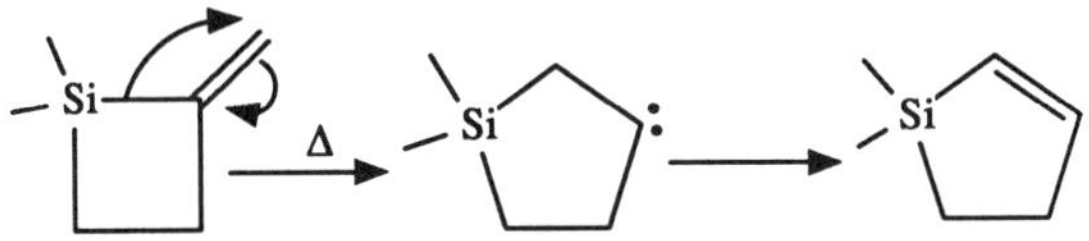

$$Me_2Si(\text{methylenecyclobutane}) \xrightarrow{\Delta} Me_2Si(\text{cyclopentylidene carbene}) \longrightarrow Me_2Si(\text{cyclopentene})$$

Free Radical.

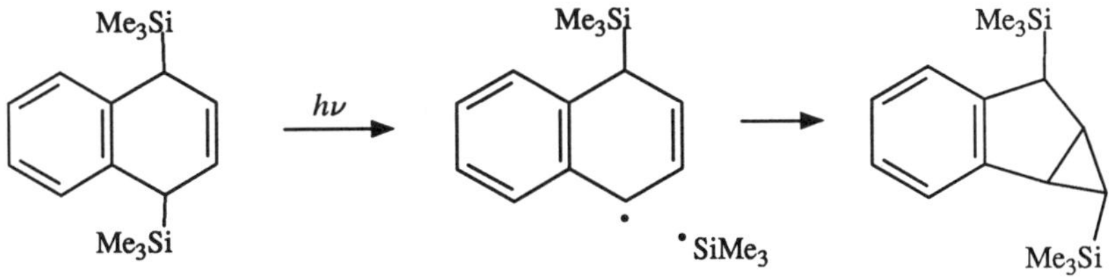

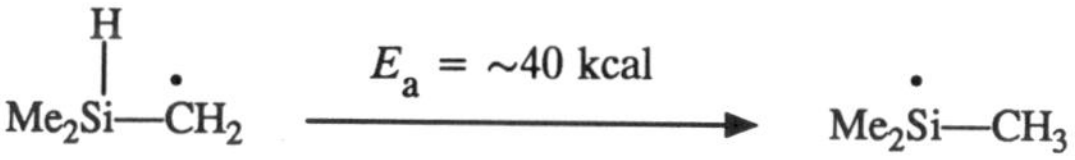

$$\text{1,4-bis}(Me_3Si)\text{-1,4-dihydronaphthalene} \xrightarrow{h\nu} \text{radical} + {}^{\bullet}SiMe_3 \longrightarrow \text{Me}_3\text{Si-substituted benzobicyclic product}$$

$$Me_2Si(H)\text{—}\dot{C}H_2 \xrightarrow{E_a\ =\ \sim 40\ \text{kcal}} Me_2\dot{S}i\text{—}CH_3$$

Miscellaneous.

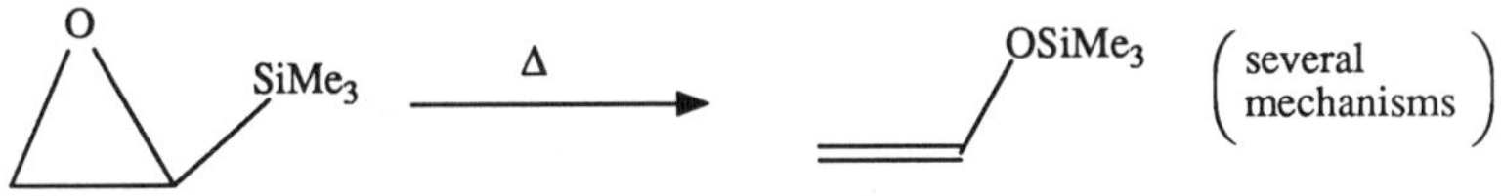

$$\text{(trimethylsilyl)oxirane} \xrightarrow{\Delta} CH_2{=}CH\text{—}OSiMe_3 \quad \left(\text{several mechanisms}\right)$$

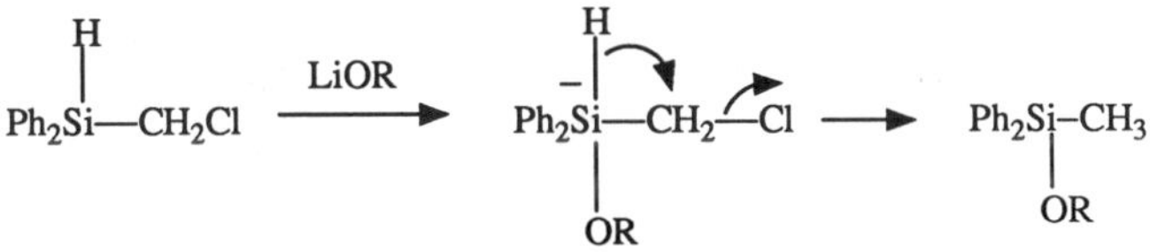

$$Ph_2Si(H)\text{—}CH_2Cl \xrightarrow{LiOR} Ph_2\overset{-}{Si}(H)(OR)\text{—}CH_2\text{—}Cl \longrightarrow Ph_2Si(OR)\text{—}CH_3$$

$$CpFe(CO)_2\text{—}CH_2SiMe_2H \xrightarrow{h\nu} CpFe(H)(CO)(\eta^2\text{-}CH_2SiMe_2) \longrightarrow CpFe(CO)_2\text{—}SiMe_3$$

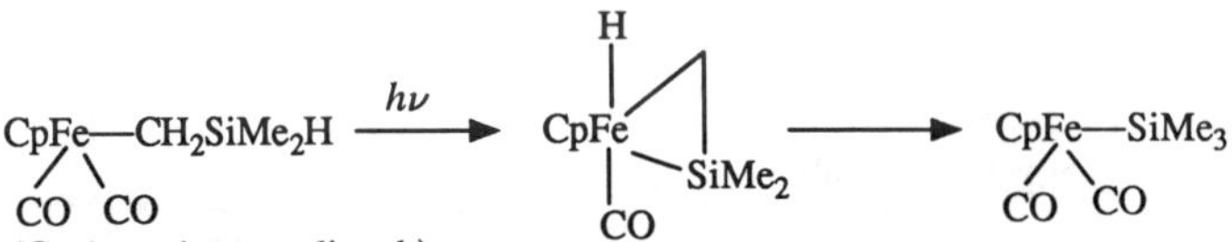

(Cp is cyclopentadienyl.)

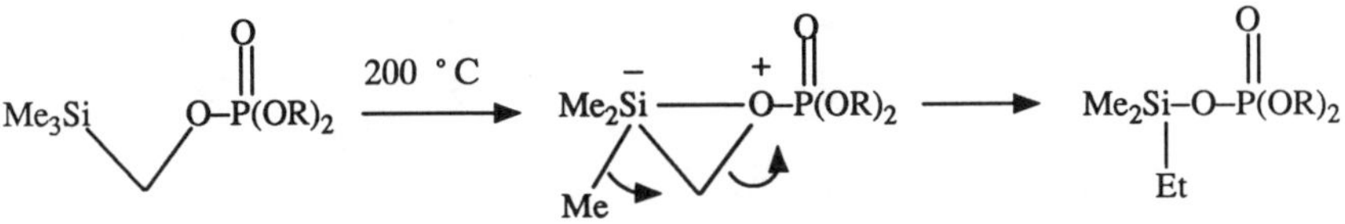

$$Me_3Si\text{—}CH_2CH_2\text{—}O\text{—}P(O)(OR)_2 \xrightarrow{200\ ^\circ C} Me_2\overset{-}{Si}(Me)\text{—}\overset{+}{O}\text{—}P(O)(OR)_2 \longrightarrow Me_2Si(Et)\text{—}O\text{—}P(O)(OR)_2$$

1,3 Rearrangements. ***Between C and O.***

R_3Si CH$_2$—C(=O)—R′ —80–175 °C→ $R_3\bar{S}i$—O (four-membered ring, $^+$ R′) → R_3Si–O–C(R′)=CH$_2$

H_2C=C(OMe)(OSiMeCl$_2$) —70 °C→ Cl_2MeSi–CH_2–C(=O)–OMe

(without Cl on Si requires R_3SiI catalyst)

$(Me_3Si)_3C$—Si(R)(F)—O$^-$ —(1) 1,3 migration; (2) H_2O→ $(Me_3Si)_2C(H)$—Si(R)(F)—$OSiMe_3$

Ph$_4$ (bicyclic Si, O, O) —1,3 migration→ (O, Si, O) —2+2→ Ph, Ph, Ph, Ph (cyclobutene fused with O–SiMe$_2$–O)

(vinyl)Si(Me$_2$)–OSiMe$_3$ —t-BuLi→ Bu—CH$_2$–$\bar{C}$H–Si(Me$_2$)–O–SiMe$_3$ —H_2O→ Bu–CH$_2$–CH(SiMe$_3$)–SiMe$_2$OH

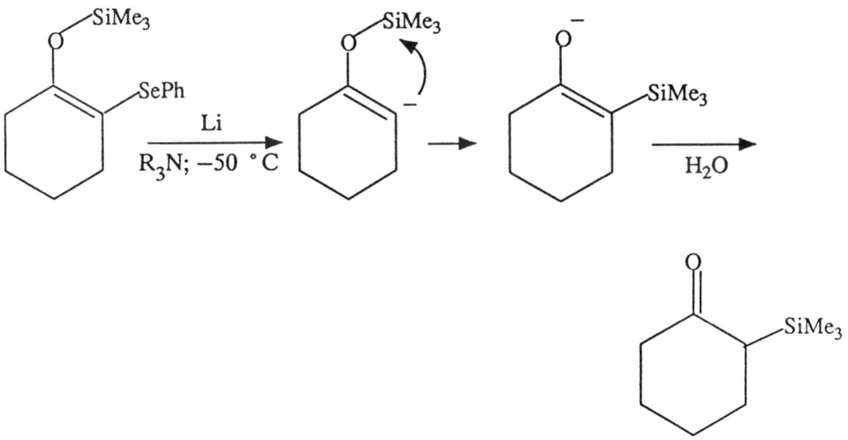

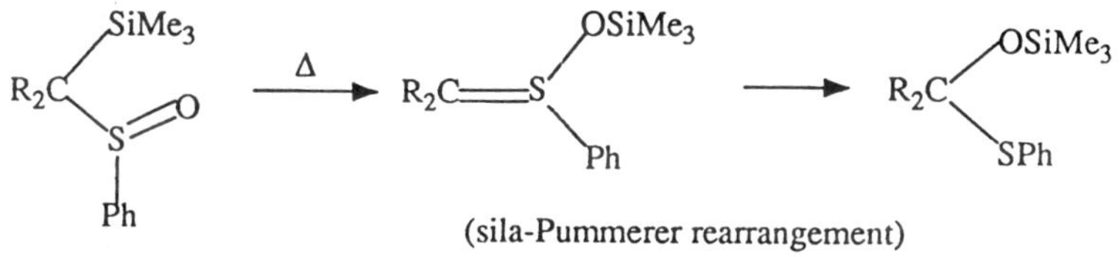

SiMe3
O
SePh
Li
R3N; –50 °C
O
SiMe3
O
SiMe3
H2O
O
SiMe3

SiMe3
R2C
S
O
Ph
Δ
OSiMe3
R2C
S
Ph
R2C
OSiMe3
SPh

(sila-Pummerer rearrangement)

Between N and O.

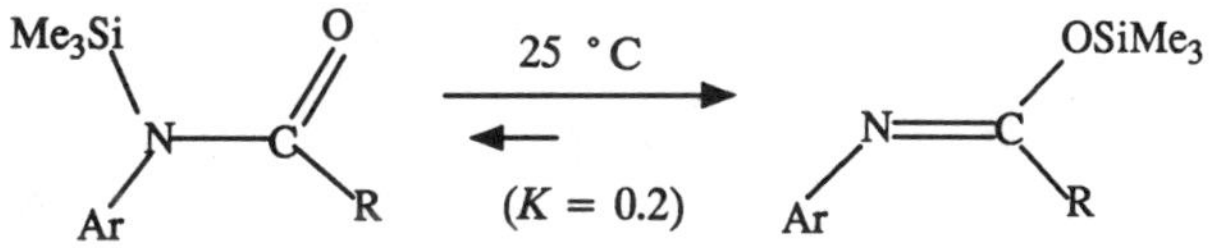

Me3Si
O
N
C
Ar
R
25 °C
(K = 0.2)
OSiMe3
N
C
Ar
R

Between C and N.

R′
R3Si
N
N
0–100 °C
K dependent on R′
R′ = Me, K < 0.02
R′ = Ph, K > 50
R′
N
N
SiR3
(Me3Si)3C–C≡N
150 °C
K > 50
(Me3Si)2C=C=N
SiMe3

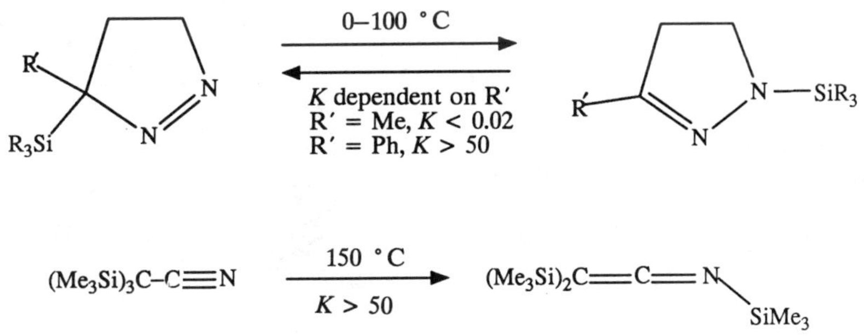

Between C and C.

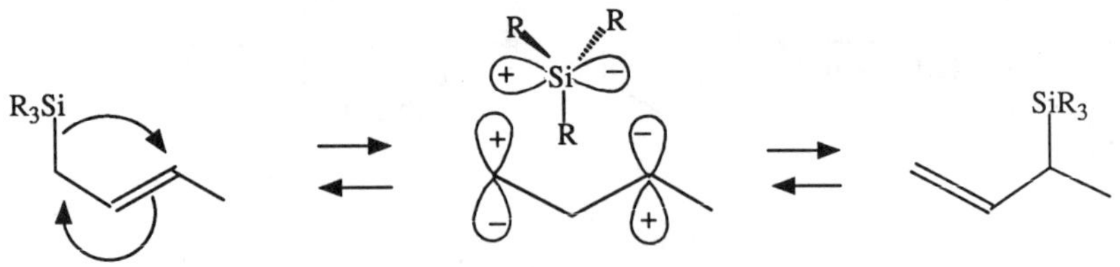

R_3Si R R Si R SiR_3

(1,3- C to C is the only rearrangement proceeding with inversion on Si)

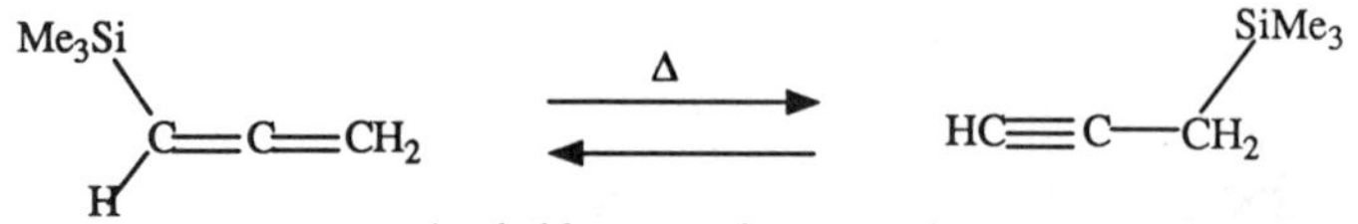

Me_3Si $SiMe_3$

C=C=CH_2 Δ HC≡C—CH_2

H

(probably a stepwise process)

1,4 Rearrangements. ***Between C and O.***

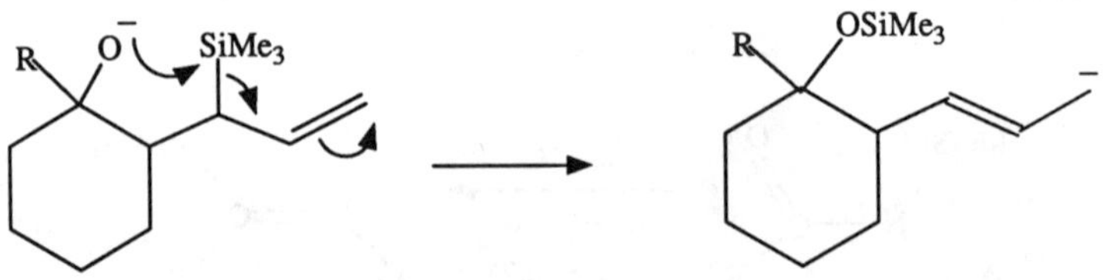

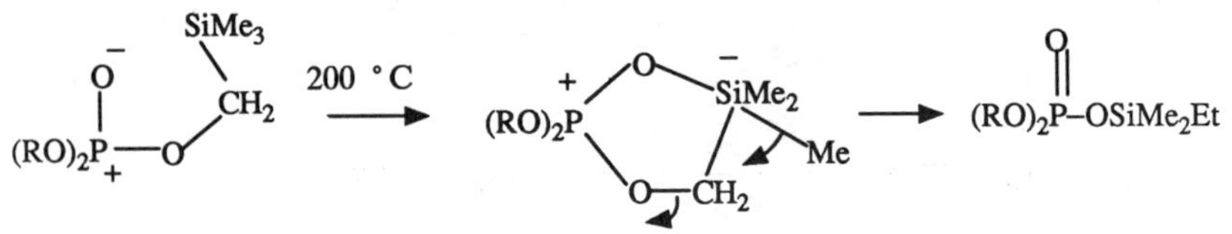

R O $SiMe_3$ R $OSiMe_3$

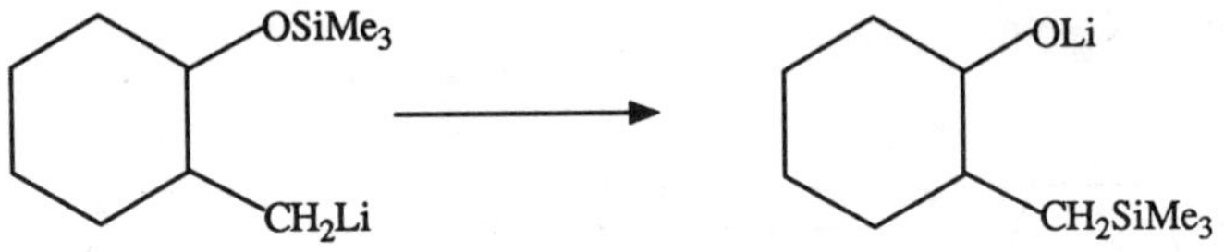

$OSiMe_3$ OLi

CH_2Li CH_2SiMe_3

$SiMe_3$

O CH_2 200 °C (RO)$_2$P O $SiMe_2$ Me O (RO)$_2$P–$OSiMe_2Et$

(RO)$_2$P—O O—CH_2

Between O and O.

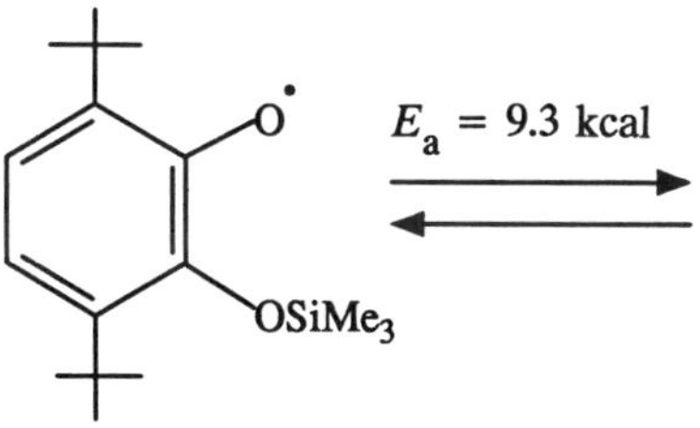

E_a = 9.3 kcal

OSiMe$_3$

OSiMe$_3$

Between C and C.

SiMe$_3$

S

S

Bu

SiMe$_3$

S

S

Bu

MeI

SiMe$_3$

S

S

Me

Bu

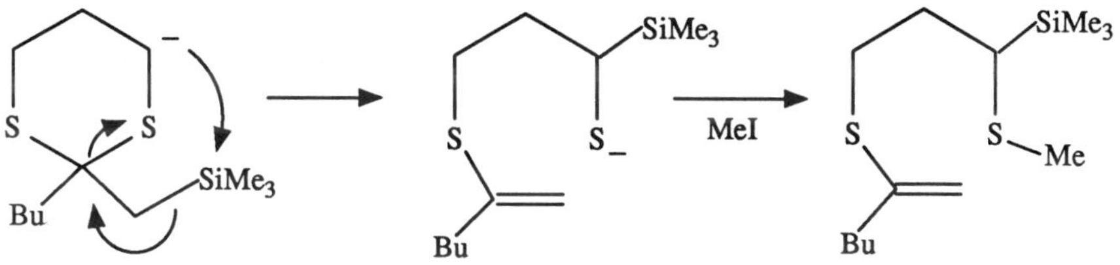

1,5 Rearrangements. *Between O and O.*

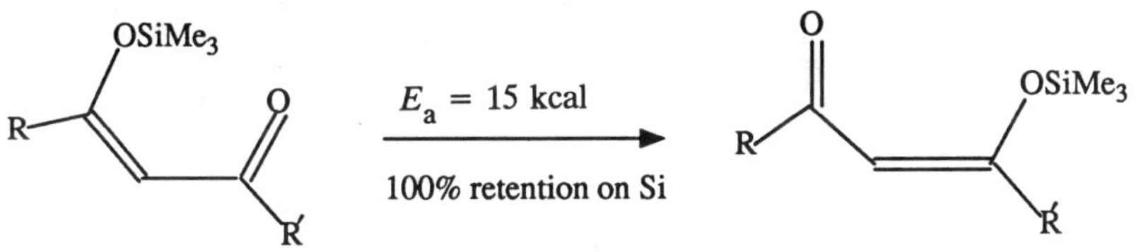

OSiMe$_3$

R

O

R′

E_a = 15 kcal

100% retention on Si

O

R

OSiMe$_3$

R′

Between Si and C.

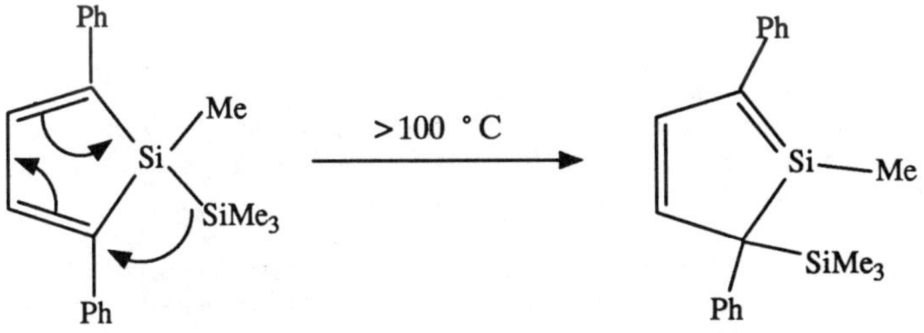

Ph

Me

Si

SiMe$_3$

Ph

>100 °C

Ph

Si—Me

SiMe$_3$

Ph

Between C and C.

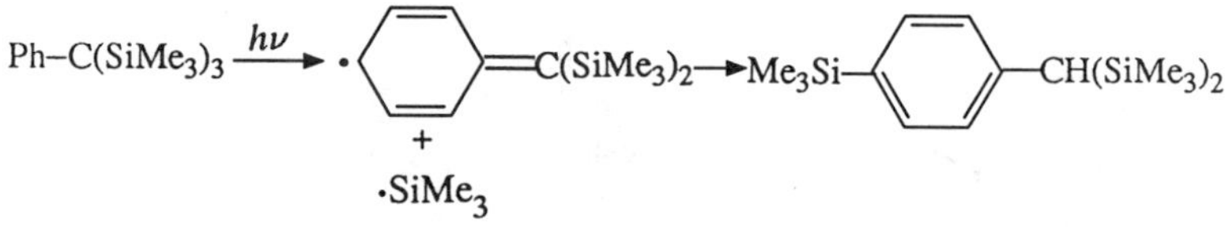

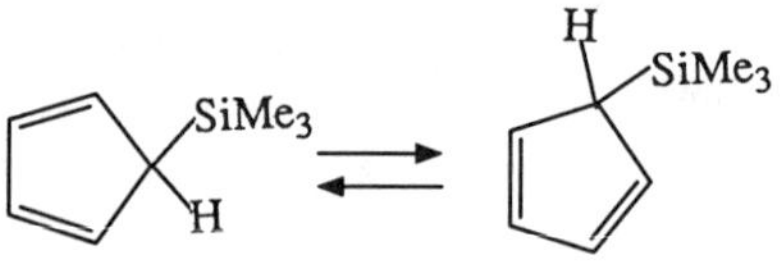

The fact that Me_3Si migrates in a 1,5-sigmatropic fashion on the cyclopentadiene ring 10^6 times faster than does hydrogen is often cited as dramatic evidence for its superior migratory aptitude. However, on the cycloheptatriene ring, 1,5-migration of H is faster than that of Me_3Si. The explanation lies in the nature of the polar transition state.

Acknowledgment

The support of this work by the National Science Foundation is gratefully acknowledged by Barton.

References

1. Onopchenko, A.; Sabourin, E. T. *J. Org. Chem.* **1987,** *52,* 4118.
2. Sheldon, J. C.; Hayes, R. N.; Bowie, J. H. *J. Am. Chem. Soc.* **1984,** *106,* 7711.
3. Dieters, J. A.; Holmes, R. R. *J. Am. Chem. Soc.* **1987,** *109*, 1692.

Bibliography

General Sources for Organosilicon Chemistry.

1. Eaborn, C. *Organosilicon Compounds;* Xerox Microfilms: Ann Arbor, MI, 1976. This is the original bible of organosilicon chemistry.
2. MacDiarmid, A. G. *The Bond to Carbon,* Part I; and MacDiarmid, A. G. *Organometallic Compounds of the Group IV Elements,* Vol. 1, Part I. These excellent comprehensive books are no longer available. I have lost mine, so if you see an extra copy lying around . . . .
3. *Comprehensive Organometallic Chemistry;* Pergamon: New York, 1982, Vol. 2. Chapters on "Organopolysilanes", Carbocyclic Silanes", "Organosilanes", and "Silicon Compounds in Organic Synthesis" (Vol. 7) by R. West, T. J. Barton, D. A. Armitage, and P. D. Magnus, respectively.
4. Bazant, V.; Chvalovsky, V.; Rathousky, J. *Organosilicon Compounds.* This extremely valuable tabulation of pertinent literature references and physical properties of every known organosilicon compound has appeared four times since 1965. It is not published or available outside of the Soviet block. I (T. J. B.) have not found a copy of the fourth serial (1980).

5. *Silicon Compounds—Register and Review*. This reference is actually the catalog of Petrarch Systems (Bartram Road, Bristol, PA 19007), which has the largest commercial offerings of organosilicon compounds. However, much of this book has about as much resemblance to a catalog as the Sears–Roebuck volume does to *Kama Sutra*. Since 1979, each issue has contained a variety of excellent, timely reviews of various aspects of organosilicon chemistry. Eminently readable!

The Nature of Silicon Bonding.

• $(p\text{–}d)\pi$ and $\sigma\text{–}\pi$ Bonding

1. Egorochkin, A. N. *Russ. Chem. Rev. (Engl. Transl.)* **1984,** *53*(5), 445.
2. Janes, N.; Oldfield, E. *J. Am. Chem. Soc.* **1986,** *108*, 5743.
3. Kwart, H.; King, K. *d-Orbitals in the Chemistry of Silicon, Phosphorus, and Sulfur;* Springer: New York, 1977.
4. Sakurai, H. *J. Organomet. Chem.* **1980,** *200*, 261.

• β-Silyl Stabilization of Cations

1. Lambert, J. B.; Wang, G. T.; Finzel, R. B.; Teramura, D. H. *J. Am. Chem. Soc.* **1987,** *109*, 7838.
2. Apeloig, Y.; Arad, D. *J. Am. Chem. Soc.* **1985,** *107*, 5285.

• α,β-Silyl Stabilization of Radicals

1. Jackson, R. A.; Ingold, K. U.; Griller, D.; Nazran, A. S. *J. Am. Chem. Soc.* **1985,** *107*, 208.
2. Davidson, I. M. T. et al. *Organometallics* **1987,** *6*, 644.

• Bond Strengths

1. Walsh, R. *Acc. Chem. Res.* **1981,** *14*, 246. This is the current bible of silicon bond strengths.
2. Kanabus-Kaminska, J. M.; Hawari, J. A.; Griller, D.; Chatgilialoglu, C. *J. Am. Chem. Soc.* **1987,** *109*, 5267. The authors use a promising photoacoustic technique to get some surprising results.

Hydrosilation.

1. Speier, J. L. *Adv. Organomet. Chem.* **1979,** *17*, 407. This review of hydrosilation is written by the master himself.
2. Ojima, I.; Kogure, T. *Reviews on Silicon, Germanium, Tin, and Lead Compounds;* Vol. 5, No. 1, 7-66, 1981. This reference gives a review of hydrosilation with metal complexes other than Speier's catalyst.

Reactive Intermediates.

1. Raabe, G.; Michl, J. *Chem. Rev.* **1985,** *85*, 419. This is the most recent comprehensive review of π-bonded silicon. We understand that an updated version is in the works.
2. Gaspar, P. P. In *Reactive Intermediates;* Jones, M.; Moss, R., Eds.; Wiley: New York. Keeping up with silylene chemistry is made easy by these critical comprehensive reviews every two or three years by Gaspar.
3. Sakurai, H. In *Free Radicals;* Kochi, J., Ed.; Wiley: New York, 1973; Vol. 2. This paper is a truly excellent review of silyl radical chemistry.

4. Wilt, J. W. In *Reactive Intermediates;* Abramovitch, R. A., Ed.; Plenum: New York, 1983; Vol. 3. This excellent review of silyl radicals is, to our knowledge, the most recent.
5. Davis, D. D.; Gray, C. E. *Organomet. Chem. Rev. A* **1970,** *6,* 283. This article is a good, but obviously dated, review of silyl anions through about 1969.
6. Colvin, E. *Silicon in Organic Synthesis;* Butterworth: London, 1980; Chapter 11. The article is only six pages but gives a good update on silyl anions.

Molecular Rearrangements.

1. Brook, A. G.; Bassindale, A. R. In *Organic Chemistry;* DeMayo, P., Ed.; Academic: New York, 1980; Vol. 2, Essay No. 9. This article is the only comprehensive review of organosilicon rearrangements and suffers only from being a bit dated and its total exclusion of the rearrangements of organosilicon reactive intermediates.

RECEIVED for review May 27, 1988. ACCEPTED revised manuscript March 20, 1989.

2

Silicon-Containing Polymers

James E. Mark

Department of Chemistry and the Polymer Research Center, University of Cincinnati, Cincinnati, OH 45221

The major categories of homopolymers and copolymers are discussed. These include (1) linear siloxane polymers, $[-SiRR'O-]_n$ (with various alkyl and aryl R and R' side groups); (2) sesquisiloxane polymers possibly having a ladder structure; (3) siloxane–silarylene polymers, $[-Si(CH_3)_2OSi(CH_3)_2(C_6H_4)_m-]_n$ (in which the phenylenes are either meta *or* para *substituted); (4) silalkylene polymers, $[-Si(CH_3)_2(CH_2)_m-]_n$; (5) random and block copolymers and blends of some of polymers 1–4; (6) polysilanes and polysilylenes, $[-SiRR'-]_n$; and (7) polysilazanes, $[-SiRR'NR''-]_n$. The structure, flexibility, transition temperatures, permeability, and other physical properties are reviewed. Applications, including uses as high-performance fluids, elastomers, coatings, surface modifiers, separation membranes, photoresists, soft contact lenses, body implants, and controlled-release systems, are discussed. Also of interest are the conversions of silicon-containing materials to novel reinforcing fillers, to ceramics by the sol–gel technique, and to high-performance fibers by controlled thermolyses.*

SEMIINORGANIC POLYMERS typically have inorganic elements making up at least part of the chain backbone, with organic groups as side chains (*1–3*). In the polymers discussed in this chapter, silicon atoms contribute the inorganic character and are present either alone in the backbone (silanes) or with atoms of oxygen (siloxanes), carbon (silalkylenes and silarylenes), or nitrogen (silazanes).

Of these, the siloxanes or silicone polymers have been studied the most and are also of the greatest commercial importance (*3–8*). Increasingly important, at least for some specialized applications, are the polysilanes (*3, 9*),

0065–2393/90/0224–0047$06.50/0

and the polysilazanes (*10, 11*) are probably not far behind. In the more general category of semiinorganic polymers, the only other commercially important class at the present time are the non-silicon-containing polyphosphazenes $[-PRR'N-]_n$ (*3, 12*).

This chapter is restricted to the silicon-containing polymers, a very broad field in its own right, and emphasizes the physical properties and applications of these materials.

Siloxane-Type Polymers

Preparation. Siloxane-type polymers are generally prepared by a ring-opening polymerization of a trimer or tetramer (*11, 13*)

$$\overline{(SiR_2O)_{3 \text{ or } 4}} \rightarrow [-SiR_2O-]_n \quad (1)$$

in which R can be alkyl or aryl and n is the degree of polymerization. In this reaction, macrocyclic species are formed to the extent of 10–15 wt %. The lower molecular weight products are generally stripped from the polymer before it is used in a commercial application. Low-molecular-weight macrocyclic species are also of interest from a more fundamental point of view in two respects. First, the extent to which they occur can be used as a measure of chain flexibility (*14*). Second, the separated species can be used to test theoretical predictions of the differences between otherwise identical cyclic and linear molecules (*15*).

In some cases, an end blocker such as $YR'SiR_2OSiR_2R'Y$ is used to give reactive $-OSiR_2R'Y$ chain ends (*16*). Some of the uses of these materials are described in the section on reactive homopolymers.

Polymerization of nonsymmetrical cyclic molecules gives stereochemically variable polymers, $[-SiRR'O-]_n$, analogous to the totally organic vinyl and vinylidene polymers $[-CRR'CH_2-]_n$. In principle, these polymers can be prepared in the same stereoregular forms (isotactic and syndiotactic) that have been achieved for some of their organic counterparts (*1, 17*). Unfortunately, the stereoregular forms have not been prepared, and only the stereoirregular form (atactic) has been obtained. Unlike the other two stereochemical forms, the atactic form is inherently noncrystallizable.

Polymerization of mixtures of monomers, such as $\overline{(SiR_2O)_m}$ with $\overline{(SiR'_2O)_m}$, can be used to obtain random copolymers. The copolymers are also generally highly irregular but in the chemical rather than stereochemical sense. Correspondingly, they also show little, if any, crystallizability. Homopolymerizations and copolymerizations of this type are discussed in detail elsewhere (*13, 17*).

Homopolymers. ***Flexibility.*** The most important siloxane polymer is poly(dimethylsiloxane) (PDMS), $[-Si(CH_3)_2O-]_n$ (*8*). PDMS is also one of the most flexible chain molecules known, both in the dynamic sense and in the equilibrium sense. Dynamic flexibility refers to a molecule's ability to change spatial arrangements by rotations around its skeletal bonds. The more flexible a chain is in this sense, the more it can be cooled before the chains lose their flexibility and mobility and the polymer becomes glassy. Thus, chains with high dynamic flexibility generally have very low glass transition temperatures (T_gs) (*18*). Because exposing a polymer to a temperature below its T_g generally causes it to become brittle, low values of T_g can be very advantageous, particularly in the case of fluids and elastomers.

The T_g of PDMS, ~−125 °C (*19*), is the lowest recorded for any polymer. The reasons for this extraordinary flexibility can be seen from Figure 1.

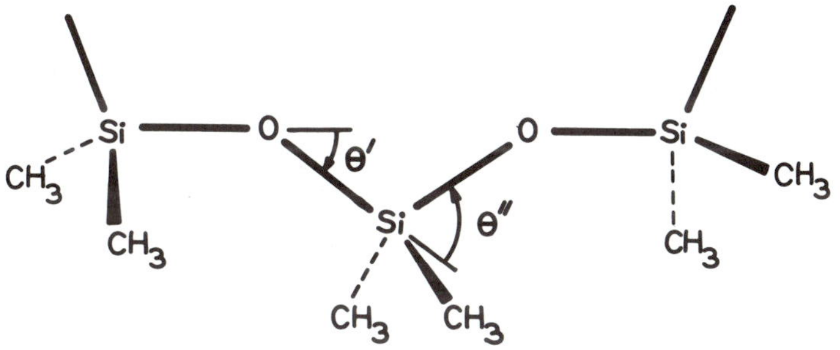

Figure 1. PDMS chain showing some structural information relevant to its high flexibility. (Reproduced with permission from reference 14. Copyright 1969 Wiley.)

First, the Si–O skeletal bond has a length (1.64 Å) that is significantly longer than that of the C–C bond (1.53 Å) central to all of organic chemistry. As a result, steric interferences or intramolecular congestion is diminished (*14*). This circumstance is true for inorganic and semiinorganic polymers in general. Almost any single bond between a pair of inorganic atoms (Si–Si, Si–C, Si–N, P–N, etc.) is longer than the C–C bond. Second, the oxygen skeletal atoms are not only unencumbered by side groups; they are as small as an atom can be and still have the multivalency needed to continue a chain structure. Third, the Si–O–Si bond angle (180 − θ′) of ~143° is much more open than the usual tetrahedral bond angle of ~110°. These three structural features have the effect of increasing the dynamic flexibility of the chain (*14*) and the equilibrium flexibility, which is the ability of a chain to be compact when in the form of a random coil. Equilibrium flexibility is generally measured indirectly by the mean-square end-to-end distance or the radius of gyration of the chain in the absence of excluded-volume effects. This type of flexibility has a profound effect on the melting point (T_m) of a polymer.

Only a polymer that is crystallizable can have a melting point, and crystallinity in a polymer has advantages. In the case of thermoplastics, the crystallites can provide the rigidity required in most molded objects. Such polymers are generally only partially crystalline (*20*), and the amorphous regions in which the crystallites are embedded have a different role. The chains in these regions have enough mobility to absorb impact energy by increasing their molecular motions. They degrade this energy harmlessly into heat and thereby increase the impact resistance of the polymer. In the case of elastomers, crystallites can act as reinforcing agents, particularly if they are strain induced, as will be described later.

Examples of ways of making a polymer less flexible, or more rigid, are shown in Figure 2 (*21*). Polymers can be made less flexible by combination of two chains into a ladder structure, insertion of rigid units such as *p*-phenylene groups into the chain backbone, or the addition of bulky side groups. The sesquisiloxane polymer shown in Figure 2 (top right) would be in the last category (*22, 23*). Attempts have been made to prepare the sesquisiloxane polymer by using the reaction shown in Scheme I (*24*), but the structure shown may not have been obtained.

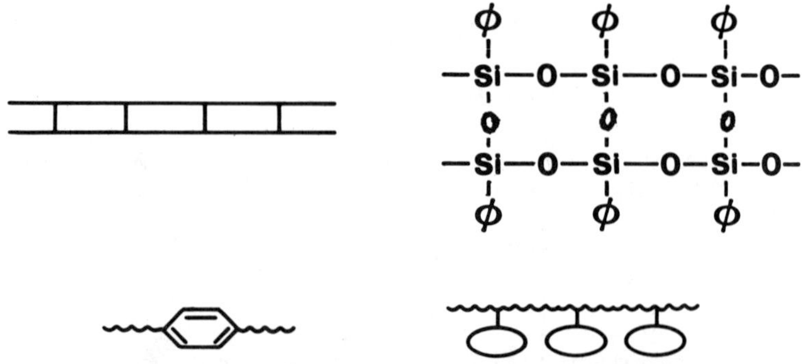

Figure 2. Ways of making a polymer more rigid. (Reproduced from reference 21. Copyright 1984 American Chemical Society.)

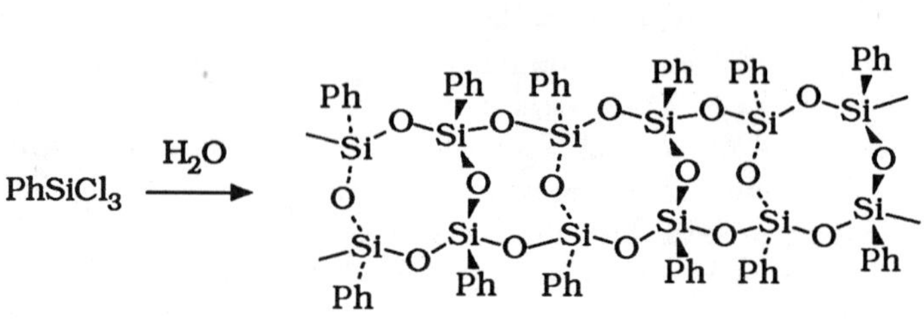

$PhSiCl_3 \xrightarrow{H_2O}$

Scheme I. Possible preparation of a sesquisiloxane ladder polymer. Ph is C_6H_5. (Reproduced with permission from reference 24. Copyright 1981 Prentice Hall.)

Figure 3 (*21*) explains the effect of rigidity on T_m. Because the Gibbs free energy change (ΔG_m) for the melting process must be zero, the melting point must be the ratio of the enthalpy of fusion to the entropy of fusion ($\Delta H_m/\Delta S_m$). When the polymer chains depart from the crystalline lattice in the melting process (Figure 3, top), they become disordered into random coils and entropy increases ($\Delta S > 0$). However, if the chains are combined into a ladder structure (Figure 3, bottom), they cannot become disordered as much. Thus, the entropy change $\Delta S' < \Delta S$ diminishes, and $T_m = \Delta H_m/\Delta S_m$ increases correspondingly. The same argument holds for the two other methods of increasing T_m.

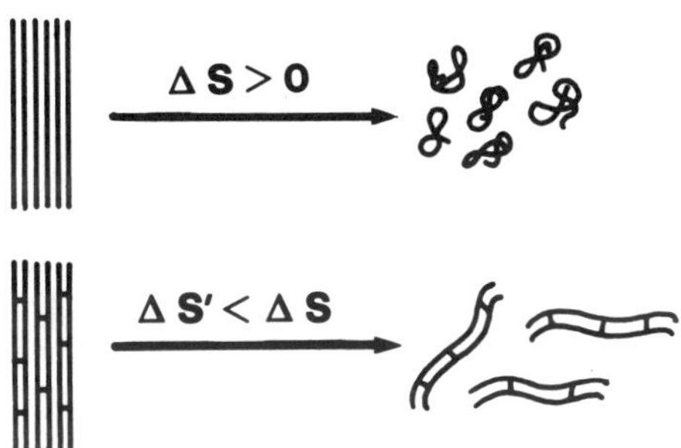

Figure 3. Increase in melting point with increase in chain rigidity. (Reproduced from reference 21. Copyright 1984 American Chemical Society.)

Another advantage of the ladder structure is its resistance to degradative chain scission. The structure will not be degraded into two shorter ladder structures except in the very unlikely event that two single-chain scissions occur directly across from one another (*25*).

Generally, structural changes that increase a chain's equilibrium rigidity also increase its dynamic rigidity and thus increase T_g. Conversely, the very high flexibility of PDMS is the origin of its low T_m (–40 °C) (*19*), as well as its very low T_g. Thus, the general effect of increased rigidity is to increase a polymer's "softening temperature", which is T_m if the polymer is crystalline and T_g (typically ~⅔T_m in kelvins) if the polymer is not crystalline (*25*).

Insertion of a silphenylene group, $-Si(CH_3)_2C_6H_4-$, into the backbone of the PDMS repeat unit yields the siloxane *meta* and *para* silphenylene polymers (Chart I) (*26*). The T_g of the *meta* polymer increases to –48 °C, but no crystallinity has been observed to date (*27*). Because the repeat unit is symmetric, crystallinity may be induced by stretching (Figure 4). The explanation for this case is the same as that given for the effect of rigidity on T_m (Figure 3), except that the chains are prevented from being completely disordered by the stretching force rather than by the structural changes in

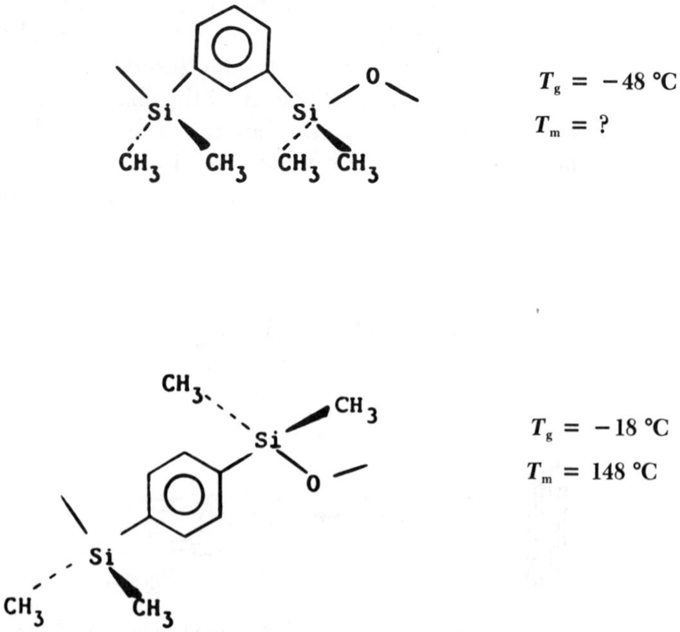

T_g = −48 °C
T_m = ?

T_g = −18 °C
T_m = 148 °C

Chart I. meta *(top) and* para *(bottom) silphenylene polymers and their transition temperatures. (Reproduced with permission from reference 26. Copyright 1988 Wiley.)*

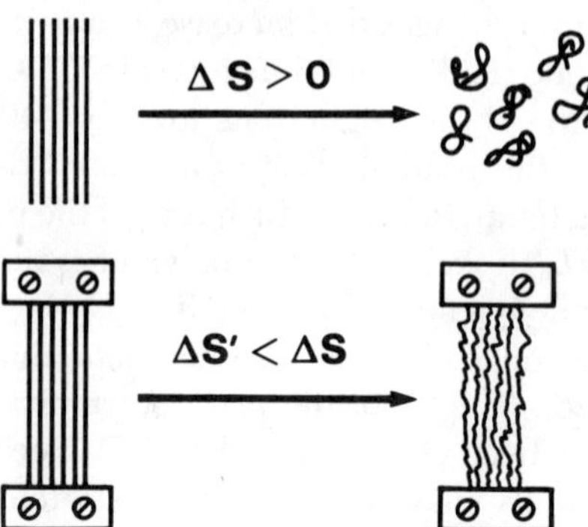

Figure 4. Increase in melting point with stretching. (Reproduced from reference 21. Copyright 1984 American Chemical Society.)

the chains. As expected, the *p*-silphenylene group has a larger rigidifying effect; the group increases T_g to −18 °C and gives rise to crystallinity, with a T_m of 148 °C. The resulting polymer is a thermoplastic siloxane. Apparently, *o*-silphenylene units have not been introduced in this way; they are probably much harder to incorporate because of steric problems. And even if *o*-silphenylene units could be incorporated, they would not be expected to have as large a rigidifying effect on the chain.

Silarylene polymers with more than one phenylene group in the repeat unit could be of considerable interest because of the various *meta* and *para* combinations that could be synthesized.

Even some flexible siloxane polymers form mesomorphic (liquid-crystalline) phases. Some illustrative data are given in Table I (*27–29*). Both poly(diethylsiloxane) and poly(di-*n*-propylsiloxane) show two crystalline modifications, as well as a mesomorphic phase. The other major class of semi-inorganic polymers, the polyphosphazenes, are also relatively flexible and show similarly interesting behavior (*12, 30*).

Table I. Transition Characteristics of Liquid-Crystalline Siloxane Polymers

Polymer	~T *(°C)*	*Nature*
$[-Si(CH_3)_2O-]_n$	−40	crystalline → isotropic
$[-Si(C_2H_5)_2O-]_n$	−60	crystalline → crystalline′
	0	crystalline′ → mesomorphic
	40	mesomorphic → isotropic
$[-Si(n\text{-}C_3H_7)O-]_n$	−55	crystalline → crystalline′
	60	crystalline′ → mesomorphic
	205	mesomorphic → isotropic

Permeability. Siloxane polymers have much higher permeability to gases than most other elastomeric materials. Therefore, these polymers have long been of interest for gas separation membranes, the goal being to vary the basic siloxane structure to improve selectivity without decreasing permeability. Some of the polymers that have been investigated in a major project (*31*) of this type are $[-Si(CH_3)RO-]_n$, $[-Si(CH_3)XO-]_n$, $[-Si(C_6H_5)RO-]_n$, $[-Si(CH_3)_2(CH_2)_m-]_n$, $[-Si(CH_3)_2(CH_2)_mSi(CH_3)_2O-]_n$, and $[-Si(CH_3)_2(C_6H_4)_mSi(CH_3)_2O-]_n$, in which R is typically an *n*-alkyl group and X is an *n*-propyl group made polar by substitution of atoms such as Cl or N. Unfortunately, structural changes that increase the selectivity generally decrease the permeability and vice versa.

In contrast to the polysiloxanes, the polysilanes $[-SiRR'-]_n$ have low permeability (*32*). This difference raises the interesting question of whether the decreased permeability is due to the absence of skeletal oxygen atoms or to the fact that polysilanes tend to be glassy rather than elastomeric at room temperature. An important comparison would therefore be between a polysilane and polysiloxane having the same side groups, at a temperature high enough for both polymers to be elastomeric or low enough for both to be glassy.

Also of great interest is the recent observation that the presence of a trimethylsilyl group, $-Si(CH_3)_3$, as a side chain in an acetylene repeat unit increases the permeability of the polymer to a value above that for PDMS! The specific polymer is poly(1-trimethylsilyl-1-propyne), and some comparisons between it and PDMS are given in Table II (*33*). Remarkably, its permeability coefficient (*P*) is about an order of magnitude higher than that of PDMS, but its selectivity (as measured by the ratio of the *P* values for oxygen and nitrogen) is not much less. The greatly increased values of *P* are due apparently to the unusually high solubility of gases in this polymer (*31*, *34*). Studies of the effects of substituting the trimethylsilyl group onto other polymer backbones are in progress.

Another type of membrane designed as an artificial skin coating for burns also exploits the high permeability of siloxane polymers (*35*, *36*). The inner layer of the membrane consists primarily of protein and serves as a template for the regenerative growth of new tissue. The outer layer is a sheet of silicone polymer that not only provides mechanical support but also permits the outward escape of excess moisture while preventing the ingress of harmful bacteria.

Soft contact lenses prepared from PDMS provide a final example (Figure 5). The oxygen required by the eye for its metabolic processes must be

Table II. Permeability of Two Si-Containing Polymers

Polymer	*Gas*	$P\ (\times 10^8)$[a]	P_{O_2}/P_{N_2}
$[-Si(CH_3)_2O-]_n$	O_2	6.0	1.9
	N_2	3.1	
$[-C(Si(CH_3)_3)]=C(CH_3)-]_n$	O_2	72	1.7
	N_2	42	

[a]The permeability coefficient is expressed in units of cubic centimeters (at standard temperature and pressure) × centimeters per square centimeter per second per centimeter of mercury.

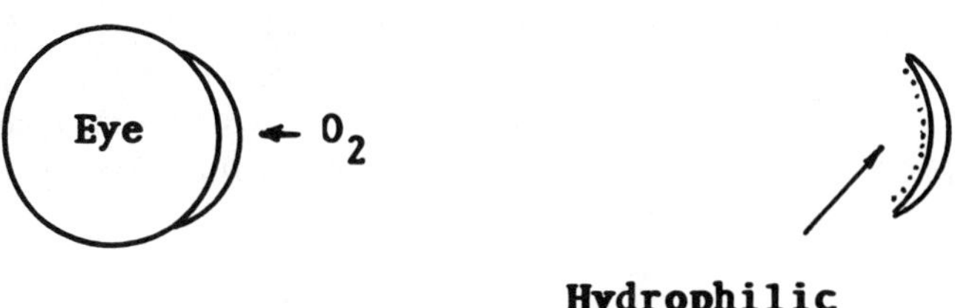

Figure 5. Use of grafting to change only the surface properties of a polymeric material.

obtained by inward diffusion from the air rather than through blood vessels. PDMS is ideal for such lenses (*36*) because of its high oxygen permeability, but it is too hydrophobic to be adequately wetted by the tears covering the eye. This hydrophobicity prevents the lens from feeling right and can also cause very serious adhesion of the lens to the eye itself. One way to remedy this problem is to graft a thin layer of a hydrophilic polymer to the inner surface of the lens. Because of the thinness of the coating, the high permeability of the PDMS is essentially unaffected.

Unusual Properties of PDMS. Some of the unusual physical properties exhibited by PDMS are summarized in List I. Atypically low values are exhibited for the characteristic pressure (a corrected internal pressure, which is much used in the study of liquids) (*37*), the bulk viscosity η, and the temperature coefficient of η (*4*). Also, entropies of dilution and excess volumes on mixing PDMS with solvents are much lower than can be accounted for by the Flory equation of state theory (*37*). Finally, as has already been mentioned, PDMS has a surprisingly high permeability.

List I. Unusual Properties of PDMS

Characteristic pressure, unusually low
Bulk viscosity (η), unusually low
Temperature coefficient of η, unusually low
Entropies of dilution, significantly lower than predicted by theory
Excess volumes on mixing, significantly lower than predicted by theory
Permeability, unusually large

Although the molecular origin of these unusual properties is still not known definitively, a number of suggestions have been put forward. One explanation involves low intermolecular interactions, and another involves the very high rotational and oscillatory freedom of the methyl side groups on the polymer (*8*). Still another theory focuses on the chain's very irregular cross section, which is very large at the substituted Si atom and very small at the unsubstituted O atom (*37*).

Reactive Homopolymers. ***Types of Reactions.*** In the typical ring-opening polymerization mentioned previously (*see* Preparation of Siloxane-Type Polymers), reactive hydroxyl groups are automatically placed at the ends of the chains (*7, 13*). Substitution reactions carried out on these chain ends can then be used to convert them into other functional groups. These functionalized polymers can undergo a variety of subsequent reactions (Table III).

Table III. Reactive Siloxane Polymers

X	*Reactant*
OH	Alkoxysilanes [e.g., $Si(OC_2H_5)_4$]
H	Unsaturated groups
$CH=CH_2$	Active H atoms and free radicals

NOTE: Polymers of the type $XSi(CH_3)_2O[-Si(CH_3)_2O-]_nSi(CH_3)_2X$ are considered.

Hydroxyl-terminated chains, for example, can undergo condensation reactions with alkoxysilanes (also known as orthosilicates) (*7, 38*). A difunctional alkoxysilane leads to chain extension, and a tri- or tetrafunctional alkoxysilane leads to network formation. Corresponding addition reactions with di- or triisocyanates represent other possibilities. Similarly, hydrogen-terminated chains can be reacted with end-linking molecules containing unsaturated groups, and conversely, vinyl-terminated chains can be reacted with molecules having active hydrogen atoms (*7, 38*).

A pair of vinyl or other unsaturated groups could also be joined by their direct reactions with free radicals. Similar end groups can be placed on siloxane chains by the use of an end blocker during polymerization (*16*). Reactive groups such as vinyl groups can be introduced as side chains by random copolymerizations involving, for example, methylvinylsiloxane trimers or tetramers (*7*).

Block Copolymers. One of the most important uses of end-functionalized polymers is in the preparation of block copolymers (*16*). The reactions are identical to the chain extensions already mentioned, except that the sequences being joined are chemically different. In the case of the $-OSiR_2R'Y$ chain ends mentioned previously, R′ is typically $(CH_2)_{3-5}$ and Y can be NH_2, OH, COOH, or $CH=CH_2$. The siloxane sequences containing these ends have been joined to other polymeric sequences such as carbonates, ureas, urethanes, amides, and imides.

Elastomeric Networks. Elastomeric networks are formed by reacting functionally terminated siloxane chains with an end linker with three or more functionalities and have been used extensively to study the molecular aspects of rubberlike elasticity (*26, 38–42*). These networks are preferred for this purpose, because relatively few complications from side reactions occur during their preparation. They are model networks in that a great deal is known about their structure by virtue of the very specific chemical reactions used to synthesize them. For example, in the case of a stoichiometric balance between chain ends and functional groups on the end linker, the critically important molecular weight M_c between cross-links is equal to the molecular weight of the chains prior to their end linking. Also, the functionality of the cross-links (the number of chains emanating from one of them) is simply the functionality of the end-linking agent. Finally, the molecular weight distribution of the network chains is the same as that of the starting polymer, and few, if any, dangling-chain irregularities are formed.

Because these networks have a known degree of cross-linking (as inversely measured by M_c), they can be used to test the molecular theories of rubberlike elasticity, particularly with regard to the possible effects of interchain entanglements (*26, 38–42*). Intentionally imperfect networks can also be prepared (Figure 6) (*38*). Such networks have known numbers and

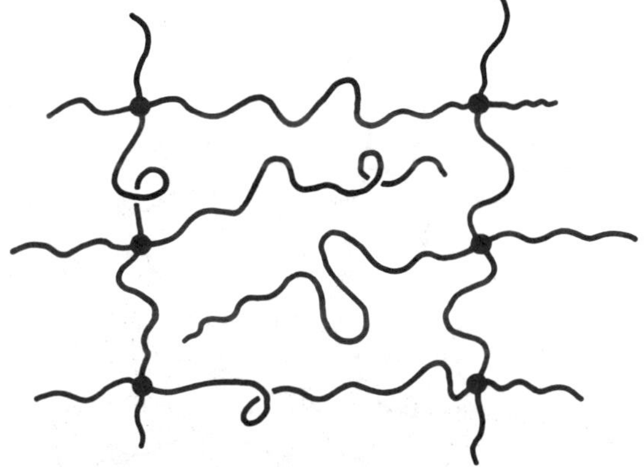

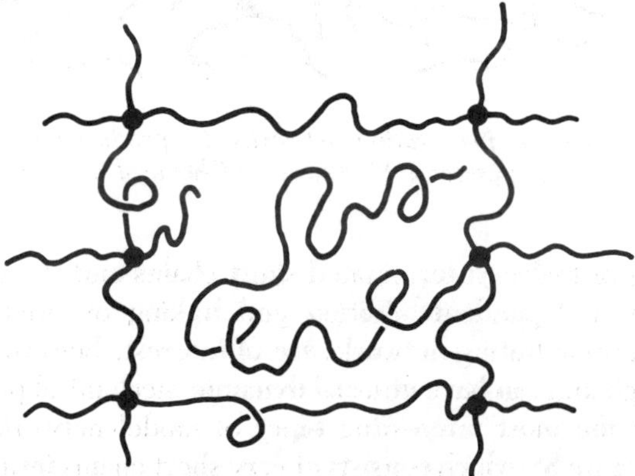

Figure 6. Two end-linking techniques for preparing networks with known numbers and lengths of dangling chains. (Reproduced from reference 38. Copyright 1985 American Chemical Society.)

lengths of dangling chains (those chains attached to the network at only one end), and thus the effects of these irregularities on elastomeric properties can be determined. In the first method (Figure 6a), the stoichiometry is unbalanced, so that there is an excess of chain ends over functional groups on the end-linking molecules. The limitation of this method is the fact that the dangling chains must have the same average length as the elastically effective chains (those chains attached to the network at both ends). The second method (Figure 6b) avoids this limitation by the use of monofunctionally terminated chains of the desired length.

These reactions can be used also to form networks that interpenetrate (Figure 7) (*38*). For example, one network could be formed by a condensation

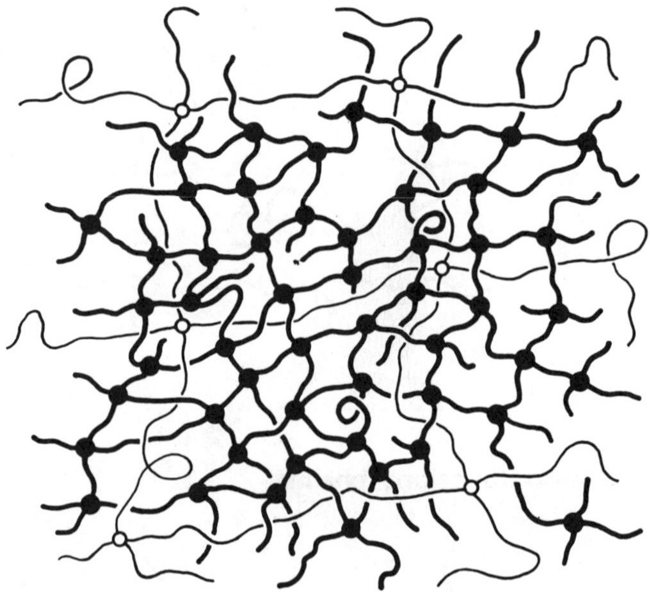

Figure 7. Two interpenetrating networks. (Reproduced from reference 38. Copyright 1985 American Chemical Society.)

end linking of hydroxyl-terminated short chains and the other by a simultaneous but independent addition end linking of vinyl-terminated long chains. Interpenetrating networks are of interest, because they can be unusually tough and can have unusual dynamic mechanical properties (*36*, *43*).

One of the most interesting types of model networks is the bimodal network (Figure 8), which consists of very short chains intimately end linked with the much longer chains that are representative of elastomeric materials (*38–40*, *44*). In Figure 8 (*44*), the short chains are drawn arbitrarily thicker

Figure 8. Bimodal network. (Reproduced with permission from reference 44. Copyright 1979 Hüthig and Wepf Verlag.)

than the long chains. These materials have unusually good elastomeric properties, specifically the large values of both the ultimate strength and maximum extensibility. Possibly, the short chains contribute primarily to the ultimate strength, and the long chains contribute primarily to the maximum extensibility. Also, not only do short chains improve the ultimate properties of elastomers, but long chains also improve the impact resistance of the much more heavily cross-linked thermosets (*45*).

Cyclic Trapping. If relatively large PDMS cyclic molecules are present when linear PDMS chains are end linked, then some of the cyclic molecules will be permanently trapped by one or more network chains threading through them (*46*), as illustrated by cyclic molecules B, C, and D in Figure 9 (*47*). Interpretation of the fraction trapped as a function of ring size by using rotational isomeric theory and Monte Carlo simulations provides very useful information about the spatial configurations of cyclic molecules (*47*).

This technique can be used to form a network with no cross-links whatsoever. Mixing linear chains with large amounts of cyclic molecules and then

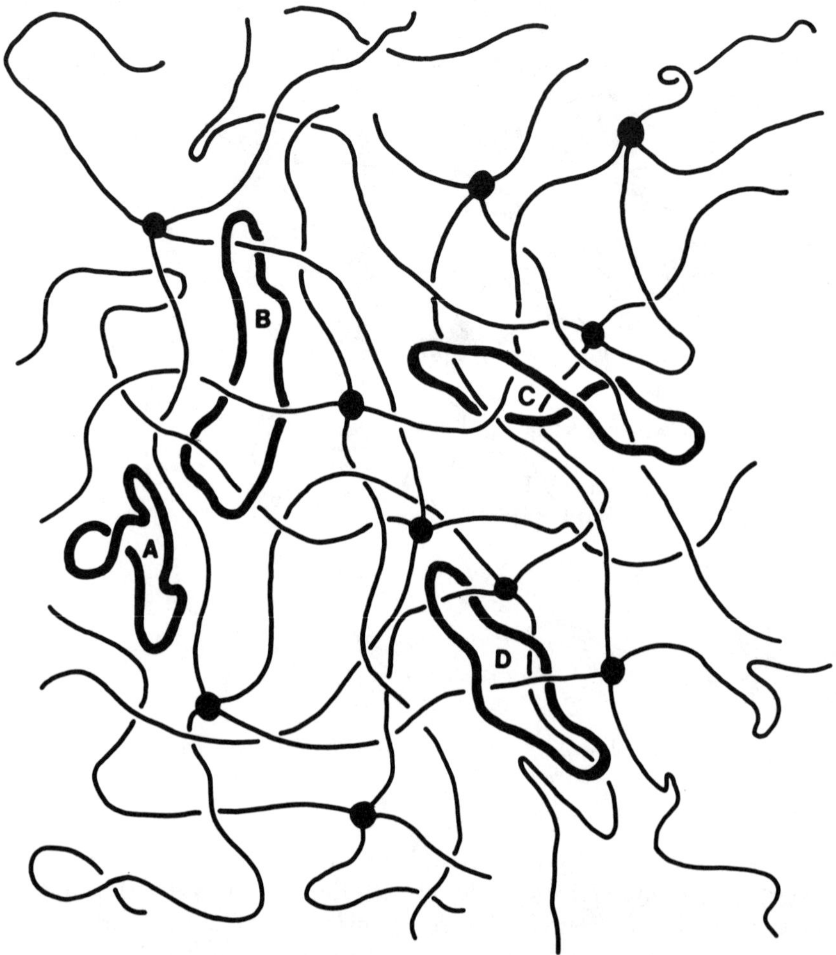

Figure 9. Trapping of cyclic molecules during end-linking preparation of a network. (Reproduced from reference 47. Copyright 1987 American Chemical Society.)

difunctionally end linking them can give sufficient cyclic interlinking to yield a "chain mail" or an "Olympic" network (*48, 49*) (Figure 10) (*50*). Such materials could have highly unusual equilibrium and dynamic mechanical properties.

Copolymers. ***Random Copolymers.*** Random copolymers may be prepared by the copolymerization of a mixture of monomers rather than the homopolymerization of a single type of monomer (*7, 13*). One reason for preparing random copolymers is to introduce functional species, such as

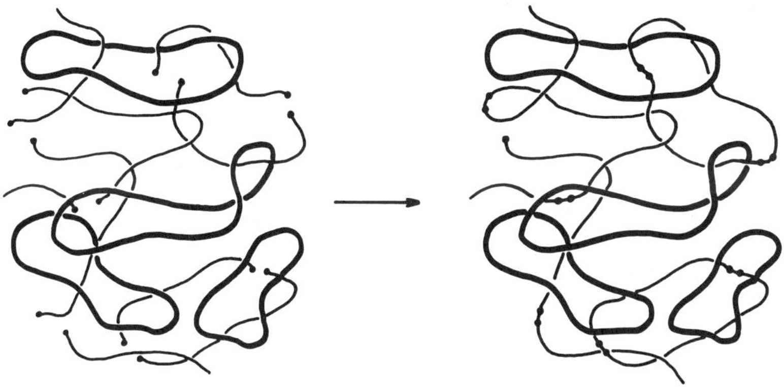

Figure 10. Preparation of a "chain mail" or "Olympic" network consisting entirely of interlooped cyclic molecules. (Reproduced with permission from reference 50. Copyright 1985 Butterworth.)

vinyl groups or hydrogens, along the chain backbone to facilitate cross-linking. Another reason is to introduce sufficient chain irregularity to make the polymer inherently noncrystallizable.

Block Copolymers. As already mentioned, the sequential coupling of functionally terminated chains of different chemical structure can be used to make block copolymers, including those in which one or more of the blocks is a polysiloxane (*16, 51, 52*). If the blocks are relatively long, separation into a two-phase system almost invariably occurs. Frequently, one type of block will be in a continuous phase, and the other block will be dispersed in the continuous phase in domains having an average size in the order of a few hundred angstroms. Such materials can have unique mechanical properties not available from either species when present simply in homopolymeric form. Sometimes, similar properties can be obtained by the simple blending of two or more polymers (*53*).

Applications. ***Medical Applications.*** The medical applications of siloxane polymers are numerous (*7, 8, 36*). For example, prostheses, artificial organs, facial reconstruction, and catheters take advantage of the inertness, stability, and pliability of the polysiloxanes. Artificial skin, contact lenses, and drug delivery systems take advantage of the high permeability of these polymers, as well.

Nonmedical Applications. Illustrative nonmedical applications are their uses as high-performance elastomers; membranes; electrical insulators; water repellents; antifoaming agents; mold release agents; adhesives; protective coatings; release control agents for agricultural chemicals; encapsulation media; and hydraulic, heat-transfer, and dielectric fluids (*7, 8*). These applications are based on the same properties of polysiloxanes just mentioned and also on their ability to modify surfaces and interfaces (for example, as water repellents, antifoaming agents, and mold release agents).

Silica-Type Materials

Sol–Gel Ceramics. A relatively new area involving silicon-containing materials is the hydrolysis of alkoxysilanes or silicates to give silica (*54*). The process, which involves polymerization and branching, is complicated, but a typical overall reaction may be written:

$$Si(OR)_4 + 2H_2O \rightarrow SiO_2 + 4ROH \qquad (2)$$

Production of ceramics by this novel route has a variety of advantages. First, much lower temperatures can be used, and higher purity products are obtained. Second, the microstructure of the ceramic can be better controlled, and it is relatively simple to form very thin ceramic coatings. Third, ceramic "alloys" are much easier to form by the hydrolysis of a mixture of organometallic materials, for example, silicates and titanates to give a SiO_2–TiO_2 alloy.

In Situ Precipitations. The same hydrolyses can be carried out within a polymer to generate particles of the ceramic material, typically with an average size of a few hundred angstroms (*38, 40*). Considerable reinforcement of elastomers, including PDMS, can be achieved in this way. Because of the nature of this in situ precipitation, the particles are well dispersed and essentially unagglomerated. A typical transmission electron micrograph of such a filled material is shown in Figure 11 (*55*). The particles are also relatively monodisperse, with almost all of them having diameters in the range 100–200 Å.

Polymer-Modified Glasses. If the hydrolyses in silane–polymer systems are carried out with relatively large amounts of silane, then the silica generated can become the continuous phase, with the elastomeric polysiloxane dispersed in it (*56–58*). The resulting composite is a polymer-modified glass or ceramic, and its hardness can be varied by control of the molar ratio of alkyl R groups to Si atoms (Figure 12) (*58*). Low values of the R/Si ratio yield a brittle ceramic, and high values yield a relatively hard elastomer. The most interesting range of values, R/Si ~ 1, can yield a relatively tough ceramic of reduced brittleness.

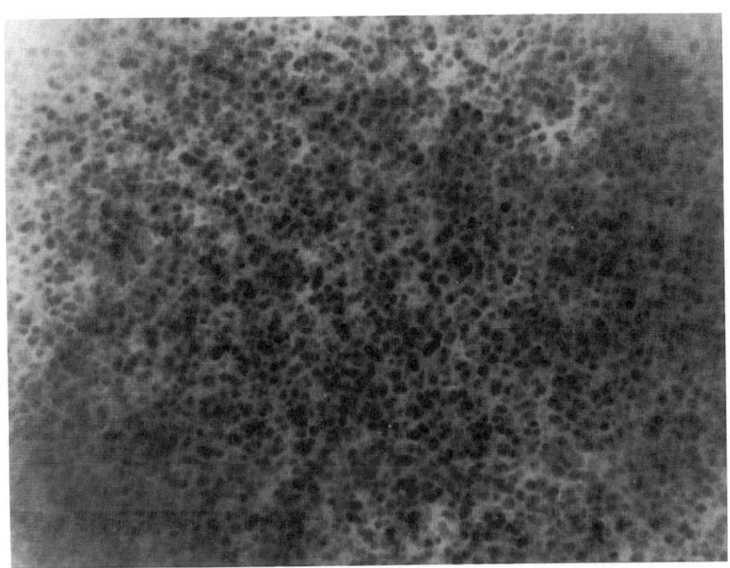

Figure 11. Electron micrograph of a PDMS network containing in situ precipitated silica particles. (Reproduced with permission from reference 55. Copyright 1984 Wiley.)

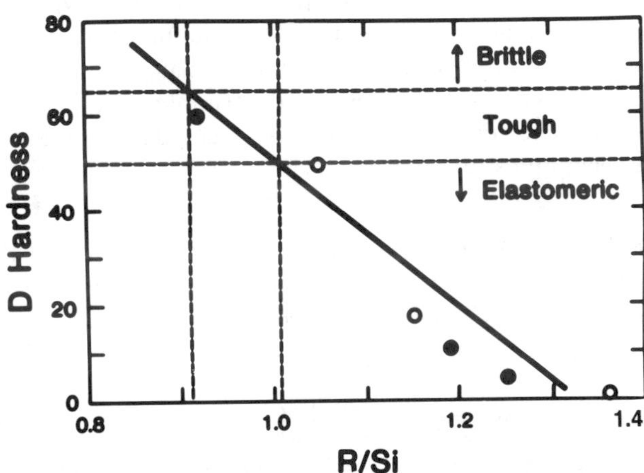

Figure 12. Hardness of a silica–PDMS composite as a function of the relative numbers of alkyl groups and silicon atoms. (Reproduced with permission from reference 58. Copyright 1987 Springer Verlag.)

Polysilanes

Synthesis. The polysilanes can be produced from silyl dichlorides by the Wurtz-type coupling (*9*):

$$SiRR'Cl_2 \xrightarrow{Na} [-SiRR'-]_n \tag{3}$$

In this reaction, R and R′ are CH_3, C_6H_5, C_4H_9, C_6H_{13}, etc. The resulting polymers can be linear or cross-linked, and glassy, elastomeric, or partially crystalline.

Flexibility. One of the interesting differences between polysilanes and their organic counterparts involves chain flexibility. For example, relevant calculations of conformational energy have been carried out recently on polysilane, $[-SiH_2-]_n$, itself (*59, 60*). Some resulting energies, shown as a function of two consecutive skeletal rotation angles ϕ, are presented in Figure 13 (*60*). The results suggest that the lowest energy conformation should be a sequence of *gauche* states ($\phi = \pm 120°$) of the same sign (*59, 60*). This result is in contrast to polyethylene, $[-CH_2-]_n$, which has the opposite preference for *trans* states ($\phi = 0°$). Such preferences generally dictate the regular conformation chosen by a polymer chain when it crystallizes. Polyethylene does in fact crystallize in the all-*trans* planar zig-zag conformation (*14*). Whether polysilane crystallizes in the predicted helical form generated by placing all of its skeletal bonds in *gauche* states of the same sign remains to be determined.

The calculations also predict that polysilane should have a higher equilibrium flexibility than polyethylene (*60*). Solution characterization techniques could be used to test this prediction. Dynamic flexibility can also be estimated from such energy maps, by determining the barriers between energy minima. Relevant experimental results could be obtained by a variety of dynamic techniques (*61*).

Applications. Some applications of polysilanes include their use as semiconductors, photoresists, photoinitiators, nonlinear optical materials, and ceramic precursors (*9*). Their use as ceramic precursors can be illustrated by the following reactions:

$$[-Si(CH_3)_2-]_n \xrightarrow{\sim 400\ °C} [-SiH(CH_3)CH_2-]_n \tag{4}$$

$$[-SiH(CH_3)CH_2-]_n \xrightarrow{\sim 1200\ °C} n\mathrm{SiC} + n\mathrm{H}_2 + n\mathrm{CH}_4 \tag{5}$$

These reactions are used to prepare high-performance silicon carbide fibers by the controlled thermolysis of poly(dimethylsilane).

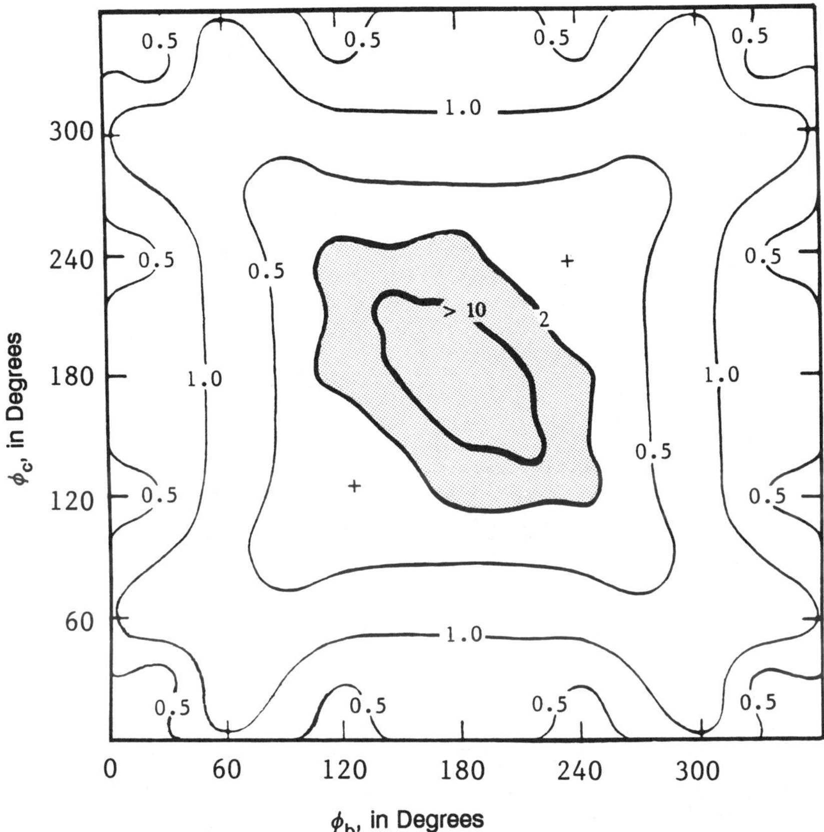

Figure 13. Map of conformational energy for rotations about the Si–Si skeletal bonds in polysilane, $[-SiH_2-]_n$. The energies (in kilocalories per mole) relative to the minima (designated by the plus signs) are shown as contour lines. (Reproduced from reference 60. Copyright 1986 American Chemical Society.)

Polysilazanes

Synthesis. Polysilazanes, $[-SiRR'NR''-]_n$, can be prepared by a variety of techniques. A typical preparative reaction is the ammonolysis of methyldichlorosilane (*10*):

$$n\mathrm{CH_3SiHCl_2} + n\mathrm{NH_3} \rightarrow [-\mathrm{SiHCH_3NH}-]_n + 2n\mathrm{HCl} \qquad (6)$$

Applications. The polysilazanes have not been studied as extensively as the polysilanes, and the applications of these polymers are just beginning to be developed. One of the most important applications is thermolysis to yield silicon nitride (Si_3N_4) ceramics (*10*) by using reactions similar to those used to prepare silicon carbide ceramics from polysilanes.

Acknowledgments

It is my pleasure to acknowledge that my work in some of these areas has been supported by the National Science Foundation, the Air Force Office of Scientific Research, the Army Research Office, and the Gas Research Institute.

References

1. Elias, H. G. *Macromolecules;* Plenum: New York, 1977; Vol. 2.
2. Allcock, H. R. *Chem. Eng. News* March 18, 1985, 22.
3. *Inorganic and Organometallic Polymers;* Zeldin, M.; Wynne, K. J.; Allcock, H. R., Eds.; ACS Symposium Series 360; American Chemical Society: Washington, DC, 1988.
4. Noll, W. *Chemistry and Technology of the Silicones;* Academic Press: New York, 1968.
5. Bobear, W. J. In *Rubber Technology;* Morton, M., Ed.; Van Nostrand Reinhold: New York, 1973.
6. *Analysis of Silicones;* Smith, A. L., Ed.; Krieger Publishing Co.: Malabar, FL, 1983.
7. Warrick, E. L.; Pierce, O. R.; Polmanteer, K. E.; Saam, J. C. *Rubber Chem. Technol.* **1979,** *52,* 437.
8. Rochow, E. G. *Silicon and Silicones;* Springer-Verlag, Berlin, 1987.
9. West, R. W. *J. Organomet. Chem.* **1986,** *300,* 327.
10. Seyferth, D.; Wiseman, G. H. In *Ultrastructure Processing of Ceramics, Glasses, and Composites;* Hench, L. L.; Ulrich, D. R., Eds.; Wiley: New York, 1984.
11. Bostick, E. E. In *Ring-Opening Polymerization;* Frisch, K. C.; Reegen, S. L., Eds.; Dekker: New York, 1969.
12. Allcock, H. R. *Angew. Chem. Int. Ed. Engl.* **1977,** *16,* 147.
13. McGrath, J. E.; Riffle, J. S.; Banthia, A. K.; Yilgor, I.; Wilkes, G. L. In *Initiation of Polymerization;* Bailey, F. E., Jr., Ed.; American Chemical Society: Washington, DC, 1983.
14. Flory, P. J. *Statistical Mechanics of Chain Molecules;* Wiley-Interscience: New York, 1969.
15. *Cyclic Polymers;* Semlyen, J. A., Ed.; Elsevier: London, 1986.
16. Yilgor, I.; Riffle, J. S.; McGrath, J. E. In *Reactive Oligomers;* Harris, F. W.; Spinelli, H. J., Eds.; ACS Symposium Series 282; American Chemical Society: Washington, DC, 1985.
17. Odian, G. *Principles of Polymerization,* 2nd ed.; Wiley-Interscience: New York, 1981.
18. Heimenz, P. C. *Polymer Chemistry: The Basic Concepts;* Dekker: New York, 1984.
19. *Polymer Handbook,* 2nd ed.; Brandrup, J.; Immergut, E. H., Eds.; Wiley-Interscience: New York, 1975.
20. Mandelkern, L. *Crystallization of Polymers;* McGraw-Hill: New York, 1964.

21. Mark, J. E. In ACS Short Course Manual *Polymer Chemistry;* Mark, J. E.; Odian, G., Eds.; American Chemical Society: Washington, DC, 1984.
22. Brown, J. F., Jr. *J. Polym. Sci., Part C* **1963,** *1,* 83.
23. Helminiak, T. E.; Berry, G. C. *J. Polym. Sci., Polym. Symp.* **1978,** *65,* 107.
24. Allcock, H. R.; Lampe, F. W. *Contemporary Polymer Chemistry;* Prentice Hall: Englewood Cliffs, NJ, 1981.
25. Billmeyer, F. W., Jr. *Textbook of Polymer Science,* 3rd ed.; Wiley-Interscience: New York, 1984.
26. Mark, J. E.; Erman, B. *Rubberlike Elasticity: A Molecular Primer;* Wiley-Interscience: New York, 1988.
27. Lee, C. L., Dow Corning Corporation, Midland, MI, personal communications.
28. Godovsky, Yu. K.; Makarova, N. N.; Papkov, V. S.; Kuzmin, N. N. *Makromol. Chem.* **1985,** *6,* 443.
29. Friedrich, J.; Rabolt, J. F. *Macromolecules* **1987,** *20,* 1975.
30. Schneider, N. S.; Desper, C. R.; Beres, J. J. In *Liquid-Crystalline Order in Polymers;* Blumstein, A., Ed.; Academic Press: New York, 1978.
31. Reports from Dow Corning Corporation, Syracuse University, and the University of Cincinnati, under Contract No. 5082-260-0666 from the Gas Research Institute, Chicago, IL.
32. West, R., University of Wisconsin, personal communication.
33. Masuda, T.; Isobe, E.; Higashimura, T.; Takada, K. *J. Am. Chem. Soc.* **1983,** *105,* 7473.
34. Stern, S. A., Syracuse University, personal communications.
35. Yannas, I. V.; Burke, J. F. *J. Biomed. Mater. Res.* **1980,** *14,* 65.
36. Arkles, B. *CHEMTECH* **1983,** *13,* 542.
37. Shih, H.; Flory, P. J. *Macromolecules* **1972,** *5,* 758.
38. Mark, J. E. *Acc. Chem. Res.* **1985,** *18,* 202, and references cited therein.
39. Mark, J. E. *Polym. J.* **1985,** *17,* 265, and references cited therein.
40. Mark, J. E. *Br. Polym. J.* **1985,** *17,* 144, and references cited therein.
41. Gotlieb, M.; Macosko, C. W.; Benjamin, G. S.; Meyers, K. O.; Merrill, E. W. *Macromolecules* **1981,** *14,* 1039.
42. Opperman, W.; Rennar, N. *Prog. Coll. Polym. Sci.* **1987,** *75,* 49.
43. Sperling, L. H. *Interpenetrating Polymer Networks and Related Materials;* Plenum: New York, 1981.
44. Mark, J. E. *Makromol. Chem. Suppl.* **1979,** *2,* 87.
45. Tang, M. Y.; Letton, A.; Mark, J. E. *Colloid Polym. Sci.* **1984,** *262,* 990.
46. Clarson, S.; Mark, J. E.; Semlyen, J. A. *Polym. Commun.* **1986,** *27,* 244.
47. DeBolt, L. C.; Mark, J. E. *Macromolecules* **1987,** *20,* 2369.
48. de Gennes, P. G. *Scaling Concepts in Polymer Physics;* Cornell University Press: Ithaca, NY, 1979.
49. Rigbi, Z.; Mark, J. E. *J. Polym. Sci., Polym. Phys. Ed.* **1986,** *24,* 443.
50. Garrido, L.; Mark, J. E.; Clarson, S. J.; Semlyen, J. A. *Polym. Commun.* **1985,** *26,* 53.
51. Noshay, A.; McGrath, J. E. *Block Copolymers: Overview and Critical Survey;* Academic Press: New York, 1977.
52. Ibemesi, J.; Gvozdic, N.; Keumin, M.; Lynch, M. J.; Meier, D. J. *Polym. Prepr. (Am. Chem. Soc., Div. Polym. Chem.)***1985** *26*(2), 18.
53. Manson, J. A.; Sperling, L. H. *Polymer Blends and Composites;* Plenum: New York, 1976.
54. Hench, L. L.; Ulrich, D. R. *Ultrastructure Processing of Ceramics, Glasses, and Composites;* Wiley-Interscience: New York, 1984.
55. Ning, Y. P.; Tang, M. Y.; Jiang, C. Y.; Mark, J. E.; Roth, W. C. *J. Appl. Polym. Sci.* **1984,** *29,* 3209.

56. Schmidt, H. *Mater. Res. Soc. Symp. Proc.* **1984,** *32,* 327.
57. Huang, H.-H.; Orler, B.; Wilkes, G. L. *Polym. Bull.* **1985,** *14,* 557.
58. Mark, J. E.; Sun, C. C. *Polym. Bull.* **1987,** *18,* 259.
59. Damewood, J. R., Jr.; West, R. *Macromolecules* **1985,** *18,* 159.
60. Welsh, W. J.; DeBolt, L.; Mark, J. E. *Macromolecules* **1986,** *19,* 2978.
61. Ferry, J. D. *Viscoelastic Properties of Polymers,* 3rd ed.; Wiley: New York, 1980.

RECEIVED for review May 27, 1988. ACCEPTED revised manuscript October 25, 1988.

ONE-DIMENSIONAL TO THREE-DIMENSIONAL SILOXANES

3

Formation of Linear Siloxane Polymers

John C. Saam

Dow Corning Corporation, Midland, MI 48686–0995

Although regarded as a mature and established field, siloxane polymerization has some distinguishing features that are well described in the literature but are often overlooked. These features, as well as some of the more-recent observations and current thinking in this field, are reviewed briefly in this chapter. Two topics not usually included in other reviews, copolymerization and condensation polymerization, are also discussed.

THE HISTORY OF LINEAR POLYSILOXANES dates back at least 116 years (*1*), and research activity in this area steadily accelerated during this period as synthetic methods improved and as the fundamental nature of polymers became clear. The industrial prominence of poly(dialkylsiloxane)s was a particularly strong impetus to the development of this field (*2*). Thus, a large body of literature has accumulated, which has been extensively reviewed. The reviews by Wright (*3*), Sigwalt (*4*), and Kendrick et al. (*5*) are excellent and current, whereas that by Voronkov et al. (*6*) covers the earlier literature. Siloxane polymerization has now become sufficiently commonplace, so that it is sometimes discussed in general textbooks on polymer chemistry (*7*).

This chapter will not duplicate the already existing reviews either in their comprehensiveness or special emphasis. Instead the distinguishing features of siloxane polymerization will be highlighted briefly, and perspectives will be given on copolymerization and condensation polymerization.

Both generic classes of polymerization, chain growth and step growth, are practiced in the formation of linear polysiloxanes. Ring-opening polymerization of cyclodialkylsiloxanes, ordinarily considered a chain growth process, is initiated by either acids or bases. Initiation by free radicals is not a consideration because of the partial ionic character and high energy of the

0065–2393/90/0224–0071$06.00/0

siloxane bond (*8*). The chief step-growth processes are condensations, of which the most important is the condensation of silanol accompanied by the elimination of water.

Equilibrium Ring-Opening Polymerization

Ring-opening polymerization can be described by the following generic equation

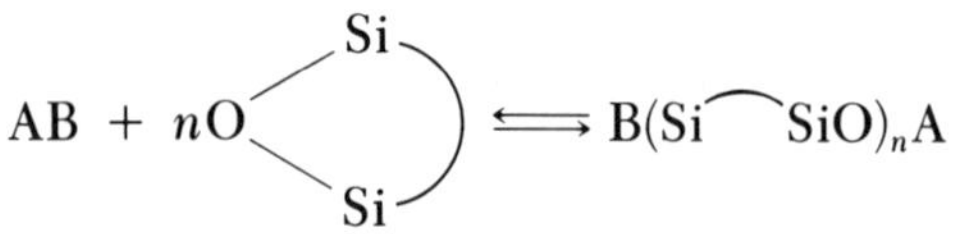

$$AB + n\,O\!\left\langle\begin{matrix}Si\frown\\ \\ Si\smile\end{matrix}\right. \rightleftharpoons B(Si\frown SiO)_nA \qquad (1)$$

in which the species AB is an acidic or basic initiator. A partial listing of initiators is given in List I. Sometimes a potent accelerating effect is seen when small amounts of Lewis bases accompany the basic catalysts. Water, alcohols, or silanols, on the other hand, will retard polymerization. At very low levels, however, water is essential for certain acid-initiated processes. A sampling of monomers consisting of strained and unstrained rings is given in List II. (Henceforth, the $(CH_3)_2SiO$ unit will be represented as D, and $(CH_3)_3SiO_{1/2}$ will be represented as M.)

List I. Typical Initiators for Cyclosiloxane Polymerization

Basic initiators

GOH[a], GOSi≡, GOR[b], GR, GSR

Acidic initiators

Acid clays, HF, HI_3, $HCl–FeCl_3$, H_2SO_4, CF_3SO_3H

[a]G can be an akali metal or a quaternary ammonium or phosphonium group.
[b]R can be alkyl, polystyryl (when G is Li, Na, or K), or poly(trimethylsilylvinyl) (when G is Li).

Reaction 1 is represented as reversible, but conditions can be adjusted so that reversibility can be greatly suppressed with the proper choice of monomer, initiator, solvent, and temperature. This is particularly true with strained rings and if lithium silanolates are chosen as initiators (*9*, *10*).

Termination is usually effected by neutralization of the acidic or basic initiators. During this step, care must be taken to ensure that traces of acidic

List II. Typical Cyclosiloxane Monomers Used in Ring-Opening Polymerization

Strained rings

$[(CH_3)_2SiO]_3$ or D_3

$[CF_3CH_2CH_2(CH_3)SiO]_3$

$[(C_6H_5)_2SiO]_3$

$CH_2\text{–}Si\text{–}(CH_3)_2$
$|\quad\quad \diagdown O$
$CH_2\text{–}Si\text{–}(CH_3)_2$

Unstrained rings

D_4, D_5, D_6, D_7

or basic residues, which might catalyze the reverse reaction, are removed. Thus, an inherent advantage of initiating with quaternary ammonium or phosphonium hydroxides is the ease of their removal by simply heating to more than 135 °C to induce their decomposition (*11*). As with most inorganic polymers (*12*), reversibility dominates the process when the rings are not strained and the initiators are strong acids or bases, such as H_2SO_4 or KOH. Even trace amounts of base can cause marked changes in the stability of the polymer at higher temperatures by inducing the formation of cyclosiloxanes and causing eventual evaporation of the polymer when conditions are chosen so that the vapor pressure of the equilibrium cyclosiloxanes exceeds the latent pressure (*13*).

Under these conditions, the system is described by two equilibria:

$$\sim D_n^* + D_x \overset{K_2}{\rightleftarrows} \sim D^*_{(n+x)} \qquad (2)$$

and

$$\sim D_n \sim + \sim D_m \sim \overset{K_3}{\rightleftarrows} \sim D_{n+y} \sim + \sim D_{m-y} \sim \qquad (3)$$

Both processes are siloxane redistributions. Reaction 2 amounts to chain growth by x units and has been studied extensively (*14–17*), whereas reaction 3, which does not give a net change in number-average molecular weight, results in an exchange of y repeat units between chains (*14–16*). The value of K_2 for poly(dimethylsiloxane) is between 0.4 to 0.6 L/mol, whereas K_3 is close to 1.0 (*14, 15*). The redistribution in reaction 3 begins early in the process (*17*), and when both equilibria are finally reached, a Gaussian distribution of molecular sizes results (*14, 15*). Thus, the system is very dynamic, and every D unit, whether located in a ring or chain, is subject to redistribution and participates in both equilibria.

Knowledge of the relative amounts of rings and chains at equilibrium permits an approximation of the thermodynamics of reaction 2 through equation 4 (*12*):

$$\ln (1 - w_p) = \frac{\Delta H_2}{RT} - \left(\frac{\Delta S_2}{R} + \ln D_o\right) \tag{4}$$

In equation 4, w_p is the weight fraction of chains, ΔH_2 is the enthalpy for equilibrium 2, ΔS_2 is the entropy for equilibrium 2, R is the gas constant, T is temperature, and D_o is the total molar fraction of D units. A number of workers (*18–20*) have reported that w_p is independent of temperature and that therefore $\Delta H_2 = 0$. These findings are consistent with the fact that the siloxane bonds on either side of the equilibrium 2 are identical in number and kind. Therefore, in the absence of ring strain or special group interactions, no net change in enthalpy can occur and temperature cannot affect equilibrium. Hence, with unstrained rings, the process is driven by entropy. The implication is that more degrees of freedom are allowed in the chain than in the rings, a conclusion that is consistent with the unrestricted rotation of the monomer units about the siloxane bond in poly(dimethylsiloxane) (*21*).

The critical conditions of temperature and dilution at equilibrium for polymer formation can be defined by setting w_p to 0 in equation 4. With $\Delta H_2 = 0$, equation 4 reduces to

$$\ln (1 - w_p) = -\left(\frac{\Delta S_2}{R} + \ln D_o\right) \tag{5}$$

and at critical conditions,

$$-\frac{\Delta S_2}{R} = \ln [D_o]_{crit} \tag{6}$$

in which $[D_o]_{crit}$ is the critical molar fraction of D units below which no polymer is present. Thus, unlike many polymers at equilibrium with their monomers, poly(dimethylsiloxane) has no definable "unzipping temperature", but it has a clearly definable "unzipping dilution". The unzipping dilution is roughly 30 vol % siloxane at equilibrium for poly(dimethylsiloxane).

The yield of polymer, w_p, is highly sensitive to the size of the substituent groups in the siloxane system (22). In the series CH_3RSiO, w_p decreases in the order R = H > CH_3 > C_2H_5 > C_3H_7 > $CF_3CH_2CH_2$. This order is not altered by dilution. Critical dilution points, as well as the equilibrium constants for the individual cyclic species, are given in reference 21 for each of the systems. Results were interpreted in terms of the Jacobson–Stockmayer theory (23), which assumes a Gaussian distribution of chain ends at juxtaposition for intramolecular bond formation, a random-walk model for chain conformation, and an entropically driven process. Application of the theory to the prediction of individual concentrations of rings at equilibrium was successful only when ring sizes exceeded 15–20 units. Details are available in the original work (*20, 22*), as well as in an excellent review (*3*).

Table I is a compilation of literature data (*24–26*) for the relative amounts of various species when reaction 2 is thought to be at equilibrium. Measurements were made at the same dilution but at different temperatures and with different solvents and catalysts. The results should be identical if the equilibrium is accurately represented by reaction 2 and if the systems are truly at equilibrium. Comparison of conditions 1 and 3 show that under quite different conditions w_ps are nearly identical, although some discrepancies are evident in the amounts of higher molecular weight cyclosiloxanes formed. Still, the similarities are remarkable when the disparity in the times of publication and analytical methods are taken into consideration.

Table I. Influence of Conditions on the Position of Equilibrium 2

Conditions	*Polymer*	D_4	D_5–D_9
1. 135 °C, toluene, KOH[a]	29	33	31.0
2. 20 °C, benzene, Li cryptate[b]	70	2.7	11.5
3. 20 °C, methylene chloride, CF_3SO_3H[c]	28	32	27
4. 20 °C, methylene chloride, CF_3SO_3H–$(CF_3SO_2)_2O$[c]	0	38	22

NOTE: All values are concentration (wt %) at equilibrium. The initial monomer was D_4 at 1.0 mol/L.
[a]Reference 24.
[b]Reference 25.
[c]Reference 26.

On the other hand, the products under conditions 2 and 4 in Table I show considerable discrepancy from what would be considered a normal distribution of equilibrium products. Although the data were taken at a point when there seemed to be little change occurring in the system, it can still be argued that equilibrium was not achieved in these examples. Later work with lithium cryptate, one of the deviants in Table I, indicates agreement in the distributions of D_4, D_5, and D_6 reported by earlier workers (*26*). Unfortunately, no data were given for w_p in that publication.

A larger counterion, such as R_4N^+ is alleged to suppress "backbiting" or the reverse reaction and reduce the concentration of cyclosiloxane byproducts (*4*). However, more convincing evidence will have to be presented to show that these data were indeed taken at equilibrium.

The molecular weight of the polymer at equilibrium during cyclosiloxane polymerizations is controlled by including a hexaalkyldisiloxane such as MM or an oligomeric species such as MDM or MD_2M. The equilibrium is represented as follows:

$$MD_nM \rightleftharpoons MD_{(n-x)}M + D_x \tag{7}$$

The conversion of end groups, p, is then correlated with molecular size. When p is defined in this way, the relation can be expressed as follows (*28, 29*).

$$p = 1 - \frac{M_o}{M_n} = 1 - \frac{1}{\overline{X}_n} \tag{8}$$

In equation 8, M_o is the monomer molecular weight, M_n is the number-average molecular weight of the chains, and $\overline{X}_n$ is the number-average degree of polymerization. Other somewhat more complex definitions for p take into account the equilibrium cyclosiloxane content (*14, 15, 24*). General relationships have also been defined that give with a reasonable degree of accuracy the distribution of chains and rings when an end-stopper is present (*15, 29, 30*).

Mechanisms of Ring-Opening Polymerization

Base-Initiated Polymerization. Although the base-initiated mechanism for reaction 2 was once thought to proceed via free ionic intermediates (*31*), little evidence supported this mechanism. Conductivities of solutions of alkali metal silanolates in moderately polar solvents are essentially nil (*32*), and most workers now agree that reactive intermediates consist of ion pairs or charge-separated ion pairs (*4–5*). The key intermediate is believed to involve coordination of the countercation of the ion pair at the chain end, for example potassium, with the cyclosiloxane in a manner analogous to the crown ethers or cryptates (*33*), as shown by structure **1**.

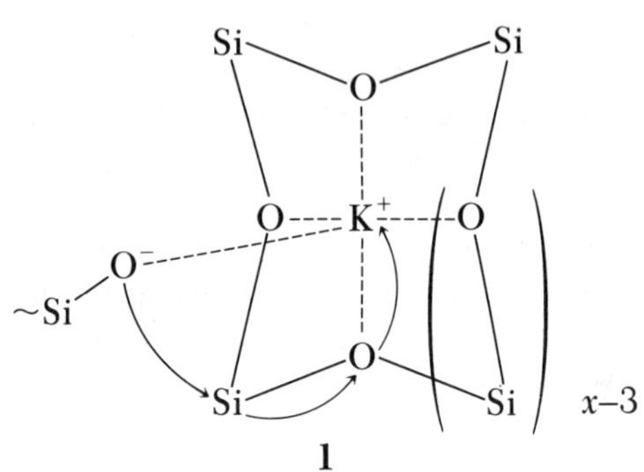

1

In **1**, x is the number of siloxane units in the ring.

Additional support for such species is found in the efficient transport of potassium ion through membranes via its coordination with D_{7-9}, which supposedly forms structures similar to **1** (*34*). The relative reactivity of D_x on the basis of rates of ring-opening polymerization initiated by potassium silanolate is $D_8 > D_7 > D_9 > D_3 > D_6 > D_4 > D_5$ (*31, 33*). This trend is consistent with the tendency of the counterion to coordinate more readily with the larger rings (*34*). The more recently observed activation energies and entropies are also consistent with a mechanism proceeding through an intermediate similar to **1** (*33*). This mechanism would require an organized transition state and, hence, a large negative activation entropy and a reduced activation energy with increased ring size because of electrophilic assistance provided by the incipient cation in **1**.

Rates of reaction (R_p) in ring opening polymerizations initiated by alkali metals are typically proportional to a fractional order of the initiator and close to first order in cyclosiloxane:

$$R_p = [CH_3Si\equiv]^{1/n}[D_x] \tag{9}$$

Values of n in equation 9 can vary from 1 for cryptate-coordinated lithium silanolates (*25, 27*) and $R_4NOSi\equiv$ (*35*) to 4 for $LiOSi\equiv$. The latter result was reported to be dependent on dilutions (*36*). In the absence of polar solvents, $n = 2$ is commonly reported for potassium silanolate initiators (*37, 38*). The fractional order is attributed to strongly associated ion pairs at the chain ends. These ion pairs must dissociate to provide a low concentration of unassociated ion pairs prior to propagation. In the case of potassium silanolate, this dissociation is pictured as the initiation of the polymerization (*37*), as shown by equations 10 and 11.

$$\underset{\mathbf{2}}{\sim Si{-}O\overset{K^+}{\diagup\diagdown}\underset{K^+}{}O{-}Si\sim} \rightleftharpoons \underset{\mathbf{3}}{2\ \sim SiO^-K^+} \tag{10}$$

$$\sim SiO^-K^+ + D_x \rightleftharpoons \mathbf{1} \rightleftharpoons \sim(Si{-}O{-})_xSiO^-K^+ \tag{11}$$

The potassium silanolate exists largely as the associated species **2**, which is considered unreactive. The molecular weight distribution inevitably shows a low-molecular-weight "tail" when these catalysts are used and is attributed to the fraction of the chain ends in the inactive form **2** (*37*).

The redistribution of siloxane units between chains via equilibrium 3 can be attributed to the existence of intermediates such as **4**, which is similar to **2**, except that siloxane units other than those at the ends become involved:

$$\sim\!Si\!-\!\left(\begin{matrix}Si\\|\\O\end{matrix}\right)_n\!\cdots K^+ \cdots O^- \cdots K^+ \cdots O\!-\!Si\!\sim\ \rightleftharpoons\ SiOSiO^-K^+ + Si(OSi)_{n-1}OK \qquad (12)$$

4

Intermediate **4** was proposed originally to explain a rapid specific redistribution in which $n = 1$ (*17*) but the intermediate is more generally represented in equation 12. Likewise, intermediates such as **5**, which would be similar in nature to **1**, could also be responsible for redistributions between chains.

$$\sim\!Si - O\cdots K^+ \left(\begin{matrix}Si\\ \\Si\end{matrix}\!>\!O\right)_{x-1}\ \rightleftharpoons\ \sim\!SiOSi + \sim\!Si(OSi)_xOK \qquad (13)$$

5

Each of the intermediates in the preceding processes (equations 10–13), **1**, **2**, **4**, and **5**, involves coordinated cyclic structures that incorporate the cationic portion of the ion pair at the chain end. Thus, the extent of reactions 12 and 13, as well as the forward and reverse of reaction 11, should be sensitive to the polar nature and the spatial requirements of the countercation. These reactions, for example, might be suppressed by the presence of species more basic than siloxane that can competitively complex more strongly with the countercation (*33*).

A large effective cation size should suppress the cation–siloxane coordination, favor the free ion pair **3** in reaction 10, and enhance charge separation. The anticipated effects would be values of n approaching 1 in the rate equation 9, greatly enhanced rates of polymerization, and suppressed formation of cyclosiloxanes. Evidence that these effects are achieved is indicated by the effects seen with R_4N^+ countercations (*35*), the lithium cryptates (*25, 27*), and the crown ether–potassium silanolate complexes (*39, 40*). Additional evidence for the influence of the countercation on the equilibria is seen in deviations of the amounts of oligomer produced in equilibrated poly(dimethylsiloxane) from the normal distribution caused by specific interactions between the potassium silanolate chain ends (*37, 38*). More definitive experiments are required, but the equilibria occurring during basic siloxane ring-opening polymerization as originally represented in equations 2 and 3 are probably too general, and the chain ends must be included somehow in the representation of these processes.

Acid-Initiated Polymerization. Acid-initiated polymerization of cyclosiloxanes has not received nearly as much attention in the published literature until recently. Consequently, the uncertainty and controversy over the mechanism is greater than that for base-initiated polymerizations. A recent paper by Sigwalt (*41*) reviews the more current activity on this subject. This focused primarily on initiation with trifluoromethanesulfonic acid in polymerizations of D_3 and D_4. Very little reliable published information exists on the most common initiator, sulfuric acid, despite a long history of continued application (*42*).

Acid-initiated polymerization resembles the base-initiated process in terms of yields of polymer and other products formed at equilibrium. The pathway to the final product, however, is quite different, because the overall rate equation and relative reactivities of the monomers differ considerably from those of the base-initiated process. The polymerization of D_4 when catalyzed by either sulfuric acid (*43*) or trifluoromethanesulfonic acid (*26*) proceeds as a step-growth polymerization as evidenced by traces of the molecular weight distribution as monomer is converted (*44*). After an initial induction period, the molecular weight of the polymer fraction increases, whereas the molecular weight distribution steadily broadens until it fits the distribution expected for a step-growth polymer (*43*). Low-molecular-weight polymer, oligomers, and cyclosiloxanes other than D_4 (D_5–D_9), on the other hand, appear quickly and then remain invariant throughout the polymerization (*26*). Relative reactivities based on observations of kinetics during acid-initiated polymerization fall into the sequence $D_3 > MM > MDM > MD_nM > D_4$, whereas the sequence for polymerizations initiated by base is $D_3 > D_4 > MD_nM > MDM > MM$ (*45*). The effect of this difference in acid-initiated polymerization is to give a continuous monotonic approach to an equilibrium molecular weight in D_4 polymerizations, whereas a maximum is often seen in base-initiated polymerizations when MM or MDM are present as end stoppers.

On the basis of initial rate studies, the rate equation for polymerization of D_4 with trifluoromethanesulfonic acid (TfOH) takes the following form (*46*):

$$R_p = [\mathrm{TfOH}]^{2.2}[D_4]^{0.7} \tag{14}$$

Studies of the kinetics at higher conversions show a first-order dependence on D_4, but the apparent first-order rate constant decreases with initial D_4 concentration according to $k \sim [D_4]_o^{-0.8}$ (*26*). In this work, the order in acidic initiator was 2.7, with either D_4 or D_6 as monomers. Polymerizations of D_3 with TfOH showed the same near-third-order dependence on acid and a strong reciprocal dependence on the initial D_3 concentration. Water had a potent effect on rate, although conflicting observations between two different schools of workers (*26, 47*) are apparent. Whereas an activating effect was seen in one case (*47*), just the opposite was seen in another (*26*). Differences between the two observations are probably due to the 10-fold differences in the concentration of TfOH used.

An unusual distribution of cyclosiloxanes appears during the TfOH-initiated polymerization of D_3 before the final equilibrium composition is reached. The cyclosiloxane byproducts had ring sizes in the series D_{3x} (in which x is an integer) with few of the other cyclosiloxanes present. One group explains the phenomena in terms of the newly formed reactive chain end preferentially reacting with its opposite chain end rather than a monomer unit in the chain. This hypothesis would explain the observed retarded propagation: Early in the polymerization, new chain ends form cyclic structures, but later, they are too far from their point of origin for ring closure to be probable (*48*). Other investigators have found fault with this argument, however, and an alternative mechanism involving ring expansion has been offered (*41*).

Although still subject to controversy and in need of more supportive evidence, a reasonable, albeit complex, mechanism for the acid-initiated process can be written that is consistent with most of the observations made so far. The mechanism consists of four parts: initiation, step growth, chain growth, and termination. The process outlined in equations 15–18 was originally proposed by Wilczek and Chojnowski (*49*), and although it differs in detail, it resembles the mechanism proposed by Sigwalt (*41, 50*). The acidic initiator, indicated by HA, represents TfOH, but a similar mechanism may be written for HA = H_2SO_4, at least in situations in which the system would be homogeneous.

initiation:

$$\left(\begin{matrix}\text{Si}\\ \\ \text{Si}\end{matrix}\right\rangle\text{O} + \text{HA} \rightleftharpoons \left(\begin{matrix}\text{Si}\\ \\ \text{Si}\end{matrix}\right\rangle\text{O HA} \rightleftharpoons \text{HOSi}\frown\text{SiA} \quad (15a)$$

$$\frown\text{SiA} + \text{HA}\cdot\text{H}_2\text{O} \rightleftharpoons \frown\text{SiOH} + 2\text{HA} \quad (15b)$$

$$\frown\text{SiA} + \text{HA} \rightleftharpoons \frown\text{SiA}\cdot\text{HA} \quad (15c)$$

step growth:

$$\frown\text{SiA} + \frown\text{SiOH} \rightleftharpoons \frown\text{SiOSi}\frown + \text{HA} \quad (16a)$$

$$\frown\text{SiOH} \overset{\text{HA}}{\rightleftharpoons} \frown\text{SiOSi}\frown + \text{H}_2\text{O} \quad (16b)$$

chain growth:

$$\frown \mathrm{SiA \cdot HA} + \left(\begin{array}{c}\mathrm{Si} \\ \mathrm{Si}\end{array}\!\!\!\!\!\!\!\!>\mathrm{O}\right) \rightleftharpoons \frown \mathrm{Si{-}^{+}O}\!\!<\!\!\left(\begin{array}{c}\mathrm{Si} \\ \mathrm{Si}\end{array}\right) \cdot \mathrm{HA_2^-} \quad (17a)$$

$$\frown \mathrm{Si{-}^{+}O}\!\!<\!\!\left(\begin{array}{c}\mathrm{Si} \\ \mathrm{Si}\end{array}\right) \cdot \mathrm{HA_2^-} + \left(\begin{array}{c}\mathrm{Si} \\ \mathrm{Si}\end{array}\!\!\!\!\!\!\!\!>\mathrm{O}\right) \rightleftharpoons$$

$$\frown \mathrm{SiOSi} \frown \mathrm{Si^{+}{-}O}\!\!<\!\!\left(\begin{array}{c}\mathrm{Si} \\ \mathrm{Si}\end{array}\right) \cdot \mathrm{HA_2^-} \quad (17b)$$

termination:

$$\frown \mathrm{Si{-}^{+}O}\!\!<\!\!\left(\begin{array}{c}\mathrm{Si} \\ \mathrm{Si}\end{array}\right) \cdot \mathrm{HA_2^-} \rightleftharpoons \frown \mathrm{SiA} + \left(\begin{array}{c}\mathrm{Si} \\ \mathrm{Si}\end{array}\!\!\!\!\!\!\!\!>\mathrm{O}\right) \cdot \mathrm{HA} \quad (18)$$

A rate equation based on such a mechanism explains the apparent −1 order in $[D_3]_0$ but not the higher negative orders sometimes observed. Likewise, the derived rate equation predicts a second-order dependence on TfOH, rather than the sometimes-observed near-third-order dependence. Formation of the unusual sequence of rings, D_{3x}, can result from intramolecular versions of reaction 16a or 16b or, alternatively, from reaction 17b leading to ring expansion rather than ring opening.

Rate retardation by water can be explained readily by destruction of the essential silyl ester via equation 15b, although the reported dramatic rate enhancement has yet to be explained satisfactorily, despite previous attempts (*41*, *47*, *49*). The intermediates advocated in the previous scheme are, at best, only partly correct and understood. This fact, combined with the complexity of the process, suggests that complete explanations will come only with considerably more effort.

Copolymerization

The classical Mayo–Lewis scheme relating comonomer feeds to relative reactivity ratios (*51*) is often applied to copolymerization of cyclosiloxanes. This scheme presumes that no depropagation of the copolymer occurs, that the copolymerization rate constants depend only on the ultimate comonomer units, and that instantaneous comonomer feed ratios and copolymer compositions are used in the analysis of data. When these assumptions hold, the Mayo–Lewis method is very useful for the analysis of copolymerization data.

Otherwise, the method can be very misleading; it failed badly when published data from certain copolymerizations were scrutinized by a specially developed analytical method for copolymerization data (*52*, *53*). For these reasons, any published data on copolymerization reactivity ratios must be approached with caution, and the reader should be assured that the criteria just mentioned have been properly met. These precautions are especially true for siloxane copolymerizations, because reversibility can be established quickly, and rapid redistribution of comonomer units between rings and chains via equilibria 2 and 3 will confound any meaningful study of comonomer reactivity ratios.

Therefore, all early studies of copolymerizations must be examined critically when the ring size of any of the comonomers exceeds three. For example, published values of r_1 and r_2 are regarded as having only qualitative significance in studies of the copolymerization of D_6 with $[CH_2{=}CH(CH_3)SiO]_4$ initiated by potassium hydroxide at 150 °C (*54*, *55*). These data failed (Saam, unpublished results) the method of analysis of copolymerization data developed by Kelen et al. (*53*). Likewise quantitative results have little meaning in the copolymerization of D_4 with $(CH_3HSiO)_4$ and $(CH_3HSiO)_5$ initiated by aluminum sulfate at an unspecified temperature (*56*). For both studies, D_6 and D_4 appeared to be qualitatively less reactive than their comonomers, but even this conclusion should be made with reservations.

More significant data are obtained from copolymerizations of strained cyclosiloxanes with initiators such as LiOSi≡, in which reversibility is minimized (*57*, *58*). Some examples are given in Table II. The potent effect of solvent is seen in the copolymerization of D_3 with $[(C_6H_5)_2SiO]_3$. The copolymer tends to be blocklike when the copolymerization is conducted in benzene with about 1% dimethyl sulfoxide as a promoter but changes to an alternating structure when only 10% of THF (tetrahydrofuran) is present. Copolymerization of the $[CH_3(C_6H_5)SiO]_3$ isomers with D_3, with no diluent or promoter but with a sodium silanolate initiator, had a strong tendency to form block polymers. This tendency was partly ameliorated when the D_3 was replaced with the co-cyclosiloxanes indicated. The results show that aromatic monomers have a strong tendency to polymerize preferentially, but this tendency changes dramatically with changes of the countercation or the solvent. These results undoubtedly reflect the importance of the nature of the ion pair at the reactive chain end and its sensitivity to changes in the type of solvent and countercation.

The lack of reliable information on kinetic parameters for the copolymerization of important cyclosiloxanes with ring sizes exceeding three is a serious deficiency. Industrial and routine laboratory syntheses rely almost entirely on such monomers, and in many cases, the processes are copolymerizations. An understanding of these constants would give valuable information on copolymer microstructure and its control. Unambiguous studies

Table II. Copolymerization of Strained Ring Cyclosiloxanes

Monomer 1	*Monomer 2*	*Initiator*	*Solvent*	*Rate Constant*	
				r_1	r_2
$[(C_6H_5)_2SiO]_3$[a]	D_3	Butyllithium	C_6H_6	2.1	0.04
$[(C_6H_5)_2SiO]_3$[a]	D_3	Butyllithium	C_6H_6–THF (0.9:0.1)	0.65	0.09
trans-$[C_6H_5(CH_3)SiO]_3$[b]	D_3	NaOSi		70.2	0.074
cis-$[C_6H_5(CH_3)SiO]_3$[b]	D_3	NaOSi		59.9	0.02
cis-$[C_6H_5(CH_3)SiO]_3$[b]	$[C_6H_5(CH_3)SiO]D_2$	NaOSi		18.6	0.22
cis-$[C_6H_5(CH_3)SiO]_3$[b]	$[CH_2{=}CH(CH_3)SiO]D_2$	NaOSi		18.2	0.11

[a]Based on data from reference 57. All runs contained 1% dimethyl sulfoxide.
[b]Based on data from reference 58.

of such copolymerizations have been hampered by reversibility until recently, when alternatives to the Mayo–Lewis model have been advocated that rigorously take into account reversibility, predict microstructure at various stages of conversion, and determine how microstructure is influenced by kinetic factors (*59–61*). Therefore, the application of these models to the kinetics of reversible cyclosiloxane copolymerizations would be useful.

Copolymerization at equilibrium appears to be an entirely random process, according to three separate studies of KOH-initiated copolymerizations of D_4 with $[CH_2{=}CH(CH_3)SiO]_4$ (*62*) and $[(C_6H_5)_2SiO]_4$ (*63*, *64*). The conclusions in each case were based on the analysis of signal intensities of pentad and triad sequences in high-resolution ^{29}Si NMR of the copolymers. Two of these studies (*62*, *64*) showed that the random sequence distribution persisted over a broad range of composition, and one (*62*) showed that this result was unaffected by temperature. The findings indicated that, as with homopolymerization, there was no enthalpic driving force. The same study also showed that the composition of the cyclosiloxanes at equilibrium and the sequencing of their comonomer units matched exactly those of the copolymer chains. The binomial distribution, which presumes random statistics, accurately described the distributions of the various equilibrium cyclosiloxanes observed by gas chromatography.

The random nature of the copolymerization equilibria can be considered a consequence of two concurrent entropically driven equilibria similar to reactions 2 and 3. These copolymerization equilibria, however, would involve the comonomers interacting reversibly with two different chain ends and the reversible transfer of the different comonomer units between chains. Expressed in another way, the equilibria could be written in a manner similar to the Mayo–Lewis model but with rate constants replaced by equilibrium constants, K_{11}, K_{12}, K_{22}, and K_{21}, and comonomer concentrations replaced by the total concentrations of the different siloxane units in the system, M_1 and M_2, regardless of their locations in the rings or chains.

$$\sim M_1^* + M_1 \xrightleftharpoons{K_{11}} M_1M_1 \quad r_1 = K_{11}/K_{12} \tag{19a}$$

$$\sim M_1^* + M_2 \xrightleftharpoons{K_{12}} M_1M_2 \tag{19b}$$

$$\sim M_2^* + M_2 \xrightleftharpoons{K_{22}} M_2M_2 \quad r_2 = K_{22}/K_{21} \tag{19c}$$

$$\sim M_2^* + M_1 \xrightleftharpoons{K_{21}} M_2M_1 \tag{19d}$$

In the preceding reactions, M_1^* and M_2^* are the active chain ends, and the copolymerization ratios r_1 and r_2 are unity in an entropically driven process (*62*).

Step-Growth Polymerization

In practice, step-growth polymerization in linear polysiloxanes consists of either homocondensations of silanol-ended siloxanes or heterocondensations of silanol-ended species with monomers containing good leaving groups, such as halogen, amino, or amido. Heterocondensations can provide convenient synthetic pathways to specifically structured polymers and copolymers (*64*, *65*). Heterocondensation between SiCl- and SiOH-ended species is important industrially in the formation of cyclosiloxanes and linear siloxane oligomers from the hydrolysis of chlorosilanes (*2*). A review of heterocondensations of chlorosilanes and silanols is contained within a broader discussion (*66*) of hydrolytic polycondensations in both linear and nonlinear organosiloxanes. The emphasis in this section will be on only silanol homocondensation.

Silanol polycondensation is catalyzed by strong acids or bases, as well as by species as mild as amines or carboxylic acids combined with their quaternary ammonium salts (*67*). With strong acids or bases, the process is highly reversible, and the equilibrium can be represented as follows:

$$n\mathrm{HOSi}\frown\mathrm{SiOH} \rightleftharpoons (n-1)\mathrm{H_2O} + \mathrm{HO(Si}\frown\mathrm{SiO)}_n\mathrm{H} \qquad (20)$$

The equilibrium constant can be defined as:

$$K = \frac{[\mathrm{SiOSi}][\mathrm{H_2O}]}{[\mathrm{SiOH}]^2} \qquad (21)$$

On the basis of this definition, the number-average degree of polymerization, X, for the polymer at equilibrium is given by the approximation

$$X \approx 2(KK_h)^{1/2}P_{\mathrm{H_2O}}^{-1/2} \qquad (23)$$

in which K_h is Henry's constant and $P_{\mathrm{H_2O}}$ is the partial pressure of water vapor. The approximation is more valid if the partial pressure and the concentration of water in the nonpolar media are sufficiently low so that pressure and concentration can replace fugacity and activity in equations 21 and 22. Equation 22 was first tested by Martellock (*68*) and found to be valid in the depolymerization of poly(dimethylsiloxane) at 150 °C at varied $P_{\mathrm{H_2O}}$ (*68*). Equation 22 can be expressed in terms of thermodynamic constants:

$$\ln X = \frac{-\Delta H + \Delta H_h}{RT} + \frac{\Delta S + \Delta S_h}{R} - \frac{1}{2}\ln P_{\mathrm{H_2O}} \qquad (23a)$$

$$\ln X = \frac{-\Delta H + \Delta H_h - \Delta H_v}{RT} + \frac{\Delta S + \Delta S_h - \Delta S_v}{R} \qquad (23b)$$

In these equations, the subscripts h and v are used to indicate terms that refer to Henry's constant and vapor pressure of water, respectively. Equation 23b closely predicted the molecular weights of poly(dimethylsiloxane) at equilibrium in emulsions stabilized by dodecylbenzenesulfonic acid. The expected reciprocal dependence of ln X on temperature was seen, and both the enthalpy and entropy terms were close to those expected from published values (*69*). Therefore, the adherence of the emulsion condensation polymerizations at equilibrium to equation 23b provides a convenient means of regulating molecular weight by adjusting temperature. Equation 23a should also be effective whenever P_{H_2O} can be adjusted independently of temperature.

The equilibrium 20 is important not only in the synthesis of linear polysiloxanes but also in their applications. The effects of water vapor on inducing chain cleavage at high temperature are not only reduced molecular weights but also a dramatic increase in the rates of chemically induced stress relaxation at 250 °C in cross-linked poly(dimethylsiloxane) networks under load (*70*). Slow hydrolytic bond cleavage in cross-linked networks is seen even in studies of stress relaxation in air at room temperature, and appreciable rates of stress relaxation in the loaded networks are measured at temperatures as low as 70 °C (*71*). The stress relaxation is greatly accelerated by ammonia but essentially eliminated when the experiments are conducted in dry air. Spontaneous adhesion of poly(dimethylsiloxane) networks to polar substrates at room temperature is also attributed to silanols formed via equilibrium 20 from the network chains (*72*). Consistent with the observations in their earlier paper, the authors saw an acceleration of adhesion in ammonia and loss of adhesion in dry air. Thus, equilibrium 20 appears to be ubiquitous and could be playing a larger role than hitherto appreciated in both synthesis and properties. Equilibrium 20 might even occur concurrently in ring opening polymerizations in which water or other hydroxylic species might be present in trace amounts (*73*). This case is seen in acid-initiated ring opening polymerization, as already mentioned (reactions 16a and 16b). Unpublished work (Saam and Stebleton) suggests that the condensation process plays a controlling role in base-catalyzed ring-opening polymerizations when carried out with KOH initiators in the presence of traces of hydroxylic species.

The rate equation (equation 24) for strong-base-catalyzed silanol condensation was based on model studies with trimethylsilanol that were intended to mimic silanol condensation polymerizations (*73*).

$$-\frac{d[\mathrm{SiOH}]}{dt} = \frac{k_a[\mathrm{MOSi}]_o[\mathrm{SiOH}]}{k_b + k_c[\mathrm{SiOH}]_o} \tag{24}$$

The catalyst in these studies was a tetramethylammonium silanolate. $[\mathrm{SiOH}]_o$ represents the initial trimethylsilanol concentration, and k_a, k_b, and k_c are rate constants. Use of alkali metal silanolates complicated the kinetics because of the self-association outlined in reaction 10. A reaction scheme consistent with the rate equation is given by equations 25–27.

$$\sim\mathrm{SiOH} + \mathrm{MOH} \rightleftharpoons \underset{\mathbf{6}}{\sim\mathrm{SiOM}\cdot\mathrm{H_2O}} \tag{25}$$

$$\sim\text{SiOH} + \mathbf{6} \rightleftharpoons \text{Si–O}\overset{\text{H}}{\cdots}\text{O–Si}\sim \quad \text{(cyclic intermediate with } \text{M} \cdot \text{H}_2\text{O}) \; \mathbf{7} \tag{26}$$

$$\mathbf{7} \rightleftharpoons \sim\text{SiOSi}\sim + \text{MOH} \cdot \text{H}_2\text{O} + \text{H}_2\text{O} \tag{27}$$

The intermediate **7** is similar in some aspects to intermediate **2** in reaction 10. The kinetics becomes second order in silanol and first order in a weak base for the polycondensation of dialkylsilanediols (*74*). The mechanism was proposed to proceed via the following reactions:

$$\text{B}' + \sim\text{SiOH} \rightleftharpoons \underset{\mathbf{8}}{\sim\text{SiO}^-\text{BH}^+} \tag{28}$$

$$\mathbf{8} + \sim\text{SiOH} \rightleftharpoons \sim\text{SiOSi}\sim + \text{B}' + \text{H}_2\text{O} \tag{29}$$

In the preceding scheme, B′ represents a weak base such as $(C_4H_9)_3N$.

The rate equation with strongly acidic catalysts is also second order in silanol and first order in catalyst (*75*). The mechanism is proposed to proceed via protonation of silanol, followed by an electrophilic attack of the conjugate acid on nonprotonated silanol. The condensation processes outlined in reactions 16a and 16b for sulfonic acids is also an applicable mechanism for the acid catalysis. The condensation polymerization in emulsion catalyzed by dodecylbenzenesulfonic acid is second order in silanol, but the rate has a complex dependence on sulfonic acid concentration (*69*). This process was most likely a surface catalysis of the oil–water interface and was complicated by self-associations of the catalyst–surfactant.

Conclusions

The present refinement in linear siloxane polymerization is a monumental achievement resulting from the astute observations and ingenuity of many chemists over the past 120 years. The workers cited in this chapter are only some of the more recent contributors. Still, much work is yet to be done, and the critical reader should be left with many questions. For example, the equilibria 2 and 3 are traditionally the basis for explaining the distribution of molecular sizes and byproducts, but they exclude any role for the reactive chain ends. Yet, the accumulating evidence of the critical role of the counterions at the reactive ends in the mechanism of the process suggests that the equilibria may have to be rewritten to include the reactive ends. Definitive experiments are needed to settle the point.

The kinetics of the rapid exchange of siloxane units between chains (equation 3) also needs better definition. The only published account giving data that defines the crucial equilibrium for condensation–cleavage of linear polysiloxanes by water (reaction 20) is old and difficult to find despite its obvious importance. In this case, the observations need to be verified, expanded, and put on sound thermodynamic ground. Mechanisms, especially for condensation polymerization in both the acid- and base-catalyzed processes, need more rigorous verification.

For some important areas, background is entirely lacking or is sparse. The kinetics of reversible copolymerization of cyclosiloxanes is a good example. Prior studies were at best qualitative because of the improper application of the Mayo–Lewis scheme and the lack of suitable alternatives. Quantitative information on copolymerization kinetics is needed to describe copolymer microstructure, which is important in determining the physical and chemical properties of copolymers. The newer models for reversible copolymerization, combined with computer and ^{29}Si NMR analyses, should be powerful tools for gathering and interpreting data unambiguously. Although work has begun in observing copolymerizations at equilibrium, the challenge will be in studies carried out prior to equilibrium. Another area where there is a deficiency is the stereoregular synthesis of polysiloxanes. Such processes should provide extremely interesting materials, and they should be achievable with the aid of knowledge gained by predecessors. Siloxane polymerization therefore is far from a stagnant field, and answers that can be found only in the laboratory are needed to some very challenging questions.

References

1. Ladenburg, A. *Ann. Chem.* **1872,** *164,* 300.
2. Noll, W. *Chemistry and Technology of Silicones;* Academic: New York, 1984.
3. Wright, P. V. In *Ring Opening Polymerization;* Ivin, K. J.; Saegusa, T., Eds.; Elsevier: New York, Vol. 2, p 324.
4. Sigwalt, P. *Actual. Chim.* **1986,** *3,* 45.
5. Kendrick, T. C.; White, J. W.; Parbhoo, B. In *Comprehensive Polymer Science;* Eastmond, G. C.; Ledwith, A.; Russo, S.; Sigwalt, P., Eds.; Pergamon: New York, 1989; Vol. 4, Part II, p 459.
6. Voronkov, M. G.; Mileshkevich, V. P.; Yuzhelevskii, Yu. A. *The Siloxane Bond;* Consultants Bureau: New York, 1978; p 159.
7. Odian, O. *Principles of Polymerization;* Wiley: New York, 1981; p 548.
8. Voronkov, M. G.; Mileshkevich, V. P.; Yuzhelevskii, Yu. A. *The Siloxane Bond;* Consultants Bureau: New York, 1978; p 77–80.
9. Lee, C. L.; Frye, C. L.; Johannson, O. K. *Polym. Prepr.* (*Am. Chem. Soc., Div. Polym. Chem.*) **1969,** *10*(2), 1361.
10. Suryanaryanan, B.; Peace, B. W.; Mayhan, K. G. *J. Polym. Sci., Polym. Chem. Ed.* **1974,** *12,* 1089.
11. Gilbert, A. R.; Kantor, S. W. *J. Polym. Sci.* **1959,** *40,* 35.
12. Gee, G. In *Inorganic Polymers, An International Symposium;* Special Publication No. 15, The Chemical Society; Sidney: London, 1961; p 67.
13. Grassie, N.; MacFarlane, I. G. *Eur. Polym. J.* **1978,** *14,* 875.
14. Carmichael, J. B.; Heffel, J. *J. Phys. Chem.* **1965,** *69,* 2213.

15. Carmichael, J. B.; Heffel, J. *J. Phys. Chem.* **1965,** *69,* 2218.
16. Brown, J. F.; Slusarczuk, G. M. *J. Am. Chem. Soc.* **1965,** *87,* 931.
17. Chojnowski, J.; Kazmerski, K.; Rubinszstein, S.; Stanczyk, W. *Macromol. Chem.* **1986,** *187,* 2039.
18. Tanaka, T. *Bull. Chem. Soc. Jpn.* **1960,** *33,* 282.
19. Lee, C. L.; Johannson, O. K. *J. Polym. Sci., Part A-1* **1966,** *4,* 3013.
20. Semlyn, J. A.; Wright, P. V. *Polymer* **1969,** *10,* 543.
21. Voronkov, M. G.; Mileshkevich, V. P.; Yuzhelevskii, Yu. A. *The Siloxane Bond;* Consultants Bureau: New York, 1978; p 51–53.
22. Wright, P. V.; Semlyen, J. A. *Polymer* **1970,** *11,* 462.
23. Jacobson, H.; Stockmayer, W. H. *J. Chem. Phys.* **1950,** *18,* 1600.
24. Carmichael, J. B.; Gordon, D. J.; Isackson, F. J. *J. Phys. Chem.* **1967,** *71,* 2011.
25. Boileau, S. In *Ring Opening Polymerization;* McGrath, J. E., Ed.; ACS Monograph 286; American Chemical Society: Washington, DC, 1985; p 23.
26. Sauvet, G.; Lebrun, J. J.; Sigwalt, P. *Cationic Polymerization and Related Processes;* Goethals, E. J., Ed.; Academic: New York, 1984; p 237.
27. Hubert, S.; Hemery, P.; Boileau, S. *Macromol. Chem., Macromol. Symp.* **1986,** *6,* 247.
28. Semlyen, J. A. *Macromol. Chem., Macromol Symp.* **1986,** *6,* 155.
29. Buese, M. A. *Macromolecules* **1987,** *20,* 694.
30. Scott, D. W. *J. Am. Chem. Soc.* **1946,** *68,* 2294.
31. Laita, Z.; Jelinek, M. *Polym. Sci. USSR (Engl. Transl.)* **1965,** *4,* 535.
32. Shinohara, M. *Polym. Prepr. (Am. Chem. Soc., Div. Polym. Chem.)* **1973,** *14*(2), 1209.
33. Mazurek, M.; Chojnowski, J. *Macromol. Chem.* **1977,** *10,* 1005.
34. Ollif, C. J.; Ladbrook, P. *J. Electroanal. Chem. Interfacial Electrochem.* **1979,** *104,* 105.
35. Andrianov, K. A.; Nogaideli, A. L.; Kopylov, V. M.; Kubulava, E. I.; Kolchina, A. G.; Khananashvili, L. M.; Sycheva, N. G. *Soobshch. Akad. Nauk Gruz. SSR* **1976,** *82,* 361.
36. Wilczek, L.; Kennedy, J. P. *Polym. J.* **1987,** *19*(5), 531.
37. Yuzhelevsky, Yu. A.; Kagan, E. G.; Timofeeva, N. P.; Doletskaya, T. D.; Klebansky, A. L. *Vysokomol. Soedin., Ser. A* **1973,** *1,* 183.
38. Mazurek, M.; Scibiorek, M.; Chojnowski, J.; Zavin, B. G.; Zdanov, A. A. *Eur. Polym. J.* **1980,** *66*(1), 67.
39. Evans, E. R. U.S. Patent 4 157 337, 1979.
40. Bargain, M.; Millet, C. U.S. Patent 4 138 543, 1979.
41. Sigwalt, P. *Polym. J.* **1987,** *19,* 567.
42. Patnode, W.; Wilcox, D. *J. Am. Chem. Soc.* **1948,** *68,* 358.
43. Kojima, K.; Tarumi, J.; Wakatsuki, Y. *Nippon Kagaku Zasshi.* **1956,** *77*(12), 1755.
44. Odian, O. *Principles of Polymerization;* Wiley: New York, 1981; p 90.
45. Kendrick, T. C.; White, J. W.; Parbhoo, B. In *Comprehensive Polymer Science;* Eastmond, G. C.; Ledwith, A.; Russo, S.; Sigwalt, P., Eds.; Pergamon: New York, 1989; Vol. 4, Part II, p 459.
46. Wilczek, L.; Rubinsztajn, S.; Chojnowski, J. *Macromol. Chem.* **1986,** *187,* 39.
47. Wilczek, L.; Chojnowski, J. *Macromolecules* **1981,** *14,* 9.
48. Chojnowski, J.; Scibiorek, M.; Kowalski, J. *Macromol. Chem.* **1977,** *178,* 1351.
49. Wilczek, L.; Chojnowski, J. *Macromol. Chem.* **1979,** *180,* 117.
50. Lebrun, J. J.; Sauvet, G.; Sigwalt, P. *Cationic Polymerization and Related Processes;* Goethals, E. J., Ed.; Academic: New York, 1984; p 237.
51. Mayo, F. R.; Lewis, F. M. *J. Am. Chem. Soc.* **1944,** *66,* 1594.
52. Kennedy, J. P.; Kelen, T.; Tudos, F. *J. Polym. Sci., Polym. Chem. Ed.* **1975,** *13,* 2277.

53. Braun, D.; Czerwinski, W.; Disselhoff, G.; Tudos, F.; Kelen, T.; Turcsanyi, B. *Angew. Macromol. Chem.* **1984,** *125,* 161.
54. Jelinek, M.; Laita, Z; Kucera, M. *J. Polym. Sci., Part C* **1967,** *16,* 431.
55. Laita, Z; Jelinek, M. *Vysokomol. Soledin.* **1963,** 5(8), 1268.
56. Reiksfeld, V. O.; Ivanova, A. G. *Vysokomol. Soledin.* **1962,** *4*(1), 30.
57. Lee, C. L.; Marko, O. W. *Polym. Prepr. (Am. Chem. Soc., Div. Polym. Chem.)* **1978,** *19*(1), 250.
58. Baratova, T. N.; Mileshkevich, V. P.; Gulari, V. E. *Polym. Sci. USSR (Engl. Transl.)* **1983,** *12,* 2899.
59. Szwarc, M.; Perrin, C. L. *Macromolecules* **1985,** *18,* 528.
60. Jensen, P. J.; Bennemann, K. H. *J. Chem. Phys.* **1985,** *83,* 6457.
61. Cai, G.; Yan, D. *J. Macromol. Sci., Chem.* **1987,** *A24*(8), 869.
62. Ziemelis, M. J.; Saam, J. C. *Macromolecules* **1989,** *22,* 2111.
63. Andolino Brandt, P. J.; Subramanian, R.; Sormani, P. M.; Ward, T. C.; McGrath, J. E. *Polym. Prepr. (Am. Chem. Soc., Div. Polym. Chem.)* **1985,** *26*(2), 213.
64. Babu, G. N.; Christopher, S. S.; Newmark, R. A. *Macromolecules* **1987,** *20,* 2654.
65. Livingston, M. E.; Dvornik, P. R.; Lenz, R. W. *J. Appl. Polym. Sci.* **1982,** *27,* 3239.
66. Kireev, V. V. In *Advances in Polymer Chemistry;* Korshak, V. V., Ed.; MIR Publishers: Moscow, 1986; p 199.
67. Bowman, S. A.; Falender, J. R.; Lipowitz, J.; Saam, J. C. U.S. Patent 4 486 567, 1984.
68. Martellock, A. C. *International Conference on Elastoplastics Technology;* Wayne State University: Detroit, MI, 1966; p 385.
69. Saam, J. C.; Huebner, D. J. *J. Polym. Sci, Polym. Chem. Ed.* **1982,** *20,* 3351.
70. Thomas, D. K. *Polymer* **1966,** 7(2), 99.
71. Gent, A. N.; Vondracek, P. *J. Appl. Polym. Sci.* **1982,** *27,* 4517.
72. Gent, A. N.; Vondracek, P. *J. Appl. Polym. Sci.* **1982,** *27,* 4357.
73. Andrianov, K. A.; Prikhod'ko, P. L.; Kopylov, V. M.; Gasanov, A. M.; Khananashvili, L. M. *J. Gen. Chem. USSR (Engl. Transl.)* **1979,** *49*(2), 109.
74. Chojnowski, J.; Chrzczonowicz, S. *Bull. Akad. Pol. Sci., Ser. Sci. Chim.* **1966,** *14*(1), 17, CA65, 586C.
75. Lasocki, Z. *Bull. Acad. Polon. Sci., Ser. Sci. Chim.* **1964,** *12*(4), 277.

RECEIVED for review May 27, 1988. ACCEPTED revised manuscript March 20, 1989.

4

Thermal and Rheological Properties of Alkyl-Substituted Polysiloxanes

Husam A. A. Rasoul, Steven M. Hurley, E. Bruce Orler, and Kevin M. O'Connor[1]

Louis Laboratory, S. C. Johnson and Son, Inc., 1525 Howe Street, Racine, WI 53403

*The thermal and rheological properties of a series of poly(dimethylsiloxane-*co*-methylalkylsiloxane) (PDM–PMAS) copolymers containing 3.5 mol % methylalkylsiloxane units of various alkyl lengths were investigated. Calorimetric results show that the alkyl side chains are crystallizable. The side-chain melting temperatures and heats of fusion normalized for side-chain weight fraction increased with increasing side-chain length. The steady-shear melt viscosity of the polymers with C_{10}, C_{12}, and C_{14} side chains decreased with increasing side-chain length. Low-strain oscillatory measurements indicated the formation of a network structure at room temperature for polymers with C_{16} and C_{18} side chains, which can be attributed to intermolecular crystallization of the paraffinic side chains.*

POLYSILOXANES WITH PENDANT SIDE CHAINS are interesting materials from both the theoretical and practical points of view. A number of polysiloxanes with various side chains, such as liquid crystals (*1*, *2*), carbazole groups (*3*), electron-donor and electron-acceptor groups (*4*), polystyrene (*5*), and functional groups (hydroxyl or carboxyl) (*6*), have been synthesized. Polysiloxanes are known for their useful properties, which include flexibility, heat resistance, water repellence, and biological inertness. These properties, combined with the ease with which a tailored polymer structure can be prepared,

[1]Current address: Research Laboratories, Eastman Kodak Company, Building 82D, Rochester, NY 14650–2116

0065–2393/90/0224–0091$06.00/0

prompted investigators to consider these types of polymers in a number of applications, including their uses as drug carriers, mold-releasing agents, and photoconductive polymers. Our own particular interest has been the thermal and rheological behavior of polysiloxanes with crystallizable side chains. In a recent work (7), we reported that the alkyl side chains of poly(methylalkylsiloxane) (PMAS) are crystallizable and exist in the hexagonal unit cell. We also observed that the melt viscosity of PMAS increased with increasing side-chain length.

In this chapter, the properties of a series of poly(dimethylsiloxane-*co*-methylalkylsiloxane) (PDM–PMAS) of various alkyl side-chain lengths are discussed. These polymers contain 3.5 mol % methylalkylsiloxane units and are prepared from the same precursor; thus, any effects due to variations in the percentage of alkyl substitution and main-chain molecular weight differences and effects due to block distribution of the alkyl substituents are eliminated.

Experimental Procedures

Synthesis. Synthesis of the copolymers was performed by a hydrosilylation reaction of poly(dimethylsiloxane-*co*-methylhydrosiloxane) (Petrarch System, Inc.) and α-olefins of various lengths (Aldrich). A round-bottomed flask equipped with a magnetic stirring bar, condenser, and calcium chloride tube was charged with a 50% solution of the reactants (up to 10% molar excess of α-olefin) in dry toluene. A solution of hydrogen hexachloroplatinate(IV) in diglyme–isopropyl alcohol (150 ppm Pt) was then added to the reaction mixture. The reaction mixture was stirred at 60 °C for 3 h. At the end of this period, the mixture was refluxed with activated charcoal for 1 h and filtered while hot. Finally the solvent and excess α-olefins were removed under reduced pressure (67 Pa at 100 °C). The reaction proceeded to completion as evidenced by the absence of the Si–H absorption at 2130 cm^{-1} in the IR spectra. Residual α-olefin in the purified polymers was determined by gas–liquid chromatography. In all polymers, residual α-olefin was less than 1.5 wt %.

Thermal Analysis. Differential scanning calorimetry (DSC) was performed with a DSC module (Du Pont 910) interfaced with a thermal analyzer (model 9900). The instrument was calibrated with indium and high-purity water. Samples ranging in size from 1 to 3 mg were initially heated to 100 °C to erase any prior thermal history. The samples were held at 100 °C for 1 min and then cooled to −20 °C at a rate of 10 °C/min. At −20 °C, the cooling rate was decreased to 2 °C/min until −100 °C and further to −150 °C at 10 °C/min. Each sample was scanned at a heating rate of 20 °C/min. The glass transition temperature (T_g) was determined as the midpoint of the change in heat capacity, and the melting temperature (T_m) was the temperature at the onset of the melting endotherm.

Rheological Experiments. Melt viscosity and low-strain oscillatory experiments were performed on a Rheometrics RDS-7700 dynamic spectrometer equipped with a 0.2–2.0-g-cm torque transducer. The samples were mounted on 25-mm-diameter parallel-plate fixtures with a gap of 0.5 mm. Prior to each scan, samples were heated to 50 °C and then cooled slowly to room temperature. Steady-shear measurements were obtained at shear rates from 0.03 to 1000 s^{-1}. Dynamic measurements were obtained by using frequencies between 0.01 to 100 rad/s at 75% strain, except for polymers with C_{16} and C_{18} side chains, for which 2% strain was used.

Results and Discussion

Thermal Analysis. As expected, PDM–PMAS polymers showed melting endotherms associated with the melting of the paraffinic side chains similar to those previously observed for PMAS polymers (7). Both side-chain melting temperatures (T_ms) and heats of fusion (ΔH_fs) normalized for side-chain weight fraction increased with increasing side-chain length. However, for a given side-chain length, T_m and ΔH_f are depressed for PDM-PMAS polymers relative to those for PMAS (Figures 1 and 2). Table I lists the side-

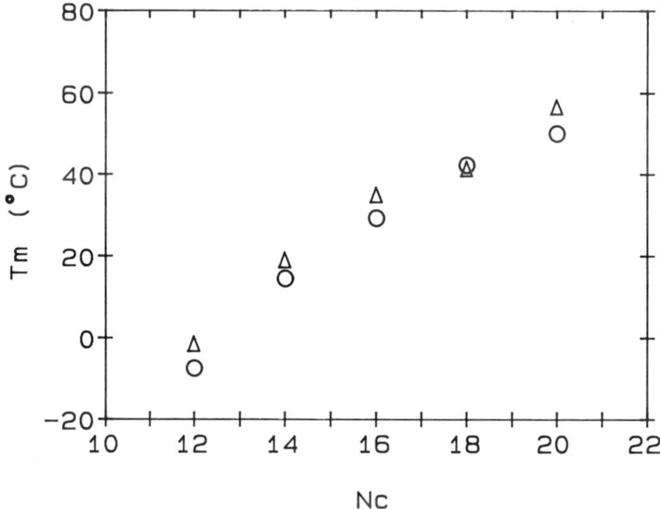

Figure 1. Side-chain melting point (T_m) *versus the number of side-chain carbon atoms* (N_c). *Key:* ○, *PDM–PMAS; and* Δ, *PMAS. The data for PMAS were obtained from reference 7.*

chain T_ms for PDM–PMAS copolymers and the corresponding ΔH_fs. The decrease in side-chain T_m and ΔH_f for PDM–PMAS compared with PMAS (7) are being analyzed at present in terms of the increased spacing between alkyl groups in the PDM–PMAS polymers and its influence on crystal thickness, lateral crystal size, and entropy of melting.

The thermal transitions associated with the backbone were investigated also. The glass transition temperatures for all PDM–PMAS copolymers and their precursor were identical (–120 °C). Other thermal events associated

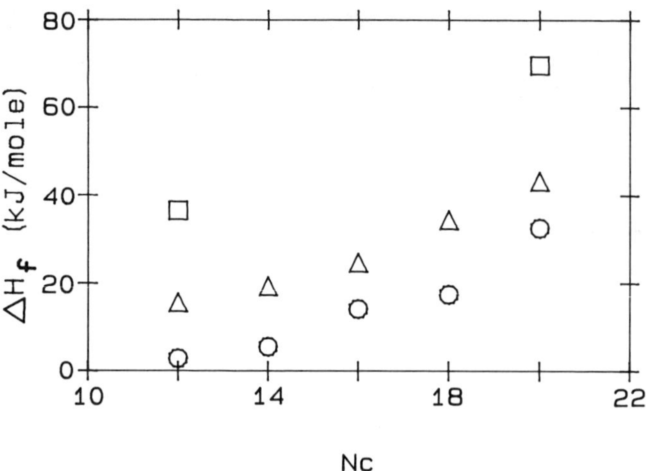

Figure 2. Side-chain heat of fusion (ΔH_f) versus the number of side-chain carbon atoms (N_c). Key: ○, PDM–PMAS; Δ, PMAS; and □, n-alkanes. The data for PMAS were obtained from reference 7.

Table I. Thermal Events for PDM–PMAS Copolymers and Precursor

	Side Chain		*Backbone*	
Material	T_m (°C)	ΔH_f (kJ/mol)	T_{mb} (°C)	ΔH_f (J/g)
Precursor	—[a]	—	−58	12.6
Copolymers[b]				
C_{10}	—	—	—	—
C_{12}	8	3.0	—	—
C_{14}	14	5.5	−68	33.5
C_{16}	29	14.2	−65	24.7
C_{18}	42	17.6	−62	22.0
C_{20}	50	32.5	−62	23.9

[a]— means material did not exhibit thermal event.
[b]The PDM–PMAS copolymers differ in the length of the alkyl side chains.

with the backbone of these polymers are summarized in Table I. The melting temperature of the backbone (T_{mb}) is not influenced by the length of the paraffinic side chain, except for polymers with C_{10} and C_{12} side chains, for which no backbone melting endotherm was observed, presumably because of the higher miscibility of shorter paraffinic chains with poly(dimethylsiloxane). The C_{10} and C_{12} side chains are probably too short to phase-separate from the backbone, and thus they disrupt backbone crystallization. Neither PMAS polymers nor their precursor studied earlier exhibit backbone melting transitions. Figure 3 shows representative DSC traces of poly(methylhexadecylsiloxane) (a C_{16} PMAS) and PDM–PMAS copolymers with C_{16} and C_{12} side chains. The precursor and most of the PDM–PMAS

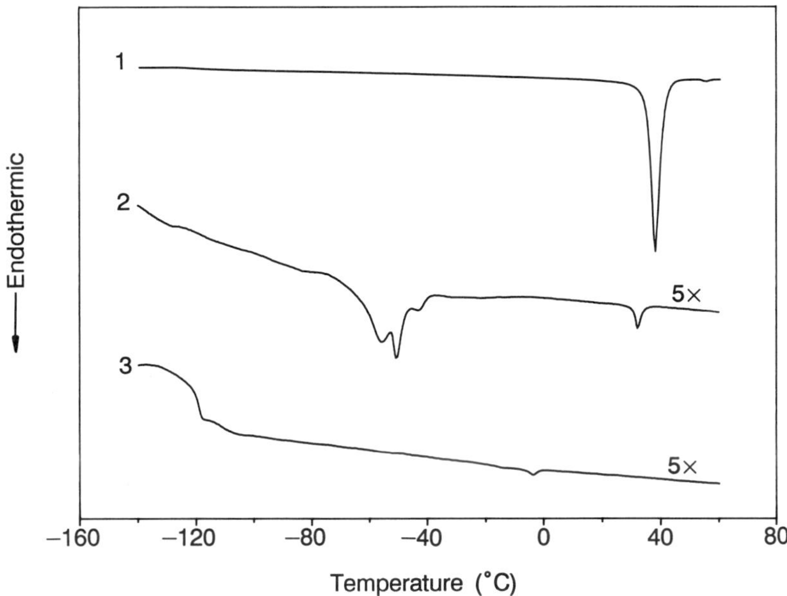

*Figure 3. Representative DSC traces: (1) poly(methylhexadecylsiloxane) (a C_{16} PMAS), (2) poly(dimethylsiloxane-*co*-methylhexadecylsiloxane) (PDM–PMAS with a C_{16} side chain), and (3) poly(dimethylsiloxane-*co*-methyldodecylsiloxane) (PDM–PMAS with a C_{12} side chain).*

copolymers show bimodal or multimodal backbone melting endotherms, when observed. A detailed study of the backbone melting behavior is reported elsewhere (*8*).

Rheological Behavior. Figure 4 shows the room-temperature steady-shear viscosity as a function of shear rate for PDM–PMAS polymers and their precursors. Polymers with C_{10}, C_{12}, and C_{14} side chains exhibit Newtonian behavior over the range of shear rates monitored.

The order of decreasing viscosity is as follows: $C_{10} > C_{12} > C_{14} >$ precursor. This order is contrary to that of fully substituted PMAS polymers (*7*). The decrease in viscosity in this order probably results from the decrease in side-chain miscibility with poly(dimethylsiloxane). Longer side chains, such as C_{14} side chains, are expected to phase-separate and form a more-ordered polymer compared with short side chains (C_{10}). However, C_{12} and C_{14} side chains are not long enough to crystallize above room temperature.

Polymers with longer side-chains (C_{16} and C_{18}) exhibit non-Newtonian viscosity at room temperature, and the viscosity decreases with increasing shear rate. Because the side-chain T_ms for C_{16} and C_{18} polymers are above room temperature, network formation via intermolecular crystallization of the paraffinic side-chains is believed to be responsible for the unusually high

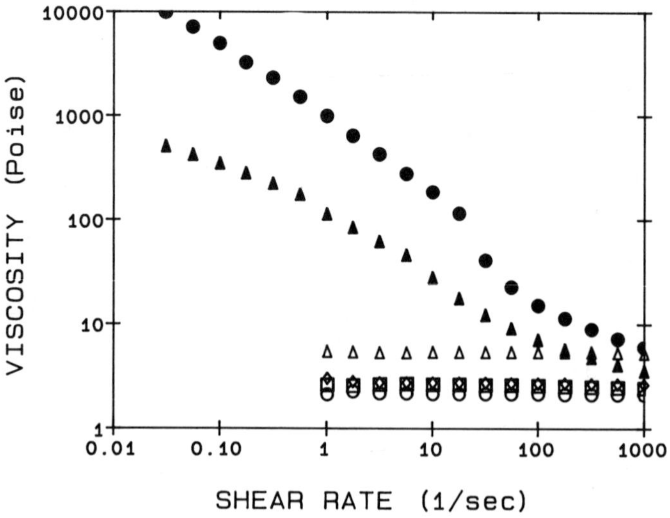

Figure 4. Steady-shear viscosity versus shear rate for PDM–PMAS. Key: ○, precursor; Δ, N_c = 10; ◇, N_c = 12; □, N_c = 14; ▲, N_c = 16; and ●, N_c = 18. N_c is the number of side-chain carbon atoms.

viscosity of these polymers at low shear rates. Indeed, low-strain oscillatory measurements indicate some type of network structure, as evident from the relatively constant value of the storage modulus G' for polymers with C_{16} and C_{18} side chains at low frequencies (Figure 5). The higher level of G' for the polymer with C_{18} side chain can be due to lower molecular weight

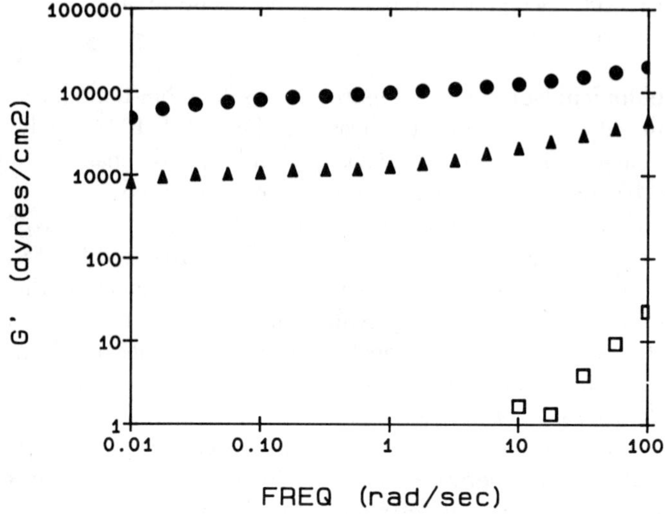

Figure 5. Storage modulus (G') versus frequency of oscillation for PDM–PMAS copolymers. Key: □, N_c = 14; ▲, N_c = 16; and ●, N_c = 18. N_c is the number of side-chain carbon atoms.

between cross-links. However, because all the polymers were prepared from the same precursor, the high level of G' is more probably due to the fact that crystals of polymers with C_{16} side chain are thermally labile (T_m is close to room temperature).

In Figure 6, the loss modulus (G'') is plotted versus the frequency of oscillation. PDM–PMAS copolymers with C_{10}, C_{12}, and C_{14} side chains and

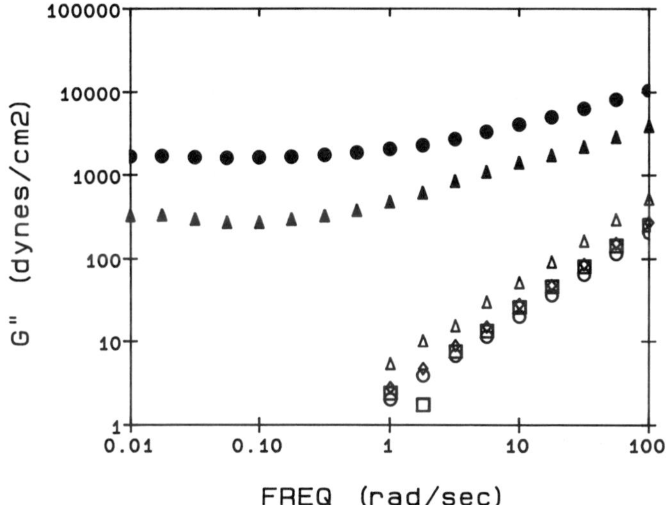

Figure 6. Loss modulus (G'') versus frequency of oscillation for PDM–PMAS polymers. Key: ○, precursor; Δ, N_c = 10; ◇, N_c = 12; □, N_c = 14; ▲, N_c = 16; and ●, N_c = 18. N_c is the number of side-chain carbon atoms.

the precursor exhibit a normal viscous liquid behavior, with G'' directly proportional to frequency and increasing in the same order as the steady-shear melt viscosity. For polymers with C_{16} and C_{18} side chains, G' is greater than G'' at any frequency because of the elastic nature of the network. G'' is closer to G' for polymers with C_{16} side chains compared with polymers with C_{18} side chains. This result indicates that PDM–PMAS copolymers with longer side chains are more structured. Figures 5 and 6 represent only part of the isothermal frequency response of these structured fluids, and we are currently investigating the use of time–temperature superposition to extend the range of frequency.

Conclusions

1. A DSC study shows that the alkyl side chains of PDM–PMAS copolymers are crystallizable and that both the melting point and the heat of fusion increase with increasing side-chain length.

2. The steady-shear melt viscosity of the same materials with C_{10}, C_{12}, and C_{14} side chains increase with decreasing side-chain length.
3. Low-strain oscillatory measurements show that PDM–PMAS copolymers with longer side-chains (C_{16} and C_{18}) form a network structure at temperatures below the side-chain T_m. Intermolecular side-chain crystallization may be responsible for this behavior.

References

1. Krueuder, W.; Ringsdorf, H. *Makromol. Chem., Rapid Commun.* **1983,** *4,* 807.
2. Mauzac, M.; Hardouin, F.; Richard, H.; Achard, M.F.; Sigaud, G.; Gasparoux, H. *Eur. Poly. J.* **1986,** *22,* 137.
3. Strohriegl, P. *Makromol. Chem., Rapid Commun.* **1986,** *7,* 771.
4. Zentel, R.; Wu, J.; Cantow, H. J. *Makromol. Chem.* **1985,** *186,* 1763.
5. Chujo, Y.; Murai, K.; Yamashita, Y.; Okumura, Y. *Makromol. Chem.* **1985,** *186,* 1203.
6. Katayama, Y.; Kato, T.; Ohyanagi, M.; Ikeda, K.; Sekine, Y. *Makromol. Chem., Rapid Commun.* **1986,** *7,* 465.
7. Rim, P. B.; Rasoul, H. A. A.; Hurley, S. M.; Orler, E. B.; Scholsky, K. M. *Macromolecules* **1987,** *20,* 208.
8. O'Connor, K. M.; Orler, E.B.; Rasoul, H.A.A.; Hurley, S. M.; Submitted for publication.

RECEIVED for review May 27, 1988. ACCEPTED revised manuscript December 15, 1988.

5

Pyridinyl- and 1-Oxypyridinyl-Substituted Silanes and Siloxanes

New Catalysts for Interfacial Transacylation Reactions

Wilmer K. Fife, Martel Zeldin, and Cheng-Xiang Tian

Department of Chemistry, Indiana University–Purdue University at Indianapolis, Indianapolis, IN 46223

*Catalysis of the synthesis of benzoic anhydride and the hydrolysis of benzoyl chloride, diphenyl phosphorochloridate (DPPC), and benzoic isobutyric anhydride in dichloromethane–water suspensions by water-insoluble silanes and siloxanes, 3- and 4-trimethylsilylpyridine 1-oxide (**3b** and **3c**, respectively), 1,3-bis(1-oxypyridin-3-yl)-1,1,3,3-tetramethyldisiloxane (**4**), and poly[methyl(1-oxypyridin-3-yl)-siloxane] (**5**) was compared with catalysis in the same systems by water-soluble pyridine 1-oxide (**3a**) and poly(4-vinylpyridine 1-oxide) (**6**). All catalysts were effective for anhydride synthesis and promoted the disproportionation of benzoic isobutyric anhydride. Hydrolysis of benzoyl chloride gave benzoic anhydride in high yield (≥80%) for all catalysts except **3a**, which gave mixtures of anhydride (52%) and benzoic acid (39%). The order of catalytic activity for DPPC hydrolysis was **5** > **4** > **3b** > **3a** > **3c** > **6**. The results suggest that hydrophobic binding between catalyst and lipophilic substrate plays an important role in these processes.*

SILANES AND SILOXANES that contain pyridinyl and 1-oxypyridinyl substituents represent an intriguing new class of catalysts for transacylation reactions (*1–4*). Interconversions of carboxylic and phosphoric acids with their derivatives (e.g., amides, anhydrides, and esters) are among the most important chemical processes observed in biological systems and used by the

NOTE: This chapter is Part 5 in a series of articles (*1–4*).

0065–2393/90/0224–0099$06.00/0

chemical industry. The reactions in vitro are usually carried out in homogeneous solution. By contrast, the corresponding processes in vivo frequently must accommodate multiple phases. For example, the transport of metabolites across the mitochrondrial membrane during fatty acid metabolism requires transacylation steps at the two faces of the membrane (*5*). In recent years, chemists have turned increasingly to studies of chemical reactions under conditions that attempt to mimic biological processes (*6–13*). The multiple-phase aspects of cellular chemistry have contributed many insights and incentives that have led to important new advances in polymer chemistry, including such areas as catalysis and synthetic methodology (*14*).

Applications of Phase-Transfer Catalysis

The advantages and possibilities associated with organic reactions carried out in mixtures of immiscible solvents were first investigated systematically by Makosza (*15*), Starks (*16, 17*), and Brandstrom (*18, 19*) during the 1960s. The methodology that brings together reagents soluble in separate liquid phases is known as phase-transfer catalysis (PTC) (Scheme I) (*20–24*). In a typical PTC process that transfers alkyl groups, R, from one nucleophilic agent (:L^-) to another (:Nuc^-), a phase-transfer catalyst (CAT^+) is required. The catalyst is usually a tetraalkylammonium or tetraalkylphosphonium salt whose lipophilic cation carries the reactant nucleophile (:Nuc^-) to the organic layer for reaction with the substrate. The process is complicated by the possibility of reaction in the interfacial region between the bulk water and organic-solvent phases (*25, 26*).

Only a few examples of PTC methodology have been reported for transacylation reactions, despite their widespread use and importance. Smalley and Suschitzky (*27*) first noted the effectiveness of pyridine 1-oxide as a catalyst for converting suspensions of benzoyl chloride and sodium benzoate in water to benzoic anhydride. More recently, Yamada and co-workers (*28*) prepared symmetrical and mixed anhydrides, even acetic benzoic anhydride, under two-phase conditions (i.e., organic solvent–water) in the presence of

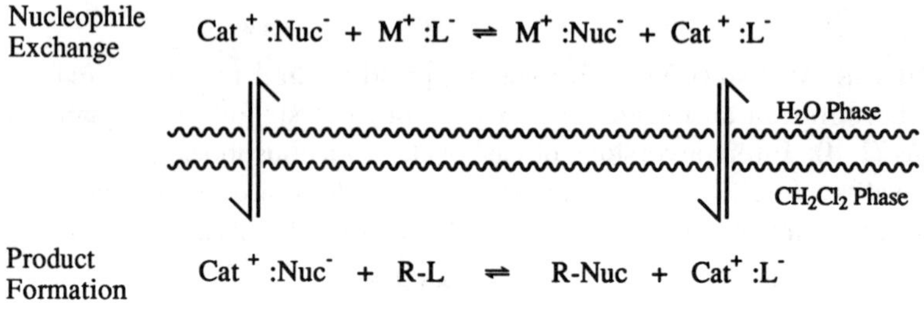

Scheme I. PTC transalkylation. (After reference 17.)

1-methyl-2(1*H*)-pyridinethione. Roulleau et al. (*29*) used conventional PTC methodology to convert acid chlorides to their corresponding symmetrical anhydrides, with tetrabutylammonium chloride as the phase-transfer catalyst in toluene–aqueous sodium hydroxide suspensions. Fife and co-workers (*30–36*) investigated the utility of multiple-phase methodologies for transacylation reactions under a wide variety of conditions. In 1986, Mathias and Vaidya (*37*) reported the first use of a polymeric tertiary amine, poly(4-diallylaminopyridine), as a phase-transfer catalyst in amide formation. Finally, Mack and co-workers (*38*) have demonstrated successful transacylation in solid–liquid, two-phase systems with crown ethers as catalysts.

Neutral nucleophilic catalysts such as tertiary amines and amine *N*-oxides represent especially attractive candidates for promoting reaction between organic-solvent-soluble acid derivatives and water-soluble carboxylate or other basic ions. Reactions (equations 1 and 2) between these catalysts and acid derivatives lead to the formation of intermediates (**1** and **2**) of greater water solubility than the reactant and enhanced reactivity toward negatively charged carboxylate or other basic ions in the aqueous phase (equation 3). Although reaction between **2** and carboxylate ion in dichloromethane solution is expected to be and is rapid, in aqueous solution, this process is also extremely fast (*35*) and perhaps diffusion controlled. Thus two important transacylation reactions that are particularly adaptable to multiple-phase methodology in immiscible organic solvent–water media are anhydride synthesis and hydrolysis of acid derivatives (equations 4 and 5).

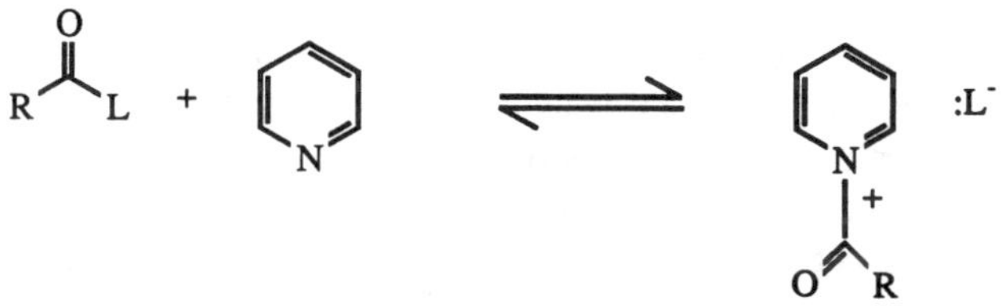

$$RC(=O)L + C_5H_5N \rightleftharpoons C_5H_5N^{+}\text{–}C(=O)R \; :L^{-} \qquad (1)$$

1

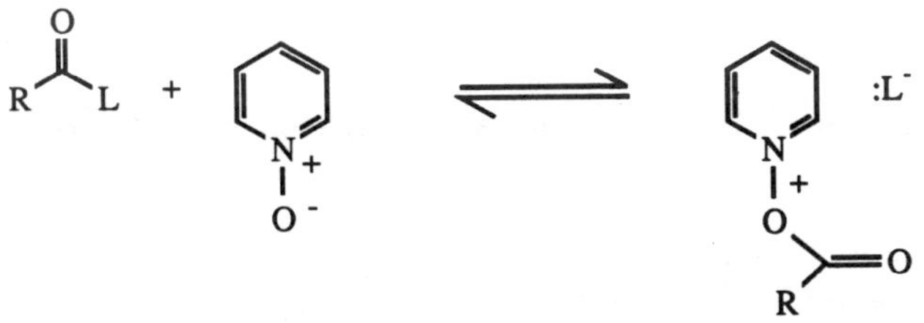

$$RC(=O)L + C_5H_5N^{+}\text{–}O^{-} \rightleftharpoons C_5H_5N^{+}\text{–}O\text{–}C(=O)R \; :L^{-} \qquad (2)$$

2

$$\text{1-(RCO)-pyridinium}^{+}\ {:}L^{-} + R'COO^{-}\ \text{or}\ OH^{-} \xrightarrow[-\ \text{pyridine}]{} R'C(O)OC(O)R\ \text{or}\ RCOO^{-}\quad {:}L^{-} \tag{3}$$

$$RC(O)L \begin{pmatrix}\textit{Organic}\\ \textit{Phase}\end{pmatrix} + R'COO^{-} \begin{pmatrix}H_2O\\ \textit{Phase}\end{pmatrix} \xrightarrow{\text{Catalyst}} RC(O)OC(O)R' \begin{pmatrix}\textit{Organic}\\ \textit{Phase}\end{pmatrix} + {:}L^{-} \begin{pmatrix}H_2O\\ \textit{Phase}\end{pmatrix} \tag{4}$$

$$RC(O)L \begin{pmatrix}\textit{Organic}\\ \textit{Phase}\end{pmatrix} + 2\,OH^{-} \begin{pmatrix}H_2O\\ \textit{Phase}\end{pmatrix} \xrightarrow{\text{Catalyst}} RCOO^{-} \begin{pmatrix}H_2O\\ \textit{Phase}\end{pmatrix} + {:}L^{-} \begin{pmatrix}H_2O\\ \textit{Phase}\end{pmatrix} \tag{5}$$

In this chapter, we describe results from a series of experiments in which phase-transfer catalysts bearing the 1-oxypyridinyl group were used in a variety of transacylation reactions. These reactions include acid anhydride synthesis, as well as hydrolysis of a mixed anhydride and two acyl chlorides. The reactions were carried out in vigorously stirred suspensions of dichloromethane and water. Particular attention was directed to molecular size (monomeric, dimeric, and oligomeric species), structural characteristics (vinyl versus siloxane chain), and solubility (water versus dichloromethane) of the catalysts.

Pyridinyl and 1-Oxypyridinyl-Substituted Phase-Transfer Catalysts

Investigations carried out in our laboratories since 1981 have demonstrated the effectiveness of PTC methodology using the 1-oxypyridinyl group in monomeric (pyridine 1-oxide, **3a**; 3-trimethylsilylpyridine 1-oxide, **3b**; and 4-trimethylsilylpyridine 1-oxide, **3c**), dimeric (1,3-bis[1-oxypyridin-3-yl]-1,1,3,3-tetramethyldisiloxane, **4**), and polymeric (poly[methyl[1-oxypyridin-

3-yl]siloxane], **5**; and poly[4-vinylpyridine 1-oxide], **6**) forms as phase-transfer nucleophilic catalysts in dichloromethane–water suspensions to effect trans-

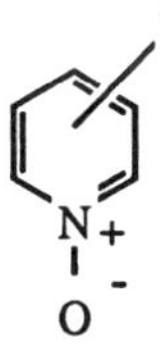

3a X = H
3b X = 3-trimethylsilyl
3c X = 4-trimethylsilyl

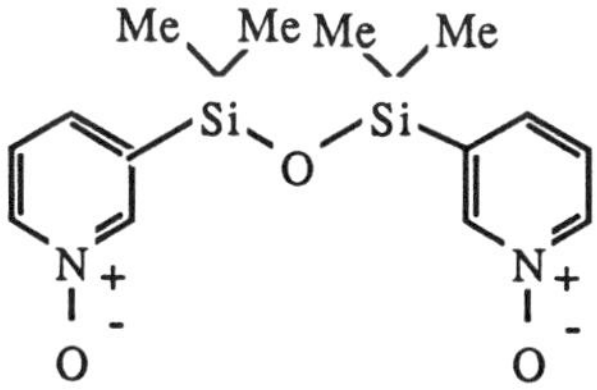

4

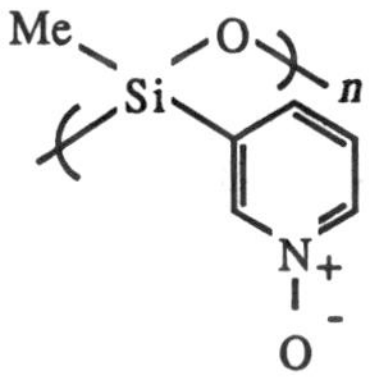

5 $n = 5 - 15$ 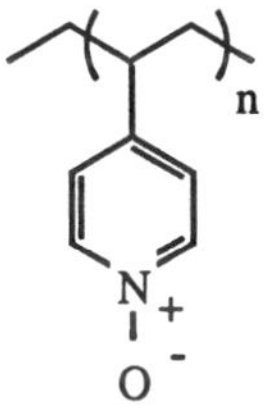 **6** $n \approx 2000$

acylation reactions rapidly, conveniently, and in high yield (*3, 4, 30–36*). Catalysts **3b**, **4**, and **5** were recently prepared and characterized as part of a continuing study of pyridinyl- and 1-oxypyridinyl-substituted silanes and siloxane in our laboratory (*1–4*).

The silicon-containing species represent a particularly interesting group of phase-transfer catalysts, because they are highly soluble in dichloromethane and virtually insoluble in water. Furthermore, their distribution behavior in dichloromethane–water suspensions is just opposite that of the well-known pyridine 1-oxide (**3a**) and its polymeric analogue, poly(4-vinylpyridine 1-oxide) (**6**), which are highly soluble in water. Polysiloxane **5** is a polymeric catalyst with a highly flexible backbone (i.e., low glass transition temperature [T_g]) in contrast to the relatively rigid vinyl-derived chain in **6**. The results from a series of investigations that compared the catalytic properties of the silicon-containing catalysts **3b**, **3c**, **4**, and **5** with those of **3a** and **6** in the synthesis of benzoic anhydride, hydrolysis of benzoyl chloride, hydrolysis of diphenyl phosphorochloridate (DPPC), and hydrolysis of benzoic isobutyric anhydride are described in the following sections.

Anhydride Synthesis. A wide variety of pyridine 1-oxides are highly selective catalysts for the conversion of acid chlorides and sodium carboxylates to symmetric and mixed carboxylic anhydrides (equation 6). Some examples of the synthetic methodology are summarized in Table I. The method produces anhydrides of high purity in excellent yield. The dichloromethane–water medium effectively compartmentalizes the reaction mixture to protect the water-sensitive acid chlorides and anhydrides in an inert, immiscible organic phase. Catalytic effectiveness, which is taken to mean

$$RC(O)Cl + R'C(O)O^- Na^+ \xrightarrow[\text{room temperature, 10 min}]{\text{Catalyst} \atop 1:1\ CH_2Cl_2/H_2O} RC(O)OC(O)R' + NaCl \qquad (6)$$

Table I. Synthesis of Benzoic Anhydride from Benzoyl Chloride and Sodium Benzoate

Catalyst	*Anhydride (%)*
None	0.0
3a	90.0
6	80.2
3b	87.9
3c	89.2
4	85.6
5	88.2

NOTE: Equimolar quantities (5 mM) of benzoyl chloride and sodium benzoate in 20 mL 1:1 (v/v) dichloromethane–water (containing 0.5 M HCO_3^-) were vigorously stirred at 22–23 °C for 10 min, and the benzoic anhydride was isolated and identified as described in reference 3.

the yield of the anhydride product, appears to be independent of catalyst structure and solubility characteristics.

A mechanism (Scheme II) similar to the generally accepted PTC process (Scheme I) has been developed to explain anhydride formation (*30*). Acid chlorides such as benzoyl chloride have a low solubility in water and, therefore, are concentrated in the dichloromethane layer. In the absence of a catalyst, benzoyl chloride is relatively inert to anhydride formation or hydrolysis under comparable experimental conditions (Table I). However, pyridine 1-oxide, which distributes relatively uniformly between the two layers, reacts quickly with the acid chloride to form 1-acyloxypyridinium chloride (**2**) in the dichloromethane layer. Intermediate **2** dissolves preferentially in the aqueous layer, where it rapidly reacts with carboxylate ion to form the anhydride and releases the catalyst (*35*).

The process is actually more complex than is suggested by Scheme II.

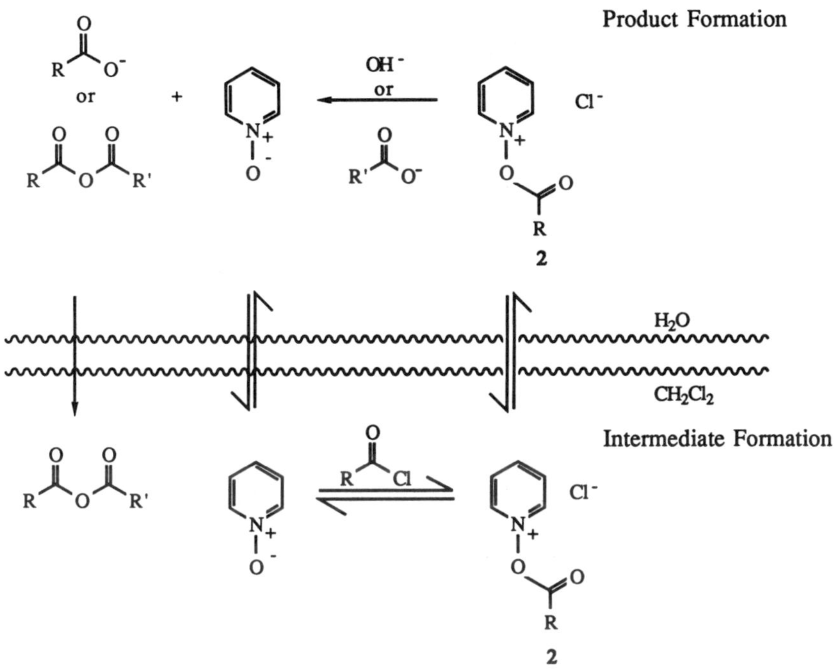

Product Formation

H_2O

CH_2Cl_2

Intermediate Formation

Scheme II. Anhydride formation via acyl group transport to the aqueous phase.

Fife and Xin (*35*) found that, under PTC conditions, the reaction between benzoyl chloride and a mixture of 1.0 equivalent of sodium *p*-toluate and 1.0 equivalent of sodium isobutyrate in the presence of 0.1 equivalent of pyridine 1-oxide gave an anhydride product mixture that contained approximately twice as much benzoic *p*-toluic anhydride as benzoic isobutyric anhydride on a molar basis (equation 7). Therefore, the process must include an important component that selects for the more lipophilic *p*-toluate ion. The differential partitioning of the competing carboxylate ions at the dichloromethane–water interface is a reasonable explanation for this observed selectivity, because product composition correlates with the Hansch (*39*) hydrophobicity parameters for competing carboxylate ions (*40*).

Alternatively, the cationic intermediate **2** may serve as a phase-transfer agent for nucleophiles as well as acyl groups. Thus, **2** could participate in an ion-pair extraction process that transports carboxylate ions to the organic phase (Scheme III) (*18*, *19*). However, an electrostatically driven rapid reaction between **2** and a carboxylate ion in the aqueous layer would minimize contributions from this pathway. To the extent that the process leading to anhydride includes components represented by Schemes II and III, anhydride formation could very well be occurring in both the bulk dichloromethane and water phases and at the interface.

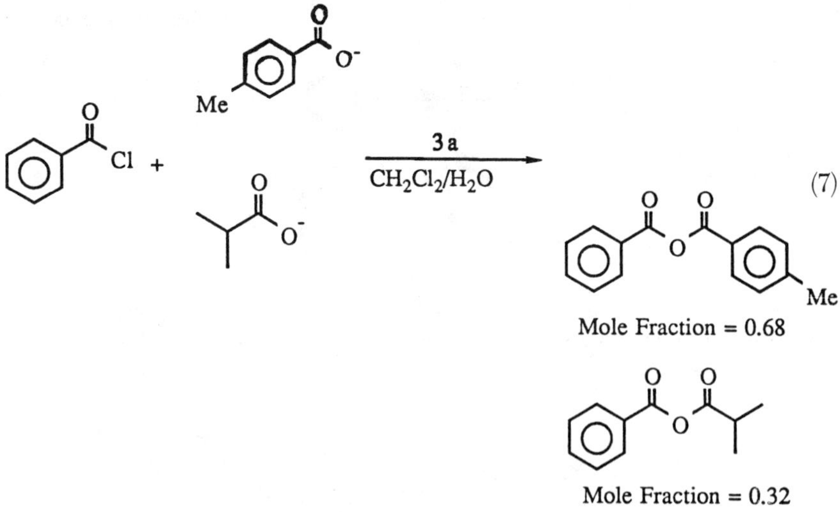

$$\text{PhCOCl} + p\text{-MeC}_6\text{H}_4\text{CO}_2^- + (\text{CH}_3)_2\text{CHCO}_2^- \xrightarrow[\text{CH}_2\text{Cl}_2/\text{H}_2\text{O}]{\textbf{3a}} \text{PhC(O)OC(O)C}_6\text{H}_4\text{-}p\text{-Me} + \text{PhC(O)OC(O)CH(CH}_3)_2 \quad (7)$$

Mole Fraction = 0.68 (benzoic p-toluic anhydride)

Mole Fraction = 0.32 (benzoic isobutyric anhydride)

Hydrolysis of Benzoyl Chloride. Early experiments demonstrated that **3b**, **3c**, **4**, and **5** are effective catalysts for the conversion of benzoyl chloride to benzoic anhydride in well-stirred suspensions of dichloromethane–0.5 M aqueous sodium bicarbonate (equation 8) (*2*, *3*). The results

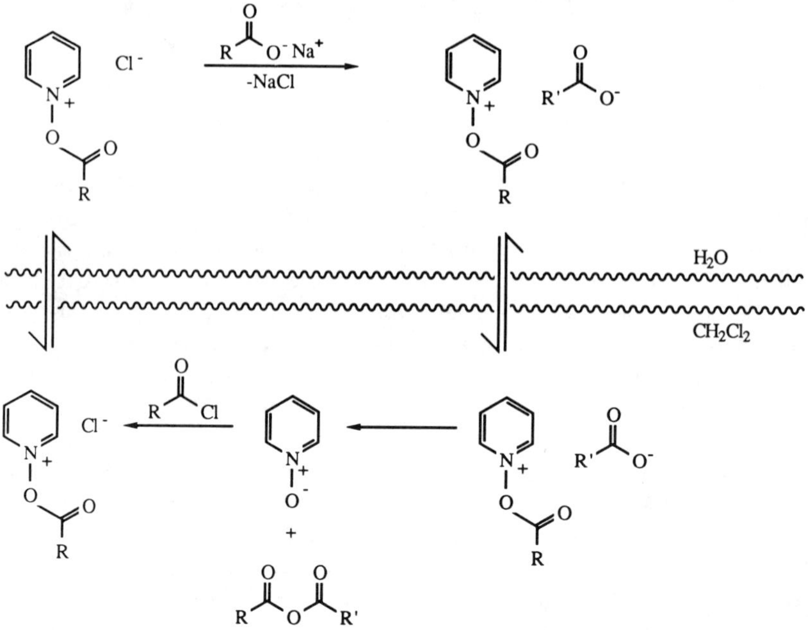

Scheme III. Anhydride formation via carboxylate ion transport by ion-pair extraction.

$$2\ C_6H_5COCl + 2\ HCO_3^- \xrightarrow[\text{room temperature, 30 min}]{\text{Catalyst, } CH_2Cl_2/H_2O} (C_6H_5CO)_2O\ (80-90\%) + 2\ CO_2 + 2\ Cl^- \qquad (8)$$

from these experiments are summarized in Table II. The hydrophilic catalyst **6** exhibited essentially the same behavior as the lipophilic catalysts **3b**, **3c**, **4**, and **5**. Thus, two products are obtained in the presence of all the catalysts: benzoic acid and benzoic anhydride. Benzoic anhydride is presumed to be formed by the reaction between sodium benzoate, the initial product, and **2** (Scheme II). A high percentage of anhydride is obtained with the organic-phase-soluble catalysts 3- and 4-trimethylsilylpyridine 1-oxide (**3b** and **3c**) and related siloxanes (**4** and **5**), as well as the water-soluble poly(4-vinylpyridine 1-oxide) (**6**). Only pyridine 1-oxide (**3a**) gives a high degree of hydrolysis.

Table II. Hydrolysis of Benzoyl Chloride

Catalyst	*Anhydride (%)*	*Acid (%)*
None	0.0	0.3
3a	52.3	39.5
6	81.0	8.3
3b	91.6	4.9
3c	80.2	2.7
4	79.1	2.2
5	89.3	5.0

NOTE: The reactions were carried out in dichloromethane–water (containing 0.5 M HCO_3^-) at 22–23 °C for 30 min. *See* reference 3 for procedural details.

Catalysts **3b**, **3c**, **4**, and **5** are restricted to the organic phase because of their limited solubility in water. Thus, these catalysts are expected to be acylated by benzoyl chloride in the organic phase to form **2**, which in the early stages of reaction is hydrolyzed to the benzoate ion (Scheme II). Because **2** is positively charged, it has enhanced access to hydroxide and other negatively charged ions. As the reaction progresses, the benzoate ion becomes an effective competitor for **2** because of its increasing concentration, as well as its greater lipophilicity relative to bicarbonate or hydroxide ions (6). The similar behavior of hydrophilic polymer **6** and the lipophilic catalysts suggests that benzoylation of 1-oxypyridinyl residues modifies the phase-distribution properties of the 1-benzoyloxypyridinium sites in **6** and in the silicon-containing catalysts **3b**, **3c**, **4** and **5** such that they become chemically indistinguishable. Alternatively, acylation of **6** may occur more slowly than for the other catalysts and, therefore, could be the rate-limiting step in the process. Thus, the benzoate ion becomes an effective competitor for **2** at an early stage of the reaction.

The markedly different results obtained with **3a** are consistent with its rapid benzoylation in the organic phase and subsequent migration into the aqueous phase for a comparatively indiscriminate reaction with the nucleophiles present (Scheme II). In any event, the results indicate that the carboxylate ion has comparable access to **2** in both **6** and the lipophilic catalysts during this process.

Hydrolysis of DPPC. The catalytic effectiveness of **3**, **4**, **5**, and **6** in the hydrolysis of DPPC was determined in well-stirred suspensions of dichloromethane–0.5 M aqueous sodium bicarbonate (equation 9; Table III).

$$(PhO)_2P(=O)Cl + 2HCO_3^- \xrightarrow[\text{1:1 } CH_2Cl_2/H_2O \text{, room temperature, 16 h}]{\text{Catalyst}} (PhO)_2P(=O)O^- + 2CO_2 + Cl^- \quad (9)$$

Table III. Hydrolysis of DPPC

Catalyst	*$(C_6H_5O)_2POCl$ (%)*	*$(C_6H_5O)_2POOH$ (%)*	$t_{1/2}$ *(min)*
None	96.0	4.3	50,000
Tetrabutylammonium chloride	20.6	69.1	500
3a	62.1	28.4	1,500
6	79.4	5.4	10,000
3b	51.3	49.8	1,000
3c	73.4	19.8	3,000
4	29.2	68.5	600
5	19.6	72.2	500

NOTE: The reactions were carried out in dichloromethane–water (containing 0.5 M HCO_3^-) at 22–23 °C for 16 h. *See* reference 3 for procedural details.

Half-lives ($t_{1/2}$) were taken as estimates of catalytic effectiveness. The order of catalytic activity was **5** > **4** > **3b** > **3a** > **3c** > **6**, which is quite similar to the order for catalysis of hydrolysis of DPPC suspended in 0.5 M aqueous sodium bicarbonate (**5** > **4** > **3b** > **6** > **3a**) (*3*). Clearly, the 1-oxypyridinyl-substituted polysiloxane **5** was the most effective catalyst of those studied under both sets of conditions: (1) DPPC and catalyst in dichloromethane–0.5 M aqueous sodium bicarbonate suspensions and (2) DPPC and catalyst in

0.5 M aqueous sodium bicarbonate. The order of catalytic activity coincides with the expected order for association between catalyst and lipophilic substrate. Therefore, catalytic activity seems to be dependent on the binding of DPPC to the catalyst prior to hydrolysis. Similar associations between lipophilic substrates and lipophilic domains in enzymes are well known. Furthermore, these so-called hydrophobic interactions are believed to play an important role in the ability of an enzyme to recognize its substrate, as well as to contribute to the catalytic process (5).

Hydrolysis of Benzoic Isobutyric Anhydride. Reactions of mixed anhydrides can be complicated by disproportionation reactions (equation 10). The process may be operative in any investigation of anhydride hydrolysis, but it would go undetected in studies with symmetrical anhydrides or in studies that just monitor reaction rates by consumption of base. To obtain a more complete understanding of the catalytic behavior of **3**, **4**, **5**, and **6** in transacylation reactions, the hydrolysis of benzoic isobutyric anhydride was investigated.

$$2\ \mathrm{RC(O)OC(O)R'} \rightleftharpoons \mathrm{RC(O)OC(O)R} + \mathrm{R'C(O)OC(O)R'} \qquad (10)$$

^{1}H NMR analysis was used to fully characterize product mixtures, because the chemical shifts of aliphatic protons are different for benzoic isobutyric anhydride and isobutyric anhydride, a possible disproportionation product (*36*). The results are given in Table IV. In all cases, disproportionation accompanied the hydrolysis of benzoic isobutyric anhydride in stirred suspensions of dichloromethane–0.5 M aqueous sodium bicarbonate. All catalysts behaved similarly, despite the major solubility differences of **3a** and **6** versus **3b**, **3c**, and **5**. For the hydrolysis of benzoic isobutyric anhydride, as with the other studies described in this report, 1-oxypyridinyl-substituted

Table IV. Hydrolysis of Benzoic Isobutyric Anhydride

Catalyst	*Recovered Starting Material*	*Disproportionation (%)*	*Hydrolysis*
None	90.0		
3a	71.0	22.5	5.7
6	40.2	39.6	6.9
3b	40.4	35.2	10.9
3c	39.9	37.0	13.6
5	36.0	40.1	9.9

NOTE: Reactions were run and worked up as described in Table II, footnote. Product analysis was by integration of ^{1}H-NMR signals for the methyl protons in the isobutyryl group of benzoic isobutyric anhydride (δ ($CDCl_3$): 1.30, d) and isobutyric anhydride (δ ($CDCl_3$): 1.24, d).

polysiloxane (**5**) is one of the most effective transacylation catalysts we have investigated.

The reaction mechanism for the hydrolysis of anhydrides is presumed to be similar to that described for acid chlorides (Scheme II). A notable difference between the reaction of acid chlorides and anhydrides, however, is the extent of reversibility in formation of **2**. The reaction between **2** and chloride ion to give acid chloride (reversal of intermediate formation; equation 2) is much less favored than the corresponding reaction between **2** and the carboxylate ion in both organic and aqueous phases.

Conclusions

The results of these studies and others reported previously demonstrate that the 1-oxypyridinyl group is an effective catalyst for the transacylation reactions of derivatives of carboxylic and phosphoric acids when incorporated in small molecules and polymers. Furthermore, this catalytic site exhibits high selectivity for acid chlorides in the presence of acid anhydrides, amides, and esters. Therefore, catalysts bearing this group as the catalytic site can be used successfully in synthetic applications that require such specificity. The results of this work suggest that functionalized polysiloxanes should be excellent candidates as catalysts for a wide variety of chemical reactions, because they combine the unique collection of chemical, physical, and dynamic–mechanical properties of siloxanes with the chemical properties of the functional group. Finally, functionalized siloxanes appear to mimic effectively enzyme–lipophilic substrate associations that contribute to the widely acknowledged selectivity and efficiency observed in enzymic catalysis.

Acknowledgment

We gratefully acknowledge financial support for this project from the Reilly Tar & Chemical Corporation.

References

1. Zeldin, M.; Xu, J. M.; Tian, C. X. *J. Organomet. Chem.* **1987,** *326*, 341.
2. Zeldin, M.; Xu, J. M.; Tian, C. X. *Polym. Prepr.* (*Am. Chem. Soc., Div. Polym. Chem.*) **1987,** *28*(1), 417.
3. Zeldin, M.; Fife, W. K.; Tian, C. X.; Xu, J. M. *Organometallics* **1988,** *7*, 470.
4. Zeldin, M.; Fife, W. K.; Tian, C. X.; Xu, J. M. In *Inorganic and Organometallic Polymers: Macromolecules Containing Silicon, Phosphorus and Other Inorganic Elements;* Zeldin, M.; Wynne, K. J.; Allcock, H. R., Eds.; ACS Symposium Series 360; American Chemical Society: Washington, DC, 1988; Chapter 15.
5. Stryer, L. *Biochemistry,* 2nd ed.; Freeman: New York, 1981; p 388.
6. Jencks, W. P. *Catalysis in Chemistry and Enzymology;* McGraw Hill: New York, 1969.
7. Klotz, I. M. *Adv. Chem. Phys.* **1978,** *39*, 109.

8. Kunitake, T.; Shinkai, S. *Adv. Phys. Org. Chem.* **1980,** *17*, 435.
9. Breslow, R. *Science (Washington, D.C.)* **1982,** *218*, 532.
10. Lehn, J. M. *Science (Washington, D.C.)* **1985,** *227*, 849.
11. Menger, F. M. *Top. Curr. Chem.* **1986,** *136* (*Biomimetic Bioorg. Chem. 3*), 1–15.
12. Wulff, G. In *Polymeric Reagents and Catalysts;* Ford, W. T., Ed.; ACS Symposium Series 308, American Chemical Society: Washington, DC, 1986; p 186.
13. D'Souza, V. T.; Bender, M. L. *Acc. Chem. Res.* **1987,** *20*, 146.
14. Fendler, J. H. *Membrane Mimetic Chemistry;* Wiley: New York, 1982.
15. Makosza, M.; Serafinowa, B. *Rocz. Chem.* **1965,** *39*, 1223.
16. Starks, C. M.; Napier, D. R. Ital. Patent 832 967, 1968; Br. Patent 1 227 144, 1971; French Patent 1 573 164, 1969; *Chem. Abstr.* **1970,** *72*, 115271.
17. Starks, C. M. *J. Am. Chem. Soc.* **1971,** *93*, 195.
18. Brandstrom, A.; Gustavii, K. *Acta Chem. Scand.* **1969,** *23*, 1215.
19. Brandstrom, A. *Kem. Tidskr.* **1970,** *Nos. 5–6*, 1.
20. Brandstrom, A. *Preparative Ion Pair Extraction: An Introduction to Theory and Practice;* Lakemodel: Apotekarsocieteten, A. B. Hassle: Stockholm, Sweden, 1974.
21. Weber, W. P.; Gokel, G. W. *Phase Transfer Catalysis in Organic Synthesis;* Springer: New York, 1977.
22. Starks, C. M.; Liotta, C. *Phase Transfer Catalysis. Principles and Techniques;* Academic: New York, 1978.
23. Dehmlow, E. V.; Dehmlow, S. S. *Phase Transfer Catalysis*, 2nd ed.; Chemie: Deerfield Beach, FL, 1983.
24. Keller, W. E. *Phase Transfer Reactions;* Georg Thieme: New York, 1986, Fluka-Compendium Vol. 1.
25. Makosza, M. *Pure Appl. Chem.* **1975,** *43*, 439.
26. Rabinovitz, M.; Cohen, Y.; Halpern, M. *Angew. Chem., Int. Ed. Engl.* **1986,** *25*, 960.
27. Smalley, R. K.; Suschitzky, H. *J. Chem. Soc.* **1964,** 755.
28. Yamada, M.; Watake, Y.; Sakakibara, T.; Sudoh, R. *J. Chem. Soc., Chem. Commun.* **1979,** 179.
29. Roulleau, R.; Plusquellec, D.; Brown, E. *Tetrahedron Lett.* **1983,** *24*, 4195.
30. Fife, W. K.; Dally, R. D. *Abstracts of Papers,* 187th National Meeting of the American Chemical Society, St. Louis, MO; American Chemical Society: Washington, DC, 1984, paper ORGN 251.
31. Fife, W. K.; Bertrand, M. A.; Nguyen, T.; Dally, R. D.; Fredrickson, W. *Abstracts of Papers,* 2nd Symposium on Pyridine Chemistry, University of Salford, UK, May 1985.
32. Fife, W. K.; Zhang, Z. D. *J. Org. Chem.* **1986,** *51*, 3744.
33. Fife, W. K.; Zhang, Z. D. *Tetrahedron Lett.* **1986,** *27*, 4933.
34. Fife, W. K.; Zhang, Z. D. *Tetrahedron Lett.* **1986,** *27*, 4937.
35. Fife, W. K.; Xin, Y. *J. Am. Chem. Soc.* **1987,** *109*, 1278.
36. Fife, W. K.; Bertrand, M. A.; Nguyen, T.; Dally, R. D.; Zhang, Z. D.; Xin, Y. manuscript in preparation.
37. Mathias, L. J.; Vaidya, R. A. *J. Am. Chem. Soc.* **1986,** *108*, 1093.
38. Mack, M. P., Dehm, D.; Boden, R.; Durst, H. D. unpublished results.
39. Hansch, C.; Leo, A. *Substituent Constants for Correlation Analysis in Chemistry and Biology;* Wiley: New York, 1979.
40. Fife, W. K.; Xin, Y. unpublished results.

RECEIVED for review May 27, 1988. ACCEPTED revised manuscript December 14, 1988.

6

Polysiloxane-Based Polymer–Electrolyte Complexes

Johannes Smid, Daryle Fish, Ishrat M. Khan, E Wu, and Guangbin Zhou

Polymer Research Institute, College of Environmental Science and Forestry, State University of New York, Syracuse, NY 13210

Comblike polysiloxanes with oligooxyethylene side chains were synthesized from poly(hydrogen methylsiloxane), and the conductivities and thermal properties of their solvent-free $LiClO_4$ *complexes were measured. The side chains were of the type* $-O(CH_2CH_2O)_7CH_3$ *(PMMS-8),* $-(CH_2)_3O(CH_2CH_2O)_nCH_3$ *(PAGS polymers with* n = 7 *or 11), and* $-(CH_2)_3OCH_2CH(OH)CH_2SO_3^-Na^+$ *($PAGSO_3^-Na^+$). Cloud points were determined in water for PMMS-8 in the presence of various salts. Maximum conductivities of the* $LiClO_4$ *complexes with PMMS-8 and PAGS polymers reached values close to* $10^{-4}/\Omega$*-cm at 25 °C, and ratios of ethylene oxide units to* Li^+ *were between 20 and 25. Mechanical properties were improved by cross-linking or by blending with high-molecular-weight poly(ethylene oxide). The conductivity of the cation conductor* $PAGSO_3^-Na^+$ *at 75 °C was* $<10^{-6}/\Omega$*-cm, but the addition of tetraethylene glycol (TEG) raised the conductivity to* $2.8 \times 10^{-5}/\Omega$*-cm when the* TEG/Na^+ *ratio was 1.*

THE SEARCH FOR PLASTIC, solvent-free electrolytes for use in solid-state batteries is being actively pursued in several laboratories (*1–4*). A number of reports have stressed the need for facile motion of the macromolecular chain in order to promote the ion conduction process in the polymer matrix, because this process occurs primarily via a free-volume mechanism (*1–4*). Comblike polymers with oligooxyethylene side chains constitute effective media for ion conduction of solubilized alkali salts (*5–8*). The low glass transition temperature (T_g) of poly(dimethylsiloxane) suggests that polysiloxane could serve as a suitable backbone for such a comb polymer, and recent studies (*9–12*) indicate this to be the case indeed.

0065–2393/90/0224–0113$06.00/0

In this chapter, we briefly review our work on the synthesis and conducting properties of complexes of lithium perchlorate and polysiloxane comb polymers with oligooxyethylene side chains. Included in the discussion are blends with high-molecular-weight poly(ethylene oxide) (PEO) and some preliminary work on cation-conducting polymers with bound sulfonate groups.

Experimental Procedures

The polymers (*see* structures **1–3**) have been designated as PMMS, PAGS, and $PAGSO_3^-Na^+$. PMMS-8 (the number refers to the average number of oxygen atoms

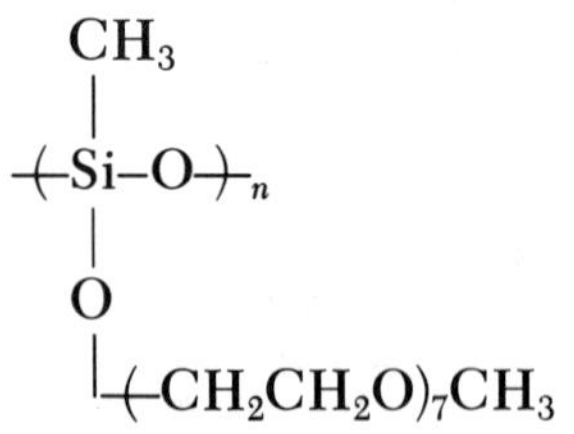

CH_3
$\{Si{-}O\}_n$
O
$(CH_2CH_2O)_7CH_3$

1: PMMS

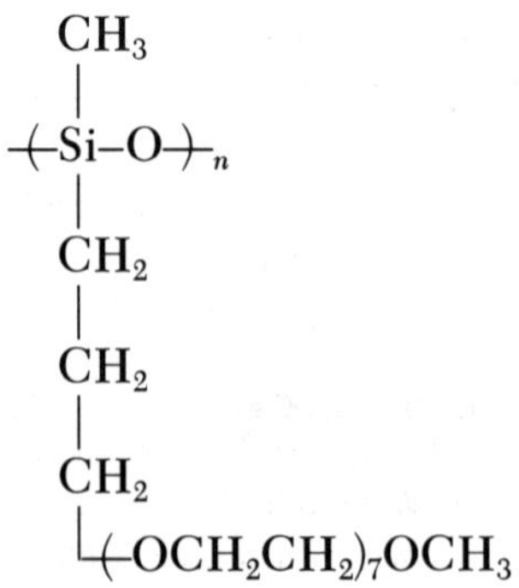

CH_3
$\{Si{-}O\}_n$
CH_2
CH_2
CH_2
$(OCH_2CH_2)_7OCH_3$

2: PAGS

CH_3 CH_3
$\{Si{-}O\}_m$ — $\{Si{-}O\}_n$
CH_2 CH_2
CH_2 CH_2
CH_2 CH_2
O $(OCH_2CH_2)_7OCH_3$
CH_2
$CHOH$
$CH_2SO_3^-Na^+$

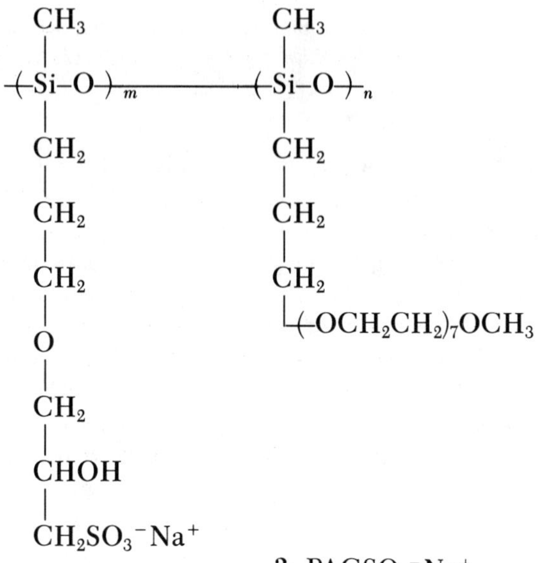

3: $PAGSO_3^-Na^+$

in the oligooxyethylene side chain) was synthesized by reacting methoxypoly(ethylene glycol) (M_w [weight-average molecular weight], 350) with poly(hydrogen methylsiloxane) (PHMS; M_w 2650) in tetrahydrofuran (THF), with zinc octanoate as catalyst. PAGS-8 was obtained by hydrosilylation of PHMS with allyl methyl poly(ethylene glycol) (AMPEG) in THF, with platinum divinyltetramethyldisiloxane as catalyst. AMPEG was synthesized from allyl chloride and a methoxypoly(ethylene glycol) with an average M_w of 350 or 500.

The single-ion conductor, $PAGSO_3^-Na^+$ (m/n = 0.4 or 1.0, in which m and n are the number of siloxy units in the structure), was made by consecutive hydrosilylation of PHMS with AMPEG and allyl glycidyl ether. The neutral polymer was sulfonated with aqueous $NaHSO_3$, which quantitatively converted the epoxy group into $-CH(OH)CH_2SO_3^-$. Excess $NaHSO_3$ was removed by repeated ultrafiltration of the polymer solution with a cutoff membrane having a M_w of 1000. The polysulfonate was azeotropically dried with benzene and finally in vacuum at 50 °C for several days.

Polymer electrolyte complexes of PMMS or PAGS with $LiClO_4$ were made by solution casting from methanol or THF. In some cases, PEO with a number-average molecular weight (M_n) of 4×10^6 was added to enhance the toughness of the films. The polymer–electrolyte mixtures were dried under vacuum at 60 °C for several days.

The polysiloxanes were characterized by Fourier transform–IR (FTIR) spectroscopy, ^1H and ^{29}Si NMR spectrometry, and by GPC. AC conductivities of the polymer electrolytes were measured under dry helium by using an automatic capacitance bridge (General Radio Corporation). Glass transition (T_g) and melt (T_m) temperatures were recorded on a differential scanning calorimeter (Perkin Elmer DSC-4). More detailed experimental procedures are published elsewhere (*9*, *12*).

Results and Discussion

The reaction of PHMS to yield PMMS-type comb polysiloxanes is essentially quantitative for methoxypoly(ethylene glycol)s of M_w up to 500. Substitution yields diminish for longer glycols, as indicated by the presence of residual Si–H groups in the IR and ^1H NMR spectra. ^{29}Si NMR spectra and GPC data revealed that PMMS-8 is contaminated with nearly 25% cyclic products and that it also contained a considerable number of branched trisiloxy units. Both cyclic products and trisiloxy units are the result of redistribution processes common in nucleophilic siloxane reactions. The hydrosilylation reaction yielding the PAGS polymers is also quantitative, but no cyclic products are formed.

The T_gs of the comb polysiloxanes increased with the oxyethylene unit content: –78 °C for PMMS-8 and –60 °C for PAGS-12. Melt endotherms resulting from side-chain crystallization were found in the DSC (differential scanning calorimetry) scans at T_m = –1 °C for PAGS-8 and at 24 °C for PAGS-12. A broad melt endotherm centered at –1 °C was found for PMMS-8. The endotherm sharpened considerably when a fractionated sample of methoxypoly(ethylene glycol) was used in the synthesis of PMMS-8.

The comb polysiloxanes are soluble in most common solvents. They can be precipitated from aliphatic hydrocarbon solvents. The expected small hydrodynamic volume of the comb polymer is demonstrated by GPC and viscosity data. For example, the intrinsic viscosity of PMMS-8 is lower than that of its precursor, PHMS. The only polymer with some water solubility is PMMS-8. Lengthening of the side chain makes the PMMS polymers less water soluble, contrary to what is found for similar polymers with a poly(methacrylate) backbone (abbreviated as PMG polymers) or polysiloxanes with poly(oxyethylene) grafts (*7, 13*).

PAGS-8 and PAGS-12 are also water insoluble. The tendency toward side-chain crystallization and the hydrophobic backbone of the polymer contributes to their poor water solubility. Cloud points could be measured in 3% aqueous solutions of PMMS-8 in the presence of various salts (Figure 1). Consistent with data reported for nonionic surfactants and polymers with oxyethylene moieties, the cloud points are sensitive to the type of salts added, especially that of the anion. Fluorides and sulfates are effective salting-out agents, whereas thiocyanates raise the cloud points.

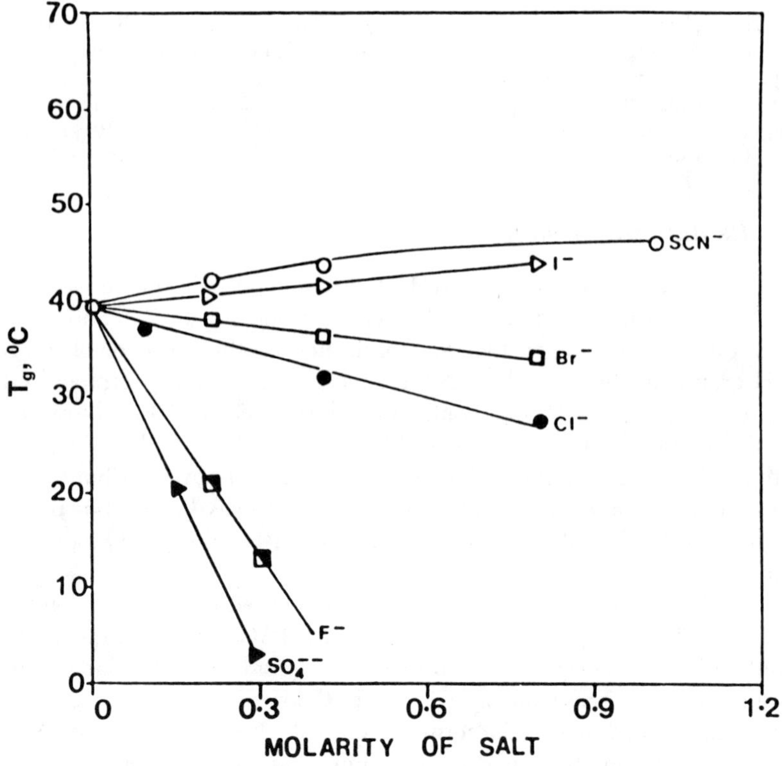

Figure 1. Change in cloud point of a 3% aqueous PMMS-8 solution on addition of sodium salts. (Reproduced from reference 19. Copyright 1988 American Chemical Society.)

Thermal and Conducting Properties of Polymer–$LiClO_4$ Mixtures. Plots of the conductivity, σ, of homogeneous, transparent mixtures of $LiClO_4$ with PMMS-8 and PAGS-12 exhibit distinct maxima at ratios of ethylene oxide units to lithium (EO/Li^+) of ~20–25 (Figure 2). This behavior is typical for amorphous polymer electrolyte complexes (*7, 8*). An increase

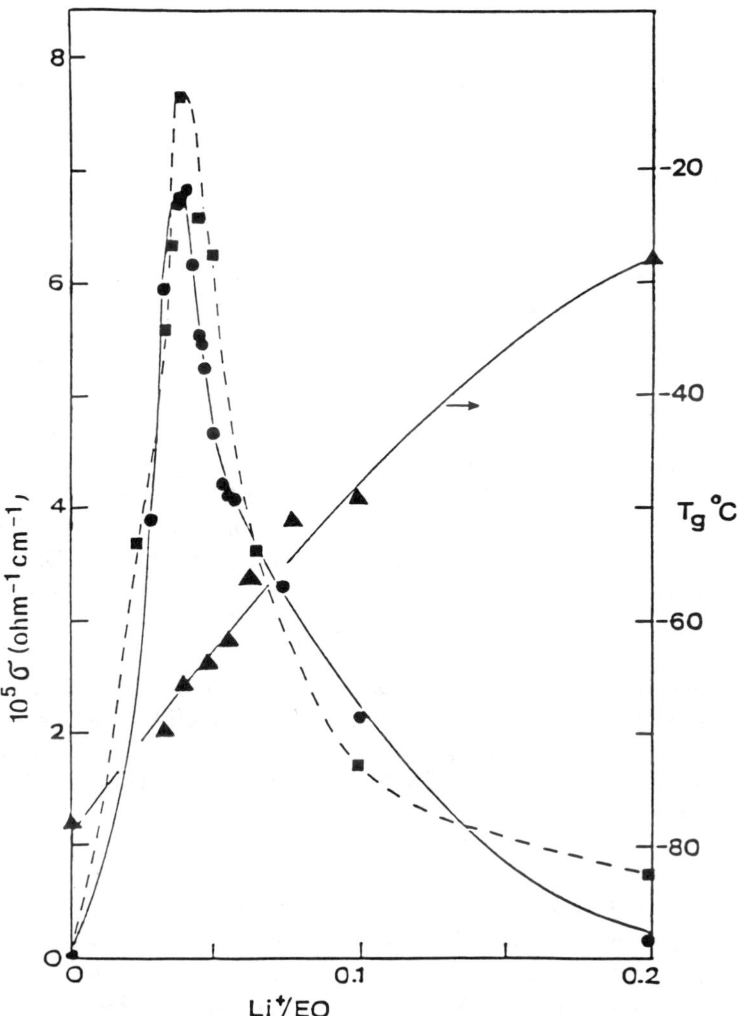

Figure 2. Plots of conductivity, σ, for $LiClO_4$ complexes with PMMS-8 (●) and PAGS-12 (■) as a function of the molar ratio of $LiClO_4$ to ethylene oxide units (Li^+/EO). Also plotted is the T_g of the PMMS-8 complex (▲) versus Li^+/EO. (Reproduced with permission from reference 12. Copyright 1988 Elsevier.)

in salt content of the plastic electrolyte raises the number of charge carriers and impedes side-chain crystallization. For example, DSC scans show a broadening of the melt endotherm of PMMS-8 when $LiClO_4$ is added; the peak disappears entirely when the EO/Li^+ ratio reaches 21. Both factors enhance the conductivity, but increase in salt content also raises the T_g (Figure 2) and lowers the conductivity. Side-chain motion is especially hindered by the chelation of Li^+ with as many as four oxygen-binding sites (*14*).

The T_g may also increase as a result of increased interactions between ion pairs and the formation of ion clusters. The increase in T_g is nearly linear with the ratio Li^+/EO, even up to a ratio of 0.5 for $LiCF_3SO_3$ solubilized in PMMS-8 or in the identical poly(methacrylate) comb polymer (Figure 3). The rise in T_g is much more rapid for the poly(methacrylate) comb polymer than for the polysiloxane. The free-volume mechanism of ion conduction is confirmed in our system by the linearity of temperature-dependent conductivity plots when the Vogel–Tammann–Fulcher (VTF) expression $\sigma = A_0T^{-1/2} \exp[-K(T - T_0)]$ (in which T_0 is the ideal T_g and A_0 and K are constants) (*1*, *2*, *9*) is applied.

The mechanical properties of the $LiClO_4$ complexes with the comb polysiloxanes improved considerably when the materials were cross-linked with TEG either by using a PMMS or PAGS polymer containing ~5–10% unreacted SiH groups or by adding 1 wt % benzoyl peroxide and heating the polymer electrolyte mixture at 100 °C for 24 h. An alternative approach that permits the plastic electrolyte to be solution cast into tough, flexible films is to add high-molecular-weight PEO (M_n 4×10^6). The DSC scans of the PMMS-8–PEO polyblends clearly reveal the presence of crystalline PEO domains. When $LiClO_4$ is added, the crystalline domains become more diffuse. When the ratio EO/Li^+ is 5, the material becomes entirely amorphous and transparent (Figure 4). At an EO/Li^+ ratio of 20–25, at which the conductivity reaches its maximum, the crystalline PEO domains impart mechanical strength to the plastic electrolyte, but they also reduce the conductivity relative to that of a PEO-free mixture of PMMS–$LiClO_4$ of comparable salt content (Figure 5).

In Table I, the conductivities, σ, for a number of polymer–$LiClO_4$ systems at 25 and 70 °C are presented. All conductivities were measured at an EO/Li^+ ratio of 25, that is, close to the expected conductivity maximum of these systems. The more hydrophobic backbone of PAGS-8 relative to PMMS-8 results in a lower value for PAGS-8, but an increase in the ethylene oxide unit content makes PAGS-12 comparable with PMMS-8. DSC scans clearly show higher T_g values for the cross-linked materials, which result in lower conductivity values for their $LiClO_4$ complexes (Table I). Blending with PEO has the same effect, because crystalline domains impede ion transport. The decrease in σ with salt content is less dramatic for the PEO blends than for the pure comb polymers (Figure 5), probably because the

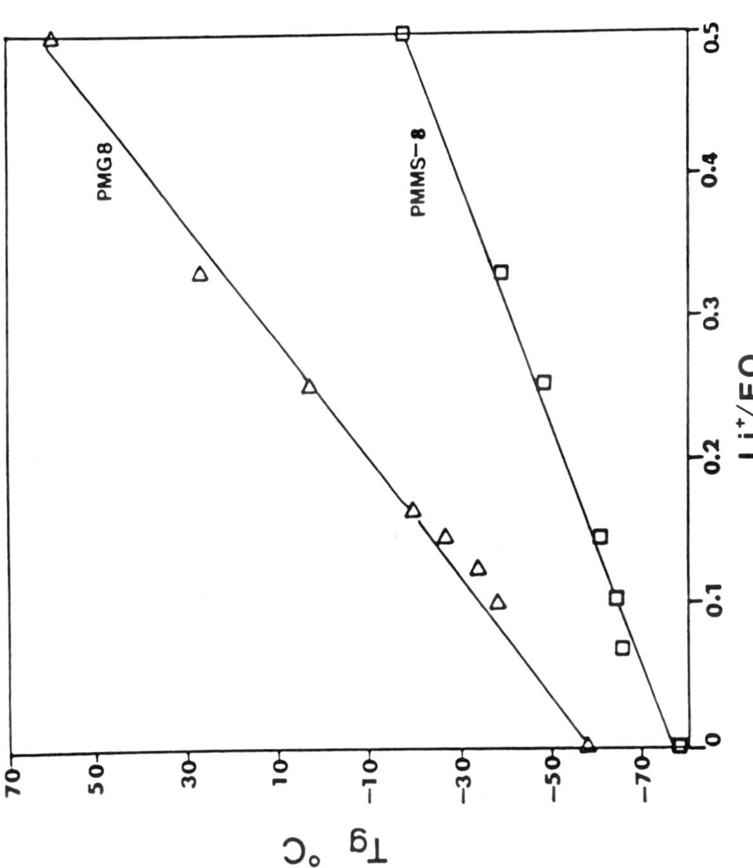

Figure 3. Plot of T_g *versus the ratio* Li^+/EO *for PMMS-8 (□) and PMG-8 (Δ), in which* $LiCF_3SO_3$ *is solubilized.*

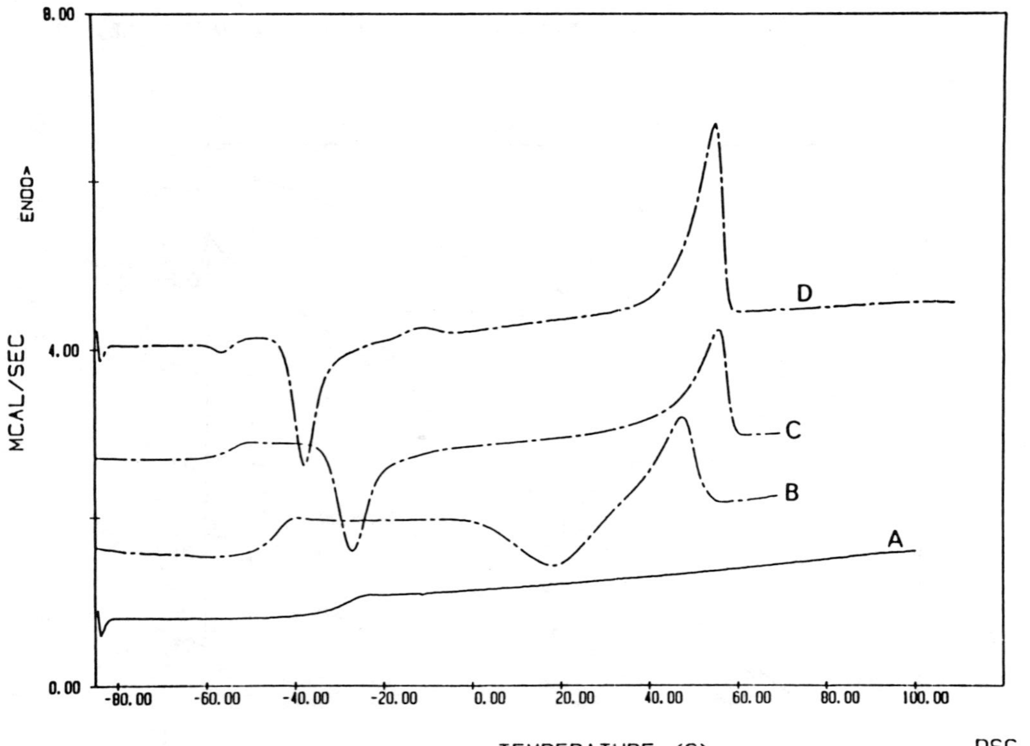

Figure 4. DSC scans of blends of PMMS-8–$LiClO_4$ with 50 wt % poly(ethylene oxide). EO/Li^+ ratios of 5 (A), 10 (B), 20 (C), and 30 (D) were used. (Reproduced with permission from reference 12. Copyright 1988 Elsevier.)

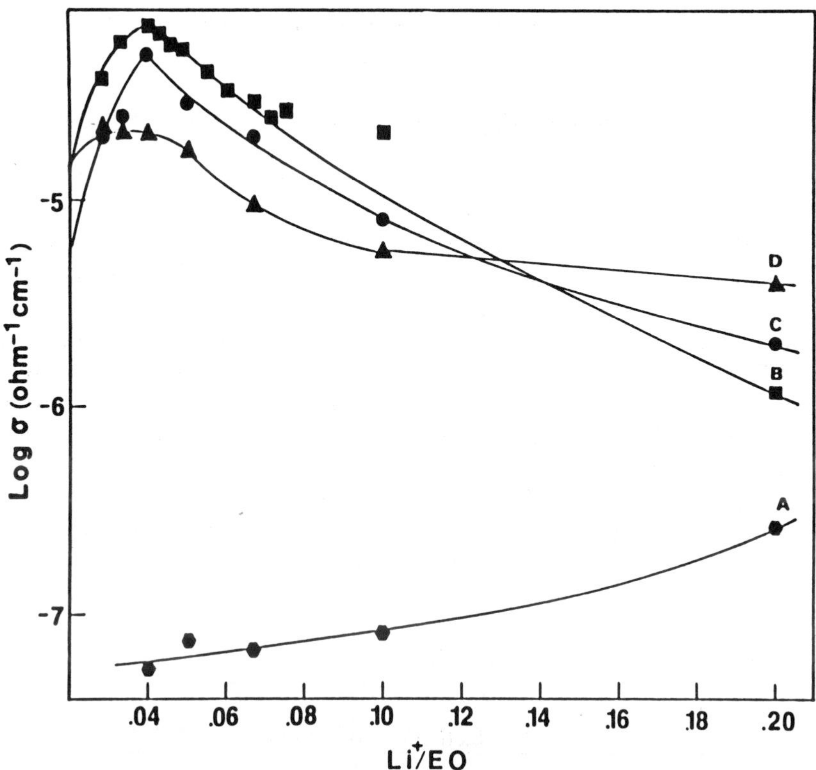

Figure 5. Plots of log conductivity versus the ratio Li^+/EO for homogeneous mixtures of $LiClO_4$ and PEO (A) and PMMS-8 (B) and blends of PEO–PMMS-8 with 30 wt % PEO (C) and 50 wt % PEO (D). (Reproduced with permission from reference 12. Copyright 1988 Elsevier.)

Table I. Conductivities (σ) of Polysiloxane–$LiClO_4$ Complexes at EO/Li^+ Ratios of 25

Polymer	*25 °C*	*75 °C*
PMMS-8	7.0	25
PMMS-8 (cross-linked)	2.0	13
PMMS-8–30 wt % PEO	3.9	—[a]
PMMS-8–50 wt % PEO	2.0	—
PAGS-8	3.0	—
PAGS-12	7.6	46
PAGS-12 (cross-linked)	4.8	30
PAGS-12–50 wt % PEO	0.54	18

NOTE: Data were obtained from reference 12. All values are multiplied by $10^5/\Omega$-cm.
[a]— means not determined.

crystalline PEO domains eventually become amorphous when the EO/Li^+ ratio approaches 5.0 (Figure 4).

Single-Ion Conductors. Determination of transference numbers in polymer–electrolyte complexes (*15*, *16*) indicates that frequently a large fraction of the current is carried by the anion. Optimization of cation transport to improve the efficiency of solid alkali batteries has been attempted by immobilizing the anion. Examples are comb poly(methacrylates) with bound $-CF_2COO^-$ substituents (*5*), poly(ethylene glycol) networks with phosphate groups (*17*), and plasticized poly(styrenesulfonate) (*18*). Recently, we carried out conductivity measurements on the polymer–electrolyte complex $PAGSO_3^-Na^+$ (Table II). For $m/n = 0.4$ ($EO/Na^+ = 17$), the conductivity σ at 25 °C is only $10^{-7}/\Omega$-cm. This value and that at 75 °C are doubled when the m/n ratio is increased to 1.0 ($EO/Na^+ = 8$). The T_gs for the two polymers ($m/n = 0.4$ and 1.0) are –67 and –60 °C, respectively. The conductivity goes up dramatically when TEG is added. The conductivity σ for $PAGSO_3^-Na^+$ with $m/n = 1$ increases at 75 °C by more than a factor of 30 when the TEG/Na^+ ratio reaches 1 (Table II). Temperature-dependent studies show that σ is $10^{-5}/\Omega$-cm at 56 °C for this system. A strong conductivity enhancement was observed also when sodium poly(styrenesulfonate) was plasticized with TEG (*18*).

Table II. Conductivities (σ) of $PAGSO_3^-Na^+$

			σ ($\times 10^6/\Omega$-cm)	
m/n	*TEG/Na*$^+$	*DMTEG/Na*$^+$	25 °C	75 °C
0.4	0	0	0.1	0.4
1.0	0	0	0.2	0.8
0.4	1.0	—[a]	1.2	7.3
0.4	—	1.0	0.12	0.7
1.0	1.0	—	6.0	28.0

NOTE: TEG is tetraethylene glycol, and DMTEG is dimethyltetraethylene glycol.
[a]— means not determined.

The effect of the dimethyl ether of TEG (DMTEG) was far less spectacular than that of TEG (Table II). The low conductivity of $PAGSO_3^-Na^+$ relative to that of salts such as sodium triflate solubilized in the neutral comb polymers is expected, because the ion pair $PAGSO_3^-Na^+$ is much tighter than $CF_3SO_3^-Na^+$. We also found that sodium methanesulfonate, $CH_3SO_3^-Na^+$, is not soluble in DMTEG. This finding suggests that the polymer-bound $-CH_2SO_3^-Na^+$ ion pairs are aggregated into larger clusters in $PAGSO_3^-Na^+$. This finding would also explain why this polymer is a solid, whereas its glycidyl precursor is a viscous melt, and why added DMTEG increases the conductivity of $PAGSO_3^-Na^+$ by a relatively small factor.

Methanol is also a poor solvent for $CH_3SO_3^-Na^+$, but the salt dissolves when TEG is added. Apparently, the multiple binding sites in TEG and its terminal hydroxyl groups are both needed to effectively chelate the bound $-CH_2SO_3^-Na^+$ ion pair. TEG breaks up the ion clusters and probably enlarges the interionic ion pair distance as the cation interacts with the oxygen-binding sites and the sulfonate ion with the hydroxyl group (*18*). The overall effect is a large increase in the conductivity. With TEG present, the $PAGSO_3^-Na^+$ polymer–electrolyte complex becomes a transparent waxy solid that can be cast into films.

Acknowledgments

The authors gratefully acknowledge the support of the Polymers Program of the National Science Foundation, Grant No. DMR 8504999, and the Petroleum Research Fund administered by the American Chemical Society.

References

1. Armand, M. B. *Annu. Rev. Mater. Sci.* **1986,** *16,* 245.
2. Papke, B. L.; Ratner, M. A.; Shriver, D. F. *J. Electrochem. Soc.* **1964,** *129,* 1982.
3. Killis, A.; LeNest, J. F.; Gandini, A.; Cheradame, H. *Macromolecules* **1984,** *17,* 63.
4. Nagoka, K.; Narusa, H.; Shinohara, I. *J. Polym. Sci., Polym. Lett. Ed.* **1984,** *22,* 659.
5. Bannister, D. J.; Davies, G. R.; Ward, I. M.; McIntire, J. F. *Polymer* **1984,** *25,* 1600.
6. Kobayashi, N.; Uchiyama, M.; Tsuchida, E. *Solid State Ionics* **1985,** *17,* 307.
7. Fish, D.; Xia, D. W.; Smid, J. *Makromol. Chem., Rapid Commun.* **1985,** *6,* 761.
8. Blonsky, P. M.; Shriver, D. F.; Austin, P.; Allcock, H. R. *J. Am. Chem. Soc.* **1984,** *106,* 6854.
9. Fish, D.; Khan, I. M.; Smid, J. *Makromol. Chem., Rapid Commun.* **1986,** *7,* 115.
10. Hall, P. G.; Davies, G. R.; McIntire, J. E.; Ward, I. M.; Bannister, D. J.; LeBrocas, K. M. F. *Polym. Commun.* **1986,** *27,* 98.
11. Bouridah, A.; Dalard, F.; Deroo, D.; Cheradame, H.; LeNest, J. F. *Solid State Ionics* **1985,** *15,* 233.
12. Khan, I.; Yuan, Y.; Fish, D.; Smid, J. *Brit. Polym. J.* **1988,** *20,* 281.
13. Nwankwo, I.; Xia, D. W.; Smid, J. *J. Polym. Sci., Polym. Phys. Ed.* **1988,** *26,* 581.
14. Xu, W. Y.; Smid, J. *J. Am. Chem. Soc.* **1984,** *106,* 3790.
15. Leveque, M.; LeNest, J. F.; Gandini, A.; Cheradame, H. *Makromol. Chem. Rapid Commun.* **1983,** *4,* 497.
16. Evans, J.; Vincent, C. A.; Bruce, P. G. *Polymer* **1987,** *28,* 2324.
17. LeNest, J. F.; Gandini, A.; Cheradame, H.; Cohen-Addad, J. P. *Polym. Commun.* **1987,** *28,* 302.
18. Hardy, L. C.; Shriver, D. F. *J. Am. Chem. Soc.* **1985,** *107,* 3823.
19. Khan, I. M.; Yuan, Y.; Fish, D.; Wu, E; Smid, J. *Macromolecules* **1988,** *21,* 2684.

RECEIVED for review May 27, 1988. ACCEPTED revised manuscript March 27, 1989.

7

Conformational Analysis of Substituted Polysiloxane Polymers

Stelian Grigoras and Thomas H. Lane

Fluids, Resins, and Process Industries Research, Dow Corning Corporation, Midland, MI 48686–0994

The nature of the siloxane bonding geometry is elucidated. When silicon is attached to oxygen, charge transfer occurs from the oxygen lone pairs to the Si–O covalent region. Therefore, the character of the lone pairs is diminished and the sp^3 *hybridization geometry is altered. The electronic charge transferred from the oxygen lone pairs creates an excess charge in the covalent region between silicon and oxygen. This excess charge shortens the Si–O bond. The conformation of substituted polysiloxanes was studied with a new method that provides a detailed description of polymer chain behavior below the glass transition temperature* (T_g) *and in the melt state by allowing partial (side chains only) or total (side chains and backbone) torsional relaxation of the bonds. The calculations corroborate the experimentally measured torsional barriers in hexamethyldisiloxane and predict a twisted conformation. The substituted polysiloxanes that are discussed contain the following pairs of substituents at Si: H,* CH_3; CH_3, CH_3; C_2H_5, CH_3; C_2H_5, C_2H_5; CH_3, C_6H_5; C_6H_5, C_6H_5. *The role of side-chain torsional freedom or rigidity in determining the crystalline forms of poly(diethylsiloxane) is explained.*

TO DESIGN THE PROPER POLYMER for the right application efficiently, the polymer chemist must establish a conceptual bridge between the structure of the polymer and its physical properties. One of the most important structural features of polymer chains is the flexibility of their backbones. Polymer chemists routinely deal intuitively with polymer chain flexibility, and they modify this flexibility through structural alterations of the polymer. How-

0065–2393/90/0224–0125$06.00/0

ever, the intuitive understanding of of chain flexibility leads to situations in which such modifications have unpredictable effects upon a large variety of polymer physical properties. If the nature of chain flexibility is well understood and properly characterized, the relationships between structure and properties can be determined, and these relationships can be used to design polymer structure for the desired changes of properties.

A promising solution for the mathematical prediction of chain flexibility is conformational analysis. Generally, conformational analysis identifies stable isomeric states for polymer chains and the energy barriers between them, which are the major elements needed to define chain flexibility in a precise manner. However for polymers, conformational analysis can become a cumbersome task, because the architecture of polymer chains allows a large number of degrees of freedom, which must be studied simultaneously. To obtain meaningful information, the analysis must be simplified, and only the most significant conformational elements must be studied.

More specifically, conformational analysis can provide information on stable isomeric states, which are defined as minima on energy–deformation plots, and on the energy barriers between these minima. Through these minima, the population of each state at different temperatures at thermodynamic equilibrium can be determined, and a description of the kinetics of the transition from one isomeric state to another can be obtained. Therefore, conformational analysis can define chain flexibility completely. Thus, conformational analysis is the key element required to establish the conceptual bridge between polymer structure and physical properties.

Among polymers, the polysiloxanes constitute an excellent family to study in terms of flexibility, because they have a high degree of flexibility. This chapter reports the results of the conformational analyses of substituted polysiloxanes by describing the isomeric states of the side chains and provides the background for a quantitative interpretation of their flexibility. In this study, the effects of adjacent-bond relaxation were analyzed when conformational analysis was performed for one bond in the substituted polysiloxane short chains. Experimental data on relaxation of substituted poly(methylsiloxane)s have been published recently (*1, 2*). These experimental data allowed us to evaluate and calibrate the theoretical results discussed in this chapter. The siloxanes used in this study have the following pairs of substituents at silicon: CH_3, H; CH_3, CH_3; CH_3, C_6H_5; CH_3, C_2H_5; C_2H_5, C_2H_5; C_6H_5, C_6H_5.

General Approach

The basic strategy in studying the conformation of polymers is to identify the principal factors that are responsible for their physical properties. Polysiloxanes constitute a suitable class of polymers, because they show two levels of flexibility in terms of molecular structure: torsional flexibility, which is common to the majority of organic polymers, and bending flexibility. *Torsional flexibility* is defined as the ability of atoms to rotate around chemical bonds. Torsional flexibility does not necessarily involve distortion of the valence geometry, that is, the bond lengths and bond angles of the chemically bonded atoms may be considered unchanged during the torsional rotation. *Bending flexibility* generally occurs when the steric hindrance between non-bonded atoms is very large for particular energetically unfavorable torsional angles. Bending flexibility is also present in most organic polymers with an angular amplitude of several degrees, but it is more prominent in polysiloxanes. The Si–O–Si bond angle can vary from 135 to 180°, with only small changes in conformational energy. In contrast, the O–Si–O bond angle is rigid, and a detailed analysis of its bending flexibility is required.

Computational Method

The conformational analysis reported herein is based on studies of the variation of nonbonding interaction for torsional deformations. The effects of bending flexibility upon these deformations were considered by repeating the conformational analysis of torsional deformations for different values of the Si–O–Si bond angle. The conformational analysis involved a systematic deformation for a specific bond and the full torsional relaxation of all other bonds in the system. This approach gives an accurate description of the long-range steric interactions. The results were obtained with the CHEMLAB II molecular-modeling package (*3*) and with an original program, SCAN, which allowed full torsional relaxation of the adjacent bonds. The molecular mechanics potentials used are described in this section.

Nonbonding Potentials. Previously (*4*), a set of molecular mechanics potentials was generated and compared with the ab initio results for organosilicon compounds. The nonbonding interactions include two terms: steric and electrostatic. The steric potential is calculated by using a Hill function with van der Waals radii approximately 20% larger than the original values (*5*). The electrostatic potential is based on net atomic charges calculated by using molecular orbital methods from a Mullikan population analysis. The net atomic charges are calculated with the extended Hückel theory method (*6*). A dielectric constant of 8.7 offers the best fit with the ab initio electrostatic interactions. This approach provides results in good agreement with the ab initio calculation of nonbonding interactions without the use of additional empirical torsional potentials.

Scanning with Full Torsional Relaxation of Adjacent Bonds. Torsional relaxation of adjacent bonds was carried out as follows. The torsional angle of one bond, which was assigned as the principal bond, was scanned systematically. In this process, the principal bond is fixed at a specific torsional angle. Then the nearest bonds are scanned successively in increments of 5°. The torsional angle corresponding to the lowest energy becomes the new value for the corresponding bond, and the next adjacent bond is scanned. This process is continued for all other bonds included in the molecular structure. To avoid variance in the final energy, this sequential scanning is performed twice. The sequential scanning is repeated two more times with the starting angle at –30° and in increments of 2° until the final point at 30°, around the value corresponding to the most stable conformation obtained. One more scan is performed from –10 to 10° in increments of 1°. The energy corresponding to this final conformation is recorded as the value associated with the torsional angle of the main bond. Then everything is repeated for a new value of the torsional angle of the main bond.

The advantage of sequential scanning versus simultaneous scanning of all bonds is the shorter computer time required for the calculations. This advantage becomes significant for molecular systems with many torsional degrees of freedom. Both methods have similar accuracies in predicting the optimum conformation. The sequential method cannot guarantee that the global minimum is achieved instead of a local minimum. However, because the bonds are scanned for a full torsional rotation, the chances to stop the minimization process in a local minimum are lower than when a regular optimization method is used. For this reason, in each case, several starting points and various torsional sequences were calculated for the relaxation of the bonds, and the conformation corresponding to the lowest energy was recorded.

Results and Discussion

Three aspects of chain flexibility in polysiloxanes will be discussed in this section: (1) the nature of the bending flexibility of the Si–O–Si angle, (2) the effects of this flexibility on the conformational analysis performed with simple scanning and with scanning that allows for torsional relaxation, and (3) the conformational analysis of various pendant groups attached to the polysiloxane bond.

Bending Flexibility. For a siloxane backbone, two different bond angles are formed: Si–O–Si and O–Si–O. The Si–O–Si bond angle is very flexible; it measures between 140 and 180° and has a small barrier of linearization (0.3 kcal/mol) (*7*). In contrast, O–Si–O is characterized by a rigid bond angle measuring between 102 and 112°, depending on the nature of the two substituents on Si. When the two substituents on Si are methyl groups, this bond angle is 112°, and when the substituents are hydrogens, this bond angle decreases to 104° (*4*). Two problems are discussed in this section: the nature of the unusual flexibility of Si–O–Si bond angle and its implications.

Flexibility of the Si–O–Si Bond Angle. Bond angles are often measured but seldom explained or interpreted, because a large number of compounds show small deviations from the standard values corresponding to the appropriate orbital hybridization. In this context, the siloxane bond is extremely interesting because of its unusually short bond length, large bond angle, and low barrier of linearization.

Disiloxane is a pertinent test compound with which to study the siloxane bond. A large body of experimental data is available for disiloxane. Its relatively small molecular size provides an attractive problem for theoreticians. When electronic effects are claimed to be responsible for a certain geometric element of a specific bond, the relative energy of the molecular orbitals and the electron occupation of these orbitals are understood to be the determining factors. Rationalization of such properties of polyatomic molecules can often be done in terms of Walsh diagrams (*8*). Yet, these diagrams do not show any interpretable relationships between the orbital energy and the angular geometry for disiloxane. However, useful qualitative information can be obtained from electronic charge distribution.

(*p–d*) π bonding is not a valid concept for the unusual bending flexibility of the Si–O–Si angle (*4*, *9*, *10*). A new basis set, 3-21G* (modified) (*9*), which uses an appropriate polarization function at oxygen and silicon atoms, accurately predicts a large variety of structural and electronic properties of the siloxane bond. The standard 3-21G* basis set does not include the polarization function at oxygen and fails to predict the correct value for the Si–O–Si angle.

The molecular wave functions of disiloxane have been generated with a Gaussian 82 program (*10*) using the two different basis sets. These two basis sets, modified and standard, were compared in terms of the difference between the electronic charge densities. The calculated wave functions were used to compute the electronic charge densities. The electronic charge density has been calculated with a modified version of the PSI77 program (*11*). The modification consisted of the inclusion of the corresponding basis sets and addition of the capability to generate the plot that shows the result of subtraction of the electronic charge density calculated with one basis set from the electronic charge density calculated with another basis set.

The plot of electronic charge density differences calculated with the two basis sets is shown in Figure 1. The blue region indicates a deficit of electronic charge when the correct basis set is used, compared with the electronic charge calculated with the standard 3-21G* basis set. The yellow regions show a surplus of electronic charge. The figure indicates that the electronic charge localized at lone pairs on oxygen is transferred to the covalent bonding region between Si and O when a correct basis set is used to calculate the charge density. The lone pairs at oxygen are diminished, and the sp^3 hybridization geometry is altered. The consequence is a wider Si–O–Si bond angle. This effect cannot be described by using a standard 3-21G* basis set.

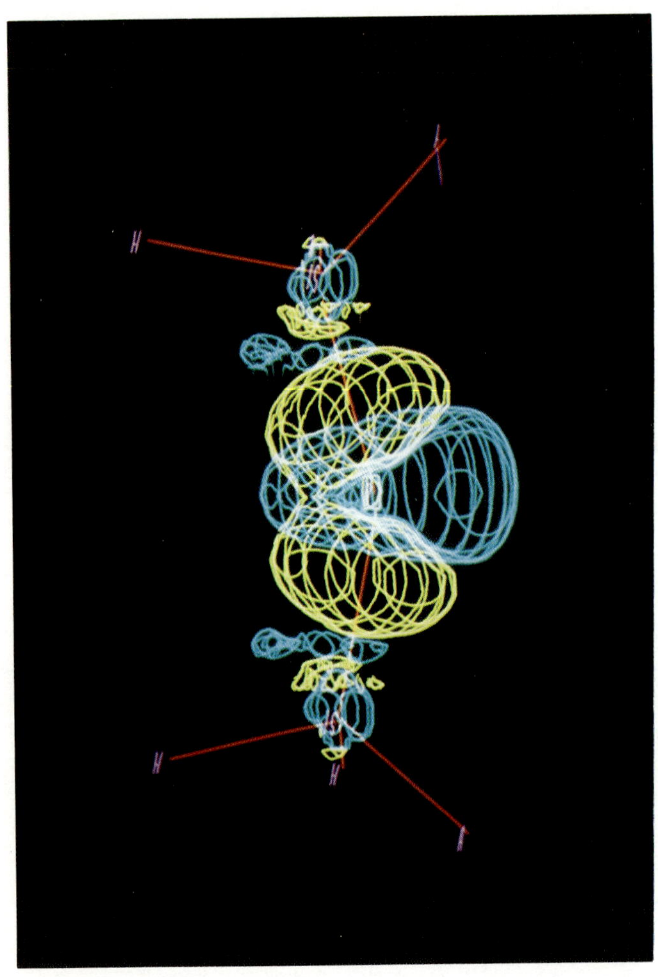

Figure 1. Differences between the electronic charge distributions as calculated with the 3-21G and the modified 3-21G* basis sets. The blue region indicates a deficit of electronic charge, and the yellow region indicates a surplus when the modified 3-21G* basis set is used with the standard 3-21G* basis set as reference.*

Therefore, charge transfer has two major effects: (1) the Si–O–Si bond angle widens, and (2) the additional charge transferred to the covalent region between Si and O shortens the bond length.

Influence of Flexible Bond Angle on Torsional Conformations. Hexamethyldisiloxane was used as a test compound for several procedures applied in conformational analysis. This compound is relatively simple, and structural data are available in the literature. The experimental value for the Si–O–Si bond angle in hexamethyldisiloxane is 148 ± 3°, as measured by electron diffraction (*12*). The electron diffraction data indicate a staggered conformation (C_{2v} symmetry), although a model with twist angles around Si–O bonds of about 30° cannot be excluded.

The torsional barrier around the Si–O bond is in the order of a few tenths of kilocalories per mole (*13*). This barrier was studied as a test for different Si–O–Si bond angles. To illustrate the influence of bending flexibility upon bond torsions, the molecule of hexamethyldisiloxane was rotated around the Si–O bond for two values of the Si–O–Si bond angle: 145 and 150°. The same compound was used to show the differences between simple scanning and the scanning with torsional relaxation.

The results of a simple scanning around the Si–O bond (Figure 2) shows that the most stable conformation corresponds to the twist conformation at 30° for both Si–O–Si bond angles considered. Two additional local minima, which have equal energies, are found at 90 and 150° about the Si–O bond. The relative energy between the absolute minimum and the local minima decreases when the Si–O–Si bond angle becomes wider. For a bond angle of 150°, a barrier of 0.2 kcal/mol exists between the absolute minimum at 30° and its symmetrical position at –30°. Then, a higher barrier of 0.5 kcal/mol separates the absolute minimum from the first local minimum, although a low barrier of 0.3 kcal exists between the local minima. For a bond angle of 145°, the energy barriers between minima increase by a factor of 2, and the energies of the local minima increase.

Scanning with torsional relaxation gives a similar profile. However, significant differences are evident. The values of the torsional angles at the absolute minimum and at the local minima are shifted by 15° to the left and to the right, respectively (Figure 3). The barrier between the absolute minimum and the local minimum next to it becomes sharper, and the second barrier between the local minima becomes wider. The barrier peaks are located at the same angle, but the barrier heights increase for the eclipsed conformation and decrease for the staggered conformation.

Despite the simplicity of this molecule and its symmetry, significant changes in the Si–O bond torsions occur when adjacent bonds are relaxed. A comparison between the results of the standard method and the new scanning method is shown in Figure 4. For this particular case, rotation of one methyl group around the Si–C bond is not sensitive to the variation of

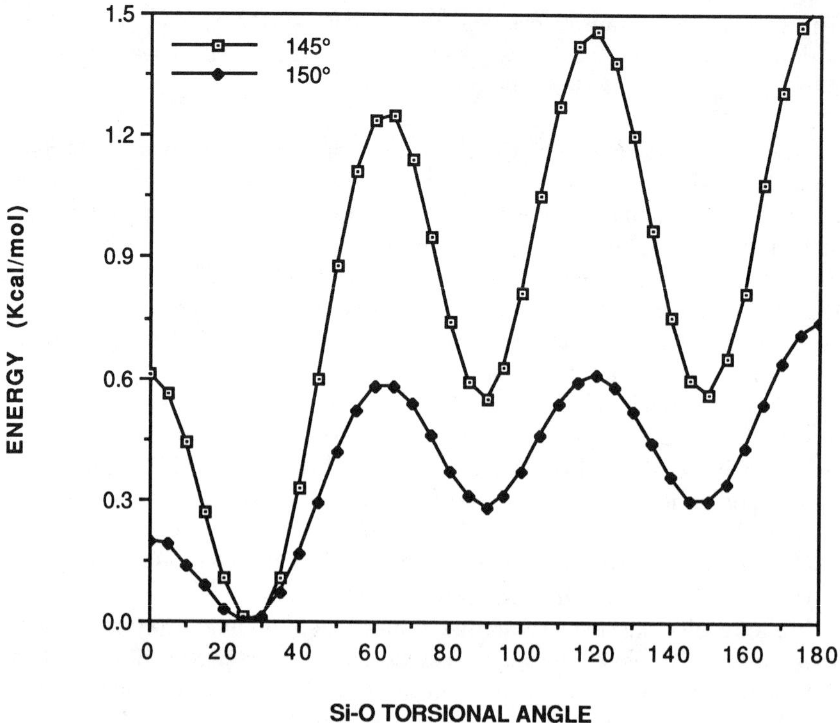

Figure 2. Plot of torsional energy of Si–O bond in hexamethyldisiloxane for two values of Si–O–Si bond angles. The data were obtained by using the simple-scanning method.

the Si–O–Si bond angle. The calculated torsional barrier is 1.6 kcal/mol, in excellent agreement with the experimental value (*13*).

These comparisons show that variation of the Si–O–Si bond angle significantly alters the values of the potential barriers between conformational minima and the relative energies of these minima. Relaxation of adjacent bonds has an important effect on bond torsions; it modifies the potential barriers, as well as the energies and positions of the minima.

Conformational Analysis of Pendant Groups. Torsional relaxation of pendant groups plays an important role in polymer phase transitions. A thorough computational study would require the simulation of several long chains by using molecular dynamics or Monte Carlo approaches. However, with some simplifying assumptions, conformational analysis of oligomers can provide useful indications of the molecular characteristics that influence the behavior of polymer chains in various phases. These phases are determined by the ability of polymer chains to order themselves. This ordering is con-

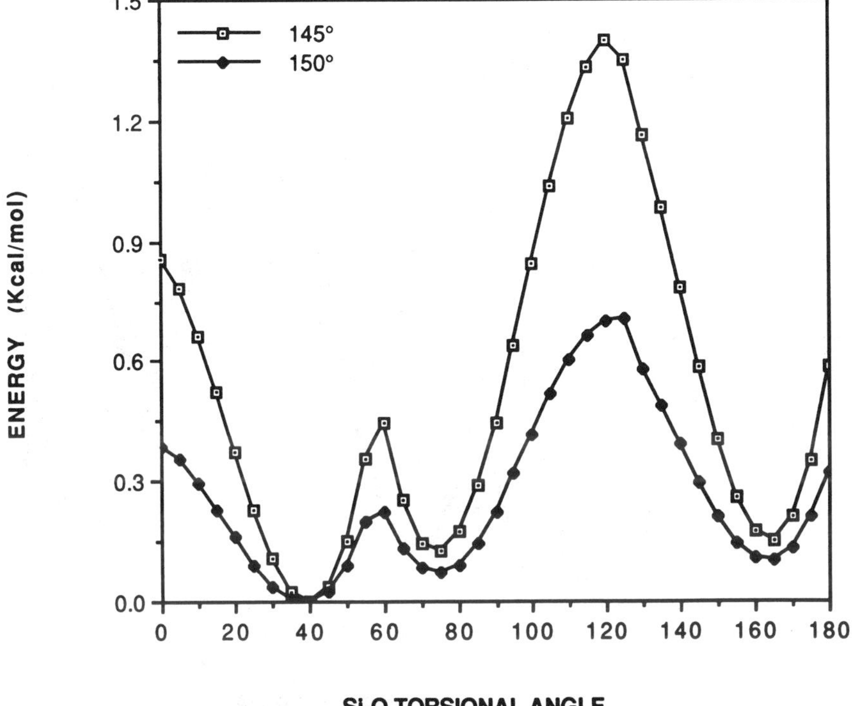

Figure 3. Plot of torsional energy of Si–O bond in hexamethyldisiloxane for two values of Si–O–Si bond angles. The data were obtained with the method using torsional relaxation of adjacent bonds.

trolled by chain conformation, chain flexibility, and the reorientation of pendant groups.

The thermodynamics of the phenomena involving polymer chain order–disorder characteristics cannot be approached by studying oligomers, because important interactions occurring in large polymeric ensembles are neglected when short chains are studied. However, conformational analysis of short chains provides a description of the possible isomeric states that can occur in the long chains and the probability of the relative occurrence of these states. The assumption in this work is that, from a conformational point of view, the main difference between the melt phase and the glassy state of polymers is that in the melt phase a complete torsional relaxation is allowed but in the glassy state the backbone is frozen and only the pendant groups can have torsional relaxation.

Torsional relaxation of pendant groups, which can occur at temperatures below or above the glass transition temperature (T_g), plays an important role in polymer chain dynamics. Even at low temperatures, despite the fact that

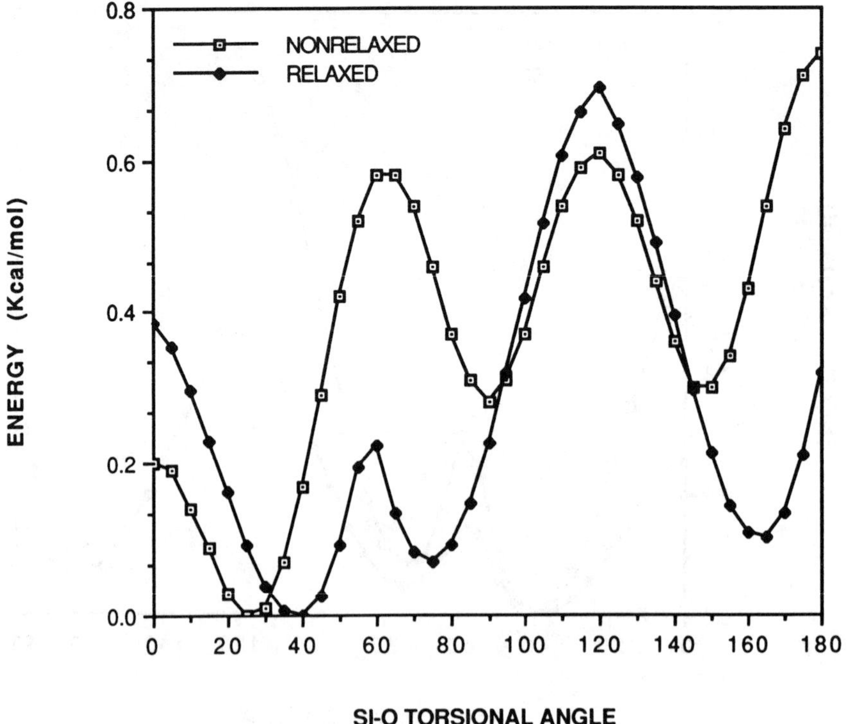

Figure 4. Comparison of results of simple scanning and method using torsional relaxation of adjacent bonds for the same bond in hexamethyldisiloxane. A Si–O–Si bond angle of 150° was used.

Boltzmann statistics predict no torsional transitions for energy barriers in the range of 2 to 4 kcal/mol, this motion has been observed experimentally. This observation has not been explained satisfactorily, but a tunneling effect may be responsible (*14*).

The torsional movement of the side chains is characterized by an activation energy that is strongly dependent on the type and size of the side chains. The bonds adjacent to the side groups belong to the backbone chain. At temperatures below T_g, the activation energy for backbone relaxation is very high (75 kcal/mol for poly(dimethylsiloxane) [PDMS; *1*]). Whether local and limited relaxation occurs for the backbone at temperatures below T_g is uncertain. At temperatures above T_g, the activation energy for backbone relaxation is much lower. For PDMS, a value of 2.75 kcal/mol was determined experimentally (*1*).

To study the relaxation of pendant groups, two conformational analysis cases were carried out. For both cases, one bond at the pendant group was scanned systematically. In the first case, the relaxation of the other bonds was restricted if the bonds belong to the backbone, that is, the backbone was frozen, and only the side chains were allowed to relax. This case simulates the molecular structure in a temperature domain below T_g. In the second case, all bonds, including the backbone and the pendant groups, were allowed to relax when one bond at the pendant group was scanned systematically. This second case simulates the conformational behavior of the polymer in the melt state.

A substituted trisiloxane structure (Figure 5) was used in this study. The initial structure has a planar *cis–trans* conformation. The terminal silicon atoms are bonded to a pair of substituents (X and R in Figure 5), a methyl group, and the oxygen of the siloxane bond. The central silicon is bonded to a pair of substituents and to the corresponding oxygens belonging to the backbone. When the substituents are not the same, they are oriented in opposite direction with respect to the orientation of the substituents at the central silicon atom. For example, in the case of methyl–phenyl substitution, if the phenyl group at the central silicon atom is oriented above the plane

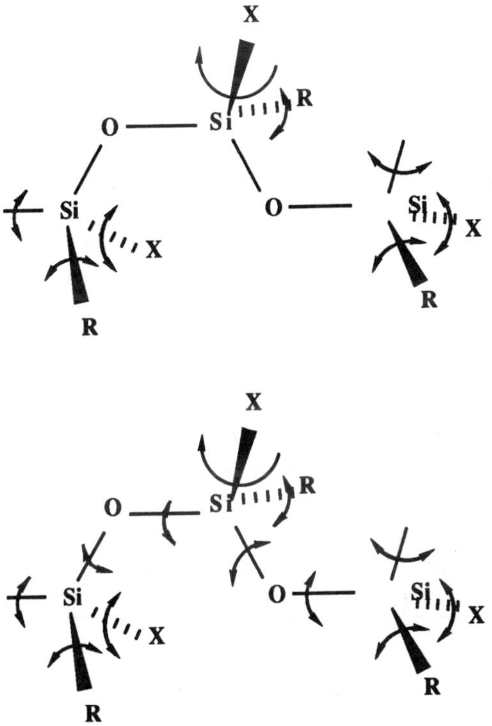

Figure 5. General structure used to analyze torsional potentials of substituted siloxanes. Top, in the glassy state (at temperatures below T_g*) and bottom, in the melt state.*

formed by the Si and O atoms, the phenyls at the terminal silicon atoms are directed below the plane of reference.

The starting structures were optimized for their torsional angles by using the sequential scanning method previously described. Then, EHT (extended Hückel theory) charges were recalculated for the optimized conformations with optimum torsional angles. With the newly optimized structures as a starting point, one bond was scanned in increments of 10°. The other bonds were relaxed with the sequential scanning procedure previously described.

With the central Si atom as reference, the reported torsional angles are defined with respect to the position of the pendant group of interest and the other silicon substituent. For example, for the methyl group in PDMS, the hydrogens were rotated by using the torsional sequence H–C–Si–C. For larger pendant groups like ethyl, the sequence H–C–C–Si is used to define the torsional angle. The convention for torsional angle measurement assigns a zero value to the *cis* conformation.

The valence geometry used in this study is given in Table I. All bond angles and bond lengths were maintained at the same values to allow the comparison of the effect of substituents on torsional behavior. Each case was studied in the solid and the melt states, by using the two relaxation modes previously described. The results (Table II) indicate two types of barriers. For the solid state (at temperatures below T_g), only the pendant groups relax (Table II, column N). For the liquid state, full relaxation is achieved during systematic scanning of pendant groups (Table II, column R). This mode is assumed to describe the polymer in the melt state.

CH_3–Si Torsion in PDMS. The torsional potential curve has three degenerate minima because of the threefold symmetry of the methyl group. For this reason, Figure 6 shows only the behavior of one-third of the full rotation. The barrier height is 1.7 kcal/mol when the backbone is allowed to relax and 2.4 kcal/mol when the backbone is frozen. The experimental value is 2.2 kcal/mol (*1*), but this value is based on results determined at two temperatures: above and below T_g. Our results indicate that at least one more experimental measurement, at T_g for the Si–CH_3 bond, is needed. These calculations predict that the new measured value will set the data

Table I. Valence Geometry of Oligomers

Bond	*Length (Å)*	*Bond*	*Angle (°)*
Si–O	1.656	Si–O–Si	148
Si–C	1.890	O–Si–O	110.7
Si–H	1.492	O–Si–C	109.47
C–C	1.540	H–C–H	109.47
C–H	1.09	Si–C–C	120
C_{ar}–C_{ar}	1.40	C_{ar}–C_{ar}–C_{ar}	120
C_{ar}–H	1.08	C_{ar}–C_{ar}–H_{ar}	120

NOTE: C_{ar} is aromatic carbon.

Table II. Potential Barriers Between Stable Isomeric States of Side Chains for Two Cases: Full Relaxation (R) and Torsional Relaxation (N) of the Side Chains

				Energy Barrier (kcal/mol)	
Substitution	*Bond*	*Angle (°)*	*Energy (kcal/mol)*	*R*	*N*
CH_3, CH_3	CH_3–Si	60	0	1.7	2.4
CH_3, H	CH_3–Si	60	0	0.7	0.7
CH_3, C_2H_5	CH_3–Si	40	0	1.9	2.2
	CH_3–C	60	0	4.1	4.2
	C_2H_5–Si	180	0		
		294	0.9	3.3	
CH_3, C_6H_5	CH_3–Si	−11	0	1.1	1.1
	C_6H_5–Si	55	0	4.9	>40
		135	1.3	4.9	
C_6H_5, C_6H_5	C_6H_5–Si	220	0	10	>40
		130	2.0	8	
C_2H_5, C_2H_5	C_2H_5–Si	150	0	19	19
		200	3.9		9.4
		−80	17		18
		−30	14.4		20.5
		80	0.9	1.6	

point above the actual line formed by two points at this time in the plot of activation energy. Therefore, two different slopes should be observed at temperature domains below and above T_g.

CH_3–Si Torsion in Poly(methylsiloxane). When one Si substituent is hydrogen and the other is methyl, the torsional barrier of the Si–CH_3 bond decreases to 0.7 kcal/mol, compared with PDMS. In this case, no significant differences were found between the bond torsions at temperatures above T_g and those below T_g. The torsional curves practically superimposed.

Side-Chain Torsions in Poly(methylethylsiloxane). Three bond rotations can be considered for the side chains of poly(methylethylsiloxane) (PMES).

CH_3–Si Torsion. Comparison of the rotations of the same bond in PDMS and PMES shows that the torsional angle corresponding to the minimum is shifted in PMES by 20°. The most stable conformation does not occur exactly at the *trans* orientation of the hydrogens in this case. This fact can be explained in terms of steric repulsions induced by the relaxation of the ethyl group vicinal to methyl. The difference between the two cases (below and above T_g) is not as large as that for PDMS. The energy barrier is 1.9 kcal/mol at temperatures above T_g and 2.2 kcal/mol at temperatures below T_g.

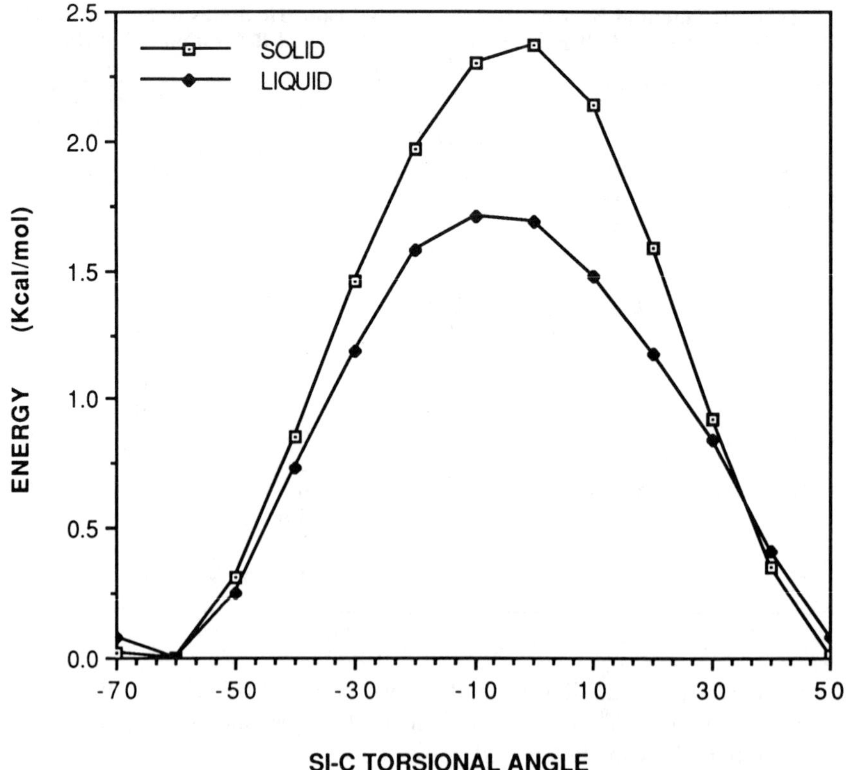

Figure 6. Torsional potential of CH_3–Si bond in poly(dimethylsiloxane) (PDMS) in solid and liquid states.

CH_3–CH_2 Torsion. The methyl group in ethane is further away from the backbone, and the relaxation of the bonds belonging to the backbone will have a minor influence on this bond. Therefore, a major difference between the two cases considered is not expected. The barrier is 4.2 kcal/mol for the solid state and 4.1 kcal/mol for the melt state.

C_2H_5–Si Torsion. The torsion of the ethyl group is more complicated. Figure 7 shows the results of conformational analysis. For clarity, the unreasonable values of energy were eliminated. In the melt state, two stable orientations exist. The most stable state corresponds to the *trans* conformation, and the local minimum corresponds to the *gauche* conformation. The energy difference between these two minima is 0.9 kcal/mol. The barrier from *trans* to *gauche* is 3.3 kcal/mol. At temperatures below T_g, only one sharp minimum is found. However, if the initial conformation of the backbone corresponds to that of the local minimum when the backbone is allowed to relax, the absolute minimum for the frozen backbone will coincide with

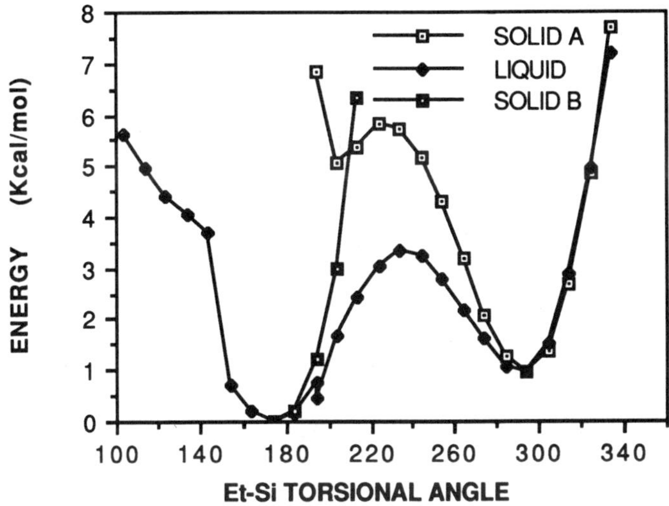

Figure 7. Torsional potential of C_2H_5–Si bond in poly(methylethylsiloxane) (PMES) in solid and liquid states. For the solid states, two conformations of the backbone were studied (see *text*).

the local minimum for the relaxed phase, which has been used as starting point. The profile of the conformational curve follows the profile of the conformational analysis that includes backbone relaxation. The barriers are significantly larger, and a local minimum occurs. This minimum corresponds to the absolute minimum in the mobile phase, but because of the inability of the backbone to relax, this additional minimum is energetically unfavored.

This case shows that the reported conformations of side chains are not unique when the the backbone is frozen. This chapter does not intend to discuss all possibilities. Rather, only the most favorable conformations that are associated with the starting structure corresponding to the most stable conformation of the backbone are covered.

Side-Chain Torsions in Poly(methylphenylsiloxane). Two bond torsions are possible for the side chains in poly(methylphenylsiloxane) (PMPS).

CH_3–Si Torsion. The CH_3–Si bond torsion in PMPS has the same features as those of the previous systems. No relevant differences were observed between the potential barriers in the solid state and the melt state. The potential barrier is low, only 1.1 kcal/mol.

C_6H_5–Si Torsion. The torsion of the phenyl group has a twofold symmetry. A big difference exists between the torsional curves corresponding to the solid and the melt states. In the solid state, when the backbone cannot relax, a very sharp minimum occurs at 240°. The potential barrier between

this minimum and its equivalent located at 60° is greater than 40 kcal/mol. Experimental measurements of the activation energy suggest a much lower value of 7.4 kcal/mol (2). However, the same problem of limited number of experimental determinations applies in this case, as for the torsional barrier of methyl groups in PDMS. When the backbone is allowed to relax, the potential wells for the minima at 240° and that at the symmetrical position at 60° become wider. A local minimum exists between them. The energy of this local minimum is 1.3 kcal/mol greater than the energy of the absolute minimum. The energy barrier between the absolute minimum and the local minimum is 4.9 kcal/mol. This value is lower than the measured value of 7.4 kcal/mol. This disagreement supports our conclusion that in some cases the activation energy of pendant groups at temperatures below T_g is different from that at temperatures above T_g. Therefore, the phenyl group prefers an orientation parallel to the Si–O bond. This orientation corresponds to the absolute minimum shown in Figure 8. Also, a stable position with higher energy is found for the phenyl ring when it is oriented between CH_3 and

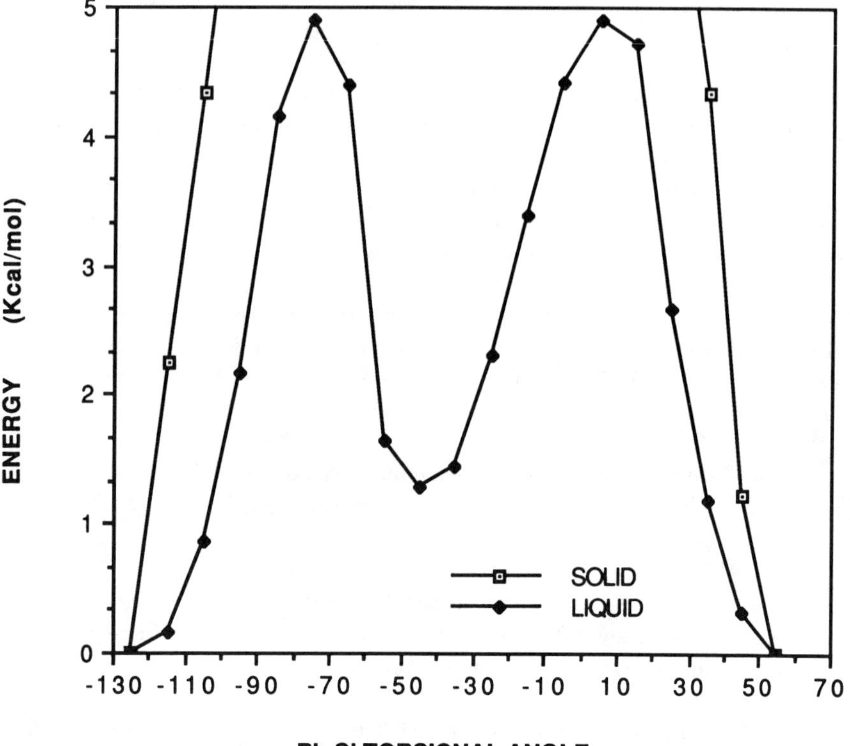

Figure 8. Torsional potential of C_6H_5–Si bond in poly(methylphenylsiloxane) (PMPS) in solid and liquid states.

O but not between the two oxygens attached to Si, as viewed through a Newmann projection.

C_6H_5–Si Bond Torsion in Poly(diphenylsiloxane). The profile of the conformational curve of the C_6H_5–Si bond in poly(diphenylsiloxane) (PDPS) (Figure 9) has features in common with that of the same bond in PMPS. The potential well is wider, but the energy barrier between the degenerate stable states is also very high when the backbone is rigid. In the case of the flexible backbone, the minimum at 130° has an energy 2 kcal/mol higher than those of the absolute minima located at 50 and 210°. The barriers between the degenerate absolute minima and the minimum at 130° are higher than the corresponding barriers in PMPS. Because of the large barriers between the stable states, the side chains in PDPS are very rigid even when the backbone is allowed to relax.

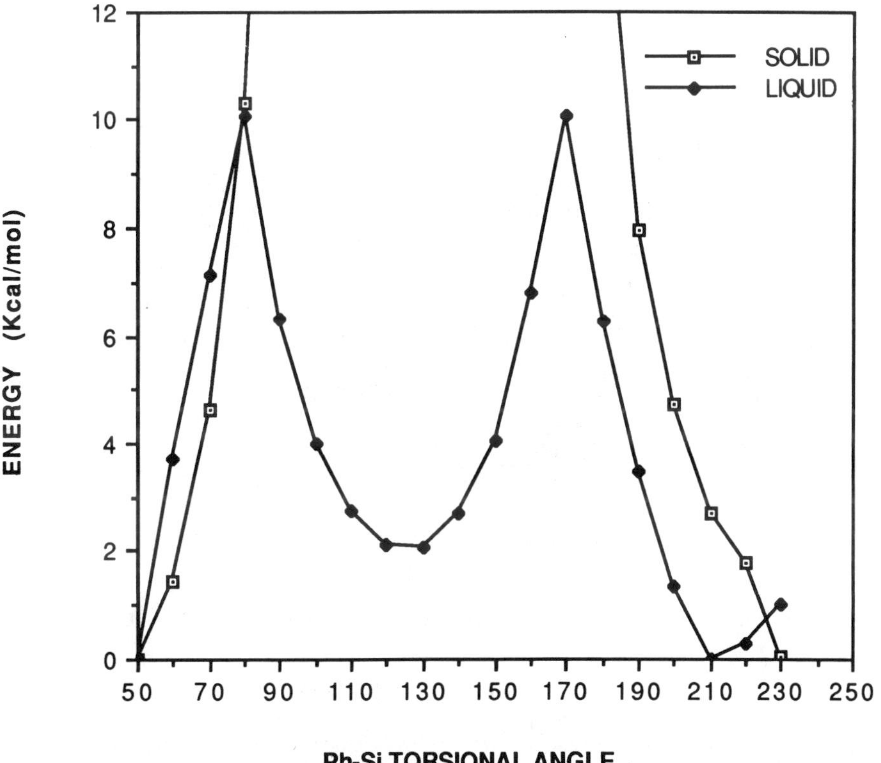

Figure 9. Torsional potential of C_6H_5–Si bond in poly(diphenylsiloxane) (PDPS) in solid and liquid states.

C_2H_5–Si Bond Torsion in Poly(diethylsiloxane). Poly(diethylsiloxane) (PDES) is a thermotropic liquid-crystal material. It exists in the mesomorphic state for a certain temperature range after melting of the crystalline phase (*15*). This property is due primarily to the side chains. This material has two crystalline forms, α and β, and these crystalline forms go through isomorphic transitions from α1 to α2 and from β1 to β2 when the temperature increases. These transitions occur because of ethyl group reorientation (*16*), which has an activation energy of 9.3 kcal/mol. The previous experimental results (*16*) did not identify the initial and final orientations. In this study, we identified these orientations by using conformational analysis.

In previous examples of substituted polysiloxanes, the relaxation of the side chains with the rigid backbone was assumed to describe the polymer chain in the amorphous glassy state. This assumption, relaxation of side chains only, can be used to study the crystalline states of PDES. Certainly, this simplification is extreme, but it can be useful to understand the available orientation of the pendant groups when the polymer chains undergo transition from one crystalline form to another. The present approach does not address the chain reorientation or the interchain interactions in the crystalline state.

Figure 10 shows the conformational behavior of the C_2H_5–Si bond. Many

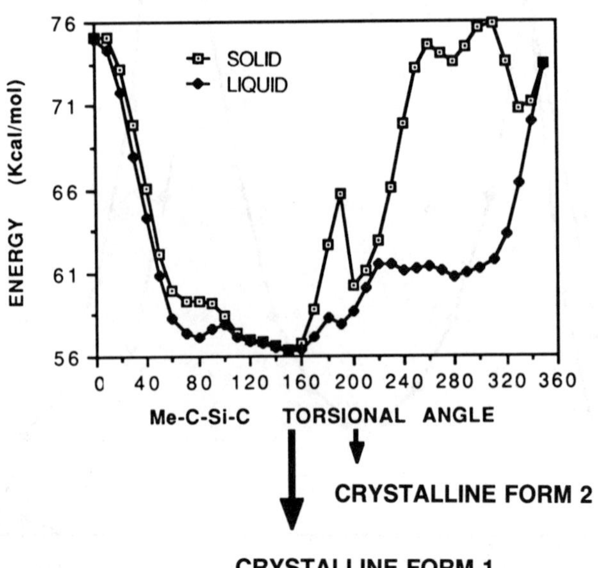

Figure 10. Torsional potential of C_2H_5–Si bond in poly(diethylsiloxane) (PDES) in solid and liquid states. PDES exists in two crystalline forms, each of which has two isomorphic forms because of ethyl group reorientation, which has an activation energy of 9.3 kcal/mol (theoretical value, 9.4 kcal/mol). Experimental values were obtained from reference 16.

local minima are present. Also, many differences between PDES with frozen backbone and PDES with relaxed backbone are evident. When the backbone is rigid, the absolute minimum corresponds to a torsional angle of 150° and is followed by a local minimum with an energy 3.9 kcal/mol higher. The barrier from the absolute minimum to the local minimum is 9.4 kcal/mol. This energy barrier is practically coincident with the activation energy of the isomorphous transformation of crystals. The local minimum corresponds to a torsional angle of 200°. Therefore, the isomorphous transition occurs when the ethyl group reorients by 50°.

Two additional local minima with very high energies (17 and 14.4 kcal/mol higher than the energy of the absolute minimum) are found. The energy decreases, a plateau is formed between 70 and 100°, and finally, the energy descends to the absolute minimum.

When backbone relaxation is included, the profile of the torsional conformational curve is different. The barrier between the absolute minimum and the first local minimum, which characterizes the isomorphic crystalline transition, becomes much lower, and this local minimum becomes insignificant. The high energy barriers, which generate the second and third local minima, disappear also, and a wide plateau for torsional angle is found between 230 and 300°. The highest energy barriers for a torsional angle of 0° are identical for the two cases. When the torsional angle approaches 80°, an additional local minimum occurs. Its energy is only 0.9 kcal/mol higher than the energy of the absolute minimum. The barrier from the absolute minimum to this local minimum is 1.6 kcal/mol.

Apparently, the structure of trisiloxane should be symmetrical when both substituents at Si are the same. However, the symmetry disappears because the optimum conformation of the backbone is *cis–trans* instead of *trans–trans*. Hence, the isomorphic transition of crystalline forms of PDES can be explained. In the liquid state, this material can form liquid crystals because various conformations for the side chain are possible. The most stable conformation of the ethyl group corresponds to a position of the methyl group in twisted *trans* orientation with respect to the ethyl group attached to the Si atom.

Conclusions

This study is the first step towards a quantitative prediction of chain flexibility based on conformational analysis. Torsional relaxation of the adjacent bonds is very important. The present approach differentiates between the polymer relaxations in the glassy state and in the melt state. It provides insight into the crystallinity of a system and succeeds in explaining the isomorphic transformations of PDES. Chain flexibility is influenced by at least two types of factors: the number of isomeric states available and the torsional freedom in a given state, which is determined by the shape of the potential well.

These results corroborate the experimentally measured data and predict torsional barriers. Furthermore, the calculations provide a detailed picture at the molecular level of the available isomeric states and show that the mobility of the side chains at temperatures below T_g should differ from that at temperatures above T_g for some cases. Additional experiments will test our conclusion. The theoretical description of the isomeric states provides grounds for predicting the crystalline and amorphous forms of the polymeric materials in the solid and liquid states.

References

1. Litvinov, V. M.; Lavrukhin, B. D.; Zhdanov, A. A. *Polym. Sci. USSR, Engl. trans.* **1985,** *27*, 2777.
2. Litvinov, V. M.; Lavrukhin, B. D.; Zhdanov, A. A. *Polym. Sci. USSR, Engl. trans.* **1985,** *27*, 2786.
3. *CHEMLAB II* Molecular Design Ltd., San Leandro, Ca.
4. Grigoras, S.; Lane, T. H. *J. Comput. Chem.* **1988,** *9*, 25.
5. Allinger, N. L.; Yuh, Y. H. Quantum Chemistry Program Exchange, Bloomington, Indiana, Program 395.
6. Howell, J; Rossi, A.; Wallace, D.; Haraki, K.; Hoffmann, R. Quantum Chemistry Program Exchange, Bloomington, Indiana, Program 469.
7. Durig, J. R.; Flanagen, M. J.; Kalasinsky, V. F. *J. Chem. Phys.* **1977,** *66*, 2775.
8. Walsh, A. D. *J. Chem. Soc.* **1953,** 2260–2331.
9. Grigoras, S; Lane, T. H. *J. Comput. Chem.* **1987,** *8*, 84.
10. Binkley, J. S.; Frisch, M. J.; Defrees, D. J.; Raghavachari, K.; Whiteside, R. A.; Schegel, H. B.; Fluder, E. M.; Pople, J. A. Gaussian 82, Carnegie–Mellon University.
11. Jorgensen, W. L. Quantum Chemistry Program Exchange, Bloomington, Indiana, Program 340.
12. Csakvari, B.; Wagner, Z.; Gomory, P.; Mijlhoff, F. C. *J. Organomet. Chem.* **1976,** *107*, 287.
13. Scott, D. W.; Messerly, J. F.; Todd, S. S.; Guthrie, G. B.; Hossenlopp, I. A., Moore, R. T.; Osborn, A.; Berg, W. T.; McCullough, J. P. *J. Phys. Chem.* **1961,** *65*, 1320.
14. Perepechko, I. *Low-Temperature Properties of Polymers;* Pergamon: Oxford, NY, 1980; p 200.
15. Bearry, C. L.; Pochan, J. M.; Froix, M. F.; Hinman, D. F. *Macromolecules* **1975,** *8*, 547.
16. Burnett, L. J.; Durret, M. G. *J. Polym. Sci., Polym. Phys. Ed.* **1980,** *18*, 227.

RECEIVED for review May 27, 1988. ACCEPTED revised manuscript March 21, 1989.

8

Synthesis and Fractionation Studies of Functionalized Organosiloxanes

Cheryl S. Elsbernd,[1] Maria Spinu, Val J. Krukonis,[2] Paul M. Gallagher,[2] Dillip K. Mohanty,[3] and James E. McGrath[4]

Department of Chemistry and NSF Science & Technology Center, Virginia Polytechnic Institute and State University, Blacksburg, VA 24061

The synthesis of difunctionalized aminopropyl-terminated poly(dimethylsiloxane) via equilibrium polymerization of the cyclic tetramer in the presence of the functionalized disiloxane was demonstrated. The use of tetramethylammonium and tetrabutylphosphonium siloxanolate anionic catalysts for these reactions was studied. The reactions of the tetramethylammonium and tetrabutylphosphonium catalysts were limited to ~80 °C because of the known transience of these species at higher temperatures. The disappearance of the cyclic tetramer and disiloxane was monitored by HPLC (high-pressure liquid chromatographic) and GC (gas chromatographic) techniques, which indicated that the tetramethylammonium and tetrabutylphosphonium catalysts were much more efficient than the potassium catalyst, even when the potassium catalyst was used at much higher (e.g., 160 °C) temperatures. This behavior may be related to the higher degree of dissociation and, possibly, the enhanced solubility of the tetramethylammonium and tetrabutylphosphonium catalysts relative to the much more studied potassium catalyst. The number-average molecular weight was independent of the catalyst concentration and was influenced only by conversion and the molar ratio of the tetramer to the disiloxane. GPC (gel permeation chromatographic) characterization was also possible by derivatizing

[1]Current address: 3M Center, St. Paul, MN 55144
[2]Current address: Phasex Corporation, 360 Merrimack Street, Lawrence, MA 01843
[3]Current address: Department of Chemistry, Central Michigan University, Mount Pleasant, MI 48858
[4]Author to whom correspondence should be addressed

0065–2393/90/0224–0145$06.00/0

the amine to ketimine functionalities. Supercritical-fluid-fractionation studies of the functionalized siloxanes yielded samples with relatively narrow molecular weight distributions, which were compared with polysiloxane standards produced in our laboratory. The number-average molecular weights determined by titration methods and those determined by GPC experiments were in excellent agreement.

THE IMPORTANCE AND UTILITY of multiphase copolymer systems have been well documented in the literature (*1–4*), with emphasis on their unique combination of properties and their potential material applications. Organosiloxane block polymers are a particularly interesting type of multiphase copolymer system because of the unusual characteristics of polysiloxanes, such as their stability to heat and UV radiation, low glass transition temperature, high gas permeability, and low surface energy (*1*, *2*, *5*). The incorporation of polysiloxanes into various engineering polymers offers an opportunity for many improvements, such as lower temperatures for the ductile-to-brittle transitions and improved impact strength.

Siloxane-containing block copolymers are often prepared by step-growth or condensation polymerization of preformed difunctionalized siloxane oligomers with other difunctionalized monomers or oligomers. Our current work (*3*, *4*, *6–8*) on siloxane chemistry includes the preparation of a number of functionalized oligomers, with emphasis on equilibration processes with the commonly available cyclic tetramer, octamethylcyclotetrasiloxane (D_4), in the presence of a functionalized disiloxane or end blocker.

Despite the importance and synthetic utility of these siloxane equilibration reactions, relatively little has been reported with respect to the detailed kinetics and mechanisms involved, especially in the presence of functionalized end blockers. A major focus of our efforts (*3*, *4*, *6–8*) is the investigation of various aspects of siloxane equilibration reactions to establish the exact nature of the active polymerization species and the effect of various reaction parameters on the preparation of well-defined difunctionalized siloxane oligomers.

The synthesis and equilibration reaction kinetics involved in the preparation of aminopropyl-terminated polysiloxanes has been studied most extensively because of the utility of the amino-terminated species as components of a large number of segmented copolymers such as imides, amides, and ureas.

The rate of disappearance of the starting materials was followed as one approach to determine the effect of catalyst type and concentration on the rate of the ring-opening polymerization. Results are presented in this chapter for the potassium-siloxanolate-catalyzed system, as well as for the analogous tetramethylammonium- and tetrabutylphosphonium-siloxanolate-catalyzed systems.

Because of the random nature of the equilibration processes, the resulting siloxane oligomers generally possess a Gaussian molecular weight distribution. In addition, a ring–chain equilibrium results in the presence of a certain amount of low-molecular-weight cyclic species at the end of the reaction. In addition to the synthesis of difunctionalized siloxane oligomers, a number of studies have been conducted to explore the usefulness of supercritical-fluid-fractionation techniques for the preparation of well-defined aminopropyl-functionalized polysiloxanes of narrow polydispersity (*9, 10*). The isolation of oligomer fractions with narrow molecular weight distributions (MWD) was interesting from a fundamental viewpoint, because incorporation of these narrow-MWD fractions into segmented copolymers would allow the determination of the effect of the polydispersity of the individual blocks on the properties of the resulting copolymer systems. Studies of the mechanical and morphological properties of such systems would result in a better understanding of structure–property relationships in multiphase copolymer systems.

Experimental Procedures

Materials. Octamethylcyclotetrasiloxane, D_4, was generously supplied by General Electric Company. 1,3-Bis(3-aminopropyl)tetramethyldisiloxane (to be referred to subsequently as aminopropyldisiloxane) was obtained from Petrarch Systems, Inc. These materials were dried over calcium hydride and vacuum distilled prior to use. Potassium hydroxide, tetramethylammonium hydroxide pentahydrate, and tetrabutylphosphonium bromide used in the preparation of the siloxanolate catalysts were used as received from Aldrich.

Catalyst Preparation. The potassium siloxanolate catalyst was prepared by charging finely crushed potassium hydroxide, D_4, and toluene to a flask equipped with an overhead stirrer and an attached Dean–Stark trap with condenser. Argon was bubbled through the solution from below the level of the liquid to promote the elimination of water via a toluene azeotrope as the reaction proceeded. Typically, a D_4/KOH molar ratio of 3:1 was used with enough toluene to form an approximately 50% (wt/vol) solution. The catalyst was allowed to form at 120 °C for 24 h, during which time the toluene–water mixture was eliminated and collected in the Dean–Stark trap. The clear catalyst was then diluted to an ~35% (wt/vol) solution with dry toluene and stored in a desiccator until use.

The tetramethylammonium siloxanolate catalyst was prepared similarly by charging tetramethylammonium hydroxide, D_4, and an azeotropic agent to the flask and heating the reaction at 80 °C for 24 h. The lower reaction temperature was necessary to avoid decomposition of the ammonium catalyst. Under most conditions, this procedure produces an active catalyst that is not completely homogeneous. Although not precisely defined, some carbonate is known to be present in addition to the siloxanolate.

The tetrabutylphosphonium siloxanolate catalyst was prepared by reacting the potassium siloxanolate catalyst with a solution of tetrabutylphosphonium bromide in toluene. The reaction resulted in a precipitate of KBr and the formation of homogeneous tetrabutylphosphonium siloxanolate. The number of moles of siloxanolate catalyst per gram of catalyst solution in all cases was determined by titration with 0.10 N HCl.

Equilibration Reactions. D_4 and the aminopropyldisiloxane were charged into a three-necked flask equipped with a magnetic stirring bar, a reflux condenser, an argon inlet, and a rubber septum for the removal of samples via a syringe during the course of the equilibration reaction. The flask was heated in a controlled-temperature bath until the temperature of the reaction mixture had stabilized, and the desired amount of siloxanolate catalyst was added. Samples were removed at various reaction times and quenched in a dry-ice–isopropyl alcohol bath for later chromatographic analysis. After 24–48 h, the equilibration reaction was terminated for analysis and isolation of the resulting siloxane oligomer. The transient tetramethylammonium and tetrabutylphosphonium siloxanolate catalysts were readily decomposed by heating for 30 min at >145 °C. The potassium siloxanolate catalyst required neutralization with acetic acid, which was followed by washing with water and drying of the oligomer. The cyclic species were removed by vacuum distillation at 100 °C and 40 Pa.

Characterization. Number-average molecular weights (M_ns) of the stripped oligomers were determined by potentiometric titration of the primary-amine end groups with 0.10 N HCl.

The D_4 content of the samples was determined by reverse-phase high-pressure liquid chromatography (HPLC) with a Varian 5500 liquid chromatograph. A DuPont Zorbax ODS (C_{18}) column was used with a Wilmad infrared detector set at 12.45 µm to monitor the Si–CH_3 vibration. The mobile phase was an 83:17 mixture of acetonitrile and acetone at a flow rate of 0.8 mL/min. A Rheodyne injector valve operating on compressed air was used with a 10-µL sample loop for reproducible injection volumes. Ethyl acetate was used to dissolve the samples for analysis.

Capillary gas chromatography (GC) was used to determine the aminopropyldisiloxane concentration. An 11-m column (0.2-mm internal diameter) coated with a dimethylsiloxane stationary phase was used. Temperature-programming techniques and a flame ionization detector were used. Tetradecane was used as an internal standard. Details of the chromatographic conditions have been reported earlier (*8*). A Varian Vista 402 data station simplified the calibration and analysis for both the HPLC and GC methods.

Gel permeation chromatographic (GPC) analysis was not possible directly on the primary-amine-functionalized oligomers because of their tendency to be adsorbed on the GPC columns rather than to elute. A method has been developed in our laboratories (*9, 10*) for derivatization of the amine end groups with benzophenone to form an imine functionality. This method allows FPC (fixed-partial-charge) analysis of the oligomers.

The titration-determined molecular weights of the individual samples were used to calculate the amount of benzophenone required to completely derivatize the amine functionalities. A 10 mol % excess of benzophenone was used to ensure complete derivatization. The reaction was done in bulk at 80 °C by heating the siloxane oligomer and benzophenone for 24 h in the presence of 3-Å molecular sieves. GPC analysis was then carried out at 30 °C with a Waters 590 GPC apparatus equipped with ultrastyragel columns with particle sizes of 500, 10^3, 10^4, and 10^4 Å. A Waters 490 programmable wavelength detector set at 218 nm was used with THF (tetrahydrofuran) as the solvent.

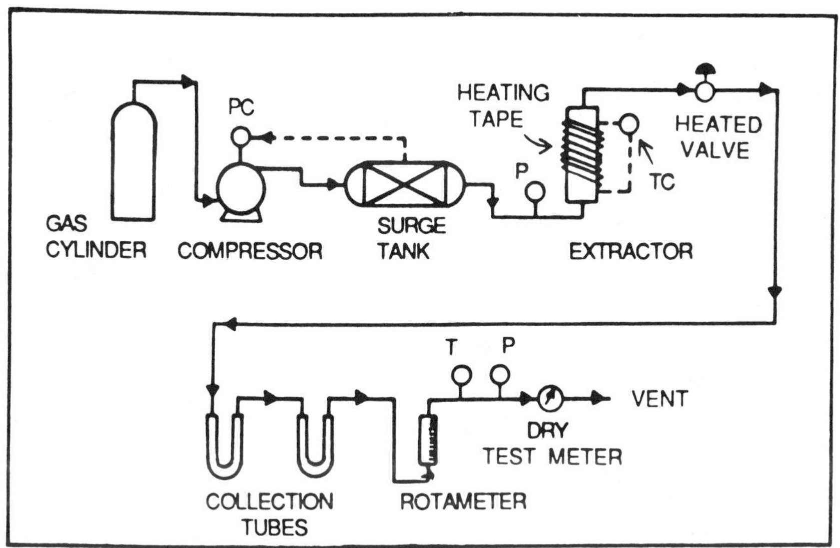

Figure 1. Setup for supercritical fluid fractionation. Abbreviations are as follows: TC, temperature controller; T, thermocouple; PC, pressure controller; and P, pressure gauge. The experimental conditions were as follows: supercritical solvent, ethane; pressure profiling, 800–4000 psi (1 psi = 6895 Pa); and temperature, 80 °C.

Supercritical Fluid Fractionation. In general, the methods previously reported (*9, 13*) were used. The setup is shown in Figure 1.

Results and Discussion

Siloxanolate Catalysts. The initial step for the study of the kinetics of base-catalyzed siloxane equilibration reactions was the preparation of a number of well-defined siloxanolate catalysts. The catalysts were prepared separately, prior to the equilibration reactions, so that a homogeneous moisture-free system with a known concentration of active centers might be obtained. The catalysts studied included potassium, tetramethylammonium, and tetrabutylphosphonium siloxanolate.

Although the hydroxide salts of these bases are effective as catalysts for siloxane equilibrations, the solid hydroxides of both potassium and tetramethylammonium are quite hygroscopic and generally contain significant amounts of water. Also, tetrabutylphosphonium hydroxide (TBPH) is available only as an aqueous solution and decomposes during attempts to concentrate the solutions to >~40% solids (*11*). Because the presence of water can complicate the kinetics of the equilibration process (*2*), steps must be taken to dehydrate the catalysts so that useful kinetic data can be obtained. The approach used in this study was to prepare the siloxanolates of the bases by reacting the hydroxides with D_4 (Scheme I). Another consideration was the fact that the hydroxide salts of the previously mentioned catalysts are not readily soluble in siloxane media, whereas preparation of their siloxanolates yields a catalyst that is homogeneous from the start of the equilibration reaction.

The catalytic solutions were characterized by potentiometric titration of the siloxanolate groups with alcoholic HCl. Titration with acid generally yielded two end points: The first end point was attributed to the siloxanolate group, and the second was attributed to the carbonate species. Previous workers (*12*) have also reported two end points for the titration of tetramethylammonium hydroxide (TMAH) with HCl, with phenolphthalein and methyl orange as indicators. They attributed the end points to hydroxide and carbonate components, because TMAH readily absorbs carbon dioxide from the air. The carbonate species was reasonably assumed to have no catalytic effect on the siloxane equilibration reactions.

Equilibration Reaction Kinetics. This study investigated the effect of various reaction parameters on the synthesis and equilibration reaction kinetics in the preparation of well-defined difunctionalized siloxane oligo-

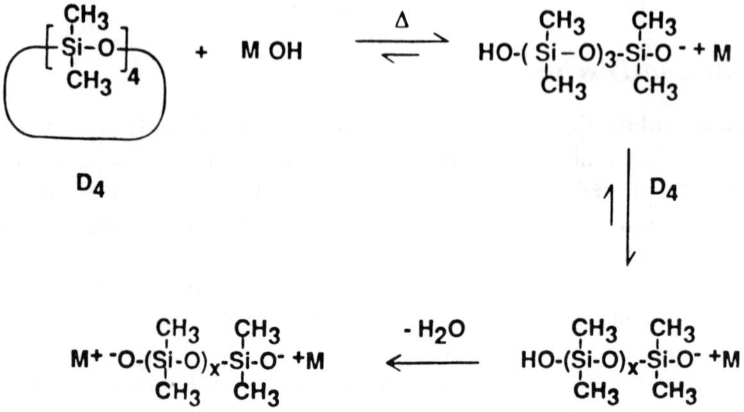

$$\underset{D_4}{\left[\text{Si}(CH_3)_2\text{-O}\right]_4} + \text{M OH} \underset{}{\overset{\Delta}{\rightleftharpoons}} \text{HO-(Si(CH}_3)_2\text{-O)}_3\text{-Si(CH}_3)_2\text{-O}^{-}\,{}^{+}\text{M}$$

$$\xrightarrow{D_4} \text{HO-(Si(CH}_3)_2\text{-O)}_x\text{-Si(CH}_3)_2\text{-O}^{-}\,{}^{+}\text{M} \xrightarrow{-H_2O} \text{M}^{+}\,{}^{-}\text{O-(Si(CH}_3)_2\text{-O)}_x\text{-Si(CH}_3)_2\text{-O}^{-}\,{}^{+}\text{M}$$

Scheme I. Preparation of siloxanolate catalyst. M^+ is K^+, $(CH_3)_4N^+$, or $(C_4H_9)_4P^+$.

mers. As pointed out earlier, the aminopropyl-terminated species was studied most extensively, because it is readily incorporated into a variety of segmented copolymer systems. A principal area of interest was the effect of catalyst type and concentration on the rate at which aminopropyl-terminated polysiloxane oligomers of well-defined end groups and controlled molecular weight are formed. Previous workers in this area have focused on the reaction kinetics either in the absence of an end blocker (*13*) or in the presence of a nonfunctionalized end blocker such as hexamethydisiloxane (*12*).

The reaction scheme for the preparation of aminopropyl-terminated difunctionalized oligomers is illustrated in Scheme II. The reaction proceeds by the anionic equilibration of the cyclic siloxane tetramer, D_4, in the presence of 1,3-bis(3-aminopropyl)tetramethyldisiloxane. The equilibration process begins immediately upon addition of the siloxanolate catalyst, and samples were removed as a function of time for the purpose of the kinetic study.

With Potassium Siloxanolate. The first set of kinetic data refers to the disappearance of D_4 as a function of time for the commonly used potassium siloxanolate catalyst. The reaction temperature was held constant at 80 °C, and the targeted molecular weight was 1500 g/mol. Several concentrations of potassium siloxanolate were studied, including 0.087, 0.134, and 0.207 mol %, on the basis of the total moles of starting material. The rate of reaction of D_4 increased, but even with 0.207 mol % potassium siloxanolate catalyst,

$$\left[\mathrm{Si(CH_3)_2{-}O}\right]_4 \;(\text{cyclic}) \quad + \quad H_2N{-}(CH_2)_3{-}Si(CH_3)_2{-}O{-}Si(CH_3)_2{-}(CH_2)_3{-}NH_2$$

D4 AMINOPROPYLDISILOXANE

80°C

SILOXANOLATE CATALYST

$$H_2N{-}(CH_2)_3{-}Si(CH_3)_2{-}O{-}\left[Si(CH_3)_2{-}O\right]_x{-}Si(CH_3)_2{-}(CH_2)_3{-}NH_2 \quad + \quad \text{CYCLIC SPECIES}$$

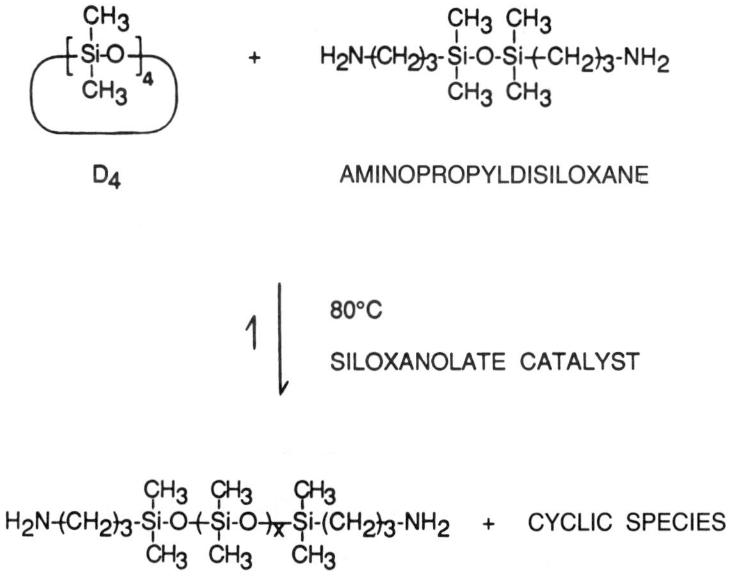

Scheme II. Synthesis of aminopropyl-terminated poly(dimethylsiloxane)s.

significant amounts of D_4 remained after 1 h of reaction at 80 °C (Figure 2). As the reactions were allowed to proceed, the level of D_4 continued to decrease until the equilibrium concentration of D_4 had been attained.

In addition to the ability of the catalyst to react rapidly with the cyclic tetramer, incorporation of the aminopropyldisiloxane was critical for the preparation of difunctionalized oligomers of controlled molecular weight. Samples were removed as a function of time and analyzed for their aminopropyldisiloxane content to determine the efficiency of the catalyst ions in this respect. For the same series of catalyst concentrations at 80 °C and a targeted molecular weight of 1500 g/mol, the aminopropyldisiloxane end blocker was not readily incorporated into the polysiloxane oligomer. For example, an initial disiloxane content of ~16.5 wt % was required to produce the 1500-g/mol oligomer with D_4. After 24 h at 80 °C and a potassium siloxanolate concentration of 0.207 mol %, only 33% of the initially charged aminopropyldisiloxane had been incorporated into the oligomer. After 40 h, the aminopropyldisiloxane level had decreased to 6.3 wt % (62% incorporation). On the basis of this value, a M_n of 2270 g/mol would be predicted. Titration of the bulk reaction mixture indicated a M_n of 1400 g/mol. The catalyst was neutralized at this point, and the oligomer was vacuum stripped

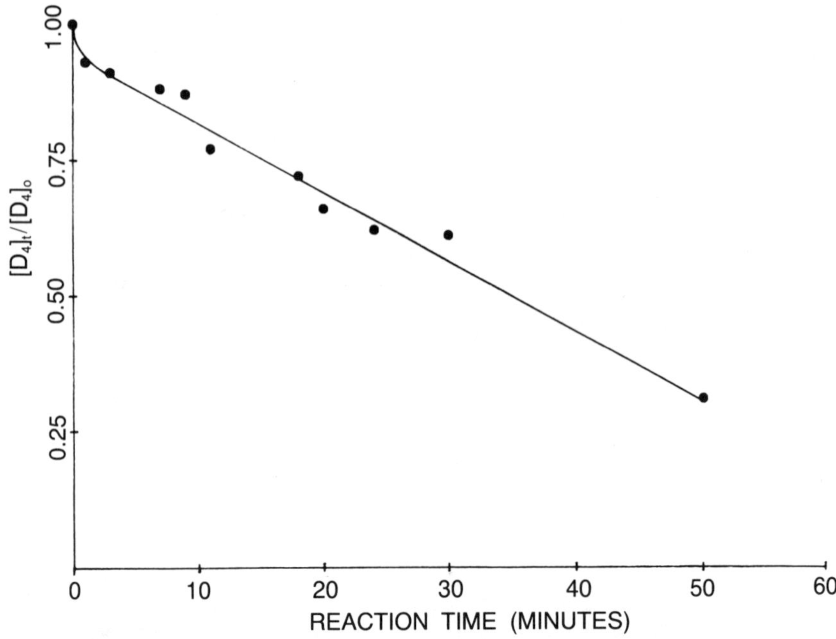

Figure 2. Disappearance of D_4 in the presence of 0.207 mol % potassium siloxanolate. The targeted M_n was 1500, and the reaction was carried out at 80 °C.

to remove any low-molecular-weight nonfunctionalized cyclic species and the unreacted aminopropyldisiloxane. The molecular weight of the vacuum-stripped oligomer, as determined by titration of the amine end groups, was 2140 g/mol. Indeed, the inefficient incorporation of the aminopropyldisiloxane end blocker had produced an oligomer with a molecular weight higher than was desired.

The lack of molecular weight control in this system was more evident at the lower concentrations of catalyst studied. For example, a potassium siloxanolate concentration of 0.087 mol % produced a 4200-g/mol oligomer after 24 h of reaction time. This value is almost 3 times the desired molecular weight. These results and previous studies (8) indicate clearly that the potassium siloxanolate catalyst did not readily incorporate the aminopropyldisiloxane at moderate catalyst concentrations.

With Tetramethylammonium Siloxanolate. The second system studied was the tetramethylammonium-siloxanolate-catalyzed equilibration process. The reaction temperature was held at 80 °C to avoid decomposition of the siloxanolate end groups. The effect of catalyst concentration on the disappearance of D_4 is shown in Figure 3. The reaction of D_4 proceeded fairly

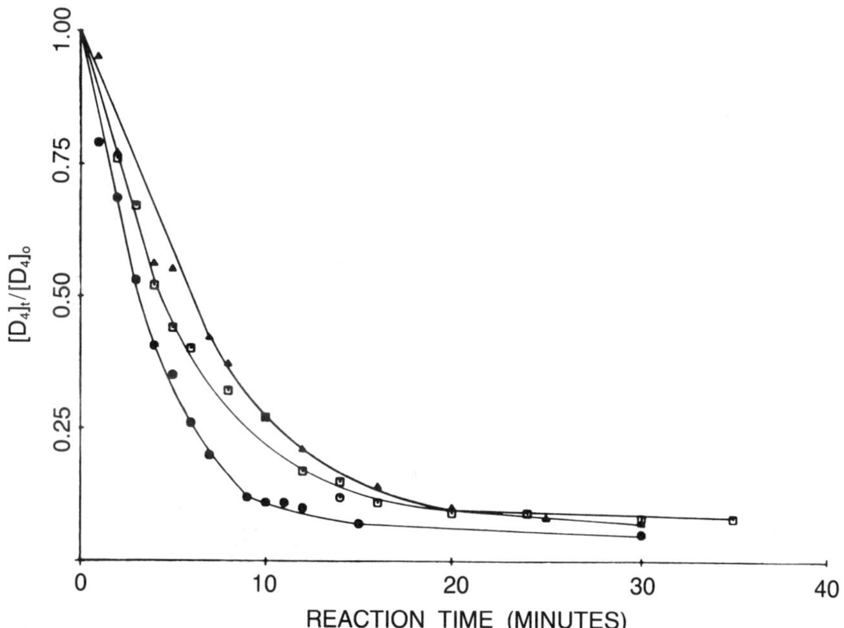

Figure 3. Effect of tetramethylammonium siloxanolate concentration on the disappearance of D_4. The targeted M_n was 1500, and the reaction was carried out at 80 °C. Key: ▲, 0.032 mol %; ◻, 0.062 mol %; and ●, 0.095 mol %.

rapidly at all three catalyst levels, and the amount of D_4 remaining after only 20 min was very close to the equilibrium value. The disappearance of the aminopropyldisiloxane for the same three reactions is shown in Figure 4. Again, the incorporation of the aminopropyldisiloxane was much slower than the incorporation of D_4, but the tetramethylammonium catalyst incorporated the aminopropyldisiloxane much more efficiently than did the potassium siloxanolate catalyst.

After 24 h of reaction time, the catalyst in these systems was decomposed by heating the reaction mixtures at >145 °C for 30 min. Then, the oligomers were analyzed to determine the equilibrium concentration of cyclic species, and the reaction mixture was subjected to vacuum stripping to remove the low-molecular-weight materials.

The M_ns of the resulting oligomers were determined by titration of the amine end groups with HCl (Table I). Oligomers with controlled molecular weights were obtained readily in 24 h at tetramethylammonium siloxanolate levels of >0.03 mol %. The generation of predictable M_ns and Gaussian distributions is consistent with the achievement of equilibrium conditions during these polymerizations.

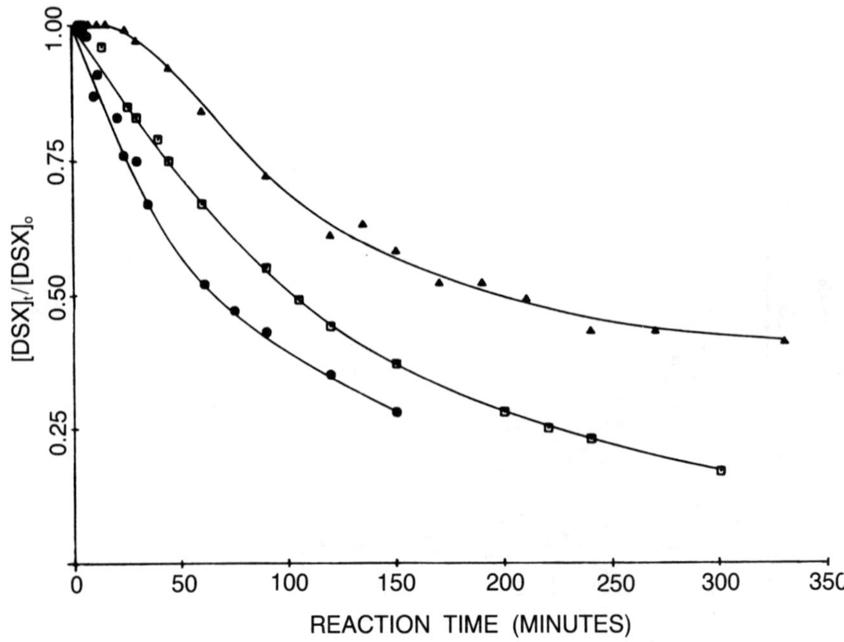

Figure 4. Effect of tetramethylammonium siloxanolate concentration on the disappearance of aminopropyldisiloxane (DSX). The targeted M_n *was 1500, and the reaction was carried out at 80 °C. Key:* ▲, *0.032 mol %;* ⊡, *0.062 mol %; and* ●, *0.095 mol %.*

Table I. Equilibration Reaction Results with Tetramethylammonium Siloxanolate

Siloxanolate (mol %)	M_n *(g/mol)*[a]
0.032	1910
0.055	1400
0.062	1540
0.095	1490

NOTE: The targeted M_n was 1500 g/mol, and the reaction was carried out at 80 °C for 24 h.
[a]Values were determined by titration.

With Tetrabutylphosphonium Siloxanolate. The final catalyst investigated in this study was tetrabutylphosphonium siloxanolate. Again, the reaction temperature was held at 80 °C because of the transience of the catalyst. The disappearance curves for D_4 and aminopropyldisiloxane are shown in Figures 5 and 6, respectively.

The disappearance of D_4 in the presence of the tetrabutylphosphonium

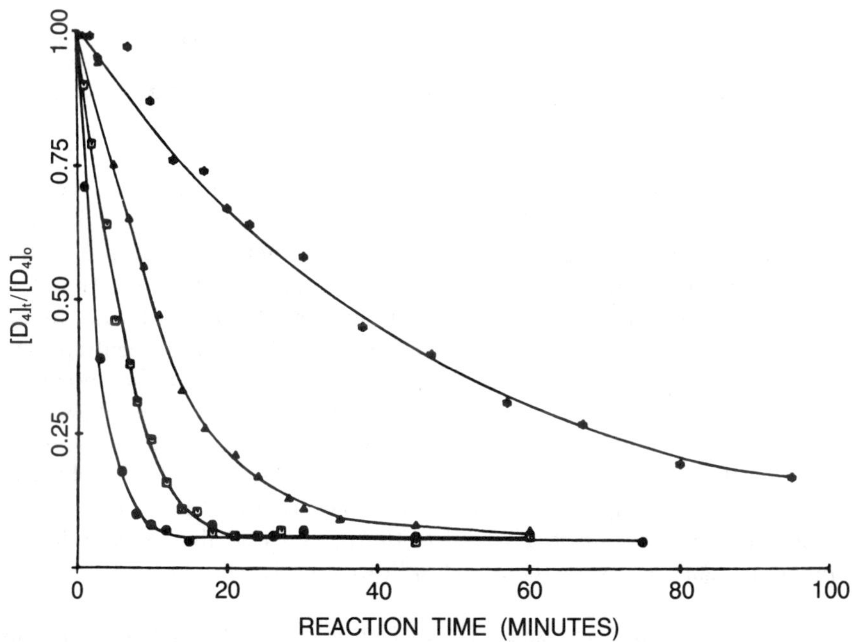

*Figure 5. Effect of tetrabutylphosphonium siloxanolate concentration on the disappearance of D_4. The targeted M_n was 1500, and the reaction was carried out at 80 °C. Key: *, 0.027 mol %; ▲, 0.038 mol %; ◻, 0.066 mol %; and ●, 0.098 mol %.*

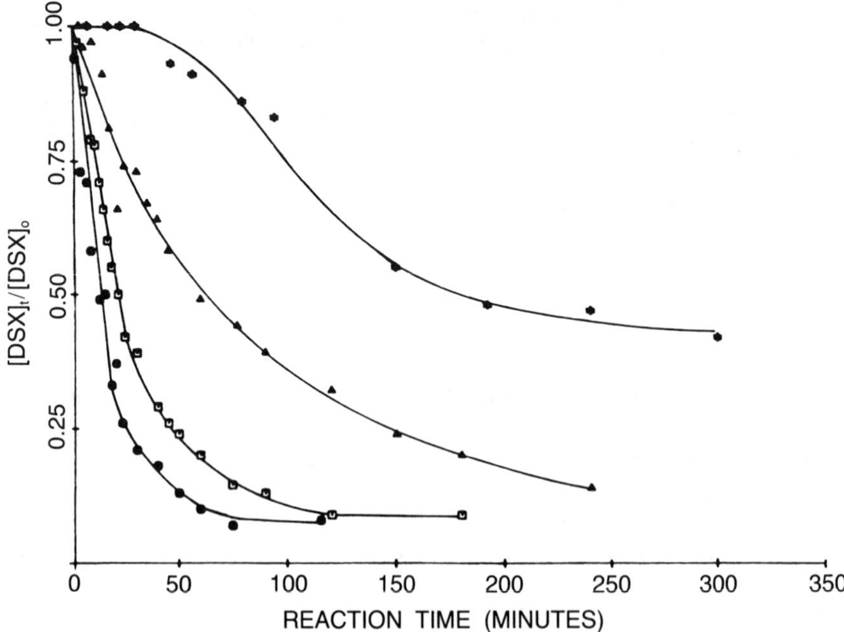

Figure 6. Effect of tetrabutylphosphonium siloxanolate concentration on the disappearance of aminopropyldisiloxane (DSX). The targeted M_n *was 1500, and the reaction was carried out at 80 °C. Key: *, 0.027 mol %; ▲, 0.038 mol %; ⊡, 0.066 mol %; and ●, 0.098 mol %.*

catalyst was again quite rapid at 80 °C, and the equilibrium level was reached in a reasonable amount of time. In addition, the disappearance of the aminopropyldisiloxane with this catalyst was extremely fast relative to the potassium and tetramethylammonium siloxanolate catalysts. At tetrabutylphosphonium siloxanolate concentrations of >0.06 mol %, the aminopropyldisiloxane content was very close to its equilibrium concentration in less than 2 h at 80 °C. Although thermodynamic equilibrium had probably been reached sooner in this system, the reactions were again allowed to proceed for 24 h, at which time the reaction temperature was increased to >145 °C for 30 min to decompose the catalyst. The molecular weight data for the vacuum-stripped oligomers are shown in Table II. The excellent molecular weight control achieved in the presence of the tetrabutylphosphonium catalyst was another good indication of its efficiency.

Comparison of Catalysts. The concentration of siloxanolate catalyst in all three systems studied ranged from 0.03 to 0.20 mol %, on the basis of the total moles of starting materials. At these concentrations of catalyst, the reactions of D_4 with the tetramethylammonium and tetrabutylphosphonium siloxanolate catalysts were fairly comparable, whereas the reaction with po-

Table II. Equilibration Reaction Results with Tetrabutylphosphonium Siloxanolate

Siloxanolate (mol %)	M_n *(g/mol)*[a]
0.027	1360
0.038	1480
0.098	1420

NOTE: The targeted M_n was 1500 g/mol. The reaction was carried out at 80 °C for 24 h.
[a]Values were determined by titration.

tassium siloxanolate proceeded more slowly, even at much higher reaction temperatures.

As a result of the method of catalyst addition and because of its fairly viscous nature even as an ~35 wt % solution in toluene, the same catalyst concentration could not be reproduced exactly from one catalyst system to another for comparison. The rate differences were sufficient, however, to compare the effects of catalyst concentrations and to examine the efficiency of the different catalysts.

The disappearance of D_4 is compared in Figure 7 for reactions conducted in the presence of 0.134 mol % potassium siloxanolate or 0.066 mol %

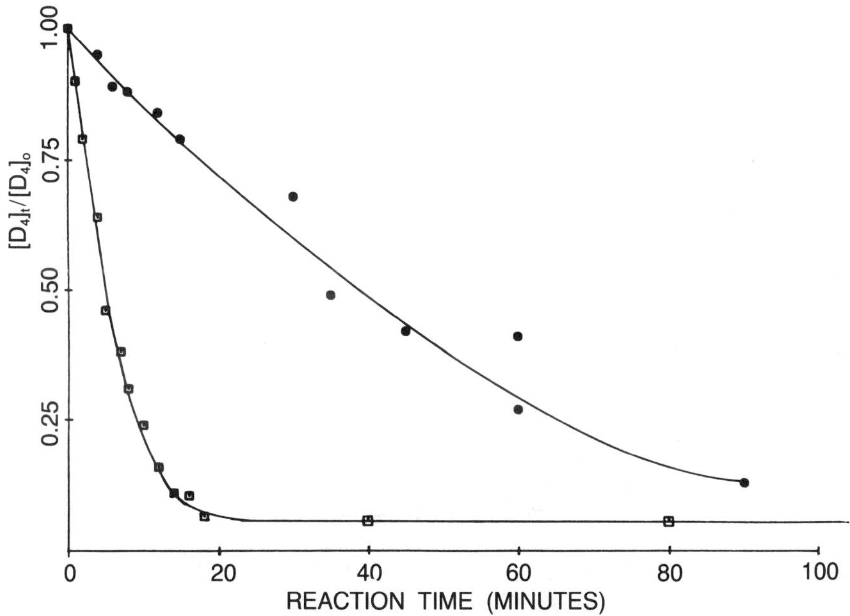

Figure 7. Effect of potassium (0.134 mol %, ●) and tetrabutylphosphonium (0.066 mol %, ▣) siloxanolate catalysts on the rate of reaction of D_4. The targeted M_n was 1500, and the reaction was carried out at 80 °C.

tetrabutylphosphonium siloxanolate. Even when the concentration of potassium siloxanolate is twice that of tetrabutylphosphonium siloxanolate, the disappearance of D_4 was still much slower in the presence of the potassium siloxanolate.

The tetramethylammonium and tetrabutylphosphonium siloxanolate catalysts are compared in Figure 8. The greater efficiency of tetrabutylphosphonium siloxanolate relative to the potassium and tetramethylammonium siloxanolate for incorporation of D_4 has been described in the past (2) for nonfunctionalized siloxane systems. Despite the rate differences, all three systems were effective in reducing the D_4 concentration to the same equilibrium level in a reasonable amount of time.

The major difference in catalytic efficiency among these three siloxanolate systems was in the rate of incorporation of the aminopropyldisiloxane. This area has not been examined in detail in the past, and the results presented in the previous tables and figures provide valuable information on the effectiveness of the various catalysts for the preparation of aminopropyl-functionalized oligomers. The data presented show that the tetramethylammonium catalyst reacts with the aminopropyldisiloxane somewhat more slowly than does the tetrabutylphosphonium catalyst. The difference is clearly observed in Figure 9, which shows that the aminopropyldisiloxane

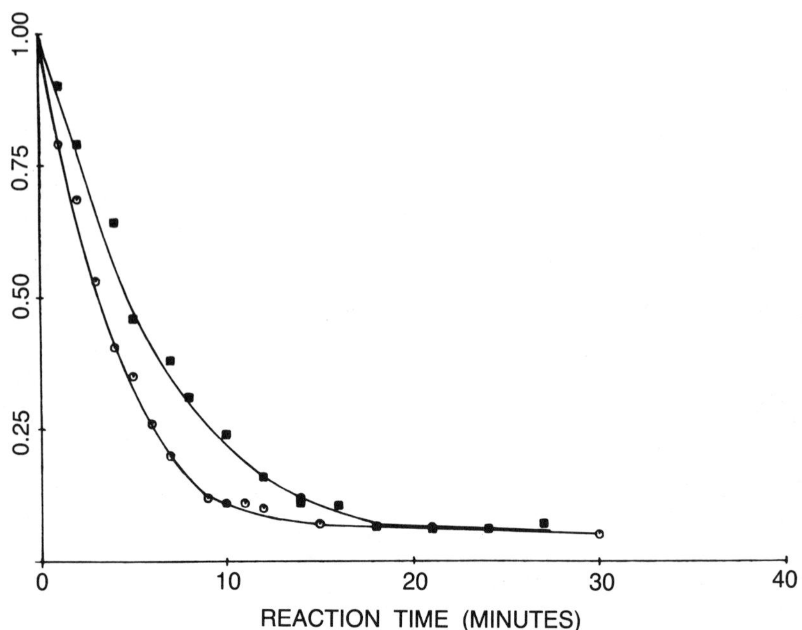

Figure 8. Effect of tetramethylammonium (0.095 mol %, ⊙) and tetrabutylphosphonium (0.066 mol %, ■) siloxanolate catalysts on the rate of reaction of D_4. The targeted M_n was 1500, and the reaction was carried out at 80 °C.

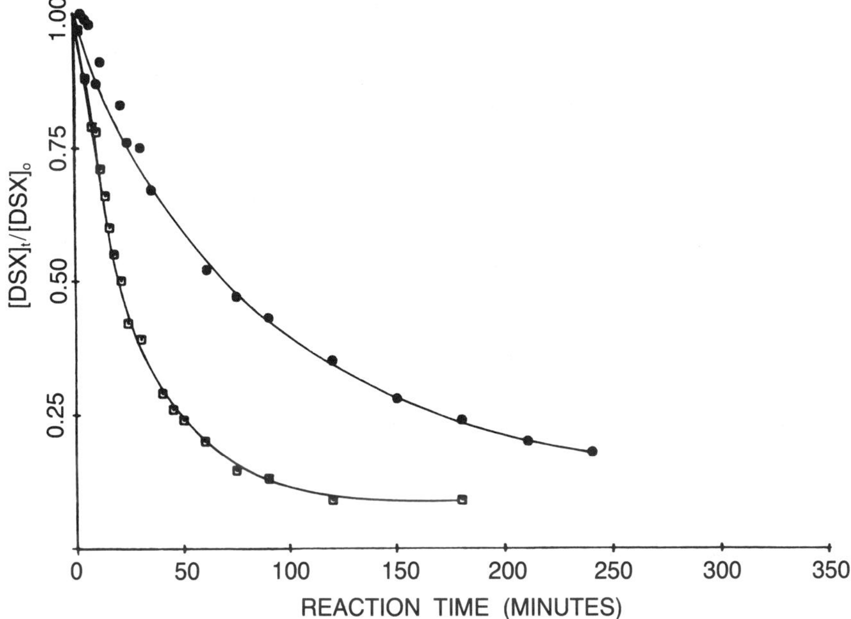

Figure 9. Effect of tetramethylammonium (0.095 mol %, ●) and tetrabutylphosphonium (0.066 mol %, ▣) siloxanolate catalysts on the rate of reaction of aminopropyldisiloxane (DSX). The targeted M_n *was 1500, and the reaction was carried out at 80 °C.*

disappears faster in the presence of 0.066 mol % tetrabutylphosphonium siloxanolate than in the presence of 0.095 mol % tetramethylammonium siloxanolate. However, both of the catalysts were significantly more efficient than potassium siloxanolate. Because incorporation of the aminopropyldisiloxane is necessary for the preparation of controlled-molecular-weight oligomers, the fact that the titration-determined molecular weights of the oligomers prepared with the potassium catalyst were significantly higher than what was targeted is not surprising.

Mechanism of Equilibration. The generally accepted mechanism for the base-catalyzed ring-opening polymerization of cyclosiloxanes involves attack of the basic catalyst at the silicon atom (*15*). It has been proposed, and generally accepted, that the active species is a partially dissociated siloxanolate anion (*13*). In the results presented in this chapter, significant differences in reaction rates were observed as the corresponding cation of the siloxanolate species was varied. The more rapid disappearance of D_4 and aminopropyldisiloxane in the presence of these catalysts increased in the following order:

$$[-SiO^-K^+] < [-SiO^-N(CH_3)_4{}^+] < [-SiO^-P(C_4H_9)^+]$$

Similar relative rates for the disappearance of D_4 in the presence of these catalysts have been observed in the past for nonfunctionalized systems. The order of reactivity follows the expected order of increasing dissociation of the ion pair. As the ion pair dissociates, the concentration of the more active species increases. The slower reaction rate of the aminopropyldisiloxane relative to D_4 was expected to some extent on the basis of electronegativity differences, but the dramatic differences in the efficiency of incorporation of the aminopropyldisiloxane have not been reported previously. Additional factors may be contributing to the observed results, such as the expected increased solubility of the more organic $(CH_3)_4N^+$ and $(C_4H_9)_4P^+$ ions relative to the K^+ ion. The better solubility may also lead to an increased dissociation of the siloxanolate species.

Supercritical Fluid Fractionation. The versatility of aminopropyl-terminated polysiloxane oligomers for incorporation of siloxane components into various block and segmented copolymer systems was elaborated on earlier. The presence of two functional groups in these materials is a distinct advantage, but the Gaussian molecular weight distributions produced by equilibration processes may limit the range of properties that may be exhibited in the resulting materials. Indeed, the properties of a siloxane-containing block copolymer are highly dependent on the nature of the difunctionalized precursors, as well as on the degree of microphase separation of the respective blocks in these segmented copolymer systems.

Supercritical-fluid-fractionation techniques have been explored to separate the aminopropyl-terminated oligomers into fractions of narrow polydispersity. The fractions have been fully characterized in terms of their structure, molecular weight, molecular weight distribution, and functionality. The results of selected studies are presented in this section.

The first set of fractionation data is for a 1900-g/mol oligomer, as determined by titration of the amine end groups in the vacuum-stripped oligomer. This particular sample was fractionated into 18 fractions, and the molecular weight data are presented in Table III. Although the initial fractions (1 through 3) were relatively broad because of an anomaly in the fractionation process, fractions 4 through 18 were of increasingly higher molecular weight and extremely narrow molecular weight distributions. The molecular weight distributions reported in Table III were calculated with respect to polysiloxane standards. Indeed, the molecular weight values determined by titration of the amine end groups were in good agreement with those calculated by GPC.

The GPC traces for the parent oligomer and several of the fractions derived from it are reproduced in Figure 10. The efficiency of this technique at producing difunctionalized fractions of narrow polydispersity is evident from Table III and Figure 10. The presence of two functional groups in these fractions was confirmed by comparing the M_n determined by titration

Table III. Fractionation of Aminopropyl-Terminated Poly(dimethylsiloxane)

Fraction	M_n (*g/mol*)[a]	M_w/M_n[b]	M_n (*g/mol*)[b]
Control	1900	1.58	1400
1	1700	—[c]	—
2	1300	—	—
3	1000	1.33	500
4	840	1.13	700
5	1050	1.22	600
6	1300	1.14	1000
7	1550	1.12	1300
8	1800	1.12	1700
9	1900	1.13	1800
10	2300	1.06	2100
11	2550	1.12	2600
12	2900	1.12	2900
13	3250	1.15	3500
14	3900	1.12	4000
15	4450	1.10	4600
16	5100	1.09	5300
17	5900	1.18	6700
18	—	—	—

[a]Values were determined by titration.
[b]Values were determined by GPC and were based on polysiloxane standards. M_w is weight-average molecular weight.
[c]— means not determined.

of the amine end groups with the value obtained by VPO (vapor pressure osmometric) analysis for several samples. The results are presented in Table IV. The excellent agreement of the molecular weight values confirms the the presence of two functional groups in these materials before and after the fractionation process.

Polysiloxanes are readily fractionated by supercritical-fluid-extraction techniques because of their excellent solubility characteristics. Indeed, poly(dimethylsiloxane)s with molecular weights up to 100,000 g/mol have been dissolved by supercritical fluids (*14*). In the present study on amino-terminated polysiloxanes, similar results were obtained. However, the desired molecular weight of difunctionalized precursors for segmented copolymer systems generally is less than 20,000 g/mol.

The next set of fractionation results are for a 7500-g/mol parent oligomer that was separated into 11 fractions with molecular weights ranging from 2400 to 17,700 g/mol (Table V). Because this sample had been vacuum stripped prior to the fractionation process, earlier fractions containing nonfunctionalized cyclic species were not obtained. The molecular weight distributions of these fractionated materials are significantly lower than that of the parent oligomer.

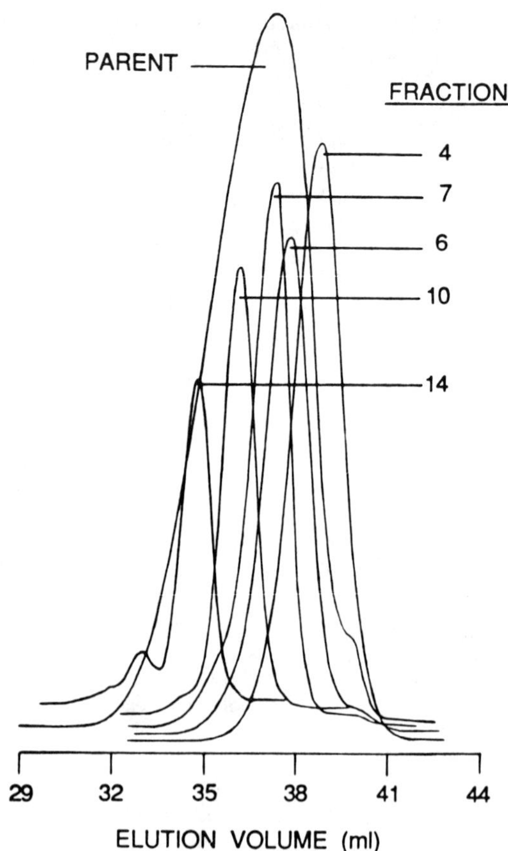

Figure 10. GPC chromatograms of aminopropyl-terminated poly(dimethylsiloxane) and its fractions.

Table IV. Molecular Weight Characterization of Aminopropyl-Terminated Poly(dimethylsiloxane) and Its Fractions

Fraction	M_n *(g/mol)*[a]	M_n *(g/mol) by VPO*[b]
Control	1900	1710
4	850	760
6	1300	1330
7	1550	1520
10	2300	2420
14	3900	3970

[a]Values were determined by titration.
[b]VPO is vapor pressure osmometry. The analyses were done in toluene at 63 °C.

Table V. Fractionation of Aminopropyl-Terminated Poly(dimethylsiloxane)

Fraction	M_n *(g/mol)*[a]	M_w/M_n[b]
Control	7,500	1.82
1	2,400	—[c]
2	2,800	1.25
3	4,560	1.18
4	6,800	1.17
5	7,000	1.13
6	6,950	1.12
7	8,650	1.12
8	12,840	1.11
9	13,600	1.08
10	14,100	—
11	17,700	—

NOTE: The supercritical solvent was ethane.
[a]Values were determined by titration.
[b]Values were determined by GPC and were based on polysiloxane standards. M_w is weight-average molecular weight.
[c]— means not determined.

Conclusions

Significant differences were observed in the rate of incorporation of D_4 and 1,3-bis(3-aminopropyl)disiloxane for similar concentrations of potassium, tetramethylammonium, and tetrabutylphosphonium siloxanolate catalysts. The rate differences affected the reaction times that were required to obtain a completely equilibrated reaction mixture with the desired molecular weight. The potassium catalyst required excessively long reaction times or high concentrations before sufficient incorporation of the aminopropyldisiloxane was realized. The tetramethylammonium and tetrabutylphosphonium catalysts were much more efficient for the preparation of controlled-molecular-weight aminopropyl-terminated polysiloxane oligomers.

The use of supercritical-fluid-extraction techniques in the fractionation of polysiloxanes has been demonstrated by the data presented. The polydispersities of the fractions were comparable with those generally attainable only by anionic-polymerization techniques, with which the incorporation of two functional groups is often difficult to attain. The ability to isolate these well-defined fractions will lead to important fundamental studies on structure–property relationships in multiphase copolymer systems.

Acknowledgments

We gratefully acknowledge the National Science Foundation and the Office of Naval Research (ONR) for their support of this research.

References

1. Noll, W. *The Chemistry of Silicones;* Academic: New York, 1968.
2. Noshay, A.; McGrath, J. E. *Block Copolymers: Overview and Critical Survey;* Academic: New York, 1977.
3. Sormani, P. M.; Minton, R. J.; McGrath, J. E. In *Ring Opening Polymerization: Kinetics, Mechanisms and Synthesis;* McGrath, J. E., Ed.; ACS Symposium Series No. 286; American Chemical Society: Washington, DC, 1985; pp 147–160.
4. Yilgor, I.; McGrath, J. E. *Adv. Polym. Sci.* **1988,** *86.*
5. Wright, P. V. In *Ring Opening Polymerization;* Ivin, K.; Saegusa, T., Eds.; Elsevier: New York, 1984; Vol. 2, p 1055.
6. McGrath, J. E.; Sormani, P. M.; Elsbernd, C. S.; Kilic, S. *Makromol. Chem., Makromol. Symp.* **1986,** *6,* 67.
7. Yilgor, I.; Riffle, J. S.; McGrath, J. E. In *Reactive Oligomers;* Harris, F., Ed.; ACS Symposium Series No. 282; American Chemical Society: Washington, DC, 1985; pp 161–174.
8. Summers, J. D.; Elsbernd, C. S.; Sormani, P. M.; Brandt, P. J. A.; Arnold, C. A.; Yilgor, I.; Riffle, J. S.; Kilic S.; McGrath, J. E. In *Inorganic and Organometallic Polymers;* Zeldin, M.; Allcock, H.; Wynne, K., Eds.; ACS Symposium Series No. 360; American Chemical Society: Washington, DC, 1988; p 180.
9. Elsbernd, C. S.; Mohanty, D. K.; McGrath, J. E.; Gallagher, P. M.; Krukonis, V. J. *Polym. Bull.*, in press.
10. Elsbernd, C. S. Ph.D. Dissertation, Virginia Polytechnic Institute & State University, 1988.
11. Gilbert, A. R.; Kantor, S. W. *J. Polym. Sci.* **1959,** *40,* 35.
12. Kantor, S. W.; Grubb, W. T.; Osthoff, R. C. *J. Am. Chem. Soc.* **1955,** *76,* 5190.
13. Grubb, W. T.; Osthoff, R. C. *J. Am. Chem. Soc.* **1955,** *77,* 1405.
14. McHugh, M.; Krukonis, V. *Supercritical Fluid Extraction: Principles and Practice;* Butterworth: Boston, 1986.
15. Hurd, D. T.; Osthoff, R. C.; Corrin, M. L. *J. Am. Chem. Soc.* **1954,** *76,* 249.

RECEIVED for review May 27, 1988. ACCEPTED revised manuscript July 27, 1989.

9

Novel Poly(imide–siloxane) Polymers and Copolymers

Frank L. Keohan and John E. Hallgren

Corporate Research and Development Center, General Electric Company, Schenectady, NY 12301

The synthesis and properties of poly(imide–siloxane) polymers and copolymers based on 5,5′-bis(1,1,3,3-tetramethyl-1,3-disiloxanediyl)norbornane dicarboxylic anhydride are described. High-molecular-weight thermoplastics and elastoplastics were prepared readily in solution from aromatic diamines, organic dianhydrides, and this unique anhydride-terminated siloxane. The thermal and mechanical properties of a variety of copolymer compositions are described. Average siloxane block length and overall siloxane content had the greatest effect on these properties.

AROMATIC POLYIMIDES comprise an important class of high-temperature polymers. Whereas their thermal and oxidative stability, physical strength, high softening temperature, and solvent resistance are highly valued, the materials are difficult to fabricate by conventional means. The physical properties of linear polyimides can be altered dramatically by the incorporation of siloxane blocks into the polymer backbone (*1–3*). Therefore, considerable effort by both industrial and academic laboratories has been focused on the synthesis and properties of siloxane–polyimide copolymers (*4*).

In principle, this approach is quite flexible and allows the tailoring of material properties for specific applications. Polymers with properties ranging from those of typical thermoplastics to those of elastoplastics can be prepared by varying the siloxane content and siloxane block sizes. Significant reductions in polymer softening temperatures, as well as solubility enhancement, have been achieved by this copolymerization approach (*5*). In addition,

0065–2393/90/0224–0165$06.00/0

dramatic improvements in adhesion to a variety of substrates have been observed (*6, 7*).

The first reported (*8*) siloxane–polyimide polymer was synthesized from 1,3-bis(aminopropyl)-1,1,3,3-tetramethyldisiloxane and pyromellitic dianhydride. This material was described as a soluble, thermoplastic polyimide. Many other siloxane–polyimide polymers have been prepared via similar methodology using other dianhydrides and diamines (*9*).

The traditional approach used in poly(imide–siloxane) synthesis is the reaction of aminopropyl-terminated dimethylsiloxane oligomers with aromatic dianhydrides and additional diamines (*9–13*). Typically, subambient temperatures and dipolar aprotic solvents are used. The resulting high-molecular-weight polyamic acid solution can be heated to effect imidization and solvent evaporation. This procedure is analogous to the synthetic method used to prepare conventional polyimides for films and coatings.

The polymers and copolymers described in this chapter were derived from the novel anhydride-terminated disiloxane, 5,5′-bis(1,1,3,3-tetramethyl-1,3-disiloxanediyl)norbornane-2,3-dicarboxylic anhydride (DiSiAn) (*14*). Both siloxane–polyimide polymers and copolymers based on DiSiAn and its polysiloxane derivatives were investigated. This chapter describes the synthesis, characterization, and physical properties of these materials.

Experimental Procedures

General Procedures. DiSiAn was prepared by the literature procedure (*15*) and purified by filtering a toluene solution through a silica gel column and recrystallizing from ether. BPADA [4,4′-bis[2,2-bis(4-hydroxyphenyl)propane]phthalic anhydride, polymer grade], *m*-phenylenediamine (mPD), 4,4′-oxydianiline (ODA), 4,4′-methylenedianiline (MDA), and 2-hydroxypyridine (Aldrich Chemical Company) were recrystallized prior to use. Octamethylcyclotetrasiloxane (D_4, GE, polymer grade) was dried over molecular sieves and distilled. Reagent-grade toluene, *o*-dichlorobenzene, and chlorobenzene (Mallinckrodt) were used as received.

The number of dimethylsiloxy repeat units, D_n, was determined by ^{1}H-NMR. Intrinsic viscosity (IV) measurements were performed in chloroform. Thermal transitions were measured by differential scanning calorimetry (DSC) at a heating rate of 40 °C/min. Thermogravimetric analysis (TGA) was performed at a heating rate of 10 °C/min. ^{1}H and ^{29}Si NMR spectra were obtained with Varian Associates XL-200 or XL-300 spectrometers in the FT (Fourier transform) mode. Dynamic mechanical analyses (DMA) were performed on a Rheometrics Dynamic spectrometer (model 7700) at a heating rate of 2 °C/min. Tensile measurements were performed on thin films cast from chloroform solution or injection-molded ASTM bars by using an Instron 4200 tensile tester at a crosshead speed of 2 in./min (5 cm/min) and an initial gauge length of 0.5 in. (1.3 cm).

Synthesis of D_{20}–DiSiAn. A 2-L, three-neck, round-bottom flask equipped with a Dean–Stark trap, water-cooled condenser fitted with a nitrogen inlet, pressure-equalizing addition funnel, and magnetic stirring bar was charged with 111.81 g (0.242 mol) of DiSiAn, 430.12 g (1.450 mol) of D_4, and 260.0 mL of chlorobenzene. The mixture was heated to reflux, and ~15 mL of solvent was removed by azeotropic distillation to dry the mixture. The solution temperature was reduced to 100 °C, and a mixture of 3 mL of concentrated sulfuric acid and 1.5 mL of fuming sulfuric acid was added. The solution became homogeneous within 10 min and was maintained at 100 °C under a nitrogen atmosphere for 48 h.

An excess of sodium bicarbonate was added to quench the equilibrium reaction. Decolorizing charcoal was added, the slurry was filtered, and the solvent was removed under reduced pressure. Small amounts of unreacted cyclic siloxanes were removed by vacuum distillation at 67 Pa and pot temperatures of up to 100 °C to obtain 486.82 g of a clear, colorless oil in 90% overall yield. ^{1}H NMR and ^{29}Si NMR analysis indicated the formula $D_{19.9}$–DiSiAn.

Synthesis of Siloxane–Polyimide Elastoplastics. In a typical polymerization, a 5-L, three-neck, round-bottom flask equipped with an overhead mechanical stirrer, a Dean–Stark trap with condenser and a nitrogen inlet, and a thermometer was charged with 484.00 g (0.2406 mol) of D_{20}–DiSiAn, 41.61 g (0.431 mol) of mPD, 19.52 g (3 wt %) of 2-hydroxypyridine, and 2 L of *o*-dichlorobenzene. The mixture was warmed to 100 °C for 1 h to dissolve the monomers and the catalyst. The polyamic acids precipitated and then redissolved when the mixture was warmed to 150 °C for 2 h. To the oligomer solution was added 99.13 g of BPADA dissolved in 200 mL of *o*-dichlorobenzene. The mixture was maintained at 150 °C for an additional 2-h period to ensure incorporation of the dianhydride and then warmed to reflux. After approximately 100 mL of a solvent–water mixture had been removed, the solution was maintained at 180 °C for 40 h. The mixture was cooled to room temperature and diluted with ~1 L of methylene chloride. Polymer was isolated from the solution by a slow addition of the polymer solution to 4 L of methanol. The resulting slurry was filtered, and the polymer was redissolved in 4 L of methylene chloride, extracted three times with 2 N aqueous HCl to remove catalyst, washed with water, dried with magnesium sulfate, reprecipitated into methanol as before, filtered, and dried in vacuo at 100 °C to obtain 522 g (85%) of a rubbery material with an IV of 0.50 dL/g. IR, ^{1}H NMR, and ^{29}Si NMR spectroscopic analysis indicated the absence of amic acid functionalities that could be present if imidization is incomplete.

Synthesis of Siloxane–Polyimide Thermoplastics. In a typical siloxane–polyimide thermoplastic preparation, a 2-L, three-neck flask equipped with an overhead mechanical stirrer, Dean–Stark trap with condenser and nitrogen inlet, and a thermometer was charged with 106.41 g (0.230 mol) of DiSiAn, 119.71 g (0.230 mol) of BPADA, 49.74 g (0.460 mol) of mPD, 8.55 g (3 wt %) of 2-hydroxypyridine, and 635 mL of *o*-dichlorobenzene. The mixture was warmed to 100 °C for 1 h to dissolve the monomers and catalyst. Polyamic acids precipitated and redissolved when the mixture was heated to 150 °C for 1 h. The solution was then warmed to reflux, and ~15 mL of water of reaction was removed by azeotropic distillation. The mixture was maintained at 180 °C for 16 h. The solution became noticeably more viscous. The polymer was isolated and purified as described previously to obtain 232 g (90%) of polymer with an IV of 0.54 dL/g. The isolated polymer was characterized spectroscopically. DSC indicated a T_g (glass transition temperature) of 196 °C.

Results and Discussion

Thermoplastic Polymers and Copolymers. One-step solution polycondensation routes were used to prepare a variety of thermoplastic DiSiAn-derived polymers and copolymers (Schemes I and II, respectively).

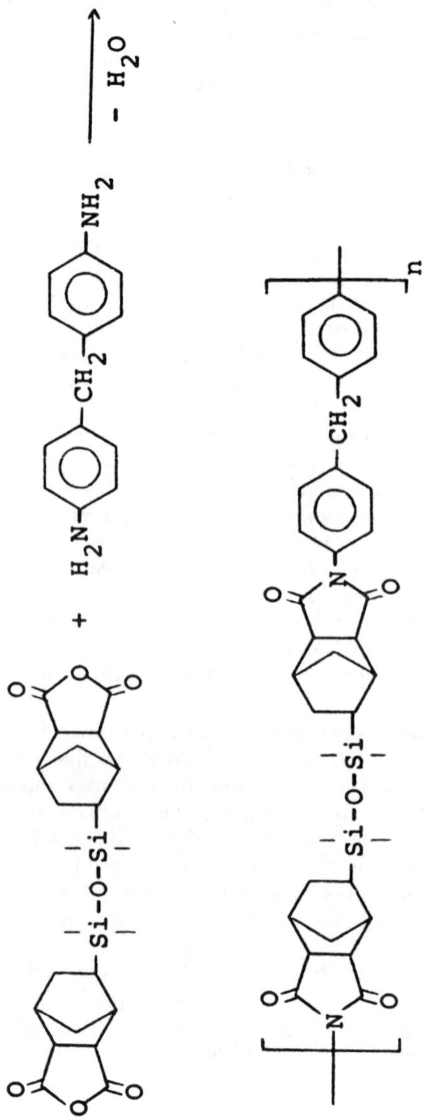

Scheme I. Synthesis of poly(imide–siloxane) thermoplastic copolymers.

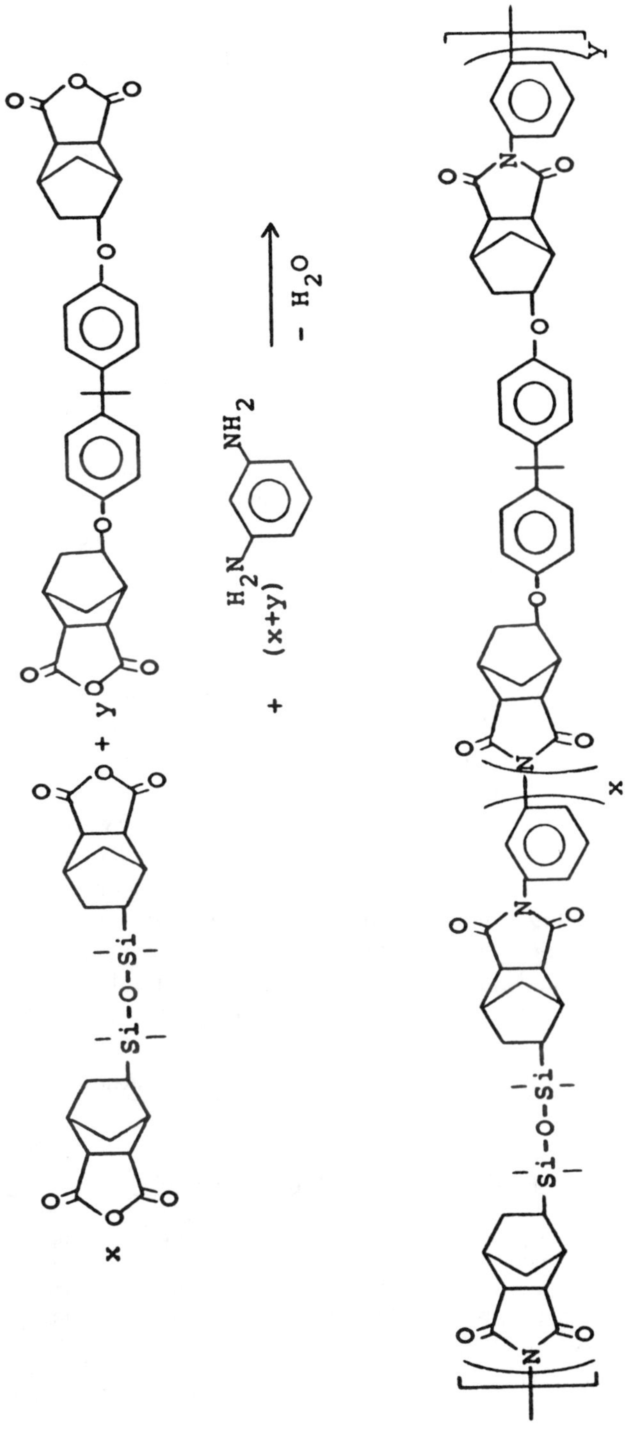

Scheme II. Synthesis of poly(ether–imide–siloxane) thermoplastic copolymers.

The polymerizations were carried out in refluxing *o*-dichlorobenzene in the presence of 2-hydroxypyridine as catalyst. In the absence of catalyst, only low-molecular-weight polymers (IV < 0.3 dL/g) could be isolated from solution. However, when 2-hydroxypyridine was used as a condensation catalyst, copolymers having IVs in excess of 0.45 dL/g could be prepared readily. The results are summarized in Table I.

High-molecular-weight polymers derived from DiSiAn and mPD, MDA, or ODA were more difficult to prepare than the corresponding copolymers containing BPADA. BPADA-containing copolymers were also difficult to synthesize when the molar ratio of DiSiAn to BPADA exceeded 50%. The addition of aromatic dianhydrides in these syntheses increased polymerization rates and yielded higher molecular weight polymers. Incorporation of organic imide blocks resulted in an ~20 °C increase in T_g in each case.

Attempts to substitute benzophenone tetracarboxylic dianhydride (BTDA) for BPADA failed to produce high-molecular-weight copolymers because of the insolubility of the BTDA–mPD oligomers. A variety of solvents and solvent combinations were tried but were unsuccessful in solubilizing the intermediate oligomers.

Both siloxane–polyimide copolymers and BPADA-derived copolymers exhibited excellent solubility in a variety of dipolar aprotic solvents, including tetrahydrofuran, *n*-methylpyrrolidone, and dimethyl sulfoxide, as well as chlorinated hydrocarbons such as *o*-dichlorobenzene and methylene chloride. The polymers and copolymers were typical thermoplastics exhibiting little elongation at failure. The tensile properties are summarized in Table II.

Films of the copolymer derived from DiSiAn and mPD were quite brittle and made tensile specimen fabrication difficult. Copolymers containing BPADA were much less brittle, and the tensile properties of the DiSiAn–BPADA–mPD-based copolymer showed a marked improvement over the DiSiAn–mPD polymer. As noted previously, the copolymers were much easier to prepare and, in general, were of higher molecular weight, which may have contributed to the improved physical properties of the copolymers.

Table I. DiSiAn–Polyimide Polymers and Copolymers

Diamine	*Dianhydride*	*IV (dL/g)*	T_g (°C)[a]	T (°C) of 5% Weight Loss[b]	
				N_2	*Air*
mPD	DiSiAn	0.52	174	490	460
MDA	DiSiAn	0.48	181	480	460
ODA	DiSiAn	0.52	184	495	480
mPD	DiSiAn–BPADA	0.60	196	470	460
MDA	DiSiAn–BPADA	0.69	198	435	400
ODA	DiSiAn–BPADA	0.76	202	460	400

[a]Data were obtained by DSC.
[b]Data were obtained by TGA.

Table II. Tensile Properties of DiSiAn–Polyimide Polymers and Copolymers

Diamine	*Dianhydride*	*Tensile Strength (kPa)*	*Elongation (%)*
mPD	DiSiAn	45,000	4.5
MDA	DiSiAn	28,000	3.0
ODA	DiSiAn	29,000	3.2
mPD	DiSiAn–BPADA	50,000	5.1
MDA	DiSiAn–BPADA	39,000	5.0
ODA	DiSiAn–BPADA	40,000	6.3

Elastoplastic Copolymers. The thermal and tensile properties of DiSiAn-based thermoplastics could be varied over a limited range by the choice of aromatic diamine and the incorporation of organic dianhydrides. More-dramatic changes in the physical properties of the poly(imide–siloxane) polymers could be effected either by increasing the siloxane block length or by increasing the overall siloxane content. Block copolymers with elastoplastic characteristics were prepared from anhydride-terminated poly(dimethylsiloxane) oligomers with an average degree of polymerization between 15 and 30.

Anhydride-terminated poly(dimethylsiloxane) oligomers were synthesized by the acid-catalyzed equilibration of DiSiAn and octamethylcyclotetrasiloxane (D_4) (Scheme III). A wide range of oligomer chain lengths (n = 1 to >100) were prepared in this manner and characterized by ^{1}H and ^{29}Si NMR spectroscopy.

The synthesis of siloxane–polyimide elastoplastics requires an approach slightly different from that used in preparing the thermoplastic materials because of differences in reactivity between the aliphatic-anhydride-terminated siloxane oligomers and the aromatic dianhydrides. A one-pot condensation of the anhydride-terminated siloxane oligomers, BPADA, and the diamine in *o*-dichlorobenzene solution in the presence of 2-hydroxypyridine as catalyst leads to a siloxane-deficient polyimide. To circumvent this deficiency, a two-step synthetic scheme was used in which the anhydride-terminated siloxane oligomers were first capped with an excess of the diamine. The aromatic dianhydride was then added to the resulting amic acid oligomeric mixture and warmed to complete imidization (Scheme IV).

The properties of three such elastoplastic copolymers derived from BPADA and D_{20}–DiSiAn in a 1.3:1 molar ratio and various diamines are summarized in Table III.

The structure of the aromatic diamines used in these elastoplastics had only a minor effect on the copolymer physical properties, as was observed for thermoplastics. The ratio of hard blocks to soft blocks in these segmented polymers could be varied, but such variations will result in a wide range of properties.

A series of copolymers based on D_{20}–DiSiAn, BPADA, and mPD were prepared in which the siloxane content was varied between 8 and 45% (by

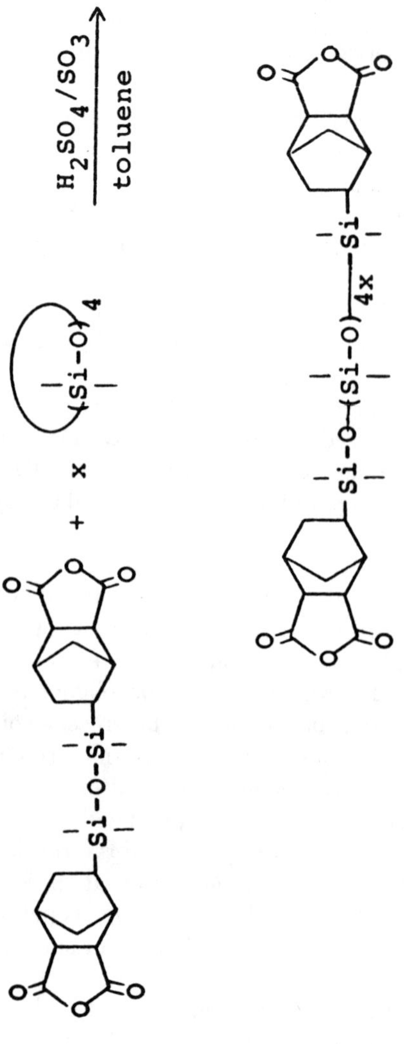

Scheme III. Equilibration of DiSiAn and octamethylcyclotetrasiloxane.

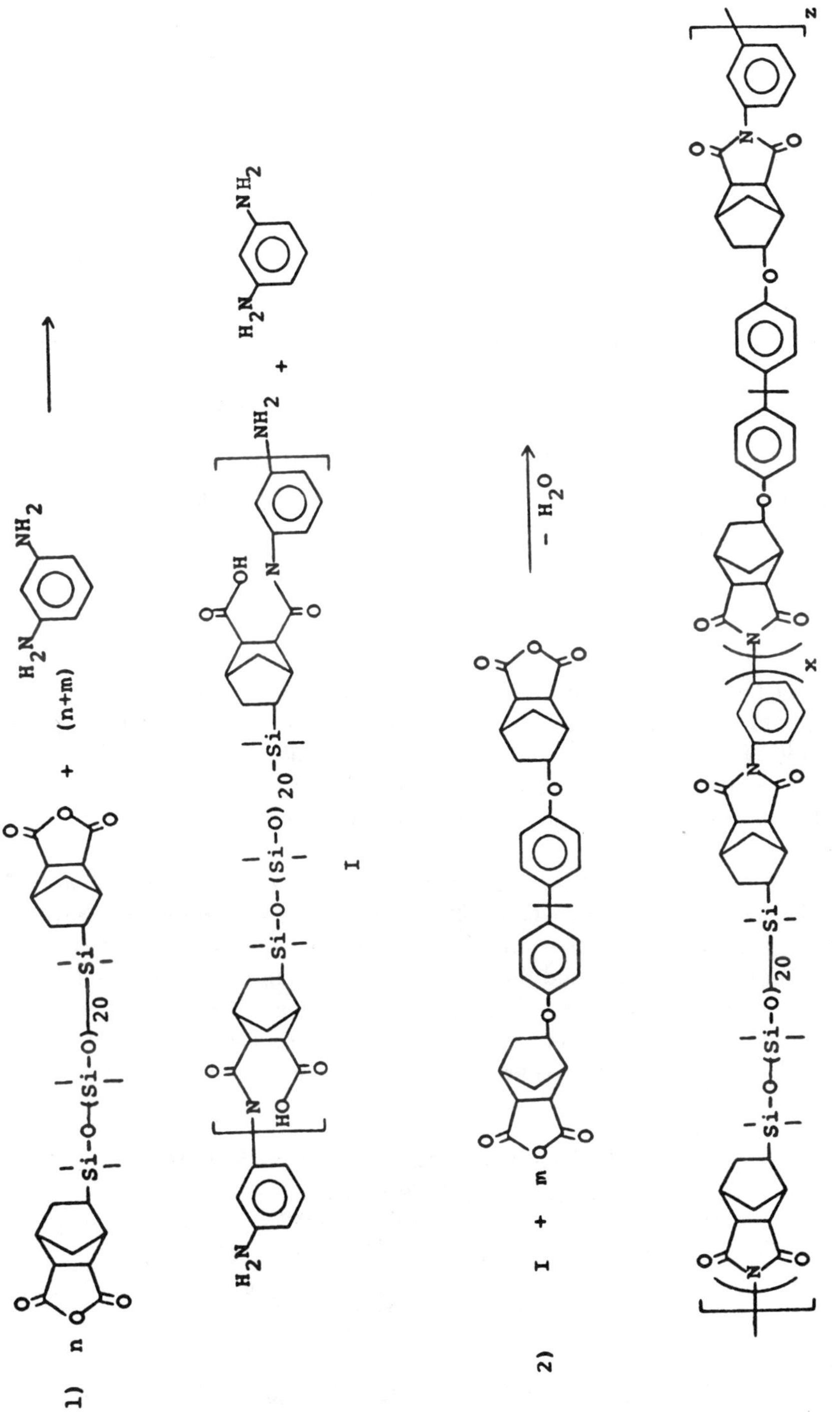

Scheme IV. Synthesis of poly(ether–imide–siloxane) elastoplastic terpolymers.

Table III. Properties of D_{20}–DiSiAn–BPADA Copolymers

Diamine	*IV (dL/g)*	*Tensile Strength (kPa)*	*Elongation (%)*
mPD	0.46	10,000	110
MDA	0.55	8,300	120
ODA	0.74	13,000	230

NOTE: BPADA and D_{20}–DiSiAn were used in a 1.3:1 molar ratio.

weight). Tough, transparent films could be cast from chloroform solutions of all of these copolymers. The properties of these materials are listed in Table IV.

A plot of elongation at break versus siloxane content (Figure 1) illustrates the transition from thermoplasticity with limited ultimate elongation to elastoplasticity. At siloxane content levels below 25%, the ether–imide blocks form the continuous phase. The result is a strong, rigid plastic. Above 25% siloxane, however, phase inversion results in a continuous siloxane phase with a much higher elongation and reduced tensile strength. Figure 2 shows the inverse relationship between siloxane content and tensile strength.

The elastoplastic materials prepared from D_n–DiSiAn in which n was 10–30 did not exhibit true elastomeric behavior. A stress–strain plot for a typical D_{20}–DiSiAn-derived elastoplastic copolymer is shown in Figure 3. This sample exhibits an initially high modulus of up to ~8% elongation with a yield of over 100%. Most of the samples exhibited only small elastic recovery after deformation. The siloxane block lengths used in this study may not have been long enough to produce the morphology necessary for good elastomeric properties.

Attempts to synthesize copolymers from D_n–DiSiAn in which n was >30 failed. Invariably, two immiscible *o*-dichlorobenzene solutions resulted. Typically, one phase contained low-molecular-weight polymer rich in poly(dimethylsiloxane), and the other contained high-molecular-weight material rich in organic block. Phase separation during polymerization was not encountered when the average siloxane blocks were less than 30 units long.

The thermal transitions of these materials could not be observed by DSC

Table IV. Properties of D_{20}–DiSiAn–BPADA–mPD Copolymers

Siloxane Content (%)	*IV (dL/g)*	*Tensile Strength (kPa)*	*Elongation (%)*
8	1.00	37,000	6
13	0.42	30,000	7
22	0.45	23,000	8
35	0.51	15,000	90
45	0.46	10,000	110

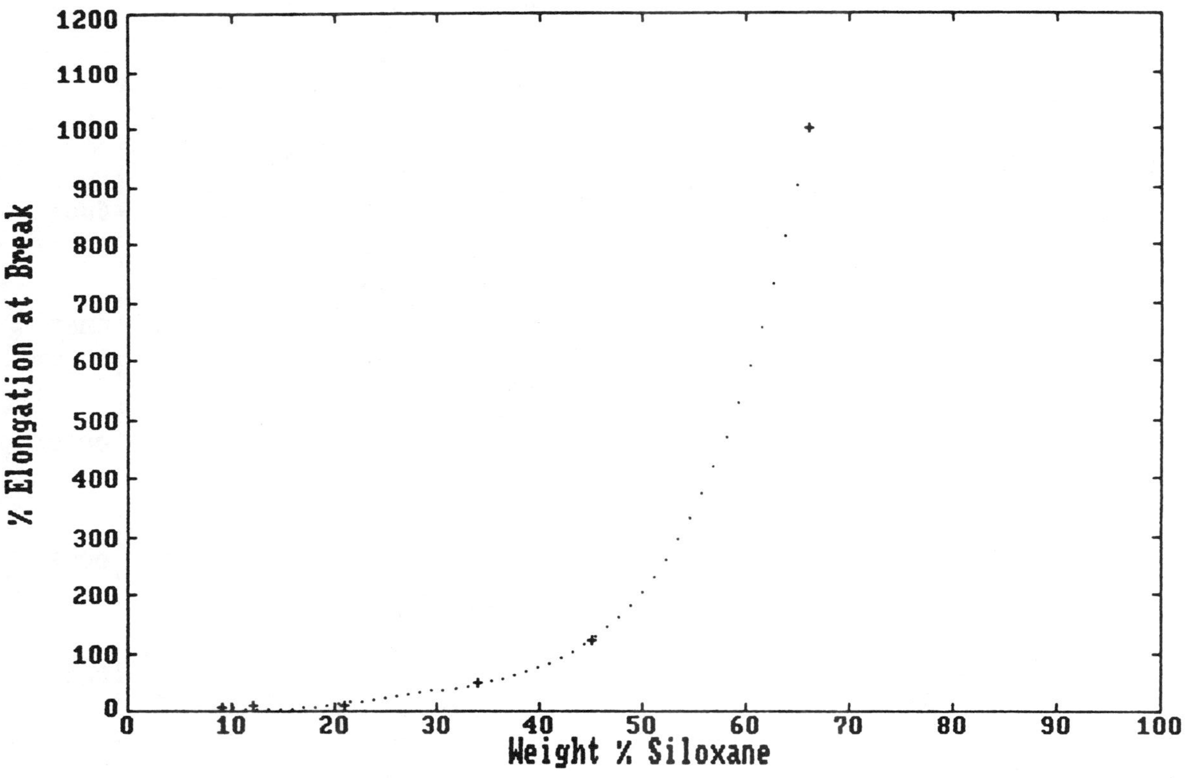

Figure 1. Elongation at break versus siloxane content.

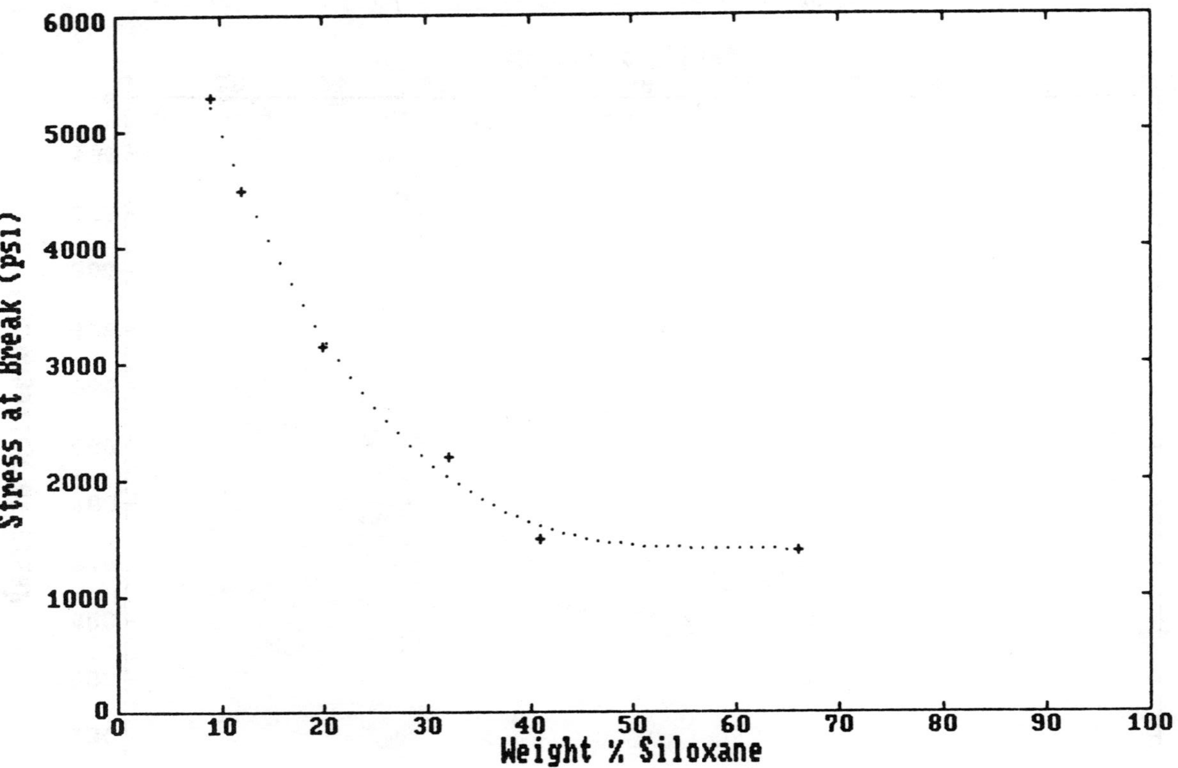

Figure 2. Tensile strength at break versus siloxane content.

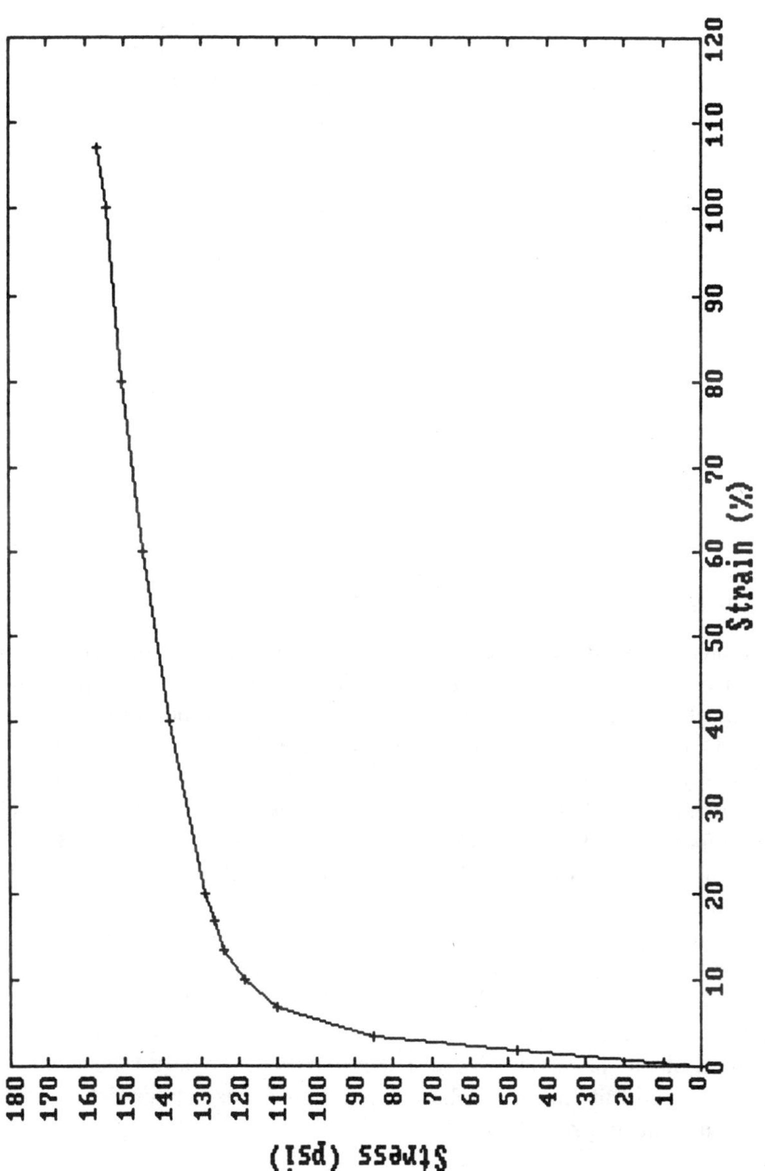

Figure 3. Stress–strain plot for a D_{20}*–DiSiAn–mPD–BPADA elastoplastic polymer containing 50% siloxane.*

analysis. However, DMA was a reliable and sensitive technique for measuring the T_gs of these copolymers.

A copolymer of D_{21}–DiSiAn, BPADA, and mPD was prepared and examined by DMA. The DMA indicated a distinct transition at 120 °C corresponding to the siloxane block and a weaker transition at 75 °C corresponding to the imide hard block. This behavior is consistent with a phase-separated material with short, well-dispersed hard blocks. The physical properties of this copolymer are shown in the following list. The elastoplastic properties of the material are also consistent with this type of microstructure.

Properties of D_{21}–DiSiAn–BPADA–mPD Copolymer

Average siloxane units	21
Siloxane content (%)	60
IV (dL/g)	0.66
Tensile strength (kPa)	13,000
Elongation at failure (%)	1,000
Temperature of 5% weight loss in N_2 (°C)	475
Temperature of 5% weight loss in air (°C)	440

Summary

High-molecular-weight poly(imide–siloxane) polymers and copolymers have been successfully synthesized from DiSiAn, its polysiloxane homologues, and a variety of organic diamines and dianhydrides. The polymers that were prepared ranged from rigid thermoplastics with T_gs in excess of 200 °C to low-modulus elastoplastics. The mechanical properties were sensitive primarily to the overall siloxane content and average siloxane block size. The copolymers prepared from DiSiAn and D_n–DiSiAn in which $n < 30$, BPADA, and various aromatic diamines had excellent solubility in many chlorinated and dipolar aprotic solvents.

Elastoplastic copolymers derived from D_n–DiSiAn in which n is between 10 and 30 exhibited high ultimate elongations. However, the materials displayed limited recoverable elongation. DMA of a typical elastoplastic revealed a weak thermal transition around 75 °C corresponding to the organic imide hard block. The small size and dispersivity of the hard blocks in these systems may inhibit the microphase separation necessary for development of true elastomeric properties.

References

1. Kuckertz, V. H. *Makromol. Chem.* **1966,** 98, 101–108.
2. General Electric Company, U.S. Patent 3 325 450, 1967.

3. General Electric Company, U.S. Patent 3 740 305, 1973.
4. Noshay, A.; McGrath, J. E. *Block Copolymers;* Academic: New York, 1977.
5. Maudgal, S.; St. Clair, T. L. *Int. J. Adhesion and Adhesives* **1984,** *4,* 1984.
6. Hitachi Ger. Offen. DE 3131613, 1981.
7. Hitachi Jpn. Kokai Tokkyo Koho 83/7473, 1983.
8. Kuckertz, V. H. *Makromol. Chem.* **1966,** *98,* 101.
9. Yilgor, I.; Yilgor, E.; Johnson, B. C.; Eberle, J.; Wilkes, G. L.; McGrath, J. E. *Polym Prepr.* **1983,** *24,* 78. *Chem. Abstr.* **1984,** *100,* 122817a.
10. Johnson, B. C.; Yilgor, I.; McGrath, J. E. *Polym. Prepr.* **1984,** *25,* 54. *Chem. Abstr.* **1984,** *101,* 152437q.
11. Arnold, C. A.; Summers, J. D.; Bott, R. H.; Taylor, L. T.; Ward, T. C.; McGrath, J. E. *Inc. SAMPE Symp. Exhib.* **1987,** *32,* 586. *Chem. Abstr.* **1987,** *107,* 59823b.
12. Arnold, C. A.; Summers, J. D.; Bott, R. H.; Taylor, L. T.; Ward, T. C.; McGrath, J. E. *Polym. Prepr.* **1987,** *28,* 217. *Chem. Abstr.* **1987,** *107,* 135020e.
13. Arnold, C. A.; Summers, J. D.; Chen, Y. P.; Chen, D. H.; Graybeal, J. D.; McGrath, J. E. *Int. SAMPE Symp. Exhib.* **1988,** *33, 960. Chem. Abstr.* **1988,** *109,* 74410z.
14. General Electric Company, U.S. Patent 4 381 396, 1983.
15. Eddy, V. J.; Hallgren, J. E. *J. Org. Chem.* **1987,** *52,* 1903.

RECEIVED for review May 27, 1988. ACCEPTED revised manuscript May 3, 1989.

10

Polysiloxane–Thermoplastic Interpenetrating Polymer Networks

Barry Arkles[1] **and Jane Crosby**[2]

[1]**Petrarch Systems, Inc., Pearl Buck Road, Bristol, PA 19007**
[2]**ICI Advanced Materials, 475 Creamery Way, Exton, PA 19341**

Polysiloxane–thermoplastic interpenetrating polymer networks (IPNs) are a unique class of polymer–polymer composites that significantly extend the process and mechanical property range of silicone elastomers and allow modification of the behavior of thermoplastics. The materials are characterized by reactive processing consistent with traditional thermoplastic technology and the development of physical properties consistent with semi-IPNs generated from polysiloxanes cross-linked in a variety of thermoplastic matrices, including polyurethanes, polyamides, and polyolefins. The semi-IPNs are prepared by melt-mixing hydride- and vinyl-functionalized polysiloxanes with thermoplastics and inducing cross-linking with a platinum catalyst in a second melt-process step. Polysiloxane composition, cross-link density, and total content affect the process and mechanical behavior of the semi-IPNs. Polysiloxane–thermoplastic IPNs may be used in biomedical materials, gears and bearings, and applications in which the dimensional stability of crystalline resins must be enhanced.

INTERPENETRATING POLYMER NETWORKS (IPNs) are a special class of polymer blends in which the polymers exist in networks that are formed when at least one of the polymers is synthesized or cross-linked in the presence of the other. Classical or true IPNs are based solely on thermosetting polymers that form chemical cross-links. More recently, two classes of thermoplastic IPNs have been developed. *Apparent IPNs* are based on combinations of physically cross-linked polymers. *Semi-IPNs* are based on combinations of

0065–2393/90/0224–0181$06.00/0

cross-linkable and nonreactive linear polymers. Polysiloxane (silicone) interpenetrating polymer networks have had a significant role in the development of thermoplastic semi-IPNs.

The first attempts to commercialize thermoplastic IPNs were based on apparent-IPN technology. In contrast to the true IPNs or semi-IPNs, in which the cross-links that stabilize the polymer networks are chemical (i.e., covalent bonds), the apparent IPNs are stabilized by physical or virtual bonds (i.e., glassy regions of block copolymers, ionic domains of ionomers, or crystalline regions in semicrystalline polymers). A series of apparent IPNs of styrene–ethylene–butylene–styrene (SEBS) block copolymers and crystalline thermoplastics (polyamides and polyesters), offered by the Shell Chemical Company, was developed by Davison and Gergen (*1*, *2*). These normally incompatible resins have been blended in combination with polypropylene or mineral oil to form co-continuous, interpenetrating phases. The IPN structure is stabilized by the physical cross-links in the styrene end blocks of the copolymers and the crystalline regions of the thermoplastic resin. The polymer networks prevent gross phase separation and lead to extended load sharing between the polymer phases. Thermoplastic ionomeric IPNs have also been reported (*3*).

The semi-IPNs are intermediate systems in which a thermosetting polymer network is formed within a thermoplastic polymer; the coexisting structures are stabilized by physical cross-links in the thermoplastic phase. Wertz and Prevorsek (*4*, *5*) have used dicyanatobisphenol A as a thermosetting monomer that is first mixed in the melt state with a thermoplastic matrix resin and later cross-linked during injection molding to develop resins offered by Allied–Signal Corporation. Silicone–thermoplastic semi-IPNs (Rimplast) (*6*, *7*) are formed by a similar reaction-molding process, but the thermosetting polymer network is formed by the addition of functionalized silicone oligomers that have been predispersed in a thermoplastic resin.

The different chemical structures of thermoset true IPNs and thermoplastic apparent and semi-IPNs can be represented conceptually as line drawings (Figures 1–3). True IPNs consisting of two, totally continuous, interlocking networks are shown in Figure 1A. Figure 1B is a geometrical abstraction of a true IPN. Neither phase can be extracted from the IPN. The semi- or pseudo-IPNs represented in Figure 2 consist of one chemically cross-linked phase within a linear thermoplastic polymer matrix. The thermoset phase cannot be extracted from the IPN. Physical cross-links in the crystalline regions of the linear thermoplastic are reversibly broken at elevated temperatures, and these breaks allow the materials to flow. Figure 3 depicts the physical cross-links of the styrene end blocks of the structured SEBS rubber as circles. Because there are no chemical cross-links in the apparent IPNs, either resin phase can be extracted independently.

The preservation of thermoplastic character in the semi-IPNs is of great commercial value. The requirements of thermoplastic processing demand

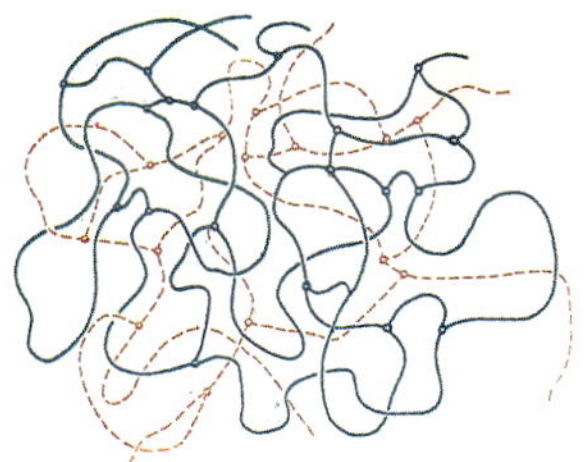

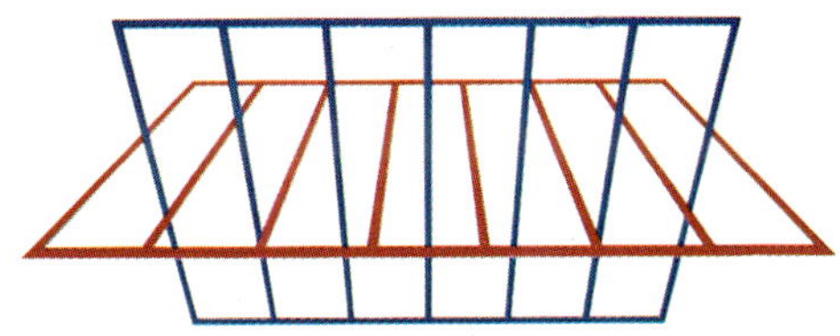

Figure 1. True IPNs. The different polymer systems are cross-linked to themselves but not to each other. The polymers thus form networks that interpenetrate each other. (A) Traditional three-dimensional depiction; (B) geometrical abstraction, in which the cross-links are depicted as junctions of vertical and horizontal lines.

Figure 2. A semi-IPN. Only one polymer is cross-linked. The other thermoplastic polymer is so long that the polymers cannot disentangle.

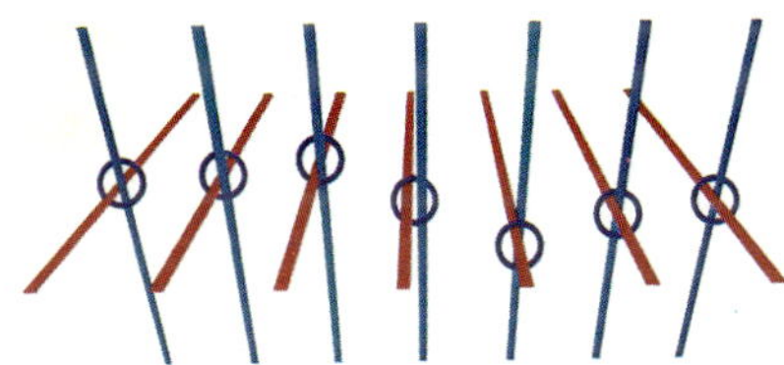

Figure 3. An apparent IPN. None of the polymers is cross-linked, but a third polymeric component, which is compatible with the other two polymers and is usually a block copolymer, holds the system together.

homogeneous rheology. At the same time, requirements for network formation, that is, the development of distinct domains, demand heterogeneous morphology. In many ways, the balance between the two requirements, homogeneous rheology and heterogeneous morphology, defines the synthetic challenge and performance parameters of the thermoplastic IPNs.

Although chemical structures are used to classify IPNs, the morphology of a polyblend defines it as an interpenetrating polymer network. The network structure imposes compatibility between otherwise incompatible resins; this compatibility allows the expression of both components. However, molecular interpenetration of the two phases in the network can lead to composite properties that are not anticipated. The properties of an IPN are not strictly additive functions of component properties; phase morphology and phase interactions that promote interfacial bonding can lead to unexpected increases in impact strength, modulus, and heat deflection temperatures. Many of these changes can be associated with changes in crystallinity.

Silicone–Thermoplastic Semi-IPNs

Chemistry and Processing. The silicone–thermoplastic semi-IPNs have two indirect antecedents: the full IPN silicone–urethane systems reported by Frisch (*8*) and Hourston (*9*), which are not amenable to thermoplastic processing, and semi-IPNs of cross-linked silicone in a silicone fluid matrix, which are used in prosthetic gels (*10*) and for which homogeneous morphology precludes the development of enhanced mechanical properties. In silicone–thermoplastic semi-IPNs, hydride- and vinyl-functionalized silicone fluids are intimately blended in the melt state with a thermoplastic resin. Examples of the silicones are given in structures **1–4.** No reaction takes place among the fluids during this extrusion step. A catalyst is then applied topically to the resin–blend pellets. When the pellets are remelted in a forming process (e.g., profile extrusion or injection molding), the heat-activated catalyst mixes with the vinyl- and hydride-functionalized silicone fluids and initiates a vinyl-addition (hydrosilylation) reaction, and a high-molecular-weight cross-linked siloxane network is developed in the finished part (Figure 4).

The addition polymerization of silicone–thermoplastic semi-IPNs is ideal

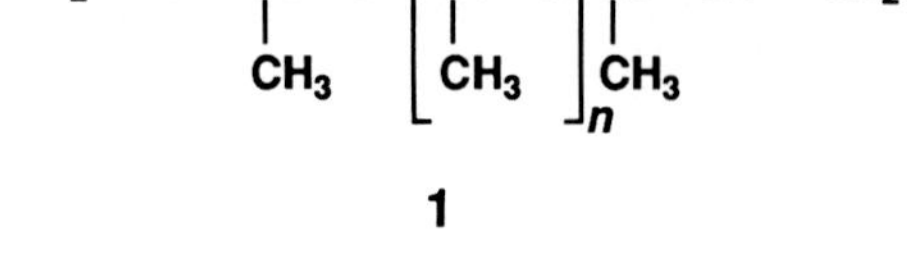

$$CH_2{=}CH{-}Si(CH_3)_2{-}O{-}\left[Si(CH_3)_2{-}O\right]_n{-}Si(CH_3)_2{-}CH{=}CH_2$$

1

$$(CH_3)_3SiO{-}\left[Si(CH_3)(H){-}O\right]_m{-}\left[Si(CH_3)_2{-}O\right]_n{-}Si(CH_3)_3$$

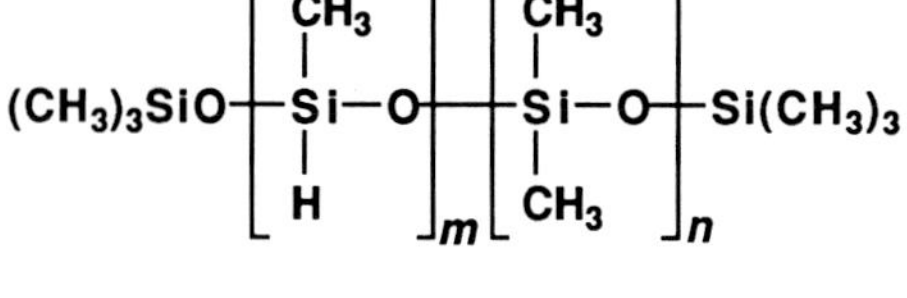

2

$$CH_2{=}CH{-}\overset{CH_3}{\underset{CH_3}{Si}}{-}O{-}\left[\overset{C_6H_5}{\underset{C_6H_5}{Si}}{-}O\right]_m\left[\overset{CH_3}{\underset{CH_3}{Si}}{-}O\right]_n\overset{CH_3}{\underset{CH_3}{Si}}{-}CH{=}CH_2$$

3

$$-\left[\overset{CH_3}{\underset{CH_3}{Si}}{-}O\right]_p\left[\overset{C_8H_{17}}{\underset{CH_3}{Si}}{-}O\right]_m\left[\overset{CH{=}CH_2}{\underset{CH_3}{Si}}{-}O\right]_n-$$

4

for a continuous reaction-molding or extrusion process. Although the reaction is initiated in the melt at elevated temperature, it will proceed in the solid state. The reaction is fast and will proceed to completion, and no chemical byproducts are formed. Conventional thermoplastic injection-molding machines or extruders can be used in forming the IPNs; no special devolatilization is needed. The rheology of a typical system is shown in three stages (Figure 5): unmodified base polymer, base polymer–silicone polyblend, and silicone semi-IPN. In general, the melt viscosity of the base polymer–silicone polyblend is lower than that of the basic polymer and that of the developed semi-IPN.

A silicone semi-IPN modified thermoplastic urethane illustrates this technology. The following composition is mixed in the melt state by pumping a preblend of the silicones into a counter-rotating twin-screw extruder with

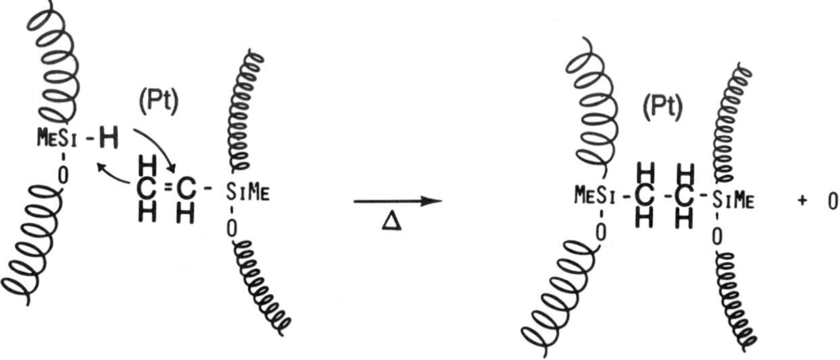

Figure 4. Addition vulcanization of polymer.

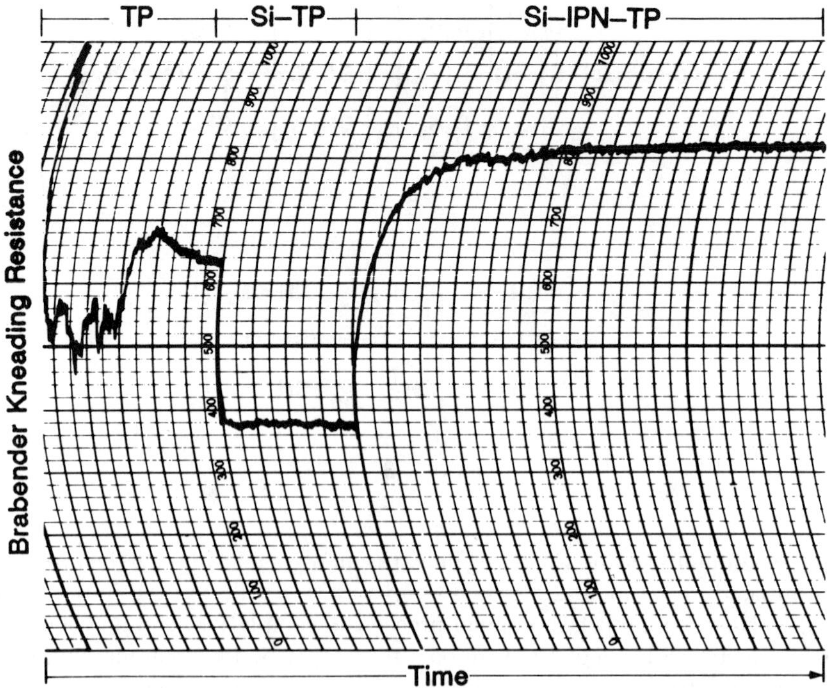

Figure 5. Typical changes in rheology accompanying the transition from a thermoplastic (TP), to a silicone–thermoplastic polyblend (Si-TP), and to a silicone-interpenetrated thermoplastic (Si-IPN-TP) for polypropylene.

thermoplastic urethane at melt temperatures of 210–215 °C: 7500 g of polyether urethane (Pellethane 2103; Dow Chemical Company), 2450 g of poly(dimethylsiloxane-*co*-diphenylsiloxane) (15–17%; vinyldimethylsiloxane terminated; 5000 ctsks.), and 50 g of poly(diphenylsiloxane-*co*-methylhydrosiloxane) (30–35% methylhydrosiloxane; copolymer terminated; 30 ctsks.)

The mixture is pelletized, and 1 g of a platinum–tetramethyltetravinylcyclotetrasiloxane complex (2.5% Pt) is tumbled onto the pellets and then deactivated by the addition of 0.05 g of 3-methylisobutynyl alcohol. The apparently homogeneous material contains two unreacted silicones in a thermoplastic urethane matrix.

The polysiloxane network is formed during part fabrication. In injection molding, the acetylenic alcohol, which acts as a fugitive inhibitor of the vinyl-addition reaction, is volatilized at low temperature as the pellets enter the feed throat. The platinum complex is activated at the process temperature of the urethane (170–185 °C). The vinyl-addition reaction is initiated by the melt state, and the parts generated demonstrate mechanical properties consistent with the formation of a silicone IPN. The fabricated parts are translucent. The physical properties of this formulation (PTUE 205) are given in Table I.

Table I. Properties of Silicone–Urethane IPNs

	Aliphatic Ether	Aromatic Esters		Aromatic Ethers				
Property	*(PTUA 102)*	*PTUE 101*	*PTUE 102*	*PTUE 202*	*PTUE 203*	*PTUE 205*	*PTUE 302*	*PTUE 332*[a]
Hardness (shore)[b]	70A	80A (30D)	55D (95A)	55D	87A	55A	60D	70D
Tear strength (kN/m)	52.8	88.0	132.0	114.4	88.0	52.8	176.0	211.2
Tensile strength (MPa)	20.7	44.8	41.4	41.4	18.6	13.8	41.4	127.6
Elongation (%)	700	520	470	425	1300	900	400	5
Flexural strength (MPa)			50.3	48.3			55.2	124.1
Flexural modulus (MPa)	27.6	41.4	110.4	103.5		13.8	186.2	5517.6
Izod impact (J/m)								
Notched	NB[c]	NB	NB	NB	NB	NB	NB	85
Unnotched	NB	NB	NB	NB	NB	NB	NB	427
Water absorption after 24 h (%)	0.3	0.3	0.3	0.4	0.4		0.5	0.4
Specific gravity	1.11	1.16	1.19	1.12	1.11	1.04	1.12	1.24
Shrinkage of a 1/8-in. (0.32-cm) section (%)	1.5–2.0	1.0–1.5	1.0–1.5	1.0–1.5	1.5–2.0		1.0–1.5	1.1–1.4
Process temperature range (°C)	165–190	165–190	165–190	205–232	205–232	190–220	205–232	215–245

[a]The urethane is an aromatic ether with carbon fiber.
[b]Shore is a standard measure of hardness for rubber recognized by ASTM. There are several scales, including A and D, which cross over.
[c]NB indicates no break.

The technology of silicone–thermoplastic semi-IPNs is highly flexible because IPN compositions and morphology can be varied. The four main factors controlling IPN morphology and properties are (1) base component composition, (2) total IPN content, (3) polymer compatibility and interfacial tension, and (4) cross-link density. Because the silicone network forms independently of the matrix resin, theoretically, an IPN can be developed in any thermoplastic polymer. Those desirable features of silicones, including physiological, release, wear, and thermal properties, can be translated into thermoplastics and thermoplastic elastomers. Some resins that have been transformed into semi-IPNs include the following:

- Polyethylene
- Polypropylene
- Polyamide (6, 6,6, 6,12, 11, 12)
- Polyacetal
- Poly(butylene)(terephthalate)
- Polyurethane
- Styrene block polymers
- Polyolefin elastomers
- Polyester elastomers
- Melt-processable fluorocarbons

Intrinsic Properties. IPN properties can also be tailored by varying silicone concentration. In some systems, dramatic changes in polymer morphology and crystallinity have been observed in loadings as low as 3%. At the other end, polysiloxane loadings of up to 55% have been achieved. Some successful semi-IPN formulations are the following:

- Poly(caprolactam-*inter*-dimethylsiloxane-*co*-vinylmethylsiloxane)
- Poly(propylene-*inter*-dimethylsiloxane-*co*-octylmethylsiloxane-*co*-vinylmethylsiloxane)
- Poly(styrene-*block*-butadiene-*block*-styrene-*inter*-dimethylsiloxane-*co*-diphenylsiloxane-*co*-vinylmethylsiloxane)

The chemical compatibility of the network and matrix polymers determines the extent of solubility of one resin in the other and, hence, the degree of molecular interpenetration. The backbone substitution groups of the silicone prepolymers are carefully chosen to achieve an appropriate degree of compatibility with a specific matrix polymer (*11*). Although most silicone oligomers are not highly soluble in high-molecular-weight thermoplastic resins, total incompatibility would lead to gross phase separation (i.e., delamination) during fabrication. At the other extreme, fully soluble silicones result in alloys in which morphology consistent with network formation is not observed. An example of a low-cross-link-density siloxane acting as a plasticizer is shown in Table II for a silicone–aliphatic urethane system. As phenyl groups are introduced into the silicone, solubility increases, network properties diminish, and physical and thermal properties decrease radically.

Various degrees of phase separation have been observed in IPN composites (*12*). The final structure consideration, cross-link density, in conjunction with the intrinsic compatibility of the polymers, determines the domain size of the reacting phase. Increasing cross-link density leads to smaller domains and greater phase continuity in the cross-linking phase and an increase in "trapped" matrix polymer chains. The greater degree of molecular interpenetration in highly cross-linked systems results in network-imposed compatibility.

Cross-link density in silicone–thermoplastic semi-IPNs is controlled by the molecular weight and number of reactive groups of the silicone oligomers. Highly branched and lightly branched siloxane networks of distinctly different behavior have been developed in the silicone–thermoplastic semi-IPN series. Although every semi-IPN base polymer system develops a distinct morphology, the case of nylon 6,6 is instructive. No dominant structures are observed for the highly branched polymers in scanning electron micrographs (SEMs). The polymer with higher cross-link density presumably forms a more rigid, extended network throughout the matrix thermoplastic. The highly branched networks exhibit greater phase separation. In SEMs of 10% of a lightly branched silicone in nylon 6,6 IPNs, the silicone networks form spherical domains ranging from 6 to 12 μm in diameter. These structures are absent in the highly cross-linked system (Figure 6.) Even in the case of the lightly branched silicone, the majority of the silicone is not involved in gross phase separation, but in the fine structure as demonstrated by the silicon EDX (X-ray dispersive microscopy).

Changes in the crystallization behavior of the thermoplastic phase of the IPN have been observed in DSC (differential scanning calorimetry) experiments (*13*). In Figure 7, a series of DSC scans on silicone–nylon 6,6 composites of various silicone contents are presented. The characteristic melt peak of the nylon matrix is observed in all of the thermograms, but a second, slightly lower temperature melt peak becomes more pronounced as the concentration or the cross-link density of silicone polymer in the IPN is

Table II. Comparison of IPN-Modified and Unmodified Aliphatic Ether Urethanes

Material	*Tensile Strength (MPa)*	*Elongation (%)*
Unmodified urethane	33.1	800
Urethane–poly(dimethylsiloxane) (10%) IPN	20.7	700
Urethane–poly(phenylmethylsiloxane) (10%) IPN	6.2	1000

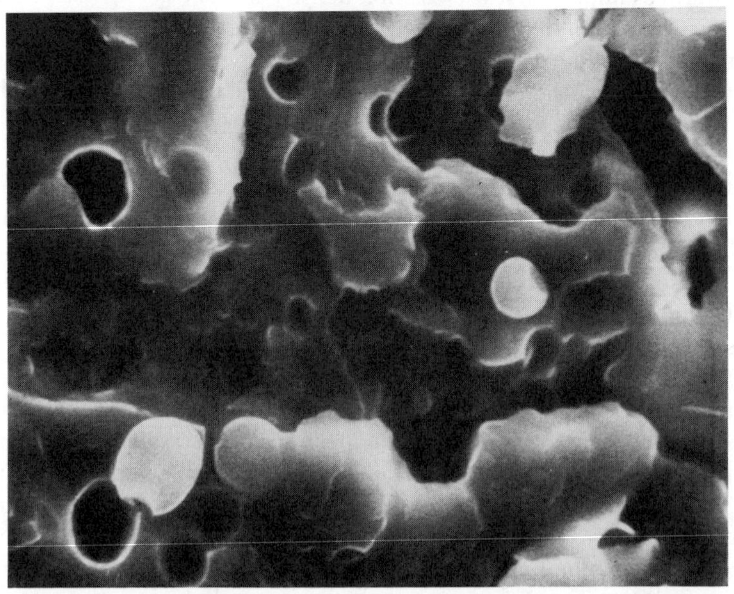

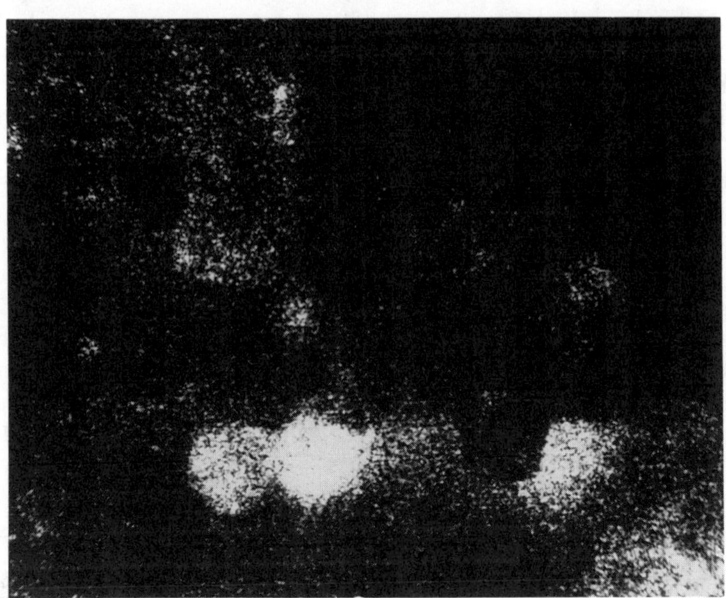

Figure 6. 6,6-Polyamide-i-low cross-link PDMSO at 1000× magnification. (Top) Scanning electron micrograph; (bottom) EDX Si.

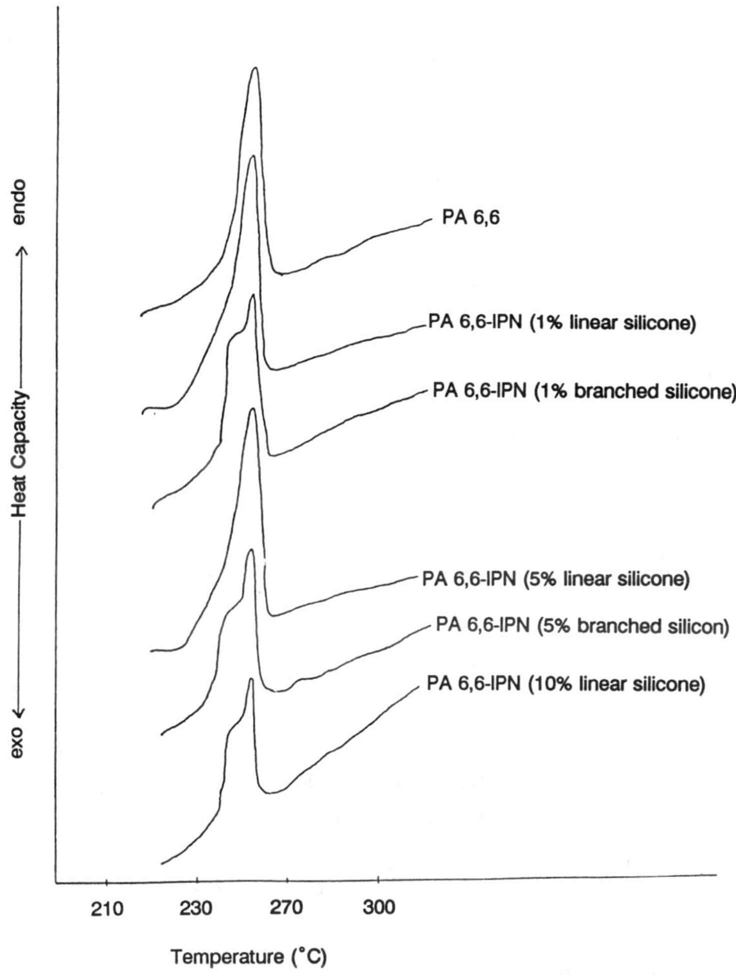

Figure 7. DSC thermograms of nylon 6,6 IPNs. PA is polyamide.

increased. Although double-melting endotherms have been observed in nylon 6,6 (*14*), they were attributed to a reorganization phenomenon during the DSC scan; the lower melting peak decreased in size and increased in temperature with increased annealing time, results suggesting that a secondary, less-ordered crystalline phase was being converted to a more favored form.

The double-melt endotherms of the silicone–nylon 6,6 IPN do not disappear with thermal cycling. Instead the peaks are further resolved in the DSC when samples are scanned to 50 °C above the melt point, cooled slowly to room temperature, and rescanned (Figures 7 and 8). The stability of the melt endotherm associated with the presence of IPN suggests the presence of an interpenetrated polymer phase of various crystalline structures.

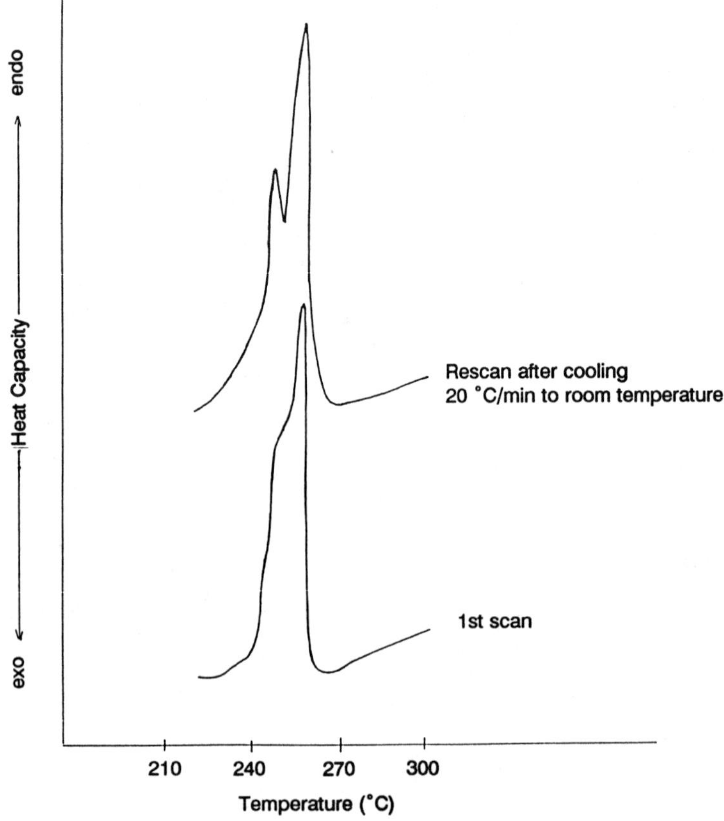

Figure 8. Multiple DSC thermograms of 30% glass fiber-reinforced, highly branched silicone–nylon 6,6 IPN.

Extrinsic and Mechanical Properties. The technology of silicone–thermoplastic semi-IPNs was originally developed with thermoplastic polyurethanes for biomedical applications (*15*). Silicone rubbers possess the good blood-contacting and low-physiological-response properties required in biomaterials but lack high tensile strength and modulus and are difficult to process. Silicone–polyurethane IPNs provide a range of favorable property combinations in easily injection-molded materials. The mechanical and physiological properties of these IPNs can be compared with those of pure silicones and base polymers:

- Compared with pure silicones
  1. Processing on conventional thermoplastic equipment with less processing variables compared with liquid silicone rubber systems

2. Greater tear, tensile, and flexural strengths
3. Lower mating surface wear (Silica-free formulation reduces abrasion.)
4. Lower oxygen permeability and reduced moisture transmission
5. Increased elastic recovery at low extensions
6. Reduced blood–polymer interaction (Filler-free silicone surface reduces thrombogenic response.)

- Compared with thermoplastic urethane
  1. Silicone release characteristics
  2. Lower wear and friction (Wear is reduced by a factor of 5–6; the coefficient of friction is reduced by 20–30%.)
  3. Improved dielectric properties
  4. Reduced blood–polymer interaction (Filler-free silicone surface reduces thrombogenic response.)
  5. Increased permeability to oxygen, which is useful in self-supporting enrichment applications
  6. Improved high-temperature performance
  7. Increased elastic recovery at high extensions

In the IPNs, the mechanical properties of the matrix resin are preserved, and wear, lubricity, and abrasion resistance are improved. In systems with high cross-link densities, the resilience and resistance of the resins to creep (set) is enhanced. The properties of a series of thermoplastic–polyurethane IPNs (TPU-IPNs) are listed in Table I.

Applications. The silicone–aromatic urethane grades demonstrate tear and tensile strengths 3 to 5 times those of silicones and wear rates that are lower by a factor 5 or 6 when compared with urethanes and by factors greater than 10 when compared with silicones. The silicone–aromatic urethanes are currently being evaluated in a wide range of medical device applications. The silicone–aliphatic urethane systems may have significant potential for long-term implant applications. Silicone–thermoplastic semi-IPNs have been extended to other thermoplastic elastomers and engineering thermoplastics for use in the medical industry.

Silicone–SEBS IPNs are the only sterilizable thermoplastic systems that resist coring and are capable of self-sealing after being pierced with hypodermic syringes. These IPNs are used in injection sites both because they allow dual-shot fabrication and because they are free of contamination from vulcanization aids found in their natural rubber predecessors.

Silicone–nylon 12 IPN is a candidate material for self-supporting medical devices such as implantable pumps, catheter guides, and connectors.

Silicone–thermoplastic semi-IPNs and fiber-reinforced IPNs based on high-modulus-engineering thermoplastic matrices have also been developed (*16*, *18*). Initially these materials were of interest as internally lubricated composites for bearing applications, in which the IPN combined the high-modulus and high temperature properties of the engineering thermoplastic with the improved tribological properties and release characteristics of a silicone. The mechanical, friction, and wear properties of nylon 6,6 and several silicone–nylon 6,6 IPNs are shown in Table III. As the volume fraction of silicone or the cross-link density increases, a higher continuity IPN is formed, and wear resistance and coefficients of friction improve dramatically. Tensile strengths are maximized, and flexural-modulus values are unchanged at low silicone concentrations. The heat deflection temperatures of all of the IPNs are are greater than that of nylon 6,6; the greatest increase is seen in the highly branched silicone networks, in which a more rigid structure is formed.

The mechanical properties of a 30% glass fiber-reinforced silicone–nylon 6,6 IPN approximate that of a conventional glass-reinforced composite (Table III). Strength and stiffness are reduced slightly in the IPN composite, but increases in elongation and Izod impact strengths indicate an increase in ductility. By contrast, low-molecular-weight silicone fluids are often used in combination with polytetrafluoroethylene (PTFE) as internal lubricants for engineering thermoplastics. Although silicone fluids are good lubricants, they tend to reduce the mechanical properties of the composite, and they migrate to the molded surface and contaminate surface mounts or trap abrasive wear debris.

The silicone–thermoplastic IPN composites exhibit superior mechanical properties (Table IV), and the network structure of the silicone prevents migration. The value of the IPN has been demonstrated in a molded journal bearing for use in high-speed paper-handling equipment. For example, part wear was reduced to a greater extent in a PTFE-lubricated, silicone–nylon 6,6 IPN bearing, compared with an uncross-linked silicone fluid, PTFE-lubricated nylon bearing, which attracted abrasive paper dust to the wearing surfaces.

Unique crystallization behavior leads to an unusual apparent property advantage of the silicone–thermoplastic semi-IPNs: low and uniform shrinkage behavior (*17*). This characteristic is of particular value in crystalline thermoplastic resins, like nylons, polyesters, polypropylene, and acetal, that exhibit high and nonuniform mold shrinkage rates. Shrinkage differential leads to part warpage, a condition that is further aggravated in fiber-reinforced composites. The IPN greatly stabilizes the mold shrinkage behavior in both fiber-reinforced and nonreinforced composites.

Results of a study of mold shrinkage and warpage are shown in Table

Table III. Properties of Silicone–Polyamide IPNs

		Modified Nylon 6,6				*Modified Nylon 12*		
Property	*Nylon 6,6*	*PTA 6601*	*PTA 6602*	*PTA 6631*	*Nylon 12*	*PTA 1201*	*PTA 1202*	*PTAE 1201*[a]
Silicone content (%)	0	5	10	3	0	5	10	5
Other materials				13% PTFE				
Tensile strength (MPa)	81.4	86.2	69.6	68.9	38.6	51.0	35.9	37.9
Elongation (%)	60	20	5	20	20	8–10	8–10	275
Flexural strength (MPa)	103.5	109.7	96.6	105.5	48.3	57.2	55.2	15.2
Flexural modulus (MPa)	2830	2830	2480	2690	1206	1489	1379	2414
Izod impact (J/m)								
Notched	48	48	43	43	107	37	32	NB
Unnotched		1121	800	694		1708	342	NB
Heat distortion temperature at 1.82 MPa (°C)	66	93	74	102	125	150	135	
Wear factor (10^{-10} in.^{3}/min)[b]	200	155	20	25		195		
Coefficient of friction								
Static (μS)	0.20	0.19	0.09	0.11		0.18		
Dynamic (μD)	0.28	0.19	0.12	0.18		0.17		
Specific gravity	1.14	1.13	1.12	1.21	1.02	1.02	1.01	1.01
Shrinkage (%)	1.2–2.0	0.5–0.7	0.4–0.6	0.5–0.7	0.7–0.8	0.5–0.7	0.5–0.7	0.5–0.7
Water absorption after 24 h (%)	1.2	0.8	0.6	0.6	0.25	0.15	0.12	
Process temperature range (°C)	245–275	235–275	235–275	235–275	180–210	180–210	180–210	180–210

[a] PTAE 1201 is an elastomer with a hardness of 60D.

[b] 10^{-10} in.^{3}/min is equivalent to 1.64×10^{-8} cm^3/min.

Table IV. Properties of Nylon 6,6 Resin and Silicone–Thermoplastic Semi-IPNs

Property	*ASTM Method*	*Nylon 6,6 (Unmodified)*	*RL-4610 (Linear Silicone)*	*RL-4620 (Linear Silicone)*	*R-5000 (Branched Silicone)*
Specific gravity	D792	1.14	1.13	1.12	1.13
Mold shrinkage of a 1/8-in. (0.32-cm) section (%)	D955	0.015	0.012	0.008	0.006
Tensile strength (MPa)	D638	81.4	86.2	69.6	76.6
Tensile elongation (%)	D638	60	20	5	20
Flexural modulus (MPa)	D790	2827	2827	2483	2827
Izod impact strength (J/m)					
Notched	D256	48	48	48	48
Unnotched	D256	>2135	1120	800	1067
Heat distortion temperature at 1.82 MPa (°C)	D648	66	71	82	93
Wear factor (10^{-10} in.^{3}/min)[a]		200	155	12	100
Coefficient of friction					
Static (μS)		0.20	0.19	0.07	0.14
Dynamic (μD)		0.28	0.19	0.08	0.16

[a]10^{-10} in.^{3}/min is equivalent to 1.64×10^{-8} cm^3/min.

Table V. Mold Shrinkage and Warpage of Nylon 6,6 and Silicone–Nylon 6,6 IPN Composites

Code	*Material*	*Mold Shrinkage (%)*	*Warpage*[a]
R-1000	Nylon 6,6	0.015	0.050
RL-4620	Nylon 6,6–linear silicone IPN	0.008	0.018
R-5000	Nylon 6,6–branched silicone IPN	0.006	0.010
RF-1006	Nylon 6,6–30% glass fibers	0.004	0.270
RF-5006	Nylon 6,6–branched silicone–30% glass fibers	0.003	0.035

[a]Warpage of an injection-molded disk (4 by 1/16 in.; 10.2 by 0.16 cm).

V. A series of 4-in. (10.2-cm)-diameter, 1/16-in. (0.16-cm)-thick edge-gated disks were injection molded under normal nylon-processing conditions. Warpage in the lightly branched and highly branched IPN parts were 2.5- and 5-fold lower, respectively, than the warpage in molded nylon 6,6 resin. The warpage in the 30% glass-reinforced nylon 6,6 IPN was 10-fold lower than that in a 30% glass-reinforced nylon 6,6 composite. The greater dimensional control afforded by the IPN enables crystalline matrix composites to be used in high-tolerance moldings.

A 30% glass-fiber-reinforced, PTFE-lubricated, nylon 6,6 IPN was molded in a high-tolerance gear designed for use in the business machine industry. Although excellent chemical resistance and high fatigue endurance limits of glass-reinforced nylon made it an attractive candidate, total composite error (TCE) for molded glass-reinforced nylon was 0.0055″, which corresponds to a class 6 American Gear Manufacturing Association (AGMA) rating. The nylon IPN exhibited lower TCE (0.0028″), produced a class 8 AGMA gear with an overall shrinkage comparable with that of 30% glass-reinforced polycarbonate. The internal lubrication and isotropic shrink characteristics of the silicone–thermoplastic semi-IPN resulted in a high-quality gear with the desired chemical resistance and fatigue endurance.

Hybrid and Secondary Cross-Linked Silicone IPNs

Hybrid versions of silicone–thermoplastic semi-IPNs have been developed (*19*). A hybrid interpenetrating network is one in which the cross-linked network is formed by the reaction of two polymers with structurally distinct backbones. Hydride-functionalized siloxanes can be reacted with organic polymers with pendant unsaturated groups such as polybutadienes (**5**) in the presence of platinum catalysts. Compared with the polysiloxane semi-IPNs discussed earlier, the hydride IPNs tend to maintain mechanical and morphologically derived properties, whereas properties associated with siloxanes are diminished. The probable importance of this technology is in cost-effective ways to induce thermoset characteristics in thermoplastic elastomers.

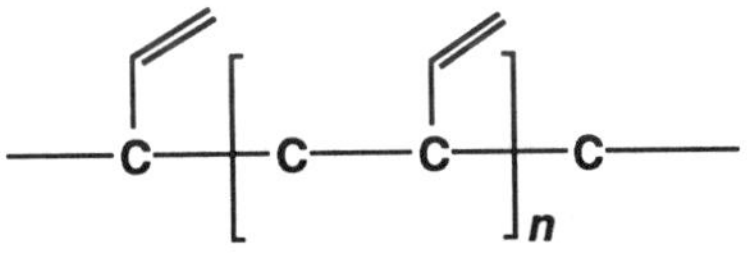

5

Another variation of silicone–thermoplastic semi-IPN technology is the use of secondary cross-linking (*20*). The incorporation of multiple alkoxysilane groups into the interpenetrating network allows a staged cross-linking reaction. The first reaction is the platinum-catalyzed vinyl-addition reaction. The second cross-linking is a hydrolysis condensation that takes place after the thermoplastic processing. This technology is anticipated to be useful in wire and cable insulation and to displace earlier technology, which utilizes peroxide grafting of vinyl silanes (Scheme I).

Conclusion

The silicone semi-IPN structure imparts unique behavior to thermoplastic composites without affecting mechanical strength and stiffness. Apart from the direct effects associated with the incorporation of thermoset character into a thermoplastic, the structure has particular value by functioning as (1) a nonmigrating internal lubricant, (2) a shrinkage and warpage modifier, (3) a flow modifier, and (4) a release agent. As a polymer–polymer composite,

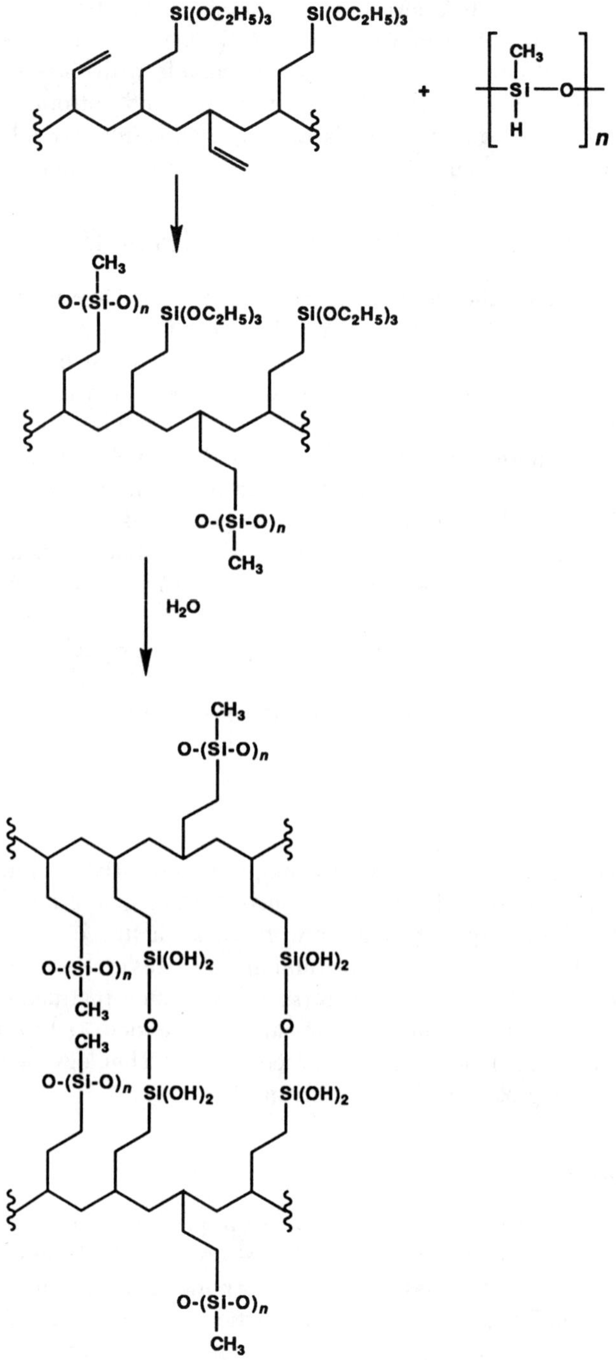

$Si(OC_2H_5)_3$ $Si(OC_2H_5)_3$ + $[-Si(CH_3)(H)-O-]_n$

H_2O

Scheme I

the IPN expresses the strengths of its components, as well as unique thermal, mechanical, and physiological characteristics. The silicone semi-IPN structures present a new avenue for developing thermoplastic elastomers with the mechanical and thermal properties of compression-molded cross-linked rubbers. Because of their synthetic flexibility, silicone IPNs have a broad range of possibilities. Their importance lies not only in these possibilities but also in the examples they provide of what the technology of reactive processes and interpenetrating polymer networks has to offer.

References

1. Davison, S.; Gergen, W. P. U.S. Patent 4 041 103, 1977.
2. Gergen, W. P.; Davison, S. U.S. Patent 4 101 605, 1978.
3. Siegfried, D. L.; Thomas, D. A.; Sperling, L. H. U.S. Patent 4 468 499, 1984.
4. Wertz, D. H.; Prevorsek, D. C. *SPE 42nd Annual Technical Conference* **1984,** p 483.
5. Prevorsek, D. C.; Chung, D. C. U.S. Patent 4 157 360, 1979.
6. Arkles, B. U.S. Patent 4 500 688, 1984.
7. Arkles, B. *Polym. Mater. Sci. Eng.* **1983,** *49,* 6.
8. Frisch, K. U.S. Patent 4 302 553, 1981.
9. Hourston, D. J.; Klein, P. G. *Polym. Mater. Sci. Eng.* **1984,** *51,* 488.
10. Jeram, E. U.S. Patent 4 072 635, 1978.
11. Arkles, B.; Carreno, C. *Polym. Mater. Sci. Eng.* **1984,** *50,* 440.
12. Sperling, L. H. *Interpenetrating Polymer Networks and Related Materials;* Plenum: New York, 1981; p 106.
13. Crosby, J. M.; Hutchins, M. K. *Proceedings of the 40th Meeting of the Society of the Plastics Industry: Reinforced Plastics/Composites;* Society of the Plastics Industry: Washington, DC, 1985; p 11-C.
14. Turi, E. A. *Thermal Characterization of Polymeric Material;* Academic: New York, 1981; p 341.
15. Ward, S. K.; Crosby, J. M. *Polym. Prepr. (Am. Chem. Soc., Div. Polym. Chem.)* **1987,** *28*(2), 173.
16. Arkles, B. *Med. Device Diagn. Ind.* **1983,** *5,* 11.
17. Naitove, H. M.; Crosby, J. M.; DeAngelis, P. J. *Plast. Technol.* **1985,** *31,* 57.
18. Crosby, J. M.; Hutchins, M. K. U.S. Patent 4 695 602.
19. Arkles, B. U.S. Patent 4 714 739, 1987.
20. Arkles, B.; Smith, R. A. U.S. Patent Applied, 1987.

RECEIVED for review May 27, 1988. ACCEPTED revised manuscript March 20, 1989.

11

Non-Gaussian Effects and Intermolecular Correlations in Bimodal Networks of Poly(dimethylsiloxane)

Vasilios Galiatsatos[1] and James E. Mark

Department of Chemistry and the Polymer Research Center, University of Cincinnati, Cincinnati, OH 45221

Strain–birefringence measurements were carried out on unswollen networks of poly(dimethylsiloxane) (PDMS) at elongation at 0–90 °C. An increase in the mole percent of short chains in the networks increased the non-Gaussian deviations from linearity of the dependence of birefringence on stress. Birefringence–temperature measurements showed the deviations to be insensitive to temperature, as would be expected for an intramolecular effect. In addition, these measurements yielded values of the optical configuration parameter and its temperature coefficient that are in good agreement with other published results for bimodal PDMS networks. The magnitudes of the intermolecular correlations were estimated from the optical anisotropies and were found to decrease as the mole percent of short chains increased. This observation is also consistent with intermolecular interactions not being the origin of the observed non-Gaussian effects.

STUDIES OF THE MECHANICAL PROPERTIES of polymer networks have yielded a great deal of important molecular information on the elasticity of polymer chains (*1–3*). For many systems, however, this approach is being supplemented increasingly by studies of the optical properties of the same

[1]Current address: Goodyear Research Laboratories, Department 410 F, 142 Goodyear Boulevard, Akron, OH 44305

0065–2393/90/0224–0201$06.00/0

networks (*2, 3*). This approach has been used for poly(dimethylsiloxane) (PDMS) $[-Si(CH_3)_2O-]_n$ (*4–8*), in part because of its superb optical properties, specifically its transparency that also makes it very attractive for applications such as contact lenses (*9*).

This chapter reports birefringence measurements carried out on PDMS networks that have a bimodal distribution of chain lengths. Such networks are of considerable interest, because their high extensibility permits them to be deformed into ranges of elongation in which both their mechanical (*3, 5–7, 10–12*) and optical (*5–7*) properties exhibit markedly non-Gaussian behavior. The desired data were obtained from networks with various short-chain–long-chain compositions in the unswollen state at 0–90 °C. The results were used to characterize (1) the non-Gaussian behavior in terms of the stress optical coefficient C (*2*) and (2) intermolecular correlations as represented by the nonintrinsic part of the optical anisotropy (*6, 13*).

Theory

The relationship between the measured relative retardation (R) and the stress-induced birefringence (Δn) is given by $R = t\Delta n$ (*2*), in which t is the sample thickness. The stress optical coefficient C is defined by $\Delta n = C\tau$ (*2*), in which τ is the true stress ($\tau = f/A$; f is the force in newtons per square millimeter, and A is the cross-sectional area of the network sample). This coefficient is thus simply the slope of the line in a plot of Δn versus τ. Finally, the optical configuration parameter Δa is defined by

$$\Delta a = \left(\frac{45kT}{2\pi}\right)\left(\frac{n}{(n^2 + 2)^2}\right)\left(\frac{\Delta n}{\tau}\right) \tag{1}$$

in which k is the Boltzmann constant and n is the refractive index of the network. This parameter may be calculated also by using rotational isomeric state theory (*13*) and is an important quantity for testing theoretical aspects of the optical behavior of networks.

The stress optical coefficient C merits special attention, because it leads directly to the parameter Γ_2 that characterizes the optical anisotropy of the network chain under strain. Γ_2 is defined by (*13*)

$$\Gamma_2 = \frac{9}{10}\sum_i \frac{\langle r^T\alpha_i r\rangle_0}{\langle r^2\rangle_0} \tag{2}$$

in which α_i is the anisotropic part of the polarizability tensor contributed by the structural unit indexed by i, r is the end-to-end distance of the chain, and T is the temperature in kelvins. The average unperturbed end-to-end distance of the chain is denoted by $<r^2>_0$. In terms of C, Γ_2 is given by

$$\Gamma_2 = \frac{27nkTC}{2\pi(n^2 + 2)^2} \tag{3}$$

in which n is the refractive index.

Experimental Details

The PDMS chains used were hydroxyl terminated and were obtained from Petrarch Systems, Inc. (Bristol, PA). Their number-average molecular weights (M_n) were 660 and 880 g/mol for the short chains and 21.3 × 10^3 g/mol for the long chains (*7*). Their polydispersity indices would be expected to be ~2.

Two sets of networks were prepared. In the first network, the M_n of the short chains was 660 g/mol (660–21.3 × 10^3), and in the second network, it was 880 g/mol (880–21.3 × 10^3). The compositions of the networks are given in Table I. The chains were end linked with tetraethoxysilane, with stannous 2-ethylhexanoate as catalyst. The reactions were carried out for 2 days under a protective atmosphere of nitrogen at room temperature. Additional details are given elsewhere (*7, 8, 14*). All sample sheets thus prepared were gently extracted to remove the approximately 3 wt % soluble material they contained.

Strips (ca. 30 mm by 80 mm by 1 mm) were cut from the extracted sheets. The thicknesses of the strips were determined from their densities by weighing them after their lateral dimensions had been determined by means of a cathetometer.

The tensile force and birefringence were simultaneously measured as functions of the strain. In brief, the samples were suspended vertically between two clamps; the lower clamp was fixed, and the upper clamp was connected to a force transducer (Statham "strain" gauge). The output of the transducer was monitored by a Hewlett–Packard chart recorder (*7, 8*). Values of the birefringence Δn were determined by using a single-frequency He–Ne laser according to well-established procedures (*2, 7, 8*). Values were calculated directly from the sample thickness and the relative retardation R, which was measured with a Babinet-type compensator. The measurements on the 660–21.3 × 10^3 samples were carried out at 0–90 °C, and those on the 880–21.3 × 10^3 samples were carried out at 25 °C.

Table I. Optical Properties of Bimodal PDMS Networks

M_n of Short Chains (g/mol)	Mol % of Short Chains in Network[a]	T (°C)	C (× 10^4)[b]	Optical Anisotropies	
				Γ_2 (Å^3)	Γ_2^i (Å^3)
660	90.8	10	1.37	0.21	0.17
	92.0		1.17	0.18	0.14
880	60.0	25	1.74	0.28	0.24
	70.0		1.52	0.25	0.21
	93.4		0.71	0.12	0.08

[a]The short chains were combined with long chains with M_n = 21.3 × 10^3 g/mol.
[b]C is the stress optical coefficient in square millimeters per newton.

Results and Discussion

Typical birefringence–stress results, for the 660–21.3 × 10^3 networks, are shown in Figure 1. The anomalous downward curvature at large α at this composition, 92.0 mol % short chains, is larger than that obtained at 90.8 mol % (7) but much less than that obtained at 93.4 mol % (8), even though the short chains were somewhat larger in this case (M_n = 880 g/mol). The curvature does not seem to depend on temperature, as it would if strain-induced crystallization was important (5). Values of C calculated from the least-squares representation of these data and from the corresponding data for the other networks are given in Table I. The decrease in C with increase in mole percent of the short chains is characteristic of this type of non-Gaussian behavior (8).

Some of the birefringence–temperature results for the same networks are given in Figure 2. Again, no changes of birefringence with temperature were observed that would suggest a significant intermolecular contribution to the non-Gaussian behavior. Values of $10^{25}\Delta a$ calculated from equation 1 for one of the 660–21.3 × 10^3 networks are shown as a function of tem-

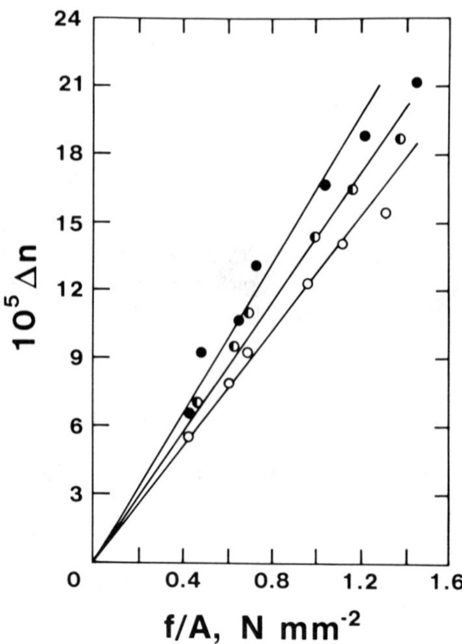

Figure 1. Dependence of birefringence on true stress of bimodal PDMS network containing 92.0 mol % short chains. The measurements were carried out at elongation in the unswollen state at 20 (○), 40 (◐), and 60 (●) °C. The slopes of the lines are C, the stress optical coefficient.

perature in Figure 3. The value at ~25 °C is 3.8 cm^3, with a temperature coefficient ($10^{25}d\Delta a/dt$) of 0.038 cm^3/K. These results are in excellent agreement with the values of 3.2 cm^3 and 0.037 cm^3/K obtained during a previous study of bimodal PDMS networks (5) but are somewhat different from the values of 5.2 cm^3 and 0.063 cm^3/K reported for unimodal PDMS networks

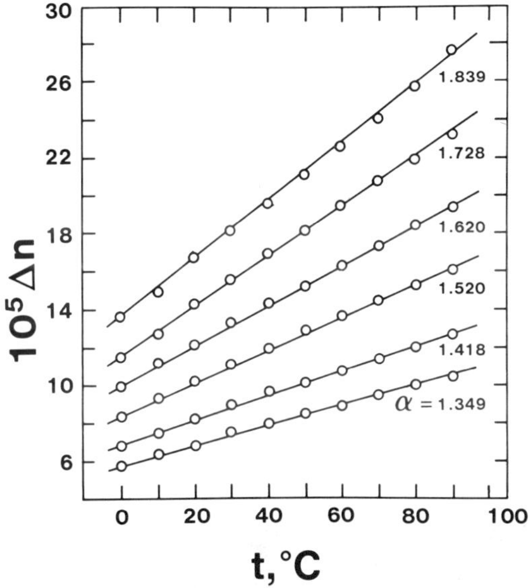

Figure 2. Temperature dependence of birefringence for PDMS network containing 90.8 mol % short chains. Each curve is labeled by the value of the elongation or relative length of the sample.

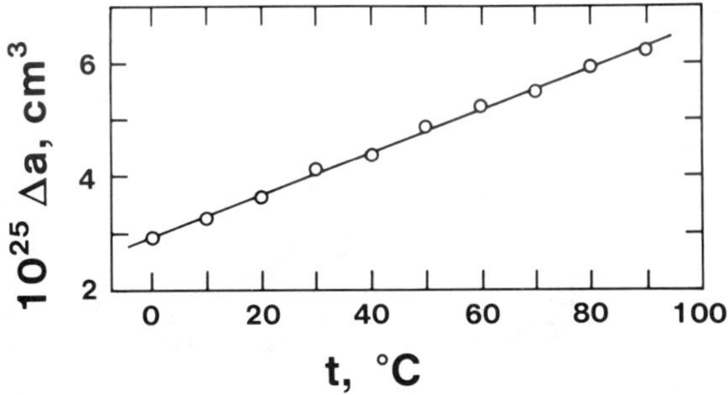

Figure 3. Temperature dependence of optical configuration parameter for PDMS network containing 90.8 mol % short chains.

(*4*). There is a large uncertainty in the intrinsic optical anisotropy of PDMS chains because of its very small value (*4, 5*).

Values of the optical anisotropy Γ_2 calculated from equation 3 are given in Table I. These values were resolved into an intrinsic contribution $\Gamma_2{}^{o}$ = 0.04 $Å^3$ (*6*) and an excess contribution $\Gamma_2{}^{i}$ due to intermolecular correlations (*6, 15*) according to the following equation:

$$\Gamma_2 = \Gamma_2{}^{o} + \Gamma_2{}^{i} \qquad (4)$$

Values of the correlational contribution are listed in Table I. The correlational contributions decrease with an increase in the mole percent of short chains in the network. This trend is also consistent with the conclusion that intermolecular interactions have little to do with the non-Gaussian behavior exhibited by bimodal PDMS networks.

Acknowledgment

It is a pleasure to acknowledge the financial support provided by the National Science Foundation through Grant DMR 84-15082 (Polymers Program, Division of Materials Research).

References

1. Flory, P. J. *Principles of Polymer Chemistry;* Cornell University Press: Ithaca, NY, 1953.
2. Treloar, L. R. G. *The Physics of Rubber Elasticity;* Clarendon: Oxford, 1975.
3. Mark, J. E.; Erman, B. *Rubberlike Elasticity: A Molecular Primer;* Wiley-Interscience: New York, 1988.
4. Liberman, M. H.; Abe, Y.; Flory, P. J. *Macromolecules* **1972,** *5,* 550.
5. Zhang, Z.-M.; Mark, J. E. *J. Polym. Sci., Polym. Phys. Ed.* **1982,** *20,* 473.
6. Erman, B.; Flory, P. J. *Macromolecules* **1983,** *16,* 1601, 1607.
7. Galiatsatos, V. Ph.D. Thesis, University of Cincinnati, 1986.
8. Galiatsatos, V.; Mark, J. E. *Macromolecules* **1987,** *20,* 2631.
9. Arkles, B. *CHEMTECH* **1983,** *September,* 542.
10. Mark, J. E. *Acc. Chem. Res.* **1985,** *18,* 202.
11. Mark, J. E. *Polym. J.* **1985,** *17,* 265.
12. Mark, J. E. *Brit. Polym. J.* **1985,** *17,* 144.
13. Flory, P. J. *Statistical Mechanics of Chain Molecules;* Wiley-Interscience: New York, 1969.
14. Llorente, M. A.; Andrady, A. L.; Mark, J. E. *J. Polym. Sci., Polym. Phys. Ed.* **1981,** *19,* 621.
15. Galiatsatos, V.; Mark, J. E. *Polym. Bull.* **1987,** *17,* 197.

RECEIVED for review May 27, 1988. ACCEPTED revised manuscript October 26, 1988.

12

New Inorganic–Organic Hybrid Materials Through the Sol–Gel Approach

Garth L. Wilkes, Hao-Hsin Huang, and Raymond H. Glaser

Polymer Materials and Interfaces Laboratory, Department of Chemical Engineering, Virginia Polytechnic Institute and State University, Blacksburg, VA 24061–6496

To meet the challenges of new material requirements, scientists must strive constantly to develop new systems. One recent approach has been to prepare inorganic–organic hybrid materials by using the sol–gel method common to ceramists. This technique allows a considerable range of chemistry to be used and may lead to the development of a whole array of systems that have potential applications (structural, optical, etc.). This chapter provides an overview of the general status of this approach, with emphasis on the work that comes from our laboratory. The chemistry discussed in this chapter involves the use of silicon-containing species in the preparation of polymeric network systems.

MANY YEARS AGO, LORD TAYLOR of the Royal Society remarked that "The development of a civilization depends on new materials." This statement remains true in today's modern world of technology and will undoubtedly continue to be valid in the future.

The three general categories of materials are metals, ceramics (inorganic materials), and polymers. This broad classification is based mainly on the nature of the chemical bonds, which are metallic, ionic, and covalent. However, these latter divisions do not have particularly distinct boundaries in most cases, because the chemical bond is often a hybrid of these three types. For example, most covalent bonds have some degree of ionic character

0065–2393/90/0224–0207$06.00/0

because of the differences in electronegativity or electron affinity of the bonded atoms, and metallic character can be present, depending on the level of electron mobility.

Through the combinations of different types of primary and secondary bonding and processing methods, materials with new properties can be produced for electrical, optical, structural, or related applications. This chapter addresses some of the principles of developing new hybrid materials by combining inorganic metal alkoxides and organic oligomers and polymers. These species are combined by using the general sol–gel reaction scheme familiar to ceramic chemists. However, in addition to the sol–gel reaction, other approaches such as ring opening and free-radical polymerization may also be incorporated. Under appropriate conditions, the combined reactions of the different precursor species can be used to develop high-molecular-weight networks, which, eventually, lead to the production of hybrid solids.

This chapter will not belabor in great detail the synthesis and structure–property behavior of those systems that have been prepared within our laboratory. Rather, some of the basic concepts involved in preparing these hybrid sol–gel materials will be reviewed. An overview of methods, including those routes taken by other researchers, will be presented. However, greater emphasis will be given to those materials that we have prepared, primarily because only limited structure–property studies have been carried out by other workers in the field. Further details of our studies on many of the hybrid materials (initially denoted as "ceramers") can be found elsewhere (*1–9*).

General Considerations of the Sol–Gel Process

The sol–gel reaction has been used to prepare inorganic–organic hybrid materials. In this general reaction, hydrolysis and condensation of a metal alkoxide species such as tetraethylorthosilicate (TEOS) take place, and a network is formed in the process. During the build-up of this inorganic network, appropriately functionalized organic (or potentially organic–inorganic) moieties that can also undergo the same condensation reaction as the hydrolyzed metal alkoxides are also incorporated. This method can lead to either an alloylike material, if molecular dispersion is obtained, or a system with more of a microphase morphological texture.

To understand the preparation of hybrid solids by the sol–gel method, the principles of the general sol–gel reaction (in the sense of inorganic constituents only) must be understood. In the last two decades, a tremendous amount of research effort has been devoted to the sol–gel area, and great progress has been made in understanding the reaction mechanisms. We will not do a thorough review of the entire sol–gel field; interested readers may refer to other studies and reviews (*10–26*). However, a general explanation of the sol–gel reaction is appropriate.

The Sol–Gel Reaction. The general sol–gel reaction scheme is composed of a series of hydrolysis steps in conjunction with condensation steps (Scheme I). Both the hydrolysis and condensation steps generate low-molecular-weight byproducts. These byproducts often cause serious complications in the preparation of thick crack-free solid materials because of the drying stresses incurred by their removal.

For example, Scheme I shows that hydrolysis of a silicon alkoxide results in the generation of an alcohol such as ethanol (in the case of TEOS). This small molecule must be removed from the system for suitable network development and solidification. Such removal would lead, in the limit, to a tetrahedral SiO_2 network. In addition, the condensation reaction of hydrolyzed silicon alkoxide results in the generation of water, which must also be removed. However, water can also promote additional hydrolysis during the reaction, as is obvious from Scheme I. Finally, Scheme I indicates that these reactions are promoted in either acidic or alkaline environments.

If these catalytic species are in sufficient quantity and are retained, the thermochemical stability and perfection of the final network may suffer. Indeed, the nature of the catalyst and, in particular, the pH of the reaction have a very pronounced effect on the hydrolysis and condensation reactions. These effects have been well reviewed in the literature (*27–31*), but the kinetics of individual reactions under acidic or basic conditions needs to be elucidated further. In fact, Chapter 13 by Keefer in this volume addresses the influence of pH on the nature of the network build-up that occurs in the hydrolysis and condensation of common metal alkoxides such as TEOS. In brief, the presence of acid tends to produce more-linear or polymerlike molecules in the initial stages. These molecules would then coalesce in the drying process and eventually form a high-density material. In contrast, an alkaline environment tends to produce more of a dense-cluster (Eden cluster) growth leading to dense particulatelike structures. These particles would later bridge to form a very inhomogeneous system in terms of density fluctuations.

$$Si(OR)_4 + H_2O \leftrightarrows (HO)Si(OR)_3 + ROH$$

$$(HO)Si(OR)_3 + H_2O \leftrightarrows (HO)_2Si(OR)_2 + ROH$$

$$(HO)_2Si(OR)_2 + H_2O \leftrightarrows (HO)_3Si(OR) + ROH$$

$$(HO)_3Si(OR) + H_2O \leftrightarrows Si(OH)_4 + ROH$$

$$\equiv Si{-}OR + HO{-}Si\equiv \leftrightarrows \equiv Si{-}O{-}Si\equiv + ROH$$

$$\equiv Si{-}OH + HO{-}Si\equiv \leftrightarrows \equiv Si{-}O{-}Si\equiv + HOH$$

Scheme I. Sol–gel reaction scheme. All reactions are catalyzed typically by the presence of an acidic or an alkaline medium.

Factors Affecting Sol–Gel Reactions. Figure 1 illustrates the general influence of pH on the hydrolysis and condensation reactions. This figure should be viewed cautiously, because it is only a general scheme and does not indicate the relative rates of hydrolysis or condensation as each individual alkoxide group in a given metal alkoxide species is removed or reacted. Further details regarding the relative rates of reaction as followed by NMR can be found elsewhere (*32*). Finally, in the later stages of the network-forming reaction, further restrictions on functional-group reactivity are controlled by steric and diffusion effects, as well as by the nature of the chemical environment (such as local pH and vitrification).

Mixed metal alkoxide systems are also of interest as a means of creating additional hybrid systems. However, recognition of the large differences in their hydrolysis and condensation rates is crucial. For example, if titanium isopropoxide is made to react under the same conditions as might be used for TEOS, hydrolysis and condensation rapidly occur and lead to particulate rather than network formation of TiO_2. Cocondensation with TEOS under these conditions does not occur because of the fast precipitation of the titanium dioxide species. Indeed, of the general metal alkoxides, those based on silicon tend to be more easily controlled because of their slower hydrolysis

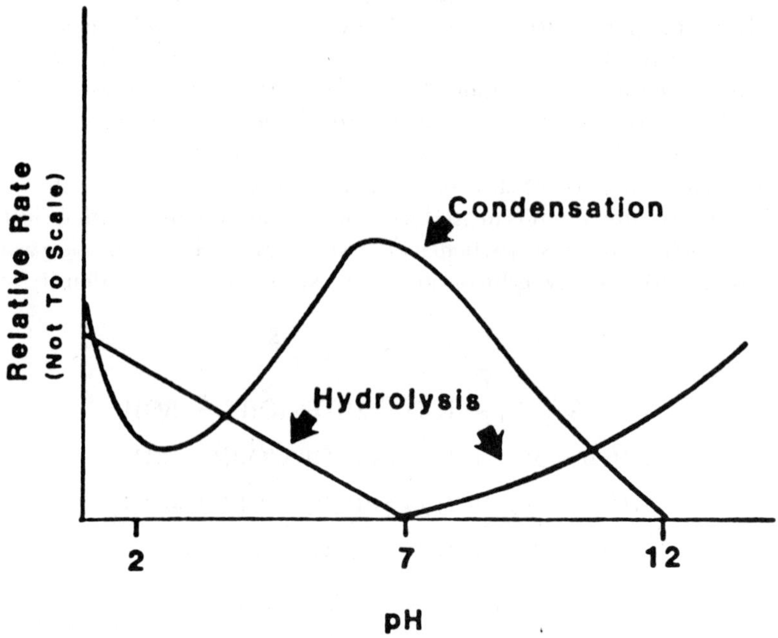

Figure 1. General effect of pH on the relative rates of hydrolysis and condensation for TEOS. (Reproduced with permission from reference 31. Copyright 1987 American Institute of Physics.)

and condensation rates relative to the majority of other metal alkoxides (*33*). However, some control of reactivity of any given metal alkoxide is possible through the choice of the alkoxy group that influences the polarizability of the metal–alkoxide bond. In addition, the size of the alkoxy group can influence the level of reactivity through steric or leaving-group-stability effects. Therefore, a species such as tetramethylorthosilicate (TMOS) tends to be more reactive than TEOS. Discussion of these points has been given by Bradley et al. (*34*).

Control of Reactivity. One way to overcome large differences in the reactivity of two metal alkoxide species is to use a chemically controlled condensation (CCC) procedure, which was first proposed by Schmidt and Seiferling (*35*). In their procedure, hydrolysis of the fast-reacting alkoxide species is slowly initiated by the controlled release of water from the esterification of an organic acid with an alcohol. The starting reaction mixture contains only the reacting alkoxides and the chosen acid–alcohol system (no water). Once the fast-reacting alkoxide has been partially hydrolyzed and condensed, water is added to complete the overall reaction and to incorporate the slower reacting alkoxide (such as TEOS).

A different method was used by Parkhurst et al. (*36*) to incorporate titanium ethoxide into TEOS-based silica gels. In their procedure, the TEOS species is allowed to hydrolyze partially and condense in the presence of an acid catalyst and water before the fast-reacting titanium ethoxide is added. Once introduced, the titanium ethoxide quickly hydrolyzes and condenses into the preexisting immature TEOS-based network rather than precipitating as titanium dioxide.

In our laboratory, we have developed a hybrid procedure that uses aspects of both of these earlier methods to incorporate titanium isopropoxide into TEOS-based gels that contain functionalized oligomeric species of either poly(tetramethylene oxide) (PTMO) or poly(dimethylsiloxane) (PDMS). In our approach (Scheme II), the TEOS moiety, as well as the triethoxysilane-end-capped PTMO or silanol-terminated PDMS species, is partially hydrolyzed and condensed by using a CCC system of glacial acetic acid and isopropyl alcohol (no water). The titanium isopropoxide is added later. Thereafter, the entire system is gently heated (refluxed) to facilitate the completion of the hydrolysis and condensation reactions. With this method, very transparent materials can be produced (*7, 9*). This result indicates a good dispersion of the titanium species, at least on a size scale distinctly smaller than the wavelength of light; otherwise, high turbidity would result because of the differences in the refractive indices of the components. Further details on the nature of these materials will be discussed later in the chapter.

The CCC route may be applied to other mixtures of metal alkoxides possessing quite different reactivities; Schmidt and Seiferling (*35*) have commented on this possibility for zirconium alkoxides. Limited data have been

hydrolysis

$$\mathrm{Si(OR)_4 + 4H_2O \xrightarrow{H+} Si(OH)_4 + 4ROH}$$

$$\mathrm{(RO)_3Si{-}(PTMO){-}Si(OR)_3 + 6H_2O \xrightarrow{H+} (HO)_3Si{-}(PTMO){-}Si(OH)_3 + 6ROH}$$

polycondensation

$$\mathrm{Si(OH)_4 + (HO)_3Si{-}(PTMO){-}Si(OH)_3 \xrightarrow{H+}}$$

$$\begin{array}{ccccccccc} & | & & | & & | & & | & \\ & \mathrm{O} & & \mathrm{O} & & \mathrm{O} & & \mathrm{O} & \\ & | & & | & & | & & | & \\ - & \mathrm{Si} & -\mathrm{O}- & \mathrm{Si} & -\mathrm{(PTMO)}- & \mathrm{Si} & -\mathrm{O}- & \mathrm{Si} & - \; + \; \mathrm{H_2O} \\ & | & & | & & | & & | & \\ & \mathrm{O} & & \mathrm{O} & & \mathrm{O} & & \mathrm{O} & \\ & | & & | & & | & & | & \end{array}$$

titanium incorporation

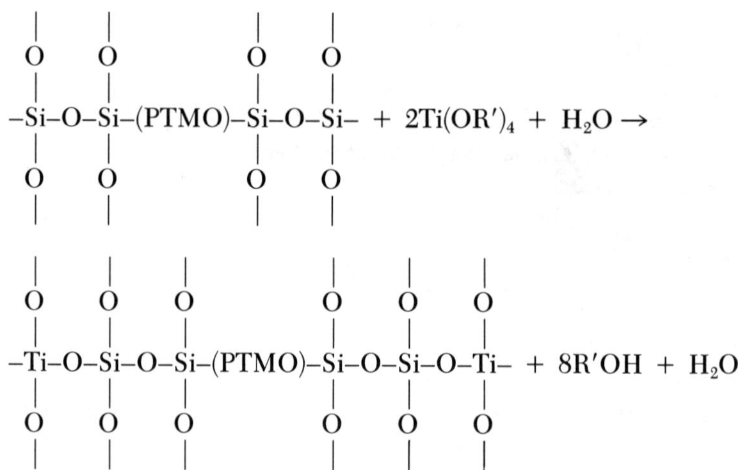

Scheme II. Titanium incorporation in TEOS-based gels containing functionalized oligomers. R is C_2H_5, and R′ is $CH(CH_3)_2$. The reactions for polycondensation and titanium incorporation are not balanced stoichiometrically.

presented on the structure of the final materials prepared. However, the data indicate that good dispersion may have been obtained. Clearly, the general CCC approach allows the building of systems with high dispersion of a variety of metal oxide species—a novel route for new materials.

Control of Morphology. An interesting modification for preparing somewhat related but different hybrid materials by using the sol–gel reaction of TEOS (although other metal alkoxides might be used) has been undertaken by Mark et al. (*37–40*). In their work, the starting material is a cross-linked PDMS elastomer network prepared by an end-linking procedure on end-capped PDMS. Such networks can be prepared by a number of methods (*39, 40*). The important feature is that once this network has been prepared, it is swollen by TEOS in the presence of alcohol. After swelling, the TEOS-imbibed network can be exposed to an alkaline or acidic environment. Because of the diffusion of the water and catalyst, the pH changes accordingly, and the reaction of TEOS takes place. This reaction will promote the development of inorganic SiO_2 (not necessarily pure SiO_2) particulates or oligomerlike species within the network. Such species, if developed into particulate form, can serve as an in situ means of preparing reinforced PDMS elastomers, in contrast to the conventional route of adding silica fillers by compounding.

As Mark et al. (*37, 38*) have shown, an alkaline catalyst such as ethylamine will promote the particulate species of condensed TEOS with a size in the order of 10 nm. These particles are well dispersed in the PDMS matrix, as has been demonstrated by transmission electron microscopy (TEM). An example of this morphology is shown in Figure 2, which illustrates the good dispersion of SiO_2 particles as promoted by alkaline catalysis. In contrast, if an acidic catalyst such as glacial acetic acid is used, the final morphology is quite different and is less well defined, as indicated by the TEM studies (*38*). This result is not surprising, in view of the fact that linear or oligomerlike species tend to form from the hydrolysis and condensation reactions of TEOS in an acidic environment. Therefore, acid catalysis is less amenable to distinct SiO_2 particulate formation in the swollen networks and, hence, leads to poor contrast in the TEM micrographs.

Hybrid Materials Incorporating Monomeric or Functionalized Oligomeric Species

The general sol–gel reaction has been modified to prepare hybrid materials (ceramers) incorporating functionalized monomers or oligomers. The functional groups can react, or partially react, with the hydrolysis products of the metal alkoxide (e.g., the silanol groups of hydrolyzed TEOS) to incorporate the oligomer or monomer of interest. With this approach, a new range of material properties can be produced by combining the features of

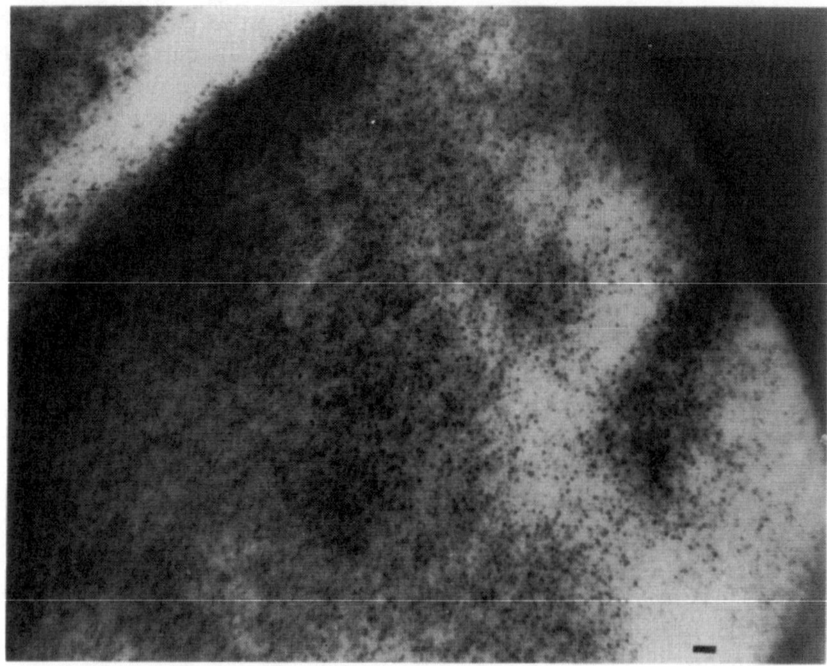

Figure 2. Transmission electron micrograph of hydrolyzed and condensed TEOS particulates dispersed in a PDMS network as promoted by an alkaline catalyst. (Reproduced with permission from reference 38. Copyright 1985 Butterworth.)

the inorganic sol–gel alkoxide moieties with those of a variety of organic or inorganic species.

Incorporation of PDMS Oligomers. One successful system prepared in our laboratory (*1–4*) is based on the incorporation of low-molecular-weight (550–1700) PDMS oligomers terminally functionalized with silanol groups into a TEOS network. In brief, if the oligomeric species can be maintained in solution with the reactive sol–gel components and if hydrolysis of the metal alkoxide is sufficiently rapid so as to provide hydrolysis products of the alkoxide for reaction with the silanol-terminated oligomers, good dispersion of the functionalized oligomer in the final network of the two components can be achieved. If, however, the hydrolysis of the alkoxide is not sufficiently rapid, self-condensation of the functionalized oligomers leads to chain extension and molecular weight build-up and finally results in poor dispersion of the oligomeric or polymeric component.

Earlier studies (*1–4*) reported from this laboratory demonstrated that functionalized PDMS can be successfully incorporated into a hybrid network by using TEOS as the metal alkoxide. When analyzed for structural features,

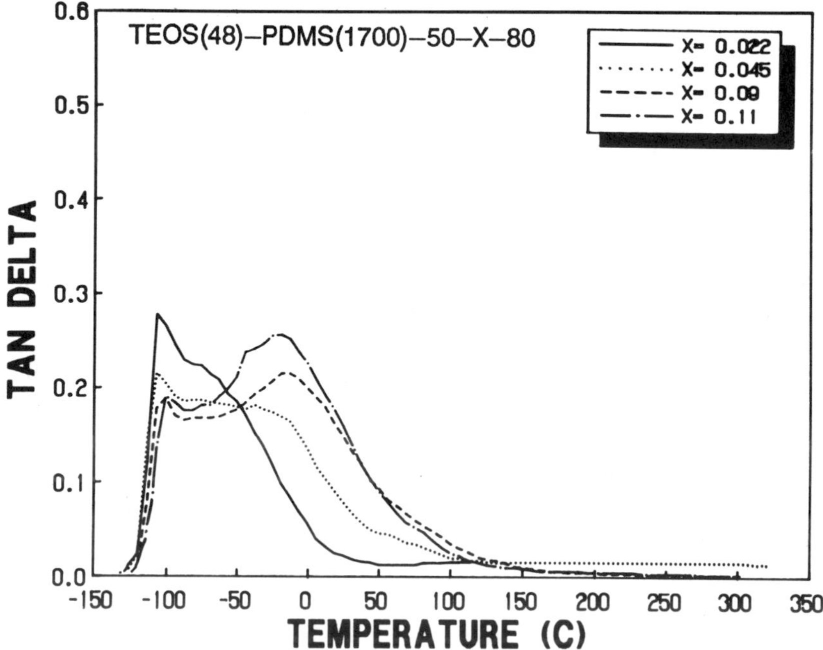

Figure 3. Effect of acid catalyst (HCl) content on the temperature dependence of the dynamic mechanical parameter tan δ for a series of TEOS(48)–PDMS(1700)–50–X–80C materials. See *text for sample nomenclature. (Reproduced from reference 4. Copyright 1987 American Chemical Society.)*

these materials showed some degree of localized phase separation of the PDMS component. However, the dispersion of the oligomer into the alkoxide-based material could be enhanced by higher levels of the acid catalyst. A higher concentration of acid catalyst would promote a higher rate of hydrolysis and provide a greater concentration of TEOS hydrolysis products to react with the silanol-functionalized oligomers. Therefore, better overall dispersion of the oligomer is achieved. Proof of this improved dispersion has been obtained and presented on the basis of small-angle X-ray-scattering (SAXS) measurements in conjunction with dynamic mechanical measurements (*3*, *4*) and, more recently, solid-state NMR investigations (Glaser, R. H.; Huang, H.; Wilkes, G. L.; Bronnimann, C. E., unpublished results).

As one example, Figure 3 shows the ratio of loss to storage modulus (tan δ) in the dynamic mechanical spectra of a series of materials made with a silanol-terminated PDMS (weight-average molecular weight [M_w] 1700) and TEOS as a function of acid catalyst content. [With regard to the sample nomenclature used in Figure 3, TEOS(48)–PDMS(1700)–50–0.045–80C indicates that the material was prepared with 48 wt % of TEOS, 52 wt % of the functionalized oligomer (PDMS in this case) with M_w of 1700, 50% of the stoichiometric amount of water for the hydrolysis reaction, an HCl/TEOS molar ratio of 0.045, and a reaction temperature of 80 °C. This nomenclature is used throughout this chapter.]

The tan δ data in Figure 3 indicate that as the acid catalyst content increases, the glass transition region (dispersion) associated with the general PDMS oligomer becomes highly broadened (The glass transition temperature [T_g] of pure PDMS is approximately −120 °C.) This broadening at considerably higher temperatures indicates restrictions on molecular mobility and indirectly suggests further oligomeric dispersion into the more-immobile cross-linked condensation products of the TEOS components.

Figure 4 shows a series of SAXS scans for the materials just discussed. The variable *s* is an angular-dependent parameter defined as $(2/\lambda)\sin(\theta/2)$, in which θ is the radial scattering angle and λ is the wavelength of X-ray (in nanometers). *I*(*s*) represents the smeared intensity. The decreasing magnitude of the scattering power (as indicated by the area under the scattering curves) suggests progressively more homogeneous materials with increasing catalyst content, that is, the scattering approaches very low levels as acid content increases. An invariant analysis of these same SAXS scans gives more positive support to this conclusion (*4*).

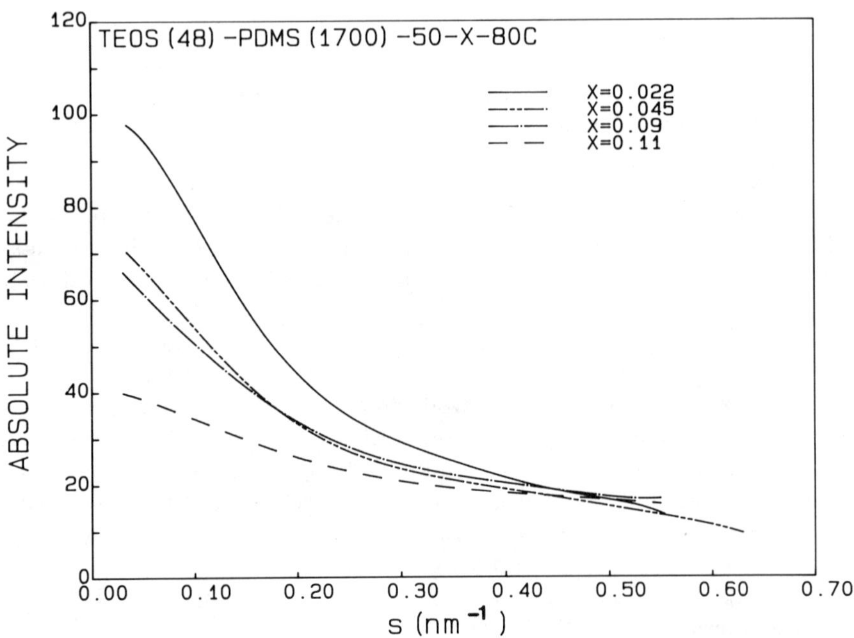

Figure 4. SAXS profiles for TEOS(48)–PDMS(1700)–50–80C materials with different levels of acid catalyst. See text for sample nomenclature. (Reproduced from reference 4. Copyright 1987 American Chemical Society.)

NMR and Raman data have indicated nearly equivalent levels of network formation for all of these systems. This result suggests that comparisons of the general SAXS intensity profiles of these materials indeed provide a comparative indication of increasing homogeneity (better mixing) of the TEOS and siloxane components as acid catalyst content increases. Although further work is needed to support this conclusion, these subtle changes in catalyst content apparently have significant effects on mechanical response. One example is the dynamic mechanical response indicated in Figure 3. Further indications of systematic effects on mechanical response with chemical composition have been presented elsewhere (*4*).

Structural Modification of Oligomers. More recently, data from our laboratory have indicated that if siloxane oligomers are end capped with triethoxysilane groups (Scheme III), oligomers with higher functionality can be obtained, and therefore, the chances of reacting with the hydrolyzed products of TEOS or other metal alkoxides are higher. Figure 5 illustrates a material prepared from such a multifunctionalized PDMS oligomer with a M_w of 1000 with TEOS. The dynamic mechanical data indicate that no significant degree of softening occurs in the temperature range expected for pure PDMS. This finding indicates that the oligomer must be highly im-

hydrolysis

$$\mathrm{Si(OR)_4 + 4H_2O \xrightarrow{H+} Si(OH)_4 + 4ROH}$$

$$\mathrm{(RO)_3Si\text{–}(PDMS)\text{–}Si(OR)_3 + 6H_2O \xrightarrow{H+} (HO)_3Si\text{–}(PDMS)\text{–}Si(OH)_3 + 6ROH}$$

polycondensation

$$\mathrm{Si(OH)_4 + (HO)_3Si\text{–}(PDMS)\text{–}Si(OH)_3 \xrightarrow{H+}}$$

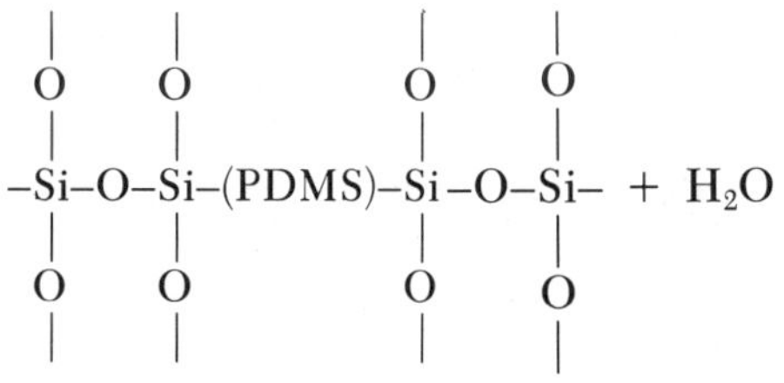

$$\begin{array}{c} | \quad\ \ | \qquad\qquad\qquad\ \ | \quad\ \ | \\ \mathrm{O} \quad \mathrm{O} \qquad\qquad\qquad \mathrm{O} \quad \mathrm{O} \\ | \quad\ \ | \qquad\qquad\qquad\ \ | \quad\ \ | \\ \mathrm{\text{–}Si\text{–}O\text{–}Si\text{–}(PDMS)\text{–}Si\text{–}O\text{–}Si\text{–}} \\ | \quad\ \ | \qquad\qquad\qquad\ \ | \quad\ \ | \\ \mathrm{O} \quad \mathrm{O} \qquad\qquad\qquad \mathrm{O} \quad \mathrm{O} \\ | \quad\ \ | \qquad\qquad\qquad\ \ | \quad\ \ | \end{array} + \mathrm{H_2O}$$

Scheme III. Use of end-capped siloxane oligomers. R is C_2H_5. The polycondensation reaction is not balanced stoichiometrically.

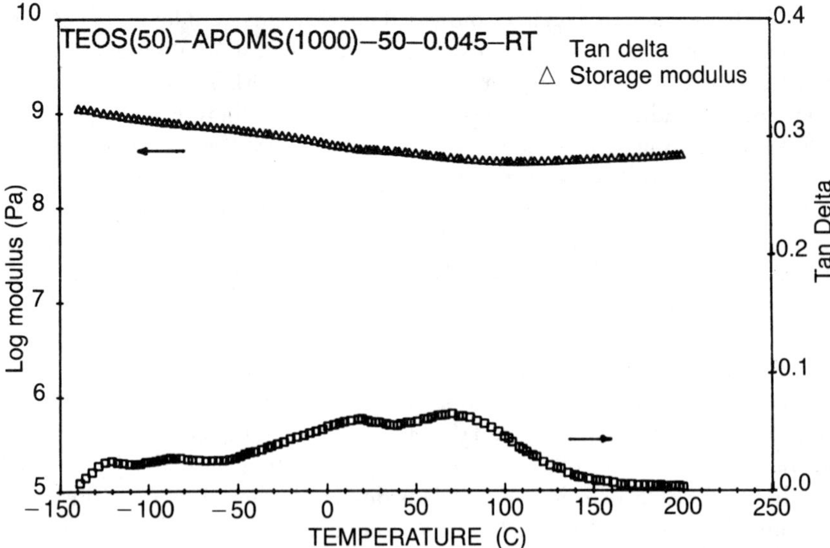

Figure 5. Dynamic mechanical behavior of a triethoxysilane-end-capped PDMS oligomer with a M_w *of 1000 reacted with TEOS: log storage modulus* (E′) *and tan* δ.

mobilized through cross-linking with the hydrolysis products of TEOS. Also, the lack of a well-defined glass transition suggests a good dispersion of the oligomer.

The use of functional groups comparable to those that exist on the alkoxide group is certainly a logical method of preparing hybrid systems. Schmidt and Seiferling (*35*) have suggested similar procedures for producing hybrid materials by using oligomeric species that contain unreacted functional groups that can be reacted (or triggered) after the sol–gel process is complete. For example, Scheme IV illustrates the coupling of an oligomeric species terminated with one functional group that can react with the metal alkoxide, with the other terminal point containing an allyl or a vinyl group that could later undergo a free-radical or ionic reaction. The reaction of such allyl or vinyl groups could lead to further chain extension, branching, or cross-linking, depending on the nature of the reactive species within the media (e.g., a species such as a divinylbenzene would provide a means of additional network formation by the free-radical approach). Further discussion of the work by Schmidt et al. (*35, 41*) is given elsewhere and will not be reviewed here, except to indicate that multiple reaction schemes (free radical, ring opening, etc.) might be utilized if the reactive medium so permits.

Incorporation of PTMO Oligomers. Another approach used in our studies is to incorporate oligomeric species whose chemistry contrasts to

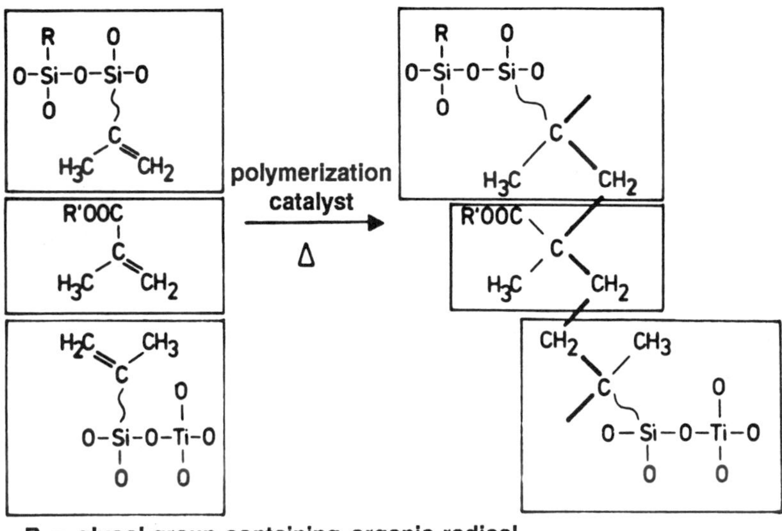

Scheme IV. Use of oligomers with unreacted functional groups. (Reproduced with permission from reference 41. Copyright 1984 Elsevier.)

that of the metal alkoxide. In particular, we have used triethoxysilane-functionalized PTMO oligomers rather than the silanol-terminated PDMS oligomers discussed earlier. These functionalized PTMO systems have been prepared with a variety of molecular weights and then reacted with TEOS, TMOS, or mixed metal alkoxide systems of TEOS and titanium isopropoxide (*5, 7–9*). This work shows that a variety of chemical reactions can be used to generate a diversity of material characteristics. Indeed, the functionalized PTMO oligomers react well with the hydrolysis products of the metal alkoxide; however, they tend to form a local phase separation or microdomain structure somewhat similar to the segmented or block polymers familiar to the polymer chemists (*42–44*). A distinct indication of this local phase separation is provided by Figure 6, which shows the SAXS profiles. In these profiles, distinct maxima (although broad because of slit smearing) strongly indicate a correlation distance over which a periodic fluctuation in electron density occurs. Such a shoulder or pseudo-Bragg peak appears in the PTMO-based systems, whereas it is absent in the PDMS-containing hybrid materials discussed earlier.

Analysis of these SAXS profiles suggests that this correlation distance (ca. $1/s$, an indicator of pseudo-Bragg spacing; Figure 6) or average periodicity in electron density fluctuations is in the order of about 10 nm. This length is not unreasonable when the general chain or coil dimensions of the

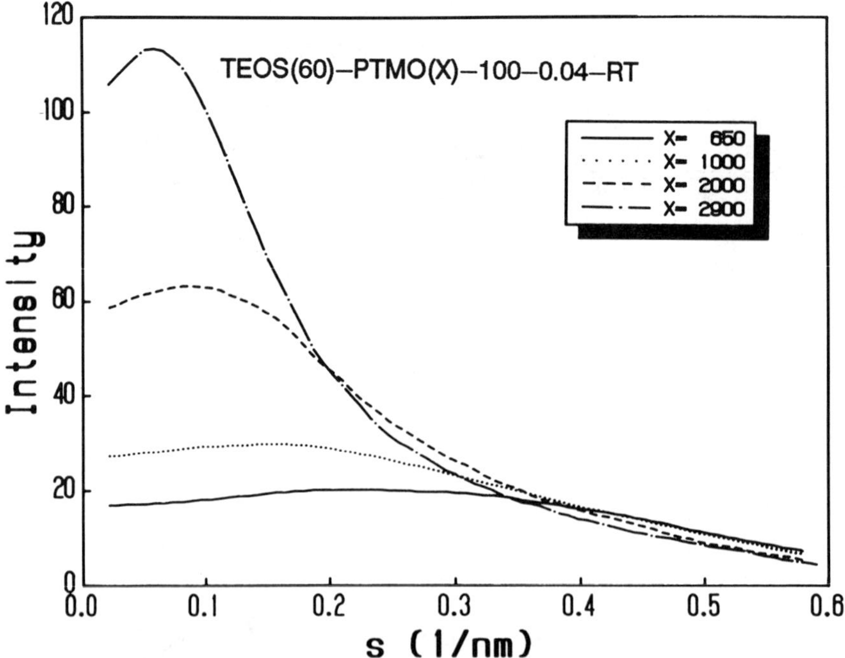

Figure 6. Effect of molecular weight of PTMO oligomer on the SAXS profiles of a TEOS(60)–PTMO(X)–100–0.04–RT system. RT is room temperature. See *text for sample nomenclature.*

PTMO oligomers are taken into account, along with the size of clusters that may form from the highly condensed hydrolysis products of TEOS (*7*).

A simplified model based on the interpretation of the mechanical data in conjunction with the SAXS results is provided in Figure 7. This highly schematic model indicates that compositional variations in the materials occur over a correlation distance given approximately by the value $1/s$. The general matrix is probably rich in PTMO, although some mixing with partially condensed TEOS is likely because of its broadened and raised glass transition response, whereas the dispersed regions are richer in TEOS condensation products. Further details regarding the justification for this model can be found elsewhere (*7*). Similar SAXS behavior is found for those systems containing titanium isopropoxide, although these materials are made by using the modified CCC method described earlier (*7*).

In contrast to the materials prepared by using PDMS oligomers, the PTMO-containing materials typically provide a much more enhanced strain-to-break mechanical response, as well as tensile strength, for an equivalent volume content (*4*, *7*). An indication of this behavior is shown in Figure 8, which illustrates three examples of the stress–strain behavior of materials prepared with TEOS and functionalized PTMO(2000) at constant water and

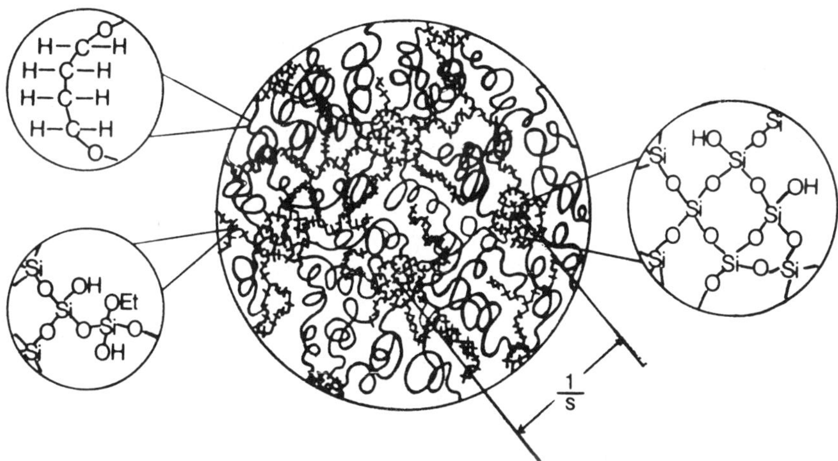

Figure 7. Proposed model of the local morphological structure in the PTMO–TEOS hybrid systems, in which 1/s is an estimate of the correlation distance. (Reproduced from reference 7. Copyright 1987 American Chemical Society.)

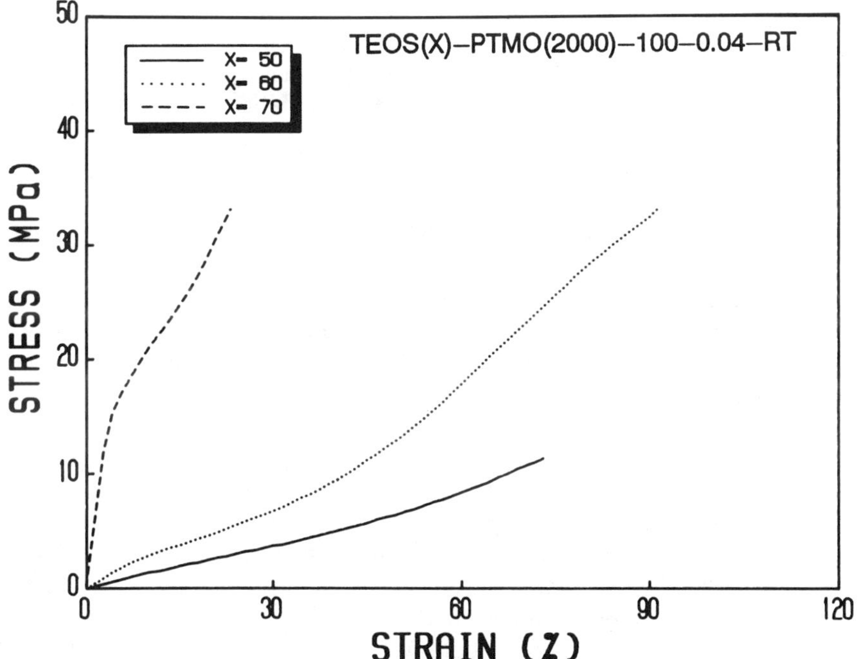

Figure 8. Effect of TEOS content (X) on the tensile behavior of a series of TEOS(X)–PTMO(2000)–100–0.04–RT materials. RT is room temperature. See *text for sample nomenclature.*

acid contents. The initial weight percentage of added TEOS ranged from 50 to 70 wt %. The data indicate clearly that the strain-at-break ranges from several percent to over 100%, whereas the tensile strength may be as high as 30 MPa. In contrast, the tensile strength measured for materials incorporating PDMS-functionalized oligomers terminated with single silanol groups never exceeds 6 MPa, and the strain-at-break typically is less than 20% (*4*). The use of triethoxysilane groups, which provides higher functionality to the organic oligomers and therefore more potential links with the hydrolysis products of TEOS, results in a considerable enhancement in mechanical properties.

The PTMO-containing materials display a glass transition dispersion that is broader and higher than that of the pure PTMO component. The T_g of pure PTMO of reasonable molecular weight is about –75 °C (*45*). Figures 9a and 9b show the dynamic mechanical behavior (log storage modulus [E'] and tan δ) of a series of PTMO-containing materials as a function of TEOS content. These materials were made with a functionalized PTMO(2000) and various TEOS contents, with the acid and water contents held constant.

The data in Figure 9a indicate a systematic change in mechanical behavior as restrictions on PTMO mobility increase as the result of a greater metal alkoxide content. As either Figure 9a or 9b indicates, the glass transition dispersion behavior of the PTMO oligomers and the general overall modulus temperature response vary quite dramatically with changing TEOS content. Indeed, with TEOS at 10 wt %, the general softening or glass transition response of the network prepared by the sol–gel approach is typical of pure PTMO. In fact, an enhancement of modulus at about –50 °C is observed. This rise is caused by partial crystallization of the oligomers, a common response of PTMO oligomers in this molecular weight range (*46*). Subsequent melting of these crystals occurs at about –25 °C. However, as the TEOS content increases above 30 wt %, not only is the glass transition dispersion broadened and shifted to higher temperatures but the inducement of crystallizability (as indicated by a rise in modulus after the glass transition) is completely eliminated. This lack of crystallizability is due, without doubt, to restrictions caused by the presence of the TEOS-based components interacting with the PTMO oligomers.

One of the unfortunate features associated with the PTMO-containing systems is slow chemical aging at ambient temperature after gelation and solidification. This aging is believed to be caused by the glass transition dispersion of the PTMO, which, in these hybrid systems, is broadened into the ambient-temperature region. This broadened dispersion restricts local mobility caused by apparent localized vitrification, which in turn limits reactivity and further network development. Unless the materials are annealed at higher temperatures, these systems will undergo continued slow network development over several weeks, which will be limited only by the diffusion and mobility restrictions in the near-glassy environment.

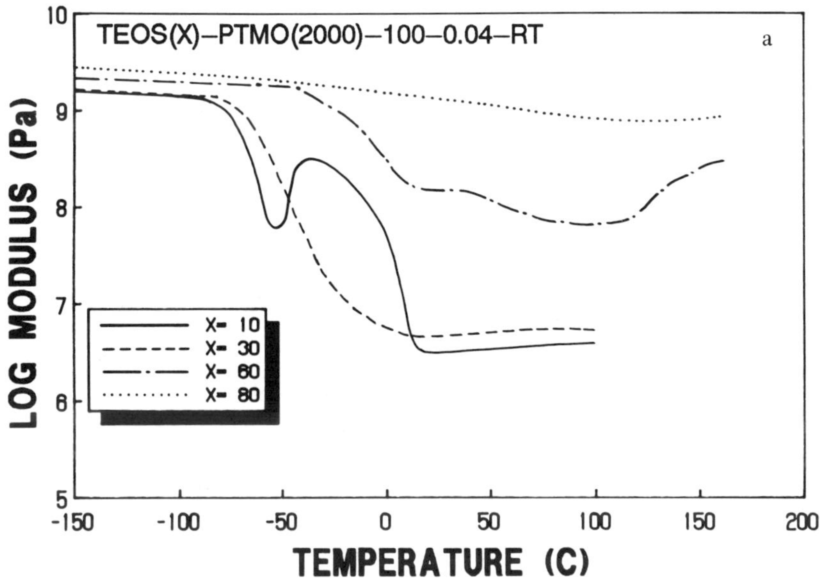

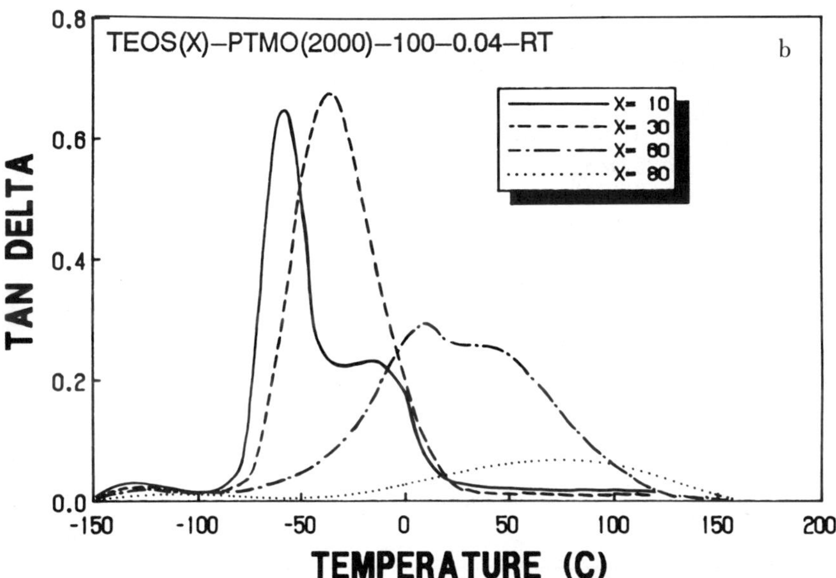

Figure 9. Effect of TEOS content (X) *on the dynamic mechanical behavior of a series of TEOS(X)–PTMO(2000)–100–0.04–RT materials: a, log storage modulus; b, tan δ. RT is room temperature.* See *text for sample nomenclature.*

An example of this effect is illustrated in Figure 10, which shows a series of stress–strain curves as a function of time for a material made with triethoxysilane-end-capped PTMO(2000) (50 wt %) in conjunction with titanium isopropoxide (30 wt %) and TEOS (20 wt %). Over a period of many days, the material clearly stiffens in terms of modulus build-up; hence, the time-dependent mechanical characteristics would have to be recognized for any application. However, such materials can be further cured thermally to achieve a more stable mechanical response with time. Therefore, if this aging phenomenon is noted, it can be dealt with accordingly.

Incorporation of Other Oligomeric Species. Oligomeric species that are glassy in pure form at ambient conditions can also be incorporated into sol–gel network systems, in addition to those possessing a more rubberlike backbone such as PDMS or the PTMO components discussed in the previous sections. Indeed, materials incorporating triethoxysilane-functionalized poly(ether ketone) (PEK) oligomers (which possess a T_g in the order of 180 °C in the pure form) have been successfully made through a sol–gel reaction with TEOS. The limitations on the extent of reaction (cross-linking) in such a system because of the more glasslike or stiffer oligomer are clearly much greater than those imposed by the broadened T_g of the PTMO dis-

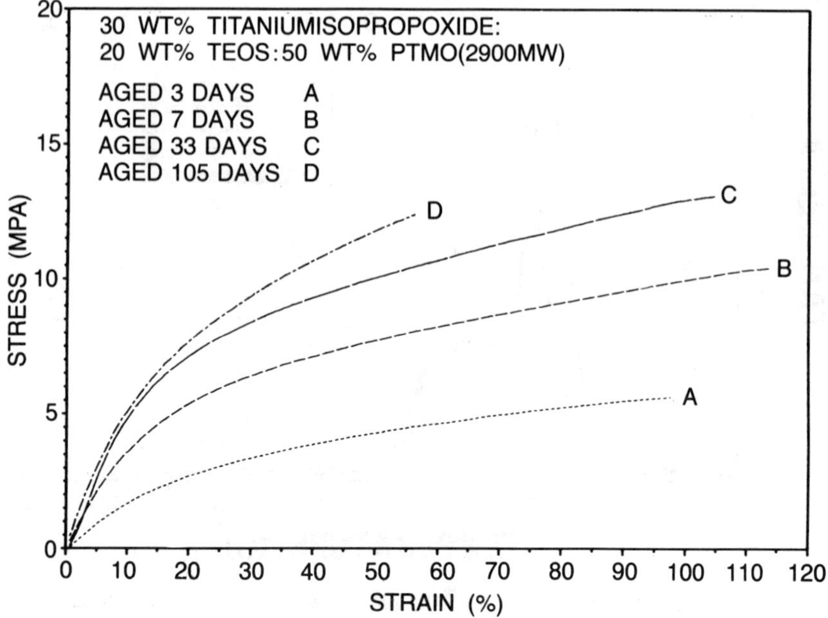

Figure 10. Effect of aging on the tensile behavior of a titanium isopropoxide–TEOS–PTMO(2900) (30:20:50 wt %) system after casting and drying at 22 °C.

cussed previously. The level of network development can be extremely restricted, because vitrification is approached through molecular weight build-up in a way similar to that explained by Gillham (*47*) for common glassy network systems such as epoxides, etc. As mentioned earlier, thermal treatment of such materials that allows further curing will result in network enhancement that will raise mechanical integrity and glass transition behavior accordingly. Studies concerning materials made by the incorporation of triethoxysilane-functionalized PEK oligomers into TEOS will be reported shortly (Noell, J. L.; Mohanty, D. K.; Wilkes, G. L. *J. Appl. Polym. Sci.*, in press). In brief, their structure has a microdomain-type morphology similar to that of the PTMO-functionalized oligomers described earlier.

Conclusion

Hybrid network materials that use the sol–gel reaction scheme familiar to ceramists and inorganic chemists can be prepared. The final properties of these materials depend on the choice of metal alkoxide and functionalized oligomer. This method of preparing new and novel materials is in its early stages and will, with time, lead to some useful applications in structural and electronic-related materials, new catalyst or porous supports for chromatography, etc. The range of chemistry available for producing such hybrid organic–inorganic systems is immense. We hope that this brief overview will stimulate new research for the design of new materials accordingly.

Acknowledgment

We acknowledge the partial financial support of this work through the Office of Naval Research.

References

1. Wilkes, G. L.; Orler, B.; Huang, H. *Polym. Prepr.* (*Am. Chem. Soc., Div. Polym. Chem.*) **1985,** *26*, 300.
2. Huang, H.; Orler, B.; Wilkes, G. L. *Polym. Bull.* **1985,** *14*, 557.
3. Huang, H.; Glaser, R. H.; Wilkes, G. L. *Polym. Prepr.* (*Am. Chem. Soc., Div. Polym. Chem.*) **1987,** *28*, 434.
4. Huang, H.; Orler, B.; Wilkes, G. L. *Macromolecules* **1987,** *20*, 1322.
5. Glaser, R. H.; Wilkes, G. L. *Polym. Prepr.* (*Am. Chem. Soc., Div. Polym. Chem.*) **1987,** *28*, 236.
6. Huang, H.; Wilkes, G. L. *Polym. Prepr.* (*Am. Chem. Soc., Div. Polym. Chem.*) **1987,** *28*, 244.
7. Huang, H.; Glaser, R. H.; Wilkes, G. L. In *Inorganic and Organometallic Polymers;* Zeldin, M.; Wynne, K. J.; Allcock, H. R., Eds.; ACS Symposium Series 360; American Chemical Society: Washington, DC, 1987; p 354.
8. Huang, H.; Wilkes, G. L. *Polym. Bull.* **1987,** *18*, 455.
9. Glaser, R. H.; Wilkes, G. L. *Polym. Bull.* **1988,** *19*, 51.

10. Mukherjee, S. P. *J. Non-Cryst. Solids* **1980,** *42,* 477.
11. Brinker, C. J.; Keefer, K. D.; Schaefer, D. W.; Ashley, C. S. *J. Non-Cryst. Solids* **1982,** *48,* 47.
12. Zarzycki, J. *J. Non-Cryst. Solids* **1982,** *48,* 105.
13. Yamane, M.; Inoue, S.; Nakazawa, K. *J. Non-Cryst. Solids* **1982,** *48,* 153.
14. Dislich, H. *J. Non-Cryst. Solids* **1983,** *57,* 371.
15. Klein, L. C.; Gallo, T. A.; Garvey, G. J. *J. Non-Cryst. Solids* **1984,** *63,* 23.
16. Yoldas, B. E. *J. Non-Cryst. Solids* **1984,** *63,* 145.
17. Wallace, S.; Hench, L. L. *Mater. Res. Soc. Symp. Proc.* **1984,** *32,* 47.
18. Prassas, M.; Phalippou, J.; Zarzycki, J. *J. Mater. Sci.* **1984,** *19,* 1656.
19. Brinker, C. J.; Scherer, G. W. *J. Non-Cryst. Solids* **1985,** *70,* 301.
20. Rabinovich, E. M. *J. Non-Cryst. Solids* **1985,** *71,* 187.
21. Gottardi, V. *J. Non-Cryst. Solids* **1985,** *73,* 625.
22. Sakka, S. *J. Non-Cryst. Solids* **1985,** *73,* 651.
23. Lacourse W. C.; Kim, S. *Mater. Res. Soc. Symp. Proc.* **1986,** *73,* 304.
24. Zerda, T. W.; Artaki, I.; Jonas, J. *J. Non-Cryst. Solids* **1986,** *81,* 365.
25. Assink, R. A.; Kay, B. D. *J. Non-Cryst. Solids* **1988,** *99,* 359.
26. Pope, E. J. A.; MacKenzie, J. D. *J. Non-Cryst. Solids* **1988,** *101,* 198.
27. Aelion, R.; Loebel, A.; Eirich, F. *J. Am. Chem. Soc.* **1950,** *72,* 5705.
28. Iler, R. K. *The Chemistry of Silica;* Wiley: New York, 1979.
29. Schmidt, H.; Scholze, H.; Kaiser, A. *J. Non-Cryst. Solids* **1984,** *63,* 1.
30. Keefer, K. D. *Mater. Res. Soc. Symp. Proc.* **1984,** *32,* 15.
31. Schaefer, D. W.; Wilcoxon, J. P.; Keefer, K. D.; Bunker, B. C.; Pearson, R. K.; Thomas, I. M.; Miller, D. E. *AIP Conf. Proc.* **1987,** *154,* 63.
32. Pouxviel, J. C.; Boilot, J. P.; Beloeil, J. C.; Lallemand, J. Y. *J. Non-Cryst. Solids* **1987,** *89,* 345.
33. Yoldas, B. E. *J. Mater. Sci.* **1979,** *14,* 1843.
34. Bradley, D. C.; Mehrotra, R. C.; Gaur, D. P. *Metal Alkoxides;* Academic: London, 1978.
35. Schmidt, H.; Seiferling, B. *Mater. Res. Soc. Symp. Proc.* **1986,** *73,* 739.
36. Parkhurst, C. S.; Doyle, L. A.; Silverman, L. A.; Singh, S.; Anderson, M. P.; McClurg, D.; Wnek, G. E.; Uhlmann, D. R. *Mater. Res. Soc. Symp. Proc.* **1986,** *73,* 769.
37. Ning, Y. P.; Tang, M. Y.; Jiang, C. Y.; Mark, J. E.; Roth, W. C. *J. Appl. Polym. Sci.* **1984,** *29,* 3209.
38. Mark, J. E.; Ning, Y. P.; Jiang, C. Y.; Tang, M. Y.; Roth, W. C. *Polymer* **1985,** *26,* 2069.
39. Mark, J. E.; Sullivan, J. L. *J. Chem. Phys.* **1977,** *66,* 1006.
40. Mark, J. E.; Jiang, C. Y.; Tant, M. Y. *Macromolecules* **1984,** *17,* 2613.
41. Philipp, G.; Schmidt, H. *J. Non-Cryst. Solids* **1984,** *63,* 283.
42. *Thermoplastic Elastomers;* Legge, N. R.; Holden, G.; Schroeder, H. E., Eds.; Hanser: New York, 1987.
43. Ophir, Z.; Wilkes, G. L. *J. Polym. Sci. Polym. Phys.* **1980,** *18,* 1469.
44. Koberstein, J. T.; Stein, R. S. *J. Polym. Sci. Polym. Phys.* **1983,** *21,* 1439.
45. Wetton, R. E.; Allen, G. *Polymer* **1966,** *7,* 331.
46. Huh, D. S.; Cooper, S. L. *Polym. Eng. Sci.* **1971,** *11,* 369.
47. Gillham, J. K. *Polym. Eng. Sci.* **1979,** *19,* 676.

RECEIVED for review May 27, 1988. ACCEPTED revised manuscript March 29, 1989.

13

Structure and Growth of Silica Condensation Polymers

Keith D. Keefer

Ceramics Development Division, Sandia National Laboratories, Albuquerque, NM 87185

Inorganic silicate polymers are prepared by the hydrolysis of silicon alkoxides and subsequent condensation of the resulting silanols. The structures can be described only in a statistical fashion by means of fractal geometry. The structures are controlled by the relative rates of the hydrolysis reactions that generate reactive silanol groups and the condensation reactions of those groups. Under alkaline conditions, hydrolysis tends to be slow with respect to condensation, and nonfractal colloidal structures form by a nucleation- and reaction-limited growth process. Under acidic conditions, hydrolysis tends to be fast with respect to condensation, and "polymeric" fractal structures with D_f *(mass fractal dimension) = 2.1 are formed by reaction-limited cluster aggregation. The exact structure in real systems may only approach these idealized cases.*

THE CONDENSATION POLYMERIZATION of tetrafunctional silicate monomers is the basis of the so-called "sol–gel" process for making glass. The goal is to produce dense, water- and carbon-free "glass" by a low-temperature, chemical route rather than by conventional high-temperature melting. These completely inorganic polymers ideally would have all of the desirable properties of fused silica: hardness, transparency, chemical durability, and thermal resistance. Low-temperature chemical processing would permit these materials to be used as protective or dielectric coatings on a variety of substrates, such as organic polymers or integrated circuits, which could not withstand contact with a molten glass.

The reactive functional groups in the polymerization reaction are sil-

0065–2393/90/0224–0227$06.00/0

anols, SiOH groups. These groups are generated concurrently with the polymerization reaction by the controlled hydrolysis of silicon compounds, typically alkoxides (equation 1).

$$Si(OR)_4 + H_2O \xrightarrow{H^+ \text{ or } ^-OH} Si(OR)_3OH + ROH \quad (1)$$

In this equation, R may be H, an alkyl group, or SiOR.

The silanols generated by the hydrolysis reaction spontaneously undergo a condensation polymerization to form a siloxane (S–O–Si) linkage and either a water molecule from the reaction of two silanol groups or an alcohol molecule from the reaction of a silanol with an alkoxide group:

$$2Si(OR)_3OH \rightarrow (RO)_3SiOSi(OR)_3 + H_2O \quad (2)$$

$$Si(OR)_3OH + Si(OR)_4 \rightarrow (RO)_3SiOSi(OR)_3 + ROH \quad (3)$$

The condensation of two alkoxides to form an ether and a siloxane bond is generally not observed.

The chemistry of these hydrolysis and condensation reactions is very complicated. Both reactions are pH sensitive. Acids and bases catalyze the hydrolysis reactions, although each to a different extent (*1*). The condensation reaction is also highly pH dependent. Water is consumed in the hydrolysis reaction but liberated in the condensation reaction. To fully activate all of the silanols initially requires twice as much water as is consumed in the net reaction. To use less water means the reactions will be closely coupled. The reactions are usually conducted in a solvent, because the alkoxides are immiscible with the water required for hydrolysis. The usual solvent is the alcohol corresponding to the alkoxide group of the silicon source, which is also a byproduct of the reactions.

Because each monomer is potentially tetrafunctional, all sorts of partially hydrolyzed and partially condensed species are possible. For example, 10 distinct dimers may be formed. Many cyclic species can occur. Many of the species in solution are metastable with respect to anhydrous, amorphous SiO_2, and rearrangements occur with time (2). The complexity of this system means that the structure and growth of these polymers can be described only in a statistical and geometric fashion rather than by a more conventional description in terms of topologies (such as chains and cross-links) and molecular weights. The most useful and intuitive description of the structure of these polymers is in terms of fractal geometry.

Fractal Structures

Random structures are inherently more difficult to describe than the ordered, periodic structures of crystals. With the exception of a few special cases, such as random-walk polymers, little of a quantitative nature could be said about random structures until the development of fractal geometry and the accompanying realization that the structures that result from many random processes are, in fact, fractal (*3–5*).

Fractal structures have the peculiar property of self-similarity; their appearance is unchanged after a change in magnification. This property is more formally described as dilational symmetry, which is analogous to the translational and rotational symmetries of a crystal. Just as the structure of a crystal appears unchanged after a vector translation of an integral number of unit cells, a fractal structure appears unchanged after a scalar multiplication. The macroscopic structure of a crystal can be generated from a description of the relative locations of its component atoms and its vector symmetry, and a fractal structure can be generated from a statistical description of its components and its scalar symmetry.

The scaling properties of fractal structures, both surface and mass structures, are quantified by their fractal dimensions. A mass fractal dimension (D_f) may be defined in the following way. The mass (M) of an object may be expressed as a power law of its radius (R).

$$M \propto R^{D_f} \tag{4}$$

For an ordinary (Euclidian) object, the exponent, D_f, is equal to the dimension of the space in which the object exists. For example, in three dimensions, the mass of a sphere scales as its radius to the third power. In contrast, for a fractal object, the exponent D_f is not equal to the dimension of space and, in general, is not an integer. However, D_f is analogous to the dimension in the equation for Euclidian objects and is called the fractal (from fractional) dimension.

A surface fractal dimension (D_s) may be defined in a similar manner:

$$S \propto R^{D_s} \tag{5}$$

In equation 5, S is the surface area.

Mass fractal dimensions are always less than the dimension of the space in which the fractal object exists. Therefore, as a fractal structure grows, its mass increases less rapidly than the volume it occupies. Therefore, the density of a fractal object is not constant but decreases with increasing size. Surface fractal dimensions, on the other hand, must lie in the range of one less than the dimension of space up to the dimension of space. The surface area of these objects increases with increasing mass at a faster rate than for Euclidian objects. As a result, the surfaces are very convoluted. Both types of structures are observed in silica polymers.

Real objects can appear fractal only over a limited range of length scales. At very short length scales, one will probe a real object's nonfractal atomic and molecular structures, and the finite size of real objects sets an upper limit to fractal behavior.

Much of the current interest in fractal geometry stems from the fact that fractal dimensions are experimentally accessible quantities. For polymers and colloids, the measurement techniques of choice are scattering experiments using X-rays, neutrons, or light. These measurements may be made on liquids or solids and can be performed readily as a function of time and temperature. Both mass and surface fractal structures yield scattering curves that are power laws, the exponents of which depend on the fractal dimension (6). For mass fractal structures the relation is

$$I(h) = Ah^{-D_f} \tag{6}$$

in which I is the observed intensity, h ($h = 4\pi \sin \Phi/\lambda$ and 2Φ is the scattering angle) is the magnitude of the scattering vector, and A is a proportionality constant that depends on the number of particles and their mass and scattering cross-section.

For surface fractal structures the analogous expression is

$$I(h) = Bh^{D_s-6} \tag{7}$$

in which B is the proportionality constant related to the total surface area in the scattering volume (7). (Equation 7 is a generalization of Porod's law, to which it reduces in the case of nonfractal, $D_s = 2$, surfaces.)

Fractal dimensions are easily extracted from log–log graphs of $I(h)$ versus h. Mass fractal structures will give slopes between –1 and –3, and surface fractal structures will give slopes between –3 and –4. If the scattering curves are recorded with a slit geometry rather than a point geometry (as is commonly the case), the integration of the observed scattering curve over an effectively infinite slit leads to a smearing that introduces a factor of h into the equations. Thus, in such experiments, mass fractal structures will give a slope between 0 and –2, and surface fractal structures will have slopes between –2 and –3.

The determination of fractal dimensions from the power law behavior of small-angle X-ray-scattering curves requires that the particles be quite large. From consideration of the effects of finite particle and monomer sizes described previously, it can be seen that power law scattering can only be observed over an angular range corresponding to $R_g >> 1/h >> a$, in which R_g is the electronic radius of gyration of the particle, and a is the dimension of a monomer.

If the particles are small, this condition may never be met. It is still possible, however, to determine a fractal dimension for such particles in a scattering experiment. This method makes use of Guinier's law, which approximates the low-angle behavior of the scattering curves according to equation 8

$$I(h) \propto N(\Delta M)^2 \exp\left(\frac{-h^2R_g^2}{3}\right) \tag{8}$$

in which N is the number of scattering particles and ΔM is the difference in the number of electrons in the particle and in an equal volume of solvent and is proportional to the molecular weight in dilute solution. The value of the scattered intensity at 0° is therefore proportional to the product of the number of scattering particles and the square of their molecular weight.

In a growth experiment in which the total scattering mass remains constant and larger particles are formed out of smaller ones, the number of particles will be inversely proportional to their molecular weight, and the intensity at 0° will increase in direct proportion to the molecular weight (8, 9). Data are analyzed with Guinier's law by graphing ln $I(h)$ versus h^2. When the Guinier approximation is valid, the data will fall on a straight line whose slope is equal to $R_g{}^2/3$ and whose intercept is $I(0)$. The particle radius and its mass are therefore determined independently from Guinier's law, and from equation 4, the slope of a log–log plot of $I(0)$ versus R_g for varying amounts of growth will be equal to the mass fractal dimension.

The concept of a fractal dimension enables the structures of silicate species to be divided into two classes: (1) those with a dimension less than 3, which may be considered to be true polymers, and (2) those with a dimension of 3, which are perhaps best described as colloidal particles (5). The colloidal class may be further subdivided into particles with fractally rough surfaces ($D_s > 2$) and particles with "smooth" surfaces ($D_s = 2$). In this chapter, the term "polymer" will be used for any product of the condensation reaction, regardless of fractal dimension, although the term "polymeric" will be used for structures with dimensions less than 3, and the term "colloidal" will be used for structures with dimensions equal to 3, in keeping with the terminology used in the literature (9).

Sol–Gel Processing

The most common process for making inorganic silica polymers is the reaction of water with silicon tetraethoxide (TEOS) in a solvent of ethanol. The major factor controlling the properties of the silica polymers is the solution pH, with water concentration being second. In acidic solution, the resulting polymers yield gels that are clear, have very fine porosity, and form fully dense silica at relatively low temperatures (<800 °C). In contrast, in alkaline solution, the resulting polymers yield gels that are porous, translucent, and become dense at higher temperatures (*10*).

Small-angle X-ray-scattering measurements have shown that the different properties of the acid- and base-catalyzed materials are related to pH-controlled differences in the structure of the macromolecules. In acidic solutions, highly ramified, low-fractal-dimension (D_f = 2.1) polymers are formed; and in alkaline solutions, non-mass fractal (D_f = 3) colloidal particles are formed (these particles may have fractally rough surfaces, as will be discussed later) (*9*, *11*, *12*). The low-fractal-dimension polymers can collapse on drying to form fairly dense materials. These materials have a high specific surface and very fine porosity. The large amount of surface water retained in these materials acts as a flux when the gel is heated, and this flux promotes the formation of a dense glass. The colloidal particles in the alkaline solutions, on the other hand, do not collapse as readily when they dry and, thus, give systems of interparticle pores that are relatively large and, hence, translucent. These gels retain less water and therefore become dense at a higher temperature.

Hydrolysis Reactions

Effect of pH. The structure of the silica polymers generated in these reaction depends primarily on the relative rates of the hydrolysis and condensation reactions (*11*). These rates, in turn, depend on the solution pH, the water concentration, and the alkoxide used. Lower molecular weight alkoxides hydrolyze faster than higher molecular weight alkoxides, and obviously, higher water concentrations favor hydrolysis. The major factor is the solution pH. The hydrolysis reaction is catalyzed by both acids and bases, and so the minimum in the hydrolysis rate occurs at a nominally neutral pH. The condensation reaction, however, is fastest at a pH of about 5–6 (*12*), as will be discussed later. Under the acidic conditions usually used in sol–gel processing (pH 3–7), the condensation reaction is the rate-limiting step, whereas at higher pH, hydrolysis turns out to be the rate-limiting step.

Mechanism. Under acidic conditions and with adequate water, the hydrolysis of silicon tetraethoxide (and tetramethoxide) is fast and complete (*13*). A plausible reaction mechanism that accounts for a number of important observations is simple acid-catalyzed nucleophilic substitution. A hydronium ion should readily attack the basic alkoxide oxygen to form an activated complex that decomposes to an alcohol and a silanol. Although silanols are fairly acidic, at low pH they should remain protonated. Such a mechanism implies that the reaction is readily reversible from a kinetic standpoint. However, orthosilicic acid, $Si(OH)_4$, is apparently much more thermodynamically stable than the alkoxide, and so the reaction goes to completion.

Base-catalyzed hydrolysis of silicon alkoxides proceeds much more slowly than acid-catalyzed hydrolysis at an equivalent catalyst concentration (*1*), the tetraalkoxides being particularly resistant to attack (*14*). Hydrolysis proceeds stepwise, with each subsequent alkoxide group more easily removed from the monomer than the previous one (*14*). Alkoxide groups are much more difficult to remove from polymers than from monomers.

These observations are easily explained by another simple reaction mechanism, nucleophilic substitution of an alkoxide on silicon (*12*). In this case, the basic alkoxide oxygens tend to repel the nucleophile, ^{-}OH, and the bulkier alkyl groups tend to crowd it. Therefore, more highly hydrolyzed silicons are more prone to attack. Because this mechanism would have a pentacoordinated silicon atom in the activated complex, hydrolysis of a polymer would be more sterically hindered than hydrolysis of a monomer. Reesterification would be much more difficult in alkaline solution than in acidic solution, because silanols are more acidic than the hydroxyl protons of alcohols and would be deprotonated and negatively charged at a pH lower than the point at which the nucleophile concentration becomes significant (*11*). Thus, although hydrolysis in alkaline solution is slow, it still tends to be complete and irreversible, if extensive polymerization does not occur first.

Polymerization

Mechanism. The polymerization of silica is thought to occur by a nucleophilic substitution mechanism as well (*2*, *11*). The nucleophile is a deprotonated silanol group, SiO^-, which attacks a neutral silicate species. The acidity of the silanol proton increases with the degree of polymerization of the silicate species to which it is attached. Thus monomer $Si(OH)_4$ is less acidic than a silanol on an end-group silicate tetrahedron and is less acidic than one on a middle group, etc. This mechanism accounts for the increase in condensation rate with pH, because the concentration of reactive SiO^- groups will increase, and it accounts for the maximum in the condensation rate at about pH 6, which occurs when the product of the concentrations of the protonated and the deprotonated silanol species is highest. This mechanism also predicts that the preferred reaction is between tetrahedra of different degrees of polymerization (*2*), because highly polymerized silicate groups are more likely to contain an SiO^-, and weakly polymerized tetrahedra are more likely to be neutral and therefore subject to attack. This mechanism was proposed for aqueous alkali silicate solutions that tend to be in the pH range of 3–12. At lower pH, other reaction mechanisms may actually occur (*2*).

A major difference between silicate polymers and most organic polymers is that the condensation of silica is reversible. Amorphous SiO_2 has a finite solubility (about 120 ppm) in water, and its solubility increases substantially with increasing pH. The solubility of SiO_2 in neutral alcohols, however, is essentially 0. Its solubility in mixed alcohol–water solutions or in acidic or alkaline alcohol solutions is unknown. The evidence is clear, however, that rearrangement of silicate polymers does occur in these solutions, and therefore, concepts such as solubility and supersaturation are valid. With the foregoing observations, the structure and growth of silica polymers can be understood in at least a qualitative fashion.

In alkaline solution, hydrolyzed silicate species are generated relatively slowly, because TEOS hydrolyzes slowly. But if enough water and catalyst are present, the degree of hydrolysis is quite complete, because the hydrolysates hydrolyze faster than the starting material. At high pH, the rate of condensation is fast, but because of the higher solubility of silica at high pH and because its rate of production is slow, the degree of supersaturation is low. The low degree of supersaturation and the reversibility of the condensation reaction mean that critical nuclei of amorphous SiO_2 will be relatively large and few in number. The preferred reaction of a monomer with highly polymerized silicate tetrahedra favors this nucleation and growth process, as does the low instantaneous monomer concentration. (The process is similar to precipitation from a homogeneous solution.) The high pH deprotonates the acidic silanols on the surface of these highly polymerized particles. The negative charge causes the polymers to repel each other and thus minimizes aggregation of the polymers while at the same time promoting reaction of the surfaces with the weakly acidic and, hence, protonated and neutral monomers.

Modeling. The process of nucleation and random growth has been studied by means of a computer simulation called the Eden model (*4, 15*). Although this model was originally designed to describe the growth of cell colonies, the growth rules are the same as those operating in alkaline solution. The value of such simulations is that the fractal dimension of the products of the growth process and their size distribution, if polydisperse, may be readily calculated (the number of monomers and the Cartesian coordinates of each are known) and compared with the results of scattering experiments.

The simulation works as follows. A lattice is defined in the computer's memory, and a site at the center is designated the "seed" or nucleus. All lattice sites immediately adjacent to the nucleus are designated "growth" sites. One of the growth sites is selected at random and occupied by a monomer. The monomer is considered to be part of the cluster, and the sites adjacent to it also become growth sites. The process is continued until the "particle" reaches any desired size. The growth sites simulate the reactive silanols, and nucleation is simulated by the rule that a monomer may attach only to the nucleus or another part of the cluster and not to another monomer. The Eden model generates clusters that have a homogeneous mass density

(Figure 1) and are therefore not fractal, just like the particles that are observed when TEOS is hydrolyzed in alkaline solution.

Figure 1. A two-dimensional cluster grown by the Eden process. The clusters in this simulation fill in, because all surface sites have equal probability of growth whether or not they have "access" to the perimeter. This artifact is more unreal in two dimensions than in three. (Reproduced with permission from reference 19. Copyright 1986 Elsevier.)

It has been observed in small-angle X-ray-scattering experiments that when silica polymers are grown in alkaline solution but under conditions of restricted water for hydrolysis, fractally rough surfaces are produced (*16*). The apparent fractal dimension of these surfaces depends on the amount of water used in the hydrolysis (Figure 2), and if the water concentration is low enough (less than 1 mol per mole of TEOS), the polymers may be mass fractal. This phenomenon has been investigated with a variation of the Eden model that simulates the effect of incomplete hydrolysis (*12*). In this variation, potential growth sites on the monomer are "poisoned" before the monomer is placed on a growth site. No monomers are allowed to occupy this poisoned site. The poisoned site simulates an unhydrolyzed alkoxide group, and the permanence of the poisoning simulates the resistance to hydrolysis of an alkoxide group on a polymer. When a distribution of monomers of different degrees of hydrolysis is used in this simulation, the resulting clusters range from being completely nonfractal, to colloidal but with fractally rough surfaces, and to true fractal polymers (Figure 3). This behavior is in qualitative agreement with the X-ray-scattering observations. The poisoned Eden model encompasses the ordinary Eden model, percolation, and even a self-avoiding random walk (the limiting case when only difunctional monomers are used).

Under acidic conditions, the relative rates of hydrolysis and condensation are reversed. The hydrolysis reaction is quite fast and produces a much higher degree of supersaturation of monomers than that produced in alkaline

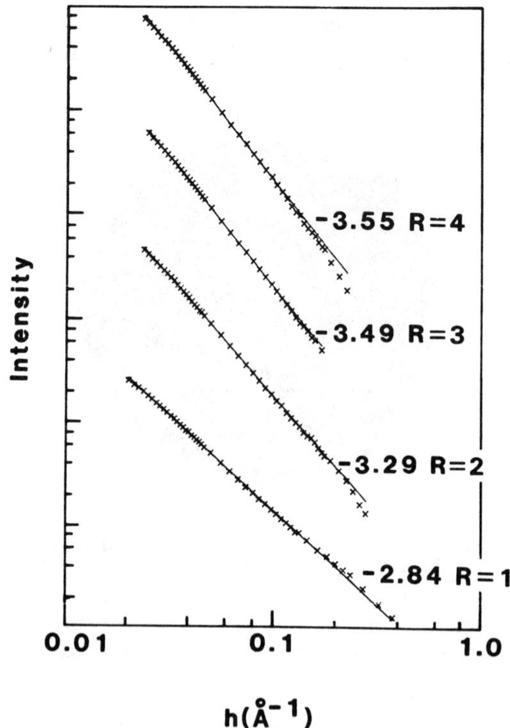

Figure 2. Small-angle X-ray-scattering curves from silica polymers grown in alkaline solution with various ratios (R) of water to TEOS. The data are slit smeared and are plotted as log I(h) versus log h, with the curves displaced vertically for clarity. The observed slopes are algebraically one greater than would be observed with pinhole geometry. The curve for R = 1 is consistent with a mass fractal structure with D_f = 2.84. The curves for R = 2–4 arise from colloids (D_f = 3) with fractally rough surfaces with D_s = 2.71, 2.51, and 2.45, respectively. (Reproduced with permission from reference 10. Copyright 1984 Elsevier.)

solution (*17*). Under these circumstances, monomer–monomer reactions are frequent initially, and the size of a critical nucleus is very small. This situation results in the formation of a large number of very small particles. Growth proceeds by the reaction of these small molecules with each other to form larger ones. Polymer–polymer reactions are more likely, because the polymers are not charged (the silanols on the weakly polymerized molecules are not very acidic and the pH is low), as they would be in an alkaline solution. The lower solubility of silica monomer at low pH means that rearrangements are less likely to occur.

Computer simulations of a similar growth process, called reaction-limited cluster–cluster aggregation, have been performed (*15*). This simulation

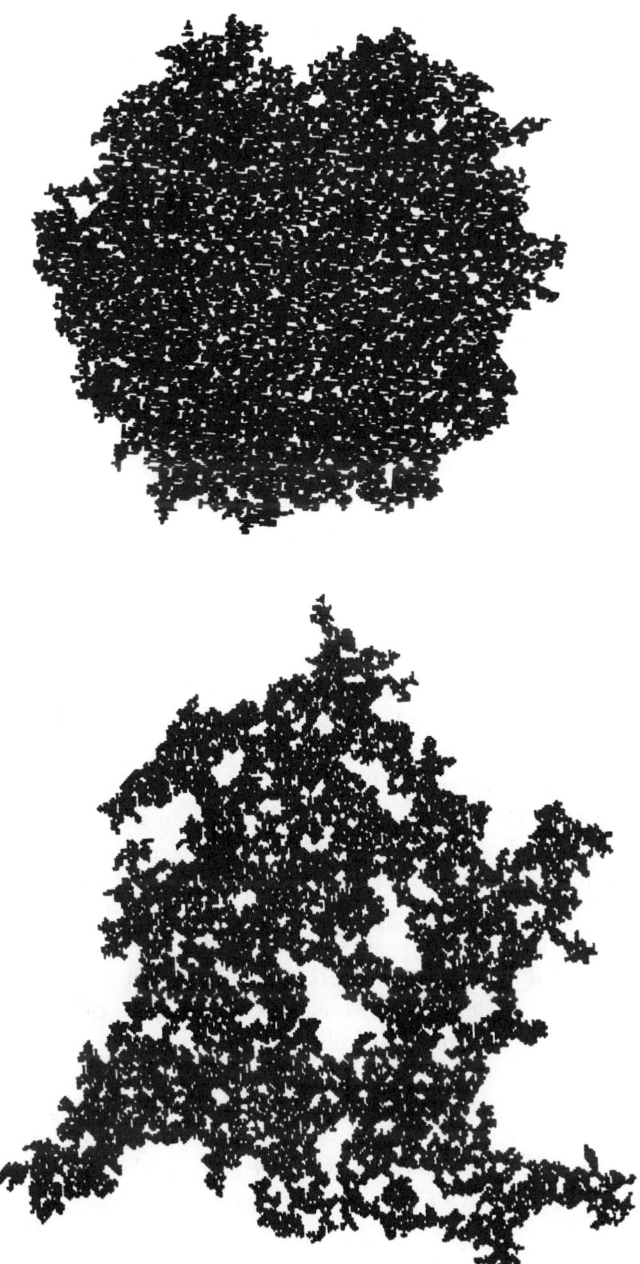

Figure 3. Two-dimensional clusters grown with the poisoned Eden model. The average functionality of the monomers in each cluster is three. The distribution of functional groups among the monomers causes the different structures.

starts with a large number of "monomers" rather than a single seed, as in the Eden model. These monomers are then moved by the computer in a random fashion to simulate Brownian diffusion. When two particles encounter one another, the probability that they will "stick" and form a larger particle is less than unity; hence the description "reaction limited" (in contrast to simulations of diffusion-limited processes, in which the sticking probability is unity [*15*]). The newly formed particles continue to diffuse and react with other particles as they are encountered. Unlike Eden growth, no new monomers are added during the growth process. The fractal dimension of the polymers formed in this simulation is about 2.1, which is very similar to the values reported in small-angle X-ray-scattering studies of hydrolysis and condensation in acidic conditions (*18*).

A simple experiment shows that, in fact, the rate of hydrolysis relative to condensation, and not the effect of the solution pH on the condensation reaction, causes the pH dependence of the polymer structure. In this experiment, the size and mass of particles grown in the same pH solution from monomers that have different degrees of hydrolysis are measured as a function of time. The growth of polymers in solutions in which the alkoxide is first hydrolyzed rapidly under acidic conditions (pH 3) and then polymerized under alkaline conditions is compared with that of polymers grown such that both hydrolysis and growth occur under alkaline conditions.

The results are shown in Figure 4, in which the scattered intensity at 0° (which is proportional to the molecular weight, as described earlier) is plotted as a function of the polymer radius of gyration. A, B, and C in Figure 4 refer to different treatments of the reaction mixture. For solution A, single-step hydrolysis of TMOS (silicon tetramethoxide) in alkaline solution (nominally pH 8.6, 4 mol of H_2O per mole of TMOS) was used. Solutions B and C were first hydrolyzed (with 12 and 4 mol, respectively, of H_2O per mole of TMOS) for 30 s, and then, the pH of the reaction mixture was adjusted (with NH_4OH) to 8.6. Slopes corresponding to a fractal dimension of 2 fit the data for the fully hydrolyzed solutions B and C, but the data for solution A are better fit by a slope of 3. Thus we can conclude that the hydrolysis pH, and not the growth pH, controls the fractal dimension.

The foregoing discussion presents a contrast of the two limiting ideal cases. Both the hydrolysis and condensation rates are affected by parameters other than pH, such as solvent, type of alkoxide, and concentration. Only the tendencies of the solution behavior are discussed in this chapter. The ideal limiting-case behavior may not always be observed.

Conclusions

Inorganic silicate polymers prepared by the hydrolysis of silicon alkoxides evolve by a complicated reaction pathway. The structures can be described only in a statistical fashion by means of fractal geometry. The structures are

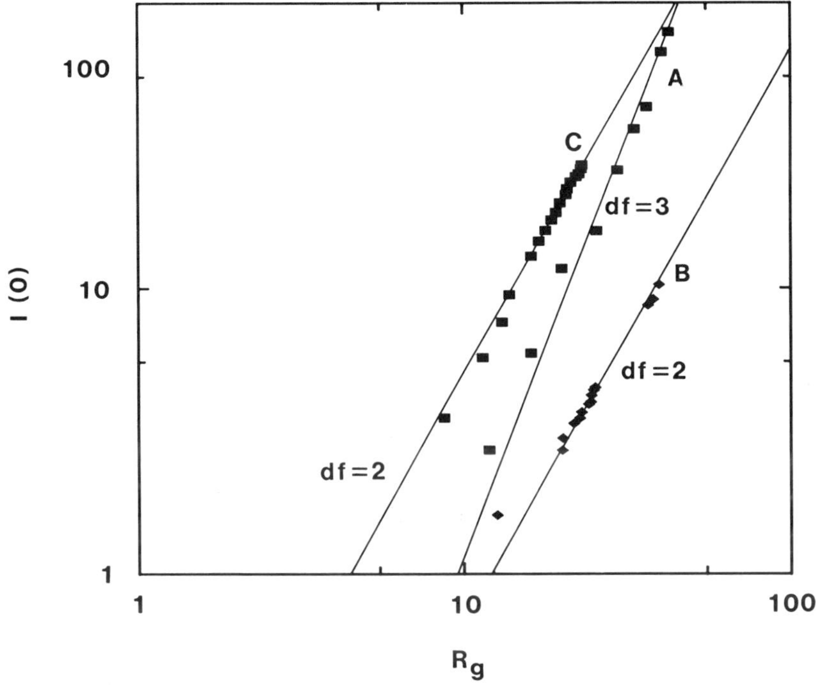

Figure 4. Results of small-angle X-ray scattering from silica polymers grown by a one-step alkaline hydrolysis and condensation (pH 8.6, 4 mol of H_2O per mole of TMOS) (A) and a two-step process consisting of hydrolysis in acid (pH 3) followed by growth at pH 8.6 with 12 (B) or 4 (C) mol of H_2O per mole of TMOS. The logarithm of I(0)*, which is proportional to the polymer molecular weight, is plotted against log* R_g*. The data for A are in good agreement with colloidal* (D_f = 3) *particles, whereas those for B and C agree well with polymers having* D_f = 2.

controlled by the relative rates of the hydrolysis reactions that generate reactive silanol groups and the condensation reactions of those groups. Under alkaline conditions, hydrolysis tends to be slow with respect to condensation, and nonfractal (D_f = 3) colloidal structures are formed by a nucleation- and a reaction-limited growth process. Under acidic condition, hydrolysis tends to be fast with respect to condensation, and polymeric structures with D_f = 2.1 are formed by a process similar to reaction-limited cluster aggregation. The exact structure in real systems may only approach these idealized cases.

Acknowledgments

Much of the work described in this chapter was performed in close collaboration with others, particularly my colleagues at Sandia. In particular, I acknowledge Dale Schaefer, who first demonstrated that the concepts of polymer physics and fractal geometry could be applied to these systems; Jim Martin, who has spent much time educating me on the intricacies of scattering and polymer theory; and Bruce Bunker, who tries to keep my chemistry honest.

This work was supported by the U.S. Department of Energy under contract number DE–AC04–76DP00789.

References

1. Aelion, R.; Loebel, A.; Erich, F. *J. Am. Chem. Soc.* **1950,** *72*, 5705.
2. Iler, R. K. *The Chemistry of Silica;* Wiley: New York, 1979.
3. Mandelbrot, B. B. *The Fractal Geometry of Nature;* Freeman: San Francisco, 1982.
4. Stanley, H. E.; Ostrowsky, N. *On Growth and Form;* Martinus-Nijhoff: Boston, 1986.
5. Martin, J. E.; Hurd, A. J. *J. Appl. Crystallogr.* **1987,** *20*, 61–78.
6. Bale, H. D.; Schmidt, P. W. *Phys. Rev. Lett.* **1984,** *53*, 596.
7. Martin, J. E., Sandia National Laboratories, personal communication.
8. Schaefer, D. W.; Keefer, K. D. In *Structure of Random Silicates: Polymers, Colloids and Porous Solids;* Proceedings of 6th International Symposium on Fractals in Physics, Trieste, Italy; Elsevier: Amsterdam, 1985.
9. Brinker, C. J.; Keefer, K. D.; Schaefer, D. W.; Assink, R. A.; Kay, B. D.; Ashley, C. S. *J. Non-Cryst. Solids* **1983,** *63*, 45–60.
10. Keefer, K. D. *Better Ceram. Chem., Mater. Res. Soc. Symp. Proc.* **1984,** *32*, 15–24.
11. Schaefer, D. W.; Keefer, K. D. *Better Ceram. Chem., Mater. Res. Soc. Symp. Proc.* **1984,** *32*, 1–14.
12. Keefer, K. D. *Better Ceram. Chem., Mater. Res. Soc. Symp. Proc.* **1986,** *73*, 245.
13. Kay, B. D.; Assink, R. A. *Better Ceram. Chem., Mater. Res. Soc. Symp. Proc.* **1986,** *73*, 157.
14. Kelts, L. W.; Effinger, N. J.; Melpolder, S. M. *J. Non-Cryst. Solids* **1986,** *83*, 353–374.
15. Eden, M. In *Proceedings of the 4th Berkeley Symposium on Mathematics Statistics and Probability;* University of California Press: Berkeley, CA, 1961; Vol. 4, p 223.
16. Keefer, K. D.; Schaefer, D. W. *Phys. Rev. Lett.* **1986,** *56*, 2376.
17. Assink, R. A.; Kay, B. D. *Better Ceram. Chem., Mater. Res. Soc.* **1984,** *32*, 301.
18. Schaefer, D. W.; Keefer, K. D. *Phys. Rev. Lett.* **1986,** *56*, 2199.
19. Schaefer, D. W.; Keefer, K. D. *Better Ceram. Chem., Mater. Res. Soc. Symp. Proc.* **1986,** *73*, 277.

RECEIVED for review May 27, 1988. ACCEPTED revised manuscript July 31, 1989.

14

Hydrolysis and Condensation Kinetics of Dimeric Sol–Gel Species by ^{29}Si NMR Spectroscopy

Daniel H. Doughty, Roger A. Assink, and Bruce D. Kay

Sandia National Laboratories, Albuquerque, NM 87185

High-resolution ^{29}Si NMR spectroscopy was used to study the hydrolysis and condensation kinetics of monomeric and dimeric species in the silicate sol–gel system. Peak assignments for the kinetics experiments were determined by comparing acid-catalyzed reaction solutions prepared with limited amounts of water with the synthetically prepared dimeric, trimeric, and tetrameric species. ^{29}Si NMR peaks were assigned for 5 of the 10 possible dimeric species. The temporal evolution of hydrolysis and condensation products has been compared with a kinetic model developed in our laboratory, and rate constants have been determined. The results indicate that the water-producing condensation of dimeric species is approximately 5 times slower than the water-producing condensation of the monomeric species. The alcohol-producing condensation of dimeric species is comparable with that of monomeric species.

SOL–GEL PROCESSING has become an active area of research (*1*, *2*), because it provides synthetic routes to novel materials. To achieve this goal, the manipulation and control of solution variables to tailor the chemical or physical properties of the product are necessary. The processing and structural evolution of sol–gel-derived materials are intimately linked with the solution chemistry used in their formation (*3*). Although many sol–gel processes have appeared in the literature, detailed kinetics studies are usually lacking. We are exploring the kinetics and mechanism of the constitutive reactions in the

0065–2393/90/0224–0241$06.00/0

silicate sol–gel system to provide the foundation necessary to understand the underlying chemistry.

At the simplest level, silicate sol–gel chemistry can be described by three reactions:

$$\equiv\text{Si–OR} + H_2O \xrightarrow{k_H} \equiv\text{Si–OH} + \text{ROH} \quad (1)$$

$$\equiv\text{Si–OH} + \equiv\text{Si–OH} \xrightarrow{\frac{1}{2} k_{cw}} \equiv\text{Si–O–Si}\equiv + H_2O \quad (2)$$

$$\equiv\text{Si–OH} + \equiv\text{Si–OR} \xrightarrow{\frac{1}{2} k_{ca}} \equiv\text{Si–O–Si}\equiv + \text{ROH} \quad (3)$$

In these equations, k_H, k_{cw}, and k_{ca} are the rate constants for hydrolysis, water-producing condensation, and alcohol-producing condensation, respectively. Equation 1 shows the hydrolysis of a Si-OR group to produce a silanol group. A silanol group can undergo condensation polymerization with another silanol (equation 2) or with an alkoxy group (equation 3) to produce a siloxane linkage plus water or alcohol, respectively. In this chapter, the rate constants discussed for these reactions are intended to be phenomenological rate constants and not meant to imply mechanistic details.

Proposed mechanisms for ester hydrolysis (*4–6*) under acidic conditions involve bimolecular nucleophilic substitution reactions that have pentacoordinate intermediates. The proposed mechanism for water-producing condensation (*5*) under acidic conditions (pH $<$ 2.5–4.5, depending on the silicon substituents) involves rapid, reversible protonation of a silanol, followed by a bimolecular reaction forming the siloxane bond and eliminating a hydronium ion.

By using these observed rate constants for hydrolysis and condensation, a detailed formalism has been developed that treats the various reactions at the functional-group level (*7–9*). The observed rate constants were reported (*8*) for the acid-catalyzed hydrolysis and condensation of monomeric species prepared from $Si(OCH_3)_4$ (tetramethoxysilane [TMOS]). In this chapter, we report the observed rate constants for condensation of the dimeric species $(CH_3O)_3Si–O–Si(OCH_3)_3$.

Experimental Procedures

Preparation of Dimer. Hexamethoxydisiloxane $[(CH_3O)_3Si]_2O$, termed "dimer" in this chapter, was synthesized by the method of Klemperer et al. (*10*) with modifications. By using Schlenk techniques, hexachlorodisiloxane (57 mL, 0.312 mol) was added dropwise (over 50 min) to a solution of freshly distilled methyl orthoformate (trimethoxymethane) (400 mL, 3.74 mol) in heptane (200 mL, dried over Na). The reaction was carried out under argon, and the temperature of the reaction mixture was maintained at 50 °C for 5 h. The mixture was stored overnight at room temperature under argon. Purification by vacuum distillation was carried out. The fraction boiling at 50 °C (43–53 Pa) was collected. Final purification to remove HCl was accomplished by passing the straw-colored liquid through neutral alumina and discarding the first cut. The purity of the colorless liquid was determined by ^{29}Si NMR spectroscopy and by gas chromatography–mass spectroscopy (GCMS). Samples were stored in the dark under argon.

NMR Spectroscopy. The ^{29}Si NMR spectra were recorded at 39.6 MHz on a Chemagnetics console interfaced to a General Electric 1280 data station and pulse programmer. The silicon-free probe and 20-mm-diameter poly(trifluorochloroethylene) sample tubes were purchased from Cryomagnetics. The 4.7-T wide-bore magnet was constructed by Nalorac Cryogenics. From 4 pulses (during the early stages of the reaction) to 64 pulses (during the latter stages of the reaction) were accumulated. The pulse repetition interval was 17.7 s.

^{1}H broad-band decoupling was applied only during data acquisition to suppress any residual negative nuclear Overhauser effect (NOE) (*11*). However, no difference was observed between spectra obtained with gated decoupling and those with continuous decoupling, a result indicating negligible NOE. This result is expected, because the observed relaxation time is at least a factor of 10 less than the dipolar relaxation time. No field lock was necessary, as the line widths of accumulated spectra were typically 0.5 Hz before a 0.3-Hz line broadening was applied.

The total integrated intensities of all the silicons in the sample were usually reproducible to within ± 3% over the course of the reaction. The ^{29}Si spin-lattice relaxation time, T_1, was 4.2 ± 1.0 s, with a slight tendency for the more highly condensed species to exhibit times toward the lower limit of the uncertainty range. The pulse repetition period of 17.7 s ensured that the magnetization has recovered to 98 ± 2% of its equilibrium value.

Kinetics Studies. The sol–gel kinetics experiments were performed on a methanolic solution of 1.12 M dimer and 1.57×10^{-2} M chromium acetylacetonate [$Cr(acac)_3$], a spin relaxation agent. Previous studies (*12, 13*) showed that $Cr(acac)_3$ concentration does not affect the product distribution or reaction rate of TMOS-derived sol–gel solutions. The solutions were acid catalyzed (1.64×10^{-3} M HCl), and various amounts of water were added. To compensate for the exothermicity caused by dimer hydrolysis when water is added to the dimer–methanol solution, the alcoholic silicate solution was chilled in a thermostatically controlled bath prior to the addition of water. By adjusting the temperature to the appropriate level, the desired reaction temperature (25 ± 1 °C) could be achieved within 60 s of mixing. At this time, the sample was removed from the thermostatically controlled bath and inserted into the spectrometer probe.

^{29}Si NMR spectra were taken immediately after the sample was mixed and then as a function of time for four water concentrations to give water/alkoxide ratios (H_2O/[Si–OR]) of 1/12, 1/6, 1/4, and 1/3. Figure 1 shows representative spectra for H_2O/[Si–OR] = 1/6 taken immediately after mixing (5 min) and at a later time (4 h). The figure shows that hydrolysis is rapid under these conditions and that condensation occurs at a much slower rate, as evidenced by inspection of the Q^1 region (*14*), which shows peaks assigned to the first (chemical shift [δ] = –84 ppm) and second (δ = –82.6 ppm) hydrolysis products of dimer within 5 min of mixing.

The ^{29}Si NMR spectra were used for two types of measurements. For the first type, the amount of dimer and the extent to which it had been hydrolyzed were determined very early in the reaction by integrating the Q^1 resonances in the –81- to –86-ppm region of the spectra. For the second type of measurement, the extent of condensation as a function of time was determined by comparing the integrals of the Q^1 and Q^2 species.

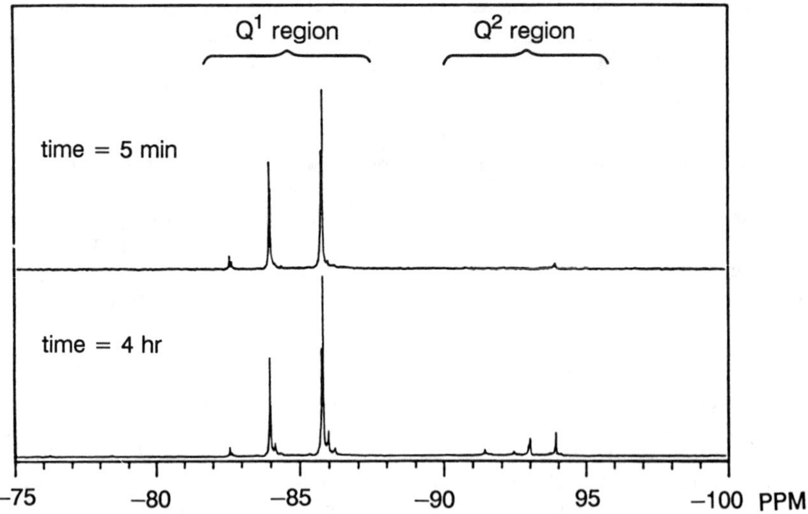

Figure 1. ^{29}*Si NMR spectra showing formation of hydrolysis and condensation products.*

The extent of condensation that occurred during the experiment was simply equal to the number of Q^2 silicons (no Q^3 silicons are observed at early times), because all silicons were initially Q^1 species. No Q^0 species were observed during the course of these experiments. Cyclic trimers consisting of Q^2 silicon atoms are expected to resonate in the Q^1 region of the spectrum (*12*). However, because low H_2O/[Si–OR] ratios were used and only the early stages of the reaction were examined, no evidence of cyclic trimers was observed. Moreover, cyclic trimers could not form without siloxane bond cleavage and, consequently, would not be expected be formed in these experiments (*12*). Cyclic tetramers have resonances in the Q^2 region (*12*) and were observed in these experiments.

Results and Discussion

^{29}Si NMR Assignments. ^{29}Si NMR spectroscopy has been used extensively to probe silicate sol–gel reactions (*8, 10, 12–23*). Peak assignments for silicon nuclei depend on the next-to-nearest neighbor of the silicon atom being observed. With this approach, the degree of hydrolysis and condensation of a given silicon nucleus can be determined.

We have confirmed these assignments for Q^1, Q^2, and a Q^3 species by recording the ^{29}Si NMR spectra of the unhydrolyzed, pure compounds of dimer, linear trimer $(CH_3O)_8Si_3O_2$, and a mixture of linear and branched tetramer $(CH_3O)_{10}Si_4O_3$ (Table I). The values are in excellent agreement with the results of Marsmann et al. (*15*) and previous work in our laboratory (*12*). We have used these assignments for the kinetics calculations discussed in the next section.

Table I. NMR Assignments for Unhydrolyzed Silicates

Silicon Environment	*This Work*	*Ref. 15*
Q^1 species		
Si–O–Si	−85.81	−85.82
Si–O–Si–O–Si	−85.99	−86.00
Si–O–Si–O–Si–O–Si	−85.98	−86.00
Si–O–Si(–O–Si)(–O–Si)	−86.19	−86.22
Q^2 species		
Si–O–*Si*–O–Si	−93.69	−93.64
Si–O–*Si*–O–Si–O–Si	−93.90	−93.86
Q^3 species		
Si–O–*Si*(–O–Si)(–O–Si)	−102.10	−102.00

NOTE: For clarity, –OR groups are not shown.
All data are chemical shifts in parts per million downfield from tetramethylsilane.

As we analyzed spectra taken at very high resolution for these experiments, we realized that additional assignments could be made. Other workers (22) have reported fine structure within a region attributed to one silicon environment (e.g., six peaks were reported for *Si*$(OCH_3)_2$(OH)(OSi), which resonates at approximately −84 ppm). The authors conjectured that the various peaks were due to different degrees of hydrolysis of the adjacent silicon. Because of the conditions of our experiments, notably the lower water concentrations and the use of a spectrometer with higher resolution, we are able to assign some of these peaks.

Assignment is aided by symmetrical relationships that exist between adjacent silicon atoms. For example, the integrated intensity of each silicon environment in the dimer $(CH_3O)_3$*Si*–O–*Si*(OH)$(OCH_3)_2$ must be equal at any given time and must have identical time dependence. A representative spectrum is given in Figure 2. Our assignments of chemical shifts for five of the possible 10 dimeric species are given in Table II.

Analysis of Condensation Kinetics. An earlier publication (*8*) explained how, under certain limiting conditions, the observed rate constants for water- and alcohol-producing condensation reactions in the TMOS sol–gel reactions can be calculated. The limiting conditions are (1) hydrolysis occurs on a time scale short compared with condensation and (2) the concentration of Si–O–Si bonds formed by condensation reactions is small compared with the initial concentration of the methoxy functional group. If these conditions

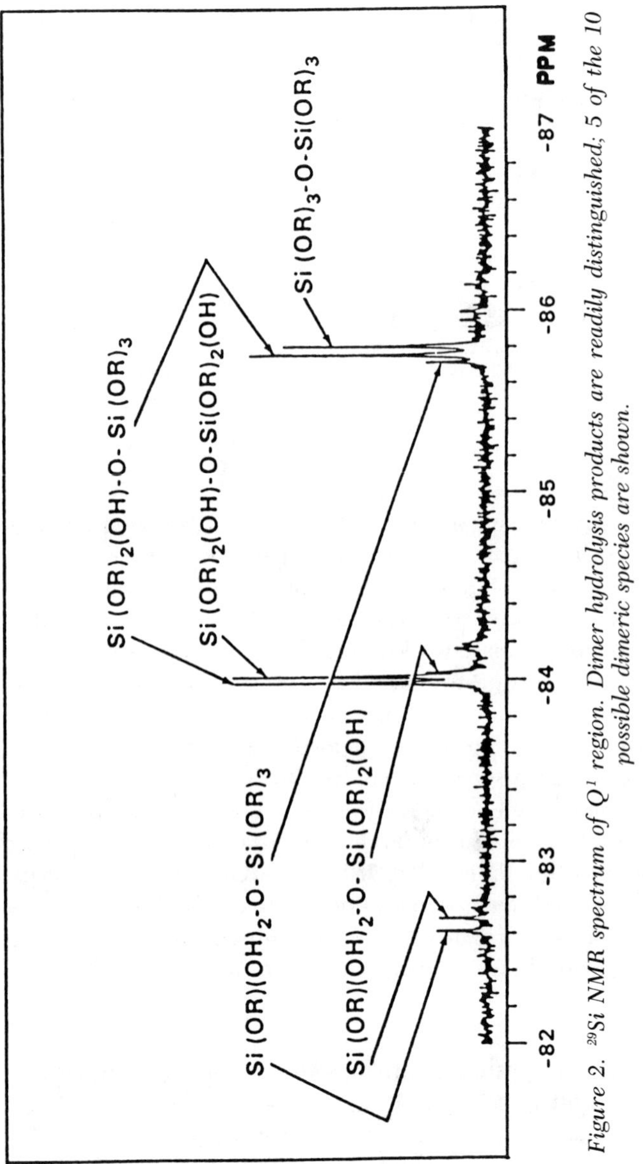

Figure 2. ^{29}*Si NMR spectrum of* Q^1 *region. Dimer hydrolysis products are readily distinguished; 5 of the 10 possible dimeric species are shown.*

Table II. Chemical Shifts of Dimeric Species

Si Environment	*δ (ppm from TMS)*
$\mathit{Si}(OCH_3)_3$–O–$Si(OCH_3)_3$	−85.81
$\mathit{Si}(OCH_3)_3$–O–$Si(OCH_3)_2OH$	−85.76
$\mathit{Si}(OCH_3)_3$–O–$Si(OCH_3)(OH)_2$	−85.72
$\mathit{Si}(OCH_3)_2(OH)$–O–$Si(OCH_3)_3$	−83.99
$\mathit{Si}(OCH_3)_2(OH)$–O–$Si(OCH_3)_2OH$	−84.02
$\mathit{Si}(OCH_3)_2(OH)$–O–$Si(OCH_3)(OH)_2$	−84.04
$\mathit{Si}(OCH_3)(OH)_2$–O–$Si(OCH_3)_3$	−82.62
$\mathit{Si}(OCH_3)(OH)_2$–O–$Si(OCH_3)_2OH$	−82.69

NOTE: The table shows the effect of different degrees of hydrolysis of the adjacent Si atom on the chemical shift of the observed Si.

are met, a good approximation of the exact kinetics expression for this system is

$$\frac{d[(\mathrm{SiO})\mathrm{Si}]/dt}{<[\mathrm{SiOH}]>} = (k_{cw} - k_{ca})<[\mathrm{SiOH}]> + k_{ca}[\mathrm{SiOCH_3}]_o \qquad (4)$$

in which $d[(\mathrm{SiO})\mathrm{Si}]/dt$ is the initial rate of formation of siloxane (Si–O–Si) bonds, $<[\mathrm{SiOH}]>$ is the average value of the silanol functional group concentration over the measurement time window, and $[\mathrm{SiOCH_3}]_o$ is the initial concentration of methoxy functional groups. Equation 4 suggests that by plotting the initial condensation rate divided by $<[\mathrm{SiOH}]>$ versus $<[\mathrm{SiOH}]>$, the value of k_{ca} can be determined from the intercept, and the value of the difference between k_{cw} and k_{ca} can be determined from the slope.

The results of this analysis applied to the condensation of dimer are shown in Figure 3. The observed rate constants for alcohol-producing condensation are approximately the same (k_{ca} = 0.001 and 0.0007 L/mol-min for monomer and dimer, respectively), whereas the rate constants for water-producing condensation are significantly different (k_{cw} = 0.006 and 0.0011 L/mol-min, for monomer and dimer, respectively).

The lower value of k_{cw} for dimers can be explained by simple arguments. The steric requirements of the bulky siloxane group should decrease the reaction rate constant of dimers compared with monomers. Inductive effects should decrease the rate as well. This latter conclusion is based on the mechanism proposed by Pohl and Osterholtz (5), which postulates that in acidic solutions ($pH < 4.5$), an equilibrium concentration of protonated silanol is rapidly established. The protonated silanol reacts with a neutral silicate species in the rate-determining step, and then, a hydronium ion is eliminated. The more electron-withdrawing siloxane group (compared with Si–OR and Si–OH) would decrease the stability of the protonated species and effectively shift the equilibrium toward the unprotonated form. Therefore, the observed reaction rate constant should decrease.

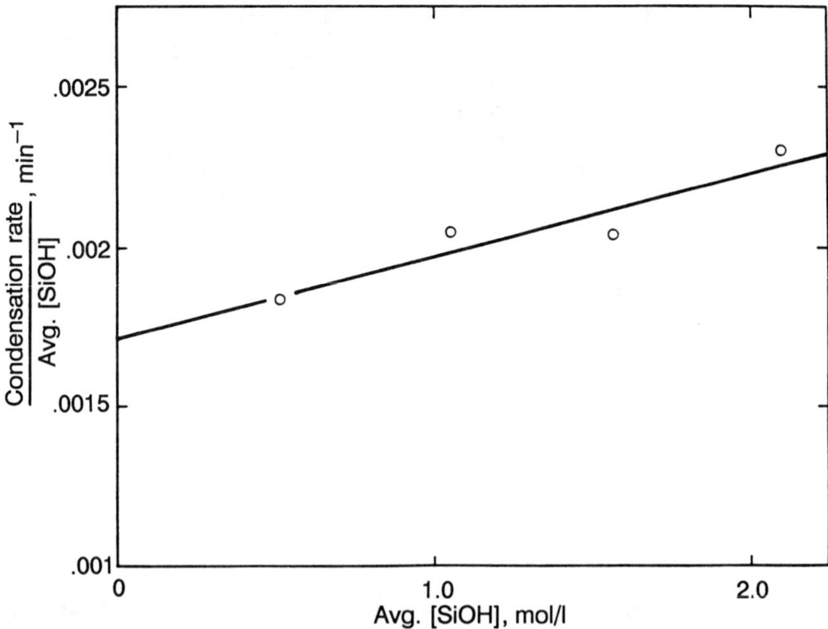

Figure 3. Graph of condensation rate divided by average [SiOH] versus average [SiOH].

The similarity of k_{ca} for monomers and dimers is unexpected. If the reaction mechanism involves a protonated silanol, as in the water-producing condensation, subsequent reaction rate constants must be proportionately larger to give an observed k_{ca} of the same magnitude. However, the likely leaving group from a protonated silanol species is water, not an alcohol.

An alternative hypothesis is that a protonated alkoxide moiety $(\equiv Si{-}O(H)R)^+$ is formed in a small steady-state concentration. Subsequently, this protonated species could react with a neutral silanol species and eliminate R–OH. The electron-donating effect of the R group would predominate over the electron-withdrawing effect of the –OSi group on silicon. These effects may explain why the measured values of k_{ca} for dimers are very similar to those of monomers.

Conclusion

Further work is needed to clarify the nature of the differences between the water- and alcohol-producing condensations of silicate species. But one conclusion is clear: The two condensation reactions proceed by different reaction pathways (or at least have different transition states) and do not necessarily follow the same reactivity trends.

A second important outcome of these data is a result of the earlier observation (8) that, depending on reaction conditions, either water- or alcohol-producing condensation reactions may be the major pathway for polymerization. If the sol–gel reaction is carried out with small amounts of added water, (i.e., [SiOH] is small and [$SiOCH_3$] is large), the alcohol-producing condensation will predominate. Conversely, at high water condensation, [SiOH] is large, and the water-producing condensation is favored. The crossover point for dimeric species will be shifted to higher water concentrations, compared with that of TMOS condensation (8). This conclusion has practical consequences for controlling the polymerization rate of silicates in sol–gel processing.

Acknowledgments

This work was supported by the U.S. Department of Energy under contract number DE–AC04–76DP00789. We gratefully acknowledge helpful discussions with Jeff Brinker, Walter Klemperer, and Vera Mainz. Walter Klemperer and Vera Mainz (Department of Chemistry, University of Illinois) provided samples of trimer and linear and branched tetramer under Sandia contract 57–3178. We also thank Sheri Martinez for synthesizing hexamethoxyldisiloxane and for other technical assistance. Likewise, the technical assistance of Duane Schneider is gratefully acknowledged.

References

1. *Better Ceramics through Chemistry;* Brinker, C. J.; Clark, D. E.; Ulrich, D. R., Eds.; Elsevier–North Holland: New York, 1984.
2. *Better Ceramics through Chemistry II;* Brinker, C. J.; Clark, D. E.; Ulrich, D. R., Eds.; Materials Research Society: Pittsburgh, PA, 1986.
3. Brinker, C. J. *J. Non-Cryst. Solids* **1988,** *100,* 31–50.
4. McNeil, K. J.; DiCaprio, J. A.; Walsh, D. A.; Pratt, R. F. *J. Am. Chem. Soc.* **1980,** *102,* 1859.
5. Pohl, E. R.; Osterholtz, F. D. In *Molecular Characterization of Composite Interfaces;* Ishida, H., Kuma, G., Eds.; Plenum: New York, 1985; p 157.
6. Pope, E. J. A.; Mackenzie, J. D. *J. Non-Cryst. Solids* **1986,** *87,* 185.
7. Kay, B. D.; Assink, R. A. In *Better Ceramics through Chemistry II;* Brinker, C. J.; Clark, D. E.; Ulrich, D. R., Eds.; Materials Research Society: Pittsburgh, PA, 1986; p 157.
8. Assink, R. A.; Kay, B. D. *J. Non-Cryst. Solids* **1988,** *99,* 359.
9. Kay, B. D.; Assink, R. A. *J. Non-Cryst. Solids* **1988,** *104,* 112.
10. Klemperer, W. G.; Mainz, V. V.; Millar, D. M. In *Better Ceramics through Chemistry II;* Brinker, C. J.; Clark, D. E.; Ulrich, D. R., Eds.; Materials Research Society: Pittsburgh, PA, 1986; p 3.
11. Horn, H.; Marsmann, H. C. *Makromol. Chem.* **1972,** *162,* 255.
12. Balfe, C. A.; Martinez, S. L. In *Better Ceramics through Chemistry II;* Brinker, C. J.; Clark, D. E.; Ulrich, D. R., Eds.; Materials Research Society: Pittsburgh, PA, 1986; p 27.
13. Kelts, L. W.; Effinger, N. J.; Melpolder, S. M. *J. Non-Cryst. Solids* **1986,** *83,* 353.

14. Q^n nomenclature is taken from that of Englehardt, G. et al., *Z. Chem.* **1974,** *14,* 109, where the superscript "n" refers to the number of O–Si groups bonded to the silicon of interest.
15. Marsmann, H. C.; Meyer, E.; Vongehr, M.; Weber, E. F. *Macromol. Chem.* **1983,** *184,* 1817.
16. Orcel, G.; Hench, L. *J. Non-Cryst. Solids* **1986,** *79,* 177.
17. Englehardt, G.; Altenburg, W.; Hoebbel, D.; Wieker, W. *Z. Anorg. Allg. Chem.* **1977,** *428,* 43.
18. Hoebbel, D.; Garzo, G.; Englehardt, G.; Till, A. *Z. Anorg. Allg. Chem.* **1979,** *450,* 5.
19. Harris, R. K.; O'Connor, M. J. *J. Magn. Reson.* **1984,** *57,* 115.
20. Harris, R. K.; Knight, C. T. G. *J. Chem. Soc. Faraday, Trans. 2* **1983,** *79,* 1525.
21. Harris, R. K.; Knight, C. T. G. *J. Chem. Soc. Faraday, Trans. 2* **1983,** *79,* 1539.
22. Artaki, I.; Bradley, M.; Zerda, T. W.; Jonas, J. *J. Phys. Chem.* **1985,** *89,* 4399.
23. Zerda, T. W.; Artaki, I.; Jonas, J. *J. Non-Cryst. Solids* **1986,** *81,* 365.

RECEIVED for review May 27, 1988. ACCEPTED revised manuscript May 5, 1989.

15

A Silicate Substitution Route to Organosilicon Compounds

George B. Goodwin[1] and Malcolm E. Kenney

Department of Chemistry, Case Western Reserve University, Cleveland, OH 44106

A substitution route to organosilicon compounds is described. In this route, a silicate ion that has the framework of the desired compound, or one similar to it, is converted to the desired compound by three successive substitutions and, if necessary, by a slight structure modification. The silicate ion is obtained by synthesizing an appropriate silicate or by acquiring one from an outside source. The substitutions involved are a protonation, an esterification, and an alkyl-de-alkoxylation. An illustration of this route is given in which the linear $(SiO_3)_n^{2n-}$ *ion is converted to the cyclic* $(SiO_3)_4^{8-}$ *ion and then to the cyclic organosilicon compound* $[(CH_3)_2SiO]_4$. *This synthesis of* $[(CH_3)_2SiO]_4$ *leads to a synthesis of* $[(CH_3)_2SiO]_n$ *that does not entail the reduction of tetravalent silicon to elemental silicon.*

THE USEFUL ROUTES TO ORGANOSILICON compounds include those in which the major identifying steps are the addition of an organic halide to silicon, the addition of a silicon hydride to an olefin, and the displacement of a halogen or an alkoxy group from a halo- or alkoxysilane with an organic group (*1–3*.) Although these three routes and some of the other known routes permit reasonable access to many organosilicon compounds, they do not permit as ready access to some of them as is desirable, and they do not permit any access to others. Because of this, alternative routes to organosilicon compounds are needed.

Of the possible alternative routes, those based on the substitution of

[1]Current address: Glass Research and Development Center, PPG Industries, Inc., Pittsburgh, PA 15238

0065–2393/90/0224–0251$06.00/0

the pendent oxygens of silicate ions appear to be particularly attractive. Recently, we have described a route of this type (*4–6*). In one variation of this substitution route, the steps can be approximated as (1) the conversion of a silicate ion, or of silica, to a silicate ion with a framework that is the same as or similar to that of the desired product; (2) the displacement of the metal ions counterbalancing this silicate ion with H^+ ions and, if necessary, a slight modification of the framework; (3) the displacement of the OH groups of the resulting silicic acid with OR groups and, if necessary, a slight modification of the framework; and (4) the displacement of the OR groups of the resulting alkyl silicate with R groups. In a second variation of the route, a silicate ion with the required framework is obtained from natural, industrial, or laboratory supply sources. The ion in this silicate is then treated as before.

Previous work (*7*) provides precedents for the type of reaction required for framework conversion in the synthetic-silicate variation of the route. Likewise, previous work provides precedents for the metal ion and OH group displacement steps (*8–10*) and for the OR group displacement step (*11–13*) of both variations of the route. In addition, work has been reported recently (*14*) on another related route to organosilicon compounds. However, no work, other than that from this laboratory, has been reported that provides a precedent for the full route.

In this chapter, a new example of the synthetic-silicate variation of the route is presented, and this variation and the available-silicate variation of the synthesis are discussed.

Illustrative Syntheses

$Ca_8(SiO_3)_4Cl_8$ from Wollastonite, $CaSiO_3$. A mixture of wollastonite (Mexico, powdered) and $CaCl_2{\cdot}2H_2O$ (1:2 mole ratio) contained in Pt dishes was placed in an inert-atmosphere furnace. Under a N_2 flow, the mixture was heated slowly to 775 °C and held at this temperature for 17 h. The resulting solid was crushed with a mortar and pestle in a N_2 atmosphere, washed with CH_3OH, and dried. The product was shown by powder X-ray diffractometry to be $Ca_8(SiO_3)_4Cl_8$ (*15*) (97% yield).

$Ca_8(SiO_3)_4Cl_8$ from α-Quartz, SiO_2. The work of Chukhlantsev (*15*) and Weiker (*16*) was used as a guide for this synthesis. A mixture of α-quartz, $CaCO_3$, and $CaCl_2{\cdot}2H_2O$ (1:1:2 mole ratio) contained in a Pt dish was placed in an inert-atmosphere furnace. The mixture was heated to 820 °C slowly and held at this temperature for 24 h and then cooled to 770 °C and held at this temperature for 48 h. The resulting solid was crushed with a mortar and pestle in a N_2 atmosphere, washed with CH_3OH, and dried. The product was shown by powder X-ray diffractometry to be $Ca_8(SiO_3)_4Cl_8$ (*15*) (99% yield).

$[(C_2H_5O)_2SiO]_4$ from $Ca_8(SiO_3)_4Cl_8$. An $HCl–C_2H_5OH$ solution was added over a 5-min period to a slurry of $Ca_8(SiO_3)_4Cl_8$ (wollastonite-based synthesis), C_2H_5OH, and toluene, and the resulting mixture (260:8:1 $C_2H_5OH:HCl:Ca_8(SiO_3)_4Cl_8$ mole ratio) was distilled until a substantial amount of distillate had been collected. The residue was filtered, and the solid was washed with pentane. The filtrate and washings were combined, and the solution formed was reduced to an oil by vacuum evaporation. This oil was mixed with C_2H_5OH, toluene, and an $HCl–C_2H_5OH$ solution, and the resulting solution was slowly distilled until a substantial amount of distillate had been collected. The residue was concentrated under vacuum, and the concentrate was separated into fractions by vacuum distillation. An appropriate fraction was examined by GC, GC–IR spectroscopy, GC–mass spectrometry, and ^{29}Si NMR spectroscopy. The results showed that the main component of the fraction (86%) was $[(C_2H_5O)_2SiO]_4$ (a known compound [*17*]). From this result, the yield of $[(C_2H_5O)_2SiO]_4$ in this component, that is, the contained yield, was calculated to be 56%.

$[(CH_3)_2SiO]_4$ from $[(C_2H_5O)_2SiO]_4$. A solution of CH_3MgCl in tetrahydrofuran was added slowly to a cooled (dry ice–acetone bath) solution of $[(C_2H_5O)_2SiO]_4$ in tetrahydrofuran. While being stirred, the resulting mixture was warmed to a somewhat higher temperature (ice bath) and held at this temperature for 48 h. The reaction product was subjected to a work-up procedure in which aqueous HCl and pentane were used, the resulting organic and aqueous phases were separated, and the organic phase was washed with aqueous NaCl and then concentrated under vacuum. The residual liquid was subjected to flash chromatography (silica gel), and the eluate was fractionally distilled. A selected fraction of the distillate was partially fractionally redistilled, and the remaining oil was retained. This oil was examined by GC, GC–mass spectrometry, and IR spectroscopy. The results showed that its main component (95%) was $[(CH_3)_2SiO]_4$. From this the contained yield of $[(CH_3)_2SiO]_4$ was calculated to be 37%.

Discussion

Synthesis of $[(CH_3)_2SiO]_4$ from Wollastonite, $CaSiO_3$. The synthesis of $[(CH_3)_2SiO]_4$ from wollastonite, $CaSiO_3$, is a clear example of the synthetic-silicate variation of the substitution route to organosilicon compounds. In this particular example, the linear $(SiO_3)_n{}^{2n-}$ ion of wollastonite is first rearranged to the cyclic $(SiO_3)_4{}^{8-}$ ion. Next, the metal ions that counterbalance the $(SiO_3)_4{}^{8-}$ ion are displaced by H^+ ions [or some of the metal ions are displaced by H^+ ions, and then simultaneously, the remaining metal ions are displaced by H^+ ions, and (formally) some of the OH groups produced in the initial or second stage of this step are displaced by OC_2H_5 groups]. After this, the OH groups of the resulting $[(HO)_2SiO]_4$ [or the resulting $(HO)_{8-x}(C_2H_5O)_xSi_4O_4]$ are (formally) displaced by OC_2H_5 groups, and finally, the OC_2H_5 groups of the $[(C_2H_5O)_2SiO]_4$ produced are displaced by CH_3 groups. (The mechanistic details of the displacement of OH by OC_2H_5 are unknown).

If the silicic acid intermediate in this synthesis is assumed to be $[(HO)_2SiO]_4$, the synthesis can be summarized as follows:

$$Ca_n(SiO_3)_n + CaCl_2 \cdot 2H_2O \xrightarrow{N_2} Ca_8(SiO_3)_4Cl_8 \quad (1)$$

$$Ca_8(SiO_3)_4Cl_8 + H^+ \xrightarrow{C_2H_5OH} [(HO)_2SiO]_4 \quad (2)$$

$$[(HO)_2SiO]_4 + C_2H_5OH \xrightarrow{H^+} [(C_2H_5O)_2SiO]_4 \quad (3)$$

$$[(C_2H_5O)_2SiO]_4 + CH_3MgCl \xrightarrow{C_4H_8O} [(CH_3)_2SiO]_4 \quad (4)$$

A low acid concentration and a low temperature were used in reaction 2 because it was believed that these conditions would help to restrict condensation and cleavage reactions and thus help to preserve the silicon–oxygen framework. [The fact that these conditions could be used results from the high susceptibility of $Ca_8(SiO_3)_4Cl_8$ to acid attack. Acid susceptibility, although not as high as that of $Ca_8(SiO_3)_4Cl_8$, is, as is apparent, a necessary feature of the silicates that are useful in the approach.] A high alcohol ratio was used in the step, because it was thought that it would lead to protection of the silicic acid through dilution and, probably, through the formation of hydrogen-bonded alcohol sheaths around the individual silicic acid molecules.

A high alcohol ratio was used in the initial part of reaction 3 partly as a result of its use in reaction 2. However, it was believed that here, too, it would yield protection for the silicic acid.

In reaction 4, a low temperature was used because it was thought that it would help limit cleavage of the silicon–oxygen ring.

The acid in reaction 3 functions as an esterification catalyst. The toluene in this reaction serves as a dehydrating agent (toluene forms an azeotrope with ethanol and water containing 12% water [*18*]). In addition, the toluene renders the reaction solution a poor solvent for the byproduct salts and thus facilitates their separation.

Synthesis of $[(CH_3)_2SiO]_4$ from α-Quartz, SiO_2. The preparation of $Ca_8(SiO_3)_4Cl_8$ from α-quartz shows that $[(CH_3)_2SiO]_4$ also can be synthesized from quartz by way of $Ca_8(SiO_3)_4Cl_8$. This synthesis is another clear example of the synthetic-silicate variation of the substitution route to organosilicon compounds. The last three steps of this synthesis are given by reactions 2–4, and the first is as follows:

$$(SiO_2)_n + CaCO_3 + CaCl_2 \cdot 2H_2O \xrightarrow{N_2} Ca_8(SiO_3)_4Cl_8 \quad (5)$$

This first step (reaction 5) clearly involves greater silicon–oxygen framework alteration than does the first step of the wollastonite-based synthesis (reaction 1). Thus, it is not surprising that, apparently, more-vigorous reaction conditions are required for this step than for the first step of the wollastonite-based synthesis.

Syntheses of $[(CH_3)_2SiO]_n$ from Wollastonite and Quartz. The wollastonite- and quartz-based syntheses of $[(CH_3)_2SiO]_4$ make possible syntheses of $[(CH_3)_2SiO]_n$ based on wollastonite and quartz, because only a known ring-opening step (*1*, *2*) needs to be added to these syntheses. Thus, a synthesis of $[(CH_3)_2SiO]_n$ from wollastonite by way of $Ca_8(SiO_3)_4Cl_8$ entails the use of reactions 1–4 and the following reaction:

$$[(CH_3)_2SiO]_4 \xrightarrow{H^+} [(CH_3)_2SiO]_n \tag{6}$$

In polymer chemistry terms, this wollastonite-based synthesis can be viewed as involving an oligomerization of a linear polymer (Scheme I, **1**) to a cyclic tetramer (**2**), three successive substitutions on the cyclic tetramer (**3–5**), and finally, a ring-opening polymerization of the last intermediate (**6**). Correspondingly, the quartz-based synthesis can be viewed as involving an oligomerization of a fully cross-linked polymer (Scheme II, **7**) to a cyclic tetramer (**8**) and then steps paralleling the last four steps of the wollastonite-based synthesis.

Syntheses of $[(CH_3)_2SiO]_n$ by Conventional and New Routes. As is well known, the conventional synthesis of $[(CH_3)_2SiO]_n$ involves the following reactions (*1–3*):

$$SiO_2 + C \longrightarrow Si + CO \tag{7}$$

$$Si + CH_3Cl \xrightarrow{Cu} (CH_3)_2SiCl_2 \tag{8}$$

$$(CH_3)_2SiCl_2 + H_2O \longrightarrow (CH_3)_2SiCl_x(OH)_{2-x} \tag{9}$$

$$(CH_3)_2SiCl_x(OH)_{2-x} \longrightarrow [(CH_3)_2SiO]_n \tag{10}$$

This conventional or reduction synthesis of $[(CH_3)_2SiO]_n$ differs in several significant ways from the wollastonite- and quartz-based syntheses of $[(CH_3)_2SiO]_n$. For example, the reduction synthesis requires the reduction of the silicon from +4 to 0 and then its reoxidation back to +4. This reduction necessitates the expenditure of considerable energy, much of which is not recovered. In the new or substitution syntheses, the reduction of silicon from +4 to 0 is not required (Figure 1).

The two types of synthesis also differ in that in the reduction synthesis the number of oxygens bound to silicon is reduced from four to zero and

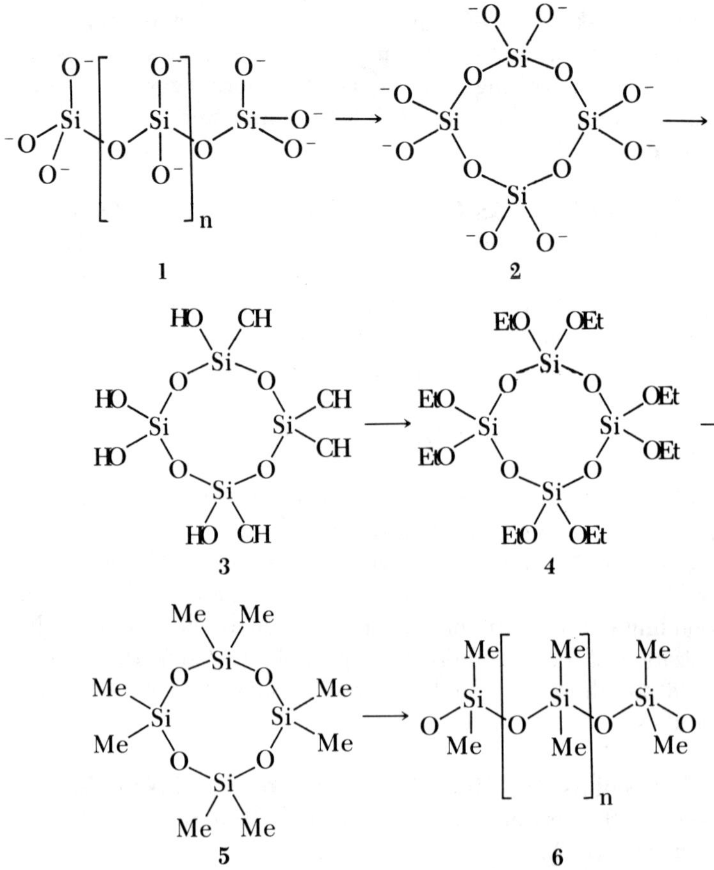

Scheme I. Species involved in wollastonite-based substitution synthesis of $[(CH_3)_2SiO]_n$.

then increased back to two. In the substitution syntheses, the number of oxygens bound to silicon is at least two throughout (Figure 2). (Because the reduction synthesis is an indirect route both in terms of structure and oxidation number, the common practice of using the adjective "direct" to describe the oxidation step of this synthesis [reaction 8] [*1–3*], and the resulting close association of this adjective with the whole synthesis leads to a substantial irony.)

Another difference between the two types of synthesis is the way in which the methyl groups are introduced. In the reduction synthesis, the methyl groups are introduced by means of an oxidation–addition reaction, whereas in the substitution syntheses, they are introduced by means of an alkyl-de-alkoxy-substitution reaction.

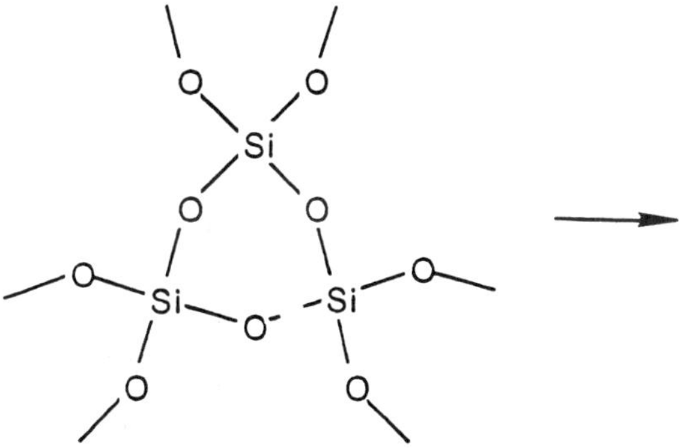

7

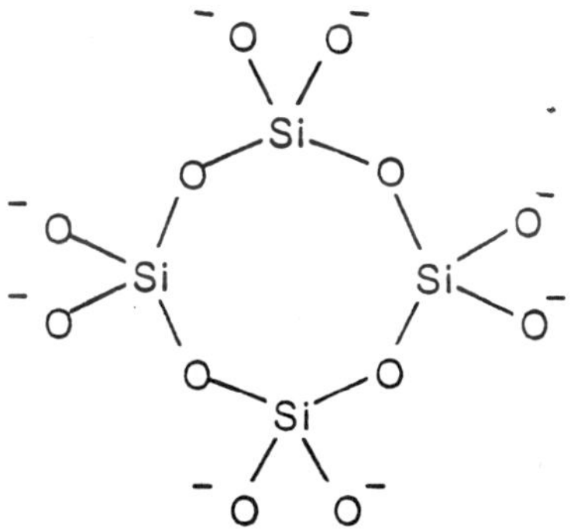

8

Scheme II. Species involved in the first step of quartz-based substitution synthesis of $[(CH_3)_2SiO]_n$.

However, it should be made clear with regard to this latter point that, whereas the alkyl groups are introduced in the new syntheses by a substitution reaction, this reaction utilizes a Grignard reagent. Thus, although the new syntheses do not require the reduction of SiO_2 to Si, they do ultimately require the reduction of $MgCl_2$ to Mg, and accordingly, they require ultimately a significant reduction reaction. Nevertheless, the existence of these substitution syntheses does show that reasonable alternatives to the reduction synthesis are possible.

Perhaps in this connection it is worth noting that the reduction synthesis

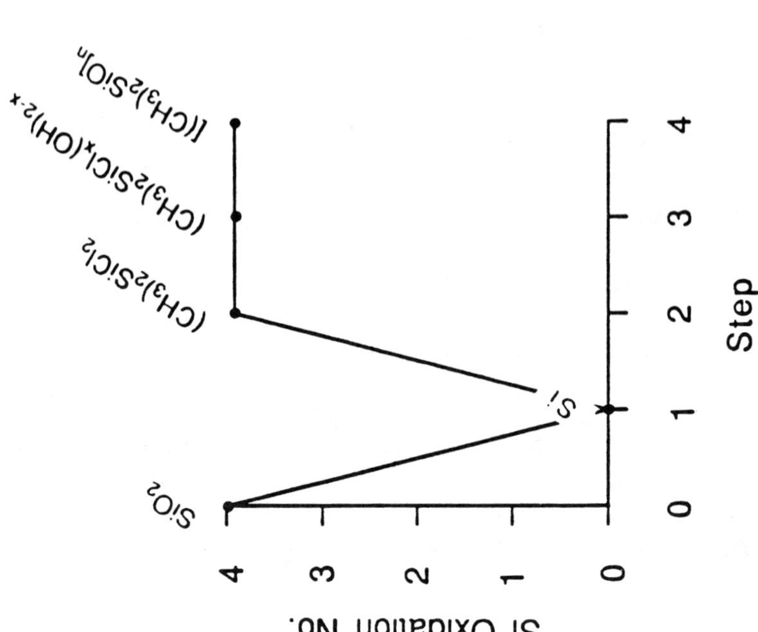

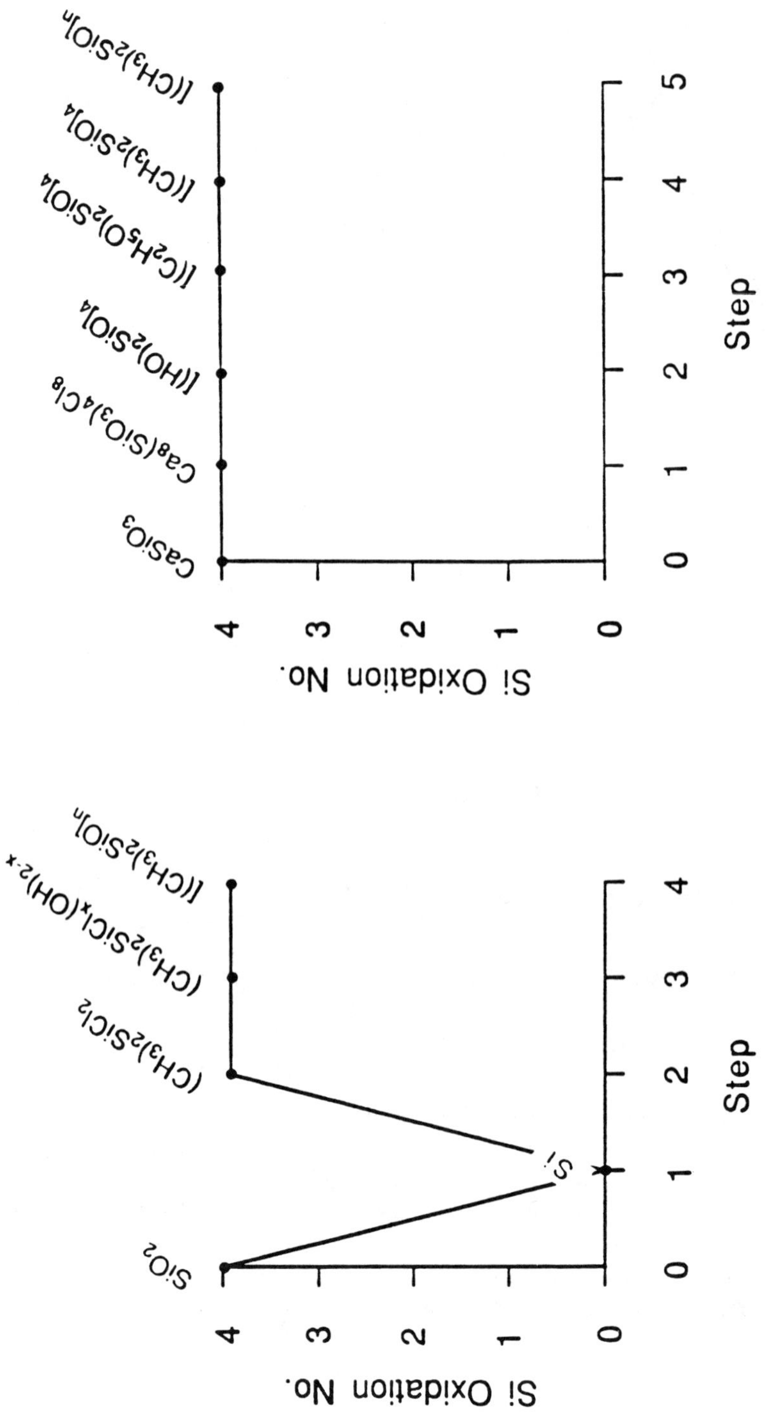

Figure 1. Variation of silicon oxidation number in reduction synthesis (top) and wollastonite-based substitution synthesis (bottom) of $[(CH_3)_2SiO]_n$.

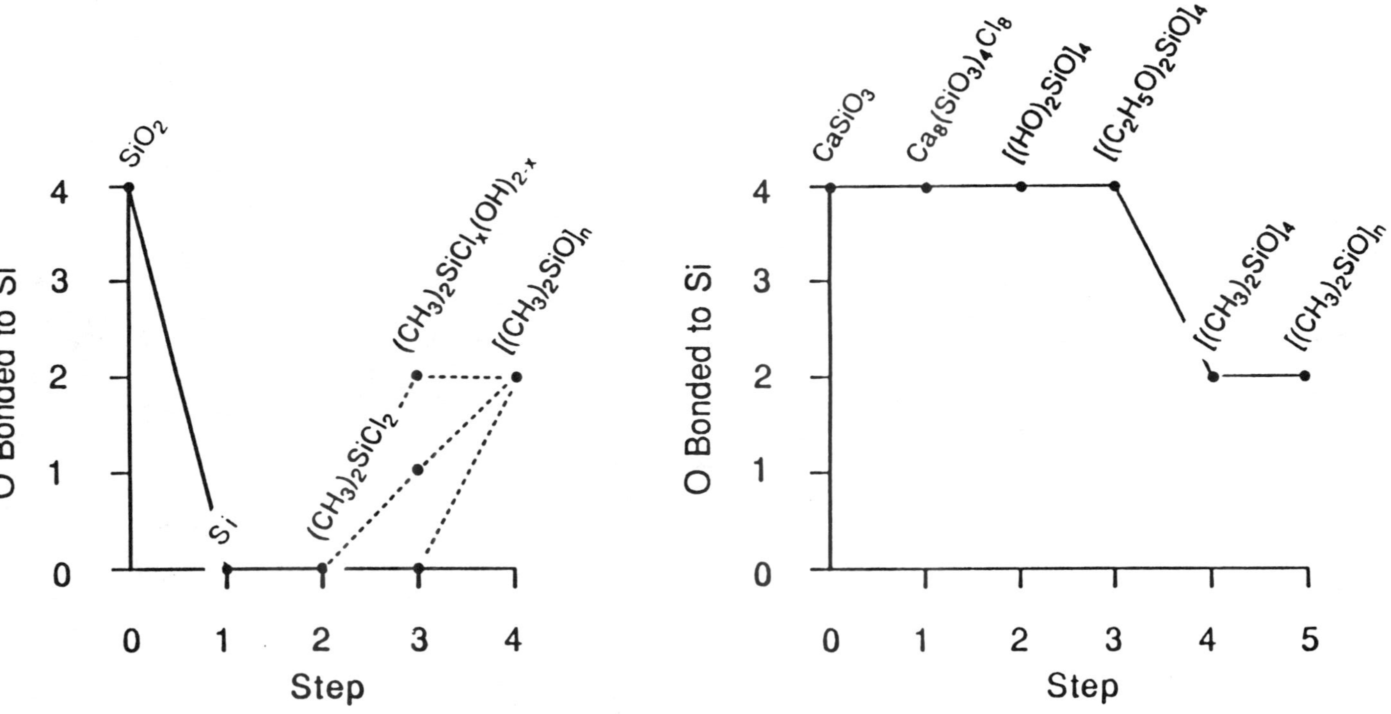

Figure 2. Variation of number of oxygens bonded to silicon in reduction synthesis (top) and wollastonite-based substitution synthesis (bottom) of $[(CH_3)_2SiO]_n$.

is now nearly 50 years old (*19*), and thus its life span is already well beyond that typical for commercial processes for organic compounds (*20*). This fact and the findings of this study suggest that further work on alternatives to the reduction synthesis could be fruitful.

Synthesis of Other Organosilicon Compounds. The preparation of the isomer pair decamethylbicyclo[5.5.1]hexasiloxane, [5.5.1] $(CH_3)_{10}Si_6O_7$ (**9**), and decamethylbicyclo[5.3.3]hexasiloxane, [5.3.3] $(CH_3)_{10}Si_6O_7$ (**10**), from SiO_2 by way of the synthetic silicate $Na_4Ca_4Si_6O_{18}$ has been reported previously (*4–6*). This synthesis provides another example of the synthetic-silicate variation of the substitution route.

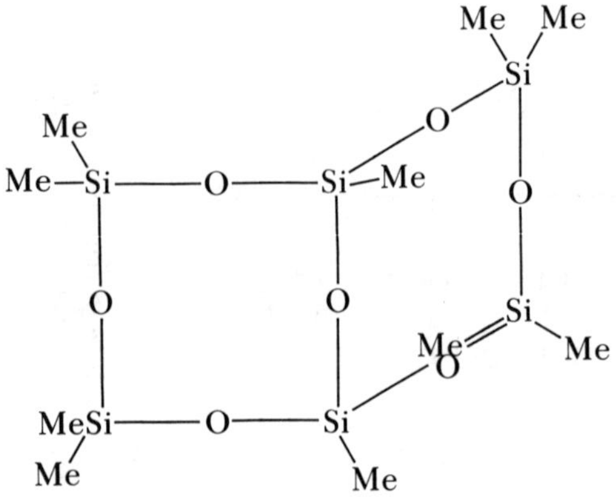

9

Decamethylbicyclo[5.5.1]hexasiloxane

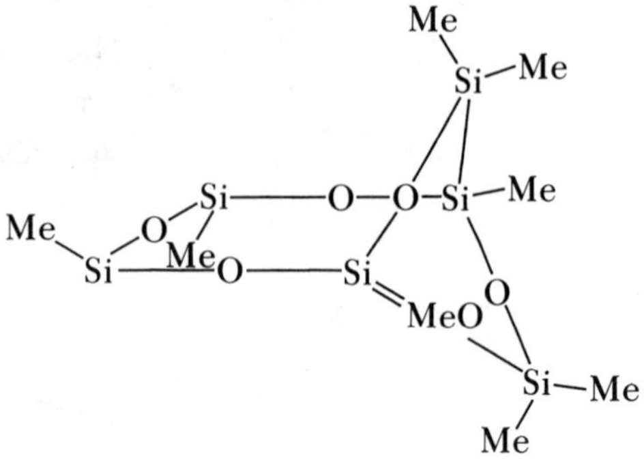

10

Decamethylbicyclo[5.3.3]hexasiloxane

The preparation of the isomer pair decaethoxybicyclo[5.5.1]-hexasiloxane, [5.5.1] $(C_2H_5O)_{10}Si_6O_7$, and decaethoxybicyclo[5.3.3]-hexasiloxane, [5.3.3] $(C_2H_5O)_{10}Si_6O_7$, from the natural silicate dioptase $(Cu_6Si_6O_{18}{\cdot}6H_2O)$ has also been reported (*4–6*). The fact that [5.5.1] $(C_2H_5O)_{10}Si_6O_7$ and [5.3.3] $(C_2H_5O)_{10}Si_6O_7$ are the alkyl silicates used in the synthesis of [5.5.1] $(CH_3)_{10}Si_6O_7$ and [5.3.3] $(CH_3)_{10}Si_6O_7$ shows that the available-silicate variation of the substitution route can be used to make the isomer pair [5.5.1] $(CH_3)_{10}Si_6O_7$ and [5.3.3] $(CH_3)_{10}Si_6O_7$ from $Cu_6Si_6O_{18}{\cdot}6H_2O$. Similarly, because $(C_2H_5O)_4Si$ can be used to make $(CH_3)_4Si$ (*21*), the synthesis of $(C_2H_5O)_4Si$ from the purchased samples of the silicates γ-Ca_2SiO_4 and Ca_3SiO_5 (*5, 6*) shows that the available-silicate variation of the route can be used to prepare $(CH_3)_4Si$ from γ-Ca_2SiO_4 and Ca_3SiO_5. Likewise, because $(C_2H_5O)_3SiOSi(OC_2H_5)_3$ can be used to make $(C_2H_5)_3SiOSi(C_2H_5)_3$ (*11*), the synthesis of $(C_2H_5O)_3SiOSi(OC_2H_5)_3$ from SiO_2 by way of hardystonite, $Ca_2ZnSi_2O_7$ (*5*), shows that the synthetic-silicate variation of the route can be used to prepare $(C_2H_5)_3SiOSi(C_2H_5)_3$ from $Ca_2ZnSi_2O_7$.

These examples show that many different ring and chain silicates can be used in the route and suggest that other silicates with more complex structures may also be candidates for use in this route. Many of these candidates occur naturally in reasonable amounts or can be synthesized easily. On the basis of the work done so far, a number of silicates having Ca^{2+}, Ba^{2+}, or Zn^{2+} ions as the cations and some silicates with water of hydration (i.e., analogues of $Cu_6Si_6O_{18} \cdot 6H_2O$) have enough acid susceptibility to be good reactants. Many silicates with Na^+, Mg^{2+}, or Al^{3+} ions as the cations have too little acid susceptibility to be good reactants.

Additional previously reported work demonstrates that alcohols other than ethanol, such as 1-propanol, 2-propanol, and 1-butanol, can be used to prepare alkoxides by the approach reported in this chapter (*4–6*). This suggests that other alcohols can be used in the route. From what is known, it appears that other acids can also be used in the route. More important, it is highly probable that a wide variety of Grignard reagents are useful. Clearly, the substitution route is flexible and can be used to make a variety of known and novel organosilicon compounds.

Synthesis of Other Silicon Compounds. The substitution route to organosilicon compounds obviously also opens up new avenues to alkyl silicates. For example, as indicated previously, the substitution route is a good avenue to the isomeric alkyl silicates [5.5.1] $(C_2H_5O)_{10}Si_6O_7$ and [5.3.3] $(C_2H_5O)_{10}Si_6O_7$. The availability of this substitution route for alkyl silicates is of considerable importance, because some of the routes that have been used to make complex alkyl silicates are not good in terms of yield or convenience (22).

Summary

A substitution route to organosilicon compounds is described. In this route, the silicon–oxygen framework of the compound wanted or a framework similar to it, is made by rearrangement of the framework in a silicate or silica or is secured by obtaining a silicate in which the framework is present from an outside source. The compound wanted is made from this framework by displacement of the pendent oxygen atoms and, where needed, by additional rearrangement of it. This route opens a way to syntheses of $[(CH_3)_2SiO]_n$ that do not entail the reduction of tetravalent silicon to elemental silicon.

Acknowledgments

We thank Ralph E. Temple, Dale R. Pulver, and Gordon Fearon for helpful discussions. We also thank Dow Corning Corporation for financial support.

References

1. Stark, F. O.; Fallender, J. R.; Wright, A. P. In *Comprehensive Organometallic Chemistry;* Wilkinson, G., Ed.; Pergamon: Oxford, 1982; Vol. 2, p 305.
2. Hardman, B. B.; Torkelson, A. In *Kirk–Othmer Encyclopedia of Chemical Technology,* 3rd ed.; Grayson, M., Ed.; Wiley: New York, 1982; Vol. 20, p 922.
3. Noll, W. *Chemistry and Technology of Silicones;* Academic: New York, 1968; Chapter 2.
4. Goodwin, G. B.; Kenney, M. E. *Polym. Prepr. (Am. Chem. Soc., Div. Polym. Chem.)* **1986,** *27,* 107.
5. Goodwin, G. B.; Kenney, M. E. In *Inorganic and Organometallic Polymers;* Zeldin, M.; Wynne, K. J.; Allcock, H. R., Eds.; ACS Symposium Series 360; American Chemical Society: Washington, DC, 1988; Chapter 18.
6. Kenney, M. E.; Goodwin, G. B. U.S. Patent 4 717 773, 1988.
7. Deer, W. A.; Howie, R. A.; Zussman, J. *An Introduction to the Rock-Forming Minerals;* John Wiley: New York, 1966.
8. Iler, R. K.; Pinkney, P. S. *Ind. Eng. Chem.* **1947,** *39,* 1379.
9. Bleiman, C.; Mercier, J. P. *Inorg. Chem.* **1975,** *14,* 2853.
10. Calhoun, H. P.; Masson, C. R. *J. Chem. Soc., Dalton Trans.* **1980,** 1282.
11. Smith, B. Ph.D. Thesis, Chalmers Technical High School, Gothenburg, Sweden, 1951, as quoted in Eaborn, C. *Organosilicon Compounds;* Academic: New York, 1960; p 14.
12. Wacker-Chemie G.m.b.H. Br. Patent 732 533, 1955.
13. Compton, R. A.; Petraitis, D. J. U.S. Patent 4 309 557, 1982.
14. Boudin, A.; Cerveau, G.; Chuit, C.; Corriu, R. J. P.; Reye, C. *Angew. Chem. Int. Ed. Engl.* **1986,** *25,* 474.
15. Chukhlantsev, V. G. *Dokl. Phys. Chem. (Engl. Transl.)* **1979,** *246,* 530; *Dokl. Akad. Nauk SSSR* **1979,** *246,* 1187.
16. Winkler, A.; Ziemer, B.; Weiker, W. Z. *Anorg. Allg. Chem.* **1983,** *504,* 89.
17. Marsmann, H. C.; Meyer, E.; Vongehr, M.; Weber, E. F. *Makromol. Chem.* **1983,** *184,* 1817.

18. *Azeotropic Data II;* Horsley, L. H., Ed.; Advances in Chemistry 35; American Chemical Society: Washington, D.C., 1962; p 61.
19. Liebhafsky, H. A. *Silicones Under the Monogram;* John Wiley: New York, 1978; p 103.
20. Achilladelis, B. G., Beckman Center for the History of Chemistry, University of Pennsylvania, Philadelphia, PA, personal communication, 1988.
21. George, P. D.; Sommer, L. H.; Whitmore, F. C. *J. Am. Chem. Soc.* **1955,** *77,* 6647.
22. Day, V. W.; Klemperer, W. G.; Mainz, V. V.; Millar, D. M. *J. Am. Chem. Soc.* **1985,** *107,* 8262.

RECEIVED for review May 27, 1988. ACCEPTED revised manuscript March 10, 1989.

16

Organosilicon Polymers for Microlithographic Applications

Elsa Reichmanis, Anthony E. Novembre, Regine G. Tarascon, Ann Shugard, and Larry F. Thompson

AT&T Bell Laboratories, Murray Hill, NJ 07974

Organosilicon polymers play an increasingly important role in the electronics industry today. These polymers are useful in multilevel-device planarization schemes, because they are soluble and may be readily spin coated to afford conformal, nominally planar surfaces for subsequent metal deposition. Lithographic applications include the use of silicon-containing resins for multilayer-resist processes. Trilevel schemes are simplified through the use of spin-on intermediate layers that act as barriers to oxygen reactive-ion etching (RIE). Alternatively, the coupling of the properties of the top imaging layer with those of the intermediate barrier material results in a bilevel scheme. The chemistry associated with these materials is outlined, and future needs are addressed.

Astonishing progress in integrated-circuit devices has been made during the last 2 decades. The cost per function has decreased by over 3 orders of magnitude, with concomitant improvements in performance. This achievement has been accomplished primarily by increasing the scale of integration. For the past several decades, the number of components per chip has increased by a factor of 10^2–10^3 per decade, and this trend will continue, although perhaps at a slower rate (Figure 1).

This increase in circuit density is made possible only by decreasing the minimum feature size on the chip. Figure 2 illustrates the decrease in minimum feature size as a function of time for dynamic random access memory (DRAM) devices. In 1975, the 4-kilobit DRAM (4×10^3 memory cells or about 8.2×10^3 transistors) had features in the 7–9-μm range, and by 1987,

0065–2393/90/0224–0265$06.00/0

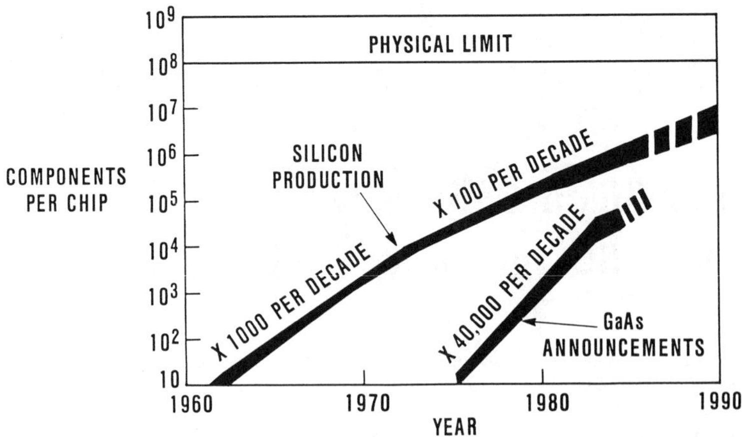

Figure 1. Number of components per chip as a function of time.

1-megabit DRAMS (1×10^6 memory cells and 2.4×10^6 transistors) with minimum features in the 0.8–1.0-μm range were in production. The size of the chip has remained nearly constant and is between 0.5 and 1.0 cm^2 in total area. This phenomenal progress is a direct result of our ability to pattern complex thin film structures with increasing precision and ever-decreasing feature sizes. In addition to finer and more-precise structures, new materials with improved performance are also needed.

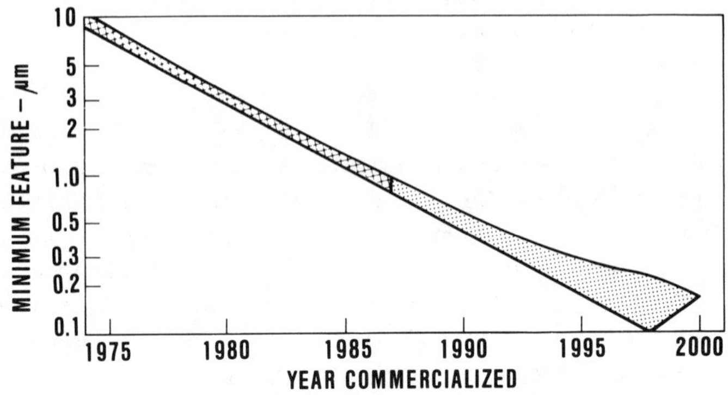

Figure 2. Minimum feature size as a function of time for DRAM devices.

Microlithography, including pattern formation and pattern transfer, is the technology in the overall manufacturing process that makes possible these improvements. Photolithography, which uses light in the 365–436-nm range, has been the dominant method for producing patterns for circuit applications. The technology has evolved over the past decades from simple

1:1 contact printing, to proximity and 1:1 projection printing, and to the current method of 5×- or 10×-reduction step-and-repeat exposure.

The basic resist systems have remained essentially the same; the positive photoresist composed of a novolac resin and a photoactive substituted diazonaphthoquinone dissolution inhibitor is the resist of choice. The current tools and resists will be able to print features as small as 0.5–0.7 μm in a production environment. These systems are almost certainly the last generation of conventional-wavelength photolithographic systems.

Alternatives for the next generation of lithographic systems include 248-nm radiation, 5×-reduction step-and-repeat systems, and electron beam (e-beam) or X-ray technologies. The 248-nm deep-UV option is most likely to become the next major technique. e-Beam and X-ray technologies will also be important in future lithographic strategies; however, when and to what extent they will be commercially significant is not well defined as yet.

All future alternatives will require new resists and processes, and for the first time, manufacturing lines will be using at least two different resists. These new materials must have satisfactory sensitivity, resolution, and process latitude. In addition, the deep-UV tools will have limited depth of focus (1–2 μm) and will be useful only with relatively planar surfaces. Multilayer-resist schemes have been proposed to overcome these limitations, and the simplest is the bilevel scheme that requires a resist that can be converted, after development, to a mask resistant to O_2 reactive ion etching (RIE). Resistance to O_2 RIE can be achieved by incorporating an element into the resist structure that easily forms a refractory oxide. Silicon performs this function very well and is relatively easy to include in a wide variety of polymer structures.

In addition to new resists, future generations of devices will require improved dielectrics that can be deposited at low temperatures and provide process flexibility. Again, organosilicon polymers have promise for this application. In this chapter, the chemistry and processes for both of these areas will be reviewed.

Device Applications

Organosilicon polymers are becoming important in many aspects of device technology. Multilevel metallization schemes require the use of a thin dielectric barrier between successive metal layers (*1*). Often, these dielectric materials are silicon oxides that are deposited by low-temperature or plasma-enhanced chemical vapor deposition (CVD) techniques. Although conformal in nature, CVD films used as intermetal dielectrics frequently result in defects that arise from the high aspect ratios of the metal lines and other device topographies (2). Several planarization schemes have been proposed to alleviate these problems, some of which involve the use of organosilicon polymers (*2–4*).

Organosilicon polymers used in device applications must satisfy a number of criteria. The polymer must have dielectric characteristics similar to those of CVD-deposited materials, exhibit good adhesion to the substrate materials, be crack resistant, and be compatible with current processes. Materials that are useful in these applications are the spin-on-glasses (SOGs) (5). SOGs are organosilicon polymers obtained upon hydrolysis of diethoxyalkyl-, triethoxyalkyl-, and tetralkoxysilanes. The polymeric precursors are soluble in a variety of organic solvents and are readily spin coated to achieve smoothing and nominal planarization of device topography. Curing at elevated temperatures leads to the formation of silicon dioxide or an SiO_2-like material. The material properties of SOGs depend upon the structure of the starting monomer. For instance, a change of substituent from methyl to phenyl on the siloxane precursor results in a change in the electrical characteristics from good to unacceptable after a 425 °C cure, whereas other properties are unaffected (5).

The use of SOG materials for intermetal dielectric applications has several advantages. A spin-on dielectric coating allows high throughput with low capital investment compared with alternative CVD-deposited materials, and process complexity is reduced. However, several material properties have to be improved before organosilicon polymers gain full acceptance by the electronics industry. Existing SOG materials are inadequate as "standalone" dielectric layers because of the extensive cracking that occurs in thick (1-μm) films. Also, adhesion loss often occurs between an SOG layer and the resists and metals used in device manufacturing. Although novel processing techniques may alleviate these problems, the ultimate solution lies with the design of new materials.

Lithographic Applications

The continuing trend toward more-complex devices with increasingly smaller feature sizes places severe demands on conventional single-layer resist processes. New lithographic technologies and processing techniques will be required to achieve the necessary improvements in resolution and line width control. The need to define smaller features, in turn, increases the effects of other problems associated with the particular lithographic strategy. For instance, standing-wave effects and reflections from the substrate surface can limit the resolution attainable with optical techniques. Alternatively, back-scattered electrons lead to proximity effects that affect line width control during e-beam exposure. These problems are intensified by the requirements of smaller features. A number of technologies that involve the use of organosilicon polymers or polymer precursors have been proposed to address these problems. Notable are the trilevel and bilevel lithographic techniques (Figure 3.) These multilayer schemes use a thick layer of organic polymer to nominally planarize the substrate and RIE image-transfer techniques.

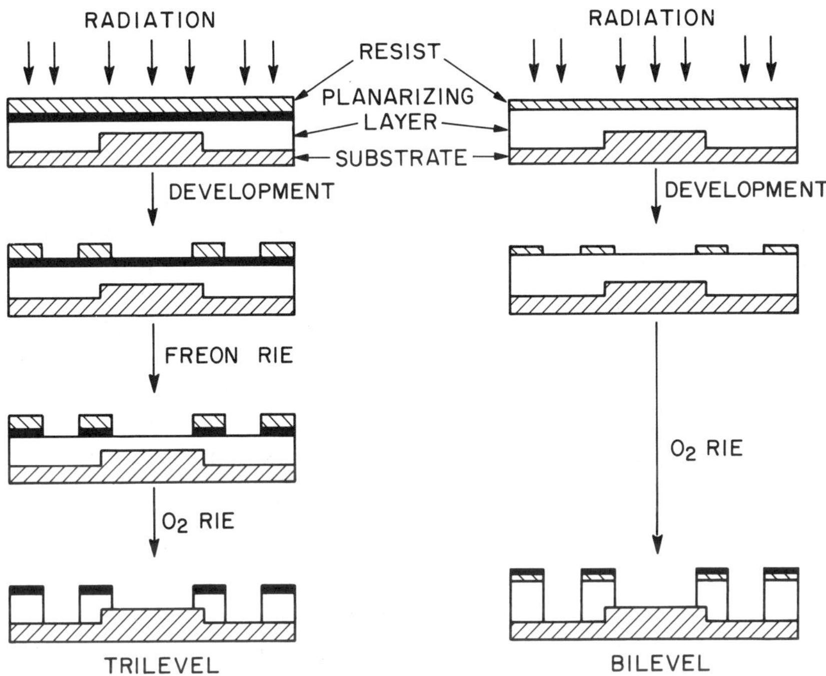

Figure 3. Schematic representation of trilevel and bilevel lithographic techniques.

Trilevel Schemes. Trilevel processing (*6, 7*) requires planarization of device topography with a thick layer of an organic polymer, such as polyimide or a positive photoresist that has been baked at elevated temperatures ("hard baked") or otherwise treated to render it insoluble in most organic solvents. An intermediate RIE barrier, such as a silicon dioxide, is deposited, and finally, this structure is coated with the desired resist material. A pattern is delineated in the top resist layer and subsequently transferred to the substrate by dry-etching techniques (Figure 3).

Whereas several variations of this process have been reported, the most common schemes use silicon dioxide as the intermediate barrier layer. This layer is generally deposited by sputtering, plasma CVD, or spin coating of an organic SiO_2 precursor (SOG material) (*8, 9*). The chemistry of the silicon-containing SOG materials is similar to that described in the previous section on device applications. Alternative SOG intermediate layers may be composed of polysiloxanes (*6*), polymers that are highly resistant to processing by oxygen RIE.

SOG layers greatly simplify trilevel processing through the elimination of expensive, low-pressure deposition steps and are thus attractive alternatives to sputtering and plasma CVD schemes. Unfortunately, these layers have higher defect densities compared with vapor-deposited films. However, proper storage and dispensing techniques can minimize this deficiency.

Bilevel Schemes. Even though the advent of SOG materials has reduced the complexity of trilevel schemes, they remain time-consuming processes that require precise control of several steps in the patterning sequence. However, multilayer techniques do improve the resolution capability of conventional resists by allowing the patterning of a thin resist to be functionally separated from the anisotropic transfer of the image to the substrate. Consequently, a further simplification of trilevel processing would be desirable. One alternative involves the incorporation of the properties of the top resist layer with those of the oxygen-RIE-resistant intermediate layer into a single imaging material. Conventional processing allows pattern definition of this upper layer, and the pattern is then transferred to the substrate by oxygen RIE techniques (Figure 3).

Organosilicon polymers are ideal candidates for use in these processes (*10*). Treatment of organometallic compounds, particularly organosilicon materials, with an oxygen discharge leads to the formation of the corresponding metal oxide. Taylor and Wolf (*11*) have shown that the incorporation of silicon into organic polymers renders the polymers resistant to erosion in oxygen plasmas. This resistance results from the formation of a protective coating of SiO_x on the polymer surface. Modeling studies predict that the thickness of this layer should be about 50 Å, a value that has been confirmed by surface analysis (*12, 13*). Knowledge that a protective layer of silicon dioxide is formed on the surface of these materials has led to considerable work in designing new polymers that incorporate silicon into the polymer structure. However, several problems with silicon-containing polymers may interfere with their lithographic performance. A decrease in glass transition temperature (T_g) often accompanies the incorporation of silicon into a polymer and may cause dimensional instability of patterns during processing. In addition, most useful silicon substituents are hydrophobic in nature. This hydrophobicity is a potentially critical problem if the use of an aqueous developer is desired. Numerous silicon-containing resist systems have been prepared and used in multilevel RIE pattern-transfer processes.

Chemistry. The first organosilicon polymers examined for use in bilevel RIE processes were the polysiloxanes (*14, 15*). These copolymers of dimethylsiloxane, methylphenylsiloxane, or methylvinylsiloxane are negative photo- and e-beam resists that exhibit erosion rates during oxygen RIE of 1–3 nm/min compared with ~80 nm/min for hard-baked positive photoresist. Although high resolution has been achieved with these materials, the imaging layer must be kept thin to avoid problems associated with image creep that arises because of the low T_g (<<30 °C) of the siloxanes.

Polysiloxanes. The initial reports of the utility of polysiloxanes for lithographic applications spurred several research groups to further investigate this class of materials. The problem of low T_g was addressed by preparing chloromethylated poly(diphenylsiloxane) (*16, 17*). More recently, poly(silsesquioxanes) (*18, 19*) have been reported as sensitive, negative, e-beam, ion-beam, and UV resists. These soluble, "ladder"-type polymers prepared by the hydrolysis of substituted chloro- and alkoxysilanes are high-T_g materials (150 °C) with high silicon contents.

Block copolymer chemistry provides a convenient means of incorporating the oxygen-RIE-resistant polysiloxane moiety into a high-T_g, radiation-sensitive polymer (*20*). The flow characteristics of the resist are determined by the unit with higher T_g, and problems associated with phase separation are minimized because of block copolymerization. Specifically, block copolymers of dimethylsiloxane and chlorinated *p*-methylstyrene exhibit good sensitivity, resolution, and thermal properties and low rates of erosion during O_2 RIE.

The synthesis of these materials is outlined in Scheme I. Transmission electron microscopy shows that the morphology of nearly equimolar compositions of the siloxane–chloromethylstyrene block copolymers is lamellar, and that the domain structure is in the order of 50–300 Å. Microphase separation is confined to domains composed of similar segments and occurs on a scale comparable with the radius of gyration of the polymer chain. Auger electron spectroscopy indicates that the surface of these films is rich in silicon and is followed by a styrene-rich layer. This phenomenon arises from the difference in surface energy of the two phases. The siloxane moiety exhibits a lower surface energy and thus forms the silicon-rich surface layer.

Polysilanes. The incorporation of Si–Si, as opposed to Si–O–Si, bonds into the polymer backbone results in resins that undergo chain scission upon exposure to UV radiation. Chain scission results in a decrease in polymer molecular weight, and proper selection of a developer allows the selective removal of the more-soluble irradiated regions. Polysilanes are high-T_g, glassy polymers that photobleach during irradiation and behave as positive resists (*21–25*). They exhibit oxygen RIE rates similar to those of the polysiloxanes and are excellent masks for RIE pattern-transfer processes. Several polysilanes are self-developing upon exposure to high doses of deep-UV radiation and require no solvent development. In related chemistry, a condensation polymer of bis(*p*-chloromethylsilylbenzene) has been reported (*26*) to undergo radiation-induced cleavage of the Si–Si linkages. This cleavage leads to a reduction in polymer molecular weight, enhanced solubility, and positive-resist behavior.

Polymers with Si-Bearing Functional Groups. An alternative to polymers with silicon in the backbone is polymeric resist systems with silicon-

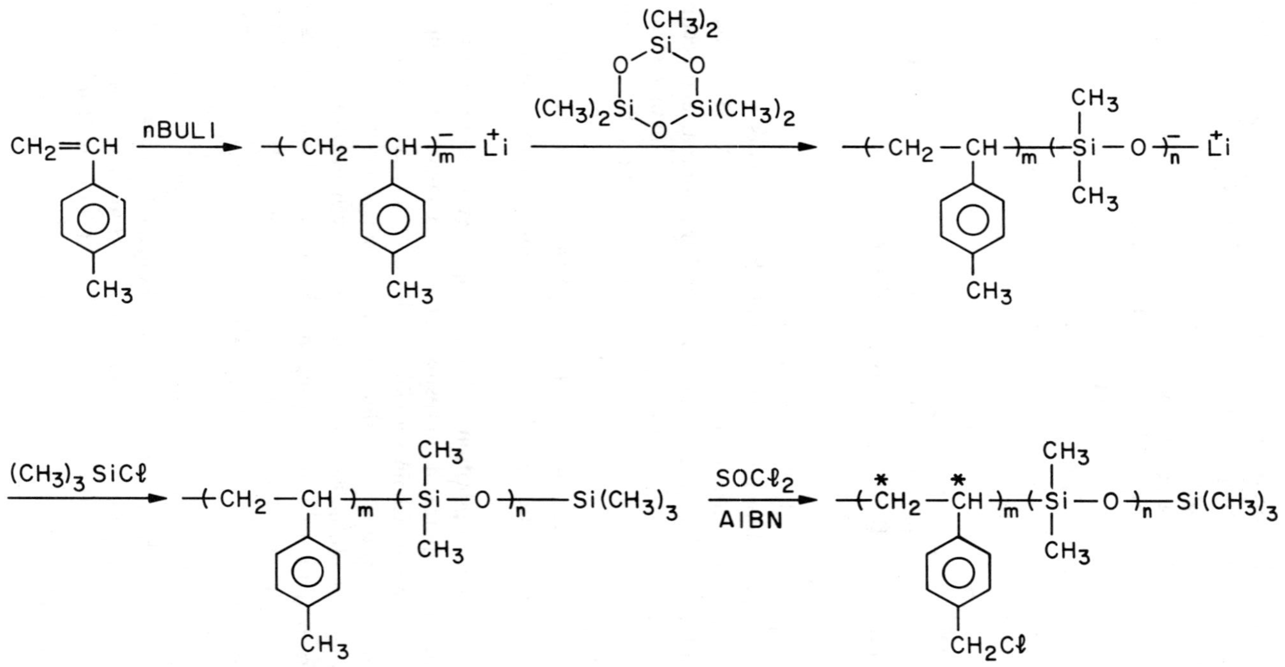

*partial chlorination

Scheme I. Schematic representation of the synthesis of a chlorinated poly(methylstyrene)-block-*poly(dimethylsiloxane).*

bearing functional groups. Negative resists based on trimethylsilylstyrene copolymers (Scheme II) have been reported by both MacDonald and coworkers (27) and Suzuki et al. (28). The silylated styrene imparts resistance to oxygen RIE and is readily copolymerized with radiation-sensitive units such as chlorostyrene and chloromethylstyrene, which undergo cross-linking reactions to impart negative-resist behavior. These random copolymers are capable of 0.5-μm resolution and have sufficient etching resistance to allow pattern transfer through a thick underlying layer of hard-baked positive photoresist.

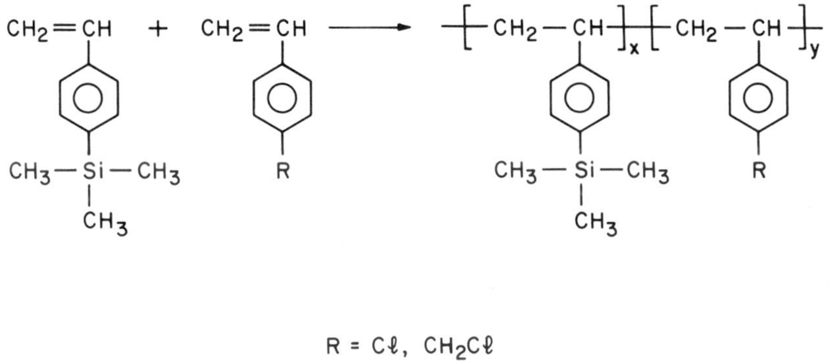

$CH_2{=}CH$ + $CH_2{=}CH$ ⟶ $-[CH_2-CH]_x-[CH_2-CH]_y-$

$CH_3-Si-CH_3$ R $CH_3-Si-CH_3$ R

CH_3 CH_3

R = Cℓ, $CH_2C\ell$

Scheme II. Synthesis of trimethylsilylstyrene negative resists.

Trimethylsilylmethyl methacrylate is an effective etching-resistant component of negative resists (29), even though methacrylates are generally positive-acting materials. Copolymerization of the methacrylate monomer with only 9 wt % chloromethylstyrene (Scheme III) affords a sensitive deep-UV (18 mJ/cm^2) and e-beam (1.95 μC/cm^2) resist capable of submicrometer resolution. High efficiency is achieved in these systems, because the chloromethylstyrene unit is highly susceptible to cross-linking. The dose

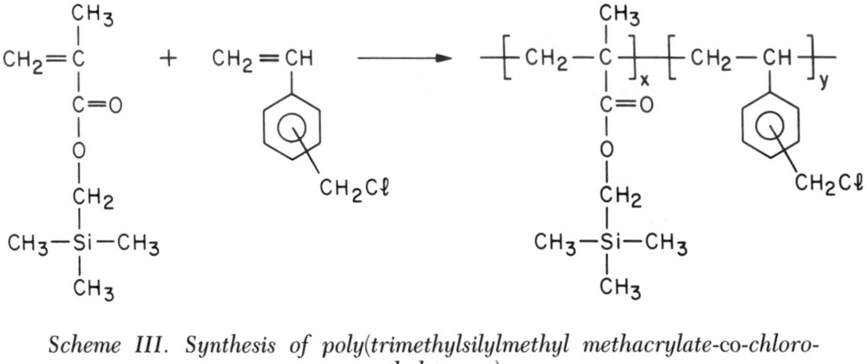

CH_3 CH_3

$CH_2{=}C$ + $CH_2{=}CH$ ⟶ $-[CH_2-C]_x-[CH_2-CH]_y-$

C=O C=O

O O

CH_2 $CH_2C\ell$ CH_2 $CH_2C\ell$

$CH_3-Si-CH_3$ $CH_3-Si-CH_3$

CH_3 CH_3

Scheme III. Synthesis of poly(trimethylsilylmethyl methacrylate-co-chloromethylstyrene).

requirements to cross-link polymers containing even small amounts of chloromethylstyrene are well below those to induce chain scission in most substituted methacrylates. Under typical oxygen RIE conditions, the etching-rate ratios of these systems relative to a hard-baked photoresist were greater than 1:10.

Silylated methacrylates are also useful in positive, chain scission, bilevel deep-UV resists (*30, 31*). However, the silyl substituent must be carefully selected to avoid an excessive decrease in T_g. For example, pentamethyldisiloxypropyl methacrylate has limited utility in resist applications, because it decreases the T_g (*30*). Copolymerization with other monomers may yield glassy, oxygen-resistant materials, but syntheses requiring controlled polymerization of three or more constituents may be unduly complicated.

Alternatively, trimethylsilylmethyl methacrylate polymers are glassy, high-T_g materials that can be readily copolymerized with other monomers to effect photosensitivity. Poly(trimethylsilylmethyl methacrylate-*co*-3-oximino-2-butanone methacrylate) (*31*) is a high-resolution, deep-UV resist capable of submicrometer imaging at a dose of ~250 mJ/cm^2. Etching-rate ratios with respect to typical planarizing layers are 1:10–12 for materials containing ~10 wt % Si. The radiation chemistry of this material involves photolytic cleavage of the N–O bond, followed by decarboxylation and main chain scission (Scheme IV).

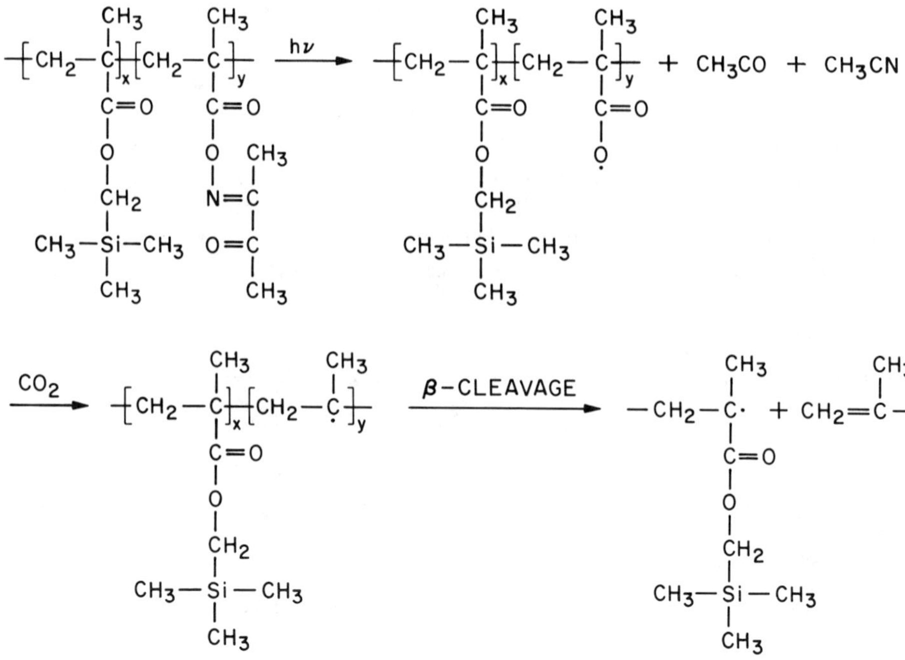

*Scheme IV. Mechanism of the photodecomposition of poly(trimethylsilylmethyl methacrylate-*co*-3-oximino-2-butanone methacrylate).*

Side-chain silicon may also be used to impart resistance to oxygen etching in olefin–sulfone polymers (*32*, *33*), a class of materials that is generally known to exhibit poor dry-etching characteristics. A resolution of 0.25 μm is readily obtained in poly(3-butenyltrimethylsilane sulfone), and formation of a thin protective layer of SiO_x during oxygen RIE allows pattern transfer through a thick underlying layer of planarizing polymer.

Brominated poly(1-trimethylsilylpropyne) is an example of a substituted polyacetylene that is suitable for bilevel-resist processes (*34*). Requiring both exposure and postexposure bake (PEB) steps, samples of the polypropyne having a mole fraction of bromine from 0.1 to 0.2 per monomer unit exhibit sensitivities in the order of 25 mJ/cm^2. Submicrometer resolution has been demonstrated, and etching-rate ratios relative to hard-baked photoresist planarizing layers are ~1:25.

Knowledge that silyl substituents may be incorporated into standard resist chemistry to effect etching resistance has prompted several workers to evaluate silylated novolacs as matrix resins for conventional positive-photoresist formulations. Typically, these resists operate via a dissolution inhibition mechanism whereby the matrix material is rendered insoluble in aqueous base through addition of a diazonaphthoquinone. Irradiation of the composite induces a Wolff rearrangement to yield an indenecarboxylic acid (Figure 4), which allows dissolution of the exposed areas in an aqueous-base developer (*35*).

The first attempt to design a conventional photoresist resistant to RIE made use of trimethylsilylphenol (*36*). However, efforts to prepare an aqueous-base-soluble novolac from this monomer were frustrated by the hydrolytic instability of the bond between the aromatic carbon and the silicon atom. These problems were overcome by insertion of a methylene spacer between the aromatic ring and the silyl substituent. Thus, trimethylsilylmethylphenol may be terpolymerized with cresol and formaldehyde to afford stable, etching-resistant, aqueous-base-soluble resins (*see* structure) (*37*). Formulation with a diazonaphthoquinone inhibitor affords a UV-sensitive resist (120 mJ/cm^2 at 405 nm) that acts as an etching mask for subsequent

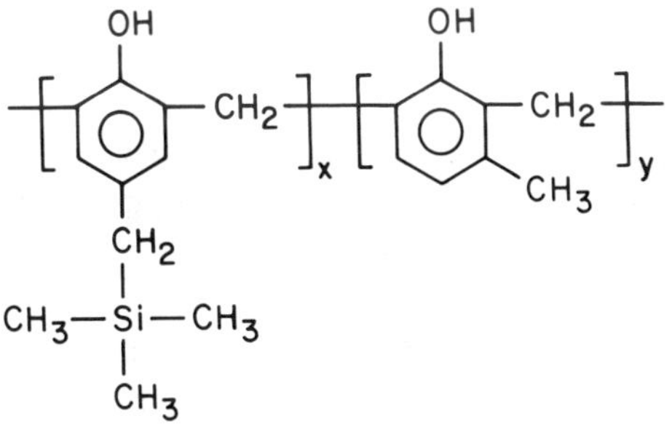

Trimethylsilylmethyl-substituted novolac resin

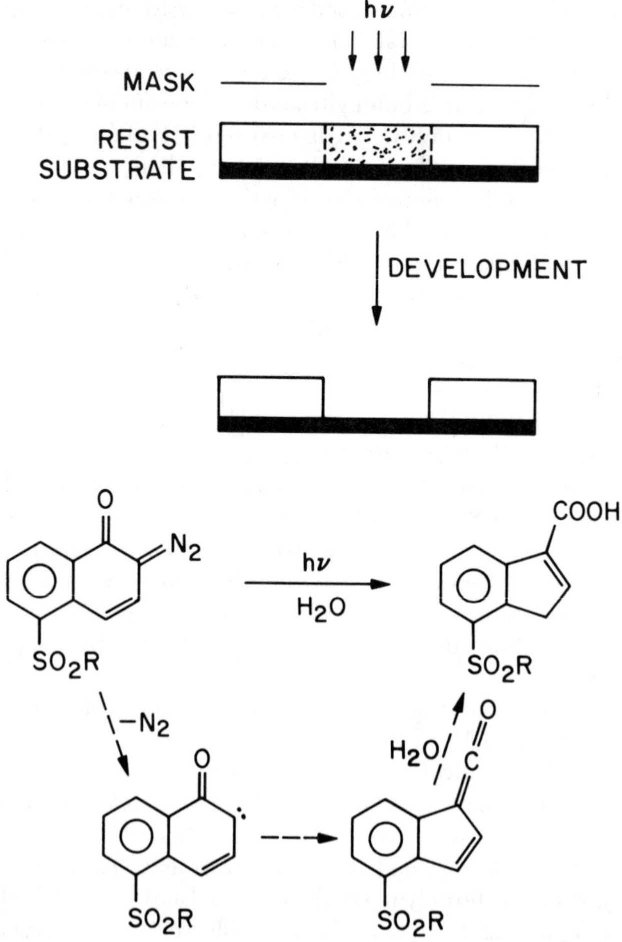

Figure 4. Schematic representation of the mechanism of conventional positive photoresists.

bilevel RIE pattern-transfer processes. Typical patterns that may be obtained are shown in Figure 5.

Poly(2-methyl-1-pentene sulfone) may be used as a dissolution inhibitor to effect e-beam sensitivity (*38*). Trimethylsilylalkoxyphenol is another monomer that has been used in the preparation of oxygen-etching-resistant novolacs for resist applications (*39*). For all of the novolac-based systems studied to date, the hydrophobic nature of the silicon moiety limits the incorporation of silicon to ~10 wt %. However, this level is sufficient to allow use of these resins as oxygen RIE masks.

Other alkali-soluble organosilicon polymers have been investigated for

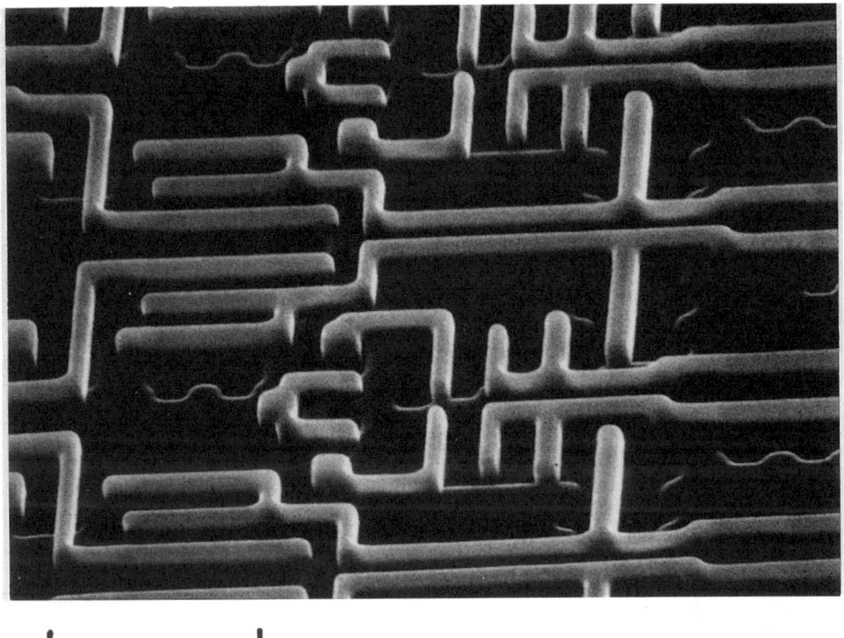

Figure 5. Scanning electron micrograph depicting 1.0-μm images obtained in a silylated novolac–diazoquinone resist formulation after conventional exposure and development followed by pattern transfer by oxygen RIE.

use as matrix resins in positive-photoresist applications. Examples include silylated polyvinylphenols (*40*) and substituted polysilsesquioxanes (*41, 42*). In particular, hydroxyl-substituted poly(benzylsilsesquioxanes) exhibit high resistance to oxygen RIE and may be formulated with diazonaphthoquinone dissolution inhibitors to afford sensitive, high-resolution photoresists (*40, 41*).

Etching Characteristics. The oxygen RIE rate of organosilicon polymers is nonlinear with respect to silicon content, and the incorporation of 10–15 wt % silicon leads to a significant reduction in etching rate (*10*). Figure 6a shows a curve of etching rate versus silicon content for poly(trimethylsilylmethyl methacrylate-*co*-chloromethylsytrene) that is typical of most organosilicon resist systems. Whereas additional incremental increases in the weight percent of Si in a given polymer do not result in a significant decrease in etching rate, the etching selectivity of a bilevel resist with respect to an organic planarizing layer is improved (Figure 6b). The oxygen-etching rates of silicon-containing polymers are related to the mass balance of silicon

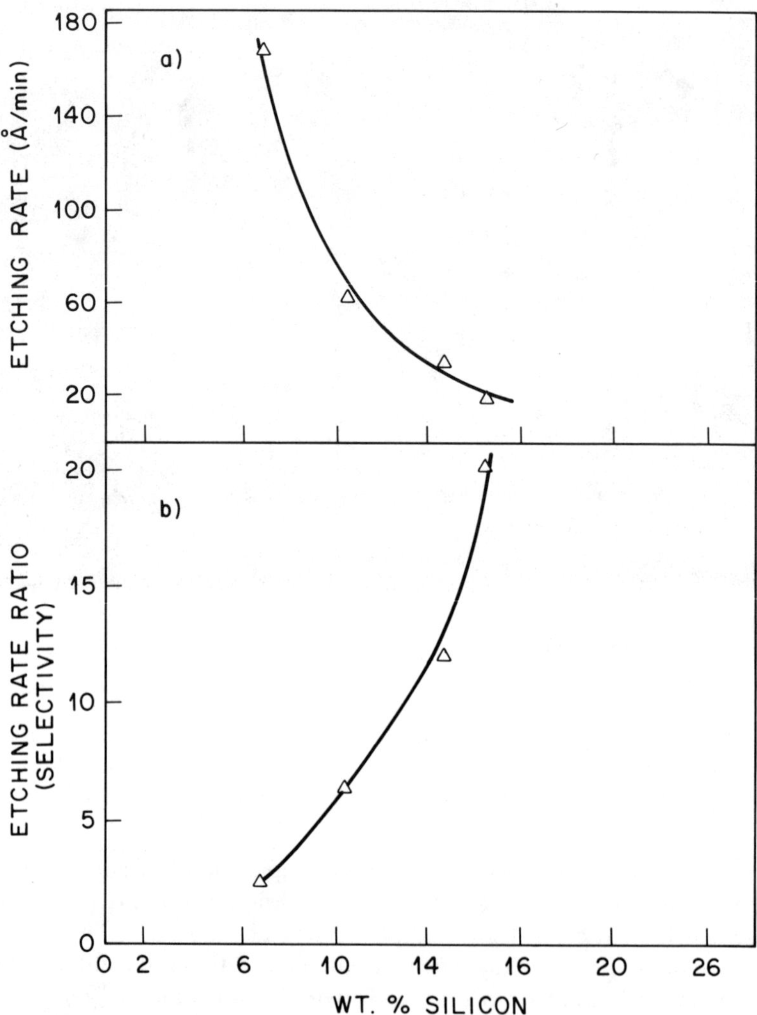

*Figure 6. (a) Etching rate versus silicon content and (b) etching selectivity versus silicon content for poly(trimethylsilylmethyl methacrylate-*co*-chloromethylstyrene).*

in the material when high ion-bombardment-energy conditions, which are typical of RIE processing, are used (*13*). In this case, polymer structure has little, if any, effect on the etching behavior of a given resin, and silicon content is the key variable. Under low-energy conditions, polymer structure plays an increasingly important role. The propensity of a given material to undergo radical-induced degradation reactions appears particularly significant, that is, the etching rate increases relative to predicted values, with increasing ease of radical reactions. Further understanding of the relationship between polymer structure and composition and etching conditions is required to achieve a viable bilevel-resist system.

Summary

Organosilicon polymers are playing an increasingly important role in the electronics industry. These materials are useful in multilevel-device planarization schemes, because they are soluble and may be readily spin coated to afford conformal, approximately planar surfaces for subsequent metal deposition. Lithographic applications include the use of silicon-containing resins for multilayer-resist processes. Trilevel planarization schemes are simplified through the use of spin-on intermediate coatings that act as barrier layers to oxygen RIE. Alternatively, coupling of the properties of the top imaging layer with those of the intermediate barrier materials affords bilevel-resist systems that require fewer processing steps.

Knowledge of the structure and properties of organosilicon polymers is a key to the development of effective materials for integrated-circuit applications. The thermochemistry and plasma chemistry that these polymers undergo during processing change markedly with what may appear to be small changes in polymer structure. These effects must be evaluated and understood to design organosilicon materials appropriate for electronic applications.

References

1. Levin, R. M.; Evans, K. *J. Vac. Sci. Technol., B.* **1983,** *1,* 54.
2. Chu, J. K.; Multani, J. S.; Mittal, S. K.; Orton, J. T.; Jecmen, R. *Proceedings of the Third International IEEE VLSI Multilevel Interconnection Conference;* IEEE: Santa Clara, CA, 1986; p 474.
3. Gupta, S.; Chin, R.; Ferguson, S. *Electrochem. Soc. Ext. Abstr.* **1984,** 2, 448.
4. Kanamori, S. *Semicond. World* **1984,** *10,* 129–133.
5. Allied Signal Corporation, Milpitas, CA, technical information.
6. Havas, J. *Electrochem. Soc. Ext. Abstr.* **1976,** *2,* 743.
7. Moran, J. M.; Maydan, D. *J. Vac. Sci. Technol. A* **1979,** *16,* 1620.
8. Ray, G. W.; Peng, S.; Burriesci, D.; O'Toole, M. M.; Lui, E. D. *J. Electrochem. Soc.* **1982,** *129,* 2152.
9. Ting, C. H.; Liauw, K. L. *J. Vac. Sci. Technol., B.* **1983,** *1,* 1225.
10. Reichmanis, E.; Smolinsky, G.; Wilkins, C. W., Jr. *Solid State Technol.* **1985,** *28*(8), 130.
11. Taylor, G. N.; Wolf, T. M. *Polym. Eng. Sci.* **1980,** *20,* 1087.
12. Watanabe, F.; Ohnishi, Y. *J. Vac. Sci. Technol., B* **1986,** *4,* 422.
13. Jurgensen, C. W.; Shugard, A.; Dudash, N.; Reichmanis, E.; Vasile, M. J. *Proc. SPIE Conf., Adv. Resist Technol. Process.* **1988,** *920,* 253.
14. Hatzakis, M.; Paraszczak, J.; Shaw, J. M. In *Microcircuit Engineering 81* Oosenbrug, A., Ed.; Swiss Federal Institute of Technology, Lausanne, 1981; pp 386–396.

15. Shaw, J. M.; Hatzakis, M.; Paraszczak, J.; Liutkus, J.; Babich, E. *Proceedings of the Regional Technical Conference on Photopolymers: Principles, Processes and Materials;* Mid-Hudson SPE: Ellenville, NY, 1982; pp 285–295.
16. Tanaka, A.; Morita, M.; Imamura, S.; Tamamura, T.; Kogure, O. *Jpn. J. Appl. Phys.* **1983,** *22,* 2659.
17. Tamamura, T.; Tanaka, A. In *Polymers for High Technology: Electronics and Photonics;* Bowden, M. J.; Turner, S. R., Eds.; ACS Symposium Series 346; American Chemical Society: Washington, DC, 1987; pp 67–76.
18. Morita, M.; Tanaka, A.; Onose, K. *J. Vac. Sci. Technol., B* **1986,** *4,* 414.
19. Brault, R. G.; Kubena, R. L.; Metzger, R. A. *Proc. SPIE Conf., Adv. Resist Technol.* **1985,** *539,* 70.
20. Hartney, M. A.; Novembre, A. E.; Bates, F. S. *J. Vac. Sci. Technol., B* **1985,** *3,* 1346.
21. Hofer, D. C.; Miller, R. D.; Willson, C. G. *Proc. SPIE Conf., Adv. Resist Technol.* **1984,** *469,* 16.
22. Miller, R. D.; Hofer, D. C.; Willson, C. G.; West, R.; Trefonas, D. T. III In *Materials for Microlithography;* Bowden, M. J.; Turner, S. R., Eds.; ACS Symposium Series 266; American Chemical Society: Washington, DC, 1984; pp 293–310.
23. Zeigler, J. M.; Harrah, L. A.; Johnson, A. W. *Proc. SPIE Conf., Adv. Resist Technol. Process. II* **1985,** *539,* 166.
24. Miller, R. D.; Hofer, D.; Rabolt, J.; Sooriyakumaran, R.; Willson, C. G.; Fickes, G. N.; Guillet, J. E.; Moore, J. In *Polymers for High Technology: Electronics and Photonics;* Bowden, M. J.; Turner, S. R., Eds.; ACS Symposium Series 346; American Chemical Society: Washington, DC, 1987; pp 170–187.
25. Taylor, G. N.; Hellman, M. Y.; Wolf, T. M.; Zeigler, J. M. *Proc. SPIE Conf., Adv. Resist Technol. Process.* **1988,** *920,* 274.
26. Nate, K.; Sugiyama, H.; Inoue, T. *Electrochem. Soc. Ext. Abstr.* **1984,** *84,* 530.
27. MacDonald, S. A.; Steinmann, A. S.; Ito, H.; Hatzakis, M.; Lee, W.; Hiraoka, H.; Willson, C. G. *Proc. Polym. Mat. Sci. Eng.* **1983,** *49,* 104.
28. Suzuki, M.; Saigo, K.; Gokan, H.; Ohnishi, Y. *J. Electrochem. Soc.* **1983,** *130,* 1962.
29. Novembre, A. E.; Reichmanis, E.; Davis, M. A. *Proc. SPIE Conf., Adv. Resist Technol. Process. III* **1986,** *631,* 14.
30. Reichmanis, E.; Smolinsky, G. *Proc. SPIE Conf., Adv. Resist Technol. Process.* **1984,** *469,* 38.
31. Reichmanis, E.; Smolinsky, G. *J. Electrochem. Soc.* **1985,** *132,* 1178.
32. Gozdz, A. S. *Solid State Technol.* **1987,** *30*(6), 75.
33. Gozdz, A. S.; Dijkkamp, D.; Schubert, R.; Wu, X. D.; Klausner, C.; Bowden, M. J. In *Polymers for High Technology: Electronics and Photonics;* Bowden, M. J.; Turner, S. R., Eds.; ACS Symposium Series 346; American Chemical Society: Washington, DC, 1987; pp 334–349.
34. Gozdz, A. S.; Baker, G. L.; Klausner, C.; Bowden, M. J. *Proc. SPIE Conf., Adv. Resist Technol. Process.* **1987,** *771,* 18.
35. Willson, C. G. In *Introduction to Microlithography;* Thompson, L. F.; Willson, C. G.; Bowden, M. J., Eds.; ACS Symposium Series 219; American Chemical Society: Washington, DC, 1983; pp 111–117.
36. Wilkins, C. W., Jr.; Reichmanis, E.; Wolf, T. M.; Smith, B. C. *J. Vac. Sci. Technol., B* **1985,** *3*(1), 306.
37. Tarascon, R. G.; Shugard, A.; Reichmanis, E. *Proc. SPIE Conf., Adv. Resist Technol. Process. III* **1986,** *631,* 40.
38. Reichmanis, E.; Novembre, A. E.; Tarascon, R. G.; Shugard, A. *Polym. Mat. Sci. Eng. Prepr.* (*Am. Chem. Soc., Div. Polym. Mat.: Sci. Eng.*) **1986,** 55, 294.

39. Saotome, Y.; Gokan, H.; Saigo, K.; Suzuki, M.; Ohnishi, Y. *J. Electrochem. Soc.* **1985,** *132*, 909.
40. Cunningham, W. C., Jr.; Park, C. E. *Proc. SPIE Conf., Adv. Resist Technol. Process.* **1987,** *771*, 32.
41. Sugiyama, H.; Inoue, T.; Mizushima, A.; Nate, K. *Proc. SPIE Conf., Adv. Resist Technol. Process.* **1988,** *920*, 268.
42. Imamura, S.; Tanaka, A.; Onose, K. *Proc. SPIE Conf., Adv. Resist Technol. Process.* **1988,** *920*, 291.

RECEIVED for review May 27, 1988. ACCEPTED revised manuscript October 27, 1988.

POLYSILANES AND RELATED POLYMERS

17

Modifications of Well-Defined Polysilanes

Krzysztof Matyjaszewski, Jeffrey Hrkach, Hwan-Kyu Kim, and Karen Ruehl

Department of Chemistry, Carnegie Mellon University, Pittsburgh, PA 15213

Different homo- and copolysilanes were prepared by sonochemical reductive coupling of disubstituted dichlorosilanes with sodium. Polymers formed in the presence of ultrasound were monomodal, with polydispersities (ratio of weight-average molecular weight to number-average molecular weight, $\overline{M}_w/\overline{M}_n$) as low as <1.2. Polysilanes that contain aryl groups were dearylated upon treatment with strong protonic acids such as trifluoromethanesulfonic (triflic) acid. The resulting triflated polysilanes reacted readily with different nucleophiles to yield alkoxy- and amino-substituted polysilanes. Triflated polymers reacted with monomers that could be polymerized cationically to form graft copolymers.

HIGH-MOLECULAR-WEIGHT POLYMERS that contain Si–Si bonds in the main chain have been known for more than 30 years, but because of the insolubility of the initially prepared poly(dimethylsilylene), they were not studied in detail until recently (*1–4*). The successful conversion of poly(dimethylsilylene) to silicon carbide fibers and the subsequent preparation of soluble polysilanes (*1–4*) stimulated the renewed study of these materials. Photochemical and photophysical studies of high-molecular-weight polysilanes confirmed earlier theoretical predictions concerning the conjugation of catenated Si–Si bonds in linear polymers (*5–12*). Very recently, thermochromism of polysilanes with long alkyl substituents has been observed and explained by the preferential formation of long (with at least 10 consecutive Si atoms) all-*trans* conformations.

Polysilanes are prepared by the reductive coupling of dichlorosilanes in

0065–2393/90/0224–0285$06.00/0

the presence of sodium (*1–4*). The coupling process used at temperatures >100 °C usually leads to a mixture of high-molecular-weight polymers ($\overline{M}_n > 100{,}000$), low-molecular-weight polymers ($\overline{M}_n < 3000$), and small cyclopolysilanes (Si_4, Si_5, and Si_6). We have obtained monomodal polysilanes (composed only of high-molecular-weight polymers) by the application of ultrasound at ambient temperatures (*13*).

The severe reaction conditions of the coupling reaction discourage the use of dichlorosilanes with substituents other than alkyl or aryl groups. Therefore, up to now, the availability of polysilanes has been limited to polymers with alkyl or aryl substituents (*1–4*). Recently, we discovered the rapid and quantitative displacement of phenyl groups from silanes and disilanes by the action of strong protonic acids such as trifluoromethanesulfonic (triflic) acid (*14*). The resulting triflates react rapidly with different nucleophilic reagents to yield polysilanes with pendant alkoxy and amino groups. The reactivity of silyl triflates is sufficient to initiate the cationic polymerization of different monomers. The polymeric initiation system yields comb-like graft copolymers.

Experimental Procedures

Details of this work have been described elsewhere (*13, 14*). All reagents were distilled and dried directly before use. Reactions were carried out either in an inert atmosphere or in a vacuum. NMR spectra were recorded with 300- and 80-MHz spectrometers. Gel permeation chromatography was performed with Ultrastyragel columns (Waters).

Sonochemical Synthesis of Well-Defined Polysilanes

Recently, ultrasound has been very successfully applied to different organometallic and catalytic reactions (*15*). Rates are strongly accelerated, and reactions are often much more selective than those carried out under typical thermal conditions. Sonication of sterically hindered dichlorosilanes in the presence of lithium yields disilenes and cyclotrisilanes (*16, 17*). We have applied ultrasound in the synthesis of poly(phenylmethylsilylene) in the presence of dispersed sodium, which was prepared directly from small pieces of sodium by sonication.

Thermal polymerization typically results in the formation of high- and low-molecular-weight polymers with a bimodal distribution. Bimodality was ascribed previously to diffusional phenomena (*4*), but an alternative explanation involving the presence of two (or more) chain carriers that exchange slowly enough to build an entire macromolecule or the presence of linear and macrocyclic polymers is possible. We expected that a lower overall temperature could favor one mechanism of propagation and yield monomodal polymers. Indeed, at ambient temperatures and by using simple ultrasonic cleaning baths or an immersion-type probe, we obtained monomodal poly(phenylmethylsilylene) with number-average molecular weights ($\overline{M}_n$) in the range of 100,000 and polydispersities ($\overline{M}_w/\overline{M}_n$; $\overline{M}_w$ is weight-average molecular weight) of $\geq$1.2.

Three phenomena may be related to the formation of monomodal polymers. The first phenomenon is the preferential contribution of intermediates of one type in the sonochemical reductive coupling. The second phenomenon accounts for the formation of high-quality sodium dispersion, which is continuously regenerated during the coupling process. The third phenomenon is related to the selective degradation of polysilanes with higher molecular weights.

Sonochemical homopolymerization of dichlorosilanes in the presence of sodium is successful at ambient temperatures in nonpolar aromatic solvents (toluene or xylenes) only for monomers with α-aryl substituents. Dialkyldichlorosilanes do not react with dispersed sodium under these conditions, but they can be copolymerized with phenylmethyldichlorosilane. Copolymers with a 30–45% content of dialkylsilanes were formed from equimolar mixtures of the corresponding comonomers. Copolymerization might indicate anionic intermediates. A chloroterminated chain end in the polymerization of phenylmethyldichlorosilane can participate in a two-electron-transfer process with sodium (or rather two subsequent steps separated by a low-energy barrier). The resulting silyl anion can react with both dichlorosilanes. The presence of a phenyl group in either α or β position in chloroterminated polysilane allows reductive coupling, in contrast to peralkyl species, which do not allow the reaction. Therefore, dialkyl monomers can copolymerize, but they cannot homopolymerize under sonochemical conditions.

Growth via radical intermediates should lead to the homopolymer of phenylmethylsilane in the presence of unreacted dialkyldichlorosilane, unless an extensive transfer process operates. Radicals can be considered as intermediates that rapidly participate in the second electron transfer that leads to the corresponding anions. High local temperatures (>2000 K) might enable the second electron transfer to proceed despite the short lifetime (<1 μs) of "hot spots" (*15*). Anions can react with both monomeric dichlorides in an S_N2-type reaction that yields a copolymer (Scheme I).

The presence of intermediate radicals was confirmed by trapping the pendant alkenyl groups during the polymerization of phenylallyldichlorosilane and phenylhexenyldichlorosilane. In the reductive coupling of phenylallyldichlorosilane, ~50% of the alkenyl groups were consumed, whereas for phenylhexenyldichlorosilane, ~25% of alkenyl groups reacted, and 75% were present in the final oligosilanes. Phenylallyldichlorosilane yields 7% of a bimodal polymer with molecular weights ($\overline{M}_{peak}$s) of 31,000 and 9100. Copolymerization of phenylallyldichlorosilane with phenylmethyldichlorosilane yields 13% of a copolymer in which 50% of the allyl groups were consumed. Again, this copolymer is a bimodal polymer with $\overline{M}_{peak}$s of 58,000

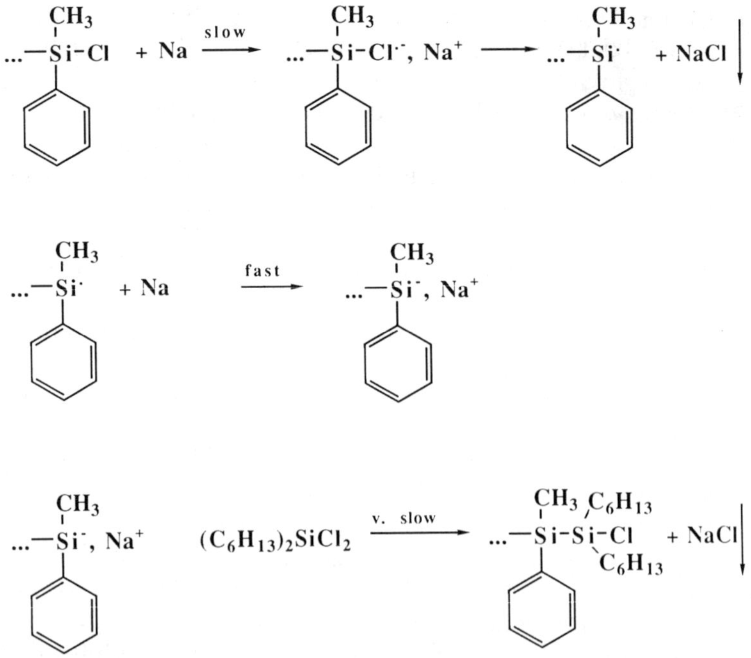

$$\ldots\!-\!\mathrm{Si}(CH_3)(C_6H_5)\!-\!Cl + Na \xrightarrow{\text{slow}} \ldots\!-\!\mathrm{Si}(CH_3)(C_6H_5)\!-\!Cl^{\cdot -}, Na^{+} \longrightarrow \ldots\!-\!\mathrm{Si}^{\cdot}(CH_3)(C_6H_5) + NaCl\downarrow$$

$$\ldots\!-\!\mathrm{Si}^{\cdot}(CH_3)(C_6H_5) + Na \xrightarrow{\text{fast}} \ldots\!-\!\mathrm{Si}^{-}(CH_3)(C_6H_5), Na^{+}$$

$$\ldots\!-\!\mathrm{Si}^{-}(CH_3)(C_6H_5), Na^{+} \quad (C_6H_{13})_2SiCl_2 \xrightarrow{\text{v. slow}} \ldots\!-\!\mathrm{Si}(CH_3)(C_6H_5)\!-\!\mathrm{Si}(C_6H_{13})_2\!-\!Cl + NaCl\downarrow$$

Scheme I. Copolymerization by reductive coupling.

and 6800. The long-lived radicals can be rapidly trapped by a hexenyl pendant group at the monomer (or terminal-unit) stage and also by an allyl group at the dimer or trimer stage. However, the formation of radicals can be considered only as the first step in the reduction of chloroterminated chains to polymeric silyl anions.

Intramolecular reactions are much faster than intermolecular reactions, and the anchimeric assistance, measured by the ratio of the rate constant of the unimolecular reaction to that of the analogous bimolecular reaction, depends strongly on the size of the ring that should be formed in the unimolecular process. The ratio of these constants can reach values of up to 10^7 mol/L. Thus, intramolecular trapping confirms the presence of radicals as short-lived intermediates but does not prove that chain growth proceeds via radical coupling.

In the reductive coupling of phenylhexenyldichlorosilane, 75% of the alkenyl groups did not react with radicals, although cyclization was highly favored in the strainless ring. This result means that the lifetime of radicals must be very short and that they can be further reduced to silyl anions prior to intramolecular cyclization. The presence of anionic intermediates is additionally supported by faster reactions in ethereal solvents, first-order kinetics of the monomer, and some model reactions. The formation of silyl anions is especially favored for monomers bearing a phenyl group, which stabilizes anions. For example, 1,1,1-triphenyl-2,2,2-trimethyldisilane is practically the only product of the reductive coupling of triphenylsilyl chloride with trimethylsilyl chloride; the radical pathway would lead to the mixture of perphenyl- and permethyldisilanes and 1,1,1-triphenyl-2,2,2-trimethyldisilane.

Polymerization by reductive coupling must start at the slow reaction between sodium and monomer and is followed by much faster reactions involving polymeric species. This sequence is synonymous to having a much faster propagation compared with initiation, as usually happens in the chain-growth process. Otherwise, no high-molecular-weight polymer could be formed in the presence of excess sodium or dichlorosilane.

The macromolecular silyl chloride reacts with sodium in a two-electron-transfer reaction to form macromolecular silyl anion. The two-electron-transfer process consists of two (or three) discrete steps: formation of radical anion, precipitation of sodium chloride and generation of the macromolecular silyl radical (whose presence was proved by trapping experiments), and the very rapid second electron transfer, that is, reduction to the macromolecular silyl anion. Some preliminary kinetic results indicate that the monomer is consumed with an internal first-order-reaction rate. This result supports the theory that a monomer participates in the rate-limiting step. Thus, the slowest step should be a nucleophilic displacement at a monomer by macromolecular silyl anion. This anion will react faster with the more electrophilic dichlorosilane than with a macromolecular silyl chloride. Therefore, polymerization would resemble a chain growth process with a slow initiation step and a rapid multistep propagation (the first and rate-limiting step is the reaction of an anion with degree of polymerization $n[\overline{DP}_n]$ to form macromolecular silyl chloride $[\overline{DP}_{n+1}]$, and the chloride is reduced subsequently to the anion).

Interchain reactions may also be involved in polymerization, and they can be responsible for very high molecular weights. The intramolecular reaction between macromolecular silyl anion and macromolecular silyl chloride is favored for small rings such as five- or six-membered rings. At that stage, anchimeric assistance is very high, and an oligosilane with a terminal chloro group can successfully compete with a more electrophilic monomer molecule. This assistance explains the large proportion of cyclic structures formed during reductive coupling. The radical mechanism can hardly explain the formation of cyclooligosilanes. When a few chains exceed the critical dimension of six or seven units, they escape the intramolecular trap, and cyclization becomes much less probable. The probability of cyclization reappears at higher degrees of polymerization, in agreement with the Jacobson–Stockmayer theory (18). The low-molecular-weight polymer might have a macrocyclic structure. Studies of the end-group structures in low-molecular-weight polysilanes are being undertaken currently.

The reactions that limit chain growth and initiate polymerization have not been defined. Is polymerization terminated by impurities? Does solvent participate in transfer? Does polymerization start by the reaction with a monomer or with impurities in the system? These questions must be answered. For example, we noticed that successful formation of the monomodal high-molecular-weight polysilane requires the addition of a few drops of monomer to sonicated sodium dispersion prior to the addition of the main part of disubstituted dichlorosilane.

Dialkyldichlorosilanes do not undergo reductive coupling with sodium at temperatures below 60 °C in toluene because of thermodynamic or kinetic reasons. The equilibrium constant for the reduction process may be unfavorably shifted at lower temperatures. Dialkyldichlorosilanes could be polymerized at ambient temperatures in ethereal solvents (*19*). For example, we prepared poly(di-*n*-hexylsilylene) with $\overline{M}_n$ = 45,000 in toluene–diglyme (1:1). The presence of diglyme may accelerate electron transfer, but it may also shift the equilibrium by the formation of complexes with sodium cations (Table I).

The active surface of dispersed sodium increases during ultrasonication because of the cavitational erosion of sodium, which is malleable at this temperature. Ultrasonic treatment also ensures a local excess of sodium by the continuous regeneration of the metal surface.

High-molecular-weight polysilanes are rapidly degraded in the presence of ultrasound (Table II). A similar effect has been observed previously for polystyrene, poly(methyl methacrylate), dextran, and other polymers (*20–22*). Selective degradation is a mechanical process caused by frictional forces between macromolecules and solvent molecules during the cavitation process. Larger molecules are more resistant to flow, have larger shear forces, and rupture more frequently than shorter macromolecules. Beyond a certain molecular weight, shear forces are smaller than bond strengths, and polymers cannot be degraded. This selective degradation reduces molecular weights to a certain value but also decreases polydispersity.

Model Dearylation Reactions

Successful reactions involving polymers should proceed with high rates and high selectivities to enable reactions with functional groups that are usually present at low concentrations. Having in mind this limitation, we searched for a rapid and selective reaction that could convert relatively inert silicon alkyl or aryl groups into intermediates capable of incorporating different functionalities. Reductive coupling in the presence of alkali metal allows only alkyl and aryl groups at silicon. Alkoxy and amino groups, and even hydrogen atoms, react rapidly with molten sodium, and the reaction leads to crosslinking and other side reactions.

On the other hand, some electrophilic silyl compounds react rapidly with a large variety of nucleophiles. The relative order of reactivity of si-

Table I. Sonochemical Synthesis of Polysilanes

Monomer	*Solvent*	*Temperature (°C)*	*Amount of Polymer (%)*	*HP/LP*	$\overline{M}_n$ *HP* $(\times 10^5)$[a]	$\overline{M}_n$ *LP* $(\times 10^3)$	$\overline{M}_w/\overline{M}_n$
Phenylmethyldi-chlorosilane	Toluene[b]	110	55	1/3 (1/9)	1.07	3.3	1.81
	Toluene[b]	60	12	1/0	1.04		1.50
	Toluene[c]	40	11.1	1/0	1.3		1.20
Di-*n*-hexyldi-chlorosilane	Toluene[b]	60	0				
	Toluene–diglyme (4:1)[b]	60	24	1/0	0.45		1.73
Phenylmethyldi-chlorosilane–di-*n*-hexyldichloro-silane (1:1 mixture)	Toluene[b]	60	12	1/0[d]	1.75		1.66

NOTE: HP is high-molecular-weight polymer, and LP is low-molecular-weight polymer. The initial concentration of monomer $[M]_o$ was 0.32 mol/L, and $[Na]_o/[SiCl]_o = 1.2$.

[a]Molecular weights were based on polystyrene standards. Vapor pressure osmometry and light-scattering measurements indicate that the true molecular weights are approximately 2 times larger.

[b]The reaction was carried out for 60 min under thermal conditions and with an immersion probe.

[c]The reaction was carried out for 180 min with a cleaning bath.

[d]The ratio of the concentration of phenylmethylsilylene to that of di-*n*-hexylsilylene units was 1.5/1.0 for this copolymer.

Table II. Effect of Sonication Time on Molecular Weights and Polydispersities of Poly(phenylmethylsilylene)

Sonication Time (min)	$\overline{M}_n$ ($\times 10^5$)	$\overline{M}_w$ ($\times 10^5$)	$\overline{M}_w/\overline{M}_n$
5	3.8	17.3	4.5
10	2.24	6.68	2.98
15	2.30	6.35	2.71
30	1.82	3.73	2.05
60	1.48	2.57	1.73
80	1.06	1.57	1.48
120	0.40	0.47	1.17

NOTES: The reaction was carried out in toluene with an immersion-type probe and the following conditions: $[M]_o$ = 0.32 mol/L; $[Na]_o/[SiCl]_o$ = 1.2; and temperature = 60 °C. Polymerization was exothermic and was completed just after the addition of monomer (15 min). Polysilane separated from the reaction mixture and sonicated in the presence of 0.2 equivalent of Na was degraded in a similar way.

lylating reagents has been established by studies of the silylation of cyclic and linear ketones (*23*) as follows:

$$\equiv Si\text{–}Cl\ (1) \ll \equiv Si\text{–}O_3SCH_3\ (40) \ll \equiv Si\text{–}Br\ (8 \times 10^4)$$
$$\ll \equiv Si\text{–}O_3SCF_3\ (7 \times 10^8) < \equiv Si\text{–}I\ (7 \times 10^9)$$

Some qualitative information indicates that perchlorates might be more reactive than iodides (*24*). Thus, three silylating compounds are much more reactive than the others: iodide, perchlorate, and triflate (trifluoromethanesulfonate). Iodides and perchlorates have considerable disadvantages as light-sensitive and explosive reagents, respectively. Triflates, however, are stable for long periods in the absence of nucleophiles (moisture included).

Four synthetic routes lead to silyl triflates: (1) reaction of silyl chlorides with silver triflate (very expensive), (2) reaction of silyl chlorides with triflic acid, (3) reaction of tetraalkylsilanes with triflic acid, and (4) reaction of arylsilanes with triflic (trifluoromethane sulfonic) acid (*24–26*). The displacement of a chlorine atom and a phenyl group from trimethylsilyl chloride and trimethylphenylsilane indicate that these groups are much more reactive than the alkyl substituents for the synthesis of triflated silanes. We found that the aryl group is over 200 times more reactive than chloride in these reactions (*14*). We were able to displace two phenyl groups from diphenyldimethylsilane in a stepwise manner:

$$CF_3SO_2OH + C_6H_5\text{–}Si(CH_3)_2\text{–}C_6H_5 \xrightarrow[-C_6H_6]{k_1}$$
$$CF_3SO_2O\text{–}Si(CH_3)_2\text{–}C_6H_5\ (>94\%) + CF_3SO_2O\text{–}Si(CH_3)_2\text{–}OSO_2CF_3 \quad (1)$$

Phenyldimethyltrifluoromethanesulfonyloxysilane was formed after the reaction with one equivalent of the acid, and the corresponding ditriflate was formed after the second equivalent of the acid was added. No ditriflate was found until the starting material was completely consumed.

A similar reaction with 1,2-diphenyltetramethyldisilane shows a much lower selectivity. Reaction with the equimolar amount of the acid results in 13% of the starting disilane, 13% of ditriflate, and 74% of monotriflate:

$$CF_3SO_2OH + C_6H_5\text{–}Si(CH_3)_2\text{–}Si(CH_3)_2\text{–}C_6H_5 \xrightarrow{k_1} CF_3SO_2O\text{–}Si(CH_3)_2\text{–}Si(CH_3)_2\text{–}C_6H_5\ (74\%) + C_6H_6 \quad (2a)$$

$$CF_3SO_2OH + C_6H_5\text{–}Si(CH_3)_2\text{–}Si(CH_3)_2\text{–}C_6H_5 \xrightarrow{k_2} CF_3SO_2O\text{–}Si(CH_3)_2\text{–}Si(CH_3)_2\text{–}OSO_2CF_3 + C_6H_6 \quad (2b)$$

These results indicate that the second rate constant is approximately eight times lower than the first rate constant, or if a statistical factor is accounted for, the phenyl group is 4 times less reactive when a strong electron-withdrawing triflate group is at the adjacent silicon atom. The presence of triflate at the same Si atom leads to much larger differences in reactivities and to a stepwise dearylation. (We prefer to use the term "dearylation", although in organosilicon chemistry, "desilylation" is used, because a large number of aryl groups are removed from these polymeric systems although the silicon backbone [polysilane] remains intact.)

1,2-Bis(trifluoromethanesulfonyloxy)tetramethyldisilane is formed in preparative yields, which exceed 75%. The reaction between two equivalents of the acid and disilane shows only one desired product when carried out directly in an NMR tube.

The ditriflates react rapidly with various alcohols

$$CF_3SO_2O\text{–}Si(CH_3)_2\text{–}Si(CH_3)_2\text{–}OSO_2CF_3 + 2ROH + 2B\text{:} \rightarrow RO\text{–}Si(CH_3)_2\text{–}Si(CH_3)_2\text{–}OR + 2(BH^+)(OSO_2CF_3^-) \quad (3)$$

in which R is CH_3–, C_2H_5–, $(CH_3)_3C$–, $CH_2{=}CH\text{–}CH_2$–, or CF_3CH_2–.

We also observed quantitative reactions with other nucleophiles such as amines and organometallics.

Reactions on Polymers

The rapid and quantitative displacement of phenyl groups from model silanes suggests that similar reactions with aryl-substituted polysilanes should lead to the triflated polymer. We have carried out this displacement under different conditions. Approximately the first 80% of the phenyl groups were removed rapidly from poly(phenylmethylsilylene). No free acid was observed in the NMR spectra at 2 min after the addition of acid. Later, displacement became more difficult, and a small amount of phenyl groups (<10%) remained in the presence of excess acid (Figure 1).

The broad NMR signals of the starting material are due to the differences in the chemical shifts of phenyl and methyl groups in isotactic, syndiotactic, and atactic triads. Tacticity remains in a partially triflated polymer in which methyl groups are strongly deshielded because they are in the neighborhood of strong electron-withdrawing triflate moieties. A small amount of unreacted phenyl groups can be ascribed to some triads in which a phenyl group is placed between two bulky triflate groups, and steric hindrance prevents the displacement of the final aryl group:

$$
\begin{array}{ccccccccc}
 & H_3C & H_5C_6 & H_3C & H_3C & H_3C & H_5C_6 & H_3C & \\
\ldots- & Si- & Si- & Si- & Si- & Si- & Si- & Si & -\ldots \\
 & C_6H_5 & CH_3 & C_6H_5 & C_6H_5 & C_6H_5 & CH_3 & C_6H_5 &
\end{array}
\quad + \text{ n } HO_3SCF_3
$$

$$\Downarrow$$

$$
\begin{array}{ccccccccc}
 & H_3C & F_3CSO_3 & H_3C & H_3C & H_3C & F_3CSO_3 & H_3C & \\
\ldots- & Si- & Si- & Si- & Si- & Si- & Si- & Si & -\ldots \\
 & O_3SCF_3 & CH_3 & O_3SCF_3 & C_6H_5 & O_3SCF_3 & CH_3 & O_3SCF_3 &
\end{array}
\quad + \text{ n } C_6H_6 \qquad (4)
$$

Strong electrophilic reagents can induce dearylation, but they can also decrease the molecular weights of polysilanes by cleavage of Si–Si bonds. We have found that the selectivity of dearylation with triflic acid in CH_2Cl_2 solutions is approximately 90%, and at higher degrees of modifications, we also observed polymer degradation. Thus, under the present conditions, dearylation with triflic acid is best suited for partial modifications and attachment of groups that can influence the physical and chemical properties of polysilanes, even those with a low content of these groups.

Polymers that contain 10–30% of methoxy, ethoxy, or *tert*-butoxy groups instead of phenyl groups have absorption maxima lower than that of the initial polymer (320 versus 340 nm). This difference can be ascribed to the decrease of a σ–π^* excitation in the modified polymer (27).

At present, we are studying the optimal conditions for the displacement reactions by using different solvents, temperatures, and acids. We have found that variation in the aryl groups also increases the selectivity of dearylation. For example, the *p*-methoxyphenyl group is displaced by triflic acid much more rapidly than the unsubstituted phenyl ring.

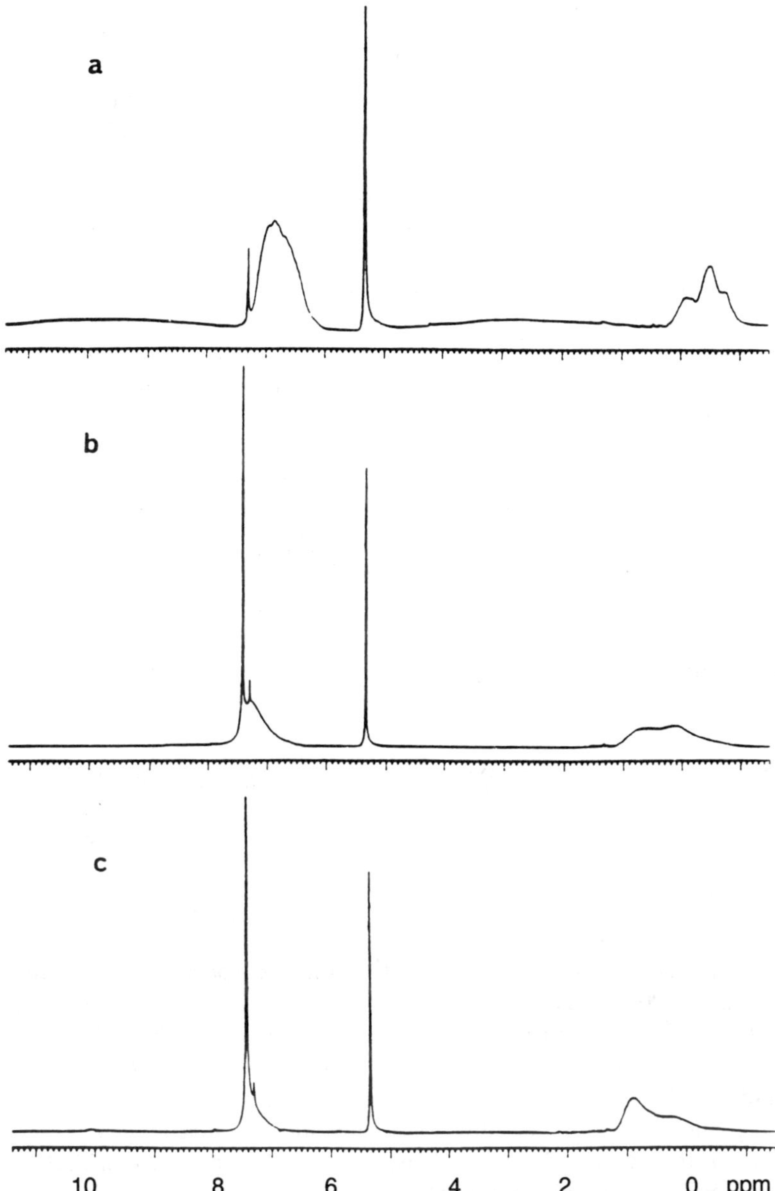

Figure 1. ^{1}H NMR spectra of poly(phenylmethylsilylene) modified by trifluoromethanesulfonic acid (HA). (a) $[-Si(C_6H_5)(CH_3)-]_0$ = 0.42 mol/L; (b) after reaction with $[HA]_0$ = 0.17 mol/L; and (c) after reaction with $[HA]_0$ = 0.34 mol/L. CH_2Cl_2 was used as internal standard (5.35 ppm). Spectra were taken in $CDCl_3$ solvent at 25 °C.

Grafting from Polysilanes

Trimethylsilyl triflate can initiate the cationic polymerization of different vinyl and heterocyclic monomers (*28*). The triflated polysilane can react in the same way and can induce growth of a large number of chains. In this way, comblike graft copolymers can be prepared.

The successful preparation of pure graft copolymer requires a transferless process; otherwise, in addition to the desired graft copolymer, a mixture of homopolymers will also be formed. The correct control of the structure of the graft copolymer can be achieved in the living system when initiation is at a comparatively higher rate than propagation and when termination is absent. A few monomers can be polymerized cationically in living systems (*29*). For example, cyclic ethers, cyclic imino ethers, and some cyclic amines give living systems. Recently, some vinyl ethers were also polymerized in transferless and terminationless processes (*30, 31*). Quite often, special features are required of the structure of the initiator. Triflate derivatives were used successfully in the living polymerization of cyclic ethers, particularly in the polymerization of tetrahydrofuran (THF) (*32*). In the polymerization of THF in the presence of an anion capable of covalent bonding, two types of active species are present: oxonium ions and covalent esters. These species isomerize with rates comparable with the rate of ionic propagation $k_p^{\,i}$, and the equilibrium position depends on the polarity of solvent:

$$\text{....-}CH_2\text{-}O\text{-}CH_2CH_2CH_2CH_2\text{-}O\text{-}SO_2\text{-}CF_3 \rightleftharpoons \text{....-}CH_2\text{-}\overset{+}{O}\langle\text{ring}\rangle,\ CF_3SO_3^- \qquad \downarrow k_p^{\,c} \ (\text{THF}) \qquad \downarrow k_p^{\,i} \ (\text{THF}) \tag{5}$$

Polymerization proceeds predominantly with ionic species, because they are more than 1000 times more reactive than esters ($k_p^{\,i} >> k_p^{\,c}$, in which $k_p^{\,c}$ is the rate of covalent propagation). Rapid isomerization allows simultaneous growth on all chains. Typically, during the polymerization of THF, polydispersities ($\overline{M}_w/\overline{M}_n$) are >1.5, because depropagation occurs simultaneously with chain growth. The equilibrium monomer concentration, $[M]_e$, at room temperature is around 5 mol/L. Thus, the initial concentration of THF under these conditions should exceed 50 vol %.

We succeeded in grafting poly(tetrahydrofuran) on partially triflated poly(phenylmethylsilylene) in CH_2Cl_2. Using a polysilane with $\overline{M}_w$ = 3600, in which 30% of phenyl groups were substituted by triflates (nine groups per chain), we observed the formation of a graft copolymer with $\overline{M}_w$ = 600,000 and $\overline{M}_n$ = 300,000. After UV irradiation and degradation of polysilane backbone, the molecular weight of this polymer decreased to $\overline{M}_w$ = 30,000. This change suggests that poly(tetrahydrofuran) was successfully grafted onto the polysilane and that the amount of homopoly(tetrahydrofuran) that could be formed by transfer or by direct initiation by the remaining triflic acid is very small. Broad molecular weight distribution is due to slow initiation.

Silyl triflates are very powerful silylating agents, and they react directly with alkyl esters to form the corresponding silyl ketene acetals, which are known as initiators of the group-transfer polymerization of methyl methacrylate (33). We prepared the initiators in situ from trimethylsilyl triflate (or premixed aryl silanes and equimolar amounts of triflic acid), methyl propionate, and triethylamine. Amines or pyridines were used in order to trap triflic acid, which was formed under these conditions. Tris(dimethylamino)sulfonium bifluoride was used as the catalyst. Typically, poly(methyl methacrylate) with higher than theoretical molecular weights (degrees of polymerization should be equal to the ratio of monomer to initiator) were formed under these conditions. This result can be due to the lower activity of silyl ketene acetals based on propionate compared with isobutyrate esters, which have higher "O/C" silylation ratios (33).

A similar reaction was used for poly(phenylmethylsilylene) with $\overline{M}_n$ = 3000, in which, statistically, three phenyl groups per polymer chain were displaced by triflic acid. Partially triflated polysilane was used as macromolecular silylating reagent. In this case, poly(methyl methacrylate)-g-poly(phenylmethylsilylene) was formed. The graft copolymer was degraded by UV light from $\overline{M}_n$ = 125,000 to 47,000 in 24 h. No further degradation was observed at longer irradiation times. Three poly(methyl methacrylate) branches per polysilane chain were expected from the ratio of initial concentrations ($[\text{polysilane}]_0/[\text{HA}]_0$). Degradation indicates that each polysilane chain contains 2.7 poly(methyl methacrylate) grafts, which suggests that nearly all triflated sites were active.

The poly(methyl methacrylate)s prepared in this experiment, as well as polymers formed in model reactions with silanes that contain bulky substituents (such as phenyldimethyl and diphenylmethyl groups), have predominantly syndiotactic structures identical to polymers prepared by the conventional group-transfer process. This result supports a two-step dissociative mechanism for a group-transfer process, because steric hindrance from bulky silyl groups should increase the proportion of isotactic triads in the hypothetical associative concerted propagation step.

The grafting process is being attempted for other monomers and for different counterions. The synthesis of true graft copolymers may lead to the formation of new morphologies, if the size of components in the copolymer allows correct microphase separation. These studies are being continued in our laboratory.

Acknowledgment

These studies were partially supported by the Office of Naval Research.

References

1. West, R. J. *Organomet. Chem.* **1986,** *300,* 327.
2. West, R. In *Comprehensive Organometallic Chemistry;* Wilkinson, G. A.; Stone, F. G. A.; Abel, E. W., Eds.; Pergamon: Oxford, 1983; Vol. 9, p 365.
3. Trefonas, P., III; West, R.; Miller, R. D.; Hofer, D. *J. Polym. Sci., Polym. Lett. Ed.* **1983,** *21,* 283.
4. Zeigler, J. M. *Polym. Prepr.* (*Am. Chem. Soc., Div. Polym. Chem.*) **1986,** *27*(1), 109.
5. Miller, R. D.; Hofer, D.; Rabolt, J.; Fickes, G. N. *J. Am. Chem. Soc.* **1985,** *107,* 2172.
6. Harrah, L. A.; Zeigler, J. M. *J. Polym. Sci., Polym. Lett. Ed.* **1985,** *23,* 209.
7. Kajzar, F.; Messier, J. *Polym. J.* (*Tokyo*) **1987,** *19,* 275.
8. West, R.; David, L. D.; Djurovich, P. I.; Stearley, K. L.; Srinivasam, K. S. V.; Yu, H. *J. Am. Chem. Soc.* **1981,** *103,* 7352.
9. Zeigler, J. M.; Harrah, L. A.; Johnson, A. W. *Polym. Prepr.* (*Am. Chem. Soc., Div. Polym. Chem.*) **1987,** *28*(1), 424.
10. Boberski, W. G.; Allred, A. L. *J. Organomet. Chem.* **1975,** *88,* 65.
11. Stolka, M.; Yuh, H. J.; McGrane, K.; Pai, D. M. *J. Polym. Sci., Polym. Chem. Ed.* **1987,** *25,* 823.
12. Pitt, C. G. In *Homoatomic Rings, Chains, and Macromolecules of Main Group Elements;* Rheingold, A. C., Ed.; Elsevier: Amsterdam, 1977.
13. Matyjaszewski, K.; Chen, Y. L.; Kim, H. K. In *Inorganic and Organometallic Polymers;* Zeldin, M.; Wynne, K. J.; Allcock, H. R., Eds.; ACS Symposium Series 360; American Chemical Society: Washington, DC, 1988; p 78.
14. Matyjaszewski, K.; Chen, Y. L. *J. Organomet. Chem.* **1988,** *340,* 7.
15. Suslick, K. S. *Adv. Organomet. Chem.* **1986,** *25,* 73.
16. Boudjouk, P.; Han, B. H.; Anderson, K. R. *J. Am. Chem. Soc.* **1982,** *104,* 4992.
17. Masamune, S.; Murakami, S.; Lobita, H. *Organometallics* **1983,** *2,* 1464.
18. Jacobson, W.; Stockmayer, W. H. *J. Chem. Phys.* **1950,** *18,* 1600.
19. Kim, H. K.; Matyjaszewski, K. *J. Am. Chem. Soc.* **1988,** *110,* 3321.
20. Basedow, A. M.; Ebert, K. H. *Adv. Polym. Sci.* **1977,** *22,* 83.
21. Smith, W. B.; Temple, H. W. *J. Phys. Chem.* **1967,** *72,* 4613.
22. Basedow, A. M.; Ebert, K. H. *Makromol. Chem.* **1975,** *176,* 745.
23. Hergott, H. H.; Simchen, G. *Justus Liebigs Ann. Chem.* **1980,** 1781.
24. Emde, H.; Domsch, D.; Feger, H.; Frick, U.; Gotz, A.; Hergott, H. H.; Hofmann, K.; Kober, W.; Krageloh, K.; Oesterle, T.; Steppan, W.; West, W.; Simchen, G. *Synthesis* **1982,** 1.
25. Schmeiser, M.; Sartori, P.; Lippsmeier, B. *Chem. Ber.* **1970,** *103,* 868.
26. Habich, D.; Effenberger, F. *Synthesis* **1978,** 755.
27. Harrah, L. A.; Zeigler, J. M. *Macromolecules* **1987,** *20,* 601.
28. Gong, M. S.; Hall, H. K. Jr. *Macromolecules* **1986,** *19,* 3011.
29. Penczek, S.; Kubisa, P.; Matyjaszewski, K. *Adv. Polym. Sci.* **1985,** *67/68,* 1.
30. Miyamoto, M.; Sawamoto, M.; Higashimura, T. *Macromolecules* **1984,** *17,* 265.
31. Sawamoto, M.; Fujimara, J.; Higashimura, T. *Macromolecules* **1987,** *20,* 916.
32. Penczek, S.; Matyjaszewski, K. *J. Polym. Sci., Polym. Symp.* **1976,** *56,* 255.
33. Sogah, D. Y.; Hertler, W. R.; Webster, O. W.; Cohen, G. M. *Macromolecules* **1987,** *20,* 1473.

RECEIVED for review May 27, 1988. ACCEPTED revised manuscript December 20, 1988.

18

Mechanistic Studies of Polysilane Polymerization

Sylvie Gauthier and Denis J. Worsfold

Division of Chemistry, National Research Council of Canada, Ottawa K1A 0R6, Canada

The copolymerization of dialkyldichlorosilanes was investigated under conditions of concurrent and consecutive monomer addition. The reactivities of the monomers in initiation and propagation reactions were different. Block copolymers may be formed both by sequential addition to existing active centers and by reactivation of existing polymer. These results were combined with previous findings to suggest general schemes for the pathways taken by the reaction.

SOLUBLE POLYSILANE POLYMERS of high molecular weight were first prepared several years ago (*1–3*). The method commonly used is the reductive coupling of dialkyldichlorosilanes with sodium in refluxing toluene. Relatively few studies of the mechanism of the reaction or discussions of the probable reaction intermediates have been reported.

The molecular equation describing the reaction,

$$n\mathrm{RR'SiCl_2} + 2n\mathrm{Na} \leftrightarrows (\mathrm{RR'Si})_n + 2n\mathrm{NaCl}$$

suggests that the reaction is the silicon equivalent of a Wurtz condensation and that the reaction would follow a condensation polymerization mechanism. The characteristics of the reaction depart considerably from that expected of a condensation polymerization, such as the formation of polyesters. In a condensation polymerization, the monomer is consumed in the formation of the first dimers with some trimers, and only late in the reaction are high-molecular-weight polymers formed. The formation of high-molec-

0065–2393/90/0224–0299$06.00/0

ular-weight polymers depends on the exact stoichiometry of the reagents. A prime condition for these characteristics is the independence of the reactivity of the active chain end groups on the molecular weight of the polymer chain.

In polysilane formation by the previously mentioned route, the products were complex and contained high-molecular-weight materials (molecular weight > 10^5) after only 10% of the dichloride had reacted. In the presence of a twofold excess or a 10% deficit of sodium, an appreciable amount of high-molecular-weight polymer was formed. Attempts to isolate dimers after 50% of the dichloride had reacted failed. The product with the lowest molecular weight was a cyclic pentamer. The polymerization, in fact, demonstrates some of the characteristics of a chain reaction as found in addition polymerization.

West et al. (*4, 5*) have suggested a number of possible reaction intermediates, which include anions, radical anions, radicals, and diradicals. Zeigler (*6, 7*) has proposed, on the basis of some radical-trapping experiments, that the intermediate is, at least at some point, a radical. He showed that the diradical silylene was not an intermediate, and he also stressed the importance of bulk solvent composition on the course of the reaction. The bulk solvent composition determines the expansion of the polymer coil as it interacts with the sodium surface. Miller et al. (*8*) initially suggested that the reaction, which is promoted by the addition of diethylene glycol dimethyl ether, proceeds by an anionic process, although they later (*9*) accepted Zeigler's bulk-solvent model.

The products of the reaction of the dichloride with melted sodium are complex. Except for sodium chloride, the products are polymeric and display a wide range of molecular weights, which fall into three groups when analyzed by size exclusion chromatography: high molecular weight (usually $>10^5$), intermediate molecular weight (10^3–10^4), and low molecular weight. The low-molecular-weight products probably consist mostly of cyclic polysilanes, primarily of pentamers starting from phenylmethyl- or hexylmethyldichlorosilane (*10*).

The multiplicity of these products suggests that two or more mechanisms are operating. Interestingly, the distribution of the molecular weights strongly resembles that found in the formation of polysiloxanes by the anionic polymerization of cyclic trisiloxanes and tetrasiloxanes (*11*). In these anionic polymerizations, an equilibrium exists between linear high-molecular-weight chains, macrocyclic products, and small ring trimers and tetramers. In the polysilane case, however, the thermodynamically stable material in solution at low temperature is probably the low-molecular-weight cyclic product. In fact, if the high-molecular-weight material is refluxed with potassium in THF (tetrahydrofuran) solution, the polymer rapidly forms the low-molecular-weight product. Also, in the reaction of dialkyldichlorosilanes with potassium in THF, considerable high-molecular-weight product is formed during the initial 1–2 min, but this product equally rapidly reverts to the cyclic material.

In the polymerization of dialkyldichlorosilanes with sodium in refluxing toluene, propagation and a concurrent back-biting reaction to cyclic material could give the range of products found if the products are kinetically, instead of thermodynamically, determined (*12*). No evidence for depolymerization has been found for the reaction in toluene solution.

Reaction rate studies (*10*) have shown that the reaction curves are sigmoidal, with an initial period of increasing rate and a final falling off in rate as the monomer is exhausted. At the maximum central part, the rate is relatively independent of the surface area of sodium, a fact suggesting that the rate-determining propagation step does not take place on the sodium surface. The mechanism suggested to explain these findings involves a two-stage reaction whereby initiation occurs at the sodium surface by a slow reaction of the dichloride to form sodium-ended chains. Propagation is by a comparatively rapid reaction of the dichloride with this chain end to re-form a chlorine-ended chain, which would interact very rapidly with the sodium surface in a non-rate-determining step to regenerate the sodium end. Copolymerization can often give insights into polymerization mechanisms. Consequently, a number of copolymerizations have been studied. The comonomers were added either concurrently or consecutively.

Experimental Procedures

The experimental techniques followed are described elsewhere (*10*). Sodium block was added to refluxing toluene under nitrogen and stirred to form small particles of the required size by controlling the stirrer speed. The monomer was added by syringe as rapidly as possible, with cooling if necessary to control the reaction. This step takes less than a minute, except for phenylmethyldichlorosilane, which was too reactive. In the sequential copolymerizations, enough time was left between the addition of the two monomers for >90% of the first monomer to react. The disappearance of the monomers was monitored by gas chromatography.

Results

Concurrent Copolymerization. Difficulties are encountered in studying the kinetics of these reactions because of reaction-to-reaction variation probably caused by the heterogeneous nature of the reaction and its sensitivity to the sodium surface area. Also, some batch-to-batch variation in the dichloride is encountered, despite redistillation. For the polymerizations, the relative rates of monomer consumption are compared under comparable conditions and are probably valid. The variations in the reaction rates of dichlorides during homopolymerization are large enough for semiquantitative conclusions to be drawn. Nevertheless, because of the sigmoidal reaction curves found for all reactions slow enough to follow, extracting a rate constant would be rather dubious. Qualitative comparisons of rate, or comparisons of half-lives at best, are all that can be done reliably at present.

On a qualitative scale, in which >> represents about a 10-fold factor, the relative rates of monomer consumption during homopolymerization are PMDS >> HMDS > DMDS >> VMDS, in which PMDS, HMDS, DMDS, and VMDS are phenylmethyl- hexylmethyl-, dimethyl-, and vinylmethyldichlorosilane, respectively. HMDS and DMDS have half-lives of 20 and 90 min, respectively. The reactions of PMDS and VMDS are, respectively, too fast and too slow to measure with any precision by our methods.

The relative rates of monomer consumption during copolymerization are in a different order. For the following pairs, the relative rates are PMDS > HMDS, PMDS > DMDS, PMDS > VMDS but VMDS > HMDS, VMDS > DMDS, DMDS > HMDS. For a terpolymerization, the relative rates are PMDS > VMDS > HMDS.

In these copolymerizations, the polymers with the vinylmethylsilane group are of interest, because the presence of the vinyl group would assist cross-linking if the polymers are used as precursors for SiC. Both IR and NMR data indicated that the vinyl content in the polymer is lower than expected and that some carbosilane may form. The yields of SiC were good (Table I).

Table I. Copolymerization with VMDS

Starting Monomer Ratio[a]	*Polymer Yield (%)*[b]	*SiC Yield (%)*[c]
1:9:0	69	75
1:9:0[d]	77	73
0.5:9.5:0	61	79
2:8:0	48	81
1:7:2	48	62
1:5:4	30	57
1:5:5	24	67

[a]The data are the starting VMDS/PMDS/HMDS ratios.
[b]The data are the yields of toluene-soluble polymer precipitated in isopropyl alcohol.
[c]The data are percentages of the theoretical yield.
[d]Diphenylmercury (1%) was added at the start of the reaction to restrict the polymer molecular weight.

Sequential Polymerization. The sigmoidal reaction curves, which indicate a tendency for the molecular weight to increase during the course of the reaction, and other considerations led to the suggestion that the polymerizing chains had long lifetimes (9), similar to the chains in a "living" polymerization. If this is the case, sequential addition of different monomers would lead to block copolymer formation. To check this hypothesis, PMDS and HMDS were polymerized sequentially with sufficient time between additions for the first monomer to be consumed. In one experiment, PMDS was the first monomer, and in another experiment, HMDS was the first. In two other experiments, diphenylmercury was added to PMDS in the first addition to limit the molecular weight of this part of the polymer (9).

Differential solubility was used to check for the presence of block copolymer. First, the higher molecular weight polymer was isolated from the reaction mixture by precipitation in isopropyl alcohol. Neither poly(hexylmethylsilane) nor poly(phenylmethylsilane) are soluble in this alcohol. Poly(hexylmethylsilane) is soluble in hexane, but poly(phenylmethylsilane) is not. The precipitated polymer was then dissolved in methylene dichloride and added to hexane. The polymer that precipitated was recovered by filtration; the rest was recovered by evaporation of the solution.

NMR analysis showed the presence of phenyl and hexyl groups in both the hexane-soluble and the hexane-insoluble fractions, a result indicating the presence of copolymers, presumably block copolymers (Table II). Because the phenyl content of the hexane-insoluble polymer was considerably higher than that of the hexane-soluble polymer (Table II), the composition distribution must have been fairly wide. Fractional precipitation of both fractions from toluene solution by methanol was attempted, and some homopoly(phenylmethylsilane) was isolated from the hexane-insoluble fraction. Most fractions contained both phenyl and hexyl groups.

Table II. Composition of Polymer from Sequential Polymerization

Monomer Sequence[a]	*Hexane-Soluble Polymer (%)*	*Phenyl Groups in Polymer (%)*	
		Hexane Soluble	*Hexane Insoluble*
HMDS, PMDS	62	66	81
PMDS, HMDS	58	60	85
PMDS, HMDS[b]	53	60	93
PMDS, HMDS[b,c]	100	25	0

[a]Equal volumes of monomers were used.
[b]Diphenylmercury (1%) was added at the start of the reaction.
[c]The reaction was analyzed at the end of the PMDS reaction. The reaction mixture was washed, dried, and restarted with fresh Na and the second monomer, HMDS.

Further sequential polymerizations were performed to determine whether the incorporated polymer had in fact to be living, as in an anionic polymerization. Previous studies (*10*) have shown that even after all the dichloride has been consumed in a PMDS polymerization, increases in molecular weight of the middle-range fraction of the polymer (10^3–10^4) occur on continued reflux with excess sodium. Also, even if the reaction solution is washed with water and dried, these changes occur on reflux with sodium.

This middle-range fraction was maximized by the addition of a little diphenylmercury at the beginning of the reaction. The reaction solution was again washed with water to deactivate any chlorine or sodium chain ends, dried, and refluxed with sodium, and HMDS was added. By analyzing for block copolymers, the final reaction product was compared with one from a reaction mixture in which the washing process had been omitted. The washed reaction mixture gave only polymer that was hexane soluble, but the highest molecular weight fraction contained 25% phenyl groups (Table II). The molecular weight of this product was also several times that of the product from the first stage. Undoubtedly, block copolymers were formed in this reaction but not as much as in the comparable reaction that was not washed. In the unwashed reaction, hexane-insoluble high-molecular-weight polymer with a higher phenyl content was formed (Table II). Some type of reactivation is possible.

Discussion

General Reaction Scheme. The polysilane reaction products from the reaction of dialkyldichlorosilanes with sodium in refluxing toluene are complex, with molecular weights spanning the three groups indicated previously. Any mechanism proposed for this reaction should take into account all the three products described at the start of this chapter, even if more than one mechanism coexist.

Before the reaction can be considered at the atomic level, the general pathways for the formation of the three products must be known. The pathway for polysiloxane formation has been mentioned. When applied to polysilanes, this pathway could imply an initiation reaction with a much faster propagation reaction whereby the silane units would add in a stepwise fashion, possibly by a two-stage process as discussed earlier. The concurrent and end-biting and back-biting reactions would chop off the low-molecular-weight cyclic materials and, less frequently, the larger cyclic materials to give the intermediate fraction. The distribution would then be determined kinetically by the relative rates of the three processes until final deactivation. Earlier, West et al. (*12*) described such a kinetic determination of products for DMDS.

An alternative scheme that does not involve end biting is possible. As before, reaction of the initial dichloride with sodium is necessary for the initiation step in pure systems. After the first one or two additions of the monomer, the growing short chain would, by the two-step addition process, have alternately sodium or chlorine ends at each end of the short chain. When the chain reaches 5 or 6 units, cyclization can take place ideally, if the two chain ends are different and can react. Only that fraction of chains that does not cyclize at this point could grow to high molecular weight. The intermediate fraction could be composed of chains of moderate length, which would still have a reasonable chance to cyclize once the chain length is greater than 8–12 units; ring formation by chains of 8–12 units is difficult. Also, condensation reactions could still occur. Chain propagation reactions may accelerate as the chain lengthens, and increased electron delocalization (*13*) may alter the activity of the chain end groups.

The middle-range polymer fraction may contain an appreciable amount of cyclic material. A fraction of PMDS polymer with a molecular weight of 2000 was isolated, and the phenyl and methyl regions were analyzed by ^{1}H NMR spectroscopy. The phenyl/methyl ratio was very close to 1:1 even if $(CH_3)_3SiCl$ was used to terminate the reaction. This result suggests that this material is cyclic. The ^{29}Si NMR spectrum showed a small extra peak at 8 ppm (TMS [tetramethylsilane] peak at 0 ppm), as well as the usual signal for Si–Si–Si at –40 ppm, corresponding to 4% Si–Si–$(CH_3)_3$. This finding casts doubt on the totally cyclic nature of this material.

Copolymerization Rates. Both of the overall schemes described rely upon a comparatively slow reaction of the initial dichloride with sodium, and this requirement is confirmed by the kinetics of the reactions. These reactions all have an initial period of accelerating rate, suggesting that the initial dichloride itself does not react rapidly with the sodium surface but that some intermediate in the reaction will react more rapidly with the dichloride. Presumably, this intermediate is the growing chain end. The homopolymerization rate is a function of the initiation rate, as well as the rates of subsequent propagation and other steps. Thus, the apparently anomalously rapid incorporation of VMDS or DMDS during copolymerization with HMDS suggests that these two monomers have fairly rapid propagation steps. Because of their very low rates of reaction with sodium in the initiation reaction, their overall homopolymerization rates are slower than their rates of copolymerization, in which initiation may be via the other monomer and faster.

Block Copolymer Formation. The formation of block copolymers during sequential polymerization suggests that chain ends do remain reactive even after all the initial dichloride is consumed. The incorporation of the first polymer into block copolymers with the second monomer is, however, compatible with the suggestion that chain extension is by monomer addition alone or by a combination of monomer addition and condensation reactions to form the higher molecular weight fractions. With chain extension by monomer addition alone, triblocks with the first monomer in the middle can be formed. Condensation reactions could give multiblock copolymers. NMR analysis has indicated block copolymer formation, but the block length has not been determined yet. The distribution of block lengths is probably large because of the broad molecular weight distribution of the first polymer before the second addition.

The inclusion of apparently "dead" polymer into block copolymer is of interest (Table II). Certainly less of the first PMDS polymer was incorporated in this system than in the corresponding living system, because all the copolymer formed in the system with dead polymer was hexane soluble. The results suggest that the active new chain end reacts with the existing polymer and reactivates it. This reaction is necessary in the mechanism that suggests back biting to eliminate cyclic materials. This type of reaction is widely postulated to account for the molecular weight distributions found in cyclic oxide polymerization, for example, THF (*14*) and oxetane (*15*).

Conclusions

If the reactions in THF and the inclusion of dead material in block copolymers are taken into account, the balance of the evidence at present seems to support the back-biting mechanism. However, the mechanisms are not really mutually exclusive, because they both suggest an alternation between a chlorine-ended chain and a sodium-ended chain. The sodium ended-chain reacts first with the initial dichloride in the rate-determining step, and then the chlorine-ended chain reacts with the sodium surface. The difference lies in the mode of formation of the cyclic materials and the importance of any condensation reactions. Both mechanisms might be operative, although one may dominate, depending on the conditions or substituents.

The details of the reaction at the atomic level are even more difficult to unravel. The reaction of a chlorine-ended chain with a sodium surface may involve two single-electron-transfer steps that may at first form a chain radical and Cl or a delocalized radical anion on the chain. The transfer of a second electron from sodium to form a delocalized anion–Na^+ ion pair may follow. The ion pair could leave the Na surface, probably as an aggregate of ion pairs, when the chain is long enough to solubilize it. The increasing ability of longer chains to delocalize the charge would account for the great difference between the activities of the initial dichloride and the growing chain in reacting with the sodium surface.

In reacting with the initial dichloride or with another chlorine-ended chain, the polysilane sodium chain end could again involve a radical intermediate, as suggested for the reaction of alkyl halides and sodium alkyls. The radical traps in the experiments of Zeigler (*6*, *7*) could operate as chain breakers at both these stages. The promoting effect of the ethers found by Miller (*8*) could assist by breaking down the ion-pair aggregates in toluene, as happens during the anionic polymerization of carbon-based polymers (*16*), or accelerate the formation of radical anions.

References

1. Wesson, J. P.; Williams, T. C. *J. Polym. Sci., Polym. Chem. Ed.* **1980,** *18,* 959.
2. Trujillo, R. E. *J. Organometal. Chem.* **1980,** *198,* C 27.
3. West, R.; David, L. D.; Djurovich, P. I.; Stearley, D. L.; Srinivasan, K. S. V.; Yu, H. *J. Am. Chem. Soc.* **1981,** *103,* 1352.

4. West, R. *J. Organomet. Chem.* **1986,** *300,* 327.
5. Zhang, Y-H.; West, R. *J. Polym. Sci., Polym. Chem. Ed.* **1984,** *22,* 225.
6. Zeigler, J. M. *Polym. Prepr.* (*Am. Chem. Soc., Div. Polym. Chem.*) **1986,** *27,* 109.
7. Zeigler, J. M. *Polym. Prepr.* (*Am. Chem. Soc., Div. Polym. Chem.*) **1987,** *28,* 424.
8. Miller, R. D.; Hofer, D.; McKean, D. R.; Willson, C. G.; West, R.; Trefonas, P. T. In *Materials for Microlithography;* Thompson, L. F.; Willson, C. G.; Fréchet, J. M. J., Eds.; ACS Symposium Series 266; American Chemical Society: Washington, DC, 1984; p 293.
9. Miller, R. D.; Rabolt, J. F.; Sooriyakumaran, R.; Fleming, W.; Fickes, G. M.; Farmer, B. L.; Kuzmany, H. In *Inorganic and Organometallic Polymers: Macromolecules Containing Silicon, Phosphorus, and Other Inorganic Elements;* Zeldin, M.; Wynne, K. J.; Allcock, H. R., Eds.; ACS Symposium Series 360; American Chemical Society: Washington, DC, 1988.
10. Worsfold, D. J. In *Inorganic and Organometallic Polymers: Macromolecules Containing Silicon, Phosphorus, and Other Inorganic Elements;* Zeldin, M.; Wynne, K. J.; Allcock, H. R., Eds.; ACS Symposium Series 360; American Chemical Society: Washington, DC, 1988; pp 101–111.
11. Brown, J. F.; Slusarczuk, G. M. J. *J. Am. Chem. Soc.* **1965,** *87,* 931.
12. Carberry, E.; West, R. *J. Am. Chem. Soc.* **1969,** *91,* 5440.
13. Trefonas, P.; West, R.; Miller, R. D.; Hofer, D. C. *J. Polym. Sci., Polym. Lett. Ed.* **1983,** *21,* 823.
14. Dreyfuss, M. P. *J. Macromol. Sci., Chem.* **1975,** *9,* 729.
15. Black, P. E.; Worsfold, D. J. *Can. J. Chem.* **1976,** *54,* 3325.
16. Bywater, S.; Worsfold, D. J. *Can. J. Chem.* **1962,** *40,* 1564.

RECEIVED for review May 27, 1988. ACCEPTED revised manuscript March 27, 1989.

19

Polymers and Copolymers of Disilacyclohexadiene

A Non-Alkali-Metal Route to Polysilanes

Robert A. Rhein

Naval Weapons Center, China Lake, CA 93555–6001

1,2-Dimethyltetramethoxydisilane was reacted with acetylene in a tube reactor heated to 400–425 °C. A mixture that is liquid at ambient temperature was formed. Careful vacuum distillation of this liquid mixture yielded crude 1,4-dimethyl-1,4-dimethoxy-1,4-disilacyclohexadiene. The crude product was reacted with potassium hydroxide, and recrystallization in isopropyl alcohol produced 1,4-dimethyl-1,4-disilacyclohexadiene-1,4-di(potassium silanoate). Upon acidification of the disilanoate, the thermally stable siloxane homopolymer was produced. Upon reaction of the disilanoate with phenylmethyldichlorosilane, the thermally stable, exactly alternating methylphenylsiloxane copolymer was obtained. Thermogravimetric analysis showed that these polymers are thermally stable. Vigorous distillation of the crude reaction mixture yielded a glassy transparent reaction product. IR and ^{1}H NMR spectroscopic data suggest that the product is poly(methylmethoxysilane). The reaction of acetylene and 1,1,2,2-tetrachlorodisilane in a heated tube at 500 °C produced a product that is liquid at ambient temperature and that produced a glassy transparent solid upon vigorous distillation. IR, ^{1}H NMR, and UV spectroscopic data suggest that the product is poly(methylchlorosilane).

THERMALLY STABLE ELASTOMERS AND POLYMERS include fluorocarbon elastomers, polyorganophosphazenes, silicone elastomers, siloxane block copolymers (including the dodecacarborane block copolymers [Dexsil elastomers] and the silphenylene–siloxane polymers), and polysilmethylenes (*1*). Of these elastomers, the polyorganophosphazenes and silicone elastomers have good thermal stability and can be produced at moderate cost, but in general, they are thermally stable only up to less than 300 °C. The other elastomers are of higher thermal stability but are expensive to manufacture.

This chapter describes the preparation of a polymer that is curable to an elastomer and should be inexpensive to manufacture. Also described is a non-alkali-metal route to polysilanes.

The polymers described in this chapter are derived from the residue of the manufacture of methylchlorosilanes by the direct process. Approximately 10% of the reaction product of methylchlorosilane manufacture is a residue (called the direct-process residue) (*2*). The fraction of direct-process residue boiling between 150 and 160 °C consists primarily of 1,2-dimethyltetrachlorodisilane and 1,1,2-trimethyltrichlorodisilane (*3*).

The preparation of 1,2-dimethyltetrachlorodisilane and 1,2-dimethyltetramethoxydisilane from this mixture of 1,2-dimethyltetrachlorodisilane and 1,1,2-trimethyltrichlorodisilane is described elsewhere (*4*), and the preparation of 1,4-dimethyl-1,4-dimethoxy-1,4-disilacyclohexadiene from 1,2-dimethyltetramethoxydisilane is shown in Scheme I (*5*).

The preparation of a thermally stable elastomer derived from 1,4-dimethyl-1,4-dimethoxy-1,4-disilacyclohexadiene and the preparation of the polymer intermediate, 1,4-dimethyl-1,4-disilacyclohexadiene-1,4-di(potassium silanoate), are discussed. Also described is an unexpected polymer resulting from the attempted distillation of 1,4-dimethyl-1,4-dimethoxy-1,4-disilacyclohexadiene. This thermally stable polymer is believed to be poly(methylmethoxysilane). The polymer resulting from the distillation of the products of the reaction of acetylene and 1,2-dimethyltetrachlorodisilane at 500 °C is also described; this polymer is believed to be poly(methylchlorosilane).

Preparation of 1,4-Dimethyl-1,4-disilacyclohexadiene-1,4-di(potassium silanoate)

1,2-Dimethyltetrachlorodisilane. To prepare 1,2-dimethyltetrachlorodisilane from commercially available direct-process residue, 754.0 g of the fraction boiling at 150–152 °C (at an elevation of 660 m) was refluxed with 54.9 g of aluminum chloride; dry hydrogen chloride was bubbled through this refluxing mixture for 34 h and 12 min. Upon cooling, the liquid portion of this mixture was decanted into another vessel, and 50 mL of reagent-grade acetone was added to the liquid. This mixture was distilled, and 652.3 g of the fraction boiling at 150–152 °C was collected. Gas chromatographic analysis of this product indicated a purity of >97%.

The IR spectrum of this sample (data are wavenumbers in reciprocal centimeters) showed medium-weak (mw) absorptions at 2960, 2900, and

Scheme I. Preparation of 1,4-dimethyl-1,4-dimethoxy-1,4-disilacyclohexadiene.

1983; a medium (m) absorption at 1716; strong (s) absorptions at 1400 and 1256; a weak (w) absorption at 1220; a medium and broad (mb) absorption at 1073; and very strong (vs) absorptions at 750–800 and 720–750. The following absorption bands are reported for 1,2-dimethyltetrachlorodisilane (*6*): 2960, 2900 (w); 1400 (m); 1256 (s); 850, 810 (w); 765 (s); and 734 (s) cm^{-1}. The ^1H NMR spectrum of this sample showed one peak at 1.00 ppm (all peaks were downfield with reference to tetramethylsilane). A single peak at 1.03 ppm is reported (*7*) in the literature. This substance was identified as 1,2-dimethyltetrachlorodisilane by boiling point, gas chromatographic data, and IR and ^1H NMR spectral data.

1,2-Dimethyltetramethoxydisilane. The next step, the conversion to 1,2-dimethyltetramethoxydisilane, is shown in Scheme II. To 466.0 g of 1,2-dimethyltetrachlorodisilane was added slowly, at 68 °C, 572 g of methyl orthoformate (trimethoxymethane). Upon completion of addition, this mixture remained at 68 °C for 14 h and 21 min; it was then distilled under vacuum at a pressure of 3724 Pa, and 246.0 g of the fraction boiling at 84 °C was collected. The reported (*4*) boiling point of 1,2-dimethyltetramethoxydisilane is 86.2–87 °C at 3990 Pa. Gas chromatographic analysis showed that this material is at least 97% pure.

The IR spectrum showed peaks at 2950 (vs), 2830 (s), 1460 (m), 1400 (m), 1253 (s), 1160 (s), 1060 (vs), 800–830 (vs), 750–760 (vs), and 640 (m) cm^{-1}. 1,2-Dimethyltetramethoxydisilane is reported (*8*) to have IR absorptions at 2960 and 2940 (vs), 2837 (vs), 1464 (mb), 1253 (s), 1190 (s), 1080 (vs), 820 (vs), 762 (vs), 730 (s), and 643 (s) cm^{-1}. The ^1H NMR spectrum of this material (referenced to tetramethylsilane) showed a a strong singlet at 0.13 ppm and a strong peak at 3.40 ppm; 1,2-dimethyltetramethoxydisilane is reported (*9*) to have ^1H NMR signals at 0.15 and 3.46 ppm. Hence, the prepared compound was identified as 1,2-dimethyltetramethoxydisilane by gas chromatographic data, boiling point, and IR and ^1H NMR spectral data.

1,4-Dimethyl-1,4-dimethoxy-1,4-disilacyclohexadiene. The reaction between 1,2-dimethyltetramethoxydisilane and acetylene (Scheme I) was carried out several times, and the following description of the procedure is typical. In a Pyrex tube (22 mm in outside diameter and about 48 cm long) heated to 400–425 °C, with nitrogen flowing through at a rate of <10 mL/min, acetylene was passed through at a rate of 43 mL/min (1.75 mmol/min). The liquid reaction product was collected in a receiving flask with a reflux condenser attached.

A gas chromatographic analysis of the liquid product indicated primarily two constituents, one of which was identified by boiling point and peak retention time as methyltrimethoxysilane. Distillation of this mixture was very difficult. Although methyltrimethoxysilane could be removed readily, the other constituent had a tendency toward polymerization. A small sample

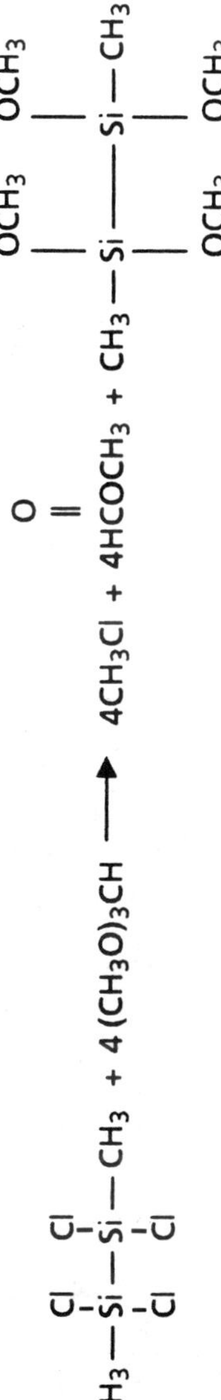

Scheme II. Preparation of 1,2-dimethyltetramethoxydisilane.

of the other constituent was analyzed and found to have a a boiling point of 70 °C at 400–530 Pa; the IR spectrum showed absorption peaks at 2937 (vs), 2815 (s), 1460 (m), 1400 (m), 1340 (s), 1256 (s), 1190 (s), 1060–1080 (vs), and 740–800 (vs) cm^{-1}. The ^{1}H NMR spectrum showed singlets at 0.12, 3.44, and 6.92 ppm. 1,4-Dimethyl-1,4-dimethoxy-1,4-disilacyclohexadiene is reported (*10*) to have IR absorptions at 2840, 1187, 1085, and 1250 cm^{-1} and ^{1}H NMR peaks at 0.17, 3.37, and 6.92 ppm. Consequently, the reaction between acetylene and 1,2-dimethyltetramethoxydisilane produced methyltrimethoxysilane and 1,4-dimethyl-1,4-dimethoxy-1,4-disilacyclohexadiene.

Because 1,4-dimethyl-1,4-dimethoxy-1,4-disilacyclohexadiene was very difficult to distill, the mixture of methyltrimethoxysilane and 1,4-dimethyl-1,4-dimethoxy-1,4-disilacyclohexadiene was distilled under vacuum at a pressure of <133 Pa and heated to a temperature not to exceed 50 °C. Gas chromatographic analysis of the residue after distillation (after removal of most of the methyltrimethoxysilane) showed that the residue contained ~6% methyltrimethoxysilane, ~67% 1,4-dimethyl-1,4-dimethoxy-1,4-disilacyclohexadiene, and smaller amounts of unknown impurities.

1,4-Dimethyl-1,4-disilacyclohexadiene-1,4-di(potassium silanoate). A typical preparation of the 1,4-di(potassium silanoate) of 1,4-dimethyl-1,4-disilacyclohexadiene is shown in Scheme III. Crude 1,4-dimethyl-1,4-dimethoxy-1,4-disilacyclohexadiene (16 mL) was added slowly to 83.2 mL of 3.835 M KOH (dissolved in 90% methanol and 10% water). The resulting mixture was filtered, and the filtrate was evaporated under vacuum (in a water bath held at ~60 °C) until most of the solvent was removed. Addition of 150 mL of tetrahydrofuran resulted in a precipitate and a two-phase liquid mixture. The precipitate was removed by filtration and washed with a small amount of tetrahydrofuran and diethyl ether. The precipitate was dissolved in 100 mL of boiling isopropyl alcohol. Upon cooling of the isopropyl alcohol solution, a crystalline mass separated. This crystalline mass was washed, first with 60 mL of isopropyl alcohol and then with 50 mL of pentane; then the solid was subjected to vacuum evaporation at ambient temperature to remove residual solvent. The result was 5.76 g of solid product.

The IR spectrum of the solid showed absorption bands at 2890 (w) and 810 (sb) cm^{-1} for Si–CH_3 and at 1340 (ms) and 930 (sb) cm^{-1}, which are probably due to the silane ring band. The ^{1}H NMR spectrum of the material dissolved in a mixture of deuterium oxide and perdeuteroacetone showed peaks close to 0.0 ppm (Si–CH_3) and 6.95 ppm (–CH=). The solid was soluble in water, methanol, and ethanol.

In anticipation of using the 1,4-di(potassium silanoate) of 1,4-dimethyl-1,4-disilacyclohexadiene in the preparation of polymers, the model reaction shown in Scheme IV was conducted. To 0.81 g of the salt believed to be

Scheme III. Preparation of 1,4-dimethyl-1,4-disilacyclohexadiene-1,4-di(potassium silanoate).

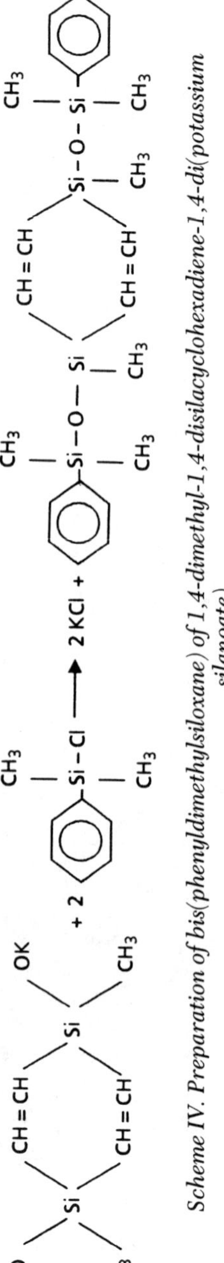

Scheme IV. Preparation of bis(phenyldimethylsiloxane) of 1,4-dimethyl-1,4-disilacyclohexadiene-1,4-di(potassium silanoate).

the 1,4-di(potassium silanoate) of 1,4-dimethyl-1,4-disilacyclohexadiene was added 8 mL of ethanol; the salt dissolved. Twice the required stoichiometric amount of phenyldimethylchlorosilane was added slowly, and a precipitate was formed. Water (15 mL) and tetrahydrofuran (5 mL) were added, and a transparent two-phase solution resulted. A gas chromatographic analysis of the organic phase indicated essentially one component at a retention time expected for the anticipated bis(phenyldimethylsiloxane) of 1,4-dimethyl-1,4-disilacyclohexadiene. This model reaction plus the IR and ^{1}H NMR spectra of the salt suggest that the solid is the 1,4-di(potassium silanoate) of 1,4-dimethyl-1,4-disilacyclohexadiene.

Preparation of 1,4-Disiloxy-1,4-dimethyl-1,4-disilacyclohexadiene Homopolymer

The homopolymer of 1,4-disiloxy-1,4-dimethyl-1,4-disilacyclohexadiene was prepared as shown in Scheme V. In this case, 3.89 g of 1,4-dimethyl-1,4-disilacyclohexadiene-1,4-di(potassium silanoate) was dissolved in 10 mL of water and titrated to the phenolphthalein end point with glacial acetic acid. A precipitate was formed; ether was added until two transparent liquid phases resulted, and the upper organic phase was removed. Removal of the ether by evaporation under vacuum yielded 1.59 g of a tan, cloudy, viscous product. The solution viscosity (tetrahydrofuran, 30 °C) was 0.142 dL/g. The IR spectrum showed absorption bands at 2933 (mw), 1255 (m), and 822 (m) cm^{-1} (Si–CH_3) and at 1035 (s, Si–O–Si) and 1340 (m, associated with the ring) cm^{-1}. The ^{1}H NMR spectrum showed a singlet at 6.86 ppm (–CH=) and very close doublets at 0.18 and 0.20 (Si–CH_3). The expected area ratio was 40/60, and the experimentally determined ratio was 38.5/60. TGA (thermogravimetric analysis) of this polymer (Figure 1), which shows a weight loss of approximately 5% at ~200 °C but no further weight loss up to 600 °C, indicates that the polymer has considerable thermal stability.

Preparation of Exactly Alternating Copolymer of 1,4-Disiloxy-1,4-dimethyl-1,4-disilacyclohexadiene and Methylphenylsiloxane

To prepare the alternating copolymer of 1,4-dimethyl-1,4-disilacyclohexadiene-1,4-siloxane and methylphenylsiloxane (Scheme VI), 5.76 g of the 1,4-di(potassium silanoate) of 1,4-dimethyl-1,4-disilacyclohexadiene was dissolved in 35 mL of ethanol, and an equimolar amount of phenylmethyldichlorosilane was added dropwise while the mixture was kept in an ice bath. A precipitate, probably potassium chloride, formed during the addition of the reactants. Upon completion of addition, 20 mL of water was added to the mixture. Cyclohexane (20 mL) and some tetrahydrofuran were added

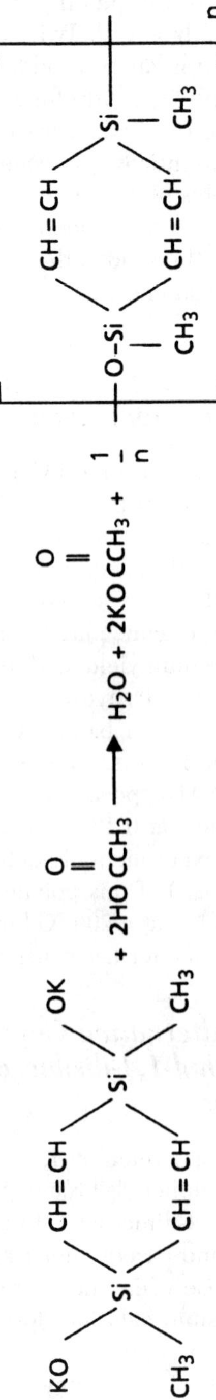

Scheme V. Preparation of 1,4-disiloxy-1,4-dimethyl-1,4-disilacyclohexadiene homopolymer.

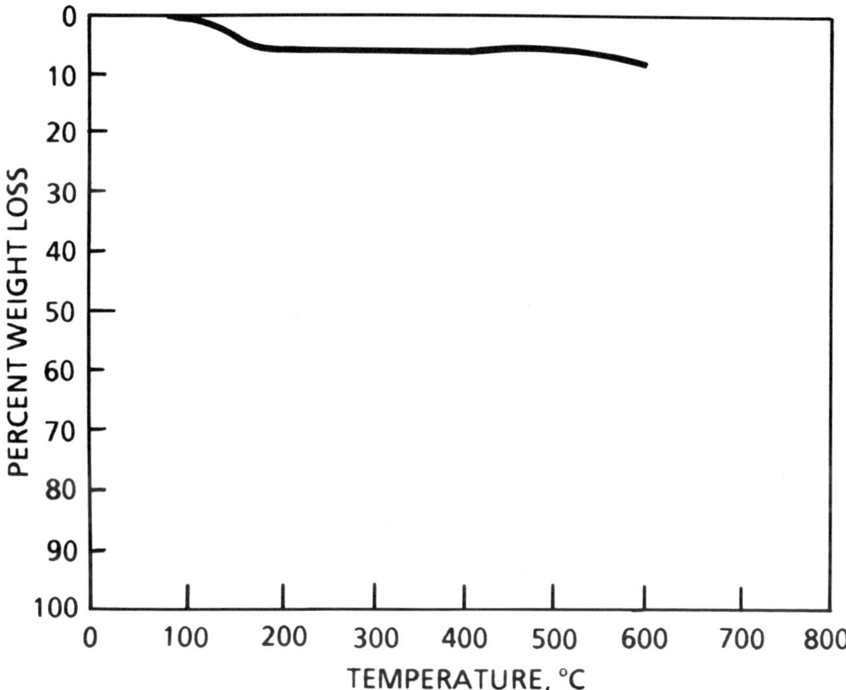

Figure 1. TGA of $\left[-\underset{CH_3}{Si}\langle{}^{CH=CH}_{CH=CH}\rangle Si \langle{}^{O-}_{CH_3} \right]_n$

until two transparent phases resulted. The organic phase was separated, and the solvent was removed by vacuum evaporation to yield 5.48 g of a tan, viscous product. This product was dissolved in 10 mL of cyclohexane, and 30 mL of methanol was added to precipitate the polymer. The solvent was removed from the polymer to yield 4.04 g of product.

The solution viscosity of this polymer (tetrahydrofuran, 30 °C) was 0.097 dL/g. The IR spectrum showed absorption bands at 3090, 3070 (w), and 1592 (m) cm^{-1} (phenyl–H); 2928 (m), 1406 (w), and 1255 (s) cm^{-1}

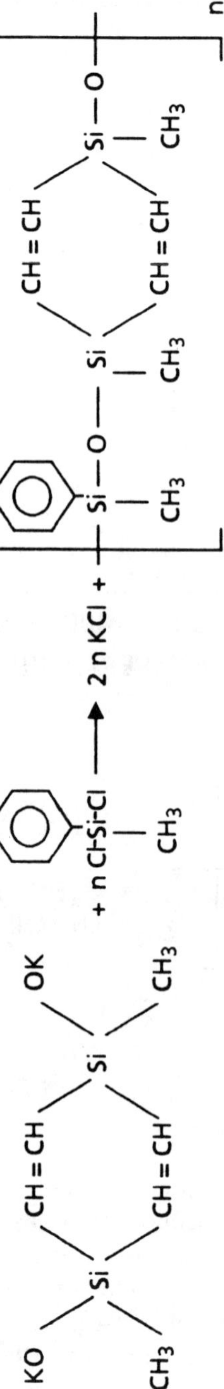

Scheme VI. Preparation of exactly alternating polymer of 1,4-disiloxy-1,4-dimethyl-1,4-disilacyclohexadiene and methylphenylsiloxane.

($Si{-}CH_3$); 1430 (m), 1120 (s), and 824 (s) cm^{-1} (Si–phenyl); 1010–1070 (s) cm^{-1} (Si–O–Si); and 1340 (m) cm^{-1} (Si ring). The ^1H NMR spectrum showed a singlet at 5.97 ppm (–CH=), a broad peak centered at 7.4 ppm (phenyl–H), and a sharp doublet at 0.26 and 0.36 ppm ($Si{-}CH_3$). As expected, the peak area of the first two peaks was equal to the peak area of the $Si{-}CH_3$ doublet. The TGA of this polymer (Figure 2), which shows a weight loss of approx-

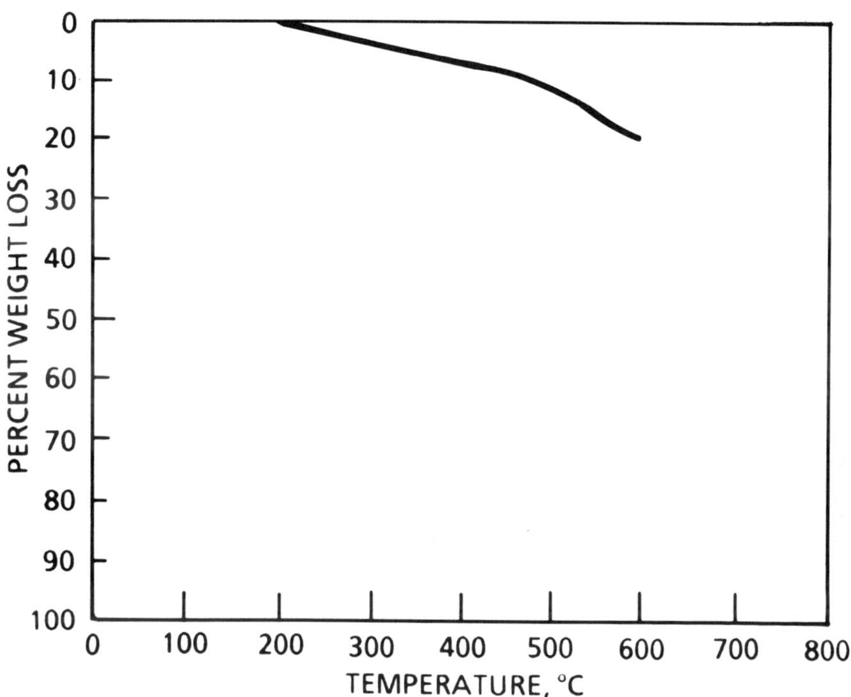

Figure 2. TGA of $\left[-Si(C_6H_5)(CH_3)-O-Si(CH_3)\langle CH{=}CH, CH{=}CH \rangle Si(CH_3)-O- \right]_n$

imately 20% at 600 °C, indicates that this polymer is thermally stable. The TGA of a sample of the known thermally stable silphenylene–methylphenylsiloxane polymer is shown in Figure 3. A comparison with the TGA of the exactly alternating copolymer of 1,4-disiloxy-1,4-dimethyl-1,4-disilacyclohexadiene and methylphenylsiloxane shows that thermal stability of the exactly alternating polymer prepared in my laboratory is superior compared with that of the known thermally stable silphenylene–methylphenylsiloxane polymer.

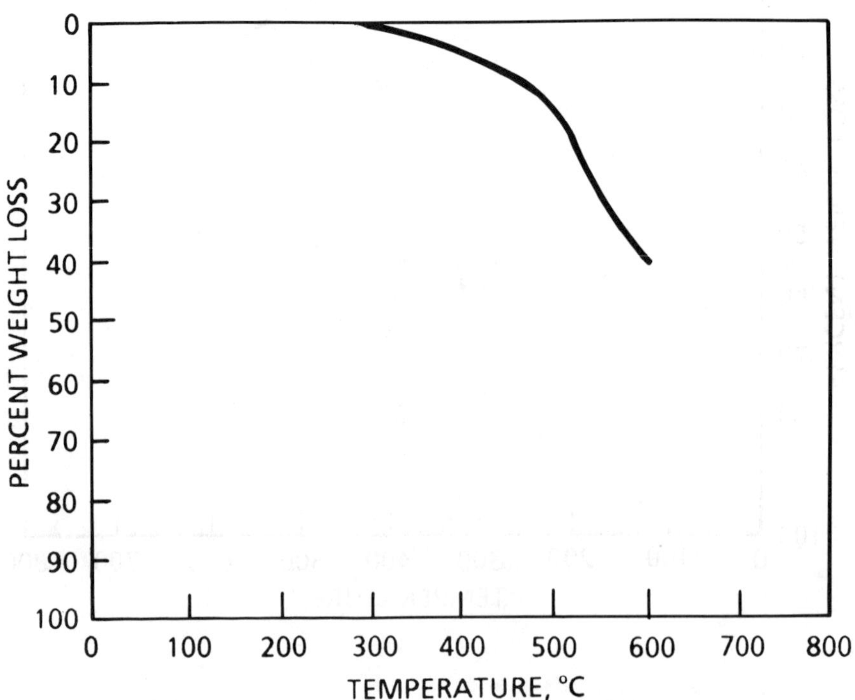

Figure 3. TGA of $\left[-Si(C_6H_5)(CH_3)-O-Si(CH_3)_2-C_6H_4-Si(CH_3)_2-O- \right]_n$

Polymer from Attempted Distillation of 1,4-Dimethyl-1,4-dimethoxy-1,4-disilacyclohexadiene

As previously described, in a heated tube held at 450 °C over a 73.8-min interval, 0.13 mol of 1,2-dimethyltetramethoxydisilane was reacted with 0.33 mol of acetylene to produce 23.1 g of liquid reaction product. Distillation of this product at atmospheric pressure yielded 11.3 g of distillate (boiling point 100–105 °C), presumably methyltrimethoxysilane (boiling point 102–103 °C). The attempted vacuum distillation of the remainder produced a polymeric mass.

In another experiment, acetylene was passed through at a rate of 1.75 mmol/min, and 1,2-dimethyltetramethoxydisilane was passed through at a rate of 0.670 mmol/min at a tube temperature of 500–525 °C. A total of 25 mL of reaction product was collected. Distillation resulted in 10.8 g of polymeric residue.

In a third experiment, acetylene was passed through at a rate of 1.75 mmol/min, and 1,2-dimethyltetramethoxydisilane was passed through at a rate of 0.670 mmol/min at a tube temperature of 400–425 °C. Use of 25 mL of 1,2-dimethyltetramethoxydisilane resulted in 22.41 g of reaction product. Distillation resulted in 3.54 g of polymeric residue, which was a light-tan glassy solid, which dissolved in hexane to produce a viscous solution. The solution viscosity (tetrahydrofuran, 30 °C) was 0.142 dL/g. The IR spectrum showed absorption bands at 2958 (m) cm^{-1} (C–H); 1408 (w), 1256 (s), and 770–830 (sb) cm^{-1} ($Si–CH_3$); 2836 (mw), 1450 (w), and 1190 (w) cm^{-1} ($Si–OCH_3$); 1020–1070 (vs) cm^{-1} (Si–O–C); and 2120 (mwb), 1340 (vs), and 930 (vw) cm^{-1}. The ^1H NMR spectrum showed broad peaks at 0.13 ppm ($Si–CH_3$) and 3.45 ppm ($Si–OCH_3$); the area ratio was 3/1. The IR and ^1H NMR spectra of this polymer are consistent with the proposal that this polymer is poly(methylmethoxysilane). The TGA of this polymer (Figure 4) shows two regimes of thermal decomposition: at 200–300 and at 400–500 °C. The weight loss at 800 °C (about 30%) indicates that this polymer is fairly stable at high temperatures.

Polymer from Reaction of Acetylene with 1,2-Dimethyltetrachlorodisilane

In a heated tube held at 500–525 °C, the following materials were passed through at the indicated rates: acetylene, 1.75 mmol/min; nitrogen, ~50 mL/min; and 50 mL of 1,2-dimethyltetrachlorodisilane, ~0.67 mmol/min. The liquid reaction product was subjected to vigorous distillation, and 14.25 g of a glassy polymer resulted.

The ^1H NMR spectrum of the product showed a broad peak centered at ~0.5 ppm (with reference to tetramethylsilane) indicating trimethylsilyl groups. The IR spectrum showed absorption bands at 2960, 2900, 1400, and

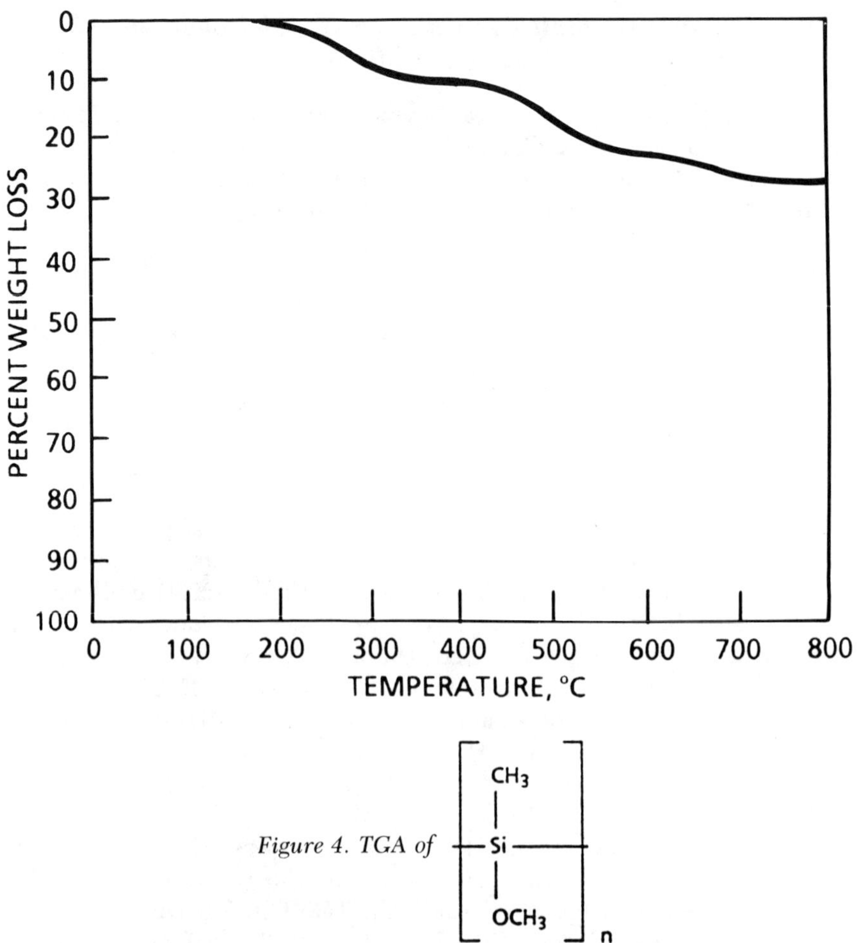

Figure 4. TGA of $\left[\mathrm{Si(CH_3)(OCH_3)}\right]_n$

1250–1260 cm^{-1}, all corresponding to Si–CH_3 bonding. Absorption bands were also observed at 509 and 537 cm^{-1} (Figure 5), indicating Si–Cl bonding. The UV spectrum (Figure 6) showed a minimum at 280 nm; polysilanes typically show minima at 300–350 nm (*11*). These results indicate that the polymer is poly(methylchlorosilane).

Summary

1. 1,4-Dimethyl-1,4-disilacyclohexadiene-1,4-di(potassium silanoate) was prepared from the reaction of 1,4-dimethyl-1,4-

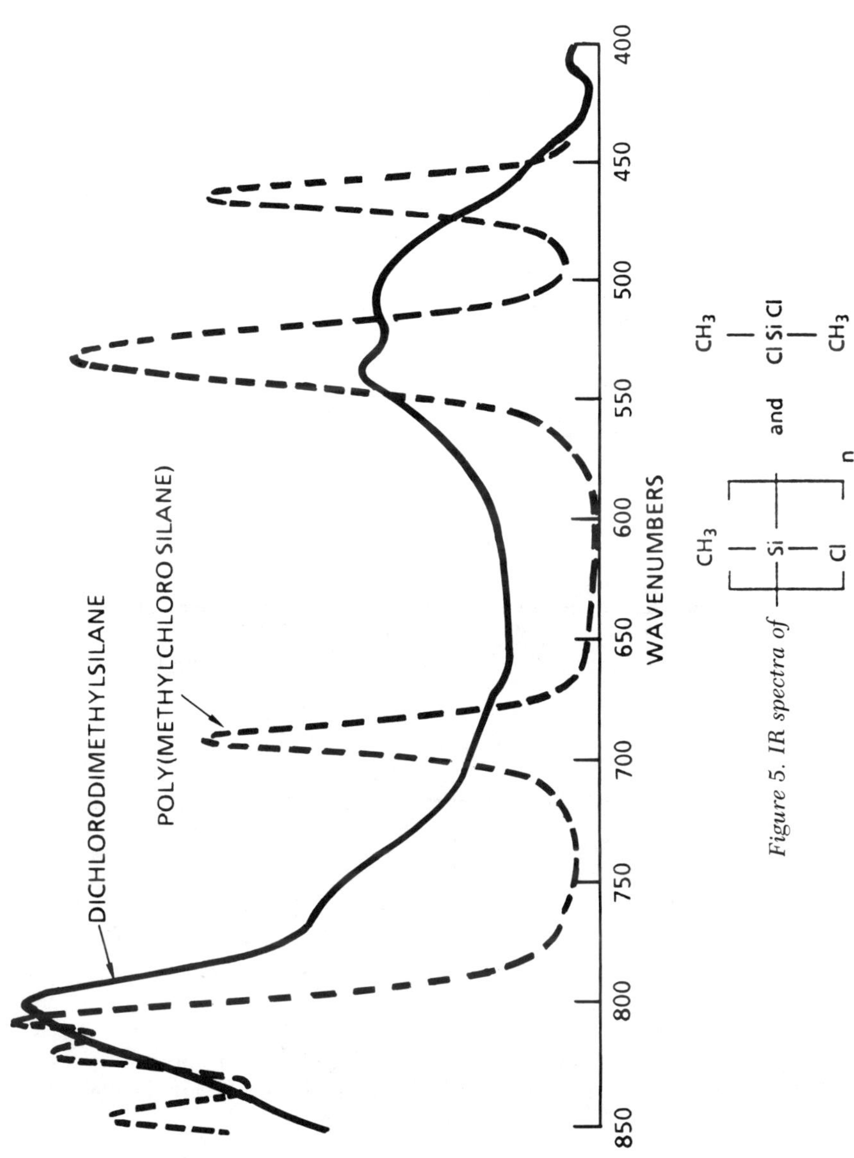

Figure 5. IR spectra of $\left[-\overset{CH_3}{\underset{Cl}{Si}}- \right]_n$ and $Cl\overset{CH_3}{\underset{CH_3}{Si}}Cl$

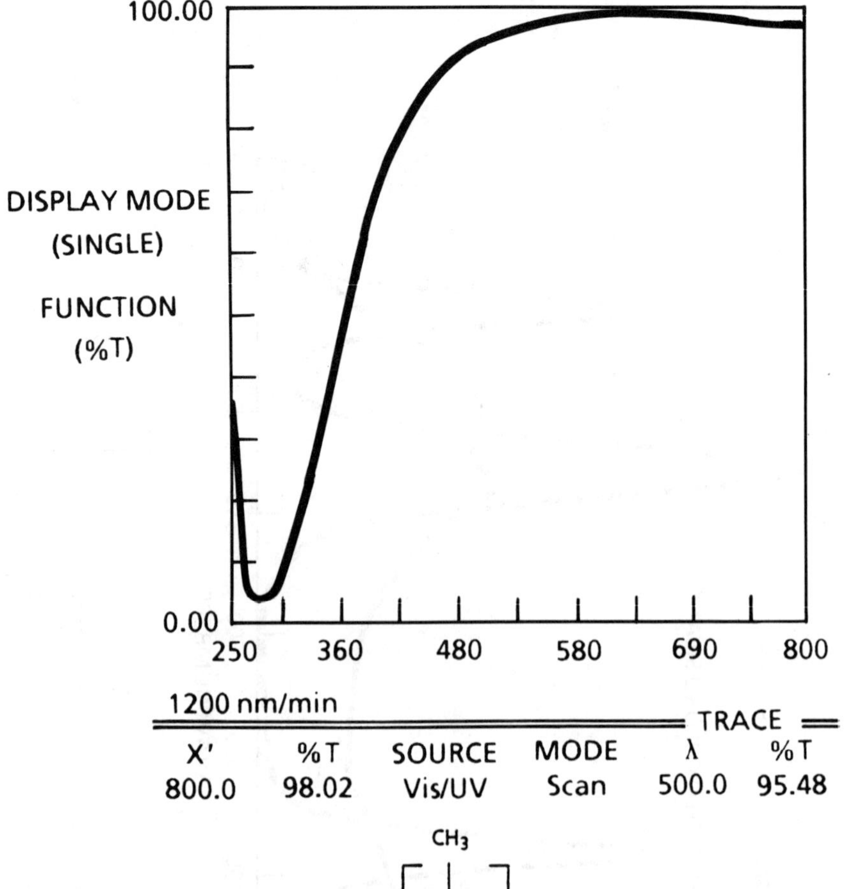

Figure 6. UV spectrum of $\left[-\mathrm{Si}(\mathrm{CH_3})(\mathrm{Cl})- \right]_n$ *in tetrahydrofuran.*

dimethoxy-1,4-disilacyclohexadiene with potassium hydroxide in solution and purified by recrystallization in isopropyl alcohol.

2. The homopolymer of 1,4-disiloxy-1,4-dimethyl-1,4-disilacyclohexadiene was prepared by acidic neutralization of the corresponding dipotassium salt.

3. The alternating copolymer of 1,4-disiloxy-1,4-dimethyl-1,4-disilacyclohexadiene and methylphenylsiloxane was prepared from the reaction of phenylmethyldichlorosilane and 1,4-di-

methyl-1,4-disilacyclohexadiene-1,4-di(potassium silanoate) in ethanol.

4. Poly(methylmethoxysilane) was prepared from the thermolysis of the reaction products of acetylene and 1,2-dimethyltetrachlorodisilane at 500 °C.

Acknowledgments

Funds for this effort were provided by NWC (Naval Weapons Center) Independent Research (reviewed by the Office of Naval Research) and technology base support (Office of Naval Technology).

References

1. Rhein, R. A. *Thermally Stable Elastomers: A Review;* Naval Weapons Center: China Lake, CA, October 1983 (NWC TP 6372, document, unclassified).
2. Barry, A. J.; Gilkey, J. W. U.S. Patent 2 681 355, 1954.
3. Eaborn, C. *Organosilicon Compounds;* Academic Press: New York, 1960; pp 353, 354.
4. Watanabe, H., et al. *J. Organomet. Chem.* **1977,** *128,* 173–175.
5. Atwell, W. H. U.S. Patent 3 465 018, 1969.
6. Hofler, F.; Hengge, E. *Monatsh. Chem.* **1972,** *103,* 1506–1511.
7. Reedy, D.; Urry, G. *Inorg. Chem.* **1967,** *6,* 169–173.
8. Hofler, F.; Hengge, E. *Monatsh. Chem.* **1972,** *103,* 1513–1522.
9. Childs, M. E.; Weber, W. P. *J. Organomet. Chem.* **1975,** *86,* 169–173.
10. Atwell, W. H. German Patent 1 768 896, 1971.
11. Burks, R. E., et al. *J. Polym. Sci., Polym. Chem. Ed.* **1973,** *11,* 319–326.
12. Pittman, C. U., et al. *J. Polym. Sci., Polym. Chem. Ed.* **1976,** *143,* 1715–1734.
13. Lai, Y. C., et al. *J. Polym. Sci., Polym. Chem. Ed.,* **1982,** *20,* 2277–2288.
14. West, R. *J. Organomet. Chem.* **1986,** *300,* 327–346.

RECEIVED for review May 27, 1988. ACCEPTED revised manuscript April 25, 1989.

20

Transition-Metal-Substituted Oligo- and Polysilanes

Keith H. Pannell, James M. Rozell, Jr., and Steven Vincenti

Department of Chemistry, University of Texas, El Paso, TX 79968–0513

New oligo- and polysilanes containing transition metals have been synthesized and studied. Oligosilanes readily deoligomerize upon photochemical irradiation via metal–silyl(silylene) intermediates. The mechanism was studied in detail via the use of variously substituted disilyl–metal complexes. New high-molecular-weight polysilanes have been synthesized as copolymers from the monomers $(C_6H_5)(CH_3)SiCl_2$ *and* $LM(CH_3)SiCl_2$*, in which LM (ligand–metal) may be* $[(\eta^5\text{-}C_5H_5)\text{-}Fe(\eta^5\text{-}C_5H_4)]$ *(Fc) or* $[(\eta^5\text{-}C_5H_5)Fe(CO)_2]$ *(Fp). The new polymers indicate that the metal substituents Fc and Fp cause a photochemical stabilization of the polymers with respect to depolymerization in a manner proportional to the metal content.*

TRANSITION METAL SUBSTITUENTS in organosilicon compounds create complexes in which the chemical and physical properties of the organosilyl radical are changed significantly (*1–3*). For example, the presence of the transition metal center allows skeletal rearrangements (*4–8*), migrations (*9–12*), and deoligomerizations (*12*). We investigated the capacity of transition metals to alter the chemical, photochemical, and physical properties of oligo- and polysilane materials.

Monosilane Complexes

The synthesis of transition metal complexes containing direct metal–silicon bonds involves three main reaction processes: salt elimination (equation 1)

0065–2393/90/0224–0329$06.00/0

(*13*), oxidative addition (equation 2) (*14*), and elimination of small molecules (equation 3) (*15*).

$$(\eta^5\text{-}C_5H_5)Fe(CO)_2{}^-Na^+ + (CH_3)_3SiCl \rightarrow (\eta^5\text{-}C_5H_5)Fe(CO)_2Si(CH_3)_3 + NaCl \quad (1)$$

$$Fe(CO)_5 + R_3SiH \xrightarrow{h\nu} cis\text{-}R_3SiFe(CO)_4H + CO \quad (2)$$

$$cis\text{-}(R_3P)_2PtCl_2 + (C_2H_5)_3SiH \rightarrow trans\text{-}(R_3P)_2PtCl[Si(C_2H_5)_3] + HCl \quad (3)$$

The chemistry of the metal–silicon bond is distinguished by greater thermal and oxidative stability compared with the related metal–carbon analogues. Thus, for example, whereas the *tert*-butyliron complex $[(\eta^5\text{-}C_5H_5)Fe(CO_2)C(CH_3)_3]$ (Fp–CMe$_3$) readily undergoes the well-established alkyl migration reaction to form acyl complexes in the presence of ligands (equation 4) (*16*), the related trimethylsilyl complex does not exhibit this type of chemistry. Only under the influence of UV irradiation will a chemical reaction occur, and this reaction leads to CO substitution (equation 5) (*17*).

$$(\eta^5\text{-}C_5H_5)Fe(CO)_2\text{–}C(CH_3)_3 + P(C_6H_5)_3 \xrightarrow{\Delta} (\eta^5\text{-}C_5H_5)Fe(CO)[P(C_6H_5)_3]\text{–}CO\text{–}C(CH_3)_3 \quad (4)$$

$$(\eta^5\text{-}C_5H_5)Fe(CO)_2\text{–}Si(CH_3)_3 + P(C_6H_5)_3 \xrightarrow{h\nu} (\eta^5\text{-}C_5H_5)Fe(CO)[P(C_6H_5)_3]\text{–}Si(CH_3)_3 \quad (5)$$

Early transition metal complexes containing a silicon–metal bond, for example, $[(\eta^5\text{-}C_5H_5)_2Zr[Si(CH_3)_3]Cl]$, exhibit both CO and O_2 insertion reactions (*18, 19*).

Oligosilane Complexes

Several examples of transition-metal-substituted oligosilanes have been reported (*20–23*). These materials have been prepared usually via salt elimination (equation 6).

$$LM^-Na^+ + R_3Si(SiR_2)_nCl \rightarrow LM\text{–}(SiR_2)_n\text{–}SiR_3 \quad (6)$$

In the previous reaction, LM (ligand–metal) is $[M(CO)_5]$ (in which M can be Re or Mn) or $[(\eta^5\text{-}C_5H_5)M(CO)_m]$ (in which M is Fe or Ru when $m = 2$ and M is Mo when $m = 3$), and $n = 1$–6.

Like the monosilane complexes, the oligosilanes form thermally and oxidatively stable complexes with the Re, Mn, Ru, and Fe systems, whereas the Mo complexes are significantly less stable. As yet such complexes have not been shown to exhibit any insertion-type reactions. The iron complexes of the type $(\eta^5\text{-}C_5H_5)Fe(CO)_2\text{–}[Si(CH_3)_2]_n\text{–}Si(CH_3)_3$ (Fp–Si_n) are interesting, because they are labile with respect to photochemical irradiation. Under such reaction conditions they readily deoligomerize to form the monosilane complex and siloxane polymers (equation 7) (*12*).

$$FpSi(CH_3)_2[Si(CH_3)_2]_nSi(CH_3)_3 \xrightarrow{h\nu} FpSi(CH_3)_3 + \text{siloxane polymers} \quad (7)$$

for $n = 1$ or 2.

Experimental Procedures

In this section, representative experimental details of the synthesis and photochemical study of the $[(\eta^5\text{-}C_5H_5)Fe(CO)_2]$ (Fp) disilanes and the Fp- and $[(\eta^5\text{-}C_5H_5)Fe(\eta^5\text{-}C_5H_4)]$ (Fc)-substituted polysilanes are presented. All manipulations were performed under dry nitrogen or argon atmosphere in oven-dried glassware with dry solvents.

Synthesis of $[(\eta^5\text{-}C_5H_5)Fe(CO)_2\text{–}Si(CH_3)(C_6H_5)\text{–}Si(CH_3)_3]$. A 250-mL, round-bottom flask equipped with a side arm was charged with 0.12 g (5.2 mmol) of Na, which was amalgamated with 15 g of Hg. To this mixture, 1.5 g (4.2 mmol) of $[(\eta^5\text{-}C_5H_5)Fe(CO)_2]_2$ (Strem Chemicals) in 60 mL of tetrahydrofuran (THF) was added. This mixture was stirred until formation of the Fp salt (*24*) was complete (1 h). Excess Hg was removed through the side arm, and the solution was cooled to 0 °C. Then, 0.95 g (4.2 mmol) of $(CH_3)_3SiSi(CH_3)(C_6H_5)Cl$ (*25*) was added. The solution was permitted to warm to room temperature and stirred for 1.5 h. At this time, the solvent was removed under reduced pressure to yield a viscous oil, which was extracted into hexane, filtered, concentrated, and applied to a silica gel column (2.5 by 25 cm). Elution with hexane yielded a bright yellow band, which upon removal of the solvent under reduced pressure yielded 1.1 g (70%) of the required complex as an orange oil. Analysis (Galbraith Laboratories, Inc.) of the complex gave the following results.

1. Elemental analysis [calculated (found), in percent]: C, 55.13 (55.60); H, 5.99 (6.18)
2. ^{1}H NMR spectrum: δ 0.15 $[(CH_3)_3Si]$; δ 0.65 (CH_3Si); δ 4.60 (C_5H_5); δ 7.26, 7.51 (C_6H_5)
3. ^{13}C NMR spectrum: δ −0.32, 1.18 (CH_3); δ 83.9 (C_5H_5); δ 127.4, 133.7, 145.8 (C_6H_5); δ 215.3 (CO)
4. ^{29}Si NMR spectrum: δ −11.8 (Fe–Si–*Si*); δ 12.6 (Fe–*Si*)
5. IR spectrum (in hexane): 2003, 1998, 1946 cm^{-1} (CO absorption band)

Photochemical Treatment of $FpSi(CH_3)(C_6H_5)Si(CH_3)_3$. Photolysis was performed at room temperature under nitrogen at 88.6 kPa with a Hanovia 450-W medium-pressure mercury lamp. Degassed cyclohexane solutions of the complex in Pyrex 9820 test tubes were illuminated for 1.5 h. Analysis of the product distribution was made with an internal standard (toluene) on a high-pressure liquid chromatographic system (Beckman Instruments, model 332) with UV detection at 270 nm. A C_{18} reverse-phase column was used with $CH_3CN\text{–}H_2O$ (65:35, vol/vol) as solvent.

Synthesis of $[(C_6H_5)(CH_3)Si_{25}Fc(CH_3)Si]_n$. A mixture of 0.5 g (1.7 mmol) of $Fc(CH_3)SiCl_2$ and 8.0 g (42 mmol) of $(C_6H_5)(CH_3)SiCl_2$ in 50 mL of toluene was heated to reflux temperature in a foil-covered three-necked flask. To this refluxing solution was added dropwise 5.8 g (101 mmol) of a 40% dispersion of Na in light mineral oil (Aldrich Chemicals). The dark-blue solution was refluxed for 2 h and cooled to room temperature. A few drops of methanol were added to remove excess Na, and the resulting solution was treated with saturated $NaHCO_3$ solution (40 mL) and toluene (40 mL). After separation and drying with magnesium sulfate, the solvent was removed under vacuum to yield an orange oil. Two precipitations from toluene with hexane and two from methanol produced 384 mg of a cream-colored powder. The ^1H NMR spectrum indicated a statistical incorporation of the two monomers. Whether the product is made up of block polymers has yet to be determined. Size exclusion chromatography using a μ-styragel 10,000-Å column (Waters Associates) with polystyrene standards indicated a bimodal molecular weight distribution at 200,000 and 11,000 (modal maxima) at a ratio of 1.4:1 (wavelength of maximum absorption [λ_{max}] 340 nm).

Synthesis of $[(C_6H_5)(CH_3)SiFp(CH_3)Si]_n$. The procedure is identical to that described in the previous section, with $Fp(CH_3)SiCl_2$ (*26*) instead of $Fc(CH_3)SiCl_2$ and two minor modifications. After removal of the toluene solvent, the product was treated initially with hexane to remove the low-molecular-weight fractions. The residue was dissolved in THF and filtered, and the THF was removed under vacuum. The residue was dissolved in toluene, washed with a saturated $NaHCO_3$ solution, separated, and dried. Two precipitations from THF with methanol produced 4% of material with a monomodal molecular weight of 8000. Spectral analysis gave the following results: IR spectrum (in THF), 1998, 1940 cm^{-1} (CO frequency); UV spectrum, λ_{max} = 332 nm; ^1H NMR spectrum (in $CDCl_3$), δ 7.2 bd (broad doublet) (C_6H_5); δ 3.8 bd (C_5H_5); δ 0.6 bd (CH_3).

Photolysis of $[[(C_6H_5)(CH_3)Si]_nLM(CH_3)Si]_m$. THF solutions of the polymers, which were obtained from reactions such as those described earlier, in quantities sufficient to produce an absorbance of 1 at λ_{max} under nitrogen at 88.6 kPa were irradiated in quartz cuvettes in a merry-go-round apparatus at 300 nm for periods of 1–60 s. A sample of authentic $[(C_6H_5)(CH_3)Si]_n$ of almost identical molecular weight was also irradiated at the same time for a comparative study. Typical results are illustrated in Figures 1 and 2.

Photochemical Deoligomerization of Fp–Si_2 Complexes

Mechanism. We have investigated in detail the mechanism of the photochemical deoligomerization for the Fp–Si_2 system by using variously substituted disilanes in place of the permethylated species. The product distribution obtained upon irradiation of $FpSi(CH_3)_2Si(C_6H_5)_3$ suggests a mechanism involving initial CO loss followed by Si–Si bond cleavage and migration to form metal–silyl(silylene) intermediates (*12*). These intermediates rapidly establish equilibrium via a series of 1,3-alkyl or -aryl shifts, and subsequent recombination of the CO yields the monosilyl complexes (Scheme I).

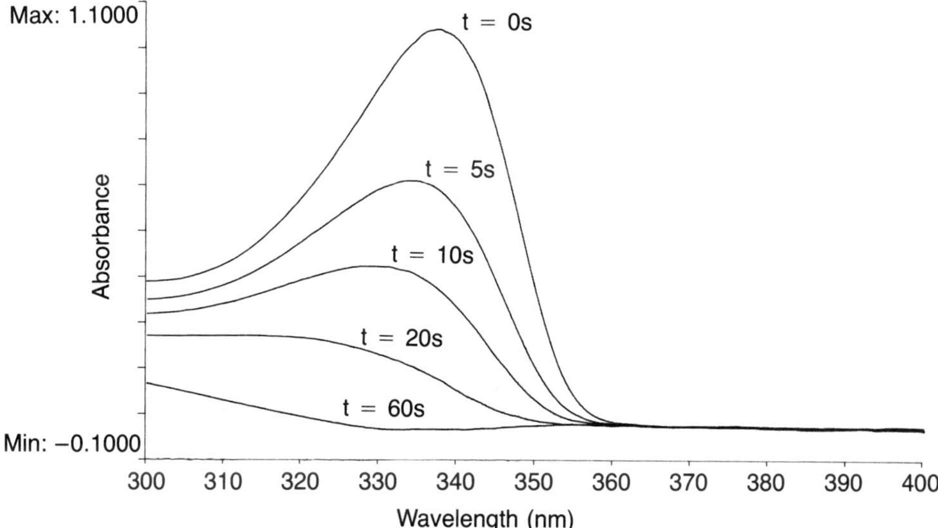

Figure 1a. Photolysis of THF solutions of $[(C_6H_5)(CH_3)Si]_n$ (molecular weight 400,000) at 270 nm.

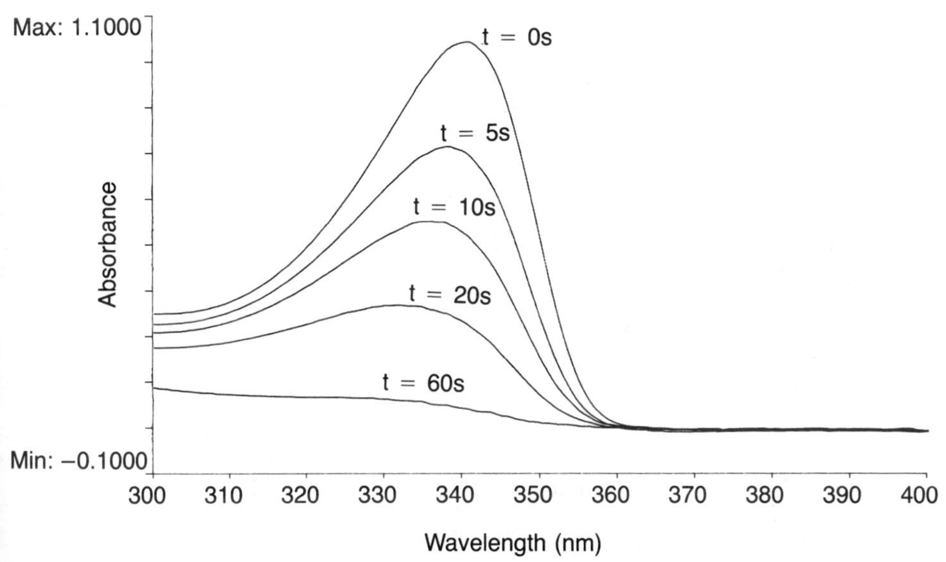

Figure 1b. Photolysis of THF solutions of $[[(C_6H_5)(CH_3)Si]_{10}Fc(CH_3)Si]_n$ (molecular weight 390,000) at 270 nm.

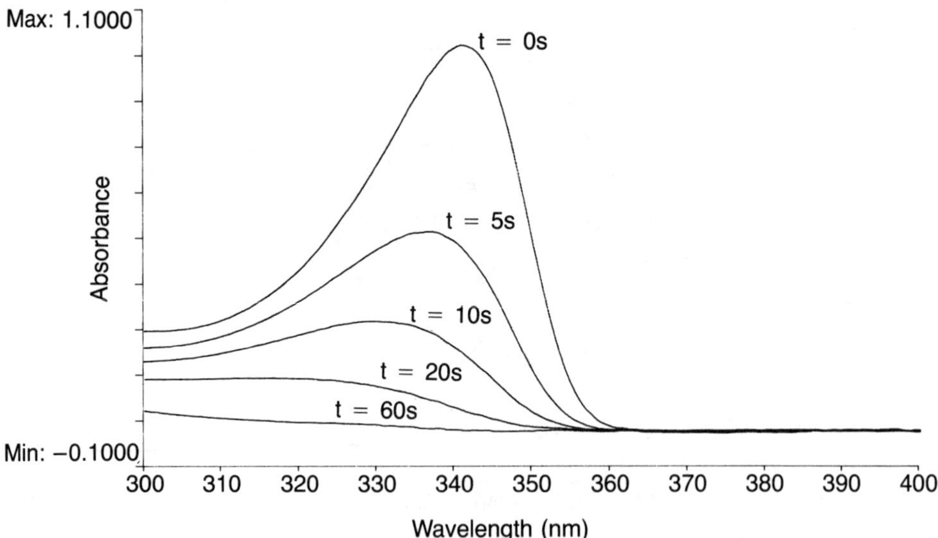

Figure 2a. Photolysis of THF solutions of $[(C_6H_5)(CH_3)Si]_n$ (molecular weight 4000) at 270 nm.

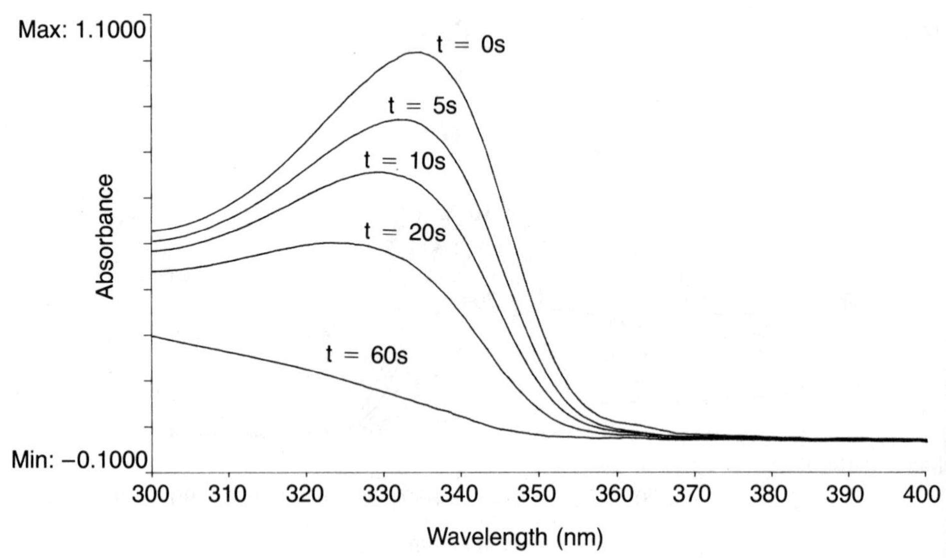

Figure 2b. Photolysis of THF solutions of $[[(C_6H_5)(CH_3)Si]_xFp(CH_3)Si]_n$ (molecular weight 8000) at 270 nm.

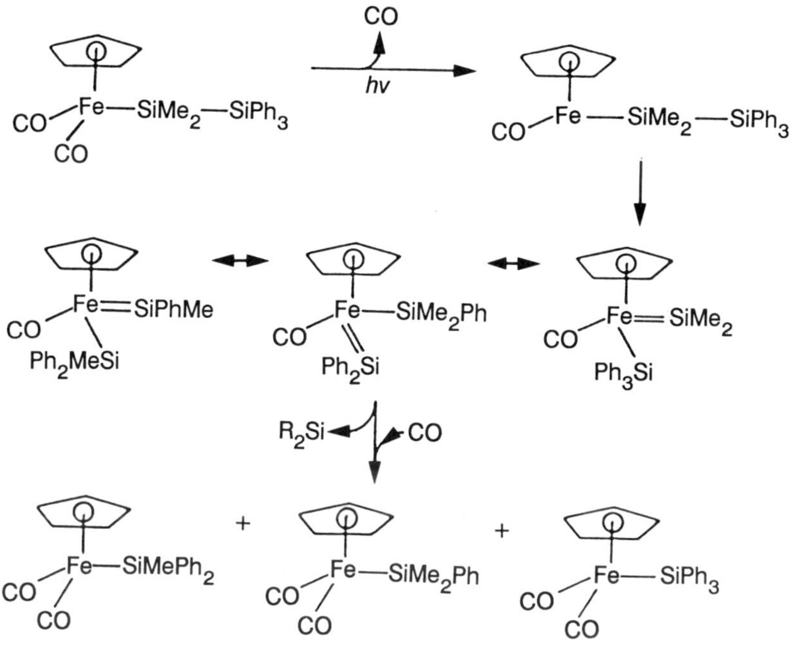

Scheme I. Deoligomerization of Fp-containing polysilanes. (Reproduced from reference 12. Copyright 1986 American Chemical Society.)

In subsequent studies, we extended this initial investigation to a series of isomeric pairs of disilyl complexes: (1) $FpSi(C_6H_5)(CH_3)Si(CH_3)_3$ and $FpSi(CH_3)_2Si(C_6H_5)(CH_3)_2$, (2) $FpSi(C_6H_5)(CH_3)Si(CH_3)_2(C_6H_5)$ and $FpSi(CH_3)_2Si(CH_3)(C_6H_5)_2$, and (3) $FpSi(C_6H_5)(CH_3)Si(CH_3)(C_6H_5)_2$ and $FpSi(CH_3)_2Si(C_6H_5)_3$.

If the mechanism outlined in Scheme I is correct, then regardless of which isomer of a given pair is irradiated, the rapid equilibrium established by the 1,3 shifts should provide the same ratio of final products. Photochemical treatment of the various isomeric pairs resulted in the product distribution data in Table I.

The data are unambiguous. Regardless of which isomer of an isomeric pair is irradiated, the product distributions are identical within experimental error. Such results clearly suggest that the mechanism outlined in Scheme I is an accurate description of the processes occurring during irradiation of the disilane complexes. Irradiation of $FpSi(CH_3)_2Si(CH_3)_2Si(CH_3)_3$ yields not only the ultimate product $FpSi(CH_3)_3$ but also transient yields of $FpSi(CH_3)_2Si(CH_3)_2$.

Table I. Product Distribution from Photolysis of Fp-Containing Disilanes

Starting Complex	$FpSi(CH_3)_3$	$FpSi(CH_3)_2(C_6H_5)$	$FpSi(CH_3)(C_6H_5)_2$	$FpSi(C_6H_5)_3$
$FpSi(CH_3)(C_6H_5)Si(CH_3)_3$	8	92		
$FpSi(CH_3)_2Si(CH_3)_2(C_6H_5)$	7	93		
$FpSi(CH_3)(C_6H_5)Si(CH_3)_2(C_6H_5)$	2	51	47	
$FpSi(CH_3)_2Si(CH_3)(C_6H_5)_2$	1	46	54	
$FpSi(CH_3)(C_6H_5)Si(CH_3)(C_6H_5)_2$		5	89	6
$FpSi(CH_3)_2Si(C_6H_5)_3$		7	86	7

$$FpSi(CH_3)_2Si(CH_3)_2Si(CH_3)_3 \xrightarrow{h\nu} FpSi(CH_3)_2Si(CH_3)_3 \xrightarrow{h\nu} FpSi(CH_3)_3 \quad (8)$$

Longer chain oligosilane complexes undergo photochemical redistribution reactions (*34*).

Although our experiments have not ruled out the possibility of a parallel reaction involving direct formation of $FpSi(CH_3)_3$ from $FpSi(CH_3)_2Si(CH_3)_2Si(CH_3)_3$ via FeSi(Si=Si) intermediates, the results are consistent with the proposed mechanism.

Photolysis of bulky permethylated Fp complexes such as $FpSi[Si(CH_3)_3]_3$ does not cause deoligomerization, possibly because stable intermediate iron–silyl(silylene) complexes are not formed (*27*). Other less bulky transition-metal–oligosilane complexes are also unreactive under the photolysis conditions. For example, the ruthenium analogues of the iron complexes, $[(\eta^5\text{-}C_5H_5)Ru(CO)_2\text{–}Si_n]$, are essentially photostable (*23*). Whether this behavior is due to the strength of the Ru–CO bond or to the enhanced stability of the Si–Si bond is not clear, and this problem is currently under investigation.

Finally, for the Fe complexes, a direct Fe–Si bond is required to activate the oligomers. When the silyl group is bonded to the complex via the cyclopentadienyl ring, as in $[(\eta^5\text{-}(CH_3)_3Si(CH_3)_2Si(CH_3)_2Si(C_5H_4)\text{-}Fe(CO)_2CH_3]$, photolysis does not lead to deoligomerization (*12*).

The mode of formation of the siloxane polymers from the photolysis of the Fp–Si_n complexes is unclear at present, and we are currently characterizing these materials. However, certain transition metal carbonylates react with silyl halides to form siloxanes, presumably via metal–silicon-bonded complexes (*28*). Furthermore, $(CH_3)_3SiSi(CH_3)_3$ is an effective reducing agent for $C^{18}O$ in the presence of H_2 and Ni catalysts; the reaction leads ultimately to the formation of CH_4 and $(CH_3)_3Si^{18}OSi(CH_3)_3$ (*29*). Such results clearly implicate CO as an important reagent for the transformation Si–Si → Si–O–Si. The inevitable presence of CO during the photochemical treatment of Fp–Si_n complexes thus seems to be responsible for siloxane polymer formation. Attempts to trap SiR_2 species by using such standard reagents as $(C_2H_5)_3SiH$, $(CH_3)_2Si(OCH_3)_2$, acetylenes, and CH_3OH have not yielded conclusive evidence regarding the fate of the lost SiR_2 fragments.

Polysilane Complexes. The photolability of the Si–Si bonds in the Fp complexes just described suggests that the incorporation of a metal center into high-molecular-weight polysilanes might affect the photodepolymerization of these photoresist materials (*30, 31*). Incorporation of the metals into the polymers was expected to permit new redox chemistry. Polysilanes are irreversibly oxidized when polysilane-coated electrodes are used in cyclic voltammetry (*32*); thus the presence of a reversible low-oxidation-state metal atom in the polymer might introduce a stabilizing effect upon the silane chain and allow switching processes to become available.

To investigate these possibilities, we have synthesized two types of metal-substituted polysilane materials incorporating both Fp and Fc groups. Fp was synthesized to investigate, inter alia, the effect on photodepolymerization, whereas Fc seemed to be an excellent model for introducing a reversible low-oxidation-potential metal center. A preliminary communication describing our initial attempts to incorporate Fc has appeared recently (*33*).

The new copolymeric materials were synthesized via a Wurtz-type coupling reaction (equation 9)

$$n\mathrm{LM(CH_3)SiCl_2} + m\mathrm{(C_6H_5)(CH_3)SiCl_2} \xrightarrow{\text{Na, toluene}} [(\mathrm{LMCH_3Si})_n(\mathrm{C_6H_5(CH_3)Si})_m]_x \quad (9)$$

in which LM is Fp or Fc. As determined by NMR spectroscopy, the copolymers formed exhibited a statistical incorporation of monomers in a manner reflecting the initial ratio of starting monomers. For the Fp-containing copolymers, molecular weights were generally low (10^3–10^4), whereas for the Fc-containing copolymers, high molecular weights (10^5–10^6) were obtained routinely. Yields were low (2–7%).

Initial attempts to make homopolymers with Fc-containing monomer were unsuccessful; however, with the Fp-containing monomer, homopolymers with molecular weights of 10^3–10^4 were prepared readily. This difference in behavior may be a feature of the particular reaction conditions used, and we are currently studying alternative synthetic procedures.

Our initial studies on the new polysilanes have concentrated on their photochemical depolymerization. The results of a study comparing the new polymers with the related homopolymers $[(C_6H_5)(CH_3)Si)]_n$ are illustrated in Figures 1 and 2. The photolyses were performed in a merry-go-round apparatus such that both samples were subjected to identical irradiation conditions. The incorporation of both Fc and Fp substituents retarded the depolymerization process. With Fp polymers, a small amount of Si–Fe bond cleavage occurs, as denoted by the appearance of the bridging CO stretching frequency at 1780 cm^{-1}, indicating the formation of $[(\eta\text{-}C_5H_5)Fe(CO)_2]_2$. The degree of photostabilization introduced by the metal substituent is proportional to the amount of metal incorporated. Thus a 5:1 $(C_6H_5)(CH_3)$-Si–Fc(CH_3)Si copolymer is markedly more photostable than a 10:1 copolymer, which in turn is more stable than the 25:1 copolymer.

The homopolymer $[FpSi(CH_3)_3]_n$ (molecular weight 8000) exhibits depolymerization, but as yet, the needed quantum yields, etc., are not available to describe accurately the effect of metal substitution on the polysilane.

Acknowledgments

This research has been supported by the Robert A. Welch Foundation, Houston, TX, and a Texas Advanced Technology research award. Financial assistance from the Dow Corning Corporation is also gratefully acknowledged.

References

1. Aylett, B. J. *Adv. Inorg. Chem. Radiochem.* **1982,** *25,* 1.
2. Cundy, C. S.; Kingston, B. M.; Lappert, M. F. *Adv. Organomet. Chem.* **1973,** *11,* 253.
3. Pannell, K. H. In *Silicon Compounds: Registry and Review;* Anderson, R.; Arkles, B. C.; Larson, G. L., Eds.; Petrarch Systems: Bristol, PA, 1987; p 32.
4. Pannell, K. H. *J. Organomet. Chem.* **1970,** *21,* C17.
5. Pannell, K. H. *Transition Met. Chem.* (*Weinheim, Ger.*) **1975/6,** *1,* 32.
6. Lewis, C.; Wrighton, M. S. *J. Am. Chem. Soc.* **1983,** *101,* 279.
7. Windus, C.; Giering, W. P. *J. Organomet. Chem.* **1975,** *101,* 279.
8. Pannell, K. H.; Rice, J. R. *J. Organomet. Chem.* **1974,** *78,* C35.
9. Thum, G.; Ries, W.; Greissinger, D.; Malisch, W. *J. Organomet. Chem.* **1983,** *252,* C67.
10. Berryhill, S. R.; Clevenger, G. L.; Burdurli, Yu, P. *Organometallics* **1984,** *4,* 1509.
11. Pannell, K. H.; Vincenti, S. P.; Scott, R. C., III *Organometallics* **1987,** *6,* 1593.
12. Pannell, K. H.; Cervantes, J.; Hernandez, C.; Cassias, J.; Vincenti, S. *Organometallics* **1986,** *5,* 1056.
13. Piper, T. S.; Lemal, D.; Wilkinson, G. *Naturwissenschaften* **1956,** *43,* 129.
14. Smid, G.; Balk, H. J. *Chem. Ber.* **1970,** *103,* 2240.
15. Chalk, A. J.; Harrod, J. F. *J. Am. Chem. Soc.* **1965,** *87,* 16.
16. Pannell, K. H.; Kapoor, R. N.; Giassoli, A. *J. Organomet. Chem.* **1983,** *316,* 127.
17. King, R. B.; Pannell, K. H. *Inorg. Chem.* **1968,** *7,* 1510.
18. Tilley, T. D. *J. Am. Chem. Soc.* **1985,** *107,* 4084.
19. Arnold, J.; Tilley, T. D. *J. Am. Chem. Soc.* **1985,** *107,* 6409.
20. King, R. B.; Pannell, K. H.; Ishaq, M.; Bennett, C. J. *J Organomet. Chem.* **1969,** *19,* 327.
21. Nicholson, B. K.; Simpson, J. *J. Organomet. Chem.* **1974,** *72,* 211.
22. Malisch, W. *J. Organomet. Chem.* **1972,** *39,* C28.
23. Pannell, K. H.; Rozell, J. M.; Tsai, W. M. *Organometallics* **1987,** *6,* 2085.
24. Pannell, K. H.; Jackson, D. *J. Am. Chem. Soc.* **1976,** *98,* 4443.
25. Tamao, K.; Kumada, M. *J. Organomet. Chem.* **1971,** *30,* 329.
26. Malisch, W.; Kuhn, M. *Chem. Ber.* **1974,** *107,* 979.
27. Pannell, K. H.; Rozell, J. M.; Wang, L. J. *Organometallics* **1989,** *8,* 550.
28. Curtis, M. D. *Inorg. Nucl. Chem. Lett.* **1970,** *6,* 859.
29. Vollhardt, K. P. C.; Yang, Z. Y. *Angew. Chem. Int. Ed. Engl.* **1984,** *23,* 460.
30. Miller, R. D.; Hofer, D. C.; McKean, D. R.; Willson, G. C.; West, R.; Trefonas, P. T. In *Materials for Microlithography;* Thompson, L. F.; Willson, G. C.; Frechet, J. M. J., Eds.; ACS Symposium Series 266; American Chemical Society: Washington, DC, 1984.

31. Zeigler, J. M.; Harrah, L. A.; Johnson, A. W. *Proc. SPIE Conf., Adv. Resist Technol. Process.* **1985,** *539,* 166.
32. Diaz, A.; Miller, R. D. *J. Electrochem.* **1985,** *132,* 834.
33. Pannell, K. H.; Rozell, J. M.; Zeigler, J. M. *Macromolecules* **1988,** *21,* 276.
34. Pannell, K. H.; Wang, L. J.; Rozell, J. M. *Organometallics* **1989,** *8,* 550.

RECEIVED for review May 27, 1988. ACCEPTED revised manuscript April 4, 1989.

21

Structures, Phase Transitions, and Morphology of Polysilylenes

Frederic C. Schilling,[1] Frank A. Bovey,[1] Andrew J. Lovinger,[1] and John M. Zeigler[2,3]

[1]AT&T Bell Laboratories, Murray Hill, NJ 07974
[2]Sandia National Laboratories, Albuquerque, NM 87185

*The results of structural studies of polysilylenes in solution and in the solid state are reviewed. The chapter provides a detailed description of the polymer chain conformation and microstructure, crystal structure, polymer morphology, and the solid-state phase transition of poly(di-*n*-hexylsilylene). Less-complete analyses have been reported for poly(di-*n*-butylsilylene), poly(di-*n*-pentylsilylene), and other polysilylenes. A wide range of structural probes has been used, including X-ray and electron diffraction; UV, IR, Raman, and NMR spectroscopy, light-scattering techniques, and differential scanning calorimetry. Many of the interesting properties of the polysilylenes can be related to the nature of the solid-state structures. Variable-temperature studies have been especially useful in this regard. Although a large amount of work remains before a full understanding of these materials is developed, recent studies have provided significant insight into the structure–property relationships and, in particular, the relationship between chain conformation and electronic properties.*

SOLUBLE DISUBSTITUTED POLYSILYLENES are a class of polymers that recently has generated great interest. These polymers have the structure $[-SiRR'-]_n$, in which R and R′ may be aryl or alkyl groups and R may be the same as R′. The substituted polysilylenes exhibit a wide variety of physical properties, depending on the nature of R and R′. Of particular interest is their intense UV absorption at 300–400 nm both in solution and in the solid state, a property conferred by the silicon backbone and accompanied

[3]Current address: SilCHEMY, 2208 Lester Drive, NE, Albuquerque, NM 87112–2640

0065–2393/90/0224–0341$10.50/0

by a sensitivity to radiation-induced chain scission. Thus, these materials may have application as photoresists.

The polysilylenes have been intensively studied by several groups using the entire armamentarium of structural methods, including UV, Raman, and IR spectroscopy, X-ray diffraction, NMR spectroscopy, calorimetry, and light-scattering techniques. Several semiempirical and quantum-mechanical studies of conformational energies have also been made. In this chapter, the polysilylene structures will be considered at several levels. At the molecular level, both the intramolecular and intermolecular structures will be examined. *Intramolecular* structures refer to the polymer chain configuration and conformation. When R is not identical to R′, the chain may exhibit varying stereochemical configurations or tacticity. In addition, in copolymers, varied sequences of comonomer units may be present. *Intermolecular* structures refer to the packing of neighboring polymer chains in the solid state. For crystalline materials, the intermolecular structure involves a description of the unit cell.

At a higher structural level, the intercrystalline structures or the polymer morphology will be examined. These structures result from the crystal growth habit of the polymer. Most of the published work in this field relates to structures at the molecular level and, in particular, to the polymer chain conformation. The chain conformation is of particular interest, because it is strongly coupled to the electronic structures and the very interesting UV absorption characteristics of these polymers.

Chain Configuration

The configuration of the polysilylene chain can range from the simple structure of the symmetrically substituted homopolymer (in which R = R′), to the more complicated structure of the asymmetrically substituted homopolymer (in which R ≠ R′), and finally, to the very complex structure of the asymmetrically substituted copolymers, in which both stereochemical and comonomer sequences need to be considered.

Proton NMR spectroscopy has been used fruitfully for many years in the analysis of polymer chain configuration (*1*). The proton solution spectra (2) for poly(methyl-*n*-propylsilylene) (PMPS), poly(methyl-*n*-hexylsilylene) (PMHS), poly(methyl-*n*-dodecylsilylene) (PMDS), and poly(di-*n*-hexylsilylene) (PDHS) are shown in Figure 1. At this high field (11.75 T), the res-

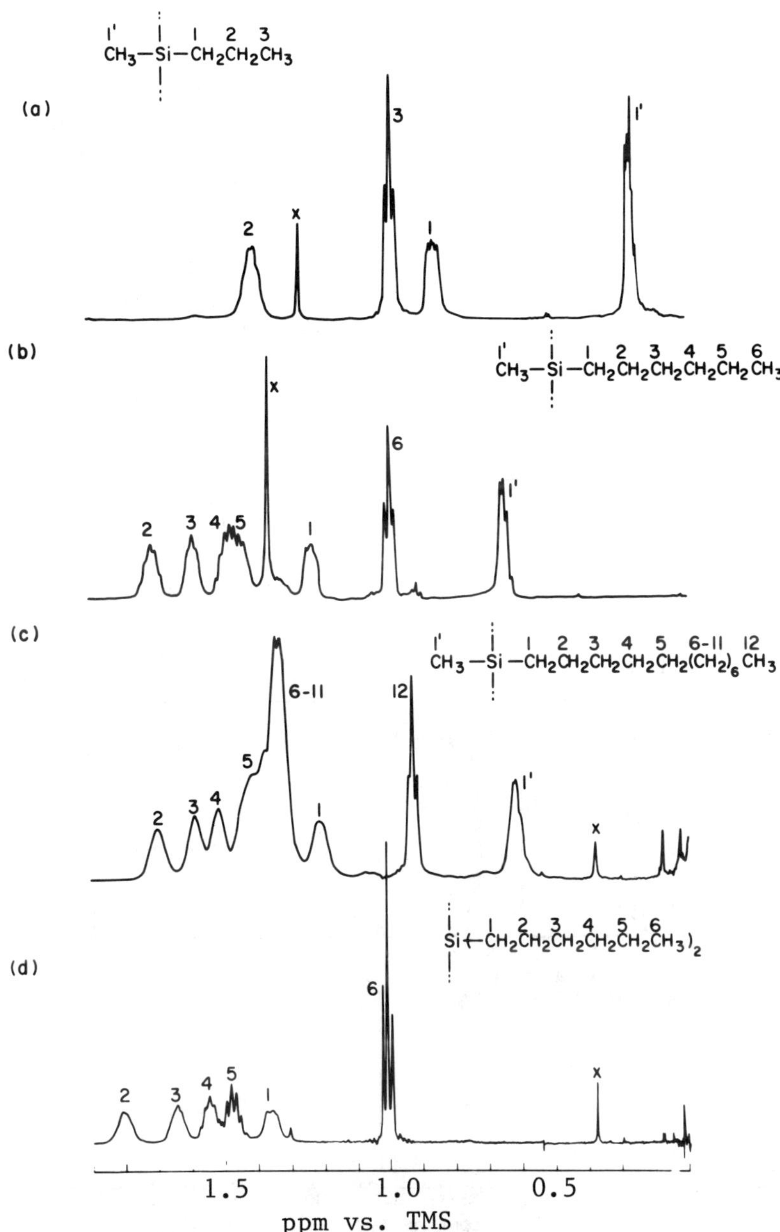

Figure 1. ^{1}H NMR (500-MHz) spectra of PMPS in methylene-d_2 *chloride at 23 °C (a), PMHS in benzene-*d_6 *at 50 °C (b), PMDS (c), and PDHS in toluene-*d_8 *at 60 °C (d). TMS is tetramethylsilane. (Reproduced from reference 2. Copyright 1986 American Chemical Society.)*

onances for the individual CH_2 and CH_3 protons in the polymer are well resolved. These resonances can be assigned specifically with the use of two-dimensional proton-NMR experiments, as has been reported for PMHS (*1*), without the need for laboriously synthesized model compounds.

^{13}C NMR spectroscopy also provides detailed information concerning polymer microstructure (*1*). Figure 2 shows the ^{13}C NMR spectra (*2*) for

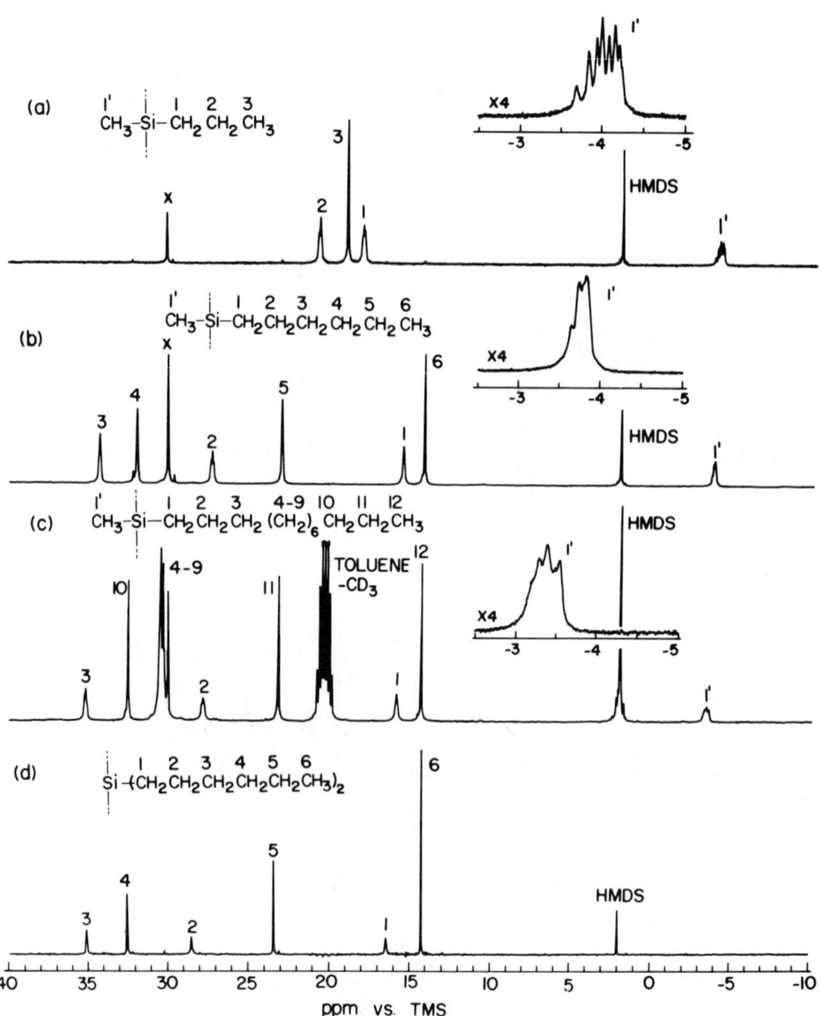

Figure 2. ^{13}C *NMR (125.7-MHz) spectra of PMPS (a) and PMHS (b) in 1,2,4-trichlorobenzene–dioxane-*d_8 *solution at 40 and 60 °C, respectively; 125.7-MHz* ^{13}C *NMR spectrum of PMDS in toluene-*d_8 *at 23 °C (c); and 50.3-MHz* ^{13}C *NMR spectrum of PDHS in toluene-*d_8 *at 100 °C (d). HMDS is hexamethyldisiloxane. (Reproduced from reference 2. Copyright 1986 American Chemical Society.)*

PMPS, PMHS, PMDS, and PDHS. Individual resonances are observed for each carbon type, except for the dodecyl polymer in which C-4 to C-9 in the side chain are not completely resolved. The carbon resonances can also be assigned without the use of model compounds by using standard two-dimensional NMR techniques (2). C-1′ and, in some cases, C-1, C-2, or both in the three asymmetrically substituted polymers show shift dispersion, which results from the many different stereochemical sequences along the polymer chain. The methyl carbon in PMPS is resolved to at least the pentad level of stereosequences. This chemical shift information can be analyzed to provide a description of the chain statistics resulting from a particular polymerization.

^{29}Si NMR spectroscopy is the technique most sensitive to the stereochemical configuration along the polymer chain. The ^{29}Si NMR spectrum for PMHS is shown in Figure 3 (2). Resonances are resolved at the pentad level of stereosequences, and the data can be fitted to Bernoullian propagation statistics with a P_m of 0.5. Thus, the addition of *meso* or racemic units along the growing chain during the polymerization is a completely random

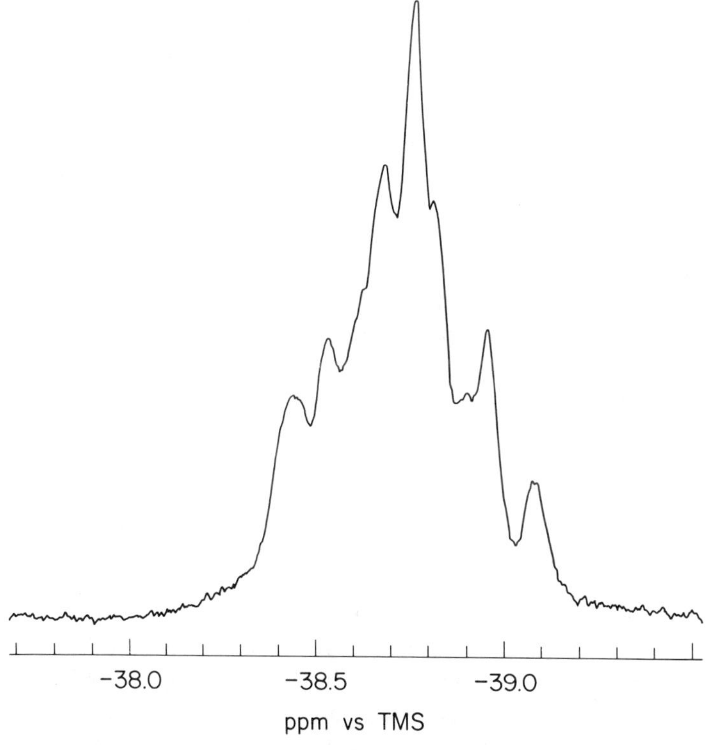

Figure 3. ^{29}*Si NMR (99.25-MHz) spectrum of PMHS in 1,2,4-trichlorobenzene–dioxane-*d_8 *at 60 °C. (Reproduced from reference 2. Copyright 1986 American Chemical Society.)*

process for this polymer. Similar results are observed for most of the asymmetrically substituted polymers that have been examined. Both the ^{13}C and the ^{29}Si NMR spectra of poly(cyclohexylmethylsilylene) (PCHMS) indicate that the polymer is atactic and probably has a random stereochemical sequence distribution.

At present, no evidence exists for long runs of either *meso* or racemic placements in the Wurtz dechlorination reaction used in the synthesis of polysilylenes. Deviation from a random stereochemical polymerization has been suggested by West et al. (*3*) in the case of poly(phenylmethylsilylene) (PPMS). An analysis of ^{29}Si NMR triads indicates that addition of monomer units with the same stereochemistry is preferred, which results in an increase in the intensity of syndiotactic and isotactic triad structures compared with a random stereochemical polymerization.

Several polysilylene copolymers have also been examined by ^{29}Si NMR. West and co-workers (*3*) report that phenylmethyldichlorosilane, when copolymerized with either dimethyldichlorosilane or methylhexyldichlorosilane, yields a copolymer with a blocklike structure. In contrast, we have observed that the copolymer of dimethylsilylene with di-*n*-hexylsilylene has random configurational structures typical of a copolymer with different comonomer reactivity ratios. These several examples indicate clearly that ^{29}Si NMR spectra can provide a complete analysis of chain microstructure for the soluble polysilylene homopolymers and copolymers.

Chain Conformation and Packing

Relationship to Electronic Properties. As a result of the close connection between bond conformation and electronic properties (*4*), the analysis of chain conformation in the polysilylenes has been of interest to researchers in this field, both from the experimental and theoretical viewpoints. As reported by Trefonas et al. (*5*), most asymmetrically substituted alkyl polysilylenes in solution at room temperature display an electronic absorption with λ_{max} ranging from 303 to 309 nm. The variable-temperature absorption spectrum of PMHS is shown in Figure 4 (*4*). At room temperature, λ_{max} is ~308 nm, and as the solution is cooled, there is a continuous red shift with the λ_{max} reaching ~328 nm at –95 °C. Some workers (*4, 6*) suggest that this observation is a reflection of an increasing population of *trans* rotational states in the silicon backbone as the temperature is lowered. This suggestion is supported by the finding that these spectra can be adequately modeled by a rotational isomeric-state treatment (*4*).

The absorption spectra reported for several symmetrically substituted polymers show contrasting behaviors. In addition to a slow red shift in λ_{max} as the temperature is lowered, the sudden appearance of a much longer wavelength band is observed. Figure 5 shows the solution data for PDHS as reported by Harrah and Zeigler (*7*). The λ_{max} shifts from ~317 nm at

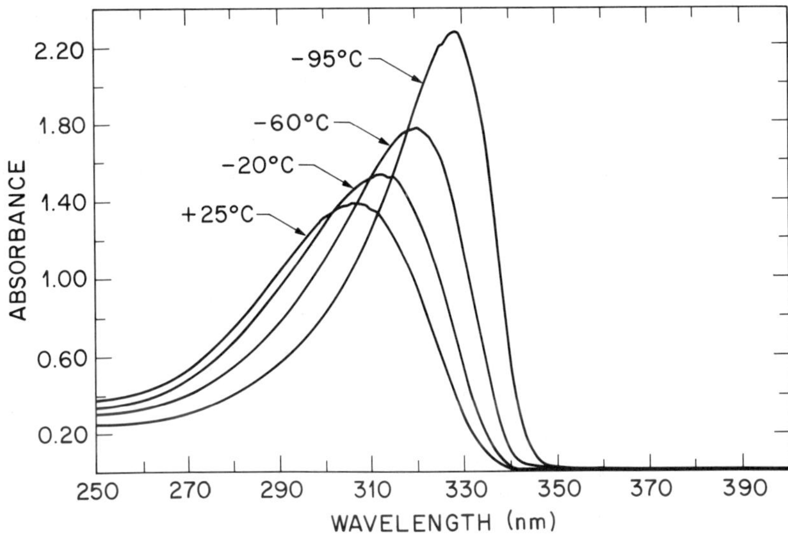

Figure 4. UV absorption spectrum of PMHS (0.004% in hexane). (Reproduced from reference 4. Copyright 1987 American Chemical Society.)

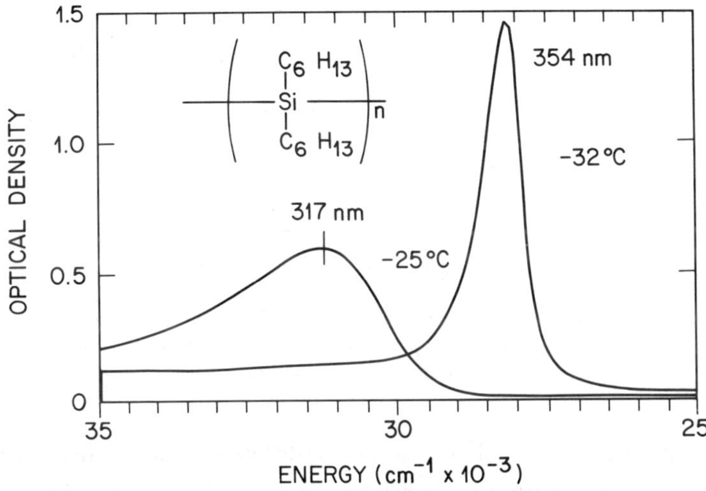

Figure 5. UV absorption spectrum of PDHS in hexamethyldisiloxane solution at 5 × 10^{-5} M. (Reproduced with permission from reference 7. Copyright 1985 Wiley.)

–25 °C to ~354 nm at –32 °C, with the transition occurring at –31 °C (±1 °C) (*4, 8*). Qualitatively similar results have been reported for poly(di-*n*-butylsilylene) (PDBS) and poly(di-*n*-pentylsilylene) (PDPS) in solution (*6*).

Absorption spectra displaying a thermochromic transition have also been reported for several of the symmetrically substituted polysilylenes in the solid state. One example is that of PDHS, as reported by Kuzmany et al. (*9*) and shown in Figure 6. At 45 °C, a single absorption at 317 nm is observed, which is very similar to the absorption maximum for this polymer in solution

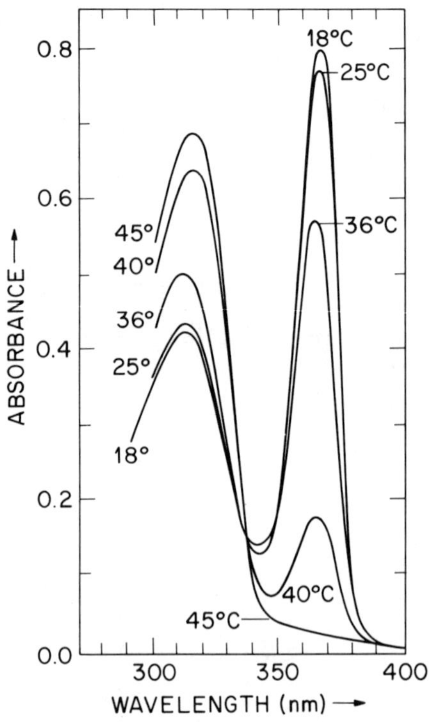

Figure 6. UV absorption spectrum of PDHS film at temperatures between 18 and 45 °C. (Reproduced with permission from reference 9. Copyright 1986 American Institute of Physics.)

(*6*). As the sample is cooled, a second absorption band is observed in the range from ~365 to 375 nm, which continues to grow upon further cooling. This behavior is completely reversible. This type of thermochromic transition is not observed in the solid-state absorption spectra of PDBS (*10*) or PDPS (*10, 11*). To understand this unusual absorption behavior of the polysilylenes in solution and in the solid state, a variety of studies have been directed toward the determination of the polymer chain conformation.

Conformational Energy Calculations. A number of theoretical studies have been conducted in an effort to establish the low-energy conformation for the symmetrically substituted polysilylenes (*12–19*). Conformational energy calculations (*13*) by the empirical force-field program MM2 (molecular mechanics) and with complete relaxation of all internal degrees of freedom (viz., bond angles, bond lengths, and torsional angles) indicate that the *gauche* conformational states are of lowest energy for unsubstituted polysilylene (PS) (in which R = R′ = H in the polysilylene structure) and for poly(dimethylsilylene) (PDMS). On the basis of the small difference in energy calculated for different conformations, these polymer chains are more flexible conformationally than polyethylene (*13*).

The calculation of a complete conformational energy map for PS and PDMS has been reported by Welsh et al. (*16*). The energy map of PDMS shows large regions of prohibitively high energy, in contrast to the energy map of PS in which nearly all regions are within 2 kcal/mol of the energy minima. This difference must be attributed to the steric bulk of the methyl groups in PDMS relative to that of the hydrogen atoms in PS. The characteristic ratio, C_∞, (i.e., the ratio of the observed mean-square end-to-end distance to that of a random-walk chain of an equal number of bonds [*20*]) ranges from 12.5 to 15.0 at 25 °C for PDMS (*16*), depending on the degree of molecular relaxation permitted for the internal degrees of freedom. These rather large values indicate a more extended chain compared with that of PS, for which the calculated C_∞ is only 4.0.

The conformational energy calculations reported by Damewood (*14*) indicate that PDHS is expected to adopt the *trans* conformation in the ground state, with the *gauche* conformation being 3.0 kcal/mol higher in energy. The increase in the differences in energy between *trans* and *gauche* in PDHS compared with PS and PDMS is attributed to the increased steric bulk of the substituent at the silicon nucleus. In contrast, the calculations of Farmer et al. (*17*) suggest that the lowest energy conformation for PDHS is one in which all Si–Si bonds are rotated 30° from *trans* and all Si–C bonds are rotated 245° from *trans*. These conformations are illustrated in the Newman projections of Figure 7, which shows the conformation along the Si–Si backbone bond (Figure 7a) and along the Si–C side chain bond (Figure 7b). The backbone conformation describes a helical structure similar to the 7/3 helix observed for PDBS and PDPS in the solid state. Both sets of calculations for PDHS involve a number of assumptions concerning the bond angles and the treatment of the proton and carbon nuclei in the side chains. Therefore, the results should be viewed as preliminary and much less certain compared with the more-complete calculations possible for PS and PDMS.

The limitations of the chosen approaches should be kept in mind when considering the results of these conformational calculations. Cooperative effects that might result from the sigma conjugation are not considered in the empirical force-field calculations, and assumptions concerning the va-

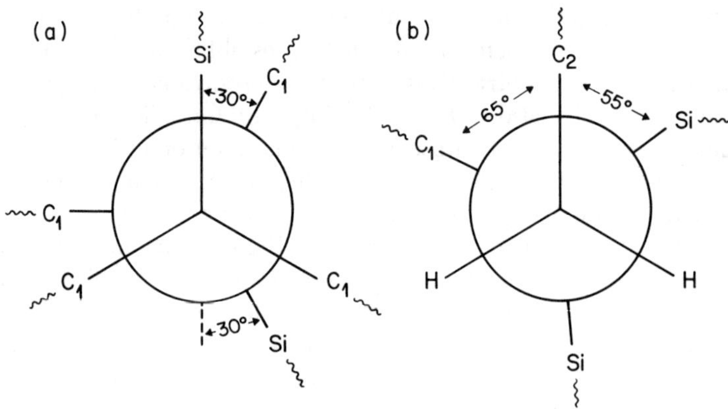

Figure 7. Newman projections of backbone (a) and side-chain (b) conformations in the 7/3 helix of PDHS.

lence angles are made in some cases (*17*). The more sophisticated ab initio calculations that have been reported (*18*) are restricted to consideration of SiH_2 and $Si(CH_3)_2$ chains with five repeat units. The results indicate small differences (<1.0 kcal/mol) between the energies of the *gauche* and *trans* conformations. These results provide qualitative validation of the less-rigorous MM2 level calculations, which are for the most part in agreement. However, the results from PS or PDMS may not necessarily be extrapolated to polymers with bulky side groups, such as butyl or hexyl groups. Complete calculations for polymers with the larger substituents will be much more difficult because of the number of bonds involved.

PDHS Structures in Solution. The determination of the chain conformation of polysilylenes in solution, particularly the conformations at temperatures just above or below the low-temperature thermochromic transition, is of great interest. NMR spectroscopy is one of the most useful techniques for probing chain conformation in solution (*21*), and ^{13}C NMR is especially effective because of the large sensitivity of the carbon chemical shift to bond conformation (*22*). Silicon nuclei are also very sensitive to chain conformation, but a good correlation between silicon chemical shift and bond conformation has not been established yet. Unfortunately, both of these nuclei suffer from low sensitivity, primarily because of their low natural abundance. In contrast, protons have an essentially 100% natural abundance, but compared with the carbon or silicon chemical shift, the proton chemical shift is not very sensitive to bond conformation. Efforts to use NMR to probe the low-temperature dilute-solution conformation of the polysilylenes have been unsuccessful thus far. The difficulty is that PDBS and PDHS precipitate from solution in 20–30 min after cooling through the thermochromic transition at –30 to –40 °C. The concentrations required to observe the transition without precipitation of polymer (0.0005–0.005 wt %) are too low for satisfactory NMR studies in the limited time available.

Light scattering is a technique that can probe polymer structure and chain dimensions in solution. The results of such measurements on PDHS and poly(di-*n*-octylsilylene) (PDOS) have been reported recently by Cotts et al. (*23*). After corrections were made for the excluded-volume effect, the light-scattering data were used to estimate the characteristic ratio, C_∞, for PDHS. A value of ~20 was found. The persistence length of PDHS was estimated to be 2.9 nm. The chain dimensions and molecular weights obtained from the data indicate that the polysilylenes are substantially more extended than hydrocarbon polymers of a similar degree of polymerization. This extension is attributed to the longer Si–Si bond and the highly substituted backbone, which provides steric hindrance to rotation. Despite the large chain dimensions, the chains have a random coil conformation in solution at room temperature. However, as the authors (*23*) have indicated, the estimation of the magnitude of the excluded-volume effect is uncertain, and conclusive statements about the flexibility of these polymers must await light-scattering measurements in a theta solvent. (In a theta solvent, the size of a macromolecule corresponds to the unperturbed dimensions.)

Light-scattering data have been reported also for solutions of PDHS at low temperature (*24*). Measurements at temperatures below –10 °C showed a tremendous increase in the scattering intensity, which reflects the precipitation of the polymer. The temperature at which precipitation occurred was dependent on the solution concentration, the rate of cooling, and the length of time the solution was held at a particular temperature. At very low concentrations in hexane (0.017 g/L), solutions could be cooled to –30 °C without precipitation. However, lowering of the temperature below –30 °C at this concentration resulted in a large increase in scattering, which was interpreted as a consequence of polymer precipitation. The root-mean-square radius of gyration, *s*, did not show a significant increase in any of the observed solutions at lower temperatures prior to the precipitation of PDHS. The failure to observe a large increase in *s* was interpreted as an indication that no conformational transition occurred (i.e., from a random coil to an extended or rodlike conformation) in solution prior to precipitation.

Two temperatures for the thermochromic transition of PDHS in solution have been reported: –24 (*6*) and –31 (*7*) °C. The difference between the two measurements is probably due to differences in material polydispersity and experimental procedures (*10*). If the light-scattering measurements were made at temperatures above the thermochromic transition and if the transition is first order, a change in *s* is not expected. Certainly, the dramatic increase in the scattering intensity observed at low temperatures is the result of the ordered phase precipitating from solution. However, whether the UV thermochromic transition occurs before the appearance of an ordered phase and whether this transition is a property of the molecules in solution are questions that remain to be answered. Additional experimental studies designed to address these questions would be extremely valuable.

PDHS Solid-State Structures. ***Differential Scanning Calorimetry.*** A significant amount of work has been reported concerning the solid-state structures of the polysilylenes. Like the conformations in the solution state, the solid-state chain conformation of these materials is interesting because of the unusual UV absorption characteristics exhibited by some of the symmetrically substituted polymers. The most thoroughly studied polysilylene is PDHS. This polymer undergoes a bathochromic transition when cooled to below ~+40 °C (Figure 6), at which the absorption maximum shifts toward the red region by 57 nm. The DSC (differential scanning calorimetry) data for PDHS, reported by Rabolt et al. (25), are shown in Figure 8. A strong endothermic transition occurs at +38 °C, with some gradual disordering evident at a few degrees lower. Upon subsequent cooling of the

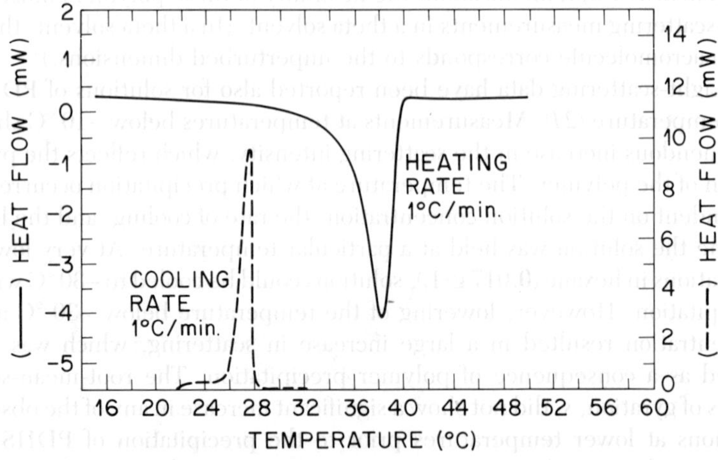

Figure 8. DSC thermograms of PDHS at the indicated heating and cooling rates. (Reproduced from reference 25. Copyright 1986 American Chemical Society.)

sample, an exothermic transition is recorded at ~28 °C. This transition of PDHS at ~+40 °C is not a melting of the polymer but, rather, a solid–solid phase transition in which the polymer is converted from a well-ordered phase to a phase with intramolecular disorder but retaining a high degree of intermolecular order. Additional heating of the sample above +40 °C does not change the DSC trace until the material decomposes at >+250 °C. First-order endothermic solid-state transitions are also observed in the DSC data for PDBS (*10*) and PDPS (*11*).

IR Spectroscopy. The observation of a solid-solid phase transition in PDHS at the temperature of the UV thermochromic transition generated interest in the nature of the polymer chain conformation at temperatures above or below this critical temperature. A number of techniques have been used to study the solid-state structures of PDHS. Rabolt et al. (25) used IR spectroscopy to monitor the conformational behavior of the alkyl side chains of PDHS and Raman spectroscopy to follow that of the backbone. The IR spectrum of PDHS at +30 °C (Figure 9) consists of sharp, intense bands. When the film of PDHS is heated to +100 °C, the sharp deformation bands typical of a highly ordered hydrocarbon chain become broad and less intense, similar to the behavior observed in the IR spectra of *n*-alkanes at temperatures above the melting temperature. The data reflect conformational disorder in the side chains at temperatures above the +40 °C transition. After

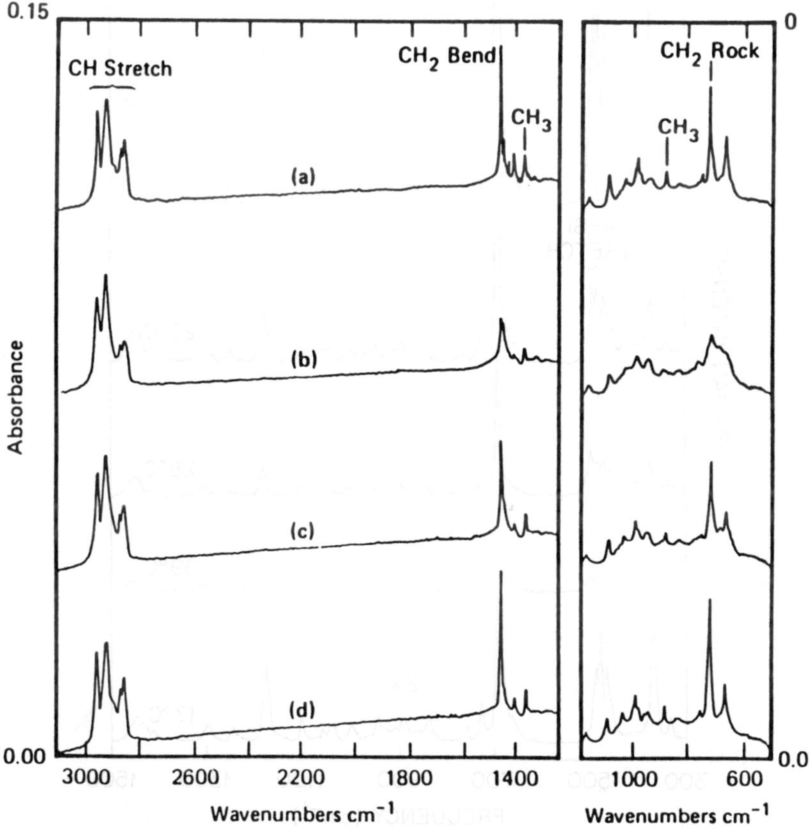

Figure 9. IR spectra of PDHS film at 30 °C (a), after heating to 100 °C (b), 50 min after cooling to room temperature (c), and 75 min after cooling to room temperature (d). (Reproduced from reference 25. Copyright 1986 American Chemical Society.)

being cooled to room temperature, the side chains again become conformationally ordered, as indicated by the return of the sharp IR bands. Similar behavior has been observed in the IR spectra of poly(di-*n*-heptylsilylene) (PDHepS) and poly(di-*n*-octylsilylene) (PDOS) (25).

Raman Spectroscopy. The conformation of the PDHS backbone can be monitored by observation of the Raman spectrum, as reported by Kuzmany et al. (9) (Figure 10). Very intense bands associated with the vibration of the silicon atoms are observed at +17 °C. The bands below 700 cm^{-1} have been assigned to the silicon backbone, whereas the weaker bands

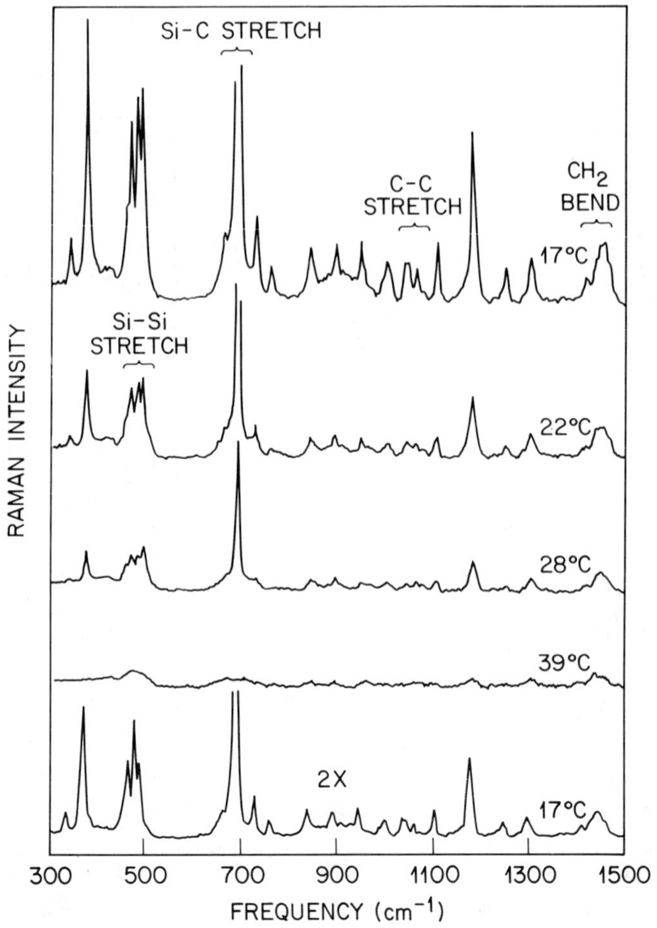

Figure 10. Raman spectra of PDHS film recorded between 17 and 39 °C and after returning to 17 °C. (Reproduced with permission from reference 9. Copyright 1986 American Institute of Physics.)

characteristic of the ordered alkyl side chain are found in the 700–1500-cm^{-1} region. Upon heating of the film, all of the sharp features of the silicon vibration bands begin to disappear as the bands broaden. Above the transition temperature, only weak bands are observed. This dramatic loss in the intensity of the Raman bands has been attributed to a preresonance-scattering process. This conclusion is based upon the observation that the Raman intensity observed in spectra taken at +17 °C is very dependent on the excitation wavelength over the range of 458–515 nm. The resonance enhancement of the Raman bands is thought to arise from the UV–visible absorption of the polymer at 374 nm, with a long-wavelength tail extending into the visible region. Upon heating the polymer to a temperature above the solid–solid phase transition, the UV absorption shifts to 317 nm, and the preresonance condition leading to the Raman enhancement no longer exists. The result is the broad, weak spectrum seen at +39 °C.

X-ray and Electron Diffraction. The molecular conformation of PDHS at temperatures below or above the transition temperature was determined by X-ray (*9, 25–26*) and electron diffraction (*26*). From uniaxially drawn samples at room temperature, molecular repeats of 0.400 (*26*) or 0.407 (*9*) nm were both interpreted as reflecting an all-*trans* Si backbone with standard (*26*) or slightly expanded (*9*) bond lengths and angles. The structure of PDHS at ambient temperature depends strongly upon crystallization and annealing conditions as seen in the diffractograms of Figure 11. In this figure, the sharp peaks at $2\theta > 7°$ correspond to the well-ordered phase I (which has the all-*trans* conformation just mentioned), whereas the broad peak at ~6.5° and the background beyond ~12° 2θ belong to the disordered phase II. The extent to which a fully amorphous phase is present is not known, because its X-ray and NMR peaks (*see* next section) are incorporated within those of phase II and because PDHS decomposes before melting. As seen in Figure 11, crystallization from methylene chloride yields predominantly the disordered phase II, whereas precipitation from THF (tetrahydrofuran) and annealing at successively higher temperatures shifts the balance increasingly toward the well-ordered phase I (Figure 11).

When PDHS is heated through its thermochromic transition (~38–43 °C), the resulting X-ray diffractometric changes (Figure 12a) indicate the progressive disappearance of all reflections from phase I, which is accompanied by a large increase in the intermolecular peak of phase II. Remarkably, this peak continues to become stronger and sharper as the temperature is raised well above that of the transition, that is, to 180 °C (Figure 12b). These changes imply that the thermochromic transition is an intramolecularly disordering transition that involves loss of the three-dimensional order of phase I but accompanied by an increase in the intermolecular packing order. When PDHS is cooled subsequently (Figure 12b), the reverse transition shows a clear hysteresis (suggestive of its first-order character), and the

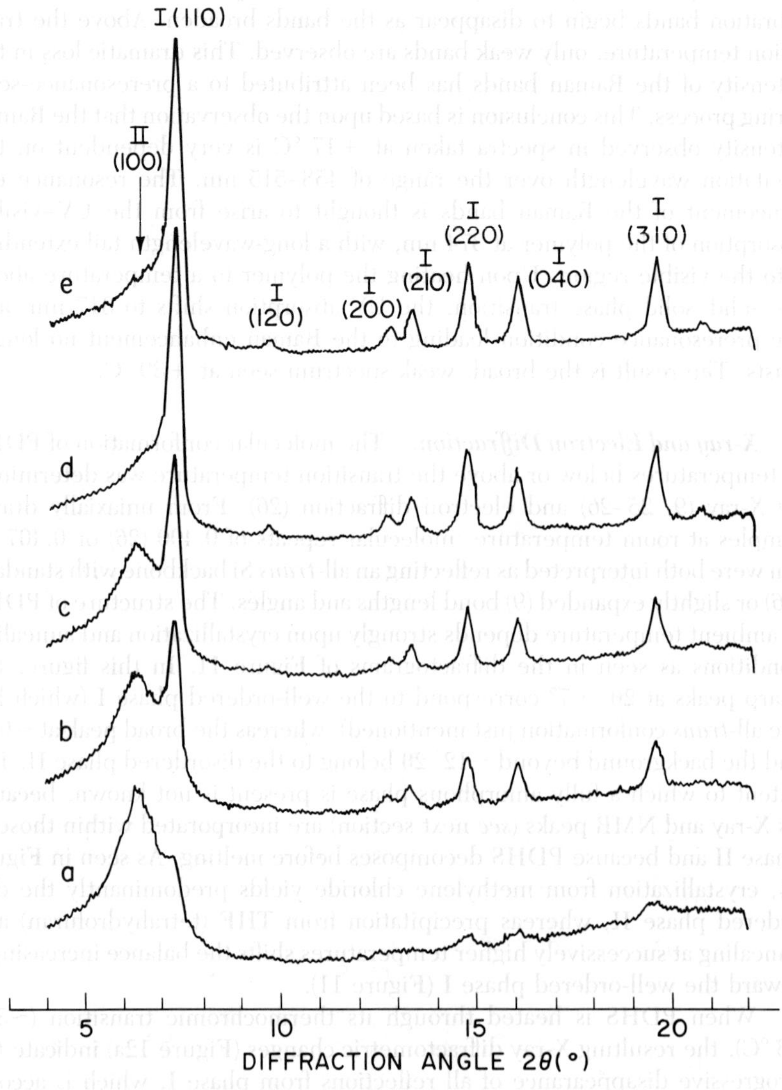

Figure 11. X-ray diffractograms (reflection mode) of PDHS samples at 25 °C after various crystallizations or annealing treatments: crystallized from a methylene chloride solution (a); crystallized from a THF solution by precipitation with methanol (b); and as in b and after heating to 60 °C (c), 150 °C (d), and 200 °C (e). Scans c–e were recorded at least 2 h after return to room temperature. (Reproduced from reference 26. Copyright 1986 American Chemical Society.)

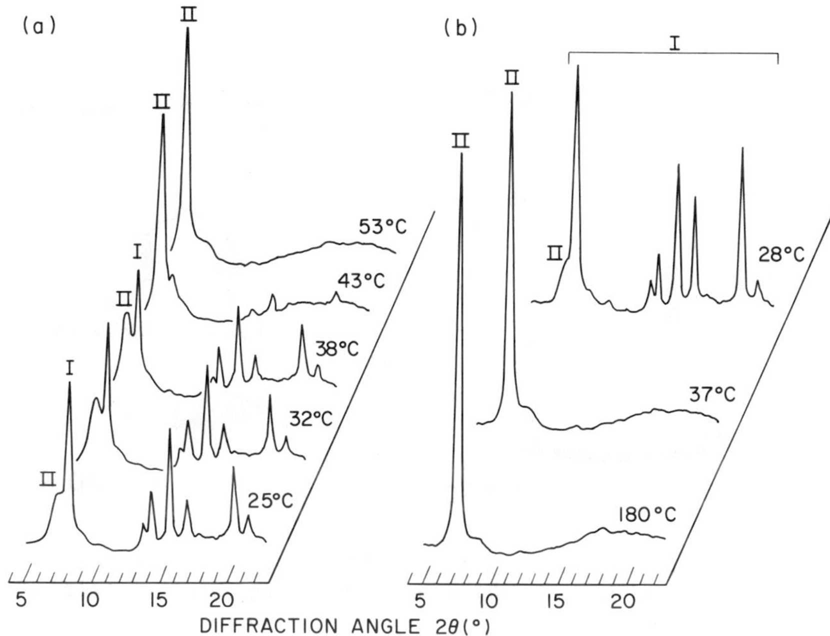

Figure 12. X-ray diffractograms (reflection mode) of PDHS during heating (a) and cooling (b) through a temperature range encompassing the disordering transition. (Reproduced from reference 26. Copyright 1986 American Chemical Society.)

resulting well-ordered phase I represents a larger fraction of the sample than it was originally (this increase is due to the intervening annealing treatment).

A better understanding of the mechanism of the thermochromic transition in PDHS is obtained by X-ray studies of fiber patterns. Comparison of such patterns from the same specimen at temperatures below or above the transition temperature (Figure 13) gives the following results:

1. The original equatorial reflections are replaced by two at 1.35 and 0.775 nm, a result indicating expansion of the intermolecular lattice and adoption of hexagonal packing.
2. Arcing of these reflections increases only slightly, a result indicating overall preservation of the main-chain orientation.
3. The outer reflections merge into an amorphous halo, a result implying intramolecular disordering, particularly of the side chains.
4. This disordering must occur preferentially within planes perpendicular to the Si-backbone axis, as evidenced by the maximized intensity of this halo at the meridional region.

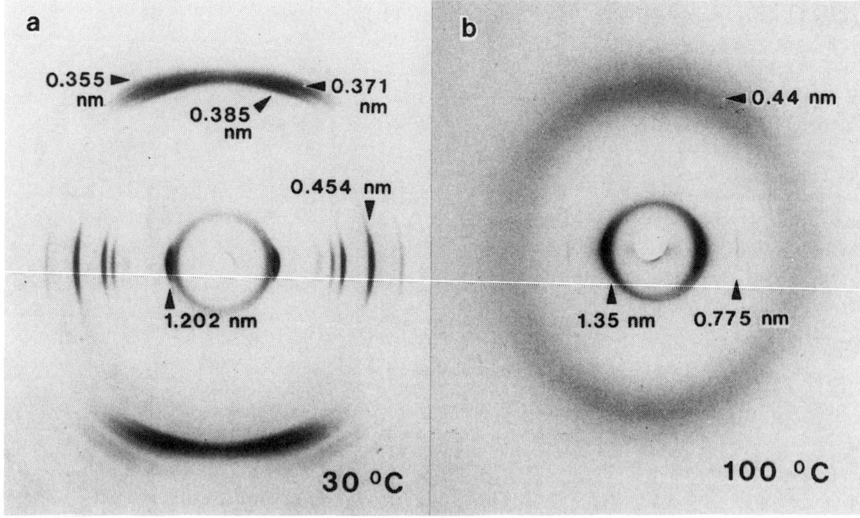

Figure 13. X-ray diffraction patterns of PDHS crystallized by slow cooling from 180 °C and then uniaxially oriented at room temperature (vertical fiber axis): (a) pattern recorded at 30 °C and (b) pattern recorded at 100 °C. (Reproduced from reference 26. Copyright 1986 American Chemical Society.)

These changes in intra- and intermolecular structures at the phase transition are consistent with those obtained by the other techniques described in this chapter.

Additional evidence for the order in phase II of PDHS can be seen in the optical birefringence, as reported by Rabolt (25). In Figure 14, the level of transmitted light is plotted as a function of temperature. The film is highly birefringent at 30 °C, but birefringence decreases dramatically as the sample is heated above the transition temperature and conformational disorder begins to be complete. However, the birefringence does not completely disappear at higher temperatures until the sample decomposes at 250 °C. In contrast, atactic PMHS exhibits no birefringence at or above room temperature. These observations suggest that despite local conformational disorder the chains of PDHS remain partially oriented at temperatures above the phase transition.

Solid-State NMR Spectroscopy. *^{13}C NMR Spectra.* Structural information can also be obtained from the solid-state NMR spectroscopic analyses of the polysilylenes. In recording the solid-state spectra under conditions of magic angle spinning and cross polarization (27, 28), the experimental parameters can be varied to permit the observation of nuclei in distinct states of motion. We have reported the ^{13}C NMR spectra of PDHS in solution (2)

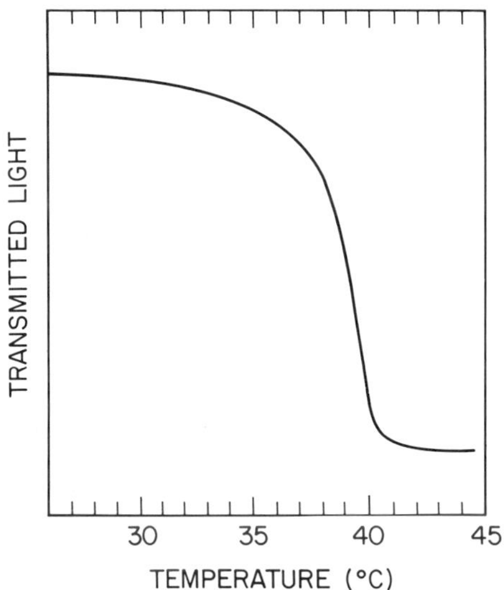

Figure 14. Optical birefringence of PDHS film as a function of temperature. (Reproduced from reference 25. Copyright 1986 American Chemical Society.)

and in the solid state (*29*). Figure 15a shows the solution spectrum of PDHS in toluene-d_8. The line width of the resonances from C-6 (3.8 Hz) to C-1 (27.0 Hz) increases continually. This trend is a result of restricted local side-chain motion near the sterically crowded polymer backbone. The reduced motion is unable to effect complete averaging of the proton–carbon dipole–dipole interactions, and line broadening is the result. This is an unusual observation for polymer solutions; however, the polysilylenes are quite different from most linear polymers in that every backbone nucleus is disubstituted with bulky groups.

Figure 15b shows a solid-state spectrum recorded under conditions such that only the mobile portions of the solid PDHS sample are observed. In this polymer (as previously indicated), the mobile portion of the sample consists of the locally disordered phase II and any amorphous material to the extent that it exists. The chemical-shift pattern for the carbons agrees very well with the solution spectrum (Figure 15a). Because carbon resonances are very sensitive to bond conformation (*22*), this result demonstrates that the phase II portion of the sample has the same average chain conformation as the polymer chains in solution. Although these NMR data permit a comparison of local bond conformations, they do not provide an indication of the more global chain dimensions. Figure 15b shows increased line widths for the carbons near the silicon backbone, with the C-1 resonance almost broadened into the baseline. This broadening reflects the severe restriction of motion near the backbone.

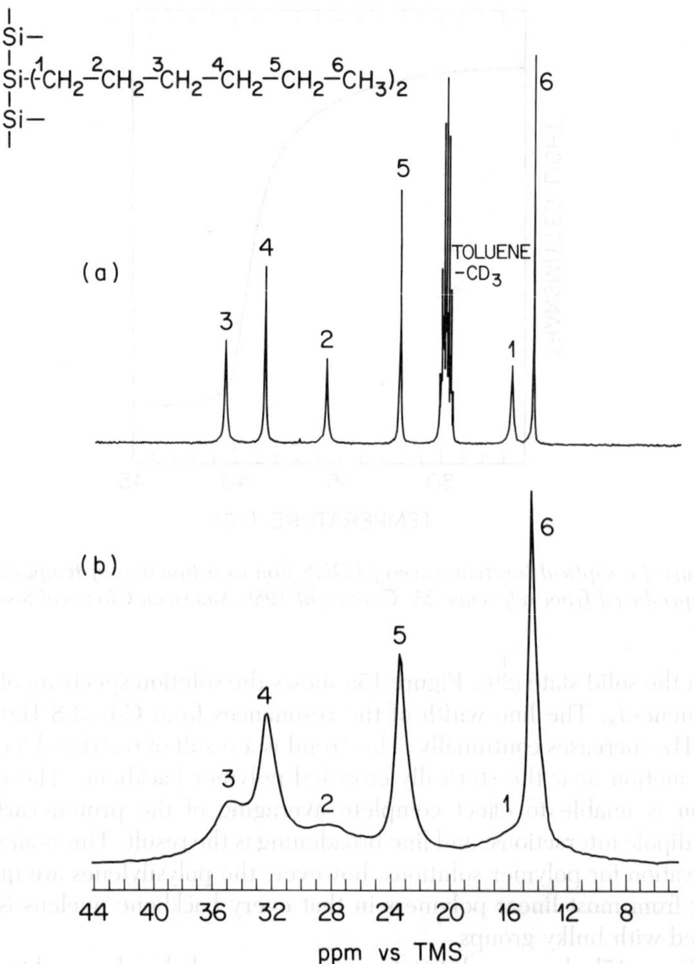

Figure 15. ^{13}C NMR (50.31-MHz) spectra of PDHS at 23 °C in toluene-d$_8$ [chemical shifts with respect to $(CH_3)_4Si$] (a) and in the solid-state with magic angle spinning (C-6 set to 14.5 ppm) (b). (Reproduced from reference 29. Copyright 1986 American Chemical Society.)

The parameters of the solid-state NMR experiment can be selected to permit observation of the carbon resonances in the entire polymer sample, including both rigid and mobile portions (29). The data from this experiment are shown in Figure 16a. The resonances of the well-ordered phase I carbons are underlined, whereas those of the mobile phase II carbons examined in Figure 15b are not underlined. The phase II resonance for each carbon is at a higher field or a more shielded position compared with the corresponding carbon of phase I. This difference reflects the presence of the more shielding

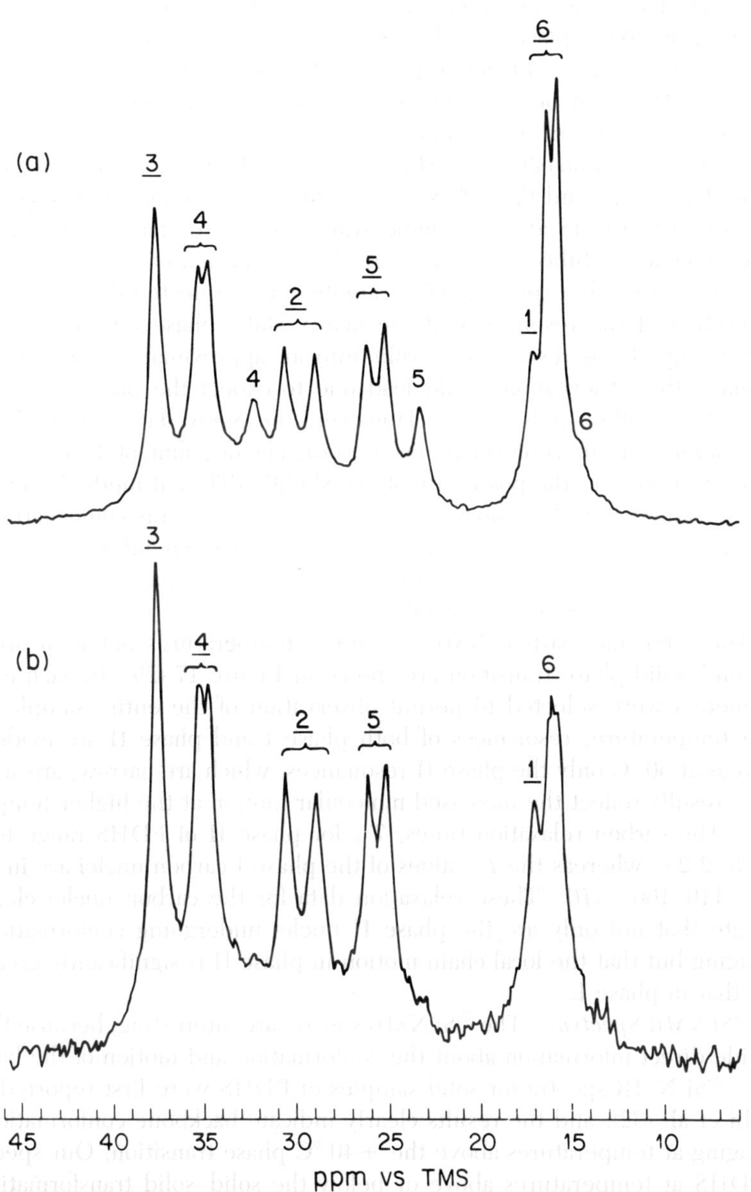

Figure 16. Solid-state ^{13}C NMR (50.31-MHz) spectra of PDHS at 23 °C in phase I and phase II (a) and in phase I only (b). Phase I resonances are underlined. (Reproduced from reference 29. Copyright 1986 American Chemical Society.)

gauche conformation in the disordered phase II side chains. The fact that single resonances are observed in phase II indicates that conformational averaging is taking place, which is consistent with the loss of reflections observed in the X-ray diffraction pattern of phase II. The peaks of phase I and phase II cannot be quantitatively compared in this type of solid-state experiment because of the significant difference in their motional behavior and cross-polarization efficiency. However, on the basis of the different chemical-shift positions and the differences in the signal intensities, two phases are present in PDHS at room temperature, and these phases differ significantly in chain mobility and in average chain conformation.

For Figure 16b, experimental parameters were selected that permit the observation of the resonances of the nearly static phase I portion of the sample only. Each resonance is split into an approximately 1:1 doublet. Normally this observation would lead one to expect the presence of two different crystalline polymorphs. However, the X-ray data indicated that only a single Si chain conformation is present. The doubling of the resonance suggests, therefore, the possibility of two slightly different methods of side-chain packing within the polymer crystal, although no X-ray confirmation of this explanation has been forthcoming. Differences in chemical shift of this magnitude attributable to alternative modes of chain packing have been found in other polymer systems (*30, 31*).

The solid-state carbon NMR spectra at temperatures below or above the solid–solid phase transition are shown in Figure 17 (*29*). In each case, parameters were selected to permit observation of the entire sample. At room temperature, resonances of both phase I and phase II are evident, whereas at 50 °C only the phase II resonances, which are narrow, are seen. These results reflect the increased molecular motion at the higher temperature. The carbon relaxation times, T_1, for phase II of PDHS range from 0.15 to 2.2 s, whereas the T_1 values of the phase I carbon nuclei are in the range 110–160 s (*10*). These relaxation data for the carbon nuclei clearly indicate that not only are the phase II nuclei undergoing conformational averaging but that the local chain motion in phase II is significantly greater than that in phase I.

^{29}Si NMR Spectra. The ^{29}Si NMR spectra are interesting, because they provide direct information about the conformation and motion of the backbone. ^{29}Si NMR spectra for solid samples of PDHS were first reported by Gobbi et al. (*32*), and the results clearly indicate backbone conformational averaging at temperatures above the +40 °C phase transition. Our spectra of PDHS at temperatures above or below the solid–solid transformations are shown in Figure 18. The vertical scaling is constant throughout the figure and permits comparison of peaks within each phase as a function of temperature. With experimental parameters selected so that nuclei of the entire sample are observed, two resonances are seen in the spectrum taken at +21 °C. As the temperature is increased, the size of the higher field res-

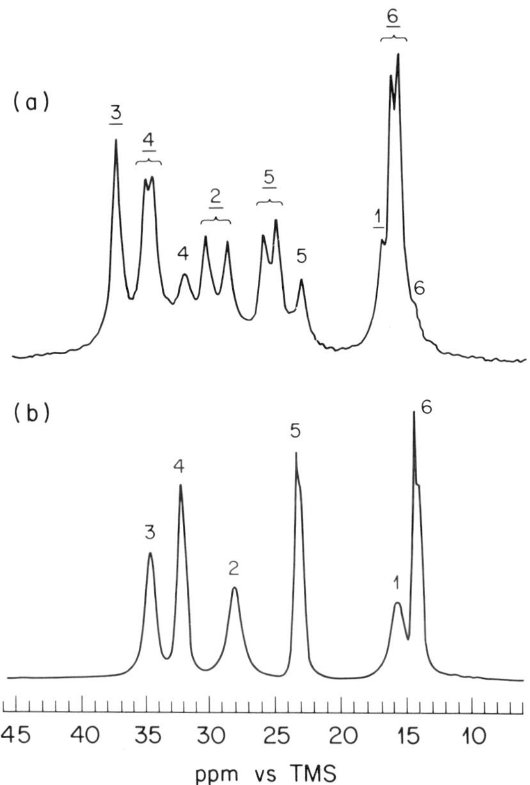

Figure 17. Solid-state ^{13}C NMR (50.31-MHz) spectra of PDHS at 23 °C (a) and 50 °C (b). (Reproduced from reference 29. Copyright 1986 American Chemical Society.)

onance increases, and the low-field resonance decreases correspondingly. Because the area of the high-field peak increases with temperature, this peak is assigned to phase II. The higher field position of this peak no doubt reflects the insertion of non-*trans*-shielding conformations into the silicon backbone, compared with the all-*trans* backbone of phase I. Changes in the bond conformations of the side chains will also effect the silicon chemical shifts. At +37 °C, the intensity of the phase I resonance was reduced by 50%, and therefore, this temperature is designated as the transition midpoint.

Measurement of the silicon spin lattice relaxation times also indicates a large difference between the local chain motions of phase I and phase II. The T_1 value is 5.5 s for phase II silicon nuclei, whereas phase I nuclei have a T_1 value of 3.2 h (*10*), one of the longest relaxation times observed for silicon nuclei. Unfortunately, as a result of these tremendous differences in motional properties, quantitative data for the entire PDHS sample cannot

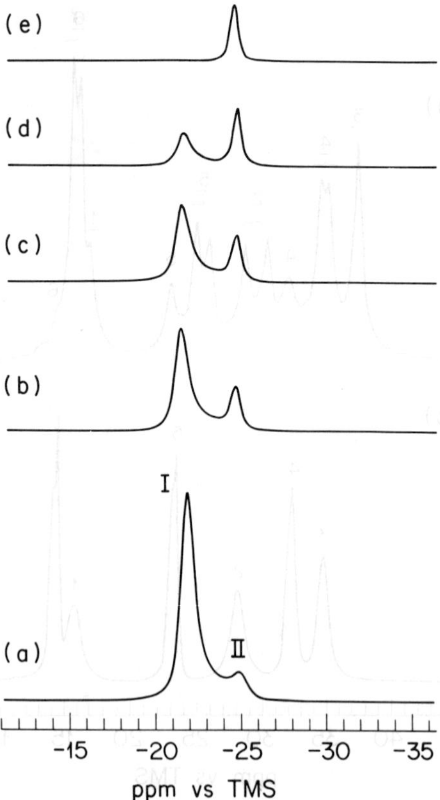

Figure 18. Solid-state ^{29}Si NMR (39.75-MHz) spectra of PDHS at 21.0 (a), 37.0 (b), 37.5 (c), 38.0 (d), and 45.0 (e) °C.

be obtained. The difference in motional properties explains the significant differences in the maximum intensities of the phase I and phase II resonances shown in Figure 18. Whereas meaningful comparisons of peak intensities can be made within a single solid phase, one should exercise caution when comparing the intensities of resonances from different motional regimes.

Summary of PDHS Structure. The information concerning the solid-state structure of PDHS that has been obtained from the studies discussed in the previous sections is summarized in the schematic drawings of Figure 19, although the complete crystal structure of PDHS has not been reported yet. The unit cell of the well-ordered phase I is monoclinic, with *a* and *b* dimensions of 1.375 and 2.182 nm, respectively, and the angle γ is 88° (*9*). The *c* dimension is 0.400 nm (*26*), which implies that the backbone has an all-*trans* conformation. The side chains are arranged in a direction nearly perpendicular to that of the backbone, with a nearly all-*trans* conformation,

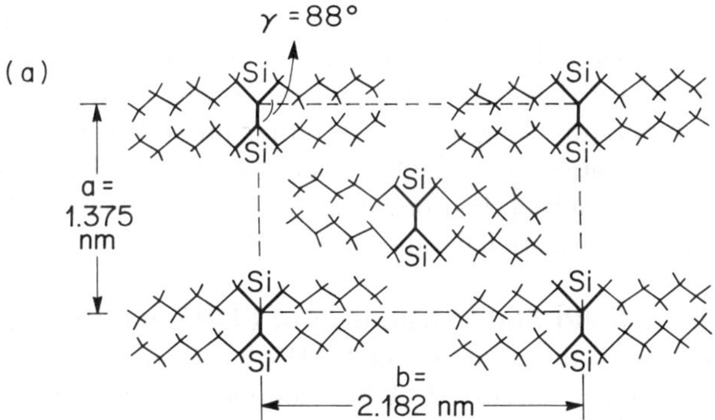

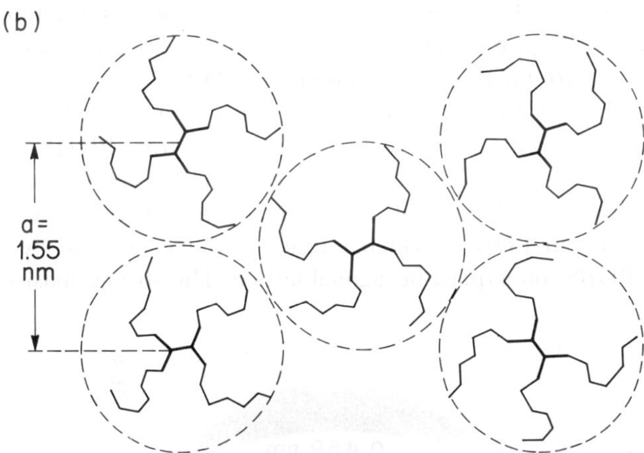

Figure 19. Schematic representations of solid-state structures of phase I (a) and phase II (b) of PDHS.

although for steric reasons, the Si–C-1 and C-1–C-2 bonds are probably rotated 30° from the *trans* conformation (9). On the basis of Raman data recorded for oriented films of PDHS and upon consideration of the interatomic distances (9), the Si–C-1 bond is probably rotated slightly away from the *trans* conformation to permit packing of the *n*-hexyl side chains.

The structure at temperatures above the transition temperature (phase II), is illustrated in the schematic drawing in Figure 19b. Both the side chains and the backbone are conformationally disordered, but the side chains remain organized preferentially on planes perpendicular to the polymer chain axis. Furthermore, significant runs of *trans* conformations remain in the backbone (26) in such a manner that the original molecular direction is preserved as the chains pack in cylinders in a hexagonal array. The resulting increase in lattice dimensions (Figure 19) renders the density of phase II less than that of phase I. The disordering transition in which phase I is converted into phase II is reversible, and by varying crystallization conditions, room-temperature samples rich in either phase I or phase II can be produced. The dramatic shift in the UV absorption observed at the transition temperature is the result of this conversion of the conformationally ordered phase I into the conformationally disordered phase II.

Solution and Solid-State Structures of PDBS and PDPS. A well-developed picture of the structure of PDHS has been obtained from the experimental work just discussed. A less complete picture is available for the structures of other polysilylenes. The structure of the dipentyl polymer, PDPS, was determined by Miller et al. (*11*) by using Raman and X-ray techniques. PDPS does not have an all-*trans* conformation, unlike phase I of PDHS; it is a 7/3 helix (*see* previous discussion of conformational energy calculations and Figure 7). The same chain conformation has been found by Schilling et al. (*10*) for the dibutyl polymer, PDBS.

Comparison of the ambient-temperature fiber diffraction pattern of PDBS (Figure 20) with the corresponding pattern from PDHS (Figure 13a) shows major differences. The equatorial region in the dibutyl homologue is dominated by the reflection at 1.098 nm (rather than by the numerous reflections seen in PDHS), which implies a much less orderly interchain packing in PDBS, on a quasi-hexagonal lattice. The strong quasi-meridional

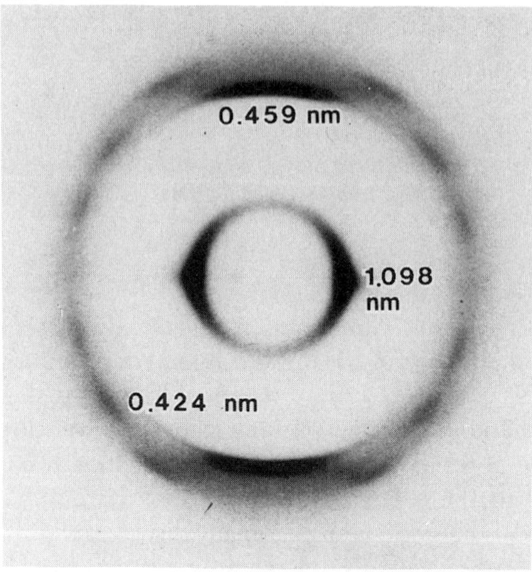

Figure 20. X-ray diffraction pattern of PDBS uniaxially drawn and recorded at ambient temperature (vertical fiber axis). (Reproduced from reference 10. Copyright 1989 American Chemical Society.)

reflection on the third layer line (0.459 nm) is actually split, with a true meridional peak seen on the seventh line in tilted fibers, which confirms the 7/3 helical conformation. Such helical structures are consistent with the energy calculations of Farmer et al. (*17*), which suggest that the critical side chain length for departure from such conformations (in favor of all-*trans* conformation) corresponds to six carbon atoms. However, recent studies have shown that the silicon bond conformation in poly(dimethylsilylene) (*33*), poly(diethylsilylene) (*34*), and poly(di-*n*-propylsilylene) (*34*) is all-*trans*. These results imply that a reexamination of the energy calculations is required.

Solid-State Transition. Evidence for a solid-state transition in PDBS can be seen from the diffractograms of Figure 21. Comparison with the corresponding X-ray evidence from PDHS (Figure 12) shows that the transition in the dibutyl polymer is characterized by much less pronounced changes. In essence, as the polymer is heated through 90 °C (which is approximately the midpoint of the transformation), the intermolecular peak (1.10 nm) becomes sharp and moves very slightly to higher angles. These changes imply an improved interchain packing on a minimally contracted lattice (0.034%), consistent with the ^{29}Si NMR evidence (*see* next section) indicating a non-*trans* average conformation at high temperatures. Randomization of side chains during heating would act in the same direction. The other clear aspect of the X-ray evidence during heating (Figure 21) is the gradual disappearance of all reflections in the 19–24° 2θ range, which implies increasing intramolecular disorder and loss of three-dimensional structure, as was also found for PDHS (*26*). Both the inter- and intramolecular manifestations of this transition are reversible upon cooling, albeit with some hysteresis (Figure 21) suggestive of a first-order process.

The structural similarities of PDBS and PDPS and their differences compared with PDHS can be clearly seen in the ^{29}Si NMR spectra (*10*) of the polymers (Figure 22). The phase I peak for PDHS occurs at ~21 ppm, which reflects the all-*trans* backbone. However, the phase II ^{29}Si resonance of this polymer is found at a higher field position, which reflects the presence of non-*trans* conformations in the backbone and side chains. These conformations are apparently more shielding than the *trans* conformations. Comparison of PDBS and PDPS shows that the silicon resonances of the disordered phase II are found in similar positions for all three polymers. However, the resonances of the ordered phase I of PDBS and PDPS are found at a field much higher than that observed for phase I of PDHS, a difference of ~5 ppm. These data indicate the presence of a more shielding conformation in the phase I structure of PDBS and PDPS polymers. These NMR results are consistent with the X-ray-determined structure of a 7/3 helix for these two polymers, a helix in which each backbone bond is rotated approximately 30° from the *trans* conformation.

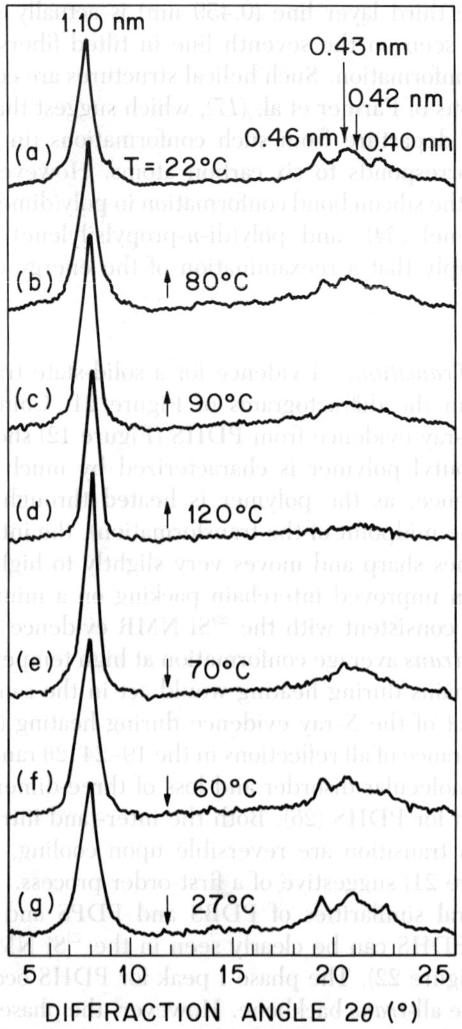

Figure 21. X-ray diffractograms of PDBS recorded at the indicated temperatures during heating and subsequent cooling. (Reproduced from reference 10. Copyright 1989 American Chemical Society.)

Thermochromic Transition. The solid-state thermochromic transition observed in PDHS is associated with conformational disordering of the polymer backbone, with the all-*trans* backbone conformation giving rise to the absorption at 374 nm. The thermochromic transition observed for PDHS in solution is also thought to result from conformational ordering, with the long-wavelength absorption of 354 nm at –40 °C associated with an all-*trans* conformation. However, no spectroscopic evidence beyond the UV results

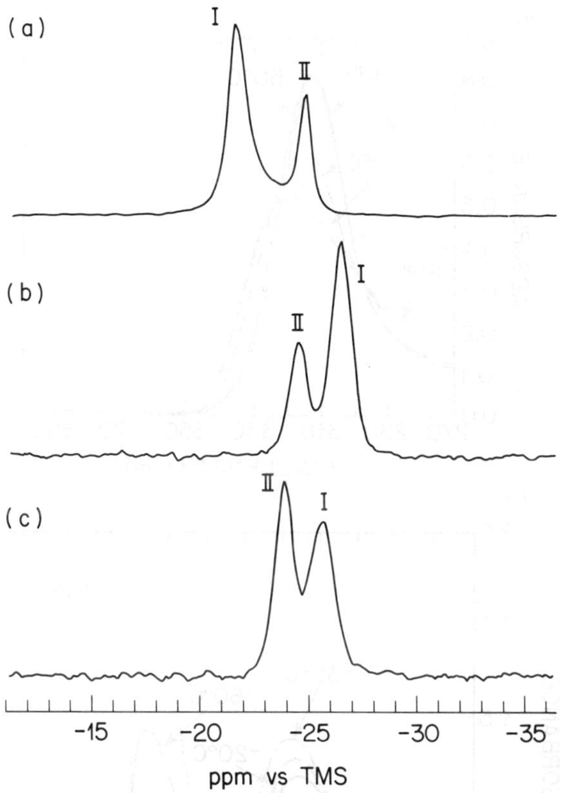

Figure 22. Solid-state ^{29}Si NMR (37.75-MHz) spectra of PDHS (a), PDPS (b), and PDBS (c) recorded near the midpoint of the phase transition for each polymer. (Reproduced from reference 10. Copyright 1989 American Chemical Society.)

previously discussed has been reported with regard to the identification of the solution conformation at this temperature, nor has there been an experimentally verified explanation for the difference in the absorption maxima between the solution and solid-state samples.

For PDBS (*10*) and PDPS (*11*), such a dramatic thermochromic transition has not been observed in the solid state. The absorption spectrum of solid PDBS broadens as the film is heated, and λ_{max} shifts to the red region by about 10 nm (Figure 23a). The DSC data for PDBS (*10*) indicate that the solid–solid phase transition is at +86 °C. The failure to observe a strong thermochromic transition in PDBS or PDPS is probably the result of the predominant non-*trans* conformations in the ordered phase I structures. These conformations prohibit the runs of *trans* conformations necessary to provide the conjugation responsible for the long-wavelength absorption observed in PDHS. In contrast, cooling of the PDBS solution results in a

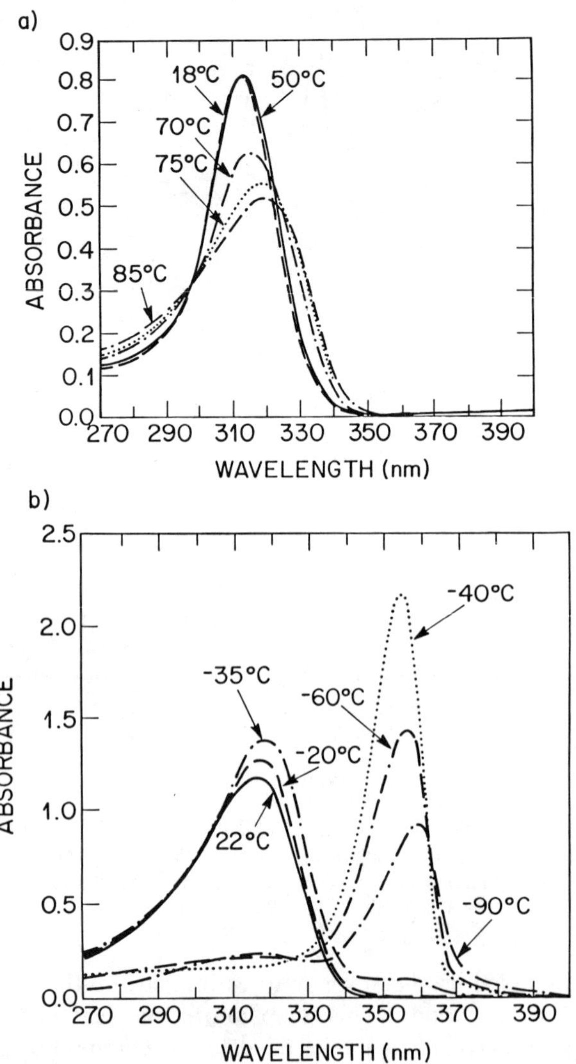

Figure 23. UV absorption spectra of PDBS as film (a) and in hexane at 5 × 10^{-5} M (b). (Reproduced from reference 10. Copyright 1989 American Chemical Society.)

dramatic change between –35 and –40 °C (Figure 23b), with λ_{max} shifting to 354 nm at –40 °C. The absorption characteristics of PDBS suggest that the driving force responsible for ordering the backbone into an all-*trans* conformation in solution may differ from that in the solid state. Crystallization of the alkyl side chains, thought to be the driving force for the generation of the all-*trans* backbone conformation in PDHS, evidently does not occur in PDBS. Therefore, something other than side-chain organization may be responsible for the conformational ordering apparent in PDBS solutions at low temperature. This may also be the case for PDPS and PDHS in solution. Additional experimental work is needed to understand fully the solution structures of the symmetrically substituted polysilylenes.

Other Polysilylenes. The symmetrically substituted poly(di-*n*-tetradecylsilylene) is reported to have a TGTG′ (*trans–gauche–trans–gauche′*) conformation, and a bathochromic shift is observed in the UV spectrum at 54 °C with λ_{max} shifting from 322 to 350 nm (*35*). The structures of many polysilylenes may be more complicated than what have been discussed thus far. As an example, the DSC data for PMHS are shown in Figure 24. The figure shows that T_g (glass transition temperature) is ~220 K and that two

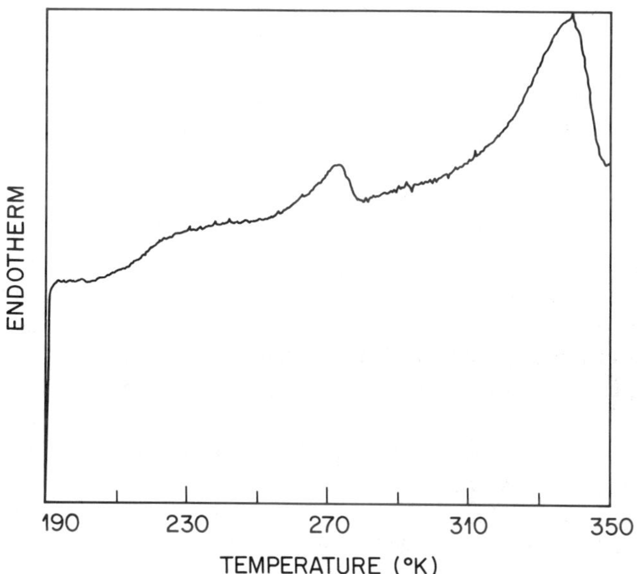

Figure 24. DSC thermogram of PMHS at a heating rate of 10 °C/min.

melting endotherms exist at higher temperatures. Slow cooling of this sample and rescanning shows the same thermal behavior. Other asymmetrically substituted polysilylenes exhibit similarly complex structures. The more-complex thermograms may reflect several polymorphs or separate transitions of the side chains and backbone.

The characterization of soluble diarylsilylene homopolymers was recently reported by Miller et al. (*36*). These materials absorb in the UV at ~400 nm (e.g., λ_{max} = 395 nm for poly[di-*p*-ethylphenylsilylene]), and the absorption bands (11–16 nm) are very narrow for polymeric materials. Most interestingly, little difference is reported in the spectra of the materials in solution and those of the solid state. The unusually long wavelength absorption characteristic of the poly(diarylsilylenes) may be conformational in origin. Such absorption would require that runs of the *trans* conformation of sufficient length to absorb at 400 nm be present in solution. If 40–50 *trans* units were sufficient to produce such an absorption, then a high-molecular-weight polymer could still exhibit the solution dimensions of a coillike polymer (*37*). However, the relationships between the chain conformation and absorption properties in solution and in the solid state have not been established experimentally for these polymers.

Polymer Morphology

The morphology of polysilylenes has only been recently investigated. Thin films of PDHS and PDBS have been prepared by evaporation of dilute solutions at ~100 °C and then examined by electron microscopy and diffraction after slow cooling to ambient temperature. The characteristic appearance of such films of the di-*n*-hexyl polymer is seen in Figure 25. The crystals have a ribbonlike morphology, with typical widths of 100–120 nm and lengths in the micrometer range. Faint striations perpendicular to the ribbon length are evident. The accompanying electron diffraction pattern shows that these striations correspond to the axis of the Si chain and that the ribbons are parallel to the crystallographic a axis. This morphology is interpreted as indicating unusually thick lamellae grown on edge.

A clearer image of the striations parallel to the molecular axis is seen in Figure 26, in which distinct growth steps are observed. Such unusually thick lamellae, with striations parallel to the chain direction, were revealed by fracture of poly(tetrafluoroethylene) (PTFE) (*38*) crystallized at atmospheric pressure, as well as of polyethylene (*39*) and poly(chlorotrifluoroethylene) (*40*) at high pressures (in all of these polymers, however, typical thicknesses are even larger than that of PDHS, i.e., 0.5–2.0 μm). Thus, PDHS appears to be only the second polymer that forms such partly extended chain crystals during normal growth from the melt without application of high pressures. As for PTFE, the origin of these extraordinary thicknesses is expected to lie in the inherent conformational and steric characteristics of the chains at the growth temperatures. Both PDHS and PTFE have conformations of low flexibility, and both (together with polyethylene at high pressures [41]) are characterized by intramolecular disorder, high mobilities, and hexagonal packing.

The molecular orientation of PDHS on a substrate, as deduced from electron micrographs such as that of Figure 25, is seen in Figure 27a. The Si-backbone chains are oriented parallel to the substrate, whereas the alkyl side chains are disposed essentially perpendicular to it. A different arrange-

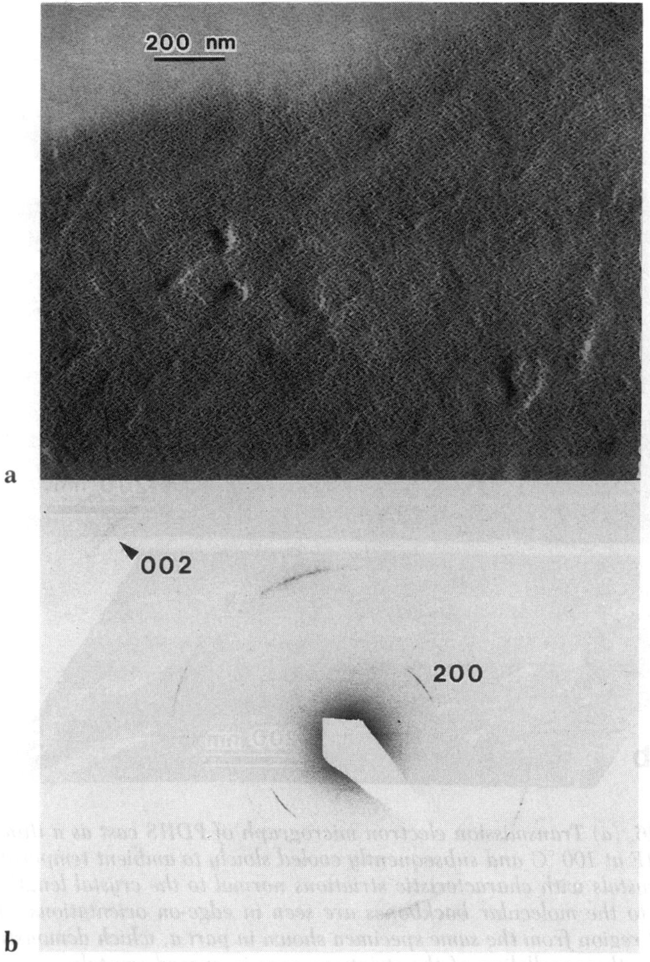

Figure 25. (a) Transmission electron micrograph of PDHS cast as a thin film from THF at 100 °C and then slowly cooled to ambient temperature. (b) Corresponding electron diffraction pattern of the region shown in part a (shown in correct orientation).

ment (Figure 27b) was reported by Miller et al. (*42*), who used grazing IR techniques on films deposited by spin coating. In this case, the preparative technique may have tipped the highly anisometric molecules to their sides.

The only other polysilylene for which morphological information has been obtained so far is PDBS. Under similar conditions of thin-film growth as described for PDHS, much smaller crystallites are obtained (Figure 28a). Their lengths are in the order of 0.2 μm, and their thicknesses are 40–70 nm. This reduced thickness (compared with that of PDHS) is consistent

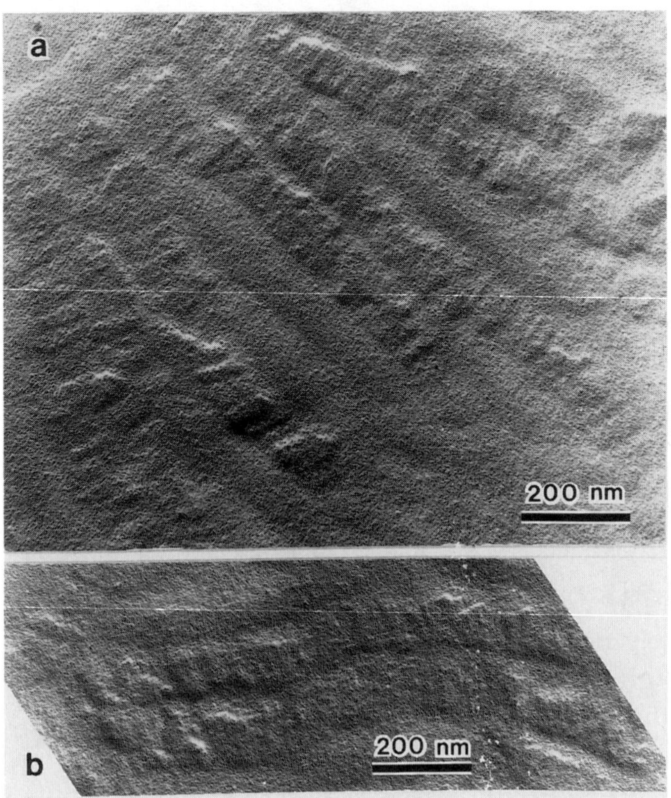

Figure 26. (a) Transmission electron micrograph of PDHS cast as a thin film from THF at 100 °C and subsequently cooled slowly to ambient temperature. Thick crystals with characteristic striations normal to the crystal length and parallel to the molecular backbones are seen in edge-on orientations. (b) A different region from the same specimen shown in part a, which demonstrates the parallelism of the striations even in curved crystals.

with their more flexible, 7/3 helical structure. At high resolution, striations normal to the length of these crystals can be discerned (although not as clearly and consistently as in PDHS). However, because of the much smaller dimensions of the PDBS crystals, confirmation of the *c*-axis parallelism to these striations has not been possible by electron diffraction.

Nevertheless, Figure 28b shows clearly that the strongest reflection is at 0.462 nm, which primarily represents an intramolecular order. On the contrary, the reflection at 1.110 nm, which represents intermolecular packing and which in bulk specimens is by far the strongest (Figure 21), is now virtually absent. These electron diffraction findings indicate that the backbone chains are disposed nearly parallel to the substrate.

Research on single-crystal growth of PDBS and PDHS is also in progress.

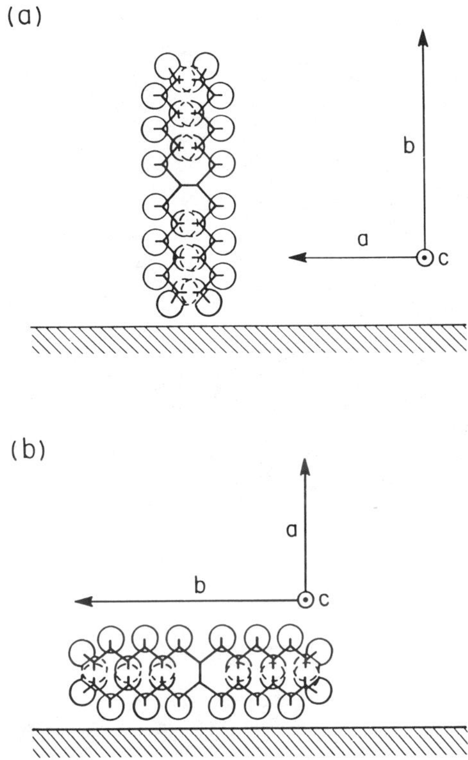

Figure 27. Schematic representations of the molecular orientation of PDHS on a substrate.

Preliminary results of morphological studies of PDBS (F. J. Padden, Jr., private communications) indicate the growth of highly ordered lamellar crystals possessing rectangular and diamondlike habits and suggest the existence of polymorphism.

Summary

Although much has been learned about the structures of polysilylenes, a tremendous amount of work remains before a full understanding of these materials is developed. The microstructure of the polymers can be studied directly by solution NMR spectroscopic techniques. The determination of the chain conformation in solution is difficult, particularly at low temperature. Light-scattering techniques may be able to establish the solution dimensions of the polysilylenes through the low-temperature thermochromic transition. The chain conformation in the solid state can be established by X-ray and electron diffraction methods. Solid-state ^{29}Si NMR spectroscopy can become

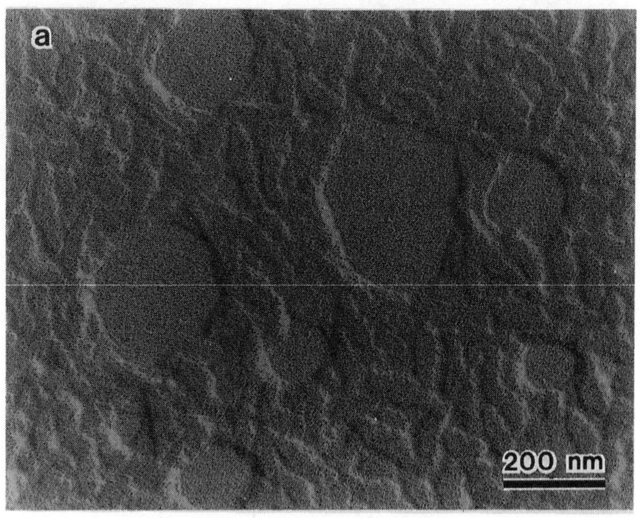

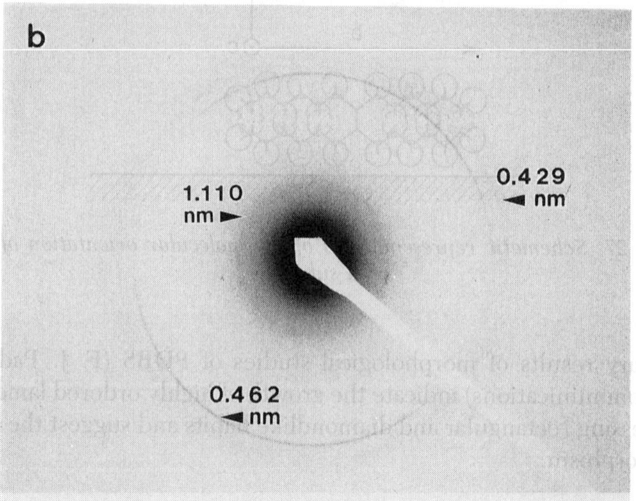

Figure 28. (a) Transmission electron micrograph of PDBS crystals grown in a thin film by evaporation of THF solution at 100 °C followed by slow cooling to ambient temperature. (b) Typical electron diffraction pattern from regions such as that shown in part a.

a quantitative probe of backbone conformation, provided that the relationship between conformation and silicon chemical shift can be established in a manner similar to that developed for the carbon chemical shift. X-ray, Raman, NMR, and IR techniques are useful in studying the solid-state order–disorder transitions in the polysilylenes. Conformational modeling faces the difficulty of developing sophisticated models that can consider all of the contributions to the conformational energies and yet deal with the large number of atoms present in the side chains and backbones of the more-interesting polysilylenes. The morphological studies of these materials are only now beginning, and the early results, which are very interesting, should inspire additional studies in this area.

Acknowledgments

We express our appreciation to the many researchers in this field who made available preprints of current work and for many stimulating technical discussions.

Abbreviations

PCHMS	poly(cyclohexylmethylsilylene)
PDBS	poly(di-*n*-butylsilylene)
PDHepS	poly(di-*n*-heptylsilylene)
PDHS	poly(di-*n*-hexylsilylene)
PDMS	poly(dimethylsilylene)
PDOS	poly(di-*n*-octylsilylene)
PDPS	poly(di-*n*-pentylsilylene)
PMDS	poly(methyl-*n*-dodecylsilylene)
PMHS	poly(methyl-*n*-hexylsilylene)
PMPS	poly(methyl-*n*-propylsilylene)
PPMS	poly(phenylmethylsilylene)
PS	polysilylene

References

1. Bovey, F. A. *Chain Structure and Conformation of Macromolecules;* Academic: New York, 1982; Chapters 3–6.
2. Schilling, F. C.; Bovey, F. A.; Zeigler, J. M. *Macromolecules* **1986,** *19,* 2309.
3. Wolff, A. R.; Nozue, I.; Maxka, J.; West, R. *J. Polym. Sci., Polym. Chem. Ed.* **1988,** *26,* 701.
4. Harrah, L. A.; Zeigler, J. M. In *Photophysics of Polymers;* Hoyle, C. E.; Torkelson, J. M., Eds.; ACS Symposium Series 358; American Chemical Society, Washington, DC, 1987; pp 482–498.
5. Trefonas, P., III; West, R.; Miller, R. D.; Hofer, D. *J. Polym. Sci., Polym. Lett. Ed.* **1983,** *21,* 823.
6. Trefonas, P., III; Damewood, J. R., Jr.; West, R.; Miller, R. D. *Organometallics* **1985,** *4,* 1318.
7. Harrah, L. A.; Zeigler, J. M. *J. Polym. Sci., Polym. Lett. Ed.* **1985,** *23,* 209.
8. Harrah, L. A.; Zeigler, J. M. *Macromolecules* **1987,** *20,* 601.
9. Kuzmany, H.; Rabolt, J. F.; Farmer, B. L.; Miller, R. D. *J. Chem. Phys.* **1986,** *85,* 7413.
10. Schilling, F. C.; Lovinger, A. J.; Zeigler, J. M.; Davis, D. D.; Bovey, F. A. *Macromolecules* **1989,** *22,* 3055.

11. Miller, R. D.; Farmer, B. L.; Fleming, W.; Sooriyakumaran, R.; Rabolt, J. F. *J. Am. Chem. Soc.* **1987,** *109,* 2509.
12. Welsh, W. J.; Beshah, K.; Ackerman, J. L.; Mark, J. E.; David, L. D.; West, R. *Polym. Prepr.* (*Am. Chem. Soc., Div. Polym. Chem.*) **1983,** *24*(1), 131.
13. Damewood, J. R., Jr.; West, R. *Macromolecules* **1985,** *18,* 159.
14. Damewood, J. R., Jr. *Macromolecules* **1985,** *18,* 1793.
15. Bigelow, R. W.; McGrane, K. M. *J. Polym. Sci., Polym. Phys. Ed:* **1986,** *24,* 1233.
16. Welsh, W. J.; DeBolt, L.; Mark, J. E. *Macromolecules* **1986,** *19,* 2978.
17. Farmer, B. L.; Rabolt, J. F.; Miller, R. D. *Macromolecules* **1987,** *20,* 1167.
18. Mintmire, J. W.; Ortiz, J. V. *J. Am. Chem. Soc.* **1988,** *110,* 4522.
19. Johnson, W. D.; Welsh, W. J. *Abstracts of Papers,* Advances in Silicon-Based Polymer Science Workshop, Honolulu, Hawaii; American Chemical Society: Washington, DC, 1987.
20. Flory, P. J. In *Statistical Mechanics of Chain Molecules;* Interscience: New York, 1969; p 11.
21. Bovey, F. A. *Chain Structure and Conformation of Macromolecules;* Academic: New York, 1982; Chapter 7.
22. Tonelli, A. E.; Schilling, F. C. *Acc. Chem. Res.* **1981,** *14,* 233.
23. Cotts, P. M.; Miller, R. D.; Trefonas, P. T., III; West, R.; Fickes, G. N. *Macromolecules* **1987,** *20,* 1046.
24. Cotts, P. M.; Maxka, J.; Miller, R. D.; West, R. *Polym. Prepr.* (*Am. Chem. Soc., Div. Polym. Chem.*) **1987,** *28*(1), 450.
25. Rabolt, J. F.; Hofer, D.; Miller, R. D., Fickes, G. N. *Macromolecules* **1986,** *19,* 611.
26. Lovinger, A. J.; Schilling, F. C.; Bovey, F. A.; Zeigler, J. M. *Macromolecules* **1986,** *19,* 2657.
27. Komoroski, R. A. In *High Resolution NMR Spectroscopy of Synthetic Polymers in Bulk;* VCH Publishers: Deerfield Beach, FL, 1986; pp 28–46.
28. Jelinski, L. W. In *Chain Structure and Conformation of Macromolecules;* Bovey, F. A., Ed.; Academic: New York, 1982; pp 232–238.
29. Schilling, F. C.; Bovey, F. A.; Lovinger, A. J.; Zeigler, J. M. *Macromolecules* **1986,** *19,* 2660.
30. VanderHart, D. L. *J. Magn. Reson.* **1981,** *44,* 117.
31. Gómez, M. A.; Tanaka, H. A.; Tonelli, A. E. *Polymer* **1987,** *28,* 2227.
32. Gobbi, G. C.; Fleming, W. W.; Sooriyakumaran, R.; Miller, R. D. *J. Am. Chem. Soc.* **1986,** *108,* 5624.
33. Lovinger, A. J.; Davis, D. D.; Schilling, F. C.; Bovey, F. A.; Zeigler, J. M. *Macromolecules,* in press.
34. Lovinger, A. J.; Davis, D. D.; Schilling, F. C.; Bovey, F. A.; Zeigler, J. M. *Polym. Commun.* **1989,** *30,* 356.
35. Farmer, B. L.; Miller, R. D.; Rabolt, J. F.; Fleming, W. W.; Ficks, G. N. *Bull. Am. Phys. Soc.* **1988,** *33,* 657.
36. Miller, R. D.; Sooriyakumaran, R. *J. Polym. Sci., Polym. Lett. Ed.* **1987,** *25,* 321.
37. Tonelli, A. E. *Macromolecules* **1974,** *7,* 628.
38. Melillo, L.; Wunderlich, B. *Kolloid Z. Z. Polym.* **1972,** *250,* 417.
39. Wunderlich, B.; Melillo, L. *Makromol. Chem.* **1968,** *118,* 250.
40. Miyamoto, Y.; Nakafuku, C.; Takemura, T. *Polym. J.* **1972,** *3,* 120.
41. Bassett, D. C.; Block, S.; Piermarini, G. J. *J. Appl. Phys.* **1974,** *45,* 4146.
42. Miller, R. D.; Sooriyakumaran, R.; Rabolt, J. F. *Bull. Am. Phys. Soc.* **1987,** *32,* 886.

RECEIVED for review May 27, 1988. ACCEPTED revised manuscript January 26, 1989.

Order–Disorder Transitions and Thermochromism of Polysilylenes in Solution

Theory and Experiment

Kenneth S. Schweizer, Larry A. Harrah[1], and John M. Zeigler[2]

Sandia National Laboratories, Albuquerque, NM 87185

The thermochromic and order–disorder transition behaviors of symmetrical and unsymmetrical alkyl-substituted polysilylenes in dilute solution are reported. The dependence of these phenomena on solvent, substituent structure, and degree of polymerization were systematically investigated. A microscopic statistical mechanical theory of order–disorder transitions of flexible conjugated polymers in solution proposed by Schweizer was used to interpret the experimental results. The theory predicts that the conformational transition is induced by attractive dispersion interactions between the delocalized electrons of the polymer backbone and the surrounding polarizable medium. Nearly quantitative agreement between the numerous theoretical predictions and the observations is found. The results lend significant support to both the single-chain mechanism of solution thermochromic transition and the role of conformation-dependent polymer–solvent dispersion interactions as the primary physical driving force. The corresponding solid-state phase transitions and a simple model to estimate rotational defect formation energies are discussed briefly.

SOLUBLE CONJUGATED POLYMERS have been the subject of increasing scientific and technological interest over the past decade. Both soluble π-conjugated (*1*, *2*) (polydiacetylenes and polythiophenes) and σ-conjugated

[1]Current address: PDA Engineering, 3754 Hawkins Street, Albuquerque, NM 87109
[2]Current address: SilCHEMY, 2208 Lester Drive, NE, Albuquerque, NM 87112–2640

0065–2393/90/0224–0379$06.00/0

(*3, 4*) (polysilylenes and polygermanes) polymers have been synthesized. The combination of extensive, one-dimensional electron delocalization along the conjugated-polymer backbone and inherent chain flexibility results in an unusually strong electronic–conformational coupling. This coupling is at the heart of a variety of fascinating phenomena, such as electronic thermochromism, solvatochromism, and coupled electronic–conformational order–disorder phase transitions. Despite the many detailed structural differences between the various flexible conjugated polymers, their electronic properties and the phenomena they exhibit are strikingly similar, both in dilute solution and in the solid state.

Order–disorder, or "rod-to-coil", transitions in dilute solution have been reported for polydiacetylenes (*2, 5–11*), polysilylenes (*12–15*), and alkyl-substituted polythiophenes (*16*). The interpretation of the experimental observations has been the subject of considerable controversy with respect to whether the observations represent a single-polymer-molecule phenomenon or a many-chain aggregation or precipitation process (*3–16*). Our own experimental evidence (*12, 13*) and that of others (*5–8, 10, 16*) weigh heavily in favor of the single-chain interpretation. In our theoretical interpretation, we will assume that the order–disorder transitions observed in dilute polysilylene solutions represent equilibrium, single-chain phenomena.

The fundamental questions are: What is the microscopic mechanism or driving force for the transition, and what physical factors are important? Two distinct possibilities have been advanced: side-chain crystallization (*5, 6, 17–19*), which is postulated to induce polymer backbone ordering, and conformation-dependent polymer–solvent interactions that arise explicitly from electron delocalization and that stabilize an ordered rodlike conformation (*20–24*). Side-chain crystallization remains a qualitative suggestion that has not been developed to the point where it has predictive power and can be critically tested. However, in the solid state, the enhanced importance of packing effects makes such a mechanism more plausible (*18, 19*).

In this chapter, the theory of conformation-dependent polymer–solvent interactions, which was developed in detail by Schweizer (*20–22*) for soluble π-conjugated polymers, will be used to explain both qualitatively and quantitatively a large body of observations on the polysilylenes (*23, 24*). The same theory has been used recently to interpret qualitatively order–disorder phenomena and the electronic thermochromism of π-conjugated-polymer solutions and films (*25, 26*). The study presented in this chapter represents part of an ongoing effort to understand in a unified fashion both the optical properties (*27–30*) and order–disorder transitions (*20–24*) of flexible, conjugated-polymer solutions.

Theory

The statistical mechanical theory of Schweizer (*20–24*) has been presented in detail elsewhere. In this section, the applicability of the theory to the σ-

conjugated polysilylenes, the relevant molecular parameters, and the set of experimentally testable theoretical predictions are discussed.

General Formulation. The remarkable qualitative similarity of the order–disorder transition phenomena of different soluble π- and σ-conjugated polymers in dilute solution suggests an underlying universal physical mechanism (*20–24*). This possibility motivated the construction of a simple, zeroth-order statistical mechanical model that contains only the most generic aspects of a monodisperse, nonpolar, conjugated-polymer solution. Two basic aspects are included: the configurational statistics of a single chain described at the level of a three-state (*trans*, *gauche*+, and *gauche*–) rotational isomeric state (RIS) model (*31*) and conformation-dependent attractive interactions (modeled as simple induced-dipole London dispersion or van der Waals forces) between the delocalized electrons of the polymer backbone and the surrounding medium, which includes both the solvent molecules and the electronically decoupled side groups. The non-*trans* ("defect") conformers are assumed to be noninteracting, require a local free energy ϵ to be created, and correspond to a $\pm\phi_D$ dihedral-angle rotation. Mathematical solution of this model (*20–22*) at the mean-field level results in an effective free-energy surface that is a function of only one "order parameter": the number of rotational defects, N_D, or, equivalently, the mean conjugation length ξ_M ($\xi_M = N/[N_D + 1]$, in which N is the number of backbone atoms).

The simple theory is completely characterized by two dimensionless parameters: $\beta\epsilon$ and V_D/ϵ. β is $(k_BT)^{-1}$, in which T is absolute temperature and k_B is the Boltzmann constant. V_D is defined as the backbone delocalized-electron contribution to the attractive polymer–solvent dispersion interaction per site when the polymer is in the fully ordered all-*trans* (rod) conformation. The ratio V_D/ϵ (in which ϵ is the mean free energy of defect formation), referred to as the coupling constant, is both a polymer- and a solvent-dependent quantity and plays a special role in the theory.

The theory yields numerical predictions upon the adoption of a specific model for rotational defects and the electronic structure of the delocalized electrons. For mathematical simplicity, previous work (*20–24*) used $\phi_D = \pm 90°$ for defects associated with rotations about the carbon single bonds of a polyene. Such defects are modeled as interruptions that completely break the π conjugation and reduce the polymer to a sequence of electronically decoupled, all-*trans* π-electron segments. By using either the Hückel theory or the Parr–Pople–Pariser valence bond results (*32*), the dependence of the polymer backbone polarizability on the order parameter ξ_M can be computed (*27*) and is found to be a strongly decreasing function of the number of defects N_D. This trend will remain qualitatively valid for any type of bond rotational defect that interrupts the conjugation and introduces electronic disorder. The theory predicts that the enhanced solvation energy of the backbone in the rod conformation is the physical driving force for an order–disorder transition.

Our inadequate understanding of rotational defects and electronic–conformational coupling in polysilylenes has precluded a quantitative formulation of the theory for σ-conjugated polymers. However, recent theoretical studies (*33–36*) and spectroscopic measurements (*3, 4, 34, 37, 38*) have repeatedly documented the similarities in the electronic behaviors of polysilylenes and π-conjugated systems. For example, the oscillator strength per bond for the low-lying $\sigma \rightarrow \sigma^*$ transition (*39, 40*) in polysilylenes is very close to the polyene value of 0.12. Quantum chemical calculations (*41*) reveal an extremely polarizable silicon backbone, which is surprisingly even more polarizable than the polyene backbone. Both spectroscopic measurements (*3, 4, 34, 37, 38*) and quantum chemical calculations (*34, 35, 41, 42*) suggest that the electronic energy levels and polarizability depend strongly on conformation. *gauche*-like defects result in larger excitation energies and smaller backbone polarizability and appear to interrupt almost completely the σ-conjugation via 1–4 hyperconjugative-type interactions (*35, 39*) and/or conformation-dependent silicon valence angles (*29*). Taken as a whole, the available experimental and theoretical results strongly suggest that the situation for polysilylenes is qualitatively very similar to that of the flexible π-conjugated polymers and that the polysilylenes can be modeled by the zeroth-order theory. Consequently, this theory (*20–24*) was applied to polysilylenes, with the obvious caveat that strict quantitative accuracy cannot be expected because of the simplicity of the polyene-based statistical mechanical model and the possible quantitative differences between π- and σ-conjugated polymers.

Theoretical Predictions and Molecular Parameters. As previously emphasized (*21, 22*), a major and unique virtue of the simple zeroth-order theory is that it makes many experimentally testable predictions. More-sophisticated versions of the theory, which include physical effects such as rotational defect interactions, solvent-induced polymer swelling, and polydispersity, have been constructed (*21*). However, the basic features of the theory are not very sensitive to these complications (*21, 22*). This encourages us to believe that the predictions extracted from the simplest version of the theory may have wide validity. In this section, we explicitly enumerate and illustrate, with our polysilylene results, the various predictions.

General Phase Behavior. The most general prediction of the theory is that the qualitative phase behavior and thermochromism in solution are controlled by the magnitude of the coupling constant V_D/ϵ. Numerical results (*22*) for the most probable conjugation length as a function of temperature are shown in Figure 1 for a 1000-carbon polyene model. Two fundamentally distinct situations are found. For $V_D/\epsilon < (V_D/\epsilon)_c$, in which $(V_D/\epsilon)_c$ is the critical or cutoff value of the coupling constant, only unimodal

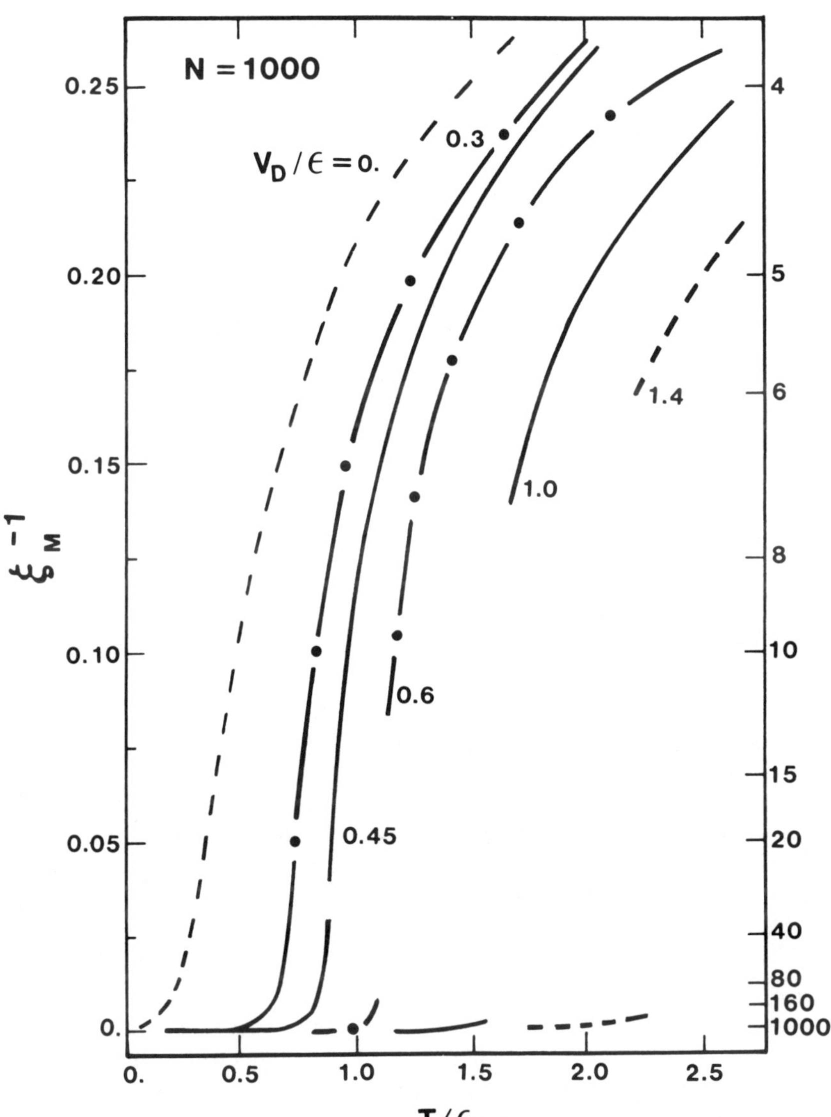

Figure 1. Most probable conjugation length and its inverse for a 1000-carbon polyene model as a function of scaled temperature and for several values of the coupling constant.

free-energy surfaces occur at all temperatures. $(V_D/\epsilon)_c$ is ~0.37 and is polymer and model dependent.

Physically, a subcritical value of the coupling constant can be due to either a very stiff polymer backbone (large ϵ) or a weak, delocalized electron–solvent interaction (small V_D arising, for example, from a relatively small conjugated-backbone polarizability). For coupling constants greater than the critical value, bimodal free-energy surfaces are found, and a pseudo-first-order phase transition occurs with a transition temperature, T_C, signaling a conformational change from a relatively disordered (smaller ξ_M) structure to a relatively ordered backbone. Crude but realistic estimates (*21*) of V_D/ϵ for polyacetylene, polydiacetylene, and polysilylene molecules yield coupling constants of order unity and realistic transition temperatures.

Even below the critical value of V_D/ϵ, quantitatively different types of behavior can occur (*21, 22*). The first case is weak coupling, for which $V_D/\epsilon << (V_D/\epsilon)_c$. In this case, ξ_M increases slowly and smoothly with cooling. The second case is intermediate coupling, for which $V_D/\epsilon < (V_D/\epsilon)_c$. In this case, ξ_M can increase very rapidly but continuously over a narrow temperature interval. Ultimately, this behavior will lead to a nearly fully ordered conformation at a nonzero temperature. Finally, the third case is strong coupling, for which $V_D/\epsilon > (V_D/\epsilon)_c$. In this case, a transition temperature T_C can be defined precisely as the point where the coexisting small- and large-ξ_M structures are present with equal probability.

The numerical analysis (*21, 22*) can be summarized by an approximate scaling formula

$$T_C \propto V_D{}^{\theta}\epsilon^{(1-\theta)} \tag{1}$$

in which θ is ~0.75 ± 0.1. This result is in accord with qualitative expectations, because larger values of V_D or ϵ stabilize the ordered conformations. T_C increases as the backbone polarizability becomes a more rapidly increasing function of ξ_M. This trend reflects the enhanced relative stability of rodlike structures. Finally, the random substitutional disorder associated with atactic polymers has not been included explicitly, and hence, its implications for phase behavior remain an open question.

For the substituted polysilylenes, $(SiRR')_n$, the coupling constant can be varied systematically by changing the side groups (this change affects ϵ and V_D via the backbone polarizability) or the solvent (this change affects V_D via the London dispersion forces; ϵ is expected to be only weakly solvent dependent for nonpolar systems). Therefore, in principle, the three distinct phase behaviors predicted by the theory may be observed by judicious choice of polymer–solvent pairs.

Solvent Dependence. The solvent dependence of T_C for a particular polymer is explicitly predicted by the theory (22). If the solvent dependences of the free energy of defect formation and the electronic structure of the all-*trans* structure are assumed to be negligible, then

$$T_C \propto \left(1 + \frac{C_S}{C_{SC}}\right)^{\theta} \quad (2)$$

in which the solvent contribution is

$$C_S = \frac{3}{4\pi}\left(\frac{n^2 - 1}{n^2 + 2}\right)\left(\frac{I_P I_S}{I_P + I_S}\right) <R^{-6}>_S \quad (3)$$

and the side-group contribution is

$$C_{SC} = \rho_{SC}\alpha_{SC}\left(\frac{I_P I_{SC}}{I_P + I_{SC}}\right) <R^{-6}>_{SC} \quad (4)$$

In these equations, the subscripts P, S, and SC refer to the polymer, solvent, and side chain, respectively. I is the first-ionization potential, n is the solvent refractive index, $\rho_{SC}\alpha_{SC}$ is the side-chain polarizability density, and $<R^{-6}>$ is a geometrical factor related to the manner in which dispersion interaction varies with the spatial separation between the polymer backbone and the surrounding medium (22).

Side-Chain Shielding of Solvent Dependence. Equation 2 has two important limiting cases. If the side chains are very small and do not appreciably shield the polymer backbone from the solvent, then equation 2 reduces to

$$T_C \propto C_S^{\theta} \quad \text{(unscreened)} \quad (5)$$

On the other hand, with large substituents, the solvent contributes perturbatively, and equation 2 takes the form

$$T_C \propto 1 + \theta C_S C_{SC} \quad \text{(highly screened)} \quad (6)$$

in which the second term is small compared with unity. In all cases, T_C increases monotonically with C_S, but because the side chains shield the backbone electrons from the solvent, the magnitude of the solvent dependence of T_C is predicted to be larger for polysilylenes with small substituents.

Dependence on Substituent Structure. The molecular nature of the side groups has two effects: (1) it influences and, presumably, largely determines the mean free energy of defect formation (ϵ) via steric effects on the enthalpy associated with defect creation and possible entropic contributions reflecting side-chain conformational changes induced by backbone isomeric-state changes and (2) it influences the extent of delocalization and, hence, polarizability along the backbone in the rod conformation (stereorandomness is especially potentially important in this case, because it introduces electronic disorder and, hence, inhibits delocalization via fluctuations in 1–4 hyperconjugative interactions or in silicon valence angles and bond lengths). These considerations are incorporated in the following theoretical expression for T_C

$$T_C \propto (\Delta\alpha_{rod})^{\theta}\epsilon^{(1-\theta)}(C_S + C_{SC})^{\theta} \qquad (7)$$

in which $\Delta\alpha_{rod}$ is the σ-conjugation component of the backbone polarizability per silicon. Standard quantum mechanical connections between polarizability and electronic structure yield the following result (*28, 29*)

$$\Delta\alpha_{rod} \propto \lambda_{max}{}^{4} \qquad (8)$$

in which λ_{max} can be identified approximately with the limiting wavelength of electronic absorption corresponding to the fully ordered polymer. Therefore, the dependence of T_C on the nature of the polymer molecule in a fixed solvent is given by equations 7 and 8. For alkane-substituted polysilylenes in simple alkane solvents, $C_S \sim C_{SC}$, and hence, the last factor in equation 7 is constant for different side-chain choices. However, in nonalkane solvents (e.g., toluene), this relation is not true, and the various factors in equation 7 can act in opposite directions when the size of the side groups is varied.

Breadth of Phase Transition. The abruptness or width of the coexistence region of the order–disorder transition has also been estimated theoretically. For a monodisperse solution of a relatively high molecular weight polymer, an intrinsic coexistence region, ΔT_C, exists because of standard finite-size fluctuation effects. Numerical calculations yield the result (*22*)

$$\frac{\Delta T_C}{T_C} \simeq \frac{10A}{N} \qquad (9)$$

in which N is the number of main-chain atoms; A is a numerical prefactor that is approximately 2 and is a weakly decreasing function of V_D/ϵ. Polydispersity effects are essentially negligible for high-molecular-weight systems, but on general theoretical grounds, stereorandomness is expected to broaden the transition (*22*).

Dependence on Molecular Weight. For a fixed V_D/ϵ and $N > 10^2$, T_C is essentially independent of N (*22*) (within a few tenths of a percent) because of a compensation of energetic and entropic factors. However, the electronic structure of conjugated polymers (e.g., limiting band gap) is sensitive to N at sufficiently low molecular weights (*3, 4, 34, 37, 38*). This sensitivity directly influences the polarizability of the all-*trans* polymer and, hence, explicitly modifies the transition temperature as follows:

$$T_C \propto (\lambda_{max})^{4\theta} \tag{10}$$

Finally, the theory makes explicit predictions (*22*) for the dependence of the transition temperature on external pressure.

Experimental Results and Comparison with Theory

Electronic Absorption Spectra. The limiting wavelength of UV absorption in dilute solution (λ_{max}) can be determined experimentally by judicious choice of solvent and cooling below the order–disorder transition temperature. Experimental details have been reported elsewhere (*12, 13, 15, 38, 43*). Results for λ_{max} of several high-molecular-weight dialkyl-substituted and atactic alkyl- and methyl-substituted polysilylenes are listed in Table I, along with the corresponding room-temperature values. The maxima for both the room-temperature wavelength of peak absorption and the limiting wavelength (when observed) were at most weakly solvent dependent; typically, the variations were only a couple of nanometers. The shorter λ_{max} for unsymmetrically substituted polysilylenes and the continual decrease of λ_{max} with substituent size asymmetry are consistent with random, stereochemically induced residual electronic disorder in the low-temperature conformation (Schweizer, K. S., unpublished results).

Table I. Spectroscopic Data and Theoretically Determined Defect Parameters

Polymer	λ_{RT} *(nm)*[a]	λ_{max} *(nm)*[b]	ϵ *(kcal/mol)*[c]	X_D[d]
Poly(di-*n*-butylsilylene)	314	355	0.98	0.18
Poly(di-*n*-pentylsilylene)	315	355		
Poly(di-*n*-hexylsilylene)	316	355	1.08	0.162
Poly(di-*n*-heptylsilylene)	318	360		
Poly(*n*-propylmethylsilylene)	310	339	1.33	0.128
Poly(*n*-hexylmethylsilylene)	307	335	1.31	0.131
Poly(*n*-dodecylmethylsilylene)	311	328	1.63	0.091
Poly(*n*-propylmethylsilylene-*co*-isopropylmethylsilylene)	311	329		

[a]Data are wavelengths of peak absorption at room temperature in dilute *n*-hexane (dialkyl-substituted polymers) and toluene (atactic polymers).
[b]Data are limiting wavelengths at low temperature.
[c]Values are free energies of defect formation from theoretical calculations and fittings.
[d]Values are defect densities at room temperature from theoretical calculations and fittings.

A fully microscopic interpretation of the temperature dependence of the absorption maximum, even well above any order–disorder transition temperature, is a formidable task because of the potential importance of many complicated physical factors (*27–30*). As a first attack on this problem, we have adopted a simple mean-field (or effective-medium) approach (*28–30*) with the assumption that the absorption peak (ω) is linearly perturbed from its limiting all-*trans* value (ω_{rod}) by the presence of bond rotational defects (free energy of formation, ϵ)

$$\omega = \omega_{rod} + \Delta\omega X_D \tag{11}$$

in which X_D is a temperature- and polymer-dependent defect density or concentration calculated from a simple three-state RIS theory, and $\Delta\omega$ is an adjustable constant indicative of the strength of the electronic–conformational coupling. For high-molecular-weight polymers, X_D is equal to the inverse of the mean conjugation length, ξ_M. By fitting the experimental absorption peak data as a function of temperature (above any order–disorder regime) in dilute *n*-hexane solution, the two positive parameters $\Delta\omega$ and ϵ can be extracted (Table I). Our values of ϵ and X_D at room temperature for poly(di-*n*-hexylpolysilylene) are in good qualitative agreement with those obtained from molecular mechanics calculations (*44*) and light-scattering measurements (*45*).

Order–Disorder Transitions. ***General Features.*** Experimental data are summarized in Table II, and representative thermochromic behaviors are shown in Figure 2. For the dialkyl-substituted polysilylenes the transition is very sharp, with a barely discernible coexistence region and an approximate isosbestic point. On the other hand, the asymmetrically substituted polymers, except poly(*n*-dodecylmethylsilylene), display very smooth behavior only in *n*-hexane solution and a broad but clearly discernible transition in dilute toluene solution. The transition width (ΔT_C) in toluene solution was taken to be the interval between departure from the extrapolated, smooth, high-temperature behavior and the onset of peak absorption wavelength saturation at low temperature. The transition temperature (T_C) is defined arbitrarily as the midpoint of this region.

The experimental results in Figure 2 and Table II clearly show three qualitatively different behaviors: an abrupt order–disorder transition; a relatively rapid continuous transition; and a gradual, smooth ordering of the polymer backbone. These observations are qualitatively identical to the three possible phase behaviors predicted by the theory. Moreover, a degree of quantitative understanding can be obtained.

As mentioned previously, the theory predicts that the qualitative form of the phase behavior is controlled by the magnitude of the coupling constant V_D/ϵ. With an inert solvent like *n*-hexane and by using equations 7 and 8, the polymer dependence of the coupling constant is given by

Table II. Order–Disorder Transition Temperatures and Widths of High-Molecular-Weight Polysilylenes

Polymer	*Solvent*	T_C (°C)[a]	ΔT_C (°C)
Poly(di-*n*-butylsilylene)	*n*-Hexane	−36	1–2
Poly(di-*n*-pentylsilylene)	*n*-Hexane	−39	2
	Toluene	−23 ± 2	1–2
	Tetralin	−13 ± 2	1–2
Poly(di-*n*-hexylsilylene)	*n*-Hexane	−30	1
	n-Heptane	−30	1
	Tetrahydrofuran	−30	1
	Methylene chloride	−23 ± 2	1
	Toluene	−20	1
	Tetralin	−16	1
Poly(di-*n*-hexylsilylene)[b]	*n*-Hexane	−44 ± 2	4–6
Poly(di-*n*-heptylsilylene)	*n*-Hexane	−25	1
	Toluene	−19	1
Poly(*n*-propylmethylsilylene)	*n*-Hexane	None	
	Toluene	−50	50
Poly(*n*-hexylmethylsilylene)	*n*-Hexane	None	
	Toluene	−65	40
Poly(*n*-dodecylmethylsilylene)	*n*-Hexane	−30	20
	Toluene	−25	8
Poly(*n*-propylmethylsilylene-*co*-*i*-propylmethylsilylene)	*n*-Hexane	None	
	Toluene	−70	50

[a]Experimental uncertainty is ± 1 °C for dialkyl-substituted polymers except where explicitly indicated.
[b]The low-molecular-weight polymer, with degree of polymerization (*N*) of 140, was used.

$V_D/\epsilon \propto \lambda_{max}{}^{4\theta}\epsilon^{-1}$. By using $\theta = 0.75$ and the values for λ_{max} and ϵ listed in Table I, estimates of V_D/ϵ relative to the corresponding value for poly(di-*n*-butylsilylene) were obtained: poly(di-*n*-hexylsilylene), 0.91; poly(*n*-propylmethylsilylene), 0.61; poly(*n*-hexylmethylsilylene), 0.59; and poly(*n*-dodecylmethylsilylene), 0.44. Therefore, although the absolute magnitude of V_D/ϵ cannot be calculated accurately, the theory is fully consistent with the observation of abrupt transitions (strong coupling) for the dialkyl-substituted polysilylenes and smeared or nonexistent transitions (intermediate or weak coupling) for the atactic polysilylenes.

Symmetrical Dialkyl-Substituted Polysilylenes. Because of their extremely sharp order–disorder transitions, the nonpolar, symmetrical dialkyl-substituted polysilylenes are almost ideal systems with which to test the predictions discussed earlier. The predicted solvent dependence of T_C was tested by performing a series of experiments with high-molecular-weight poly(di-*n*-hexylsilylene) in dilute solution. Experimental results for six solvents are listed in Table II, and the theoretically defined solvation coupling constants and solvent parameters are collected in Table III.

The one-to-one correspondence between the measured values of T_C and the calculated solvation coupling constants agrees with the theoretical pre-

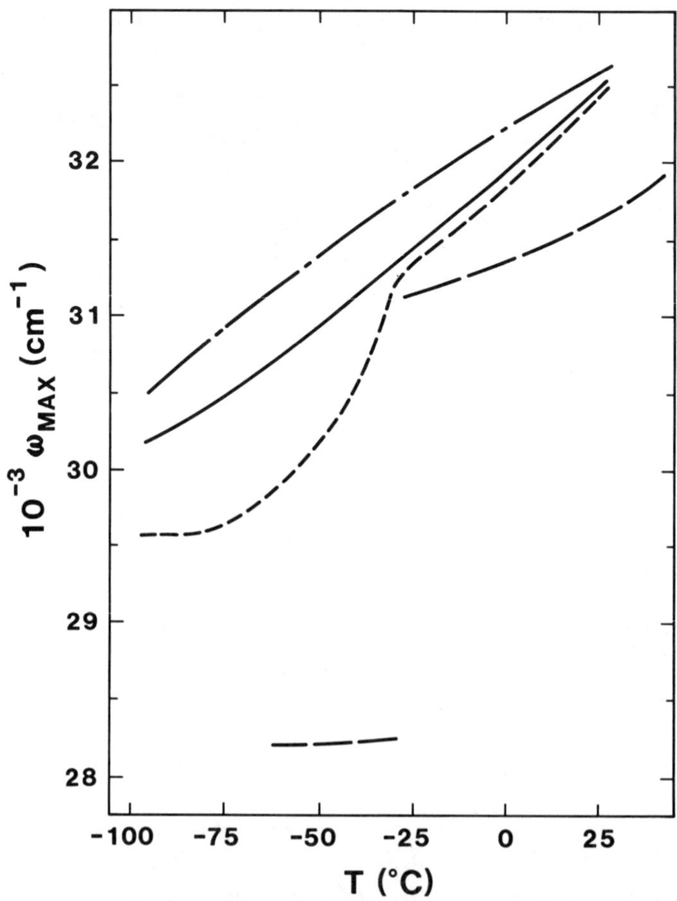

Figure 2. UV absorption maximum as a function of temperature. Key: — — —, poly(di-n-hexylsilylene) in n-hexane solution; — -, poly(n-hexylmethylsilylene) in n-hexane solution; —, poly(n-propylmethylsilylene) in n-hexane; – – –, poly(n-propylmethylsilylene) in toluene.

Table III. Room-Temperature Solvent Parameters and Solvation Coupling Constants

Solvent	n^{2a}	$I_S{}^b$ *(eV)*	$(n^2 - 1/n^2 + 2)I_{av}{}^c$
n-Hexane	1.89	10.18	0.98
n-Heptane	1.93	10.08	0.975
Tetrahydrofuran	1.98	9.42	0.99
Methylene chloride	2.02	11.35	1.10
Toluene	2.24	8.82	1.14
Tetralin	2.38	8.58	1.22

[a]Data are from reference 51.

[b]Data are from reference 52.

[c]Values are computed solvation coupling constants calculated with $I_P = 7$ eV (*34, 36*). I_{av} is $I_P I_S/(I_P + I_S)$.

diction of equation 2. The identical observed T_Cs for the nonpolar n-hexane and polar THF (tetrahydrofuran) solvents are consistent with the conformation-dependent solvation mechanism. The unimportance of the permanent (static) dipole moment of the solvent is a consequence both of the side-chain shielding effect and the highly delocalized nature of the backbone σ-conjugated electrons, which results in an averaging out of the static solvent polarization. Because the hexyl side chains are quite large, the heavily shielded limit of equation 6 is expected to be applicable, that is, the dependence of T_C on the solvation coupling constant is linear. A correlation plot (Figure 3) demonstrates the excellent agreement between experimental results and the theoretical prediction. The requirement that the experimental and theoretical slopes must be equal implies that the solvent contributes ~25% to the solvation energy of the backbone, with the hexyl side chains accounting for the remainder.

A more limited study of the solvent dependence of T_C was also performed with poly(di-n-pentylsilylene). Experimental results for three solvents are listed in Table II, and a correlation plot is presented in Figure 3. As was found for poly(di-n-hexylsilylene), an excellent agreement with the theoretically predicted linear dependence was obtained. The steeper slope for the di-n-pentyl polymer is consistent with theoretical expectations, because the pentyl group is smaller than a hexyl side chain and therefore provides less shielding of the polymer backbone from the solvent molecules.

The predicted increased solvent sensitivity of T_C with decreasing substituent size was tested further by comparing the transition temperatures of a series of di-n-alkyl polymers in n-hexane and toluene solutions. The results (Table II) are again in good qualitative agreement with the theoretical prediction. In particular, the difference between T_C in toluene and that in n-hexane solution increases monotonically with decreasing side-chain length: 6 °C, heptyl; 10 °C, hexyl; and 16 °C, pentyl.

The predicted intrinsic width of the order–disorder transition of a monodisperse, finite-molecular-weight polymer solution was also tested. The average molecular weights of dialkyl-substituted polysilylenes are in the order of 6×10^5, which implies that N is ~3000–5000 silicon atoms. With equation 9, the theory predicts that $\Delta T_C/T_C$ is ~0.004–0.006, which for T_C = –30 °C corresponds to an intrinsic width of roughly 1 or 2 °C. This result is in good agreement with the experimental observations summarized in Table II.

Finally, the predicted dependence of T_C on mean molecular weight (equation 10) was tested with poly(di-n-hexylsilylene) in n-hexane solution. With θ = 0.75, a limiting λ_{max} of 355 nm for the high-molecular-weight polymer, and the observed λ_{max} of ~346 nm for the low-molecular-weight material listed in Table II, the following is predicted: T_C (high molecular weight) – T_C (low molecular weight) = 18 °C. The experimentally observed lowering of T_C is ~14 ± 2 °C (Table II), which is in good agreement with the theoretical estimate.

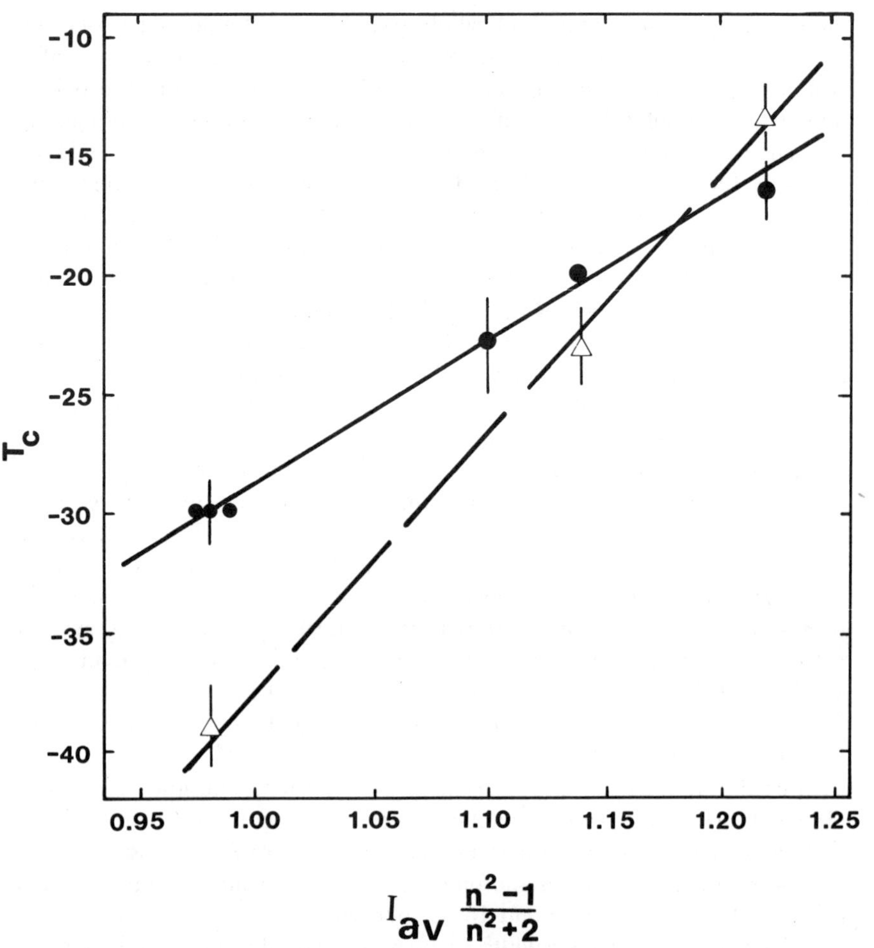

Figure 3. Correlation plots of experimental transition temperatures in different solvents versus the theoretically defined solvation coupling constant. Key: △, poly(di-n-pentylsilylene); ●, poly(di-n-hexylsilylene). The solid lines are straight-line fits. Typical experimental uncertainties are indicated by vertical error bars.

Unsymmetrical Alkyl-Substituted Polysilylenes. A detailed comparison of theoretical predictions and experimental results for the atactic polysilylenes is more difficult for several reasons: (1) the observed transitions are much broader, (2) the effects of random substitutional disorder are not included in the theory, and (3) the magnitudes of the consequences of stereochemical disorder are expected to vary for different atactic polymers. Nevertheless, for all the asymmetrically substituted polysilylenes studied, except poly(*n*-dodecylmethylsilylene), the predictions discussed earlier about solvent dependence, side-chain shielding, and dependence on substituent structure are clearly obeyed qualitatively.

The higher transition temperature of poly(*n*-propylmethylsilylene) in toluene compared with that of poly(*n*-hexylmethylsilylene) is especially significant, because these two polysilylenes have very similar free energies of defect formation (Table I), and poly(*n*-propylmethylsilylene) is characterized by a larger λ_{max} and less side-chain screening of the backbone from toluene.

Discussion

In summary, we believe that the order–disorder transition phenomena observed in dilute polysilylene solutions represent true thermodynamic, single-chain processes and not many-chain aggregation or microcrystallization. In view of the complexity of the systems and the simplicity of the theory, the degree of agreement between the experimental observations and the detailed theoretical predictions is remarkable and difficult to dismiss or explain by other models. The theoretical ideas have been demonstrated to possess considerable predictive ability and, at a minimum, point to the importance of conformation-dependent polymer–solvent interactions. To make further progress, additional careful spectroscopic, light-scattering, and magnetic-resonance experiments on a wide variety of polymer–solvent systems are desirable, along with parallel work on the soluble polygermanes (*46, 49*), polydiacetylenes, and polythiophenes. High-pressure dilute-solution experiments would be particularly useful.

Another class of soluble polysilylenes exhibits essentially no or very weak thermochromism. This class includes poly(cyclohexylmethyl- (*15, 38*), poly(phenylmethyl- (*15, 38*), (polytrimethylsilylmethyl- (*15*), and poly(diarylsilylenes) (*46*), all of which appear to be conformationally locked over a wide range of temperatures. In terms of our theoretical perspective, this behavior would arise from the steric effects of bulky substituents, which imply a large value of ϵ and, hence, a small coupling constant V_D/ϵ. For aryl-substituted polysilylenes, the proximity of an aromatic group to the backbone could also stabilize a highly ordered rodlike conformation via enhanced dispersion interactions.

Finally, solid films of some polysilylenes exhibit thermochromism and undergo true thermodynamic order–disorder phase transitions at much higher temperatures (*18, 19, 47, 48*) than in solution (typically, $T_C \sim$ 40–80 °C). In the context of the theory, a larger refractive index of the neat solid compared with that of the dilute solution results in a higher predicted T_C (*21*). However, we do not believe that the observed high T_Cs in films can be explained solely by this effect. Previous explanations have been made exclusively in terms of side-chain crystallization (*18, 19, 48*). Packing effects should be more important in the solid state, but both intramolecular and intermolecular packing effects must be carefully considered. Indeed, the fact that poly(di-*n*-butylsilylene) (*43*) and poly(di-*n*-pentylsilylene) (*50*) solids crystallize as helices and not in the all-*trans* rod form implies that solid-state phenomena are very complex and also that solution phase behavior need not be controlled by the same physical factors as those for solid films. Finally, the occurrence of thermochromic phase transitions in amorphous atactic polysilylene films (*15*) emphasizes the need to consider seriously the role played by conformation-dependent dispersion interactions in the solid state, as well as in solution.

Acknowledgment

This work was supported by the U.S. Department of Energy under contract number DE–AC04–76DP00789.

References

1. For a recent review, *see* Schoot, M.; Wegner, G. In *Nonlinear Optical Properties of Organic Molecules and Crystals;* AT&T: New York, 1987; Vol. 2, Chapter 3, p 3.
2. *Polydiacetylenes;* Bloor, D.; Chance, R. R., Eds.; NATO ASI Series E-102; Nijhoff: The Hague, Netherlands, 1985.
3. For a recent review, *see* West, R. *J. Organomet. Chem.* **1986,** *300*, 327.
4. Zeigler, J. M. *Synth. Met.* **1989,** *28*, 581.
5. Patel, G. N.; Chance, R. R.; Witt, J. D. *J. Chem. Phys.* **1979,** *70*, 4387.
6. Patel, G. N.; Witt, J. D.; Khanna, Y. P. *J. Polym. Sci., Polym. Phys. Ed.* **1980,** *18*, 1383.
7. Lim, K. C.; Fincher, C. R.; Heeger, A. J. *Phys. Rev. Lett.* **1983,** *50*, 1934.
8. Lim, K. C.; Heeger, A. J. *J. Chem. Phys.* **1985,** *82*, 522.
9. Plachetta, C.; Rau, N. O.; Schulz, R. C. *Mol. Cryst. Liq. Cryst.* **1983,** *96*, 141.
10. Rughooputh, S. D. D. V.; Phillips, D.; Bloor, D.; Ando, D. *Chem. Phys. Lett.* **1984,** *106*, 247.
11. Wenz, G.; Muller, M. A.; Schmidt, M.; Wegner, G. *Macromolecules* **1984,** *17*, 837.
12. Harrah, L. A.; Zeigler, J. M. *J. Polym. Sci., Polym. Lett. Ed.* **1985,** *23*, 209.
13. Harrah, L. A.; Zeigler, J. M. *Bull. Am. Phys. Soc.* **1985,** *30*, 540.
14. Trefonas, P.; Damewood, J. R.; West, R.; Miller, R. D. *Organometallics* **1985,** *4*, 1318.
15. Harrah, L. A.; Zeigler, J. M. In *Photophysics of Polymers;* Hoyle, C. E.; Torkelson, J. M., Eds.; ACS Symposium Series 358; American Chemical Society: Washington, DC, 1987; p 482.
16. Rughooputh, S. D. D. V.; Hotta, S.; Heeger, A. J.; Wudl, F. *J. Polym. Sci., Polym. Phys. Ed.* **1987,** *25*, 1071.
17. Berlinsky, A. J.; Wudl, F.; Lim, K. C.; Fincher, C. R.; Heeger, A. J. *J. Polym. Sci., Polym. Phys. Ed.* **1984,** *22*, 847.
18. Miller, R. D.; Hofer, D.; Rabolt, J. F.; Fickes, G. N. *J. Am. Chem. Soc.* **1985,** *107*, 2172.
19. Rabolt, J. F.; Hofer, D.; Miller, R. D.; Fickes, G. N. *Macromolecules* **1986,** *19*, 611.
20. Schweizer, K. S. *Chem. Phys. Lett.* **1986,** *125*, 118.
21. Schweizer, K. S. *J. Chem. Phys.* **1986,** *85*, 1156.
22. Schweizer, K. S. *J. Chem. Phys.* **1986,** *85*, 1176.

23. Schweizer, K. S. *Polym. Prepr. (Am. Chem. Soc., Div. Polym. Chem.)* **1986,** 27(2), 354.
24. Schweizer, K. S. *Synth. Met.* **1989,** *28,* 565.
25. Inganas, O.; Salaneck, W. R.; Osterholm, J. E.; Laakso, J. *Synth. Met.* **1988,** *22,* 395.
26. Hennecke, M.; Dull, I.; Fuhrmann, J. *Ber. Bunsenges. Phys. Chem.* **1988,** *92,* 596.
27. Schweizer, K. S. *J. Chem. Phys.* **1986,** *85,* 4181.
28. Soos, Z. G.; Schweizer, K. S. *Chem. Phys. Lett.* **1987,** *139,* 196.
29. Soos, Z. G.; Schweizer, K. S. In *Nonlinear Optical and Electroactive Polymers;* Prasad, P.; Ulrich, D., Eds.; Plenum: New York, 1988; p 331.
30. Schweizer, K. S. *Bull. Am. Phys. Soc.* **1987,** *32*(3), 885.
31. Flory, P. J. *Statistical Mechanics of Chain Molecules;* Interscience: New York, 1969.
32. Soos, Z. G.; Ramasesha, S. *Phys. Rev. B* **1984,** *29,* 5410.
33. Dewar, M. J. S. *J. Am. Chem. Soc.* **1984,** *106,* 669.
34. Pitt, C. J. In *Homatomic Rings, Chains and Macromolecules of Main Group Elements;* Rheingold, A. L., Ed.; Elsevier: Amsterdam, 1977.
35. Klingensmith, K. A.; Downing, J. W.; Miller, R. D.; Michl, J. *J. Am. Chem. Soc.* **1986,** *108,* 7438.
36. Takeda, K.; Teramae, H.; Matsumoto, N. *J. Am. Chem. Soc.* **1986,** *108,* 8186.
37. Trefonas, P.; West, R.; Miller, R. D.; Hofer, D. *J. Polym. Sci., Polym. Lett. Ed.* **1983,** *21,* 823.
38. Harrah, L. A.; Zeigler, J. M. *Macromolecules* **1987,** *20,* 601.
39. Michl, J.; Downing, J. W.; Karatsu, T.; Klingensmith, K. A.; Wallraff, G. A.; Miller, R. D. In *Inorganic and Organometallic Polymers;* Zeldin, M.; Wynne, K. J.; Allcock, H. R., Eds.; ACS Symposium Series 360; American Chemical Society: Washington, DC, 1988; p 61.
40. Naylor, P.; Whiting, M. C. *J. Chem. Soc.* **1955,** 3042.
41. Bigelow, R. W.; McGrane, K. M. *J. Polym. Sci., Polym. Phys. Ed.* **1986,** *24,* 1233.
42. Takeda, K.; Matsumoto, N.; Fukuchi, M. *Phys. Rev. B* **1984,** *30,* 5871.
43. Schilling, F. C.; Lovinger, A. J.; Zeigler, J. M.; Davis, D. D.; Bovey, F. A. *Macromolecules* **1989,** *22,* 3055.
44. Damewood, J. R. *Macromolecules* **1985,** *18,* 1793.
45. Cotts, P. M.; Miller, R. D.; Trefonas, P.; West, R.; Fickes, G. N. *Macromolecules* **1987,** *20,* 1046.
46. Miller, R. D.; Sooriyakumaran, R. *J. Polym. Sci., Polym. Lett. Ed.* **1987,** *25,* 321.
47. Gobbi, G. C.; Fleming, W. W.; Sooriyakumaran, R.; Miller, R. D. *J. Am. Chem. Soc.* **1986,** *108,* 5624.
48. Miller, R. D.; Rabolt, J. F.; Sooriyakumaran, R.; Fickes, G. N.; Farmer, B. L.; Kuzmany, H. *Polym. Prepr. (Am. Chem. Soc., Div. Polym. Chem.)* **1987,** *28*(1), 422.
49. Trefonas, P.; West, R. *J. Polym. Sci., Polym. Chem. Ed.* **1985,** *23,* 2099.
50. Miller, R. D.; Farmer, B. L.; Fleming, W. W.; Sooriyakumaran, R.; Rabolt, J. F. *J. Am. Chem. Soc.* **1987,** *109,* 2509.
51. *Handbook of Chemistry and Physics,* 51st ed.; Weast, R. C., Ed.; CRC: Boca Raton, FL, 1971.
52. Watanabe, K.; Nakayama, T.; Mottl, J. *J. Quant. Spectrosc. Radiat. Transfer* **1962,** *2,* 369.

RECEIVED for review May 27, 1988. ACCEPTED revised manuscript February 2, 1989.

23

Configurational Properties of a Stiff-Chain Diaryl-Substituted Polysilane in Dilute Solution

Patricia M. Cotts, Robert D. Miller, and Ratnasbapa Sooriyakumaran

Almaden Research Center, IBM, 650 Harry Road, San Jose, CA 95120–6099

The properties in dilute solution of a diaryl-substituted polysilane indicate that the chain exhibits substantial stiffness in solution, with a persistence length of approximately 100 Å. To estimate the persistence length, measurements of the mean-square radius of gyration, intrinsic viscosity, and hydrodynamic radius at infinite dilution for a series of molecular weights were fitted to theoretical expressions for the wormlike chain. The large persistence length is consistent with the long wavelength (395 ± 10 nm) for maximum electronic absorption observed for the diaryl-substituted polysilanes. Results are compared with those reported previously for dialkyl-substituted polysilanes.

THE EQUILIBRIUM FLEXIBILITY of a polymer chain backbone, that is, the ability of the backbone bonds to rotate, has always been of great interest to polymer scientists. Silicon-containing polymers present opportunities for the investigation of configurational properties that are not available with more-common carbon-backbone polymers.

A completely unrestricted chain consisting of n bonds of length l, with all bond angles equally probable, has a mean-square end-to-end distance $<r^2>$ given by

$$<r^2> = nl^2 \tag{1a}$$

Real polymer chains of fixed bond angles and restricted rotation have

0065–2393/90/0224–0397$06.00/0

larger values of $<r^2>$, which may be expressed in a similar form

$$<r^2> = n_k l_k^2 \tag{1b}$$

in which n_k and l_k are no longer the actual number of bonds and bond length, respectively, but are the number and length of larger imaginary bonds that yield the correct value of $<r^2>$ when the polymer backbone is treated as an unrestricted chain (*1*). The additional criterion necessary to define the length of the Kuhn statistical segment l_k and n_k is that

$$L = n_k l_k \tag{2}$$

in which L is the contour length or the length of the fully extended chain.

Adherence to equation 1b may be regarded as a definition of flexibility. Thus, chain molecules may be flexible despite considerable energy differences among rotational states of adjacent bonds. Larger hindrance to rotation necessitates larger values of l_k, and thus the ratio l_k/l may be used as a measure of chain flexibility. For example, for typical synthetic organic polymers with carbon backbones, $l_k/l \simeq 10$. On the other hand, larger values of l in the real chain can result in higher flexibility (l_k/l is small), as is observed for poly(dimethylsiloxane) (PDMS). The longer bond reduces steric hindrance to rotation. For highly hindered chains, as l_k becomes very large, the number of statistical segments n_k becomes too small for equation 1b to be valid for molecular weights of interest, and these chains are regarded as stiff. With any flexibility in the chain, equation 1b is valid at sufficiently high molecular weights, but these molecular weights are often beyond those obtainable for stiff synthetic polymers.

Polymers exhibiting substantial stiffness are often treated as a Kratky–Porod wormlike chain (*2*) or a variation thereof. The degree of stiffness is represented by the persistence length q, which may be defined as the projection of the end-to-end vector **r** in the direction of the first bond. The wormlike chain encompasses the limits of the flexible Kuhn chain discussed earlier, with $q = l_k/2$, and the completely rigid rodlike polymer, with $q = \infty$. Actual polymer chains with a persistence length equal to or greater than their fully extended end-to-end length (contour length) L will approach the behavior of rodlike chains, but again at sufficiently high molecular weights (M), they may be treated as flexible chains.

The models just discussed apply to chains in an ideal or θ solvent, in which excluded volume interactions are absent. In practice, solution properties are more often determined in good solvents in which long-range excluded volume interactions lead to an increase in $<r^2>$. These long-range interactions increase with M, so that equation 1b is no longer valid. Excluded volume interactions and chain stiffness can be difficult to distinguish without measurements in a θ solvent.

The long bonds in silicon-containing polymers can result in greater flexibility, relative to carbon-backbone polymers, as observed for PDMS. However, highly substituted all-silicon-backbone polymers (polysilanes) may be stiff despite the long Si–Si bond (2.35 Å) because of large steric hindrance caused by two bulky groups on each backbone atom.

Previously, we investigated the properties in dilute solution of dialkyl-substituted polysilanes, including poly(di-*n*-hexylsilane) (PDNHS), in good solvents (3). The global polymer chain dimensions in solution were measured by using both thermodynamic and hydrodynamic techniques. The ratio l_k/l was ~20 for this polymer; this value is twice as large as values obtained for most carbon-backbone polymers. However, because the molecular weights are large, the polymers are still flexible and are expected to obey equation 1b in the absence of long-range excluded volume interactions.

In this chapter, we report measurements on a diaryl-substituted polysilane, poly[bis(*p*-*n*-butyl)phenylsilane] (PBPNBPS), in which the bulky phenyl groups may provide more substantial steric hindrance to rotational freedom than do the dialkyl substituents. Reported values of the wavelength of maximum electronic absorption are large (395 ± 10 nm) and cannot be attributed solely to electronic substituent effects (*4*).

Experimental Procedures

Intrinsic Viscosity. Dilute-solution viscometry of samples in toluene was carried out in a Cannon–Ubbelohde semimicrodilution viscometer (size 25) in a temperature-controlled bath (25.0 ± 0.2 °C). At least three concentrations were measured, and the results were extrapolated to infinite dilution by using the Huggins and Kramers relations

$$[\eta] = \frac{\eta_{sp}}{c} + k'[\eta]^2 c + \ldots \qquad (3a)$$

$$[\eta] = \frac{\ln \eta_{rel}}{c} + \left(k' - \frac{1}{2}\right)[\eta]^2 c + \ldots \qquad (3b)$$

in which $[\eta]$ is intrinsic viscosity, k' is the Huggins coefficient, $\eta_{rel} = t_{soln}/t_{solv}$, and $\eta_{sp} = \eta_{rel} - 1$. t_{soln} and t_{solv} are the efflux times of the solution and solvent, respectively. Values of η_{rel} varied from 1.2 to 1.6, and no correction for kinetic energy was applied for $t_{solv} \geq 100$ s.

Static Light Scattering. The weight-average molecular weight (M_w), the thermodynamic second virial coefficient (A_2), and the *z*-averaged mean-square radius of gyration ($R_{G,z}^2$) were determined with a Chromatix KMX-6 low-angle light-scattering photometer and a Brookhaven BI200SM photogoniometer. Values of M_w and A_2 were obtained with the more precise KMX-6 photometer from measurements at a scattering angle of 4° at several concentrations c

$$\frac{Kc}{R_o} = \frac{1}{M_w} + 2A_2 c + \ldots \qquad (4)$$

in which the Rayleigh factor R_θ for a scattering angle of 4° was substituted for $R_o = R_{\theta\to 0}$ without correction. The constant K is given by

$$K = \frac{4\pi^2 n^2 (dn/dc)^2}{N_A \lambda_o^4} \tag{5}$$

in which n is the refractive index of the solvent (1.497 for toluene), N_A is Avogadro's number, λ_o is the wavelength of the incident light (6328 Å), and dn/dc is the differential refractive index increment (0.131 mL/g). No corrections were made for anisotropy (ρ_v), because at a scattering angle of 0°, ρ_v was small ($\rho_v[0°] = 0.0044$ for sample 4–2).

The root-mean-square z-averaged radius of gyration $R_{G,z}$ was measured with both instruments. The BI200SM photogoniometer was used with software from Brookhaven Instruments to determine Kc/R_θ for 11 angles (15° $< \theta <$ 150 °), and these values were fitted to the equation

$$\frac{Kc}{R_{\theta,c}} = \frac{Kc}{R_{\theta=0,c}}\left(1 + \frac{q^2 R_{G,z}{}^2}{3}\right) \tag{6}$$

in which $q = (1/\lambda_o)(4\pi n) \sin(\theta/2)$.

Measurement of $R_{G,z}$ with the KMX-6 photometer using the dissymmetry method described previously (3) yielded values within 5% of those determined with the Brookhaven photogoniometer. For all measurements, $R_{G,z}$ was independent of concentration in the range measured. The angular dependence of $Kc/R_{\theta,c}$ of sample 64–2 is shown in Figure 1.

Dynamic Light Scattering. The hydrodynamic radius, R_h, is defined as the Stokes radius from the mutual diffusion coefficient at infinite dilution (D_o)

$$D_o = \frac{k_B T}{6\pi\eta_o R_h} \tag{7}$$

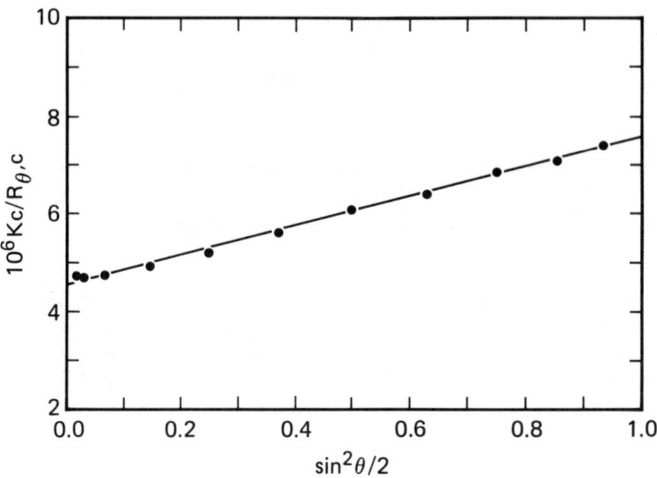

Figure 1. Angular dependence of Kc/R$_{\theta,c}$ *for sample 64–2, with* c = *1.58 mg/mL.*

in which η_0 is the solvent viscosity, k_B is the Boltzmann constant, and T is temperature.

The limiting diffusion coefficient was determined by extrapolation of D_c at four to five concentrations c to infinite dilution. D_c is defined by

$$D_c = \Gamma_c / q^2 \tag{8}$$

in which Γ_c is the first cumulant of the correlation function $C(t)$ at each concentration,

$$\ln \left[\frac{C(t)}{B} - 1 \right]^{1/2} = \ln b^{1/2} - \Gamma_c t + \frac{\mu_2 t^2}{2} + \ldots \tag{9}$$

in which B is the measured base line, b is an optical factor, t is time, and μ_2 is the second cumulant.

Software for cumulant analysis from Brookhaven Instruments was used to fit the measured correlation function $C(t)$ to equation 9 by using a weighted second-order polynomial nonlinear regression. The measured base line B of the correlation function $C(t)$ was determined from the average of four delay channels (1029–1032) multiplied by the sample time Δt. The calculated and measured base lines were within 0.1% for all runs used in the analysis.

The sample time Δt was chosen with the criterion

$$\Delta t = \frac{2}{m\Gamma_c} \tag{10}$$

in which m is the number of channels (m = 128). This criterion provides a larger number of points in the initial decay of the correlation function (for the determination of Γ_c) at the expense of less precision in the second cumulant μ_2. The normalized second cumulant, μ_2/Γ_c^2, was 0.3 ± 0.1, consistent with the polydispersity of these samples. A representative correlation function is shown in Figure 2.

Size Exclusion Chromatography. The molecular weight distribution of each sample was determined by size exclusion chromatography (SEC). A volume of 150 μL of each sample at a concentration of 1–2 mg/mL was injected onto a set of four 30-cm PLgel (cross-linked polystyrene) columns (with porosities of 10^6, 10^5, 10^4, and 10^3 Å) housed in a Waters 150C liquid chromatograph at 40 °C. The mobile phase was tetrahydrofuran, and the concentration detector was a differential refractometer.

The column set was calibrated with a series of 15–20 narrow-distribution polystyrene (PS) standards (Polymer Laboratories), and the data were fitted to a third-order polynomial

$$\log M = A + Bt + Ct^2 + Dt^3 \tag{11}$$

in which t is the peak elution time of a PS standard of molecular weight M. Calculation of molecular weight averages relative to PS was carried out with an IBM PCXT with software from Nelson Analytical. The chromatograms of all samples are shown in Figure 3.

Results

Molecular Weight Determination. Results obtained for all samples measured are listed in Table I. No unusual behavior was observed with any

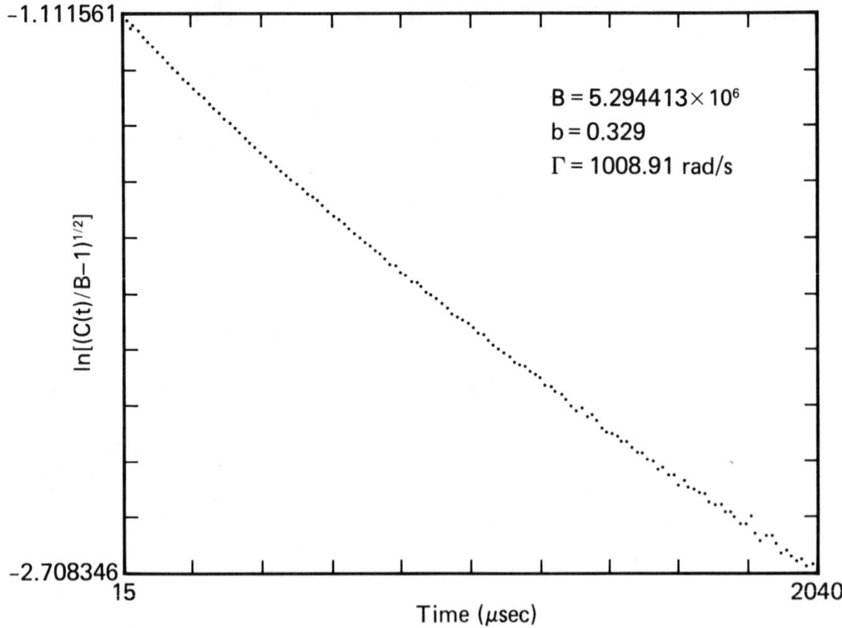

Figure 2. log of normalized correlation function C(t) *as a function of time for sample 64–2, with* c = *1.18 mg/mL.*

of the measurements, and interpretation of the data was straightforward. All reported results (other than those for SEC) were obtained in toluene, which was a thermodynamically good solvent for the polymer. However, as will be discussed later, the contribution of the excluded volume effect to the experimental parameters is expected to be small. Measurements of M_w by light scattering carried out in other solvents (tetrahydrofuran and hexane) were in agreement with results obtained in toluene. This result and the consistency of results from the variety of techniques used suggest that aggregation is not a problem for these solutions. Aggregation is often a contributing factor in solutions of stiff-chain polymers, particularly those that can crystallize.

The M_w values obtained from SEC (M_w^{SEC}) calibrated with PS standards agree quite well with M_w values from light scattering (M_w^{LS}). This unexpected result may be explained by Figure 4, which shows the $[\eta]$·M relations for PS, PBPNBPS, and PDNHS. At similar molecular weights, PS and PBPNBPS have very similar values of $[\eta]$, that is, their hydrodynamic volumes are similar. In comparison, the PDNHS curve (3) lies significantly below that for PS, so that values of M_w^{SEC} relative to PS are expected to be much smaller than the true M_w determined by light scattering. The increased stiffness of PBPNBPS relative to either PDNHS or PS may be seen in the increased slope of the curve relative to the other two in Figure 4.

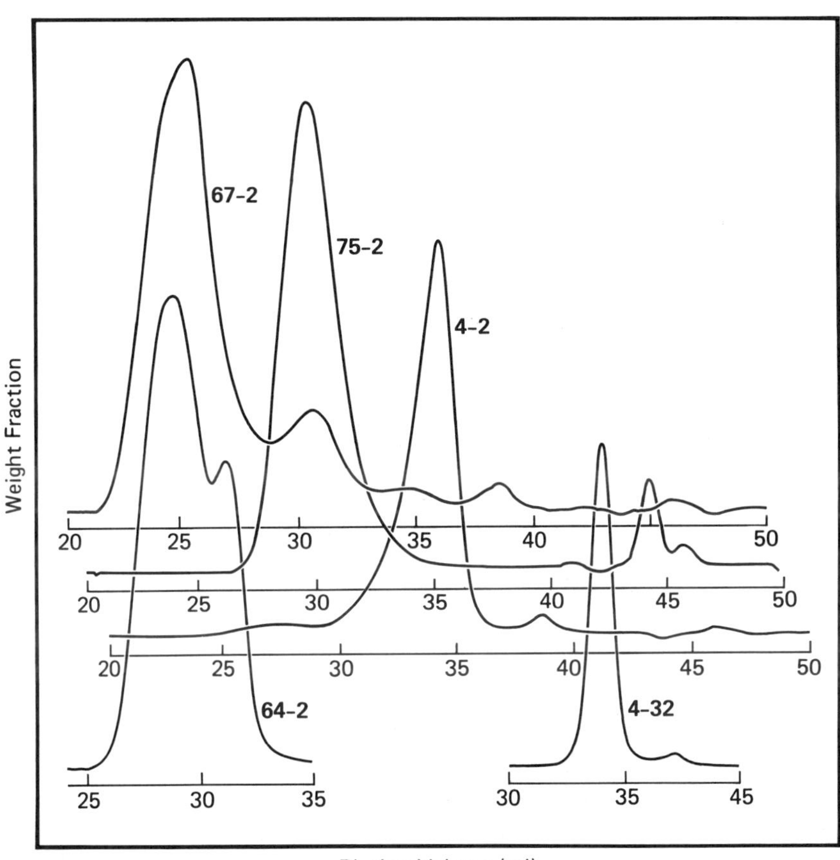

Figure 3. SEC chromatograms of samples studied.

Estimation of Persistence Length. ***Correction for Polydispersity Effects.*** Estimation of the persistence length from the experimental data by using the Kratky–Porod wormlike chain model requires correction for polydispersity effects. The unusual (and sometimes multimodal) distributions obtained for polysilanes are not amenable to an analytical form for the distribution. The samples used in this study were nearly monomodal by SEC, with only sample 4–2 containing a high-molecular-weight tail (Figure 3). For the purpose of this study, correction of the z-averaged radius of gyration ($R_{G,z}$) to a weight-averaged value ($R_{G,w}$), that is, the radius of gyration that would be measured for a monodispersed sample with $M = M_w$, may be approximated by

$$\frac{R_{G,z}}{R_{G,w}} \cong \left(\frac{M_z}{M_w}\right)^{0.7} \tag{12}$$

Table I. Summary of Experimental Data

Sample	M_w^{LS} (×10^3 g/mol)	A_2^{LS} (×10^{-4} mL • mol) g^z	$R_{G,z}^{LS}$(Å)	$R_{G,w}$ (Å)[a]	R_h (Å)	[η](mL/g)	M_w^{SEC} (×10^3 g/mol)[b]	M_z/M_w
4–32	13 ± 1	2.1 ± 0.5	—[c]	—	20 ± 2	—	9.5	1.2
4–2	49 ± 5	1.0 ± 0.4	265 ± 15	70 ± 20	50 ± 3	17.7 ± 0.5	41	10[c]
64–2	450 ± 50	3.3 ± 0.5	600 ± 20	335 ± 35	292 ± 18	169 ± 12	506	2.6
75–2	550 ± 80	—	600 ± 20	430 ± 30	268 ± 30	—	193	1.8
67–2	1800 ± 200	2.6 ± 0.5	1235	800 ± 80	510 ± 20	454 ± 20	1800	2.0

[a]Values were calculated from $R_{G,z}^{LS}$ as described in text.
[b]Values are relative to polystyrene.
[c]This sample showed a high-molecular-weight tail (*see* Figure 3).

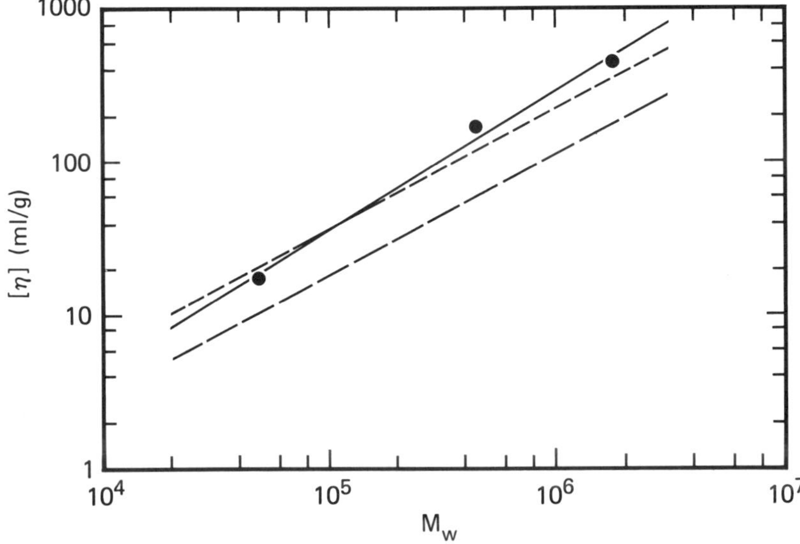

Figure 4. [η]·M relations for PS (short dashes), PDNHS (long dashes), and PBPNBPS (solid line).

with M_z/M_w determined by SEC, because R_G is approximately proportional to $M^{0.7}$ for these polymer samples. Equation 12 is correct for Gaussian coils with an exponent of 0.5 for any distribution. For coils expanded from Gaussian dimensions because of either or both long- and short-range interactions, the molecular weight distribution must be known to calculate the precise average. For the distributions observed in this study, the z average is expected to be within 15% of the correct average.

The experimental values of [η] and R_h were sufficiently close to weight-averaged quantities to be used without correction. The R_h value was obtained from the first cumulant of the autocorrelation function, which yields the z average of the diffusion coefficient. However, because the diffusion coefficient D is inversely proportional to R_h, the actual R_h value more closely reflects M_w for many synthetic polymers. Although these quantities (and the values of $R_{G,w}$ obtained previously) are not exactly equal to those that would be obtained for a monodispersed sample with $M = M_w$, further correction is within the uncertainty of the limited data.

Excluded Volume Interactions. The contribution of long-range excluded volume interactions to $<r^2>$ and R_G decreases as the polymer chain becomes stiffer. This tendency may be visualized as arising from the less frequent intramolecular segment–segment interactions in a stiff-chain polymer. A more quantitative assessment may be obtained through the interpenetration function ψ

$$\psi \equiv \frac{A_2 M^2}{4\pi^{3/2} N_A R_G^3} \tag{13}$$

or, similarly, the parameter $A_2M/[\eta]$. Both parameters vanish when excluded volume interactions are absent ($A_2 = 0$) and increase to an asymptotic value as the excluded volume parameter α increases. α is defined as

$$\alpha^2 \equiv \frac{<r^2>}{<r^2>_o} \tag{14}$$

in which $<r^2>$ is determined for the solvent and temperature of interest, and $<r^2>_o$ is determined under θ conditions, in which excluded volume effects are absent.

Inspection of the parameters ψ and $A_2M/[\eta]$ shows that each may also become small if R_G and $[\eta]$ are large, even if A_2 is larger than 0. For example, for very stiff chains, R_G and $[\eta]$ may be so large that no excluded volume effect is observed, even though A_2 may be large. For a small value of α, experimental values of ψ and $A_2M/[\eta]$ may be used to estimate α by using theoretical or empirical methods. For the PBPNBPS samples measured in this study, $\alpha \leq 1.2$, and the difference between the measured dimensions and those expected for θ conditions is at most 20%. This effect is expected only at the highest molecular weights studied. Because the values of R_G, $[\eta]$, and R_h for the highest highest molecular weight samples do not deviate from the curves in Figures 5–7, the effect appears small.

Experimental Assessment. The most direct experimental assessment of the persistence length may be obtained from the root-mean-square radius of gyration measured by light scattering. The analytical expression for the unperturbed R_G^2 of a wormlike chain has been given by Benoit and Doty (5):

$$R_G^2 = \frac{qL}{3} - q^2 + \frac{2q^3}{L}\left[1 - \left(\frac{q}{L}\right)(1 - e^{-L/q})\right] \tag{15}$$

The contour length L and molecular weight M are related by the mass per unit length, M_L, also referred to as the shift factor.

$$M_L = \frac{M}{L} \tag{16}$$

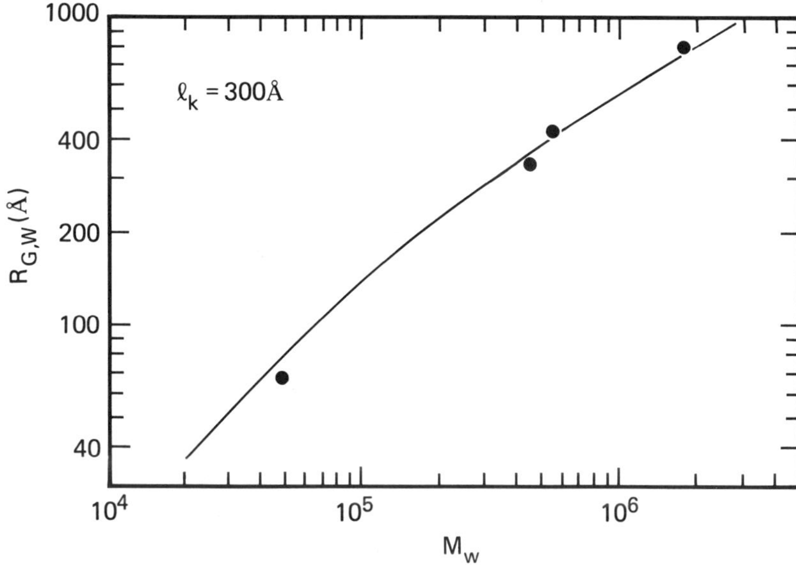

Figure 5. Comparison of $R_{G,w}$ *values (calculated from measured* $R_{G,z}^{LS}$*) with the theoretical values (solid line) calculated from equation 2, with* q = *150 Å and* M_L = *149/Å.*

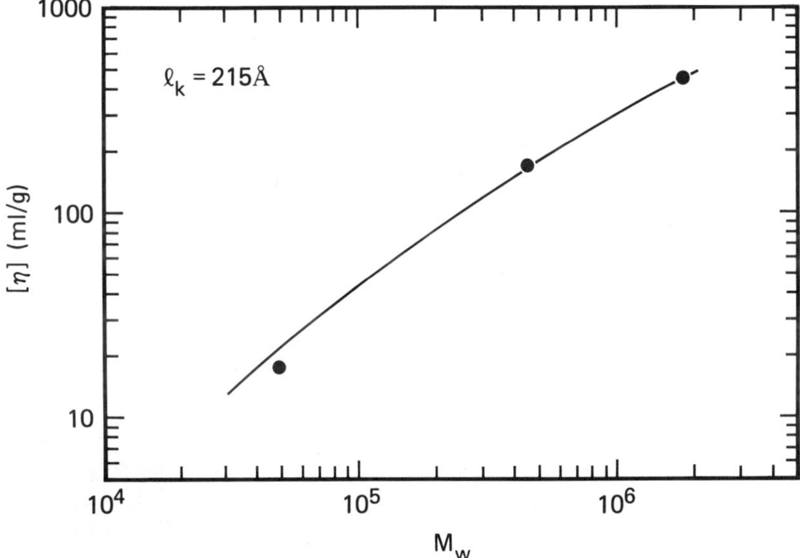

Figure 6. Comparison of measured values of [η] with theoretical values (solid line) calculated with q = *108 Å,* M_L = *149/Å, and* d = *20 Å.*

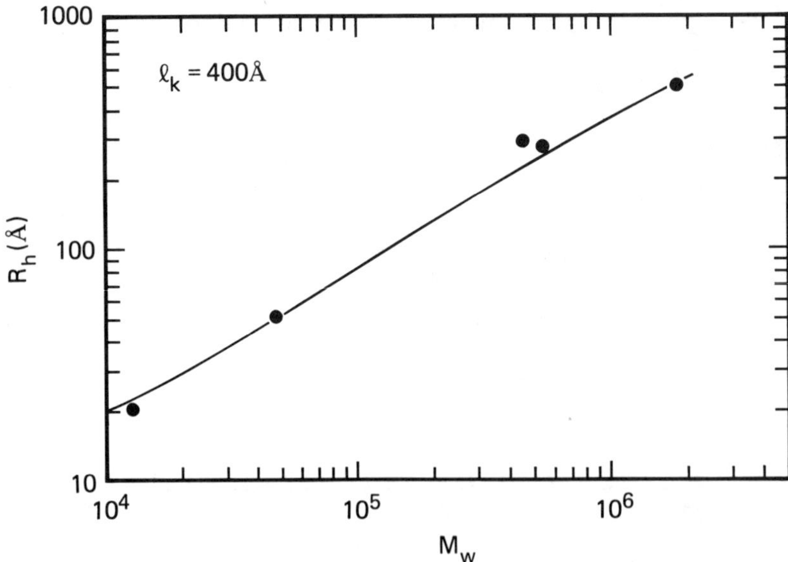

Figure 7. Comparison of experimental values of R_h *with theoretical values (solid line) calculated with* q = *200 Å*, M_L = *149/Å, and* d = *20 Å.*

Although M_L as used in the wormlike chain model may not be exactly equivalent to the mass per unit length of the fully stretched chain (2), we have used M_L = 149/Å and determined L as the sum of the monomer unit projections on the backbone for a given M. This approach leaves only the persistence length q to be determined. A comparison of the experimental values ($R_{G,w}$) with the theoretical values calculated from equation 15 with q = 150 Å (l_k = $2q$ = 300 Å) and M_L = 149/Å is shown in Figure 5. The wormlike chain model is a relatively good fit for the limited data. The theoretical line shows the curvature reflecting increased flexibility at higher M, which is also seen in the experimental data. This curvature would not be observed for a more flexible polymer in a good solvent, such as the di-*n*-hexyl-substituted polysilane reported previously.

A disadvantage in the use of R_G to estimate q is that R_G can be measured only for relatively high molecular weights with light scattering, and the lower molecular weight region, where stiffness is more apparent and excluded volume effects are minimized, is often inaccessible. The other disadvantage is that the R_G measured is a *z*-averaged quantity and is thus highly dependent on polydispersity, particularly on the high-molecular-weight portion of the distribution.

Estimation of q from hydrodynamic techniques such as intrinsic viscosity or diffusion requires an additional parameter, the hydrodynamic diameter or cross-section of the wormlike chain. This diameter d may be estimated from the partial specific volume $\bar{v}$ (*6*)

$$\bar{v} = \left(\frac{\pi N_A}{4}\right)\left(\frac{d^2}{M_L}\right) \tag{17}$$

or left as an adjustable parameter. By estimating the specific volume at 1.0 ± 0.1 mL/g on the basis of the measured film density, we obtained d = 18 ± 1 Å. We have used d = 20 Å in the hydrodynamic calculations that follow. The value of q is relatively insensitive to the value of d used, and d does not need to be known with precision for the estimation of q.

Expressions for the intrinsic viscosity $[\eta]$ and the translational friction coefficient Ξ have been given by Yamakawa and Fujii (*7, 8*). Ξ is simply related to the translational diffusion coefficient D measured by dynamic light scattering by the expression

$$D = \frac{k_B T}{\Xi} = \frac{k_B T}{6\pi\eta_o R_h} \tag{18}$$

Comparisons of the experimental values of $[\eta]$ and R_h with the theoretical values are shown in Figures 6 and 7. For the viscosity data, the best fit was obtained with q = 108 Å (l_k = 215 Å), whereas for R_h, q = 200 Å (l_k = 400 Å) yielded a better fit.

Discussion

For all three measurements, q is substantially larger than the 30 Å estimated for poly(di-*n*-hexylsilane) (*3*), a fact indicating substantially more hindrance to rotation in the diaryl-substituted polysilane than in the dialkyl-substituted one, as expected. Although a precise determination of the persistence length requires a series of narrow-distribution samples covering a wide range of molecular weights and a quantitative assessment of the excluded volume interaction, we can conclude that $q \geq 100$ Å for this polymer. Thus, the axial ratio l_k/d is expected to be ~10, and a stable ordered state in the bulk polymer is predicted (*9–11*).

Comparison with results reported previously for PDNHS and other polymers (*12–16*) is shown in Table II. Included in Table II are values of the characteristic ratio C_∞

$$C_\infty \equiv \frac{\langle r^2 \rangle_o}{nl^2} \tag{19}$$

which is often used as a stiffness parameter. The diaryl-substituted polysilane is more rigid than common carbon-backbone polymers but less rigid than

Table II. Comparison of Configurational Parameters

Polymer	q (Å)	l_k/l	C_∞
PDMS	4	5	6
PMMA[a]	7	9	7
PS	9	12	10
PDNHS[b]	30	20	20
PMDA–ODA PI[c]	30	1.7	2.9
PBPNBPS	100	85	70
PPTA[d]	290	88	88
PHIC[e]	420	600	350

NOTE: These values are approximate and are for purposes of comparison only. Estimates were obtained from references 9 and 10 unless indicated.
[a]PMMA is poly(methyl methacrylate).
[b]Data are from reference 3.
[c]Data are from reference 14.
[d]Data are from reference 15.
[e]Data are from reference 16.

some synthetic polymers such as poly(phenylene terephthalamide) (PPTA) or poly(hexylisocyanate) (PHIC), which are considered as rodlike chains, with persistence lengths of several hundred angstroms. A persistence length of ~100 Å is somewhat larger than those expected for high-temperature polyimides (e.g., PMDA–ODA PI), in which aromatic rings in the backbone form unusually long bonds connected by the ether oxygens or other groups about which nearly free rotation occurs.

In the diaryl-substituted polysilane, the global dimensions are approximated by a wormlike chain model with a persistence length of ~100 Å. Thus, for the highest molecular weight studied, $L/q \simeq 100$, and the chain dimensions are similar to those of a flexible-model Kuhn chain with bond lengths l_k of ~200 Å, that is,

$$R_G = \left(\frac{<r^2>_o}{6}\right)^{1/2} = \frac{n_k^{1/2} l_k}{\sqrt{6}} \tag{20}$$

or $R_G = 630$ Å.

The experimental value of $R_{G,w}$ (815 Å), is in good agreement with the calculated value, in view of the uncertainty caused by polydispersity and excluded volume contributions. In contrast, $L/q < 1$ for the lowest molecular weight sample, which may be treated as rodlike chains, that is,

$$R_G = \left(\frac{L^2}{12}\right)^{1/2} \tag{21}$$

or $R_G = 26$ Å.

Although this value of R_G is too small to be measured by light scattering, it is consistent with the experimental value of R_h (20 Å). These approximate calculations illustrate the strong dependence of polymer chain geometry on molecular weight, particularly for $q \sim 100$ Å. At low molecular weights, the chain may be viewed as nearly rodlike, whereas at high molecular weights, flexible chain behavior is observed.

Conclusion

The diaryl-substituted polysilanes provide a unique opportunity to investigate a wide range of backbone stiffness with the degree of stiffness accessible to experimental investigation. Unlike many other stiff-chain polymers in which the stiffness arises from long rigid groups in the backbone or from specific intramolecular interactions among backbone segments leading to a helical configuration, the stiffness in the diaryl-substituted polysilanes is due only to unusually large energy differences among rotational states and should thus be very sensitive to changes in temperature. This behavior is consistent with the large temperature dependence ($d \ln \langle r^2 \rangle / dT = -3.2 \times 10^{-3}$/K) observed for poly(di-*n*-hexylsilane) (*17*).

Acknowledgment

We appreciate the help of C. Weidner with the SEC measurements.

References

1. Kuhn, W. *Kolloid Z.* **1939,** *87*, 3.
2. Kratky, O.; Porod, G. *Rec. Trav. Chim. Pays-Bas* **1949,** *68*, 1106.
3. Cotts, P. M.; Miller, R. D.; Trefonas, P. T., III; West, R.; Fickes, G. N. *Macromolecules* **1987,** *20*, 1046.
4. Miller, R. D.; Sooriyakumaran, R. *J. Polym. Sci., Polym. Chem. Ed.* **1987,** *C25*, 321.
5. Benoit, H.; Doty, P. *J. Phys. Chem.* **1953,** *57*, 958.
6. Tsuji, T.; Norisuye, T.; Fujita, H. *Polym. J.* **1975,** *7*, 558.
7. Yamakawa, H.; Fujii, M. *Macromolecules* **1973,** *6*, 407.
8. Yamakawa, H.; Fujii, M. *Macromolecules* **1979,** *7*, 128.
9. Flory, P. J. *Macromolecules* **1978,** *11*, 1141.
10. Flory, P. J.; Ronca, G. *Mol. Cryst. Liq. Cryst.* **1979,** *54*, 289.
11. Ronca, G.; Yoon, D. Y. *J. Chem. Phys.* **1982,** *76*, 3295.
12. Kurata, M.; Tsunashima, Y.; Iwana, M.; Kanada, K. In *Polymer Handbook;* Brandup, J., Ed.; Wiley: New York, 1975; pp iv-34–iv-50.
13. Flory, P. J. *Statistical Mechanics of Chain Molecules;* John Wiley and Sons: New York, 1969; pp 40–43.

14. Birshtein, T. M. *Vysokomol. Soedin.* **1977,** *A19*, 54.
15. Ying, Q.; Chu, B. *Makromol. Chem., Rapid Commun.* **1984,** *5*, 785.
16. Murakami, H.; Norisuye, T.; Fujita, H. *Macromolecules* **1980,** *13*, 345.
17. Cotts, P. M.; Maxka, J.; Miller, R. D.; West, R. *Polym. Prepr. (Am. Chem. Soc., Div. Polym. Chem.)* **1987,** *28*, 450.

RECEIVED for review May 27, 1988. ACCEPTED revised manuscript March 13, 1989.

24

Radiation Sensitivity of Soluble Polysilane Derivatives

Robert D. Miller

Almaden Research Center, IBM Research Division, 650 Harry Road, San Jose, CA 95120–6099

Polysilane derivatives have very unusual electronic properties associated with extensive σ delocalization along the polymer backbone. Strong electronic transitions that depend on the nature of substituents, polymer molecular weight, and backbone conformation appear in the UV spectra. These materials are radiation sensitive and are degraded to lower molecular weight fragments upon exposure to light and ionizing radiation. This chapter is an overview of the nature of these radiation-induced processes and describes some potential applications, primarily in the field of microlithography.

THE FIRST DIARYL-SUBSTITUTED POLYSILANE derivative was probably prepared over 60 years ago by Kipping (*1*). The simplest dialkyl-substituted material, poly(dimethylsilane), was described in 1949 by Burkhard (*2*). These materials are highly insoluble and intractable and attracted little scientific interest until recently.

The modern era in polysilane chemistry began about 10 years ago with the synthesis of a number of soluble homo- and copolymers (*3–5*). The current interest in substituted polysilanes has resulted in a rapidly expanding list of new potential applications, which include the use of polysilanes as:

- thermal precursors to β-silicon carbide (*6–8*),
- oxygen-insensitive photoinitiators for vinyl polymerizations (*9*),
- polymeric charge conductors (*10, 11*),

0065–2393/90/0224–0413$12.50/0

- radiation-sensitive materials for microlithographic applications (*12, 13*), and
- materials with interesting nonlinear optical properties (*14–15*).

High-molecular-weight substituted polysilanes are usually prepared by a modified Wurtz coupling of the respective dichlorosilanes with sodium metal (*16*). Other procedures have been described recently (*17–21*), but these methods generally result in the production of lower molecular weight polymers or oligomers. With the modified Wurtz coupling, a large number of high-molecular-weight, soluble polysilane derivatives have been prepared (*16, 22*). The mechanism of this heterogeneous polymerization is quite complex, and significant solvent effects have been reported (*12, 16, 23*).

Electronic Spectra

One of the most interesting characteristics of the polysilanes is their unusual electronic spectra (*16*). Even though the backbone is fully σ bonded, all substituted polysilanes absorb strongly in the UV–visible region. Their absorption spectra depend to some extent on the nature of the substituents. Alkyl-substituted, atactic, amorphous materials absorb at 300–325 nm, with sterically bulky groups producing a shift to longer wavelengths (*13, 24*). Aryl substituents that are directly bonded to the silicon backbone result in significant red shifts of 25–30 nm (*24*).

The absorption spectra of polysilane derivatives (*25–31*) depend also on the conformation of the silicon backbone. This polymer backbone effect results in a curious thermochromic behavior for many polysilane derivatives both in the solid state and in solution (*32, 33*). The planar zigzag conformation results in large spectral red shifts, and the position of the absorption maximum is sensitive to small changes in backbone conformation (*31*). Soluble diaryl-substituted polysilane derivatives exhibit the greatest red shifts, and it has been proposed that this effect is probably conformational in origin (*31, 34*). Recent light-scattering studies have provided some additional support for the hypothesis that diaryl-substituted polysilanes are significantly extended even in solution (*35*).

The absorption spectra of polysilane derivatives also depend on the polymer molecular weight (*24*). The λ_{max} of the long-wavelength absorption, which moves progressively to longer wavelengths with increasing catenation for short oligomeric chains, rapidly approaches a limiting value for high-molecular-weight polymers at a degree of polymerization of ~40–50 (Figure 1a). A similar limiting effect is observed for the molar extinction coefficients as a function of molecular weight (Figure 1b).

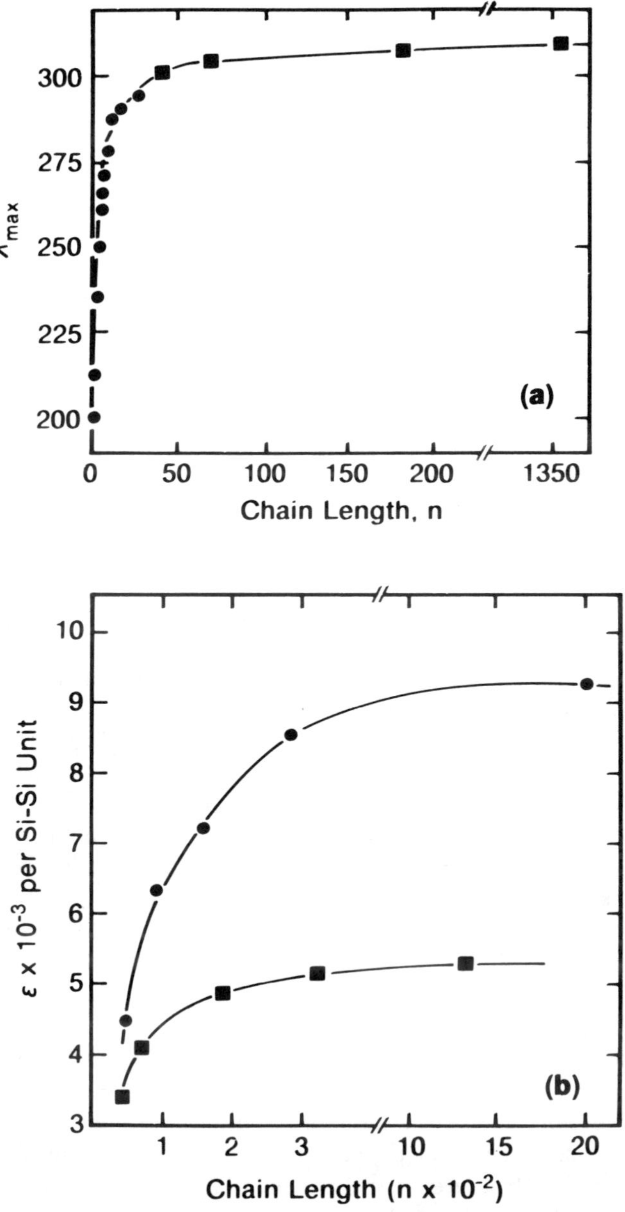

Figure 1. (a) Plots of UV absorption maxima versus chain length (n) *for poly(alkylsilane)s. Key:* ●, *poly(dimethylsilane);* ■, *poly(*n-*dodecylmethylsilane). (b) Plots of absorptivity per Si–Si bond at* λ_{max} *versus chain length* n. *Key:* ●, *poly(phenylmethylsilane);* ■, *poly(*n-*dodecylmethylsilane).*

Radiation Sensitivity of Polysilanes

The polysilanes constitute a new class of radiation-sensitive materials that are sensitive to UV light, as well as to various types of ionizing radiation (vide infra). The observed dependence of both the position of the absorption maximum and the molar extinction coefficients on polymer molecular weight suggests that any process that significantly reduces the molecular weight should result in spectral bleaching. This effect would have major lithographic consequences (vida infra).

Spectral Bleaching. The response of a typical polysilane derivative to irradiation is shown in Figure 2. The strong bleaching suggests that the molecular weight of the polymer is reduced significantly upon exposure. The bleaching phenomenon is characteristic of both alkyl- and aryl-substituted polysilanes upon exposure to UV light and is quite general. Subsequent studies have confirmed that the polymer molecular weight is reduced significantly upon exposure. This effect is demonstrated in Figure 3, which shows the decrease in the molecular weight of a typical polysilane, poly(*n*-dodecylmethylsilane), in solution upon exposure to UV light. Similar effects are observed upon irradiation of solid polysilane films.

Although spectral bleaching with molecular weight reduction occurs both in air and under high vacuum, some oxidation of the silicon backbone undoubtedly occurs in the presence of air. This reaction was first demonstrated by Zeigler and co-workers (*13*), who reported the appearance of a strong IR band at 1020 cm^{-1}, which is characteristic of the Si–O–Si functionality, upon exposure of a poly(cyclohexylmethylsilane-*co*-dimethylsilane) film at 254 nm in the presence of air. Similar changes occur in poly(di-*n*-pentylsilane) films upon irradiation in the presence of air (Figures 4a and 4b). In addition to the appearance of the characteristic band at 1021 cm^{-1}, weaker bands at 2080 and 3382 cm^{-1}, which are ascribed to Si–H and Si–OH vibrations, also are apparent. Similar observations are described in reference 13.

Effect of Air. Because most imaging processes involving polysilanes use the materials in thin-film form and are most often conducted in air, the effect of air on the photochemical degradation is important. In a study of spectral bleaching rates for a number of polysilanes in air relative to those observed in vacuum, we (*36*) noticed that the magnitude of the effect depends strongly on the structure of the polymer, although some acceleration is observed for almost all samples when irradiated in air. At this point, it is not clear whether the differences in the bleaching rates are a function of the nature of substituents or depend on the physical characteristics of the polymer, such as molecular weight and glass transition temperature (T_g). The polymer properties could influence reactivity by changing the mobility of reactive sites, the rate of oxygen diffusion, etc.

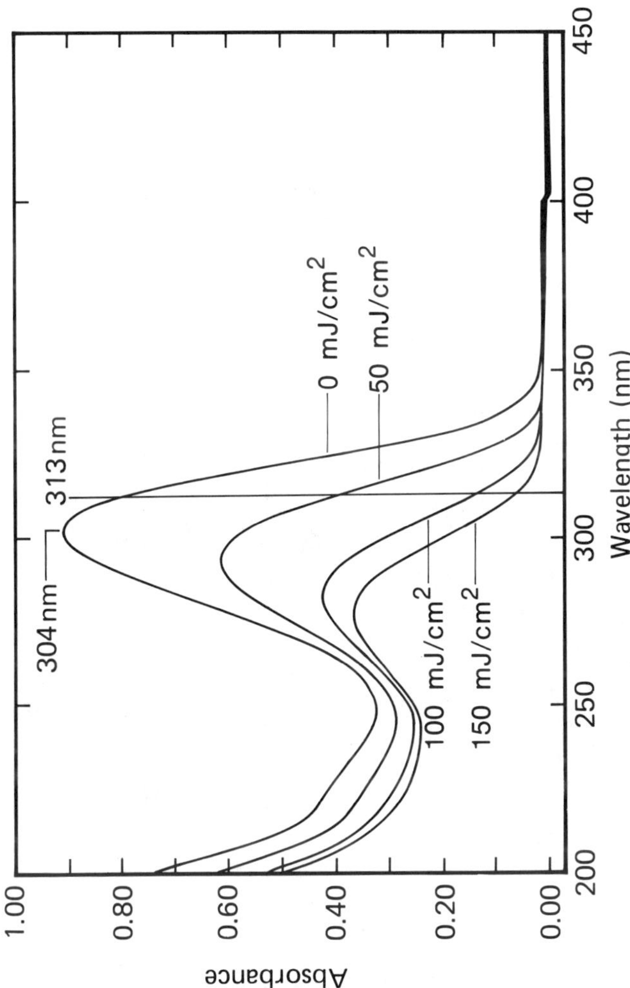

*Figure 2. Spectral bleaching of a film of poly(*n*-hexylmethylsilane) upon irradiation at 313 nm.*

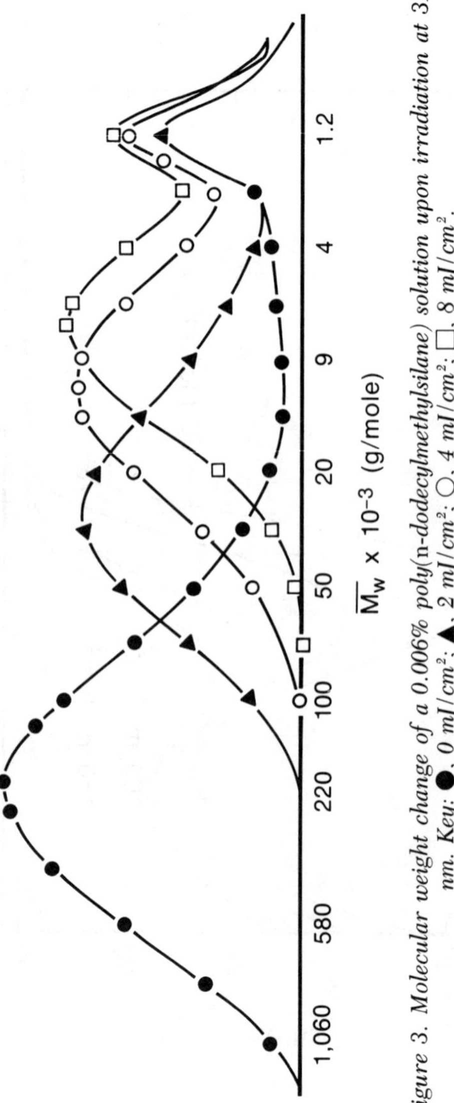

*Figure 3. Molecular weight change of a 0.006% poly(*n*-dodecylmethylsilane) solution upon irradiation at 313 nm. Key: ●, 0 mJ/cm^2; ▲, 2 mJ/cm^2; ○, 4 mJ/cm^2; □, 8 mJ/cm^2.*

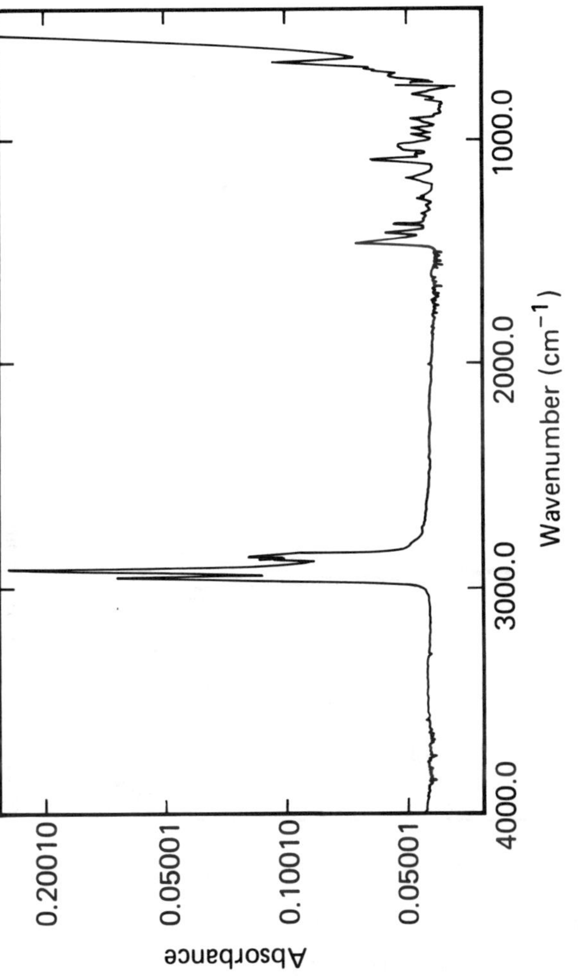

Figure 4a. IR spectrum of an 815-nm poly(di-n-pentylsilane) film.

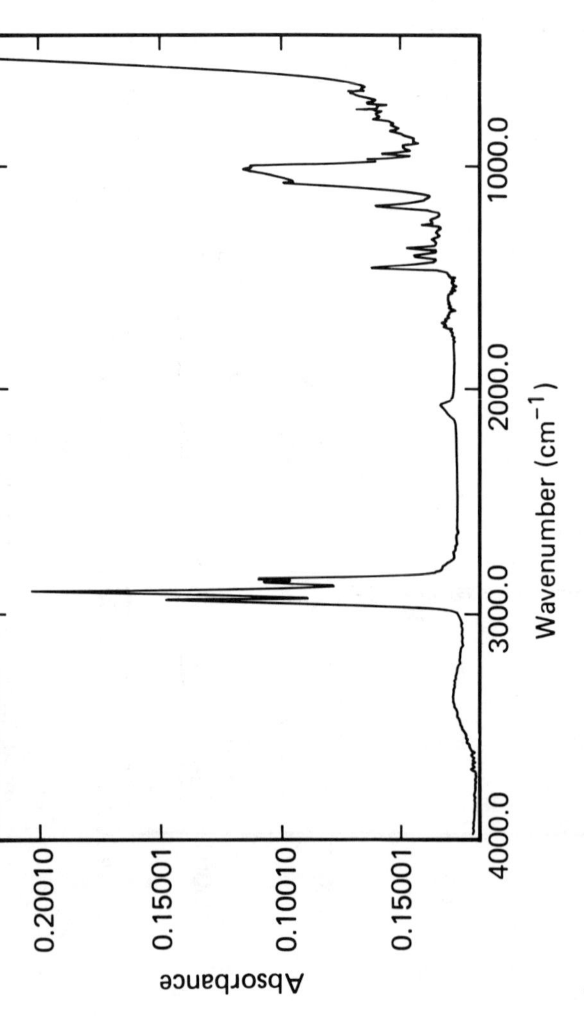

Figure 4b. IR spectrum of an 815-nm poly(di-n-pentylsilane) film after exposure at 254 nm in air (exposure dose 400 mJ/cm^2).

A detailed study of the effect of air has been reported recently by Ban and Sukegawa (*37*), who analyzed the siloxane content of irradiated films of poly(methylphenylsilane) (PMPS) and poly(*n*-propylmethylsilane) (PMPrS) by IR spectroscopy. On the basis of these experiments, the authors concluded that at 254 nm PMPS and PMPrS are almost completely oxygenated at saturation. The extent of oxygenation of both samples was considerably less at 330 nm, presumably because the spectral bleaching of the initial absorption band results in inefficient light absorption. In each case, the molecular weight of the oxygenated polymer was reduced. Subsequent comparative imaging studies on these materials irradiated in air and under vacuum suggested that photooxidation plays an important role in subsequent pattern development.

Effect of Polyhalogenated Additives. We (*36*) have also observed that the rate of bleaching of solid polysilane films upon irradiation is considerably slower than that observed for solutions at comparable optical densities. Although this result is consistent with the observed decrease in the quantum yields for scission, Φ(s), in going from solution to the solid state (*24*) (vide infra), this decreased sensitivity is inconvenient for imaging processes. For this reason, a search was made for compatible additives that might influence the bleaching rate of polysilane derivatives in the solid state.

We have found that a certain number of polyhalogenated aromatic derivatives such as compounds **1–3** greatly accelerate the rate of bleaching of a number of polysilane derivatives in the solid state (*38*). This effect is dramatically demonstrated in Figure 5 for a PMPS film doped with ~20% by weight of 1,4-bis(trichloromethyl)benzene, **1**. Similar results were obtained with substituted triazine sensitizers such as **2**. In these cases, the polysilane is the primary absorber of the incident radiation. Interestingly, when compound **3**, which absorbs at ~400 nm, was incorporated into a PMPS film and the sample was irradiated at 404 nm, where only the sen-

CCl_3 CCl_3 **1**

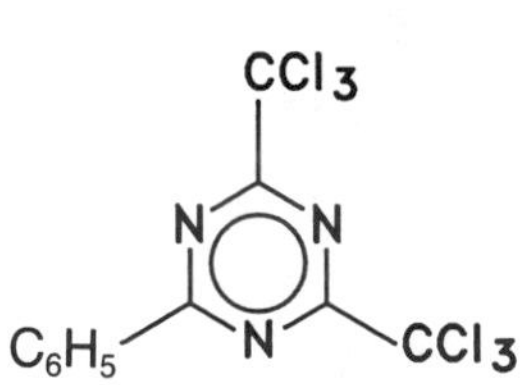

CCl_3 N N C_6H_5 N CCl_3 **2**

CCl_3 N CCl_3 N N OCH_3 OCH_3 **3**

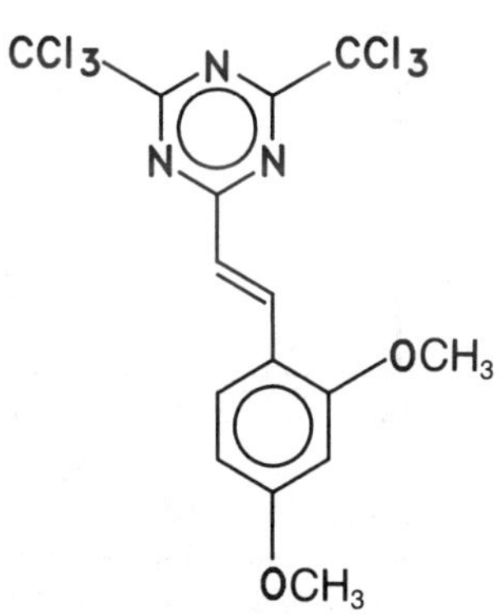

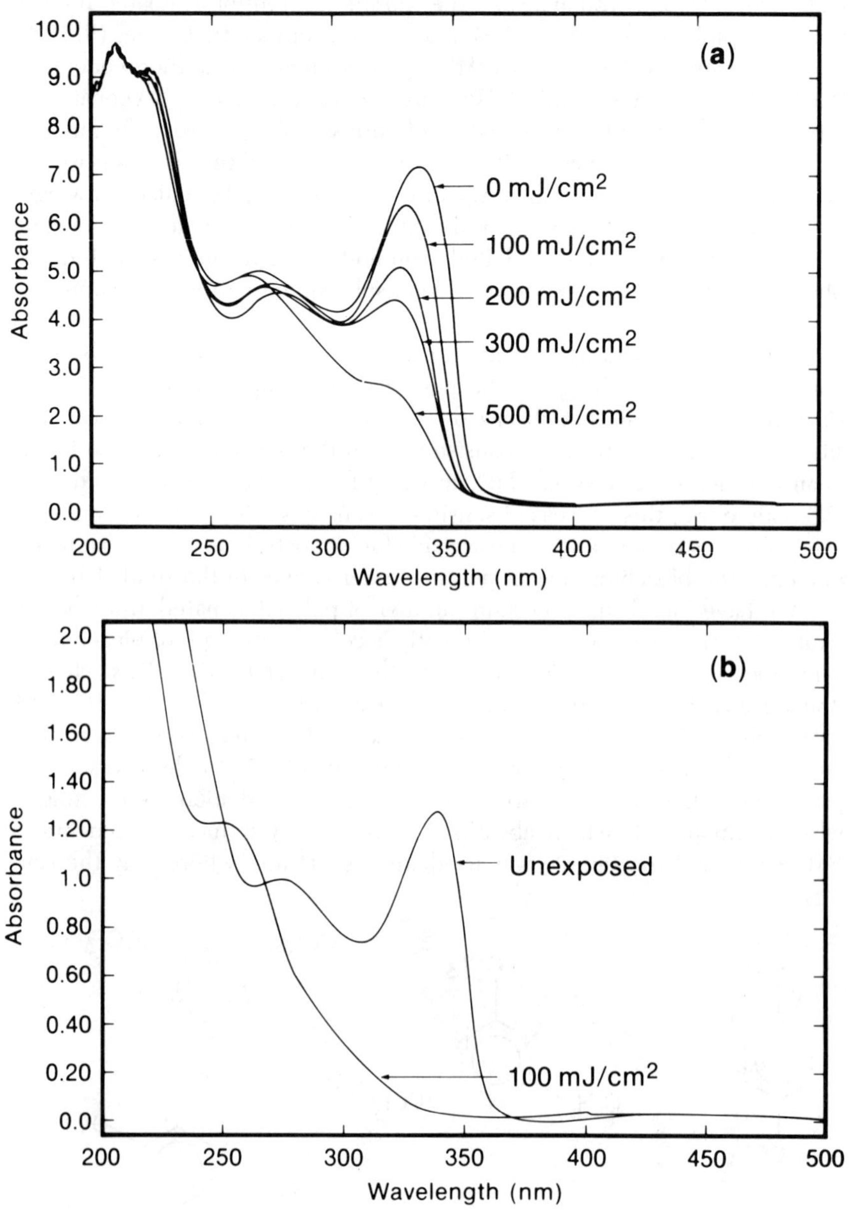

Figure 5. Irradiation of poly(phenylmethylsilane) at 313 nm in the presence or absence of polyhalogenated additives: (a) film containing no additive and (b) film containing approximately 20% by weight of 1,4-bis(trichloromethyl)benzene.

sitizer absorbs, both the polysilane and the sensitizer absorptions at 343 and 400 nm, respectively, were efficiently bleached. However, this accelerated bleaching in the solid state does not appear to be a general phenomenon and depends on the structure of the polysilane. For example, the incorporation of sensitizer **2** into a film of a typical dialkyl-substituted polysilane such as poly(cyclohexylmethylsilane) did not accelerate the bleaching of the polymer film. In fact, the presence of compound **2** actually seems to inhibit somewhat the photochemical bleaching process.

Selectivity of Polyhalogenated-Additive Effect. The mechanism by which halogenated additives promote the bleaching of PMPS but inhibit the same process in poly(cyclohexylmethylsilane) is a source of some speculation and is currently under investigation. Although halogenated sensitizers such as compounds **1–3** could serve simply as halogen-atom-transfer reagents to inhibit the recombination of incipient reactive silicon chain fragments (e.g., silyl radicals), it seems unlikely that they would be selective only for fragments derived from PMPS, unless the radicals generated from PMPS are intrinsically longer lived and radical recombination is the predominant pathway for the fragments in the solid state in the absence of additives.

The photochemistry of polysilane derivatives may occur also via the triplet state (*13, 30*), on the basis of the observation of a weak-structured phosphorescence characteristic of a localized excited state for a number of polysilane derivatives. In principle, the halogenated additives could promote intersystem crossing via an intermolecular heavy-atom effect (*39*). Again, however, why the two structurally similar polysilanes should respond so differently to the presence of the additive is unclear.

One possible explanation for the selective accelerated decomposition of PMPS in the presence of the chlorinated additives is that photochemical bleaching occurs by electron transfer. If electron transfer is involved, the two polysilane derivatives may behave differently, because the peak oxidation potential of a thin film of PMPS is lower by ~0.4 eV than that of typical dialkyl-substituted polysilane derivatives (*40*). In addition, the electrochemical oxidation of PMPS is highly irreversible, and the polymer film is removed from the electrode presumably as smaller silicon-containing fragments. Such a photochemically mediated electron-transfer reaction would also be facilitated by the high electron affinities of polyhalogenated aromatic compounds, which subsequently decompose by dissociative electron attachment (*41*). The instability of both the polysilane radical cations and the polyhalogenated radical anions should improve the overall efficiency of the process by inhibiting back electron transfer.

Photochemistry

Despite the considerable interest in the radiation sensitivity of high-molecular-weight polysilanes, relatively few detailed studies on the nature of the products and intermediates produced upon irradiation have been reported. Oligomeric polysilane derivatives, on the other hand, have been studied extensively (*42*). Because the results of these pioneering investigations form the basis for the interpretation of many subsequent photochemical studies on the high-molecular-weight polysilanes, this background will be described here.

Reactive Intermediates. The simplest radiation-sensitive polysilane contains disilane structural units. Although simple alkyl-substituted disilanes absorb in the far- and vacuum-UV regions, aromatic substituents cause a red shift to the accessible UV region. The photochemical decomposition of disilanes leads predominantly to silicon–silicon bond homolysis and the production of substituted silyl radicals. Once generated, silyl radicals may abstract hydrogen or halogen atoms from suitable donors, add to vinyl and aromatic groups, react with alcohols by abstracting hydrogen atoms or alkoxy groups, or undergo other characteristic reactions (*43*). Because of the relatively low energy of the Si–H bond (81–88 kcal/mol) (*44*), the abstraction of hydrogen atoms from typical alkanes is normally endothermic, except for the most highly activated carbon–hydrogen bonds.

Encounters between silyl radicals in solution or in the gas phase usually result in recombination and disproportionation (*45*, *46*). Disproportionation results in the production of silanes and highly reactive silenes. The disproportionation reaction is thermodynamically favorable because of the formation of a silicon–carbon double bond, which, although subsequently chemically reactive, is worth ~39 kcal/mol (*44*). For pentamethyldisilanyl radicals, disproportionation is kinetically competitive with radical dimerization (*46*). In an earlier study, Boudjouk and co-workers (*47*) demonstrated conclusively by isotopic substitution and trapping that the silyl radicals generated by photolysis undergo disproportionation, as well as, presumably, dimerization (Scheme I). In deuterated methanol, the silanes produced were predominantly undeuterated, whereas methoxymethyldiphenylsilane was extensively deuterated in the α position. The results of these experiments strongly implicated the substituted silene produced by disproportionation.

In another model study, Ishikawa and co-workers (*42*, *48*) showed that for phenylpentamethyldisilane, the phenyl-substituted silyl radicals can undergo *ortho* radical addition to produce unstable silenes that can be subsequently trapped and identified (Scheme I). Recent spectroscopic studies have identified the silene as a reactive intermediate in this process (*49*).

Si–Si Bond Homolysis. Ishikawa and co-workers (*50*, *51*) have also studied the photolysis of a number of polymeric disilane derivatives. These workers have examined both derivatives with pendant silicon substituents (*50*) and materials in which the disilane moiety is incorporated into the polymeric backbone (*51*). For derivatives with pendant silicon substituents (Scheme II), irradiation destroys the characteristic phenyldisilane absorption

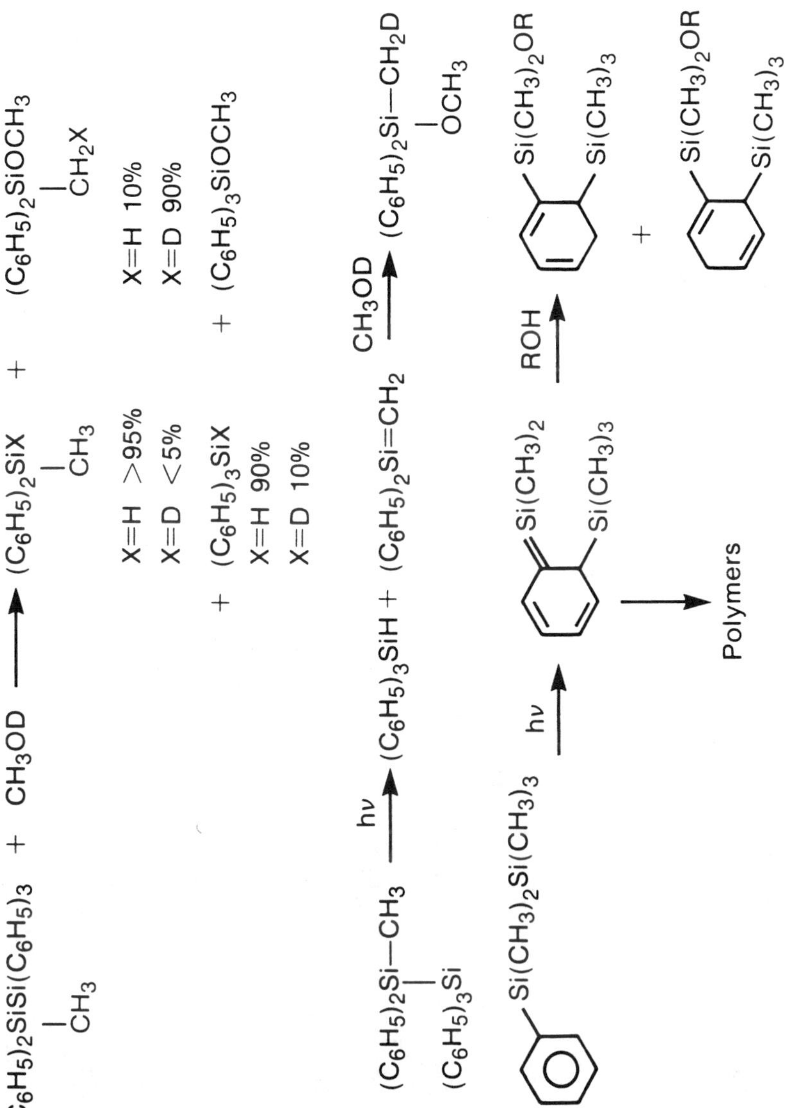

$$(C_6H_5)_2Si(CH_3)Si(C_6H_5)_3 + CH_3OD \longrightarrow (C_6H_5)_2Si(CH_3)X + (C_6H_5)_2Si(OCH_3)CH_2X + (C_6H_5)_3SiX + (C_6H_5)_3SiOCH_3$$

For $(C_6H_5)_2Si(CH_3)X$: X=H >95%; X=D <5%

For $(C_6H_5)_2Si(OCH_3)CH_2X$: X=H 10%; X=D 90%

For $(C_6H_5)_3SiX$: X=H 90%; X=D 10%

$$(C_6H_5)_2Si(CH_3){-}Si(C_6H_5)_3 \xrightarrow{h\nu} (C_6H_5)_3SiH + (C_6H_5)_2Si{=}CH_2 \xrightarrow{CH_3OD} (C_6H_5)_2Si(OCH_3){-}CH_2D$$

$$C_6H_5Si(CH_3)_2Si(CH_3)_3 \xrightarrow{h\nu} \text{[cyclohexadiene with =Si(CH}_3)_2\text{ and Si(CH}_3)_3\text{]} \xrightarrow{ROH} \text{[Si(CH}_3)_2\text{OR, Si(CH}_3)_3\text{ cyclohexadienes, two isomers]}$$

(The intermediate also → Polymers)

Scheme I. Disproportionation and dimerization of silyl radicals as demonstrated by isotopic substitution and trapping.

$$\left(\!\sim \underset{\underset{Si(CH_3)_3}{|}}{\overset{\overset{C_6H_5}{|}}{Si}}-O-Si(CH_3)_2-O\sim\!\right)_n \xrightarrow{h\nu} \text{cross-linked polymer}$$

Model Studies

$$(CH_3)_3Si-O-\underset{\underset{Si(CH_3)_3}{|}}{\overset{\overset{C_6H_5}{|}}{Si}}-O-Si(CH_3)_3 \xrightarrow[CH_3OH]{h\nu} \text{Polymer} + \underset{4\%}{(CH_3)_3SiO\underset{\underset{H}{|}}{\overset{\overset{C_6H_5}{|}}{Si}}OSi(CH_3)_3} + \underset{5\%}{(CH_3)_3SiO\underset{\underset{OCH_3}{|}}{\overset{\overset{C_6H_5}{|}}{Si}}OSi(CH_3)_3}$$

$h\nu$, cumene (isopropylbenzene) $\longrightarrow$ 4-isopropylphenyl–$C_6H_5Si[OSi(CH_3)_3]_2$ + 4-isopropylphenyl–$Si(CH_3)_3$

$$\left(\!\sim \underset{\underset{Si(CH_3)_3}{|}}{\overset{\overset{C_6H_5}{|}}{Si}}-O-Si(CH_3)_2-O\sim\!\right)_n \xrightarrow{h\nu} \left(\!\sim \underset{\bullet}{\overset{\overset{C_6H_5}{|}}{Si}}-O-Si(CH_3)_2-O\sim\!\right)_n + \bullet Si(CH_3)_3$$

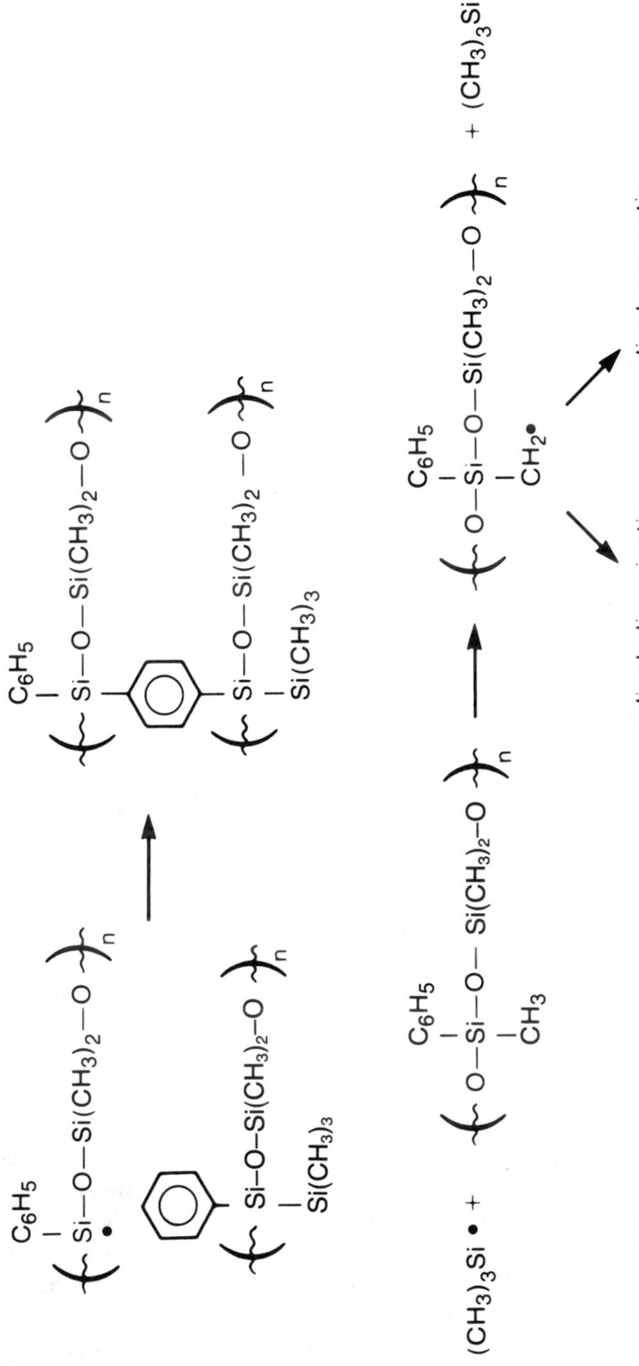

Scheme II. Si–Si bond homolysis in siloxane polymers containing pendant silicon substituents.

at 235 nm, and the polymer becomes insoluble through cross-linking. The IR spectrum of the photodegraded polymer shows a weak –SiH absorption at 2150 cm^{-1}. Model studies on 1,1-bis(trimethylsiloxy)-1-phenyltrimethyldisilane showed that Si–Si bond homolysis occurs, and products characteristic of silyl radical abstraction and substitution were isolated. On the basis of these model studies, the mechanism shown in Scheme II, which involves Si–Si bond homolysis followed by characteristic silyl radical reactions, was postulated. Furthermore, oxidation studies on the photodegraded polymer suggested that, unlike phenylpentamethyldisilane, the disiloxydisilane polymer shows little inclination toward trimethylsilyl radical addition at the *ortho* positions of the incipient phenylsilyl chain radical.

Ishikawa and co-workers (*50*, *51*) have also reported the photodegradation of backbone disilane polymers (Scheme III). In this case, as in the previous example with pendant silyl substituents, silicon–silicon bond homolysis to produce silyl radicals was the predominant process. When irradiated in toluene, the photodegraded polymer showed a substantial –SiH band at 2150 cm^{-1} in the IR spectrum. Similarly, NMR spectroscopic examination of the irradiated polymer showed evidence of silyl radical substitution into the solvent toluene. Irradiation of the polymeric disilane in deuterated methanol produced no bands due to Si–D in the IR spectrum and resulted in the incorporation of the elements of methanol into the chain ends (as revealed by NMR spectroscopy). For the phenyl-substituted polymer, the NMR evidence indicated that $<5\%$ of rearranged cyclohexadiene derivatives were formed.

On the basis of these experiments and molecular weight studies on the photodegraded samples, which confirmed that a considerable reduction in molecular weight had occurred, Ishikawa et al. (*51*) have postulated a mechanism involving silicon–silicon bond homolysis followed by radical disproportionation. Irradiation of solid films resulted in increased solubility in the exposed areas and the formation of very little SiH in the photodegraded sample (this result depends somewhat on the nature of the silicon substituents). Instead, the IR spectra of the products show strong absorption bands attributed to –SiOH and Si–O–Si groups, a fact suggesting that in the solid films the incipient silyl radicals are efficiently scavenged by oxygen to produce predominantly silanols and siloxanes.

Degradation of Larger Silicon Catenates. Although it seems clear that polymeric disilane derivatives photodecompose by bond homolysis to produce silyl radicals, model studies on larger silicon catenates indicate that their photochemistry may be more complex (Scheme IV). Cyclic silane derivatives seem to extrude monomeric silylenes upon irradiation to produce smaller cyclic silanes (*52*). The proposed silylene intermediates have been identified spectroscopically (*49*, *53*), and trapping adducts have been isolated in solution. Exhaustive irradiation ultimately results in acyclic silanes, which have been suggested to result via hydrogen abstraction by silyl radicals. The extrusion of substituted silylenes from larger cyclosilanes is interesting, and in at least one instance, the stereochemistry of the starting material was preserved. This observation suggests that extrusion and rebonding may be synchronous processes (*54*).

Similar extrusion reactions have been observed for acyclic polysilanes (*55*). The isolation of hydrogen-terminated silanes containing fewer silicon atoms than the starting materials was taken as evidence for the intermediacy of silyl radicals, and the importance of chain scission seemed to increase with increasing catenation.

Photolysis of High-Molecular-Weight Polysilanes

Photolysis Products. Similar trapping experiments have been performed with high-molecular-weight, soluble polysilanes in the presence of suitable trapping reagents (*56*). Because silyl radicals readily abstract a chlorine atom from chlorinated alkanes (*43*), a sample of high-molecular-weight poly(cyclohexylmethylsilane) was irradiated at 300 nm in carbon tetrachloride. The formation of hexachloroethane was taken as indicative of the presence of trichloromethyl radicals formed possibly via chlorine atom abstraction by silyl radicals. The yield of hexachloroethane was 33%, on the basis of the number of monomer units. This yield suggests that ~66 trichloromethyl radicals were formed for every 100 silicon atoms in the backbone. Although instructive, this experiment does not rule out the possibility that the trichloromethyl radicals could be produced by electron transfer rather than by direct halogen atom abstraction.

A number of polysilane homo- and copolymers were also irradiated in the presence of trapping reagents such as triethylsilane (TES), methanol, and *n*-propyl alcohol. The results of the exhaustive irradiation at 254 nm in the presence of TES are shown in Table I. In each case, the adduct of the dialkyl-substituted silylene and TES was the dominant product. Also produced concurrently in significant yields were the corresponding disilanes.

Table I. Product Distributions and Yields from 254-nm Photolysis of High-Molecular-Weight $(R^1R^2Si)_n$ in Solution with TES Chemical Trapping Reagent

Product	R^1 = n-C_4H_9, R^2 = n-C_4H_9	R^1 = n-C_6H_{13}, R^2 = CH_3	R^1 = c-C_6H_{11}, R^2 = CH_3
$(C_2H_5)_3SiR^1R^2SiH$	59	70	71
$HSiR^1R^2SiR^1R^2H$	11	11	14
$(C_2H_5)_3SiSiR^1R^2Si(C_2H_5)_3$	a	3	a
$HSiR^1R^2OSiR^1R^2H$	a	a	2
$(C_2H_5)_3SiOSiR^1R^2SiR^1R^2H$	a	a	3

NOTE: All values are given in percents. a indicates that no attempt was made to identify products whose yields were less than 2%.

SOURCE: Reproduced from reference 56. Copyright 1985 American Chemical Society.

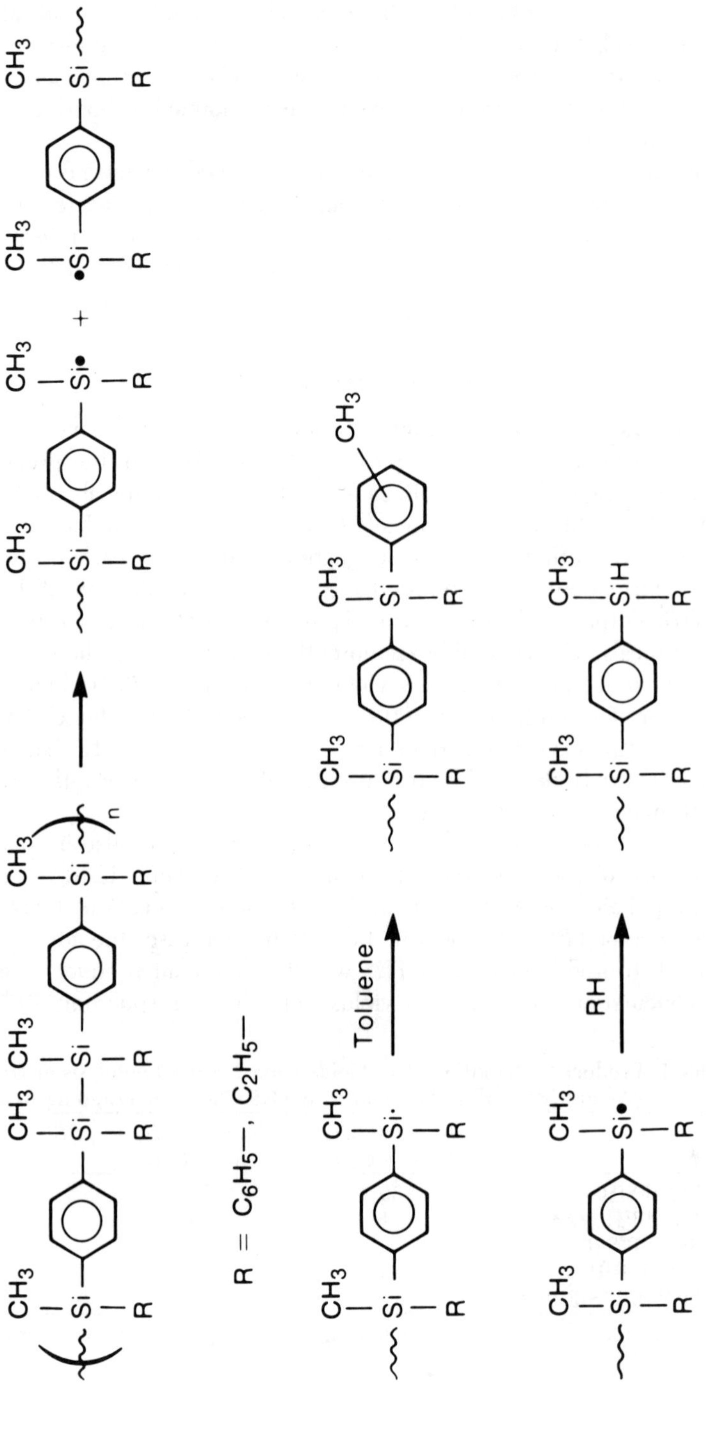

R = C_6H_5—, C_2H_5—

Toluene

RH

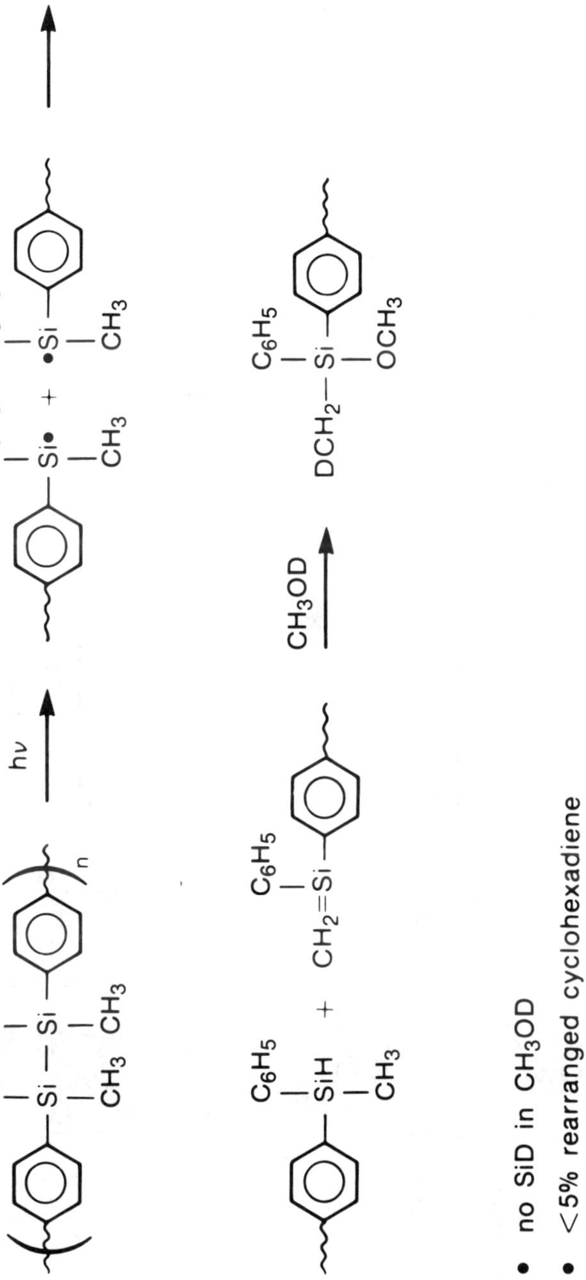

$$\left(-C_6H_4-Si(C_6H_5)(CH_3)-Si(C_6H_5)(CH_3)-C_6H_4- \right)_n \xrightarrow{h\nu} -C_6H_4-Si^{\bullet}(C_6H_5)(CH_3) + {}^{\bullet}Si(C_6H_5)(CH_3)-C_6H_4- \longrightarrow$$

$$-C_6H_4-SiH(C_6H_5)(CH_3) + CH_2{=}Si(C_6H_5)-C_6H_4- \xrightarrow{CH_3OD} DCH_2-Si(C_6H_5)(OCH_3)-C_6H_4-$$

- no SiD in CH_3OD
- <5% rearranged cyclohexadiene

Scheme III. Photodegradation of disilane backbone polymers.

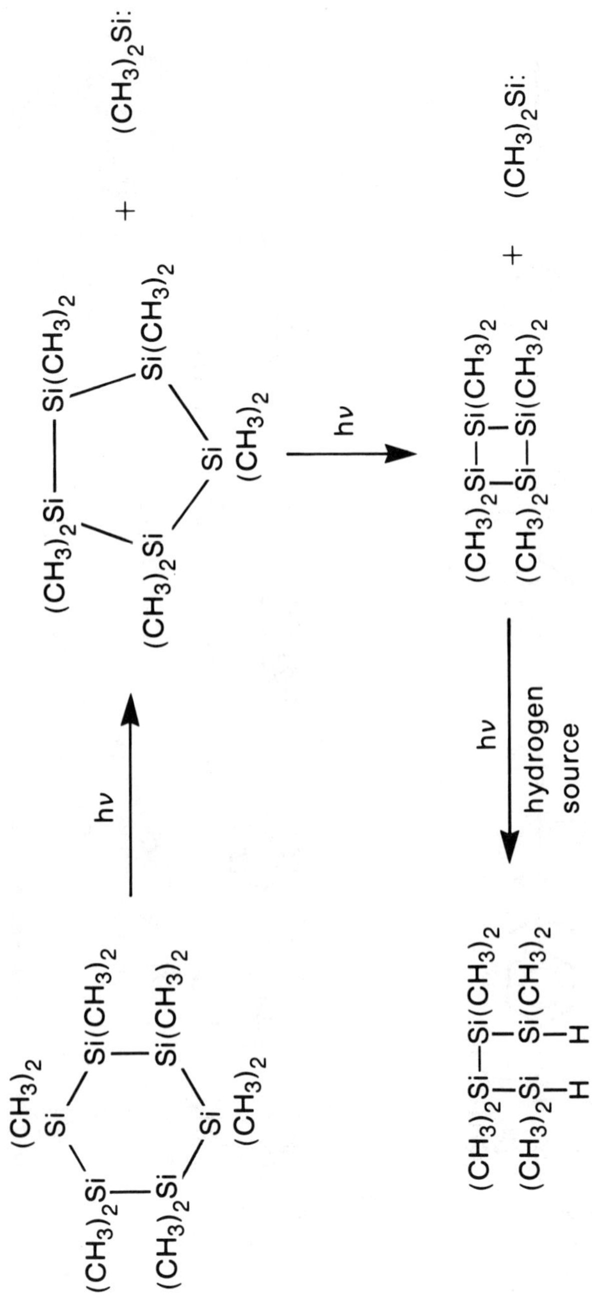
hν
+
$(CH_3)_2Si:$
hν
+
$(CH_3)_2Si:$
hν
hydrogen
source

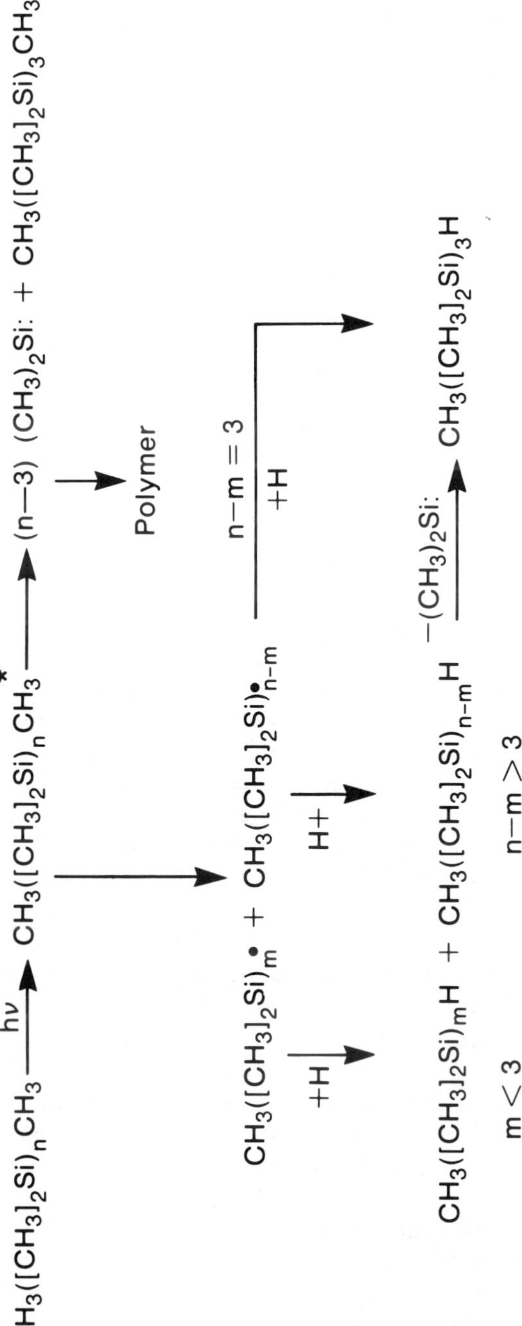

Scheme IV. Photochemical transformations in cyclic and acyclic oligosilanes.

Similar products were produced from copolymers such as poly(cyclohexylmethylsilane-*co*-dimethylsilane), and materials characteristic of the copolymer composition were obtained. Three disilanes (i.e., 1,2-dicyclohexyl-1,2-dimethyl-; 1,1,2,2-tetramethyl-; and 1-cyclohexyl-1,2,2-trimethyldisilane), as well as the two substituted silylene adducts, were produced from the predominantly random copolymer. The disilanes, which are taken to be diagnostic of silyl radical abstraction, were assumed to accumulate in the mixture, because they no longer absorbed light significantly at 254 nm. The photoinstability of a model trisilane, 2-*n*-butyl-1,1,1,2,3,3,3-heptamethyltrisilane, was demonstrated under the reaction conditions even though its absorption maximum occurred at ~215 nm.

Similar results were obtained for the photolysis of poly(*n*-hexylmethylsilane) using either methanol or *n*-propyl alcohol as the trapping reagent (Table II). The additional complexity of the disilane products was anticipated, because silyl radicals can abstract either a hydrogen atom or an alkoxy radical. Although the alkoxy disilanes could be produced by other routes (e.g., alcoholysis of Si–H bonds or addition to an intermediate disilene), these routes were considered unlikely on the basis of appropriate control experiments or the lack of a literature precedent.

Reactive Intermediates in Photolysis. ***Silylenes and Silyl Radicals.*** The results of exhaustive irradiation of substituted high-molecular-weight polysilanes in solution suggest that at least two reactive intermediates (i.e., silylenes and silyl radicals) are produced. On the basis of this information, a tentative reaction scheme (Scheme V) is proposed.

It should be emphasized that Scheme V is based only on the trapping of reactive intermediates. It does not preclude the existence of independent photochemical channels that might produce exclusively silylenes or silyl radicals or that these pathways may be occurring concurrently, as would seem likely from the thermodynamic considerations. It also does not preclude the possibility that silylenes are produced from silyl radicals by a cascade process, although there are no compelling reasons based on either thermodynamics or literature precedents (*46*) to so conclude, or that the photochemical intermediates are even necessarily the same when they are produced from the high-molecular-weight polymer or the smaller oligomeric fragments. Obviously, the elucidation of a photochemical mechanism for the photodegradation of a high-molecular-weight polymer is a complex task, and in the case of polysilanes, additional mechanistic studies are necessary.

Persistent Silicon-Centered Radicals. Although the production of substituted silylenes during the exhaustive photolysis of substituted high-molecular-weight polysilanes in solution seems to be supported by the diagnostic trapping products, the suggested intermediacy of silyl radicals rests mainly on the isolation of fragments that contain Si–H bonds, which

Table II. Product Distributions and Yields from 254-nm Photolysis of High-Molecular-Weight Poly(*n*-hexylmethylsilane) in Toluene with Excess Methanol and *n*-Propyl Alcohol

Products	$R = CH_3$	$R = n\text{-}C_3H_7$
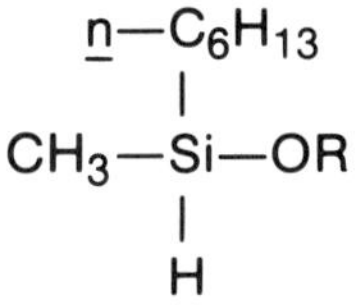 $CH_3-Si(n\text{-}C_6H_{13})(H)-OR$	65	39
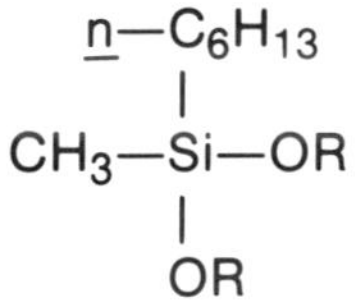 $CH_3-Si(n\text{-}C_6H_{13})(OR)-OR$	0	5
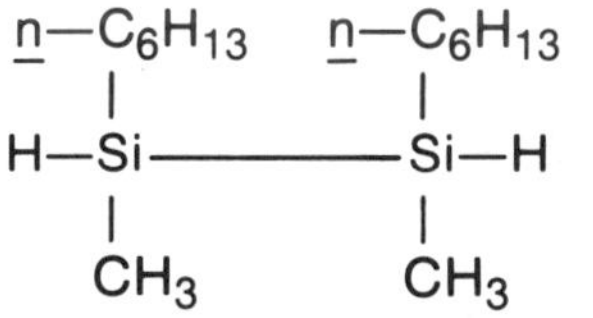 $H-Si(n\text{-}C_6H_{13})(CH_3)-Si(n\text{-}C_6H_{13})(CH_3)-H$	7	7
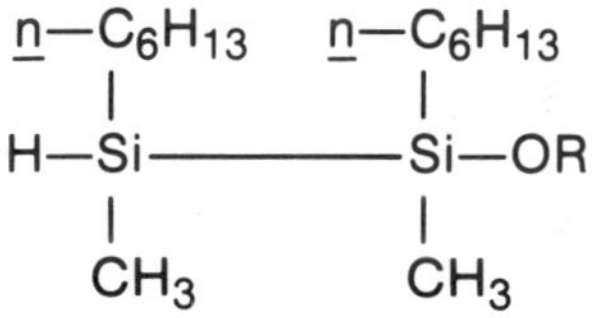 $H-Si(n\text{-}C_6H_{13})(CH_3)-Si(n\text{-}C_6H_{13})(CH_3)-OR$	15	20
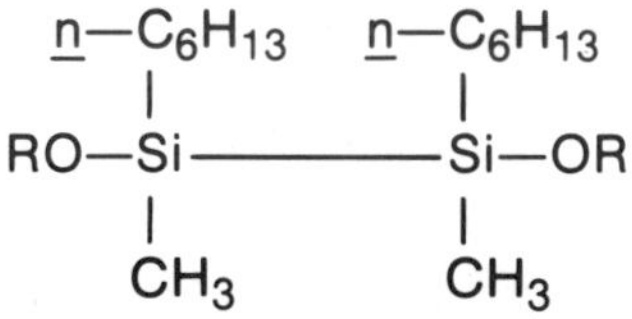 $RO-Si(n\text{-}C_6H_{13})(CH_3)-Si(n\text{-}C_6H_{13})(CH_3)-OR$	7	12

NOTE: The data were obtained from reference 56. All values are yields expressed in percent.

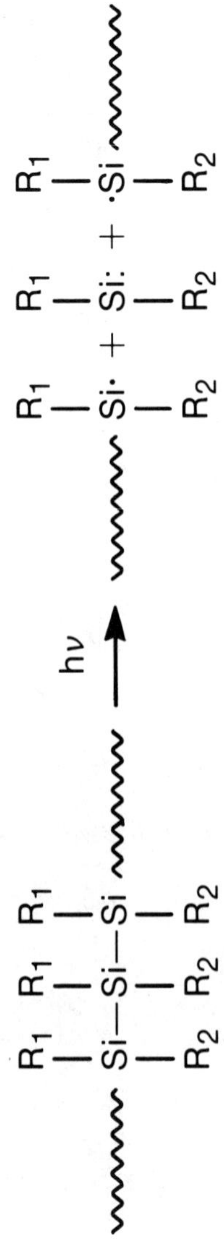

Scheme V. Photochemical decomposition of high-molecular-weight polysilanes.

are presumably formed by hydrogen abstraction. Recently, however, Michl and co-workers (57) have reported the detection by electron spin resonance (ESR) spectroscopy of persistent radicals produced upon irradiation of a number of dialkyl-substituted polysilanes in 3-methylpentane at room temperature.

The ESR signals are complex and overlapping and are probably derived from at least two radical species. The radicals have been identified as silicon centered, on the basis of their *g* values, the observation of a ^{29}Si hyperfine interaction, and their reactivity with oxygen and chlorocarbons. In no case, however, was a persistent-radical signal observed that is split by four equivalent α hydrogens, as would be expected for a radical produced by simple chain scission.

In the case of poly(di-*n*-hexylsilane) (PDHS), an early radical with a *g* value of 2.0046 was observed with a ^{29}Si hyperfine splitting of ~75G. This *g* value is consistent with those previously reported for persistent silicon-centered radicals (*43*), and the low value for the ^{29}Si hyperfine splitting is reasonable for a polysilylated silyl radical. Continued irradiation produced another persistent silicon-centered radical with no proton splittings and with a ^{29}Si hyperfine coupling of ~56G. Irradiation of PDHS in deuterated pentane produced no photodegraded materials that showed a Si–D stretch in the IR spectrum. On the other hand, irradiation at 248 nm of PDHS that had been fully deuterated in the α-carbon positions resulted in the formation of both Si–H and Si–D bonds.

Disproportionation of Silyl Radicals. Faced with the task of generating a series of sterically hindered polysilyl substituted radicals upon irradiation, the authors have proposed a tentative reaction scheme involving silyl radical formation, disproportionation to silanes and silenes, and readdition of silyl radicals to the silenes. The disproportionation of silyl radicals is a well-established process that is kinetically competitive with recombination (*46*). Repetition of this process would lead eventually to highly sterically encumbered and undoubtedly persistent silicon-based radicals carrying only silicons in the α positions. Although such a scheme would explain much of the data in this obviously very complex process, it is very tentative, and other possible routes to and structures for the persistent silyl radicals have not been ruled out (*58*).

The hypothesis that silyl radical disproportionation may play a significant role in hindering chain repair by silyl radical recombination suggests that polysilane derivatives with no α hydrogens might be more resistant to photochemical degradation, as determined by spectral bleaching. Unfortunately, fully alkyl-substituted polysilanes with no α hydrogens are very sterically hindered and synthetically inaccessible. Diaryl substitution relieves some steric congestion, but the simplest derivative, poly(diphenylsilane), is extremely insoluble and intractable.

Recently, we (*34*) described the preparation of a number of poly[*p*-bis(alkylphenyl)silane]s that are soluble and can be spin cast into films. These materials have no α hydrogens, although they do contain reactive benzylic hydrogens in the remote *para* positions. Figure 6 shows that in solution poly[*p*-bis(*n*-hexylphenyl)silane] (PBHPS) is extremely photolabile and bleaches readily upon exposure, in a manner similar to that described for the alkyl-substituted polysilane derivatives. This finding is not surprising, because the chain-end mobility in solution is such that the benzylic hydrogens might be available for abstraction. In this case, disproportionation is kinetically competitive with silyl radical recombination. In the solid state, however, where this mobility is greatly restricted, the remote benzylic hydrogens should be much less accessible to a silyl radical produced by chain scission. Consistent with this proposal is the observation that PBHPS bleaches extremely slowly in the solid state (Figure 6).

Quantum Yields of Photolysis. These photochemical studies on high-molecular-weight polysilanes suggest that these polymers constitute a new class of photosensitive materials. The bleaching studies indicate that significant molecular weight reduction occurs upon irradiation in air or in vacuo, which in turn implies that the polymers are predominantly of the scissioning type.

For polymers that undergo scissioning and cross-linking simultaneously, the relative efficiencies of each process can be estimated by monitoring the changes in the number- and weight-average molecular weights ($\overline{M}_n$ and $\overline{M}_w$, respectively) as a function of the absorbed dose (*59*). The relationships between the molecular weights and the respective quantum yields are described by equations 1 and 2.

$$\frac{1}{\overline{M}_n} = \frac{1}{\overline{M}_n^{\circ}} + \left[\Phi(s) - \Phi(x)\right]\frac{D}{N_A} \tag{1}$$

$$\frac{1}{\overline{M}_w} = \frac{1}{\overline{M}_w^{\circ}} + \left[\frac{\Phi(s) - 4\Phi(x)}{2}\right]\frac{D}{N_A} \tag{2}$$

In these equations, $\overline{M}_n^{\circ}$ and $\overline{M}_w^{\circ}$ are the initial molecular weights, $\Phi(s)$ and $\Phi(x)$ are the quantum yields for the scissioning and cross-linking reactions, respectively; D is absorbed dose; and N_A is Avogadro's number. The slopes of the respective plots of $1/\overline{M}_n$ and $1/\overline{M}_w$ versus dose produce two simultaneous equations, the solution of which yields values for $\Phi(s)$ and $\Phi(x)$. We (*60*) have analyzed the data obtained for a number of polysilane derivatives by GPC (gel permeation chromatography) to evaluate the respective molecular weights and distributions (Table III). Polystyrene standards were used for molecular weight calibration.

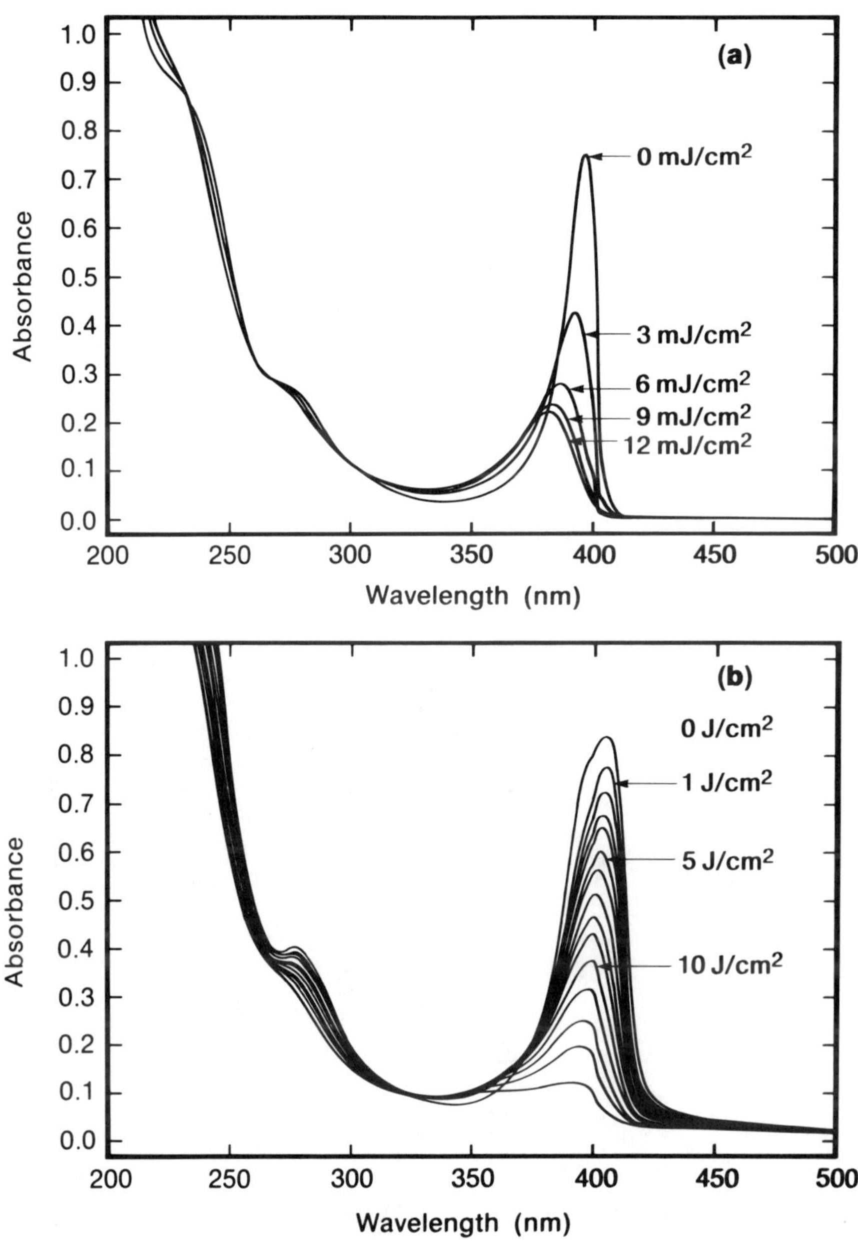

*Figure 6. Irradiation of poly[*p*-bis(*n*-hexylphenyl)silane] at 405 nm: (a) solution in cyclohexane and (b) film.*

Table III. Quantum Yields for Photodecomposition of Substituted Polysilane Derivatives

R^1	R^2	*Solvent*	$\overline{M}_w^{o}$ [a] $\times 10^3$	λ_{ex} (nm)	*Solution* $\Phi(s)$	*Solution* $\Phi(x)$	*Solid Film* $\Phi(s)$	*Solid Film* $\Phi(x)$
	—CH_3	Toluene	41.4	355	1.2	0.06	0.022	0
C_6H_{13}	C_6H_{13}	Toluene	44.6	353	0.6	0	— [b]	— [b]
CH_2CH_2—	—CH_3	Toluene	208.9	347	0.97	0.08	— [b]	— [b]

$C_{12}H_{25}$—	—CH_3	Toluene	185.1	347	0.54	0	—[b]	—[b]
C_3H_7—	—CH_3	—[c]	3900	—[c]	—[c]	—[c]	0.011	0.0013[d]
C_3H_7—	—CH_3	—[c]	190	—[c]	—[c]	—[c]	0.015	0.0013[d]
C_6H_5—	—CH_3	THF[e]	245	313	0.97	0.12	0.015	0.002
p-t-$C_4H_9C_6H_4$—	—CH_3	Toluene	633.6	367	1.00	0.18	—[b]	—[b]

[a]Molecular weights were determined by GPC and are relative to polystyrene calibration standards.
[b]The samples were studied only in solutions.
[c]The samples were studied only as solid films.
[d]Data are from reference 85.
[e]THF is tetrahydrofuran.

Although the data in Table III are approximate (uncertainties of up to 20% would be reasonable) because of variations in the molecular weight distributions of the individual samples, uncertainties in actinometry, the use of polystyrene calibration samples for molecular weight determinations, etc., certain interesting features are apparent. The quantum yields for scission in solution are high, ~0.5–1.0. This result indicates that the polymers predominantly undergo chain scission, as predicted from the spectral-bleaching studies. Cross-linking is somewhat more important for those derivatives with aromatic substituents directly bonded to the silicon backbone. One particularly striking feature is the significant decrease in the quantum yields for both processes in going from solution to the solid state. This result would be expected if species such as silyl radicals are involved in the polymer degradation because of decreased chain mobility and solid-state cage effects.

Polysilane Lithography

Polysilanes possess a number of unique characteristics that make them suitable for many lithographic applications. Specifically, these polymers are soluble, castable into high-quality optical films, thermally stable, imageable over a broad spectral range, strongly absorbing yet readily bleached upon exposure, and resistant to oxygen etching in plasma environments. Resistance to oxygen etching is particularly important for multilayer-resist schemes using oxygen reactive-ion etching (O_2-RIE) techniques for image transfer.

O_2-RIE processes have been proposed for the production of high-aspect-ratio, high-resolution images over chip topography (*61*). This image-transfer technique requires an image-forming polymer that is considerably more stable in oxygen plasma environments than normal carbon-based polymers, so as to serve as an effective etch barrier for the protection of unexposed regions during the pattern-transfer step. By virtue of their high silicon content, which leads to the formation of a thin, highly resistant silicon oxide etch barrier in an oxygen plasma (*62*), polysilanes are ideally suited. An example of a typical bilayer process for the production of high-resolution images is shown in Figure 7 (*63*). The thin top layer plays a dual role by serving both as the imaging layer and as the barrier layer in the subsequent image-transfer process.

Because of their desirable properties and radiation sensitivity, polysilanes have been used in a variety of microlithographic applications as: (1) mid-UV contrast-enhancing materials, (2) imaging layers in a variety of bilayer lithographic processes, and (3) new resist materials for ionizing radiation.

Contrast-Enhanced Lithography. Contrast-enhanced lithography is a clever technique that uses a thin layer of a bleachable contrast-enhancing layer coated over a conventional photoresist to sharpen the mask image,

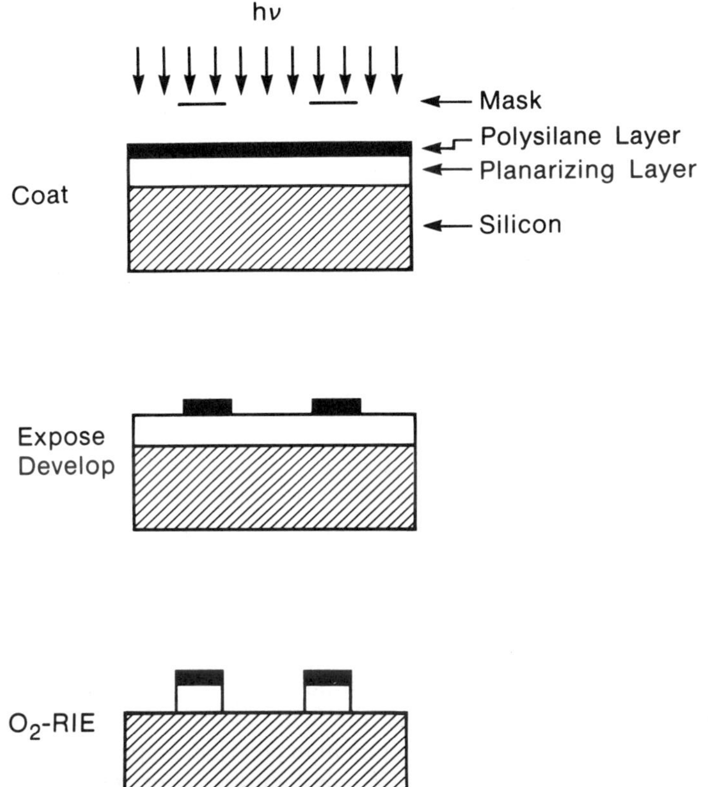

Figure 7. Schematic of a bilayer polysilane process for the production of high-resolution images using O_2-RIE techniques for image transfer.

which has been distorted somewhat by optical diffraction (*64, 65*). The actual process is beyond the scope of this chapter, and interested readers are encouraged to consult the literature cited for further details.

An effective contrast-enhancing layer must possess certain essential characteristics: It must be thermally stable, soluble in solvents that permit casting on photoresist layers with no mixing at the interface, strongly absorbing in the desired spectral region, and bleachable upon exposure. In addition, removal of this layer from the exposed photoresist prior to image development is desirable.

Certain polysilane derivatives possess all of these characteristics and have been used in the mid-UV (300–340-nm) spectral region (*66*). In this regard, it is significant that the polysilanes are readily bleached at shorter wavelengths, because this potentially provides the improved resolution intrinsic to shorter wavelength exposure sources.

Bilayer Applications. Polysilane derivatives have been used both as the imaging layer and the O_2-RIE barrier layer in a number of bilayer processes such as described in Figure 7 (*12*). The images may be developed both by classical wet-development processes and an all-dry technique involving photoablative imaging. In the classical processes, the polysilane is patterned in a conventional fashion, and the images are developed to the planarizing polymer layer using an organic developer. These images can be subsequently transferred to the substrate by an O_2-RIE treatment. Figure 8 shows some high-resolution submicrometer features generated by such a process. The images shown in Figure 8 were initially generated by exposure with a 1:1 commercial projection printer and a mid-UV exposure source. Because the etching process can be highly anisotropic, nearly vertical wall profiles can be produced for high-aspect-ratio features.

Polysilane Self-Development. A second technique that involves photoablative imaging has attracted considerable interest recently, because the procedure represents an alternative that does not require the use of organic solvents for development. For this reason, the photoablation of polysilane derivatives has been studied extensively (*12*, *13*, *63*, *67–69*) and will be discussed in some detail in this section.

In 1982, organic polymers were reported to be self-developable in an imagewise fashion by the use of high-flux light sources such as UV excimer lasers. Since these initial reports, the field has expanded rapidly (*70*, *71*).

The first report of self-development in polysilanes appeared in 1984 (*67*). This prospect was exciting because polysilanes resist oxygen plasma etching by the rapid formation of a protective coating of SiO_2. Because images created in the polysilane can be transferred through a planarizing organic polymer by O_2-RIE processing, laser imaging via self-development should constitute an all-dry (solventless), multilayer lithographic imaging process. Zeigler et al. (*13*) have also reported the imaging of thick single-layer polysilane films upon excimer laser exposure.

Photoablation of Poly(**p-tert*****-butylphenylmethylsilane).*** We (*63*, *68*) have investigated the self-developing characteristics of a typical aromatic polysilane, poly(*p-tert*-butylphenylmethylsilane) (PTBPMS), by using a quartz crystal microbalance to monitor the rate of polymer removal in real time. This technique is particularly useful for measuring thickness changes in materials that are soft and for which stylus penetration from mechanical profilometers is a problem. Changes in polymer weight are accompanied by changes in crystal oscillation frequency, which can be translated into changes in polymer thickness. This technique is sensitive to nanogram changes in weight.

Figure 9 shows the change in crystal oscillation frequency for a PTBPMS sample upon exposure to 248-nm radiation at various fluences from a KrF

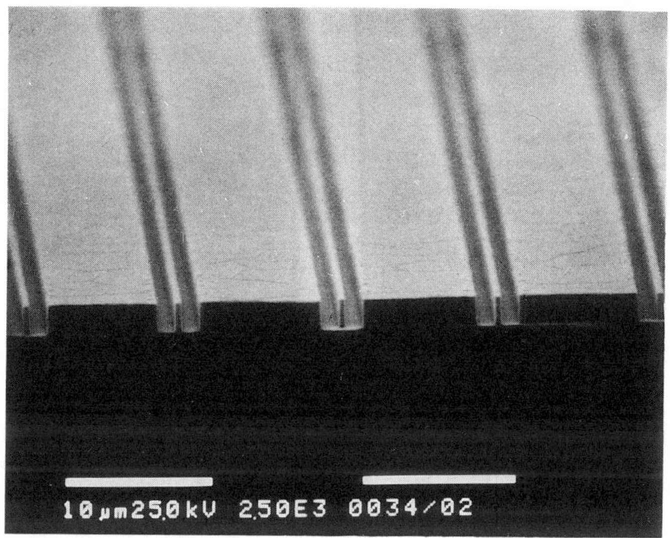

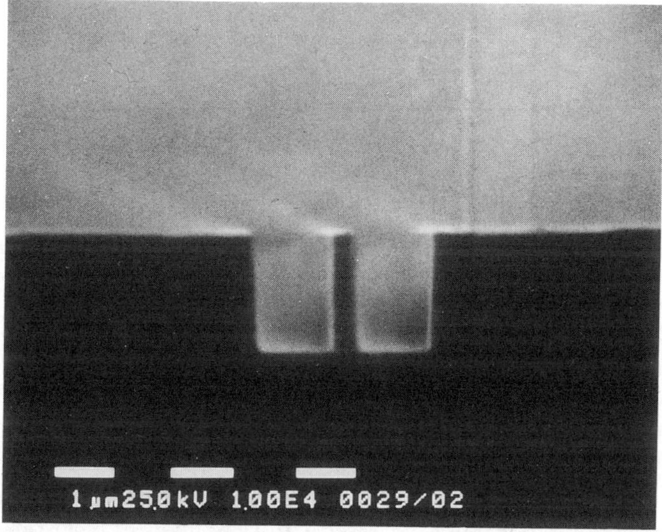

Figure 8. Submicrometer (0.75-µm) features generated in a bilayer of 0.2 µm of poly(cyclohexylmethylsilane) coated over 2.0 µm of a hard-baked AZ photoresist by mid-UV projection lithography (100 mJ/cm²). Image transfer was by O_2-RIE. (Top photo is reproduced with permission from reference 66. Copyright 1984 Society of Photo-Optical Instrumentation Engineers. Bottom photo is reproduced with permission from reference 63. Copyright 1986 Society of Plastics Engineers.)

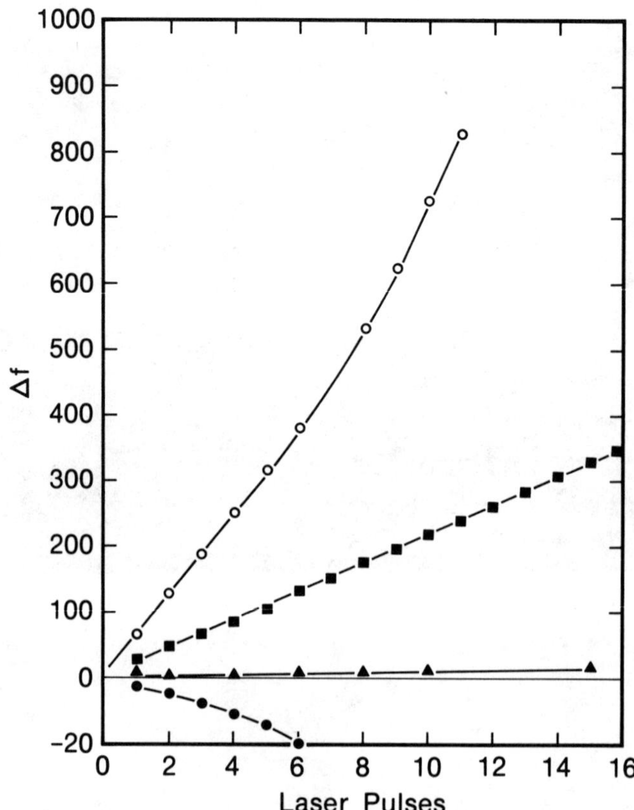

Figure 9. Self-development of a thin (0.2-μm) PTBPMS film upon irradiation at 248 nm (KrF excimer laser) in air. Mass loss is proportional to the frequency change (Δf) of the quartz crystal microbalance. Key to fluences expressed in millijoules per square centimeter per pulse: ●, 34; ▲, 44; ■, 54; ○, 95. (Reproduced with permission from reference 63. Copyright 1986 Society of Plastics Engineers.)

excimer laser. A number of features are obvious upon examination of Figure 9. A threshold fluence above which significant weight loss occurs is identifiable. Above this threshold, the loss of material per pulse is roughly constant, and the weight loss per pulse increases with laser fluence. Near the threshold, significant weight loss does not occur even though photochemistry is occurring, as evidenced by spectral bleaching. Interestingly, below the threshold, a small weight gain is observed. This weight gain would be consistent with polymer photooxidation via the reaction of oxygen with intermediates such as silyl radicals or silylenes.

Although the rate of photoablation above the threshold is not particularly sensitive to the nature of the bath gas, it does depend on the pressure. At a fluence of 76 mJ/cm^2 per pulse, the rate of weight loss of PTBMPS is ~4–6 times greater in vacuum (1.33×10^{-4} Pa) than at atmospheric pressure. In vacuum, the energy utilization efficiency for PTBPMS at 248 nm is estimated to be 1×10^4–1.2×10^4 J/cm^3.

Interestingly, a sample of poly(di-*n*-pentylsilane), which absorbs only weakly at 248 nm, was ablated very slowly at a fluence of 75 mJ/cm^2 per pulse. However, ablation increased markedly with a XeCl source (308 nm), whose spectral output is near the absorption maximum of the sample (313 nm). For this sample at 308 nm, an ablation threshold was also observed at ~50 mJ/cm^2 per pulse. This experiment suggests that the energy deposited per unit volume is an important feature of the ablation process.

We (*63*) have also demonstrated that PTBPMS can be used in a bilayer configuration in an all-dry lithographic process. Figure 10 shows submicrometer images produced initially in the bilayer upon exposure at 248 nm and subsequently transferred by O_2-RIE.

Photoablation of Copolymers. Other workers have also investigated the phenomenon of polysilane self-development. Zeigler and co-workers (*13*) have studied the self-development of a number of polysilane homo- and copolymers and found that self-development efficiencies increase with the size of substituents. They also suggested that the material removal process for alkyl-substituted polysilanes at low fluences (<50 mJ/cm^2 per pulse) is predominantly photochemical rather than photothermal. By using a 1:1 copolymer, poly(methyl-*n*-propylsilane-*co*-isopropylmethylsilane), images were generated by excimer laser exposure at 248 nm.

The same copolymer has been investigated recently in some detail by Hansen and Robitaille (*69*) using He–Ne interferometry for real-time monitoring of the film thickness changes upon exposure at both 248 and 193 nm. At 248 nm and fluences between 60 and 200 mJ/cm^2 per pulse, the energy utilization efficiencies varied from 8.0×10^3 to 20×10^3 J/cm^3. These values are similar to those observed in earlier studies and are also similar to results obtained for other organic polymers such as poly(methyl methacrylate) and poly(ethylene terephthalate). These authors (*69*), however, suggest that high-peak-power values are necessary to remove material cleanly and that the presence of oxygen is detrimental to the production of clean images.

Photoablation Mechanism. *Studies by Mass Spectrometry.* The mechanism of the photovolatilization of polysilanes derivatives has been investigated by several groups, who have probed the nature of the gaseous products by mass spectrometry. Zeigler and co-workers (*13*) studied the fragments ejected from poly(cyclohexylmethylsilane-*co*-dimethylsilane) in vacuo upon irradiation in the deep-UV region and concluded that the major fragments were primarily single monomer units (silylenes), with small amounts of chain fragments up to pentamers. The irradiation source in this

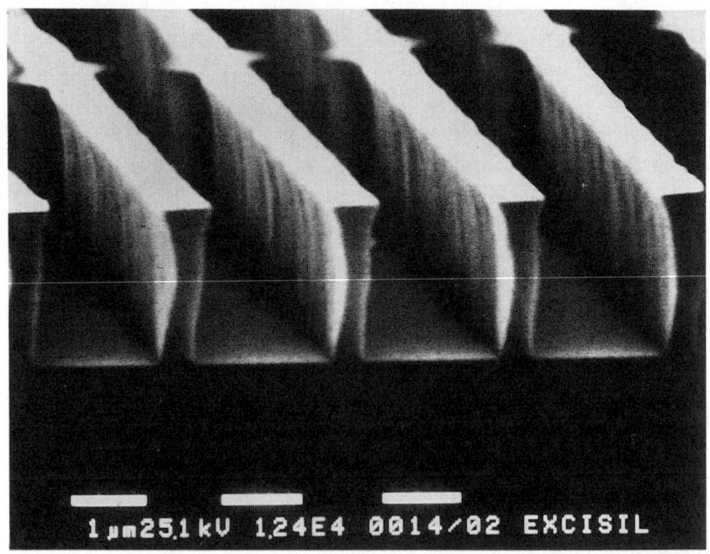

*Figure 10. Self-developed submicrometer images produced in a thin film (0.2 μm) of poly(*p-tert-*butylphenylmethylsilane) by irradiation at 248 nm (55 mJ/cm² per pulse) to a dose of 550 mJ/cm². The images were transferred into 2.0 μm of a hard-baked AZ photoresist by O_2-RIE. (Reproduced with permission from reference 63. Copyright 1986 Society of Plastics Engineers.)*

study was apparently a low-pressure mercury lamp. In the presence of oxygen, a number of mass spectral peaks ascribed to the presence of various cyclic siloxanes were identified.

We (72) have also investigated in some detail the gaseous products ejected from a wide variety of polysilane homo- and copolymers by using a triple-quadrupole mass spectrometer for mass analysis. The excitation source was a XeCl excimer laser (308 nm) operated at 10 Hz. Ionization of the fragments was induced either by electron impact (50 eV) or by multiphoton ionization (MPI). The experiments were conducted mainly in the high-fluence regime (90–250 mJ/cm^2 per pulse), considerably above the polymer ablation thresholds. Seven homopolymers, including samples of poly(di-*n*-hexylsilane) that had been specifically labeled with deuterium in either the α or β position or with ^{13}C on the α sites, as well as a number of copolymers, were investigated. A laser desorption electron-impact mass spectrum of a typical polysilane, poly(dimethylsilane), is shown in Figure 11 (72).

Features of Mechanism. As a result of these studies, a number of conclusions may be drawn about the laser-induced volatilization of polysilane derivatives at 308 nm. The organosilanes are volatilized in sizes ranging from one to at least five monomer units. The mass spectra obtained argue against the generation of significant quantities of isocyclic silane derivatives, because these materials usually show strong parent ion peaks. The observation of significant intensities for $(R_1R_2Si)_nH^+$ ions is interesting because electron-impact studies on model organosilanes indicate that the formation of these ions is relatively rare. This result suggests that SiH compounds are probably formed prior to ionization. Isotopic-labeling studies indicate that the hydrogens come from positions other than just the α- and β-side-chain positions and also that silicon atoms become attached to carbon atoms other than those originally bonded in the α positions.

No compelling evidence exists for the presence of monomeric silylenes (RR′Si) as the bulk of the ejected neutral species. This conclusion is based on a number of observations. First, in all but one case studied, the intensity of the monomeric parent ion was quite low, and considerable quantities of silicon fragments with higher masses were evident. MPI studies (at 345–505 nm) of the gaseous products from poly(dimethylsilane) failed to produce any significant enhancement in the parent ion of dimethylsilylene, and higher mass fragments were always present. Finally cryotrapping of the gaseous products in an argon matrix did not produce any evidence of dimethylsilylene absorption (73).

Another interesting observation is that significant silicon–carbon bond cleavage occurs in the high-fluence regime. This cleavage is manifested by the appearance in the mass spectra of the characteristic signals for the olefin derived from the aliphatic side chain. In the case of poly(cyclo-

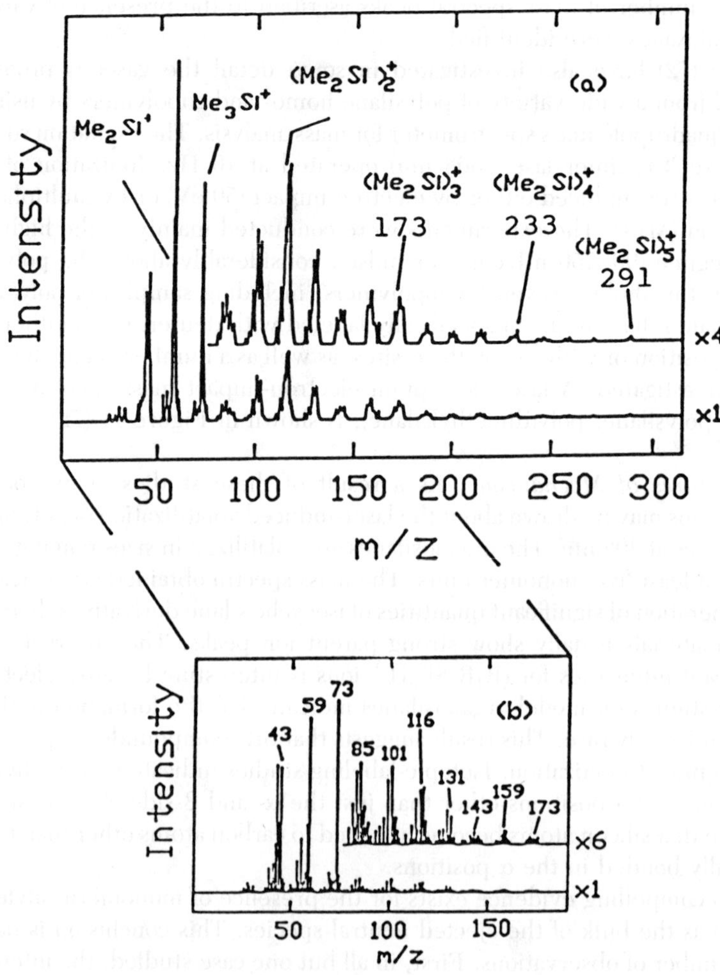

Figure 11. Laser desorption electron-impact mass spectrum of poly(dimethylsilane) under low (a) and high (b) resolution.

hexylmethylsilane) and poly(phenylmethylsilane), cyclohexene and benzene are produced. Simple homolysis of the silicon–carbon bond to generate the corresponding alkyl radicals is considered unlikely, because no alkane is produced concurrently. The corresponding olefins are produced prior to ionization, as demonstrated by trapping and analysis of the gaseous products from ablation. Furthermore, isotopic-labeling experiments indicate that the olefin is formed by the loss of one β hydrogen. Although the ratio of olefin to silicon-containing fragments decreases with decreasing fluence, the presence of olefinic products (derived from the longer alkyl substituents) persists even at low fluences.

The olefins represent the first stable products isolated from the photoablation of substituted polysilanes, and as such, any mechanistic hypothesis must include these species. The absence of literature precedents for photochemical silicon–carbon bond cleavage suggests the possibility of a thermal route. This suggestion is further strengthened by the observation that exposure of a number of polysilanes to IR radiation from a cw CO_2 laser results in gaseous mixtures with compositions (as determined by mass spectral analyses) that closely resemble those produced by irradiation at 308 nm.

The origin of the olefin product requires some interesting speculation. The only silicon-containing intermediates that are known to lose an alkene fragment thermally with the specific transfer of a β hydrogen are the alkyl silylenes (activation energy [E_a] ~ 30 kcal/mol) (*74–77*).

$$-SiCH_2CH_2- \rightarrow \overline{SiHCH_2CH}- \rightarrow -SiH + CH_2{=}CH- \quad (3)$$

Because no clear evidence of dimethylsilylene was obtained upon irradiation of poly(dimethylsilane), longer chain dialkylsilylenes are unlikely to be produced.

An attractive, although tentative, alternative would be an alkyl-substituted silylsilylene formed from the polymer chain. Two thermodynamically reasonable routes to such intermediates are possible. The first route (equation 4) involves 1,1-elimination to produce the silylsilylene directly. This route has a precedent in organosilane thermal processes (*78, 79*). The second route (equations 5a and 5b) involves rearrangement from a silene produced by the disproportionation (*46, 80, 81*) of two silyl radicals caused by bond homolysis. This type of rearrangement has also been described in the literature (*82*). The postulated silylsilylenes are also attractive intermediates to explain the rebonding of silicon to carbon atoms other than those in the original α positions (CH insertion), which is obvious from the mass spectral analysis of gaseous products from the laser ablation of isotopically labeled poly(di-*n*-hexylsilane).

$$-SiR_2{-}SiR_2{-}SiR_2{-}SiR_2- \rightarrow -SiR_2{-}\ddot{Si}{-}R + R_3Si{-}SiR_2- \quad (4)$$

$$-SiR_2{-}\dot{Si}RCH_2R' + \dot{Si}R_2- \rightarrow -SiR_2{-}SiR{=}CHR' + HSiR_2- \quad (5a)$$

$$R'CH{=}SiR{-}SiR_2- \rightarrow -SiR_2{-}CHR'{-}\ddot{Si}R \quad (5b)$$

Radiation Chemistry of Polysilanes

Whereas oxygen can play a role when polysilanes are photodegraded in air, photobleaching with attendant molecular weight reduction occurs even when polysilane derivatives are irradiated in an inert atmosphere or under vacuum.

This has prompted the investigation of the potential of substituted polysilane derivatives as electron beam resists.

Figure 12 shows images created in a bilayer composed of a thin layer of a typical polysilane coated over a thick, hard-baked layer of a typical AZ-type photoresist. After solvent development of the imaged layer, the pattern

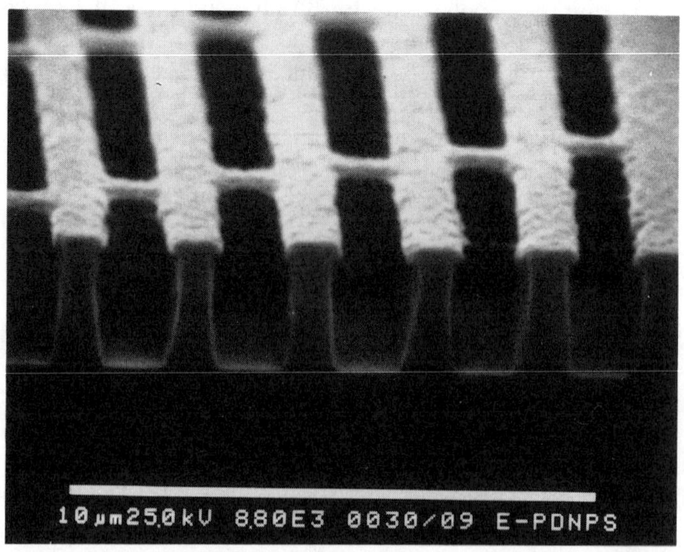

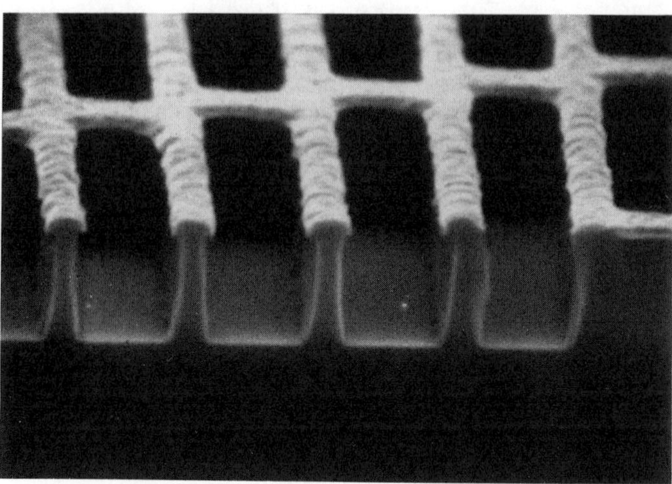

Figure 12. Electron beam imaging of poly(di-n-pentylsilane) (0.14 µm) coated over 2.0 µm of a hard-baked AZ-type photoresist exposed at 20 µC/cm² and wet developed. Pattern transfer was by O_2-RIE.

was transferred by O_2-RIE. The process described is unoptimized, but it demonstrates that polysilanes can be imaged in a positive mode even when the imaging step is conducted in vacuum with ionizing-radiation sources.

To study the structural sensitivity of polysilanes to ionizing radiation, a number of samples were irradiated with a calibrated ^{60}Co source, and the degraded materials were analyzed by GPC in a manner similar to that described for the determination of photochemical quantum yields (59). In radiation processes, the slopes of the plots of molecular weight versus absorbed dose yield the *G* values for scissioning, *G*(s), and cross-linking, *G*(x), rather than the respective quantum yields. These values, which represent the number of chain breaks or cross-links per 100 eV of absorbed dose, are indicative of the relative radiation sensitivity of the material. The data for a number of polysilanes are given in Table IV. Also included in Table IV for comparison is the value for a commercial sample of poly(methyl methacrylate) run under the same conditions. The *G*(s) value of this sample compares favorably with that reported in the literature (83).

The data in Table IV show that polysilanes undergo predominantly scission during γ radiolysis. In all cases, the *G*(s)/*G*(x) ratios were greater than 10. When aromatic substituents are directly attached to the aromatic ring, the *G*(s) value decreases significantly. This trend is consistent with the previous observation that *G* values for polystyrene derivatives are considerably lower than those observed for comparable saturated carbon-backbone polymers and suggests that aromatic substituents impart some radiation stability (83). The data in Table IV further suggest that for polysilanes this effect is limited to aromatic substituents that are directly attached to the polymer backbone, because the *G*(s) value for poly(phenethylmethylsilane) is among the highest that we have measured.

These results raise the interesting question of whether remote aromatic substitution in carbon-based polymers will still impart the radiation stability characteristic of substituted polystyrene derivatives. These data (Table IV), although preliminary and somewhat incomplete, suggest that structural variations in high-molecular-weight polysilanes can result in significant differences in their sensitivity to ionizing radiation. These preliminary trends provide some insight for the development of new polysilanes for resist studies.

Conclusion

Polysilane derivatives constitute a new class of radiation-sensitive materials with interesting physical and electronic properties. The photochemical decomposition of polymers containing disilanyl units seems adequately explained by silicon–silicon bond homolysis and subsequent radical reactions. The solution photochemistry of longer silicon catenates results in the extrusion of substituted monomeric silylenes, as well as the formation of silyl radicals produced by chain homolysis. Recent studies indicating that the

Table IV. ^{60}Co γ-Radiolysis Data for Substituted Polysilane Derivatives

Polymer	$\overline{M}_n^o \times 10^3$	G(*s*)	G(*x*)	G(*s*)/G(*x*)
$[(C_4H_9)_2Si]_n$	304.9	0.42	0.023	18
$[(C_5H_{11})_2Si]_n$	807.6	0.40	—	—
$[(C_6H_{13})_2Si]_n$	971.9	0.42	0.041	10.5
$[(C_{14}H_{29})_2Si]_n$	1472.9	0.86	0.035	24.6
$(C_6H_5SiCH_3)_n$	738.2	0.26	0.014	18.6
$(t\text{-}C_4H_9C_6H_4SiCH_3)_n$	153.9	0.14	0.004	35
$(C_6H_5CH_2CH_2SiCH_3)_n$	167.3	0.90	0.03	30
PMMA	428.7	1.40	—	—

NOTE: The symbol — indicates that the property was too small to measure for the sample. PMMA is poly(methyl methacrylate).

photolysis of high-molecular-weight polysilanes is wavelength dependent (*84*) suggest that the photodecomposition is more complex than was first indicated by initial exhaustive irradiation studies performed at 254 nm. Also, ESR studies on irradiated samples show that the persistent silicon-centered radicals generated are not those expected from simple chain homolysis. Additional studies will be necessary before the mechanism of persistent radical formation can be described with confidence.

In addition, mechanistic studies of the photochemical reactions are necessary to determine whether similar processes occur in the solid state. Polymer chain scission is usually the predominant process in the solid state, although cross-linking reactions become more important in the presence of pendant unsaturation. However, little is known about the nature of the intermediates produced in the solid state. Information of this type is important, because most of the applications of polysilane derivatives require the materials as solid films.

The usefulness of high-molecular-weight polysilane derivatives has been demonstrated for a variety of lithographic processes. Although the intense absorption of the polysilane backbone suggests that high-molecular-weight polysilanes are best suited for multilayer processes, the observed bleaching of the absorption upon irradiation, as well as the tendency of these polymers to photoablate with intense light sources, suggests that single-layer applications are not necessarily precluded. A better understanding of the radiation-induced processes, particularly the structure–reactivity relationships, is essential for the optimization of current lithographic procedures, as well as the development of new applications.

In summary, polysilane derivatives constitute a new class of radiation-sensitive materials for which a number of new applications have appeared. Much of the interest has centered around their unusual electronic properties and their sensitivity to various types of radiation. The nature of these radiation-induced processes is very complex, and although some progress has been made recently in understanding these processes, this area remains a fertile field for future study.

Acknowledgments

I acknowledge the contributions of D. Hofer and D. LeVergne of IBM in the γ-radiolysis experiments. The GPC studies were performed by C. Cole of IBM. Finally, I gratefully acknowledge the partial financial support for the work performed at IBM by the Office of Naval Research.

References

1. Kipping, F. S. *J. Chem. Soc.* **1924,** *125*, 2291.
2. Burkhard, C. A. *J. Am. Chem. Soc.* **1949,** *71*, 963.
3. Trujillo, R. E. *J. Organomet. Chem.* **1980,** *198*, C27.

4. Wesson, J. P.; Williams, T. C. *J. Polym. Sci., Polym. Chem. Ed.* **1980,** *180,* 959.
5. West, R.; David, L. D.; Djurovich, P. I.; Stearley, K. L.; Srinivasan, K. S. V.; Yu, H. G. *J. Am. Chem. Soc.* **1981,** *103,* 7352.
6. Yajima, S.; Hayashi, J.; Omori, M. *Chem. Lett.* **1975,** 931.
7. Hasegawa, Y.; Iimura, M.; Yajima, S. *J. Mater. Sci.* **1980,** *15,* 1209.
8. West, R. In *Ultrastructure Processing of Ceramics, Glasses and Composites;* Hench, L.; Ulrich, D. C., Eds.; John Wiley and Sons: New York, 1984.
9. West, R.; Wolff, A. R.; Peterson, D. J. *J. Rad. Curing* **1986,** *13,* 35.
10. Kepler, R. G.; Zeigler, J. M.; Harrah, L. A.; Kurtz, S. R. *Phys. Rev. B* **1987,** *35,* 2818.
11. Stolka, M.; Yuh, H.-J.; McGrane, K.; Pai, D. M. *J. Polym. Sci. Part A: Polym. Chem.* **1987,** *25,* 823.
12. Miller, R. D.; Hofer, D.; Rabolt, J. F.; Sooriyakumaran, R.; Willson, C. G.; Fickes, G. N.; Guillet, J. E.; Moore, J. In *Polymers for High Technology;* Bowden, M. J.; Turner, S. R., Eds.; ACS Symposium Series 346; American Chemical Society: Washington, DC, 1987; p 170 and references cited therein.
13. Zeigler, J. M.; Harrah, L. A.; Johnson, A. W. *Proc. SPIE* **1985,** *539,* 166.
14. Kajzar, K.; Messier, J.; Rosilio, C. *J. Appl. Phys.* **1986,** *60,* 3040.
15. Baumert, J.-C.; Bjorklund, G. C.; Lundt, D. H.; Jurich, M. C.; Looser, H.; Miller, R. D.; Rabolt, J. F.; Swalen, J. D.; Twieg, R. J. *Appl. Phys. Lett.* **1988,** *53,* 1147.
16. West, R. *J. Organomet. Chem.* **1986,** *300,* 327 and references cited therein. Miller, R. D.; Michl, J. *Chem. Rev.* **1989,** in press.
17. Aitken, C. T.; Harrod, J. F.; Samuel, E. *J. Am. Chem. Soc.* **1986,** *108,* 4059.
18. Aitken, C. T.; Harrod, J. F.; Samuel, E. *J. Organomet. Chem.* **1985,** *279,* C11.
19. Aitken, C. T.; Harrod, J. F.; Samuel, E. *Can. J. Chem.* **1986,** *64,* 1677.
20. Aitken, C. T.; Harrod, J. F.; Gill, U. S. *Can. J. Chem.* **1987,** *65,* 1804.
21. Becker, B.; Corriu, R.; Guerin, C.; Henner, B. *Abstracts, 8th Int. Symp. Organosilicon Chem.*, St. Louis, Missouri, 1987; 84.
22. Trefonas, P. T., III; Djurovich, P. I.; Zhang, X.-X.; West, R.; Miller, R. D.; Hofer, D. *J. Polym. Sci., Polym. Lett. Ed.* **1983,** *21,* 819.
23. Zeigler, J. M. *Polym. Prepr. (Am. Chem. Soc., Div. Polym. Chem.)* **1986,** *27,* 109.
24. Trefonas, P. T., III; West, R.; Miller, R. D.; Hofer, D. *J. Polym. Sci., Polym. Lett. Ed.* **1983,** *21,* 823.
25. Miller, R. D.; Hofer, D.; Rabolt, J.; Fickes, G. N. *J. Am. Chem. Soc.* **1985,** *107,* 2172.
26. Rabolt, J. F.; Hofer, D.; Miller, R. D.; Fickes, G. N. *Macromolecules* **1986,** *19,* 6111.
27. Kuzmany, H.; Rabolt, J. F.; Farmer, B. L.; Miller, R. D. *J. Chem. Phys.* **1986,** *85,* 7413.
28. Lovinger, A. J.; Schilling, F. C.; Bovey, F. A.; Zeigler, J. M. *Macromolecules* **1986,** *19,* 2657.
29. Miller, R. D.; Farmer, B. L.; Fleming, W.; Sooriyakumaran, R.; Rabolt, J. *J. Am. Chem. Soc.* **1987,** *109,* 2509.
30. Harrah, L. A.; Zeigler, J. M. *Macromolecules* **1987,** *20,* 601.
31. Miller, R. D.; Rabolt, J. F.; Sooriyakumaran, R.; Fleming, W.; Fickes, G. N.; Farmer, B. L.; Kuzmany, H. In *Inorganic and Organometallic Polymers;* Zeldin, M.; Wynne, K. J.; Allcock, H. R., Eds.; ACS Symposium Series 360; American Chemical Society: Washington, DC, 1987; p 43 and references cited therein.
32. Trefonas, P. T., III; Damewood, J. R.; West, R.; Miller, R. D. *Organometallics* **1985,** *4,* 1318.
33. Harrah, L. A.; Zeigler, J. M. *J. Polym. Sci., Polym. Lett. Ed.* **1985,** *23,* 209.

34. Miller, R. D.; Sooriyakumaran, R. *J. Polym. Sci., Polym. Lett. Ed.* **1987,** *25,* 321.
35. Cotts, P. M.; Miller, R. D.; Sooriyakumaran, R. *Abstracts* Advances in Silicon-Based Polymer Science, Makaha, Oahu, Hawaii, 1987, 54.
36. Miller, R. D.; Sooriyakumaran, R., unpublished results.
37. Ban, H.; Sukegawa, K. *J. Appl. Polym. Sci.* **1987,** *33,* 2787.
38. Miller, R. D.; Hofer, D.; McKean, D. R.; Willson, C. G.; West, R.; Trefonas, P. T., III In *Materials for Microlithography;* Thompson, L. F.; Willson, C. G.; Fréchet, J. M. J., Eds.; ACS Symposium Series 266; American Chemical Society: Washington, DC, 1984; p 294.
39. Turro, N. J. In *Modern Molecular Photochemistry,* Benjamin/Cumming Publishing: Menlo Park, CA, 1978; Chapter 5.
40. Diaz, A. F.; Miller, R. D. *J. Electrochem. Soc.* **1985,** *132,* 834.
41. Hamill, W. H. In *Radical Ions;* Kaiser, C. T.; Kevan, L., Eds.; Interscience: New York, 1968.
42. Ishikawa, M.; Kumada, M. *Adv. Organomet. Chem.* **1981,** *19,* 51 and references cited therein.
43. Wilt, J. W. In *Reactive Intermediates;* Abramovitch, R. A., Ed.; Plenum: New York, 1983; Chapter 3.
44. Walsh, R. *Acc. Chem. Res.* **1981,** *14,* 246.
45. Raabe, G.; Michl, J. *Chem. Rev.* **1985,** *85,* 419.
46. Hawari, J. A.; Griller, D.; Weber, W. P.; Gaspar, P. P. *J. Organomet. Chem.* **1987,** *326,* 335.
47. Boudjouk, P.; Roberts, J. R.; Golino, C. M.; Sommer, L. H. *J. Am. Chem. Soc.* **1972,** *94,* 7926.
48. Ishikawa, M.; Fuchikami, T.; Kumada, M. *J. Organomet. Chem.* **1976,** *118,* 155.
49. Gaspar, P. P.; Holter, D.; Konieczny, C.; Corey, J. Y. *Acc. Chem. Res.* **1987,** *20,* 329 and references cited therein.
50. Nate, K.; Ishikawa, M.; Imamura, N.; Murakami, V. *J. Polym. Sci., Part A, Polym. Chem.* **1986,** *24,* 1551.
51. Ishikawa, M.; Hongzhi, N.; Matsusaki, K.; Nate, K.; Inoue, T.; Yokono, H. *J. Polym. Sci., Polym. Lett. Ed.* **1984,** *22,* 669. Nate, K.; Ishikawa, M.; Ni, H.; Watanabe, H.; Saheki, Y. *Organometallics* **1987,** *6,* 1673.
52. Ishikawa, M.; Kumada, M. *J. Organomet. Chem.* **1972,** *42,* 325.
53. Drahnak, T. J.; Michl, J.; West, R. *J. Am. Chem. Soc.* **1979,** *101,* 5427.
54. Sakurai, H.; Kobayashi, Y.; Nakadana, Y. *J. Am. Chem. Soc.* **1974,** *96,* 2656.
55. Ishikawa, M.; Takaoka, T.; Kumada, M. *J. Organomet. Chem.* **1972,** *42,* 333.
56. Trefonas, P. T., III; West, R.; Miller, R. D. *J. Am. Chem. Soc.* **1985,** *107,* 2737.
57. Michl, J.; Downing, J. W.; Karatsu, T.; Klingensmith, K. A.; Wallraff, G. M.; Miller, R. D. *Inorganic and Organometallic Polymers;* Zeldin, M.; Wynne, K. J.; Allcock, H. R., Eds.; ACS Symposium Series 360; American Chemical Society: Washington, DC, 1988; Chapter 4.
58. Recent variable-temperature ESR (electron spin resonance) studies have determined that the early persistent radicals produced from symmetrical dialkylpolysilanes upon photolysis appear to have the structure $\left(SiR_2\dot{S}iRSiR_2\right)_n$. Although these radicals could conceivably be produced by simple silicon–carbon bond homolysis, supporting studies indicate that a more complex pathway to these radicals is involved. McKinley, A. J.; Karatsu, T.; Wallraff, G. M.; Miller, R. D.; Sooriyakumaran, R.; Michl, J. *Organometallics* **1988,** *7,* 2569.
59. Willson, C. G. In *Introduction to Microlithography;* Willson, C. G.; Bowden, M. J., Eds.; ACS Symposium Series 219; American Chemical Society: Washington, DC, 1983; Chapter 3 and references cited therein.
60. Miller, R. D.; Guillet, J. E.; Moore, J. *Polym. Prepr.* **1988,** *29,* 552.

61. Willson, C. G. In *Introduction to Microlithography;* Willson, C. G.; Bowden, M. J., Eds.; ACS Symposium Series 219; American Chemical Society: Washington, DC, 1983; Chapter 6.
62. Reichmanis, E.; Smolinsky, G.; Wilkins., C. W., Jr. *Solid State Technol.* **1985,** *28*(8), 130.
63. Miller, R. D.; Hofer, D.; Fickes, G. N.; Willson, C. G.; Marinero, E.; Trefonas, P. T., III; West, R. *Polym. Eng. Sci.* **1986,** *26,* 1129.
64. Griffing, B. F.; West, P. R. *Polym. Eng. Sci.* **1983,** *23,* 947.
65. West, P. R.; Griffing, B. F. *Proc. SPIE* **1984,** *33,* 394.
66. Hofer, D. C.; Miller, R. D.; Willson, C. G.; Neureuther, A. R. *Proc. SPIE* **1984,** *469,* 108.
67. Hofer, D. C.; Jain, K.; Miller, R. D. *IBM Tech. Disclosure Bull.* **1984,** *26,* 5683.
68. Marinero, E. E.; Miller, R. D. *Appl. Phys. Lett.* **1987,** *50,* 1041.
69. Hansen, S. G.; Robitaille, T. E. *J. Appl. Phys.* **1987,** *62,* 1394.
70. Srinivasan, R. *Science* **1986,** *234,* 559 and references cited therein.
71. Yeh, J. T. C. *J. Vac. Sci. Technol.,* A **1986,** *4,* 653.
72. Magnera, T. F.; Balaji, V.; Michl, J.; Miller, R. D. In *Silicon Chemistry;* Corey, J. Y.; Corey, E. R.; Gaspar, P. P., Eds.; Ellis Horwood Publishers: Chichester, England, 1988, p 491.
73. Drahnak, T. J.; Michl, J.; West, R. *J. Am. Chem. Soc.* **1979,** *101,* 5427.
74. Gusel'nikov, L. E.; Polyakov, Yu. P.; Volnina, E. A.; Nametkin, N. S. *J. Organomet. Chem.* **1985,** *292,* 189.
75. Barton, T. J.; Burns, G. T. *Organometallics* **1983,** *2,* 1.
76. Rickborn, S. F.; Ring, M. A.; O'Neal, H. E. *Int. J. Chem. Kinet.* **1984,** *16,* 1371.
77. Sawrey, B. A.; O'Neal, H. E.; Ring, M. A.; Coffey, D., Jr. *Int. J. Chem. Kinet.* **1984,** *16,* 801.
78. Davidson, I. M. T.; Howard, A. V. *J. Chem. Soc. Faraday Trans. 1* **1975,** *71,* 69.
79. Chen, Y. S.; Cohen, B. H.; Gaspar, P. P. *J. Organomet. Chem.* **1980,** *195,* C1.
80. Gammie, L.; Safarik, I.; Strausz, O. P.; Roberge, R.; Sandorfy, C. *J. Am. Chem. Soc.* **1980,** *102,* 378.
81. Doyle, D. J.; Tokach, S. K.; Gordon, M. S.; Koob, R. D. *J. Phys. Chem.* **1982,** *86,* 3626.
82. Barton, T. J.; Burns, S. A.; Burns, G. T. *Organometallics* **1982,** *1,* 210.
83. Schnabel, W. In *Aspects of Degradation and Stabilization of Polymers;* Jellinek, H. H. G., Ed.; Elsevier: Amsterdam, 1978; Chapter 4.
84. Karatsu, T.; Miller, R. D.; Sooriyakumaran, R.; Michl, J. *J. Am. Chem. Soc.* **1989,** *111,* 1140.
85. Ban, H.; Sukegawa, K. *J. Polym. Sci., Polym. Chem. Ed.* **1988,** *26,* 521.

RECEIVED for review May 27, 1988. ACCEPTED revised manuscript March 27, 1989.

25

Exciton–Exciton Annihilation in Polysilanes

R. Glen Kepler and John M. Zeigler[1]

Sandia National Laboratories, Albuquerque, NM 87185

*The exciton–exciton annihilation rate constant for singlet excitons in solid films of poly(*n*-propylmethylsilane) was determined by measuring the fluorescent intensity from the films as a function of incident-light intensity in two types of experiments: one in which the excitons were created by single-photon transitions and one in which the excitons were created by two-photon transitions. The rate constant is* $\sim 10^{-7}$ cm^3/s.

THE TRANSPORT OF BOTH CHARGE CARRIERS and neutral excited states in organic molecular crystals and polymers has been studied extensively in recent years (*1–6*). Almost all of these studies have been on molecules containing delocalized, conjugated π electrons and have focused on the role of these electrons in the transport. Very recently, silicon-backbone polymers, polysilanes, were found to exhibit many of the same electronic properties as those previously thought to arise exclusively from conjugated π electrons, even though many of the σ-bonded polysilanes contain no π electrons (*7*). In both solution and the solid state, these molecules exhibit high quantum efficiencies for fluorescence (*8–10*) and absorption spectra that depend strongly on molecular weight (*8*) and conformation (*8–15*). Solid films of polysilanes are excellent photoconductors (*16, 17*) and transport charge efficiently (*18, 19*). These properties are believed to result from σ electrons delocalized along the Si backbone, but relatively little is known presently about the theoretical description of the delocalization of these electrons and their excited states.

In this chapter, we report some measurements of energy transport by

[1]Current address: SilCHEMY, 2208 Lester Drive, NE, Albuquerque, NM 87112–2640

0065–2393/90/0224–0459$06.00/0

singlet excitons (*8, 16, 17*) in poly(*n*-propylmethylsilane). We studied exciton–exciton annihilation, a process that occurs at relatively high exciton concentrations, by using two experimental techniques to create the excitons: single-photon and two-photon transitions. We found that exciton–exciton annihilation is an easily observed phenomenon that becomes the dominant process for exciton loss at a concentration in the order of 10^{16} excitons per cubic centimeter, even though the exciton lifetime is only 0.6 ns.

Experimental Procedures

Crude poly(*n*-propylmethylsilane) was obtained by adding sodium dispersion to a solution of purified *n*-propylmethyldichlorosilane in refluxing dry toluene or in dry toluene mixed with heptane (*20, 21*) under well-controlled conditions. Two precipitations of the crude material from toluene with ethyl acetate and two from tetrahydrofuran with methanol afforded a pure white rubbery solid in 25% overall yield. The weight-average molecular weight (M_w) of this solid was found to be 2.7×10^5 daltons by gel permeation chromatography (calibrated with polystyrene), a value that corresponds to a polymer containing 3140 Si atoms in the backbone chain. The polymer was established (*22*) to be essentially atactic, that is, it has random stereochemistry about the backbone silicon atoms. Films of various thicknesses were prepared by solvent casting, spin casting for films less than 1 μm, and multiple-layer solvent casting for thick films. UV quartz was used as the substrate. The samples were mounted on a cold finger in a vacuum chamber. Most experiments were conducted at room temperature, but a few were conducted at temperatures as low as 20 K.

Excitons were created in the films in two ways: by single-photon transitions using strongly absorbed light and by two-photon transitions using photons with energies well below that required to create an exciton directly. The two-photon-transition exciton–exciton annihilation experiments were conducted with frequency-doubled light from a neodymium-doped yttrium–aluminum–garnet (YAG) laser with a pulse width at half maximum of about 5 ns. Most of the single-photon experiments were conducted at 320 nm, but some were conducted at 280 nm. In both cases, the light was obtained by pumping a dye laser with the frequency-doubled neodymium-doped YAG laser light. Excitons for this exciton lifetime experiment were created by two-photon transitions using frequency-doubled light from a neodymium-doped YAG laser with a pulse width of approximately 200 ps.

An experiment consisted of measuring the peak fluorescent light intensity during a pulse as a function of the intensity of the incident light. The incident light intensity was measured by using a pellicle to reflect a fraction of the beam into a light meter (Laser Precision Corporation energy ratiometer, model Rj-7200). The intensity was varied by using a series of filters. The intensity of the fluorescent light was measured with a gated microchannel plate photomultiplier (Hamamatsu, model R2024V). The light intensity incident on the photomultiplier was maintained in the linear range of the photomultiplier by using a series of calibrated neutral density filters. The intensity of the incident light was limited to those intensities that did not damage the sample, as evidenced by the repeatability of experimental results.

Results

Typical experimental results obtained on 50-μm-thick samples are shown in Figures 1 and 2 for single-photon excitation and two-photon excitation, re-

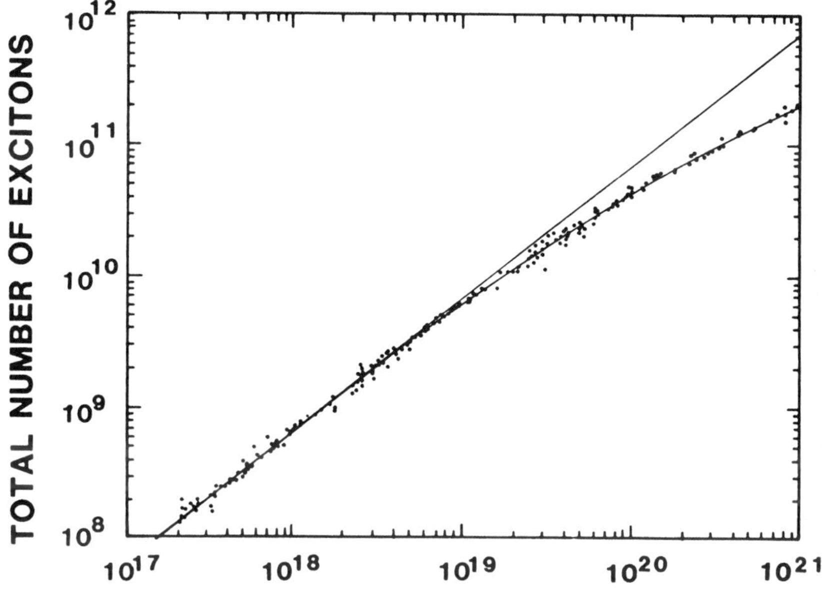

Figure 1. Fluorescence intensity versus incident laser light intensity at 320 nm. The straight line was fit to the low-intensity data and is drawn with the assumption that the fluorescence intensity is proportional to the intensity of the incident laser light. The curved solid line is the result predicted by the phenomenological equation for single-photon excitation with $\gamma = 4.2 \times 10^{-8}$ cm^3/s. The area of the exciting laser beam was about 0.1 cm^2.

spectively. The solid lines are theoretical best fits to the experimental data, which were determined by using the following phenomenological equations and the exciton–exciton annihilation rate constant as a parameter.

$$\frac{dn}{dt} = \alpha I_0 e^{-\alpha x} - \beta n - \gamma n^2 \qquad \text{(single–photon excitation)} \qquad (1)$$

$$\frac{dn}{dt} = \kappa I_0^2 - \beta n - \gamma n^2 \qquad \text{(two–photon excitation)} \qquad (2)$$

In these equations, n is the exciton concentration, t is time, α is the absorption coefficient for the incident light, I_0 is the intensity of the incident light, κ is the two-photon absorption rate constant at 532 nm, β is the reciprocal of the exciton lifetime, and γ is the exciton–exciton annihilation rate constant.

The absorption coefficient α was determined to be 2.4×10^5 cm^{-1} at 320 nm and 9.2×10^4 cm^{-1} at 280 nm by measuring the optical density and thickness of several films. The exciton lifetime t ($t = 1/\beta$) was measured by using frequency-doubled light from a 200-ps neodymium-doped YAG laser

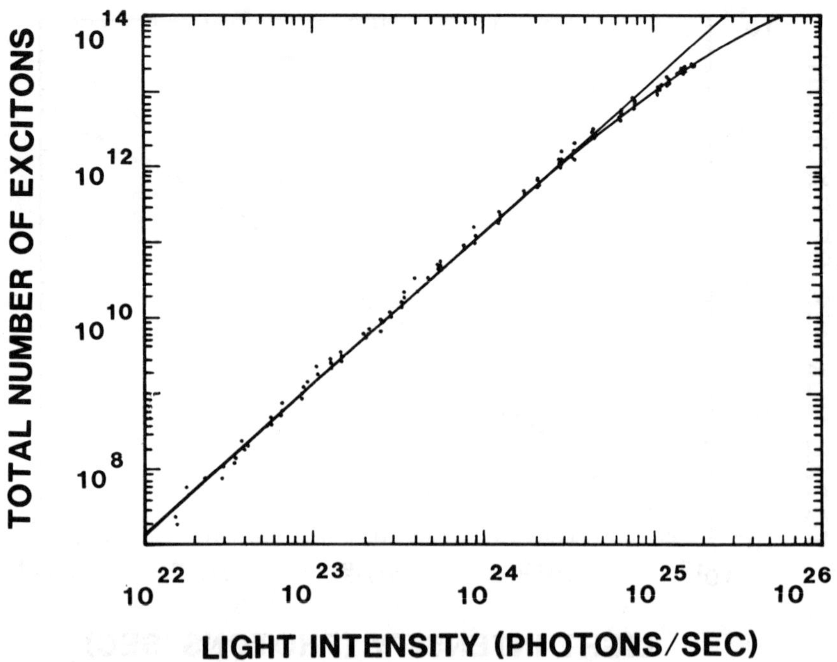

Figure 2. Fluorescence intensity versus incident laser light intensity at 532 nm. The straight line was fit to the low-intensity data and is drawn with the assumption that the fluorescence intensity is proportional to the square of the intensity of the incident laser light. The curved solid line is the result predicted by the phenomenological equation for two-photon excitation with $\gamma = 10^{-7}$ *cm³/s. The area of the exciting laser beam was about 0.1 cm².*

to create excitons in a thick, approximately 50-μm film, and the time dependence of the fluorescence was measured with a fast photodiode. We used this technique to measure the fluorescent lifetime for three types of polysilane with the following results: for poly(*n*-propylmethylsilane), $t = 0.6$ ns; for poly(di-*n*-hexylsilane), $t = 1.0$ ns; and for poly(phenylmethylsilane), $t < 0.3$ ns.

For both experiments, we assumed that $dn/dt = 0$, because the exciton lifetime is considerably shorter than the excitation light pulse width. Therefore, for the two-photon experiment at light intensities such that $\gamma n^2 << \beta n$, $\kappa = \beta n/I_0^2$. We found $k = 10^{-26}$ cm^3/s at 532 nm. The solid curve in Figure 2 was determined by using the constants just mentioned and $\gamma = 10^{-7}$ cm^3/s. For the single-photon-transition experiment, the exciton concentration n is a strong function of depth in the sample, and thus $\int ndx$ is plotted as the solid line in Figure 1 with $\gamma = 4.2 \times 10^{-8}$ cm^3/s.

Because the incident laser light intensity was not uniform over its area, the theoretical curves in Figures 1 and 2 were calculated by first measuring the incident laser beam intensity at 900 equally spaced points over its area.

We then assumed that the light intensity was uniform within each of these 900 equal-area segments. The total number of excitons present at equilibrium was then calculated for a given total number of excitons in a pulse by summing the theoretically calculated equilibrium density in each of the segments. The variation in light intensity over the area of the beam was taken into account in a similar fashion when we calculated the two-photon-absorption rate constant κ.

Preliminary experiments conducted at 20 K indicate that the value of γ at 20 K is within our experimental error of about a factor of 3 of its value at room temperature.

Discussion

The possibility that our observations are the result of some process other than exciton–exciton annihilation should be examined. The most obvious experimental observation that will establish exciton–exciton annihilation as the process would be a decrease in the exciton lifetime at high concentrations. We have been unable to carry out this experiment so far. The most obvious possible alternative process that might explain the experimental results is exciton photoionization. It is highly unlikely that our results can be explained by this process for two reasons. First, the exciton photoionization coefficient at 320 nm would have to be unrealistically high to account for the observations. If σ is the exciton photoionization cross-section, then $\sigma I = \beta$ at the intensity at which the rate of exciton loss by the intensity-dependent process equals the rate of loss at low intensities. Because the intensity at which this occurs is about 10^{22} photons per square centimeter per second, σ would have to be about 10^{-13} cm^2, an unreasonably large number. Second, the cross-section for photoionization at 532 nm would have to be about 3 orders of magnitude lower and fortuitously of just the right magnitude to allow the two different experiments to give results that are almost consistent with a single exciton–exciton annihilation rate constant.

These results are strong evidence that excitons in polysilanes are highly mobile. The exciton concentration when the rate of loss by exciton–exciton annihilation equals the rate of loss by other nonannihilation processes can be determined by setting $\beta n = \gamma n^2$ to give $n = \beta/\gamma \sim 10^{16}$ excitons per cubic centimeter. At this concentration, the excitons are, on the average, about 400 Å apart.

If excitons are assumed to be particles that diffuse with a diffusion coefficient D in three-dimensional space and that annihilate one another if they come within the distance R of each other, the rate constant for exciton–exciton annihilation is given (*23*) by $\gamma = 8\pi DR$. If R is assumed to be 20 Å, D is about 0.01 cm^2/s. This value is surprisingly large, because the diffusion coefficient for singlet excitons in anthracene single crystals is thought to be (*24*) about 10^{-3} cm^2/s. The reported values of the exciton–exciton annihilation rate constant in anthracene single crystals at room temperature (*24*) range from 2 $\times$ 10^{-9} to 4 $\times$ 10^{-8} cm^3/s. The excitons in the polysilane, however, almost certainly move quite rapidly along the chain on which they exist and then frequently hop from chain to chain. The consequences on exciton–exciton annihilation of this type of motion have not been discussed, to our knowledge.

In these experiments, the values of γ in the single-photon-excitation experiments were consistently lower than those in the two-photon-excitation experiments. We are presently investigating the possibility that neglect of exciton diffusion in the single-photon-excitation experiment might account for this discrepancy.

Finally, because polysilanes show promise as photoresists (*25, 26*) and nonlinear optical materials (*27*), the fact that the rate of loss of excitons in poly(*n*-propylmethylsilane) by exciton–exciton annihilation becomes equal to the monomolecular loss rate at laser intensities of only 10^{-6} J/cm^2 in a 5-ns pulse is significant. Because exposure levels are typically much higher for photoresist and nonlinear optical studies, exciton–exciton annihilation must be considered when attempting to understand the underlying photochemical reactions, because annihilation may dominate other photophysical and photochemical processes.

Summary

By measuring the intensity of fluorescence versus exciton concentration in solid films of poly(*n*-propylmethylsilane), the exciton–exciton annihilation rate constant was determined to be $\sim 10^{-7}$ cm^3/s. These results show that excitons are highly mobile in these materials. One theoretical interpretation of the rate constant indicated that the diffusion coefficient is about 10^{-2} cm^2/s, an exceptionally high value. These results also show that exciton–exciton annihilation is an important process that will have to be taken into account when attempting to interpret experiments designed to investigate fundamental photochemical and photophysical processes in these materials if the exposure light intensity is higher than $\sim$100 W/cm^2.

Acknowledgments

This work was supported by the U.S. Department of Energy under contract number DE–AC04–76DP00789. The technical assistance of P. M. Beeson and L. I. McLaughlin is gratefully acknowledged.

References

1. Wudl, F. *Acc. Chem. Res.* **1984,** *17*, 227.
2. Tanaka, K.; Yamabe, J. *Adv. Quantum Chem.* **1985,** *17*, 251.

3. *Handbook of Conducting Polymers;* Skotheim, A., Ed.; Marcel Dekker: New York, 1986.
4. *Polydiacetylenes;* Bloor, D.; Chance, R. R., Eds.; Maritimes Nijoff: Netherlands, 1985.
5. Chien, J. C. W. *Polyacetylene;* Academic: New York, 1984.
6. Hayes, W. *Contemp. Phys.* **1985,** *26,* 421–441.
7. West, R. *J. Organomet. Chem.* **1986,** *99,* 300.
8. Harrah, L. A.; Zeigler, J. M. *Macromolecules* **1987,** *20,* 601–608.
9. Harrah, L. A.; Zeigler, J. M. In *Photophysics of Polymers;* Hoyle, C. E.; Torkelson, J. M., Eds.; ACS Symposium Series 358; American Chemical Society: Washington, DC, 1987; pp 482–498.
10. Johnson, G. E.; McGrane, K. M. In *Photophysics of Polymers;* Hoyle, C. E.; Torkelson, J. M., Eds.; ACS Symposium Series 358; American Chemical Society: Washington, DC, 1987; pp 499–515.
11. Harrah, L. A.; Zeigler, J. M. *J. Polym. Sci., Polym. Lett. Ed.* **1985,** *23,* 209.
12. Trefonas, P.; Damewood, J. R.; West, R.; Miller, R. D. *Organometallics* **1985,** *4,* 1318–1319.
13. Miller, R. D.; Hofer, D.; Rabolt, J.; Fickes, G. N. *J. Am. Chem. Soc.* **1985,** *107,* 2172–2174.
14. Miller, R. D.; Farmer, B. L.; Fleming, W.; Sooriyakumaran, R.; Rabolt, J. *J. Am. Chem. Soc.* **1987,** *109,* 2509–2510.
15. Kuzmany, H.; Rabolt, J. F.; Farmer, B. L.; Miller, R. D. *J. Chem. Phys.* **1986,** *85,* 7413–7422.
16. Kepler, R. G.; Zeigler, J. M.; Harrah, L. A.; Kurtz, S. R. *Phys. Rev. B.: Condens. Matter* **1987,** *35,* 2818.
17. Kepler, R. G.; Zeigler, J. M.; Harrah, L. A.; Kurtz, S. R. *Bull. Am. Phys. Soc.* **1983,** *28,* 362.
18. Stolka, M.; Yu, H. J.; McGrane, K.; Pai, D. M. *J. Polym. Sci., Part A* **1987,** *25,* 823–827.
19. Abkowitz, M.; Knier, F. E.; Yu, H. J.; Weagley, R. J.; Stolka, M. *Solid State Commun.* **1987,** *62,* 547–550.
20. Zeigler, J. M. *Polym. Prepr. (Am. Chem. Soc., Div. Polym. Chem.)* **1987,** *28,* 424–425.
21. Zeigler, J. M. *Polym. Prepr. (Am. Chem. Soc., Div. Polym. Chem.)* **1986,** *27,* 109–110.
22. Schilling, F. C.; Bovey, F. A.; Zeigler, J. M. *Macromolecules* **1986,** *19,* 2309–2312.
23. Agranovich, V. M.; Galanin, F. D. *Electronic Excitation Energy Transfers in Condensed Matter;* North-Holland: New York, 1982.
24. Kepler, R. G. In *Treatise on Solid State Chemistry; Crystalline and Non-Crystalline Solids;* Hannay, N. B.; Ed., Plenum: New York, 1976; Vol. 3, p 615.
25. Zeigler, J. M.; Harrah, L. A.; Johnson, A. W. *Proc. SPIE Conf., Adv. Resist Technol. Process. II* **1985,** *537,* 166–174.
26. Hofer, D. C.; Miller, R. D.; Willson, C. G. *Proc. SPIE Conf., Adv. Resist Technol. Process. I* **1984,** *469,* 16–23.
27. Kajzar, F.; Messier, J.; Rosilio, C. *J. Appl. Phys.* **1986,** *60,* 3040–3044.

RECEIVED for review May 27, 1988. ACCEPTED revised manuscript October 20, 1988.

26

Electronic Transport in Polysilylenes

Martin A. Abkowitz, Milan Stolka, Ronald J. Weagley, Kathleen M. McGrane, and Frederick E. Knier

Xerox Corporation, 0114–39D, Webster, NY 14580

*Recent analysis of electronic-transport data in poly(methylphenylsilylene) (PMPS) obtained by the time-of-flight technique is compared in detail with similar experimental data on carbon-backbone polymeric glasses. The extensive literature on poly(*N*-vinylcarbazole) (PVK) and molecularly doped polymers (MDPs) is presented in this chapter in the form of an abbreviated historical review. Field- and temperature-dependent characteristics of transport in PMPS are strikingly similar to those of the previously studied PVK and MDPs. In PVK, holes migrate by thermally assisted hopping among carbazole side groups in a process not involving the carbon main chain. However, comparative studies of aromatic- and aliphatic-side-group-substituted polysilylenes indicate that transport in Si polymers is not significantly sensitive to side-group substituents. That transport persists even in the total absence of π electrons on the pendant groups indicates the involvement of states primarily derived from the Si main chain. Recent optical experiments suggest that carriers hop among domainlike islands of 10–15 Si repeat units.*

ELECTRONIC TRANSPORT IN AMORPHOUS SOLIDS and in random organic media in particular has been the subject of vigorous scientific activity for more than 2 decades. The approach taken in this chapter is to integrate recent studies of electronic transport in polysilanes with the extensive earlier work on poly(*N*-vinylcarbazole) (PVK) and molecularly doped polymers (MDPs) (i.e., systems in which transport-active molecular species are dispersed in an inert binder) and thereby allow the distinctive features of transport in the Si-based systems to emerge.

Interest in electronic transport in amorphous films is clearly both sci-

0065–2393/90/0224–0467$10.25/0

entific and technological. Technological interest stems principally from the need to fabricate large-area photoreceptors (drums or belts) for page-sized electrophotographic images (*1*). Large-area devices can be coated by thermal evaporation (e.g., chalcogenides), plasma-assisted chemical vapor deposition (e.g., amorphous Si [a-Si:H]), or, perhaps most conveniently, solution techniques.

In this chapter, a selective overview of technological and historical background is followed by a general discussion of the microscopic details of the transport phenomenon and experimental techniques. Key results of earlier studies on carbon-based systems are presented and then compared with corresponding data on poly(methylphenylsilylene) (PMPS), which has been taken as the prototype for studies of transport system in polymers with silicon backbones. Key points are then summarized. Those wishing to omit the extensive background section may proceed directly to the section on electronic transport in polysilylenes (page 492).

Perspective

Interest in organic polymeric electrophotographic receptors arose from a simultaneous need for flexibility and mechanical toughness. In contemporary machine architecture, the photoreceptor belt is often required to bend around small-diameter rollers. Materials commonly used in the past that are capable of both efficient generation and transport of free carriers after the absorption of light, such as amorphous Se (a-Se) and its alloys with Te and As or a-Si:H, are not flexible enough to achieve these newer design goals. On the other hand, organic polymers with good film-forming properties (at dielectric thicknesses dictated by the developer system) that are also capable of supporting surface potentials in the order of 1000 V across 20 μm are usually not efficient charge carrier generators in the visible range (400–700 nm), although in certain instances, they can transport injected charge with the required efficiency.

The two distinct requirements have led to the concept of a layered photoreceptor (*2*) (Figure 1). A very thin substrate charge generator layer (GL) absorbs the incident radiation to which the thick transport layer (TL) overcoat is transparent. The transport layer embodies the desired mechanical and adhesive properties, charge retention, and resistance to the corrosive action of corona effluents and can transport carriers of at least one sign, which are injected from the generation layer. The technological figure of merit for the transport polymer is the transit time, the time required by the photoinjected carrier pulse to complete its traversal with negligible loss of carriers to deep traps.

The electrophotographic process is initiated by deposition of a uniform surface charge from the corona on the xerographic photoreceptor belt in the

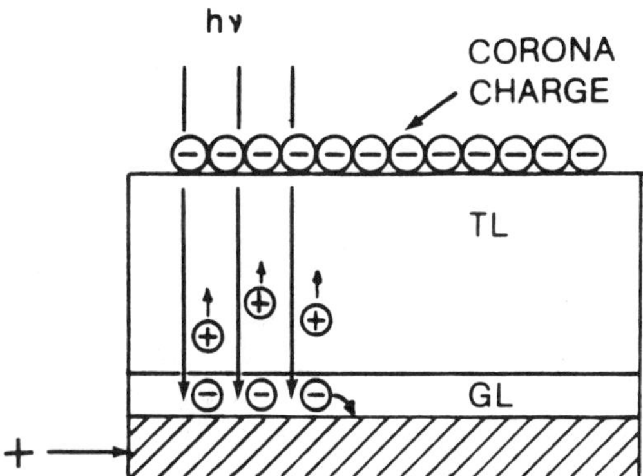

Figure 1. Scheme of layered electrophotographic photoreceptor.

dark. The charge must be retained on the photoreceptor for the duration of the xerographic cycle, which means that the polymer must, in fact, be a good insulator. An electrostatic latent image is then generated on the photoreceptor either by full (light-lens) imagewise exposure or by laser writing. The areas exposed to light become conductive and their surface charge is neutralized, whereas charge is retained in the nonexposed areas.

The latent image is developed by depositing appropriately charged toner particles on the photoreceptor. If the toner particles, for instance, carry charge of the opposite sign, they adhere to the unexposed dark areas. The toned image is transferred to oppositely charged plain paper and fused. The photoreceptor itself is then erased by light and cleaned to complete the cycle. For our present purposes, a key requirement is that the photoinjected-carrier transit time (time to traverse 20–30-μm layer) must be short compared with the interval between exposure and development of the latent image by toner.

Technological Requirements

The drift mobility (μ) of a photoinjected carrier is its mean velocity per unit field (E) and may itself be field dependent. The transit time (t_{tr}) and mobility are related to specimen thickness (L) according to equation 1:

$$t_{tr} = \frac{L}{\mu E} \tag{1}$$

The importance of μ in the electrophotographic process can be understood as follows: For an increase in the light exposure intensity (ΔF), the final decrease in surface potential (ΔV) is proportional to the number of injected carriers (supply) and the distance they travel within the TL (if the GL is negligibly thin and carriers leave it instantly). Carrier displacement or range (R) in a given field (E) is controlled by $\mu\tau$, the product of mobility (μ) and deep-trapping lifetime (τ). Thus, the following relation exists:

$$R = \mu\tau E \quad (2)$$

During xerographic discharge, a charge of CV_o (in which C is the film capacitance and V_o is the charge voltage) traverses the bulk and induces time-dependent variation in the electric field behind the leading edge of the injected carrier front. Thus, as the fastest carriers transit the TL, the electric field behind them is reduced, and the slower carriers transit at lower field, which in turn makes their velocities lower.

Xerographic discharge, which is a highly space–charge-perturbed process, is therefore characterized by significant dispersion in the arrival times of photoinjected carriers. In some cases, the transit times of the slowest carriers are 10 times or more than the transit times of the fast carriers. However, the slowest carriers, too, must exit the TL before the photoreceptor reaches the development zone, typically 0.3–1.0 s after exposure. In practice, carrier mobilities that significantly exceed 10^{-6} cm^2/V-s are desirable.

For example, a 25-μm-thick TL (in which there is no deep trapping) with $\mu = 10^{-6}$ cm^2/V-s at $E = 10^4$ V/cm in a photoreceptor charged initially to 1000 V still retains a residual voltage of 20 V after 0.3 s, or 7 V after 1 s, even when enough incident light is used to photogenerate all carriers needed (i.e., CV_o) to discharge the device. A device with $\mu = 10^{-7}$ cm^2/V-s at the same field will retain as much as ~60 V at 1 s after exposure, which is probably unacceptable.

A finite carrier range (*3*) reflects the influence of deep traps and is controlled by the mobility–lifetime product ($\mu\tau$). By using the numbers just given, $\mu\tau$ in practical polymeric TL should exceed 10^{-6} cm^2/V substantially.

Practical TL materials must also satisfy the requirements of chemical integrity (resistance to corona effluents, etc.), photochemical stability, mechanical integrity, and abrasion resistance. The TL polymer must also match the selected generation material, that is, it should accept charges photogenerated in the GL. Modern GL materials are either one-component vapor- or solvent-deposited thin layers of a-Se, a-As_2Se_3, or various azo dyes or solvent-coated dispersions of photogenerating pigments, such as trigonal (crystalline) Se, squarylium, or perylene derivatives, in a polymer binder. The overall effective thickness of the GL is 0.3–0.5 μm. In laser printers in which the latent image is generated by a solid-state laser emitting in the 780–830-nm range (near IR), the photogenerating pigment must be sensitive to near-IR radiation (*4*). The transporting polymers must be electronically compatible with the photogenerator of choice.

Historical Background

Many charge-transporting polymers have already been identified (*5*). The most thoroughly studied charge-transporting insulating polymer is poly(*N*-vinylcarbazole) (PVK; *6*). (*See* Figure 5 for the structure of PVK.)

A layered photoconductor with a PVK TL was reported by Regensburger (*2*). Many other charge-transporting polymers were then synthesized, but none was studied as extensively as PVK. Structurally, all charge-transporting polymers are derived from monomers containing an ionizable (mainly oxidizable) group with a stable oxidized form in which charge is delocalized in an aromatic unit (*5*). The most predominant structural moiety in these polymers is an aromatic amine group with all imaginable substitutions. PVK is an example of a substituted aromatic amine polymer. Polymers with complex aromatic groups such as anthracene or pyrene have also been studied as potential transport materials, but they are generally oxidatively or photolytically unstable (*5*).

Initially, the ability of PVK to transport charges was assumed to be related to the spatially crowded structure of PVK (*7*), in which the carbazole groups, because of their covalent-bonding backbone, are forced to interact with each other to the point that significant orbital overlap creates a band structure. However, the observed mobilities were low (10^{-7}–10^{-6} cm^2/V-s), with field and temperature dependences that were not characteristic of band transport alone. Moreover, it was discovered later that the same charge-transport phenomena could be achieved simply by dissolving in inert polymeric binders the functional groups in the polymers just mentioned (*8*, *9*). This observation led in turn to the concept of MDPs. Obviously, the carbon backbone in PVK played only the secondary role of holding carbazole groups together (*9*, *10*). Charge transport in amorphous σ-bonded carbon-backbone polymers is widely recognized now as a hopping process that involves electric-field-driven charge exchanges among discrete sites that are dissolved molecules or covalently bonded pendant groups.

Most practical polymer-based electrophotographic photoreceptor systems now in use are in fact solid solutions of an active species in a binder host polymer (*11*). This system concept embodies the notion of full chemical control of the transport process. Thus, the concentration of dopant molecules directly controls the drift mobility, which is in turn controlled by the overlap of wave functions between active sites. The host polymer binder is then specialized for its mechanical and adhesive properties. Understanding the key features of small-molecule transport provides guidelines for the optimization of injected-carrier range. The key point is to understand how chemically induced traps arise in such systems and how molecularly doped materials containing various contaminants in substantial quantities can still efficiently transport charge. (Even part-per-million concentrations of contaminants are extraordinarily high by conventional semiconductor standards.)

Many practical small-molecule-doped polymeric systems have been studied. Two notable examples are *N*,*N*′-diphenyl-*N*,*N*′-bis(3-methylphenyl)-1,1′-diphenyl-4,4′-diamine (TPD) (*12*) and *p*-diethylaminobenzaldehyde diphenylhydrazone (DEH) (*13*), both dispersed in polycarbonate. Most small-molecule systems that have been investigated have been complex substituted aromatic amines.

The design of single-component polymer transport materials continues to interest researchers in this field. The use of such materials will completely eliminate solvent extraction, diffusional instability, and crystallization of the small molecules. One obvious route that has not been successful to date is the design of yet another aromatic-amine-containing carbon-backbone polymer. An alternative may be to explore the large class of glassy silicon-backbone polymers, such as polysilylenes (*14*) and polyphosphazenes (*15*).

Polysilylenes have been known for many years but were not considered interesting until the mid-1970s, when Yajima (*16*) and West (*17*) used them as precursors for Si–C ceramics. In 1980, the first soluble high-molecular-weight polymers were synthesized (*18*). Photoconductivity in poly(methylphenylsilylene)–TNF (2,4,7-trinitrofluorenone) was reported by Kepler et al. (*14*). Preliminary charge-transport studies showed that at 295 K hole mobility in poly(methylphenylsilylene) (PMPS) is $\sim 10^{-4}$ cm^2/V-s at about 10^5 V/cm, quite high for a polymer (*14, 19, 20*).

Transport Measurements

Xerographic discharge involves photogeneration and injection of the mobile carrier type (transport is typically unipolar) into transport states, transport of the photoinjected carrier, and fractional loss to and immobilization in deep traps. Transport measurement techniques fall into two subclasses: steady state or transient. Transient measurements themselves are further classified according to whether they are carried out with small or large signals. For example, under conditions of intense-light-pulse-induced exposure (in which the pulse width is much less than the transit time), xerographic discharge becomes a large-signal process. In this case, the charge in transit, a full CV_0's worth, will significantly disturb the field distribution in the transport layer and cause it to become time dependent. Thus, the transiting charge is forced to become spatially dispersed even when the mobility is itself uniform and time independent for each transiting carrier.

Under small-signal conditions, on the other hand, the transiting carrier packet is constrained to contain much less than a CV_0 of charge and, consequently, moves in a uniform field. Any complicated time dependence of mobility (*21*) that becomes experimentally visible under these conditions must, we believe, reflect microscopic and fundamental features of transport in the disordered medium and should be distinguished from the macroscopic dispersion induced by carrier motion under highly space–charge-perturbed conditions (*22*).

Steady-State Measurements. The operational definition of ohmic conductivity (σ) is well known:

$$\sigma = ne\mu \tag{3}$$

In equation 3, n is the carrier density at thermal equilibrium, e is the electronic charge, and μ is the conductivity mobility, which is the mean velocity of a carrier per unit applied field under steady-state conditions. In relatively good glassy polymeric insulators, despite arguments to the contrary, unambiguous determination of the mobility directly from curves of current density (j) versus electric field (E) is very difficult. The problem is always the uncertain role of contacts, which must be neutral, in the sense that during measurement the bulk carrier density that exists under zero-field conditions is not perturbed (*23*). The latter is rarely the case. In fact, the observation of linear curves of j versus E guarantees nothing, and this behavior is often observed when currents are limited by contact emission and not bulk controlled (*24*) (i.e., controlled by the injection rate from the contact rather than by the bulk of the glassy film). Even if ohmic contact can be made, the respective contributions of n and μ must be deconvoluted from the curves of j versus E. Independent determination of n in insulating systems is difficult. Thermopower measurements (Seebeck coefficients commonly carried out to determine n in doped crystalline semiconductors) are not practical in good insulators.

Unambiguous steady-state measurements are carried out only in the trap-free space–charge-limited-current (SCLC) regime (*23*), when the average transit time (t_{tr}) of any excess injected carrier across the sample bulk is shorter than the time required for the bulk to locally neutralize the carrier (*22*). The transit time (t_{tr}) has been defined in equation 1. The time to neutralize any excess injected carrier is the bulk dielectric relaxation time, τ_r ($\tau_r = [\rho\epsilon]^{1/2}$ in which ρ is the bulk resistivity and ϵ is the bulk dielectric constant). The condition for SCLC is given by the following expression:

$$t_{tr} = \frac{L}{\mu E} < \tau_r = (\rho\epsilon)^{1/2} \tag{4}$$

The trap-free space–charge-limited current (TFSCLC) (*22*) is proportional to mobility, as shown by equation 5:

$$L_{TFSCLC} = \frac{9}{8}\mu\epsilon\frac{E^2}{L} \tag{5}$$

SCLCs require that the contacts on the polymer be able to act as an infinite carrier reservoir with respect to bulk demands; this requirement is just an alternative operational description of an ohmic contact. To reiterate, unambiguous determination of mobility directly from steady-state electrical measurements requires ohmic contacts and space–charge-limited conditions. The observation of linear curves of j versus E is usually of no significance.

Transient Measurements. Time-resolved injection photocurrents have the particular virtue, in the present context, of stimulating the conditions under which carriers transit a polymer film during xerographic discharge. However, this discussion is limited to the case in which the integrated number of charges q in the transit pulse is small enough ($q << CV_0$) so that transit occurs in a uniform field. This condition, which significantly simplifies the interpretation of data, describes the canonical small-signal time-of-flight (TOF) experiment (25). The field is such that equation 4 applies. For TOF experiments, the electrical contacts must block dark injection. This condition is relatively easy to satisfy. With the field applied, the photoreceptor is excited with a weak flash of visible light that penetrates to the generation layer. The flash, typically supplied from a nitrogen-pumped dye laser, is of short duration compared with the transit time.

A schematic of the apparatus is shown in Figure 2. The series resistor senses current. When the overall circuit RC (R and C are circuit resistance and capacitance, respectively) time constant is short compared with t_{tr}, the voltage developed on the sensing resistor is proportional to the instantaneous

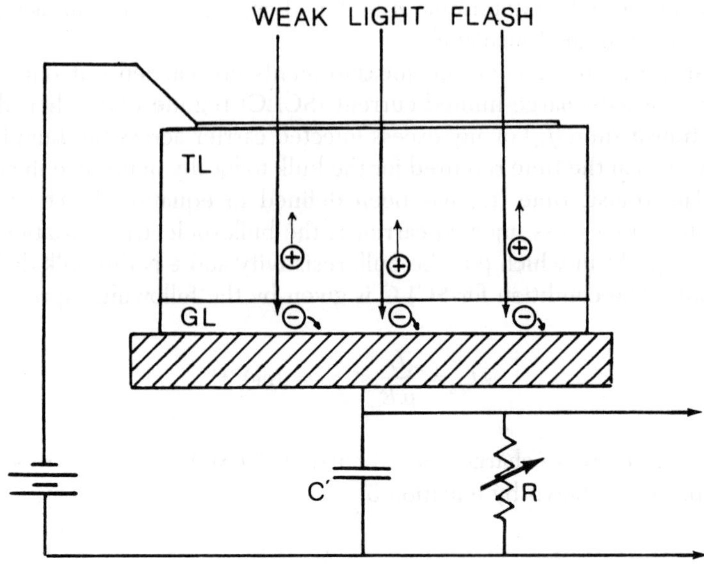

Figure 2. Schematic of a TOF apparatus. R is the current-sensing resistor. C′ is the total shunt capacitance.

current in the specimen. Ideally, the current drops to zero when the transiting carriers reach the collecting electrode as a coherent sheet. In practice, rounding of the transit pulse is observed, because some dispersion in the transit times of arriving carriers is always induced by thermal diffusion and mutual electrostatic repulsion.

Superimposed on these pervasive minimal effects are dispersion phenomena that are more fundamentally related to the degree of disorder in any given system. For example, for a disordered molecular solid, the separation of the overall effects of steric disorder into (1) a self-consistent site-variable random potential and (2) randomly varying inter- and intramolecular transition matrix elements that directly encompass the system's geometric (i.e., positional) disorder, is useful. These effects contribute to the diagonal and off-diagonal terms in the system's effective Hamiltonian and so are referred to as diagonal (energetic) and off-diagonal (positional or site) disorder, respectively.

Transport in many disordered systems can be thought of as a chain of discrete events (e.g., hopping or detrapping), in which each event is associated with a waiting time on a given site. When the waiting-time distribution function is Gaussian, the mean position of a transiting carrier packet initially injected as a thin sheet is proportional to time, whereas the width of the spread increases only as the square root of time. In a current-mode TOF experiment, the time-resolved current is constant until the leading edge of the transiting packet reaches the absorbing boundary. In effect, the subsequent decay of current maps the distribution of arriving-carrier transit times.

In units of time normalized to the transit time (t/t_{tr}) the width of this decaying tail scales as $(t_{tr})^{1/2}$ and is thus actually smaller at lower fields. However, when the waiting-time distribution, $\psi(t)$, is very much broader, it is more appropriately represented by an algebraic function (*21*) of a form encountered in phenomenological descriptions of pair-recombination luminescence in crystalline semiconductors [$\psi(t)\alpha t^{-(1+\alpha)}$], in which the dispersion parameter α varies between 0 and 1. Under these circumstances, the maximum of the carrier distribution can remain (depending on α) close to the plane of origin, even when the leading edge of the distribution has penetrated to the absorbing boundary (*21*). In a TOF experiment, the time-resolved current will acquire two algebraic branches that intersect at a statistically defined t_{tr}. For $t < t_{tr}$,

$$i(t) \sim t^{-(1-\alpha)} \tag{6a}$$

whereas for $t > t_{tr}$,

$$i(t) \sim t^{-(1+\alpha)} \tag{6b}$$

A key result is the prediction (*21*) of a unique scaling of this statistically defined transit time with applied field and sample thickness that leads to an anomalous thickness-dependent mobility. The anomalous behavior (i.e., compared with Gaussian transport) ceases only when the dispersion parameter α approaches 1. The transit time (t_{tr}) obeys the following scaling law:

$$t_{tr} = \left(\frac{L}{E}\right)^{1/\alpha} \tag{7}$$

As α decreases from a maximum of 1, the current-mode transit pulse acquired in real time becomes increasingly featureless but retains a clearly defined kink when the logarithm of current is replotted versus the logarithm of time. When the log–log plot is of the current normalized to its value at the transit time (i/i_{tr}) versus time normalized to the transit time (t/t_{tr}), then current transients collected at different fields become superimposed and display the distinctive feature of "universality".

A key feature of systems in which transport is characterized by an algebraic distribution of waiting times is the correlation between the respective slopes of these universality plots (equation 6) and the scaling law for the transit time (equation 7), both of which are determined by the degree of dispersion as manifested by α (*21*). Some confusion in terminology has arisen already regarding the dispersion in TOF experiments. For example, "relative dispersiveness", as specifically manifested in the shape of TOF transients in polymeric glasses, characteristically varies with temperature, sample purity, sample surface condition, and even the details of the initiating injection step. In some cases, dispersive-looking TOF transients cannot be identified clearly with anomalous scaling. Thus, experimentally observed dispersion in transit times may or may not be associated with the so-called anomalous features that are the direct consequence of a specific phenomenological representation of a very broad algebraic distribution of carrier waiting times. (For the case of anomalous dispersion, the distribution is broad enough to encompass the actual time of observation.)

Electronic Transport Mechanisms

In glassy polymeric media, electronic transport can be phenomenologically thought of as a series of discrete steps characterized by a distribution of waiting times $\psi(t)$. When the distribution extends into the time scale of observation, the mobility itself will always appear to be time dependent. However, when the distribution does not extend into the time scale of observation, mobility can be characterized by an averaged value for most of the transit event, even though it exhibits thermalization at early times, which may be resolvable under certain experimental conditions (*26*, *27*). Several more microscopically detailed pictures can correspond to this phenomenological description of electronic transport.

Trap-Controlled Band Transport. Carrier displacement in the field direction takes place in an effective density of extended states N_{eff} in the band in which the microscopic mobility (μ_o) is controlled by scattering processes. In disordered solids, in which mean free paths are of the order of atomic spacing, carrier motion in extended states would likely be diffusive, not ballistic, and microscopic mobilities would fall in the range 1–10 cm^2/V-s at room temperature (*28*). The observed drift mobility in a TOF experiment is much smaller, because the carrier moving in the band is frequently interdicted by a trapping event. Trapping, thermally activated release, and retrapping control the temperature dependence of the drift mobility, which is characteristically activated (*29*). For example, in the simplest case of control by a density N_t of nearly monoenergetic traps, which are displaced from the transport level by energy E_t, the drift mobility (μ) is written as:

$$\mu = \mu_o \left[1 + \frac{N_t}{N_{eff}} \exp\left(\frac{E_t}{kT}\right) \right]^{-1} \tag{8}$$

In equation 8, k is the Boltzmann constant, and T is temperature.

In this situation, there is a single waiting time, which is simply the detrapping time, and transport is nondispersive. As the trap distribution broadens, so does the distribution of trap release times. In the limiting case in which the traps are organized into a distribution falling off exponentially from a demarcation level in the band called a mobility edge, transport can become highly dispersive (*28*). The featureless transit pulses seen in amorphous As_2Se_3 conform well with this picture of multiple trapping in an exponential band tail and, in fact, manifest all the features characterizing dispersive transport (*29*).

Hopping in a Manifold of Localized States. If the hopping (i.e., quantum-mechanical-tunneling) sites are monoenergetic and the carriers do not relax with their surroundings to form a polaron during each wait at a discrete site, then transport is temperature independent, and the mobility scales experimentally with average intersite hopping distance. In glassy polymers, a random-disorder potential at each site should cause site energies to reorganize into a Gaussian distribution (*30*). The simplest description of the ensuing transport is phonon-assisted hopping involving an activation (ϵ) related to the Gaussian width. The mobility is proportional to $\rho^2 \exp(-2\rho/\rho_o) \exp(-\epsilon/kT)$, in which ρ is the average intersite distance and ρ_o is the localization radius (*31*).

Recently, the time-resolved behavior of a carrier packet injected into such a Gaussian distribution of localized states was considered analytically by Movaghar and co-workers (*32*) and was extensively simulated in computer experiments by Bässler and his co-workers (*30, 33*), who obtained the following novel results. Relaxation of the excess injected carriers to a temperature-dependent level below the average energy in the distribution occurs in a relatively few hops. Transport occurs in the long-time limit in which the drift mobility is predicted to exhibit a non-Arrhenius type of temperature dependence given by:

$$\mu(T) = \mu_o \exp\left(\frac{-T_o}{T}\right)^2 \tag{9}$$

In equation 9, T_o is the width of the Gaussian distribution expressed in temperature units and μ_o is the mobility in an analogous system with no disorder (i.e., the Gaussian distribution collapsed to a single energy level). A carrier hopping in the field direction gains energy equivalent to $eE\rho$, in which ρ is the hopping distance, and can therefore gain access to more states in the distribution. The net effect of an electric field is to reduce the effective width of the Gaussian distribution. Therefore, the combined field and temperature dependence of the drift mobility is contained in the following expression:

$$\mu(E, T) = \mu_o \exp\left(\frac{-T_o}{T}\right)^2 \exp\left(\frac{E}{E_o}\right) \tag{10}$$

The key point is that the full behavior of the drift mobility can be expressed in terms of the mobility of an ordered microscopic analogue, together with a parameter T_o that expresses the effect of disorder in terms of the energy broadening of the level into a Gaussian distribution. Because this model presumes a time-independent mobility, it makes no allowance for the observation of anomalous dispersion. In fact, anomalous dispersion in this picture must be identified with additional traps lying well outside the Gaussian manifold, that is, trap-controlled hopping.

Trap-Controlled Hopping. In trap-controlled hopping, the scenario described for trap-controlled band mobility applies. However, the microscopic mobility is associated now with carriers hopping in a manifold of localized states. Overall temperature and field dependence reflects the complicated convolution of the temperature and field dependence of both the microscopic mobility and the trap kinetic processes. Clearly, the observed behavior can now range from nondispersive to anomalously dispersive behavior as before, depending on the energy distribution of transport-interactive traps.

Small-Polaron Hopping. The microscopic hopping mobility even among identical sites in the absence of disorder acquires an extra activation associated with the relaxation of the carrier and its polarized surroundings, that is, a self-trapping process. As before, all other processes would be expected to convolute, with the extra temperature dependence associated with polaron formation.

Historical Examples

The experimental features of electronic transport in polysilylenes to be described later are best understood when contrasted with results of earlier studies in random organic media, key examples of which are highlighted in the following sections.

Poly(*N*-vinylcarbazole). The following are the most relevant characteristics of transport in poly(*N*-vinylcarbazole) (PVK) and in PVK complexed with TNF.

Transit Pulse Shapes. Figure 3 displays the current-mode hole transit pulse shapes in PVK and in Br-substituted PVK from 264 to 490 K, together with the corresponding Arrhenius plots of mobility (*34*). For brominated PVK, the transit pulses remain relatively featureless over the entire experimentally accessed range. For PVK, however, the transit pulses display a well-developed shoulder above 414 K, which is characteristic of nondispersive transport of the photoinjected carrier sheet.

A striking feature of the PVK data is the absence of any evidence of a change in the temperature dependence of mobility on passing through a range in which transit pulses clearly undergo a significant increase in dispersion. The relative increase in dispersiveness at high temperatures suggests that the chemical disorder introduced by Br substitution in effect broadens the waiting-time distribution function. Gill (*35*) demonstrated electron transport in PVK complexed with TNF. With this system, Seki (*10*) demonstrated that room-temperature electron transits involve a hopping motion among TNF molecules. At room temperature, superposition with respect to the applied field is observed for electron transit pulses (Figure 4). Superposition is a key feature of anomalous dispersion, as described earlier.

Strong Field Dependence of Drift Mobility. Hole drift mobility increases sharply with field, and with PVK, a characteristic $E^{1/2}$ dependence of log μ is exhibited, as originally reported by Pai (*36*), whose data are displayed in Figure 5. Drift mobility values are low and fall into the range 10^{-9}–10^{-6} cm^2/V-s at room temperature.

Strong Temperature Dependence of Drift Mobility. Figure 6 is an Arrhenius representation of hole drift mobility data on a 0.2:1 TNF–PVK film (*10*). The apparent activation in this representation is field dependent,

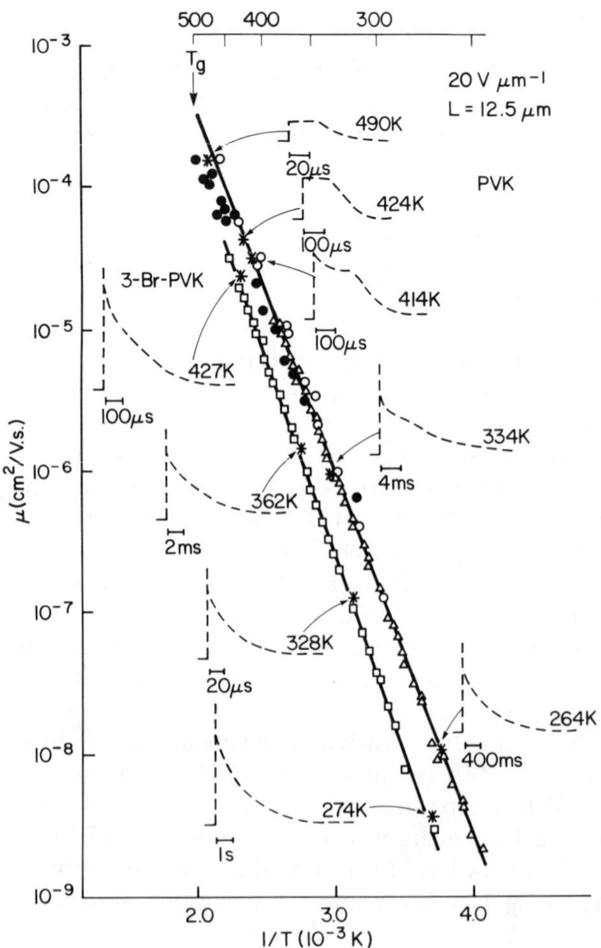

*Figure 3. μ versus 1/*T *for hole transport in PVK and poly(3-bromo-*N*-vinylcarbazole) (3-Br-PVK), with current-mode transients at the indicated temperatures. The various symbols indicate data taken at various times. (Reproduced with permission from reference 34. Copyright 1978 American Institute of Physics, Inc.)*

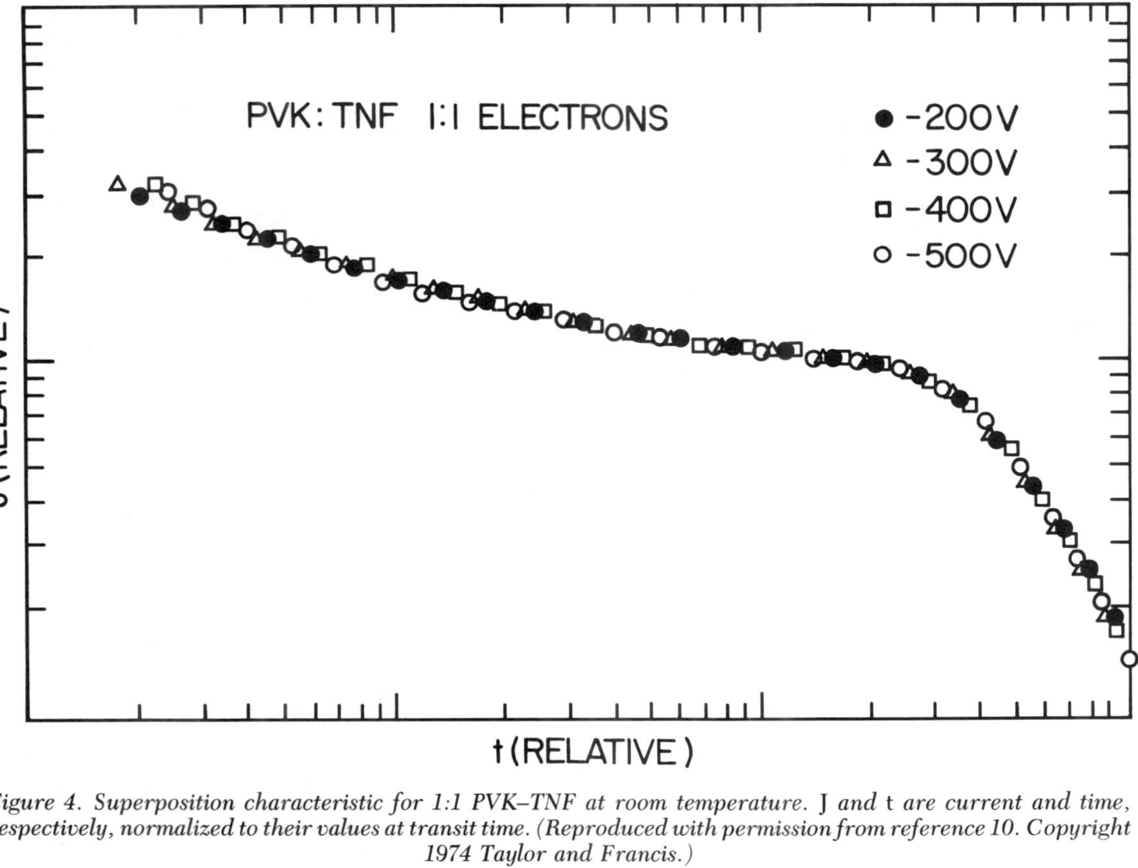

Figure 4. Superposition characteristic for 1:1 PVK–TNF at room temperature. J and t are current and time, respectively, normalized to their values at transit time. (Reproduced with permission from reference 10. Copyright 1974 Taylor and Francis.)

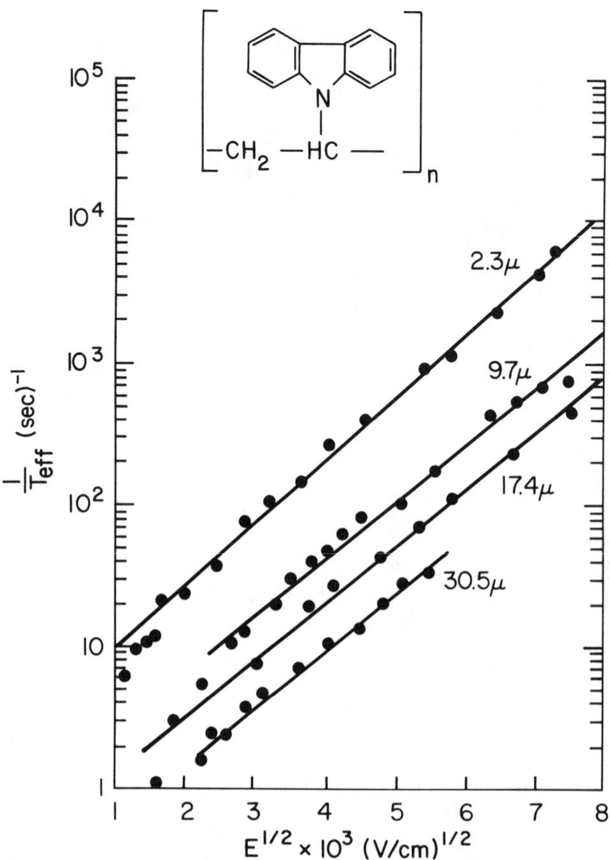

Figure 5. Inverse effective transit time ($1/T_{eff}$) versus square root of applied field ($E^{1/2}$) for four film thicknesses of PVK. (Reproduced with permission from reference 36. Copyright 1971 American Institute of Physics, Inc.)

as shown in Figure 7 for both electrons and holes. The activation in the zero-field limit is close to 0.7 eV for both carriers. In the Arrhenius representation, the family of curves of log μ versus $10^3/T$ that are parametric in field tends to converge at a finite value of $10^3/T$ corresponding to a temperature T equal to T_0. T_0 is 416 K, which is in the vicinity of T_g (glass transition temperature). The effect of small-molecule additives, which act as plasticizers in PVK, on transport has been noted by Fujino et al. (37). Their data suggest that a strong correlation exists between T_0 and the T_g of the PVK-based polymeric system.

An alternative framework for interpreting the temperature dependence of drift mobility is provided by the model suggested by Bässler and co-

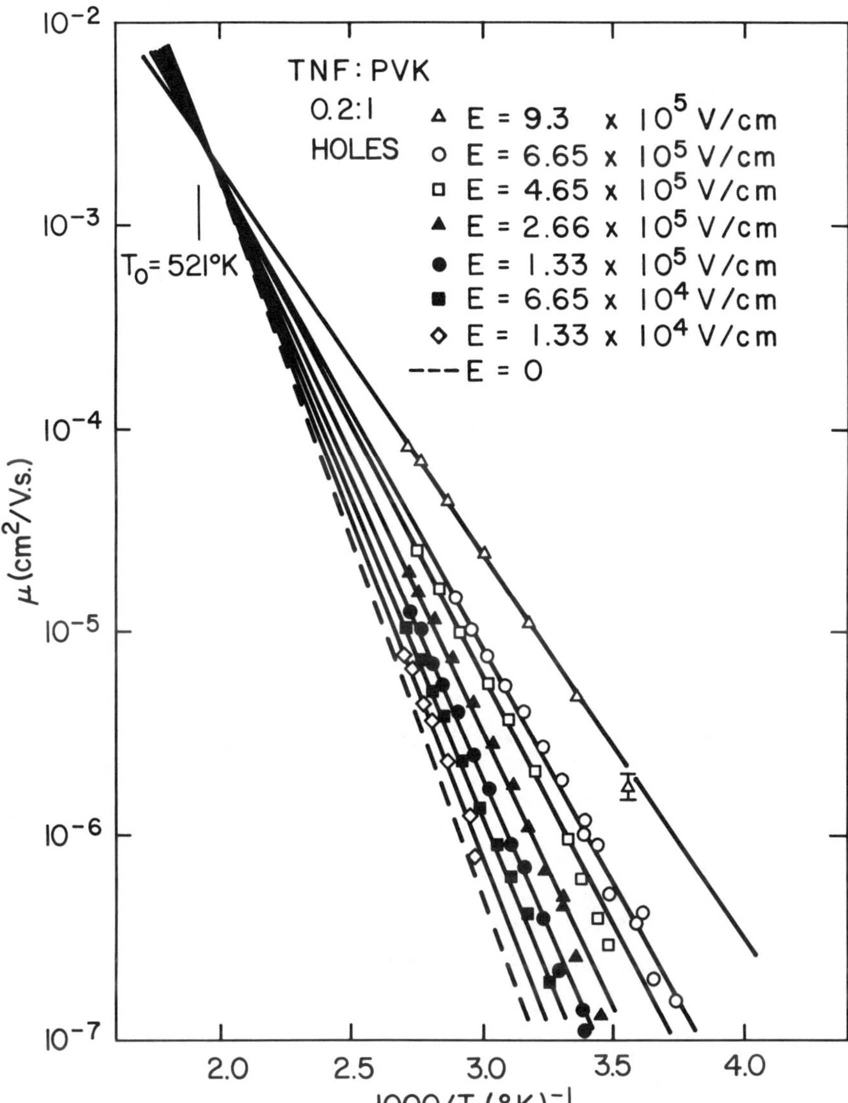

Figure 6. Arrhenius representations of hole drift mobility data on a 0.2:1 TNF–PVK film that is parametric in field. (Reproduced with permission from reference 35. Copyright 1972 American Institute of Physics, Inc.)

workers (33) and Movaghar and co-workers (32). To reiterate, the model assumes that a carrier initially injected into a Gaussian manifold of localized state of width σ (in electron volts) or equivalent temperature width T_0 establishes thermal equilibrium within a few hops, after which the mobility achieves its stationary value. The model does not treat the effect of additional states outside the Gaussian manifold and presumes that in the absence of diagonal and off-diagonal disorder the manifold would collapse into a single monoenergetic level, within which the mobile carrier would undergo activationless hops with mobility μ_0. As stated earlier, a key prediction is an inverse quadratic-temperature dependence. In a recent paper, Bässler (*30*) represented Gill's hole-transport data (*35*) in this framework (Figure 8). From the slopes of this family of Bässler plots that are parametric in field, the width of the Gaussian distribution can be estimated by extrapolating to the zero-field limit. This point will be brought up again, when overall comparisons with PMPS are made.

Mobility Scales with Average Intersite Hopping Distance. Gill identified an exponential dependence of electron drift mobility on average TNF intersite hopping distance and noted an associated decrease in hole drift mobility (*35*). Gill's subsidiary observation that TNF addition also decreased hole mobility precisely because complexed carbazole is removed as a hole transport site provided evidence for the key role of the discrete carbazole groups in hole transport.

Molecularly Doped Polymers. Work on molecularly doped polymers (MDPs), which are dispersions of transport-active molecular species in inert polymeric binders, evolved directly as a result of mechanistic insights gained from the numerous preceding studies of polymers and charge-transfer complexes. A number of hole-transport studies have been carried out on substituted aromatic amines (*13, 31, 38–42*) with polycarbonate (PC) used as an inert binder. Results on the system *N,N'*-diphenyl-*N,N,'N'*-bis(3-methylphenyl)-1,1′-diphenyl-4,4′-diamine (TPD) in bisphenol A–PC are illustrative (*12*).

Concentration Dependence of Hole Drift Mobility in TPD–PC. Figure 9 shows the concentration dependence of hole mobility expressed (*12*) in terms of the average intersite distance (ρ; expressed in angstroms) between TPD transport-active sites. The localization radius (ρ_0) is a measure of the attenuation of the molecular wave function associated with the transport state. The 1.4-Å value of ρ was similar to values (*31*) reported for other MDPs. Figure 10 demonstrates that the apparent activation energy determined from Arrhenius plots of mobility are concentration and field dependent. For most of the concentration range, μ/ρ^2 varies exponentially with ρ. The average intersite distance varies from 20 Å in a 9 wt % sample of

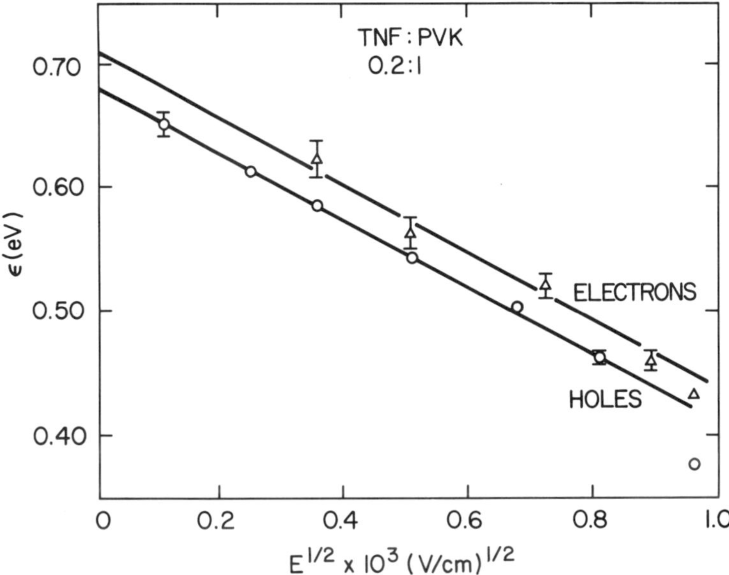

Figure 7. Activation energies (ε) of hole and electron drift mobilities versus applied field in a 0.2:1 TNF–PVK film. (Reproduced with permission from reference 35. Copyright 1972 American Institute of Physics, Inc.)

TPD–PC to 9 Å estimated for amorphous TPD. The average distance between TPD sites was computed with the following assumption:

$$\rho = \left(\frac{M}{Ad}\right)^{1/3} \tag{11}$$

In equation 11, M is the molecular weight (516), d is the density (1.2 g/cm^3), and A is Avogadro's number. Figure 10 shows the significant change between the overall activation energies of the glassy TPD film and the solid solution. For example, a slight change in intermolecular distance from 9 Å (for a-TPD) to 9.6 Å (for a film containing 80 wt % diamine) produces an approximately twofold increase in activation energy. Further dilution of TPD in the solid solution, which induces a significant change in intermolecular distance, induces only a gradual change in activation energy. A corresponding steep increase in mobility between the 80 wt % solid solution and glassy TPD is indicated clearly in Figure 9. The concentration region of maximum change corresponds to the case in which the average intersite distance approaches the maximum molecular dimensions. At these high concentrations, percolative effects may contribute significantly to transport. The large change in mobility and its apparent activation may signal an alteration in transport mechanism.

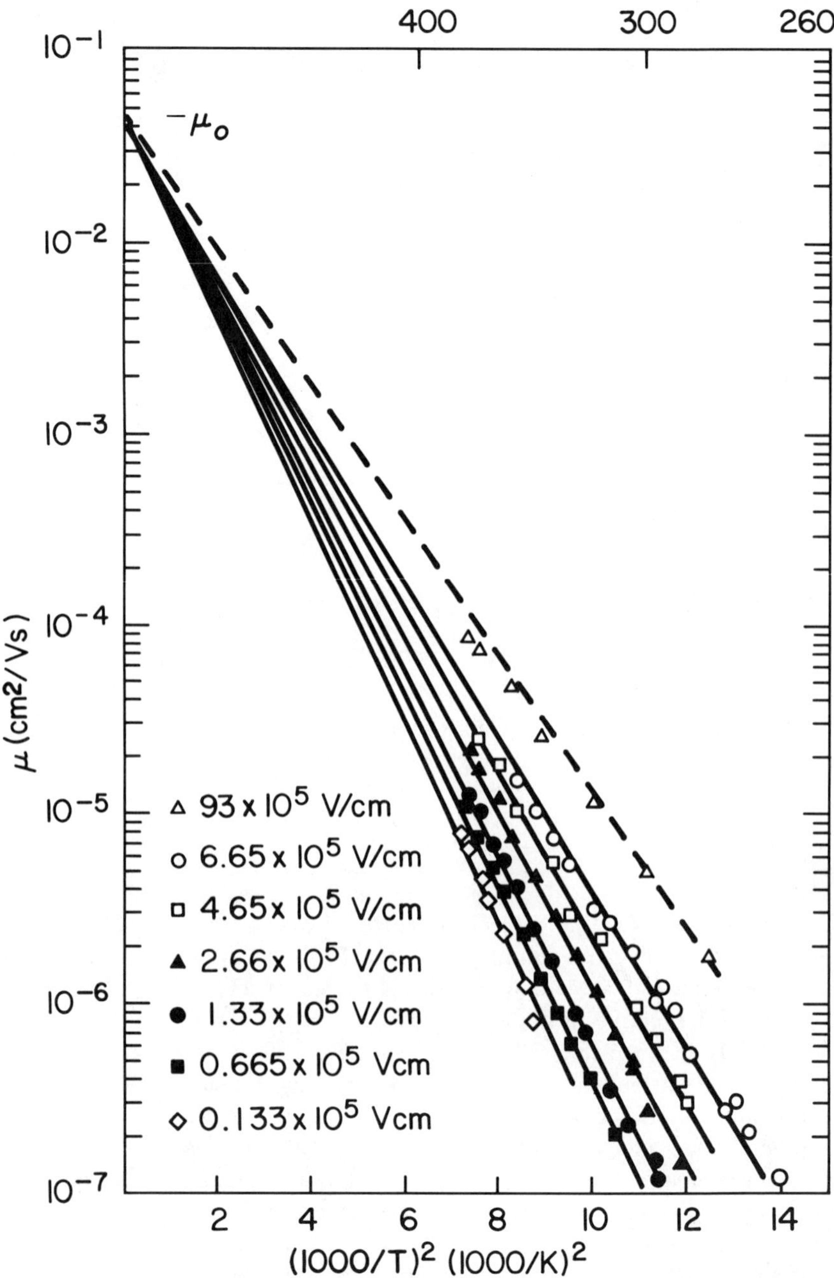

Figure 8. Data of Figure 6 replotted as μ versus $(10^3/T)^2$. (Reproduced with permission from reference 30. Copyright 1984 Taylor and Francis.)

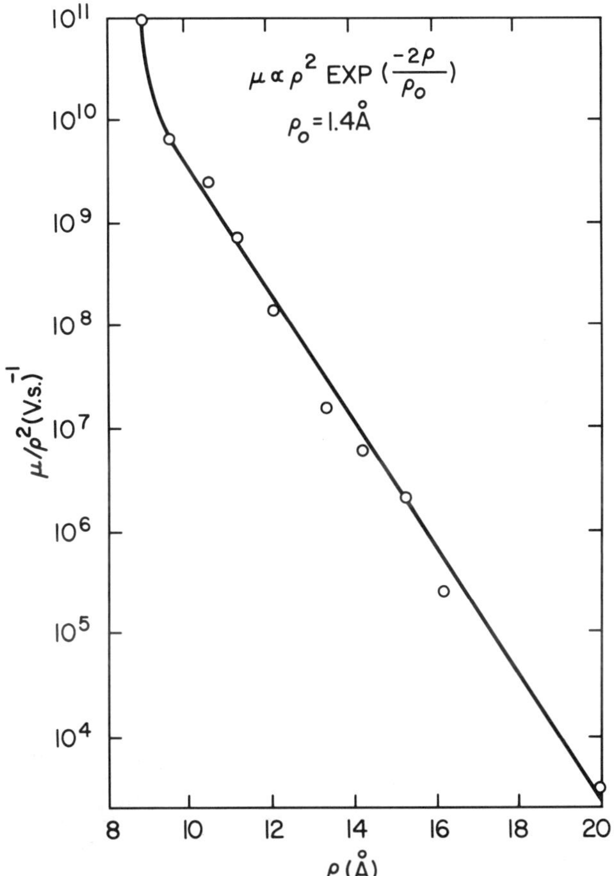

Figure 9. Concentration dependence of hole mobility (μ) in TPD–PC. ρ is the intersite distance. (Reproduced from reference 12. Copyright 1984 American Chemical Society.)

Scaling of Mobility with Thickness and Field. Figure 11 shows that the mobility in a 50:50 TPD–PC film is independent of film thickness in the range 5–90 μm. The absence of anomalous scaling is generally reflected in hole-transport behavior over a wide temperature range. The system TPD–PC was particularly free of space–charge memory effects visible in TOF and did not develop significant residual potentials (i.e., voltage remaining on photoreceptor after discharge, which is due to bulk space–charge), even after repeated xerographic photoinjection. This result indicates strongly that the system is trap free, that is, the states that can act as traps outside of the manifold of states among which carriers hop are totally absent. This behavior was clearly manifested in the shapes of space–charge-limited dark-current injection transients (*43*). Figure 12 shows that steady-state cur-

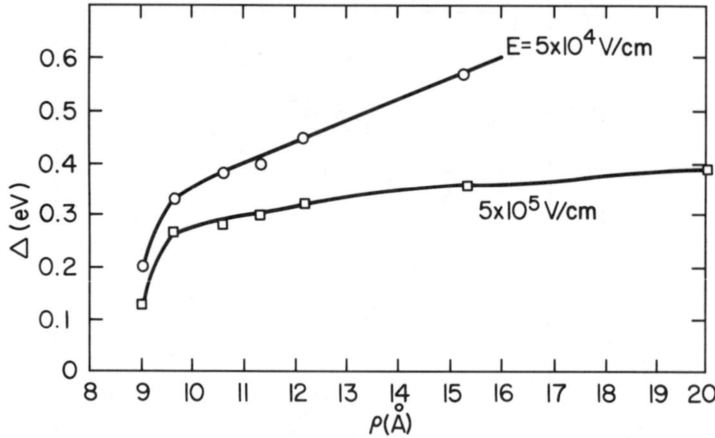

Figure 10. Concentration dependence of activation energy (Δ) at two fields in TPD–PC. ρ is the intersite distance. (Reproduced from reference 12. Copyright 1984 American Chemical Society.)

rent equilibrium is rapidly achieved (i.e., within several transit times). In trap-containing insulators, such as molecular crystals, the space–charge-limited current is expected to decay rapidly after the kink that occurs at approximately 0.8 t_{tr}. The behavior illustrated in Figure 12 is, in fact, quite extraordinary.

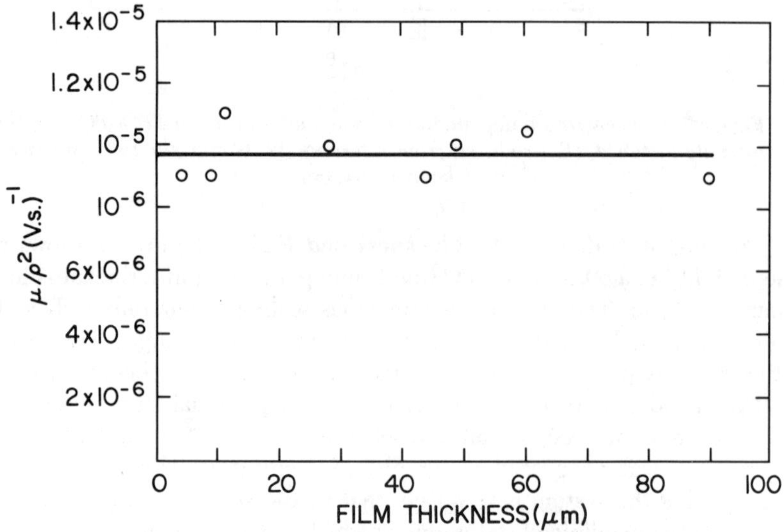

Figure 11. Thickness dependence of hole mobility in 50:50 TPD–PC film. (Reproduced from reference 12. Copyright 1984 American Chemical Society.)

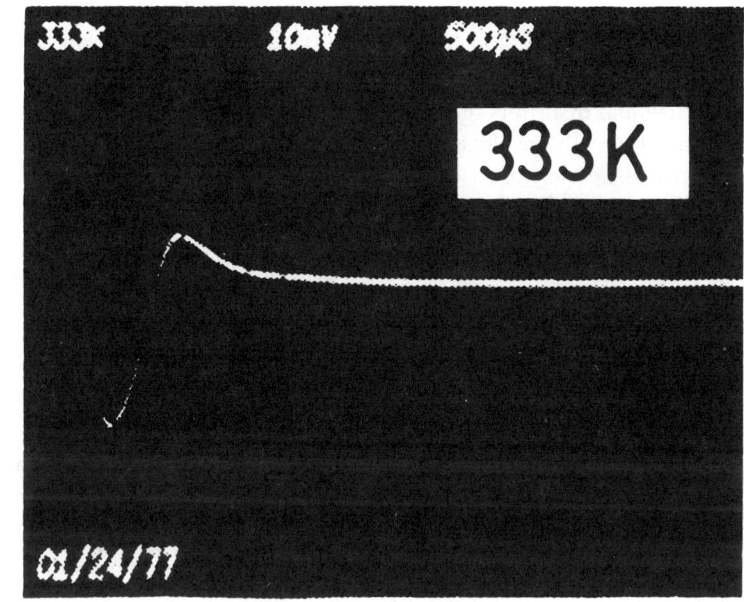

Figure 12. Dark-current response to step field excitation in 44 wt % TPD–PC film with Au (ohmic) contacts. T = *333 K, and* L = *26* μ*m. A 500-V field was applied. The cusp occurs at approximately 0.8* t_{tr}*, as determined independently with the TOF technique. The current equilibrium achieved in several transit times indicates the absence of deep trapping. (Reproduced with permission from reference 43. Copyright 1986 Taylor and Francis.)*

Temperature-parametric plots of log μ versus E for a 60:40 TPD–PC film clearly demonstrate a linear field dependence. This behavior contrasts with the square-root behavior of drift mobility on field strength, which was observed in PVK (*38*), PVK–TNF (*35*), and *N*-isopropylcarbazole (*44*) dispersed in PC and which has also been reported for other substituted aromatic amines (*40*).

Although field dependence of drift mobility in organic glasses below their respective T_gs appears to be pervasive, for reasons not yet understood, the dependence of log μ on field ranges in behavior from $E^{1/2}$ to E in otherwise closely related systems.

Trap-Controlled Hopping. Physical defects that are associated with intermolecular conformations that deviate in some specific way from the normal statistical distribution of molecular arrangements and extrinsic chemical species can generate additional localized states that lie outside of the distribution of bulk states. The bulk states control the primary transport channel, the "microscopic" mobility, whereas the additional extrinsic states can superimpose the effects of multiple trapping much as band tails do in inorganic amorphous semiconductors. According to Bässler (*30*), the distribution of primary transport states in organic glasses controlled by diagonal and off-diagonal disorders (geometric and electrostatic fluctuations from site to site) are typically insufficient to induce anomalous dispersion. Therefore, anomalous dispersion must be symptomatic of additional multiple trapping in the system and thus can vary in degree according to sample purity.

However, relative energetics are key in determining whether an impurity or structural defect can act as a trap or will be electrically inert and act as an "antitrap". For example, because the ionization potential of TPD is lower than that of PVK, TPD can introduce hole traps (*45*) into PVK. As the concentration of TPD is increased beyond a threshold level, it can proceed to provide the dominant transport channel. This situation is demonstrated clearly in Figure 13, which is a plot of log μ in a PVK host as a function of the concentration of TPD dopant molecules. At TPD concentrations below 5×10^{19} molecules per cubic centimeter, TPD traps progressively limit the transport that proceeds by hole hopping among carbazole pendant groups. At higher TPD concentrations, the TPD molecules provide first a parallel then the dominant transport channel. When TPD molecules are the dominant transport channel, PVK in effect acts as an inert binder.

Experimental Procedures

Polysilylene Synthesis. Polysilylenes are synthesized by the Wurtz coupling reaction of dichlorodialkyl- or dichloroarylalkylsilanes with molten sodium in an aromatic solvent according to the following reaction:

$$n\ \mathrm{Cl{-}SiR_1R_2{-}Cl} + 2n\ \mathrm{Na} \rightarrow 2n\ \mathrm{NaCl} + {-}({-}\mathrm{SiR_1R_2}{-})_n{-}$$

The PMPS used in this study was synthesized in refluxing toluene. Typically, the PMPS synthesis results in a mixture of three products; a cyclic oligomer ($n \simeq 5$), a low-molecular-weight linear polymer (number-averaged molecular weight [M_n] $\simeq$ 4–20,000), and a high-molecular-weight polymer ($M_n \simeq 300{,}000$ or more).

The molecular weight distribution factor of the high-molecular-weight fraction is typically 1.5–3.0. When the reaction is carried out by the method of Zeigler (*46*) (sodium dispersion steadily added to the solution of monomer in a mixture of aliphatic and aromatic solvents), the narrow-distribution low-molecular-weight fraction is so well separated from the narrow-distribution high-molecular-weight fraction that acetone or hexane precipitation easily isolates the monomodal high-molecular-weight fraction. Two different polymerization mechanisms proceeding in parallel are probably responsible for the unusual molecular weight distribution of the product.

High-molecular-weight PMPS is soluble in common solvents and is a good film former. Qualitatively, the overall handling characteristics resemble those of polystyrene. Films of PMPS were cast from toluene solution by a solvent evaporation technique. Other polysilylenes, poly(methylcyclohexylsilylene), poly(methyl-*n*-propylsilylene), poly(methyl-*n*-octylsilylene), and a copolymer of dimethyl- and methylphenylsilylene (1:1 ratio) were also prepared by Wurtz coupling by the method of Zhang and West (*47*). Films of these polymers were also cast from toluene. All polymers form clear transparent colorless films.

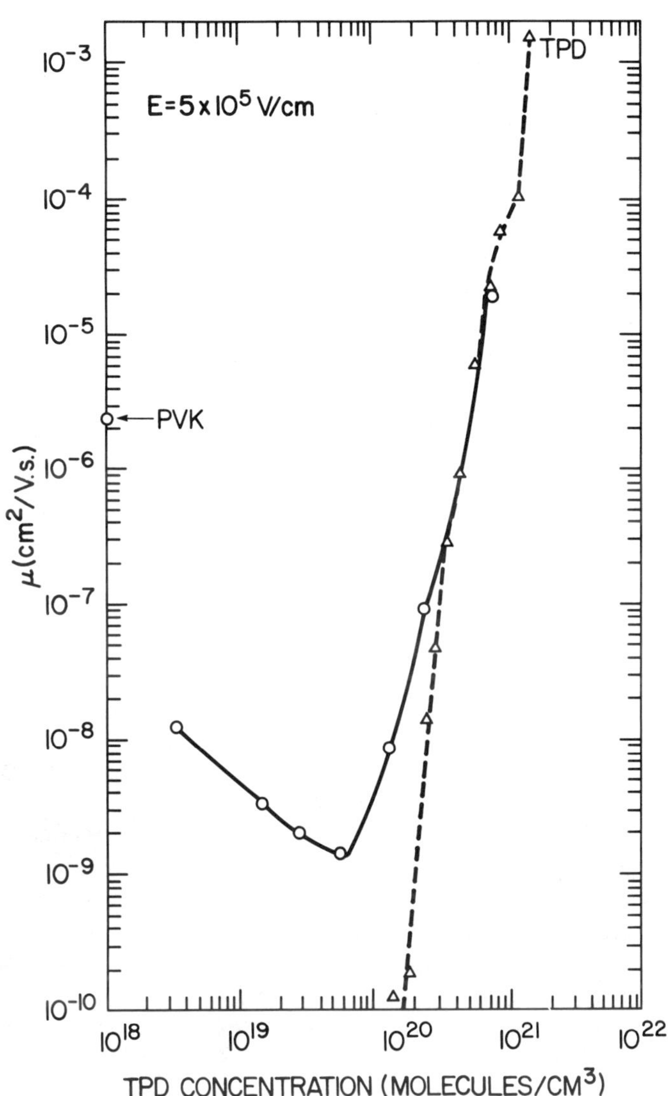

Figure 13. μ versus concentration of TPD dopant molecules at T = *295 K. The solid curve is for TPD–PVK, and the dashed curve is for TPD–PC. (Reproduced from reference 45. Copyright 1984 American Chemical Society.)*

Measurement of Drift Mobilities. Injected carrier drift mobilities were measured directly over a wide range of electric fields and temperature by the TOF technique with thin layers (0.5 μm) of a-Se or As_2Se_3 as photogenerators. The top electrodes consisted of evaporated semitransparent aluminum layers. Small-signal current-mode TOF transients were produced by hole injection from a-Se or As_2Se_3 excited at 450 or 550 nm by attenuated pulses from a nitrogen-pumped dye laser. The current transients were also generated intrinsically by using highly attenuated

337-nm pulses from the nitrogen laser. The absorption coefficient of PMPS at 337 nm is high (~10^5); consequently, most carriers were produced at the top of the film. The samples were examined for evidence of photodegradation, particularly during the UV photoexcitation experiments. No photodegradation was observed.

Electronic Transport in Polysilylenes

Certain features characteristic of the broader class of polysilylenes are revealed by the behavior of PMPS, which has been the most extensively studied material. The following sections focus on PMPS, but the findings can be regarded as representative for polysilylenes in most respects.

Conductivity in PMPS. Dark conductivity in PMPS and other investigated polysilylenes is negligible. Nonohmic (i.e., contact-controlled) direct-current (dc) steady-state dark injection currents were also infinitesimal. Correspondingly, during xerographic experiments in which the top surface is corona charged instead of being charged with the semitransparent electrode, the dark decay of the potential was extremely small, typically <<1%/s, as with good dielectric polymers.

Hole Transport in PMPS. In the experiments with layered structures (*20*) and visible excitation (to which PMPS is transparent), transient currents were observed only when the top electrode was negatively biased with respect to the substrate. The substrate was composed of a visible photoconductor (charge generation layer) overcoated aluminum ground plane. When the polymer top surface was directly (intrinsically) photoexcited with pulsed 337-nm excitation, current transit pulses were observed only when the top electrode was positively biased. Therefore, under the experimental conditions described, only hole transient transport could be directly observed. Transit pulses were nondispersive over a wide range of temperature. Figure 14 illustrates the relative increase in dispersion with decreasing temperature. In addition, no evidence for anomalous thickness dependence at the transit time was obtained, even at the lowest temperature.

Transit pulses are represented as the logarithm of current normalized to its value at the transit time. In highly dispersive transport media like a-As_2Se_3, such a plot would exhibit two algebraic branches corresponding to the power law curves $t^{-(1-\alpha)}$ when $t < t_{tr}$ and $t^{-(1+\alpha)}$ when $t > t_{tr}$, with $\alpha < 1$, as discussed earlier. In dispersive media at fixed temperature, transit pulses represented in this form show the universality exemplified in Figure 4 for a 1:1 PVK–TNF film (*10*). Such universality is not observed in PMPS (*20*). Neither was it observed in the TPD–PC system (*12*). Figure 15 shows the field dependence of hole drift mobility in PMPS at various temperatures. Field dependence grows progressively stronger as temperature is lowered. At temperatures near the T_g (determined calorimetrically), field dependence

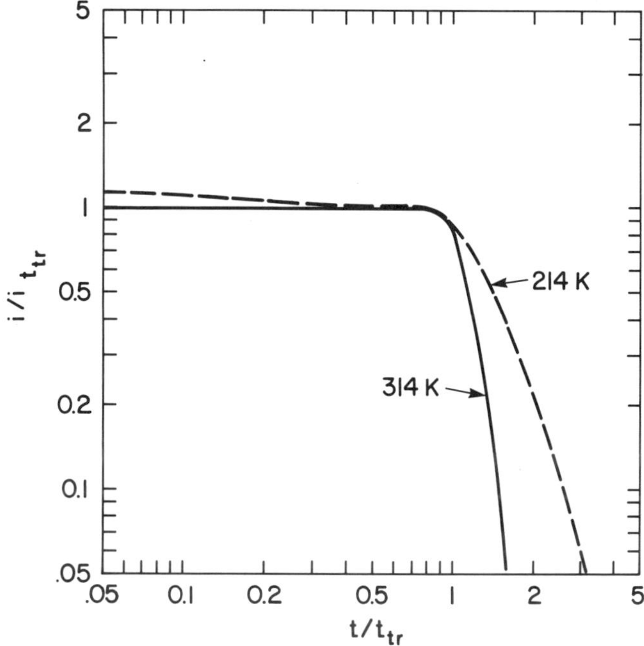

Figure 14. Logarithm of current (i) *normalized to its value at the transit time* (t_{tr}) *versus time in units of the transit time. An 8-μm PMPS film was used at 314 and 214 K. The applied field was 75 V/μm.*

(over the range indicated) of the drift mobility is negligible. These characteristics parallel the behavior described for PVK–TNF and MDPs in general.

To analyze hole-transport data in PMPS, the functional form of the field dependence of hole drift mobility must be determined first. This information is used to analyze the temperature-dependent behavior of the mobility extrapolated to the zero-field limit. In this limit, mobility data are analyzed in two phenomenological frameworks. First, the mobility is treated as simply and singly activated, after the procedure used by Pai (*36*), Gill (*35*), and others to represent transport data in PVK. The resulting activation energy is field dependent. Second, the temperature dependence of mobility is treated in the manner suggested by the numerical simulation studies of Bässler and co-workers (*33*) and analytical calculations of Movaghar and coworkers (*32*). This procedure establishes the width of the distribution of sites among which the carriers hop.

A square-root dependence of drift mobility on field was observed in PMPS (*20*) at 292 K. A family of such plots is shown in Figure 16. From the family of plots, the temperature dependence of the slopes (β) can be determined (Figure 17).

When hole mobilities are represented as a family of Arrhenius plots that are parametric in field, the apparent activation energies exhibit the char-

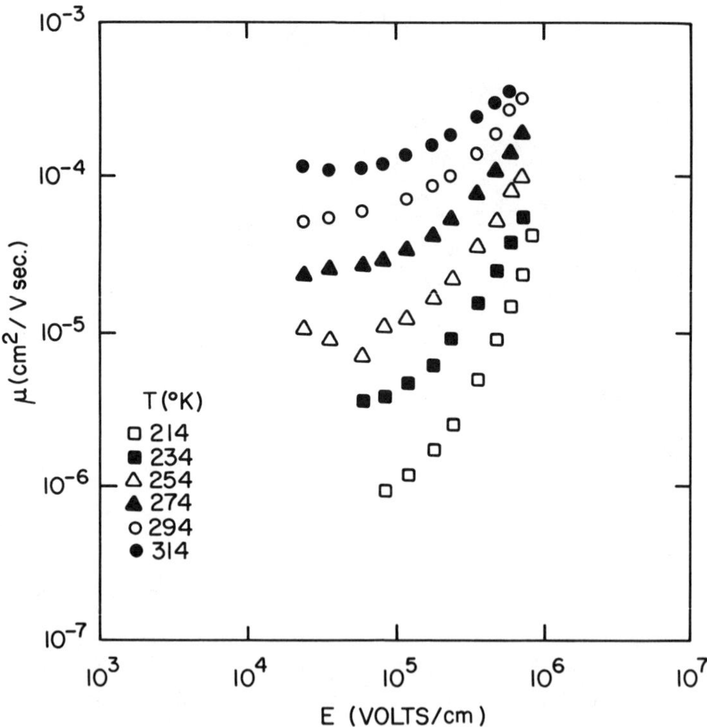

Figure 15. μ versus E for PMPS at various temperatures.

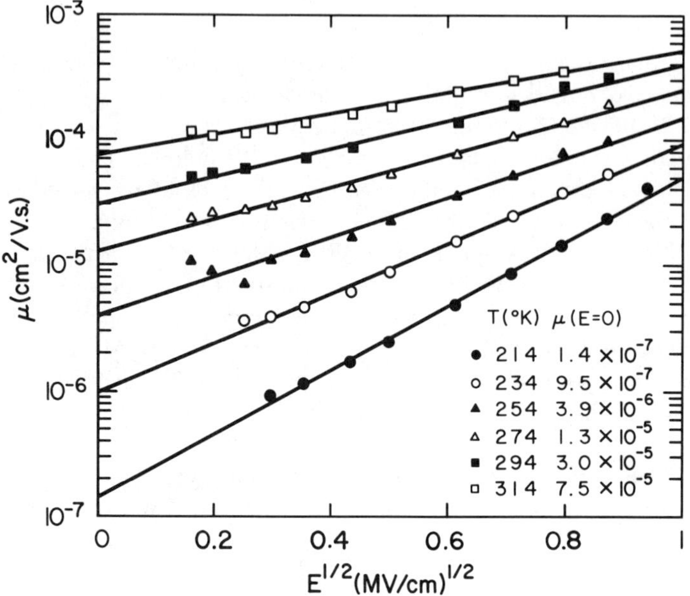

Figure 16. μ versus $E^{1/2}$ for PMPS at different temperatures.

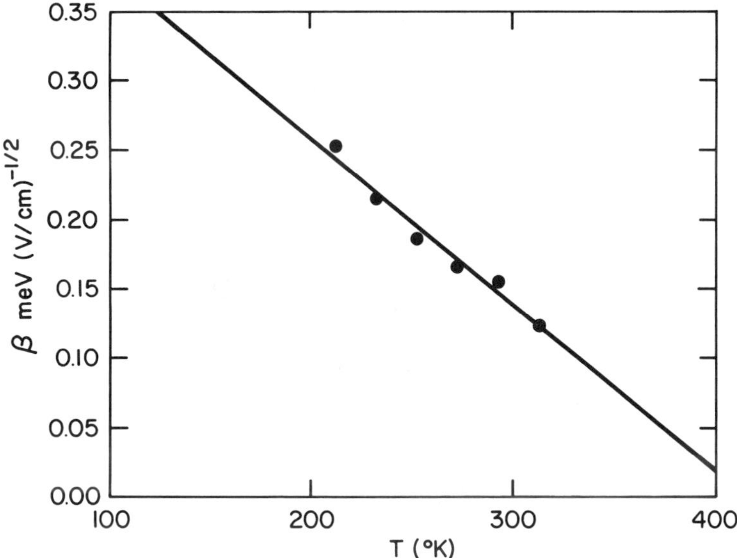

Figure 17. Slope of the curves (β) in Figure 16 versus temperature.

acteristic field dependence illustrated by the slopes of the curves in Figure 18 and more explicitly by those in Figure 19. The zero-field activation energy for a large number of sample films clusters in the 0.34–0.37-eV range; these values are substantially less than the corresponding 0.7 eV reported for PVK–TNF (Figure 7) (*35*).

For poly(methylcyclohexylsilylene) at $E = 1 \times 10^5$ V/cm, drift mobility activation is only 0.15 eV, which is approximately half that in PMPS at the same field, and yet, the drift mobilities of the two polysilylenes at room temperature are about the same (*48*).

The Arrhenius plots of log μ versus $1/T$ that are parametric in field tend to converge to $T_0 = 416$ K (Figure 18), which is in the vicinity of the T_g of PMPS. As noted earlier, a similar coincidence of T_0 with T_g was suggested in PVK plasticized with a small molecule (*37*). Both the apparent activation energy and the drift mobility become field independent at $T > T_0 = T_g$. The extrapolated value of μ_0 at T_0 is 4×10^{-3} cm^2/V-s.

The field- and temperature-dependent behavior is analogous to that described earlier for PVK and MDPs. All of these data are phenomenologically described by the following expression:

$$\mu = \mu_0 \exp\left(-[\epsilon_0 - \beta(T)]\frac{E^{1/2}}{kT_{\mathrm{eff}}}\right) \qquad (12)$$

in which $\dfrac{1}{T_{\mathrm{eff}}} = \dfrac{1}{T} - \dfrac{1}{T_0}$.

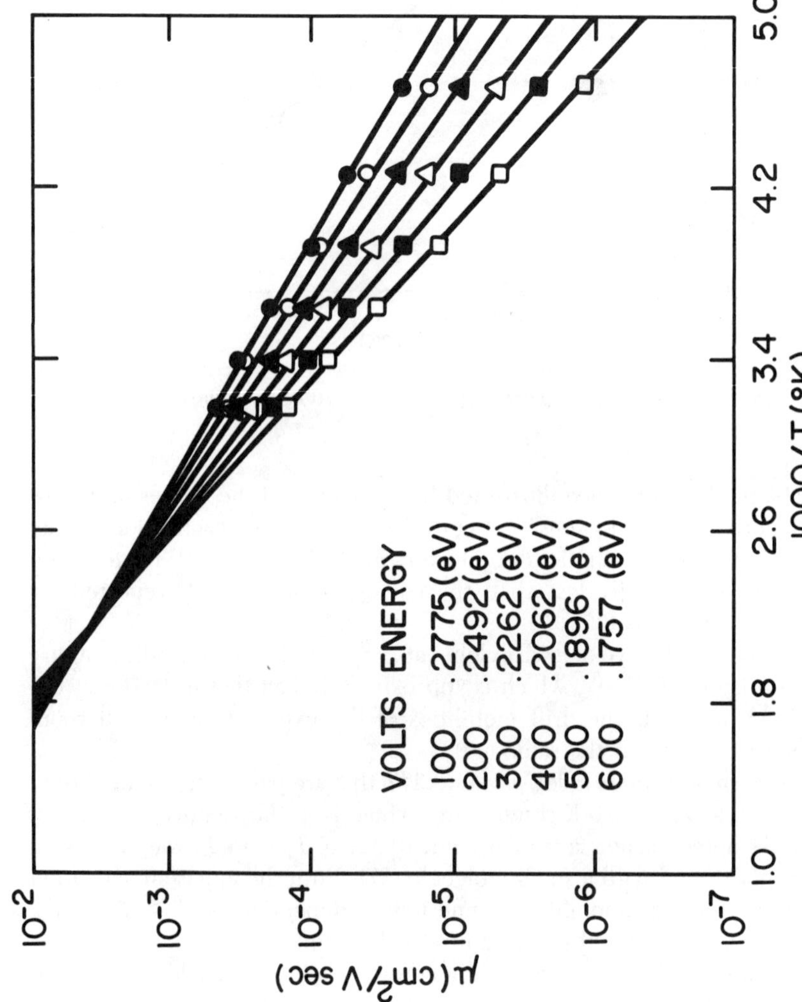

Figure 18. Arrhenius plots of μ for PMPS that are parametric in field.

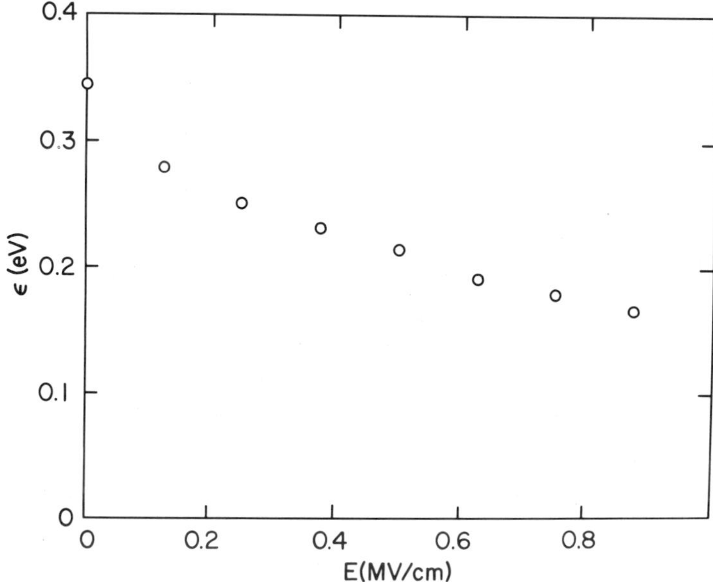

Figure 19. Field dependence of apparent activation energy of hole transport in PMPS.

Numerical simulation studies (*35*) have provided an alternative model for hopping transport in organic glasses. In this case, carriers are assumed to be injected and transported in a disorder-induced Gaussian distribution of localized states of width σ or equivalent temperature $T_0 = 7400\sigma$. This model further predicts that the temperature dependence of μ, which is non-Arrhenius in the long-time limit, takes the form given by equation 9 and, in its initial formulation, at least predicts a linear field dependence of log μ. For PMPS, the observed dependence is closer to $E^{1/2}$. Bässler's model (*30*) predicts that plots of the logarithm of mobility versus $1/T^2$ should intersect at a common point on the mobility axis corresponding to the value μ, the mobility at the limit of no disorder. In Figure 20, the temperature dependence of hole drift mobility at various fields, when replotted in conformity with the predictions of Bässler's model, leads to $\mu_0 \sim 10^{-2}$ $cm^2/V\text{-}s$.

From the analysis of the field-dependent slopes of the curves in Figure 20, the underlying disorder-induced Gaussian width (σ) of the hopping states can be calculated. The resulting field-dependent width expressed in terms of energy and equivalent temperature units is shown in Figure 21, together with the analogous data obtained for PVK.

In this analysis, the hopping-site distribution, which is represented by the value of σ extrapolated to zero field, appears to be very narrow (i.e., 0.09 eV) compared with the distribution for PVK–TNF. This result should

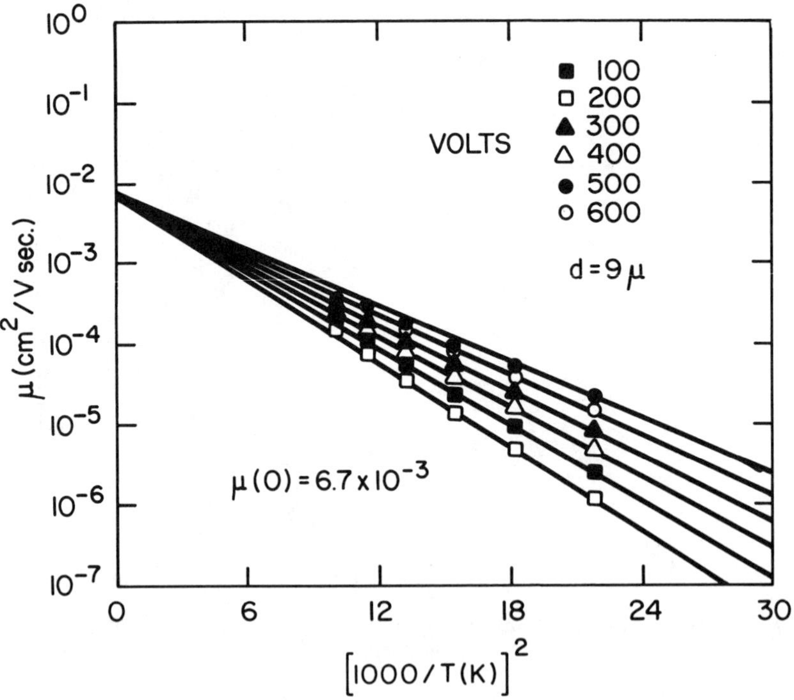

Figure 20. μ versus $(10^3/\mathrm{T})^2$ for PMPS at different fields.

be compared with the relatively nondispersive transit pulses at room temperature in PMPS (Figure 14) and the corresponding dispersive transits in PVK–TNF (Figure 3).

Transport Mechanisms in Polysilylenes

In PVK, hole transport is a thermally assisted hopping process among the pendant carbazole side groups, which are characterized by their relatively extended π electronic wave function overlap. Spectroscopic evidence indicates that the carbazole pendant groups interact only very weakly with the saturated carbon backbone of the polymer. Therefore, the primary role of the chain backbone on transport in PVK is probably limited to fixing the relative spatial orientation of the pendant groups. Extensive concentration-dependent studies of transport in various MDPs demonstrate that a similar hopping mechanism underlies the exponential dependence of drift mobility on the average intersite distance commonly observed in these systems. In PVK and MDPs, mobility, which is field dependent in the glassy state, can be represented as activated, with a field-dependent activation energy. Furthermore, the field dependence of the drift mobility is itself temperature dependent and becomes progressively weaker as the T_g is approached from

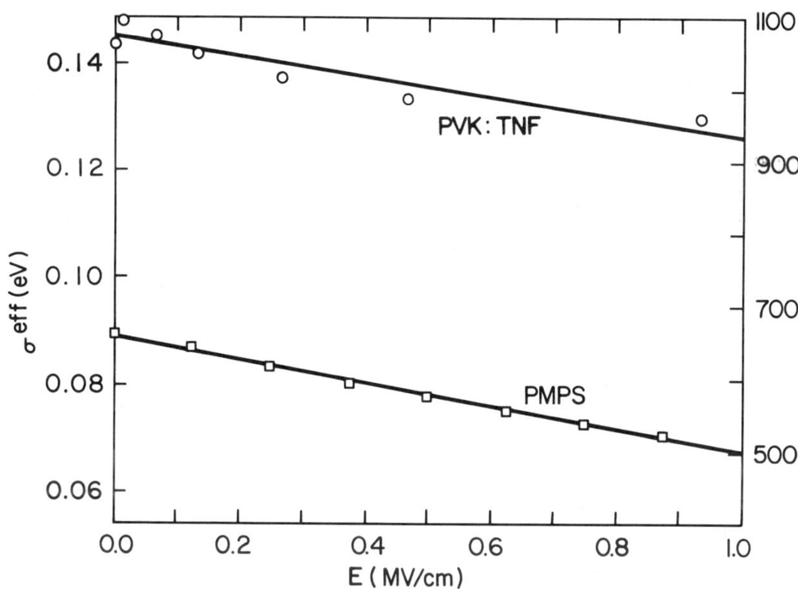

Figure 21. Width of Gaussian distribution of hopping sites expressed in terms of energy (σ) (left ordinate) and equivalent temperature (right ordinate) for PMPS and PVK.

below. Though not understood in all its microscopic details, the transport process operating in PVK and MDPs manifests a complex but clearly developed pattern of behavior, as revealed by experiments.

Transport measurements in PMPS reveal (*49, 50*) a behavior that is qualitatively similar in all important respects to that in PVK and MDPs. Does this similarity then imply a role (as hopping sites) for the π-electron-containing pendant groups on the saturated Si chain, which is analogous to the role played by carbazole on the saturated carbon backbone in PVK? In this connection, we have not been able to observe hole transport in polystyrene, the phenyl-pendant-containing carbon-backbone system most analogous to PMPS, but more compelling and more direct evidence to support another interpretation of the transport mechanism in polysilylenes exists.

We have already noted that hole transport exhibiting an overall pattern of behavior similar to PMPS is observed (*50*) in poly(methylcyclohexylsilylene), the aliphatic system most closely resembling PMPS but distinguished by the absence of π electrons. On the basis of simple chemical reasoning, one should expect no significant pendant wave function overlap in this system. From this and related comparative studies, the side groups in Si polymers do not seem to control transport critically.

Instead, transport is most likely a hopping process involving states associated directly with the silicon backbone. Trap-controlled band transport, the process believed to be operative in the three-dimensional (space polymer) network of a-Si:H, seems a less likely possibility in the disordered Si polymer systems characterized by both bulky side groups (i.e., lack of interchain intimate contact) and quasi one-dimensional network topology. Recent calculations of Rice and co-workers (*51*) suggest that extraction of a hole from a silicon polymer would, as a result of its one-dimensional character, lead to the immediate formation of a polaronic state. They suggest a mechanism of polaronic hopping, although their theory in its present form does not account for the highly correlated field- and temperature-dependent behavior observed in the drift mobility.

Finally, an interesting possibility is suggested by recent studies of polarized excitation in and fluorescence spectra of poly(di-*n*-hexylsilylene) (*52*). On the basis of these measurements, this polymer chain is believed to be organized into a domainlike series of chromophores. Associated calculations further suggest that the domains are all-*trans* segments of approximately 10–15 repeat units separated by *gauche* links. If these domains could also behave as small-molecule analogues, then transport by intra- and interchain domain hopping might be expected to display the observed characteristics. A broad program of experiments designed to explore further the origin and nature of transport states in Si-backbone polymers is currently in progress.

Summary

Interest in glassy dielectric (small dark-carrier population) polymers for application as page-size (large-area) flexible, robust, solution-coated electrophotographic receptors has been a principal motivator of research on electronic transport processes in a wide variety of such systems. In conventional inorganic photoreceptors, photogeneration and transport occur in the same film, whereas in most contemporary organic-polymer-based systems, photogeneration and transport occur in compositionally distinct layers (i.e., the generation layer [GL] and transport layer [TL]), each of which needs to satisfy specific requirements. Thus, the ability to efficiently transport holes photoinjected from adjacent generation layers makes polysilylenes interesting candidates for both scientific and technological investigations.

A key requirement of the transport layer is that charge accepted from the excited photogeneration layer must pass through a dielectric thickness, which is fixed by system developability criteria, in a short time compared with the interval between exposure and development of the latent image. This condition imposes a requirement on the drift mobility of the charge carrier. In addition, the fractional loss of transiting charge to deep traps must be minimized, because trapped space–charge in the bulk produces unacceptable image degradation. Relatively high mobility for the photoinjected carriers must be achieved simultaneously with high charge retension in the dark (negligible free-carrier density) and high dielectric strength.

The requirement on space–charge buildup during repeated xerographic photoinjection is stringent for an organic polymer. In practical terms, for typical transport layer thicknesses (10–25 μm), the trap density should not exceed $10^{14}/cm^3$, corresponding to no more than one active trap site per 10^7 monomer units. This trap density is far below the expected level of chemical contamination associated with typical synthetic and purification procedures. Proper choice of transport media renders unavoidable resident impurities electrically inactive. For example, in MDPs that transport holes, the oxidation potential of the transporting species in a given host polymer must be less than that of resident background impurities. A similar requirement applies for the carbazole groups in PVK.

PVK was an early candidate for photoreceptor application. The embodiment deployed in copiers was the complex of PVK with TNF. The mobilities in this system are not sufficient to satisfy the contemporary requirements of high-speed copiers, duplicators, and laser printers. In addition, PVK is too brittle for belt architecture. However, mechanistic studies on this system provide design criteria for closely related MDPs.

Deployment of MDPs, which are solid solutions of a transport-active species (typically in the range 30–50 wt %) in an inert host polymer, embodies the concept of full compositional control of mobility (53). Transport molecules are chosen with the aim of rendering accidental contaminants electrically inactive. The concentration of small molecules controls drift mobility. Appropriate choice of a compatible host polymer can then focus on mechanical properties and environmental stability. MDPs are limited by phase separation, crystallization, and leaching by contact with the development system, cleaning solvent, etc. These problems continue to motivate the search for single-component transport polymers.

TOF studies of glassy silicon-backbone polymers containing both aromatic and aliphatic side groups reveal the relatively high hole drift mobility at room temperature. Hole drift mobility at room temperature is not profoundly sensitive to side-group substituents and remains relatively high (10^{-4} cm^2/V-s at 1×10^5 V/cm) even in the complete absence of π electrons. Films can be prepared in which space–charge buildup even under conditions of repeated photoinjection remains small. Hole drift mobility is typically field dependent and thermally activated. The field dependence of hole drift mobility, which is itself temperature dependent, grows progressively weaker as the specimen temperature approaches the T_g from below.

The qualitative behavior of electronic transport in polysilylenes is thus very similar to that observed in PVK and MDPs. However, unlike transport in PVK, in which hole transport clearly involves hopping of photoinjected holes among carbazole side groups, transport in glassy polysilylenes appears to involve holes hopping among states which derive from the saturated Si main chain. One possibility suggested by optical measurements and associated calculations is that the silicon polymer chain is typically suborganized into domains of 10–15 repeat units, among which photoinjected holes could execute thermally assisted hopping motion, a situation resembling that prevailing in MDPs.

References

1. For comprehensive review see Schafert, R. M. *Electrophotography;* Focal Press: New York, 1975.
2. Regensburger, P. J. *Photochem. Photobiol.* **1968,** *8,* 429.
3. Rose, A. *Photoconductivity and Applied Problems;* Interscience: New York, 1963.
4. Loutfy, R. O.; Hor, A. M.; Rucklidge, A. *J. Imaging Sci.* **1987,** *31,* 31 and references therein.
5. Stolka, M.; Pai, D. M. *Adv. Polym. Sci.* **1978,** *29,* 1.
6. Pearson, J. M.; Stolka, M. *Poly(N-vinylcarbazole);* Gordon and Breach: New York, 1981.
7. Williams, D. J. *Macromolecules* **1970,** *3,* 602.
8. Regensburger, P. J. Can. Patent 932199, 1973.
9. Gill, W. D. *Proc. 5th Int. Conf. on Amorphous and Liquid Semicond.;* Taylor and Francis: London, 1974; p 901.
10. Seki, H. *Proc. 5th Int. Conf. on Amorphous and Liquid Semicond.;* Taylor and Francis: London, 1974; p 1015.
11. Murayama, T. *Denshi Shashin Gakkaishi 25,* 290.
12. Stolka, M.; Yanus, J. F.; Pai, D. M. *J. Phys. Chem.* **1984,** *88,* 4707.
13. Schein, L. B.; Rosenberg, A.; Rice, S. L. *J. Appl. Phys.* **1986,** *60,* 4287.
14. Kepler, R. G.; Zeigler, J. M.; Harrah, L. A. *Bull. Am. Phys. Soc.* **1984,** *29,* 509.
15. Fatori, V.; Bürge, G.; Ggeri, A. *J. Imaging Technol.* **1986,** *12,* 334.
16. Yajima, S.; Okamura, K.; Hayashi, T. *Chem. Lett.* **1975,** 1209.
17. Mazdiyasni, K. S.; West, R.; David, L. D. *J. Am. Ceram. Soc.* **1978,** *61,* 504.
18. Trujillo, R. E. *J. Organomet. Chem.* **1980,** *198,* C27.
19. Stolka, M.; Yuh, H.-J.; McGrane, K.; Pai, D. M. *J. Polym. Sci., Polym. Chem. Ed.* **1987,** *25,* 823.
20. Abkowitz, M. A.; Knier, F. E.; Yuh, H.-J.; Weagley, R. J.; Stolka, M. *Solid State Commun.* **1987,** *62,* 547.
21. Scher, H.; Montroll, F. W. *Phys. Rev. B* **1975,** *12,* 2455.
22. Lampert, M. A.; Mark, P. *Current Injection in Solids;* Academic Press: New York, 1970.
23. Abkowitz, M.; Scher, H. *Philos. Mag., B* **1977,** *35,* 1585.
24. Abkowitz, M. *J. Appl. Phys.* **1979,** *50,* 4009.
25. Dolezalek, F. J. In *Photoconductivity and Related Phenomena;* Mort, J.; Pai, D. M., Eds.; Elsevier: New York, 1976; p 27.
26. Muller-Horsche, E.; Haarer, D.; Scher, H. *Phys. Rev. B* **1987,** *35,* 1273.
27. Orlowski, T. E.; Abkowitz, M. *Solid State Commun.* **1986,** *59,* 665.
28. Mott, N. F.; Davis, E. A. *Electronic Processes in Non-Crystalline Materials;* Clarendon Press: Oxford, 1971.
29. Enck, R.; Pfister, G. In *Photoconductivity and Related Phenomena;* Mort, J.; Pai, D. M., Eds.; Elsevier: New York, 1976; p 27.
30. See Bässler, H. *Philos. Mag., B* **1984,** *50,* 347 and references therein.
31. Pfister, G. *Phys. Rev. B* **1977,** *16,* 3676.
32. See for example Movaghar, B.; Würtz, D.; Pohlmann, B. *Z. Phys. B: Condens. Matter* **1987,** *66,* 523.
33. Schönherr, G.; Bässler, H.; Silver, M. *Philos. Mag., B* **1981,** *44,* 47.

34. Pfister, G.; Griffiths, C. H. *Phys. Rev. Lett.* **1978,** *40,* 659.
35. Gill, W. D. *J. Appl. Phys.* **1972,** *43,* 5033.
36. Pai, D. M. *J. Chem. Phys.* **1971,** *52,* 2285.
37. Fujino, M.; Kanazawa, Y.; Mikawa, H.; Kusabayashi, S.; Yokoyama, M. *Solid State Commun.* **1984,** *49,* 575.
38. Bässler, H.; Schönherr, G.; Abkowitz, M.; Pai, D. M. *Phys. Rev. B* **1982,** *26,* 3105.
39. Mort, J.; Pfister, G.; Grammatica, S. *Solid State Commun.* **1976,** *18,* 693. For a more complete review and additional references see Mort, J.; Pfister, G. In *Polym. Plast. Technol. Eng.* **1979,** *12*(2), 89–139.
40. Pai, D. M.; Yanus, J. F.; Stolka, M.; Renfer, D.; Limburg, W. W. *Philos. Mag., B* **1983,** *48,* 505.
41. Borsenberger, P. M.; Mey, W.; Chowdry, A. *J. Appl. Phys.* **1978,** *49,* 273.
42. Tsutsumi, M.; Yamamoto, M.; Nishijiina, Y. *J. Appl. Phys.* **1986,** *59,* 1557.
43. Abkowitz, M.; Pai, D. M. *Philos. Mag., B* **1986,** *53,* 193.
44. Santos Lemus, S. J.; Hirsch, J. *Philos. Mag., B* **1986,** *53,* 25.
45. Pai, D. M.; Yanus, J. F.; Stolka, M. *J. Phys. Chem.* **1984,** *88,* 4714.
46. Zeigler, J. M. *Polym. Prepr.* **1986,** *27,* 1, 109.
47. Zhang, X.-H.; West, R. *J. Polym. Sci., Polym. Chem. Ed.* **1984,** *22,* 159, 225.
48. Yuh, H.-J.; Abkowitz, M.; McGrane, K.; Stolka, M. *Bull. Am. Phys. Soc.* **1986,** *31,* 386.
49. Kepler, R. G.; Zeigler, J. M.; Harrah, L. A.; Kurtz, S. R. *Phys. Rev. B* **1987,** *35,* 2818.
50. Abkowitz, M.; Stolka, M.; Yuh, H.-J. *Bull. Am. Phys. Soc.* **1987,** *32,* 885.
51. Rice, M. J.; Phillpot, S. R. *Phys. Rev. Lett.* **1987,** *58,* 937.
52. Klingensmith, K. A.; Downing, J. W.; Miller, R. D.; Michl, J. *J. Am. Chem. Soc.* **1986,** *108,* 7438.
53. A particularly significant early reference is Mehl, W.; Wolf, N. E. *J. Phys. Chem. Solids* **1964,** *25,* 1221.

RECEIVED for review May 27, 1987. ACCEPTED revised manuscript October 20, 1988.

27

Synthesis and Properties of Silicon-Branched Organosilicon Polymers

Yoichiro Nagai,[‡] Hamao Watanabe,[1] Hideyuki Matsumoto, Yoshitake Naoi,[2] and Naotake Sutou[2]

Department of Chemistry, Gunma University, Kiryu, Gunma 376, Japan

Two types of new silicon-branched organosilicon polymers, linear and ladder polysilane structures, were produced from dihalo- and tetrahalodisilane, respectively, via alkali-metal-mediated reactions. Further investigations disclosed that the polymers may be useful as photoresists, semiconductors, ceramic precursors, and composite materials in high-technology fields.

THE RADIATION SENSITIVITY of soluble alkyl- and aryl-substituted linear polysilanes with high molecular weights has been studied with much vigor recently (*1–9*). Because silicon branching in the silicon framework results in significant electronic stabilization (*10–12*), silicon-branched organosilicon polymers are expected to exhibit unique properties that are considerably different from those of existing organosilicon polymers. The structures of polysilanes are so far limited to the linear backbone. However, silicon-branched organosilicon polymers have been prepared (9).

Because of the significance of silicon branching on electronic properties, we became interested in preparing the following two silicon-branched organosilicon polymers according equations 1 and 2.

Linear Polysilane

Polymer **1** was obtained (*13, 14*) by reductive condensation of 1,1-dichloro-2-phenyl-1,2,2-trimethyldisilane (a silyl-substituted methyldichlorosilane)

[‡]Deceased.
[1]Author to whom correspondence should be addressed.
[2]Current address: Yuki Gosei Kogyo Company, Itabashi, Tokyo 176, Japan

0065–2393/90/0224–0505$06.00/0

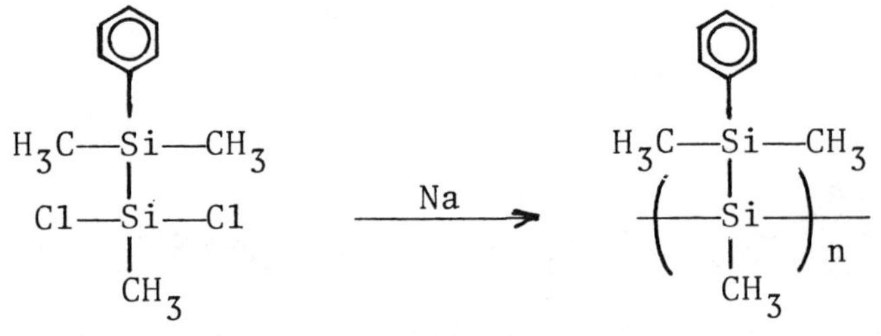

Na

1

Li

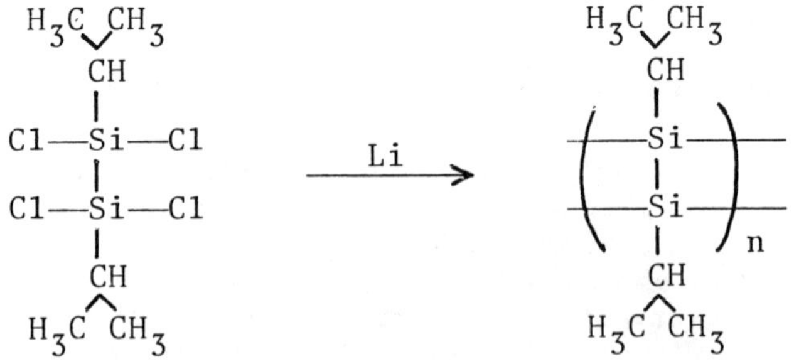

2

with sodium dispersed in toluene at 50 °C. The starting dichlorodisilane was obtained readily by the reaction of 1,1,2-trichloro-1,2,2-trimethyldisilane with phenylmagnesium bromide in the presence of copper iodide (equation 3) with more than 99% regioselectivity.

$$Cl(CH_3)_2SiSiCH_3Cl_2 \xrightarrow[CuI]{C_6H_5MgBr} C_6H_5(CH_3)_2SiSiCH_3Cl_2 \qquad (3)$$

Reductive condensation of the dichlorodisilane monomer with sodium yielded a pale-yellow semisolid crude product. Fractionation of the polymer was carried out by precipitation with ethanol from a THF (tetrahydrofuran) solution. A polymer with a weight-average molecular weight (M_w) of 4×10^4 (determined by gel permeation chromatography [GPC] with polystyrene standards) was obtained as a nice white powder that was soluble in common organic solvents, fusible, and shapable. The polymer obtained showed no absorption bands due to Si–H or Si–O–Si bonds in the IR region (Figure 1). The IR spectrum indicates that the polymer is of the expected poly[(phenyldimethylsilyl)methylsilylene] structure. The observed phenyl-to-methyl ratio (1:3) in the ^{1}H NMR spectrum also supports this structure, and obviously, the original Si–Si bond in the starting disilane was retained during the condensation reaction.

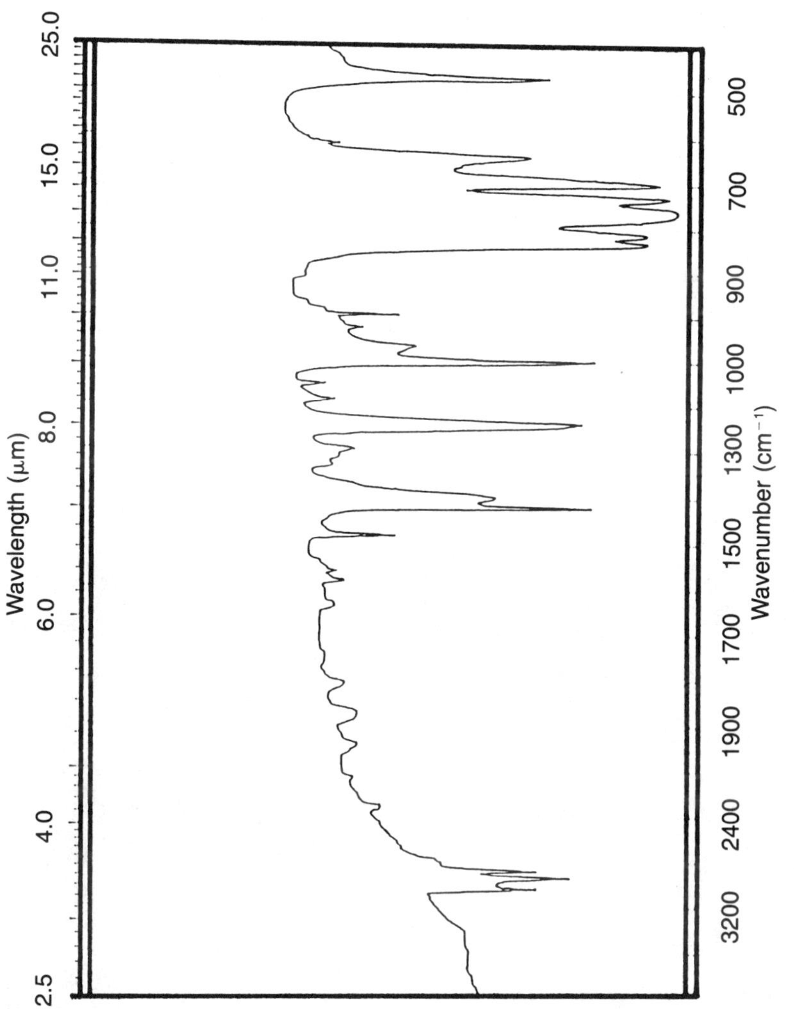

Figure 1. IR spectrum of $-[[(C_6H_5)(CH_3)_2Si]SiCH_3]_n-$ *(KBr pellet).*

The polymer exhibited UV absorption bands near 245 and 310 nm (Figure 2). Irradiation of its solution in cyclohexane with a mercury lamp resulted in the rapid decrease of the molecular weight at the early stage of the reaction, but further irradiation gave rise to a considerable increase in molecular weight (Figure 3). Presumably, initial chain cleavage followed by cross-linking occurred. The polymer can be shaped into a film by the casting method. Irradiation (with a low-pressure lamp) of the polymer film in air gave an insoluble material. The IR spectrum of the irradiated film showed a strong band due to the Si–O–Si bond.

The silicon-branched polymer also became semiconducting upon exposure to iodine vapor; it is, therefore, closely related to "polysilastyrene", for which West et al. (2) used highly toxic pentafluoroarsine as dopant. Polymer **1** is originally insulating, but an iodine-doped sample, a black solid, showed a conductivity (σ) of $\sim 10^{-3}/\Omega$-cm. The black solid was stable in air for several days and then gradually softened. The stability of the doped sample remains to be improved further.

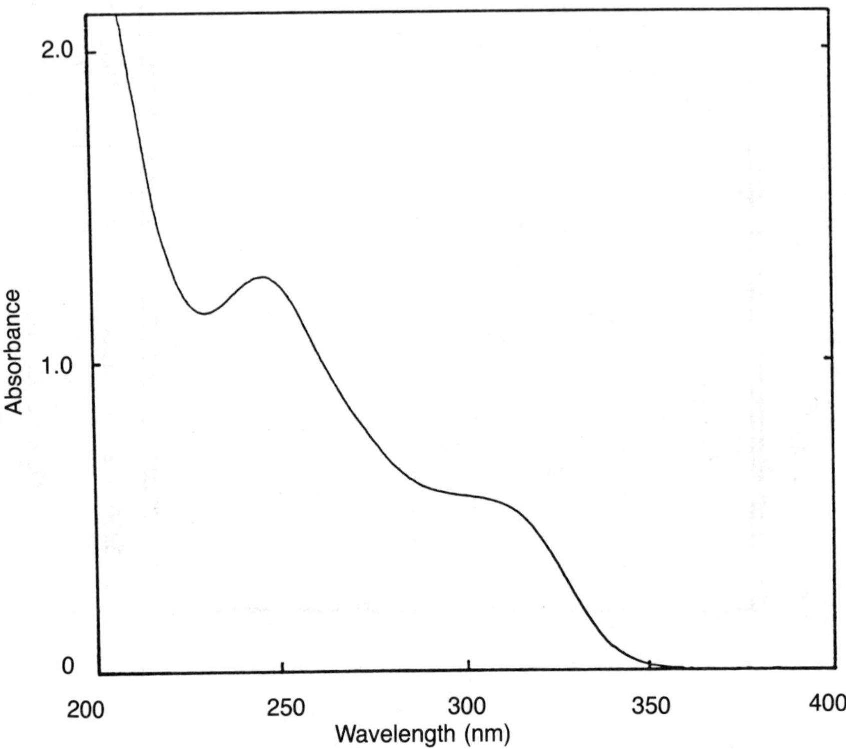

Figure 2. UV spectrum of $-[[(C_6H_5)(CH_3)_2Si]SiCH_3]_n-$ *(cyclohexane solution).*

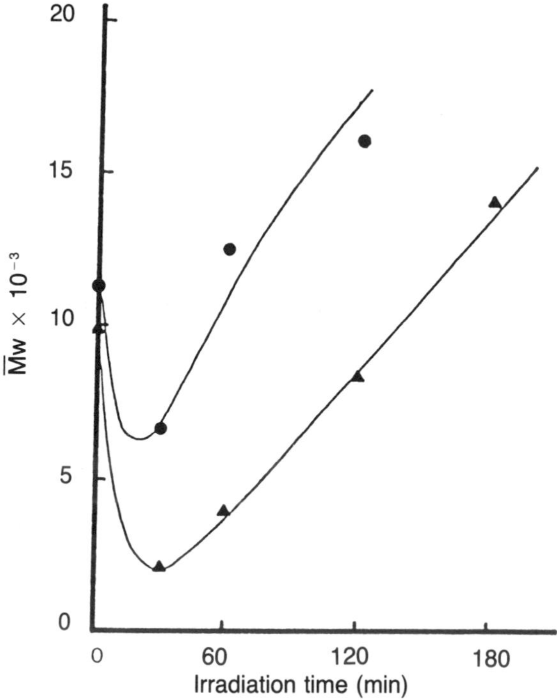

Figure 3. Photodegradation of $-[[(C_6H_5)(CH_3)_2Si]SiCH_3]_n-$. *Key:* ●, *irradiation with a high-pressure mercury lamp (100 W);* ▲, *irradiation with a low-pressure mercury lamp (30 W).*

Ladder Polysilane

Previously, a several-ring system polysilane, which is soluble in toluene, had been prepared by Baney et al. (*15*), starting from methylated polychlorodisilanes by heating. This system has a complex network structure. Our second silicon-branched organosilicon polymer, polymer **2**, was obtained by the cocondensation of 1,1,2,2-tetrachloro-1,2-diisopropyldisilane and 1,2-dichloro-1,1,2,2-tetraisopropyldisilane. We have shown previously (*16*) that the reaction of a 1:3 mixture of the tetrachlorodisilane and the dichlorodisilane with lithium in THF gives a series of annulated cyclotetrasilane ring such as structures **3–5** (Scheme I).

As is shown in Figure 4, a high-pressure liquid chromatogram (HPLC) for a typical run gave major peaks, each of which can be assigned to one of the annulated cyclotetrasilane systems. Small peaks accompanying the major peaks could be due to linear oligomeric polysilanes or to a series of persilaprismanes. Nevertheless, Figure 4 clearly shows that a major part of the product mixture consists of the desired polycyclopolysilanes of ladder struc-

$Cl_2RSiSiRCl_2$ + ClR_2SiSiR_2Cl $\xrightarrow[\text{THF}]{\text{Li}}$

$R = (CH_3)_2CH$

```
          R
R2Si—Si—SiR2
   |     |     |                          3
R2Si—Si—SiR2
          R

          +

          R     R
R2Si—Si—Si—SiR2
   |     |     |     |                    4
R2Si—Si—Si—SiR2
          R     R

          +

          R     R     R
R2Si—Si—Si—Si—SiR2
   |     |     |     |     |              5
R2Si—Si—Si—Si—SiR2
          R     R     R

          +

        Others
```

Scheme I

ture. Thus, a 10:1 mixture of the tetrachlorodisilane and the dichlorodisilane was treated with lithium or sodium in THF to give a reddish yellow semisolid in 80–100% yields. The average molecular weight of the crude polymer was 2000, but fractions of $M_w = 25{,}000$ were observed by GPC. Iodine doping converted the crude polymer to a semiconducting black solid with $\sigma = \sim 10^{-2}/\Omega\text{-cm}$.

Thermogravimetric analysis of the polymer was carried out under very nearly the same conditions as previously reported for permethylpolysilane and polysilastyrene (*17, 18*) (Figure 5). Unfractionated polymer **2** lost 35% of its weight upon heating under nitrogen at a 10 °C/min rise in temperature, up to 800 °C. The theoretical loss according to equation 4 or 5 is 27 or 40%, respectively, and the results indicate that practically no volatile silicon compounds have been driven off.

$$(C_3H_7Si) \rightarrow CH_4 + 3/2H_2 + C + SiC \quad (4)$$

$$(C_3H_7Si) \rightarrow C_2H_4 + 3/2H_2 + SiC \quad (5)$$

West et al. (*17*) have reported that permethylpolysilane is completely lost at 650 °C, whereas 70% of a sample of polysilastyrene was lost as volatile

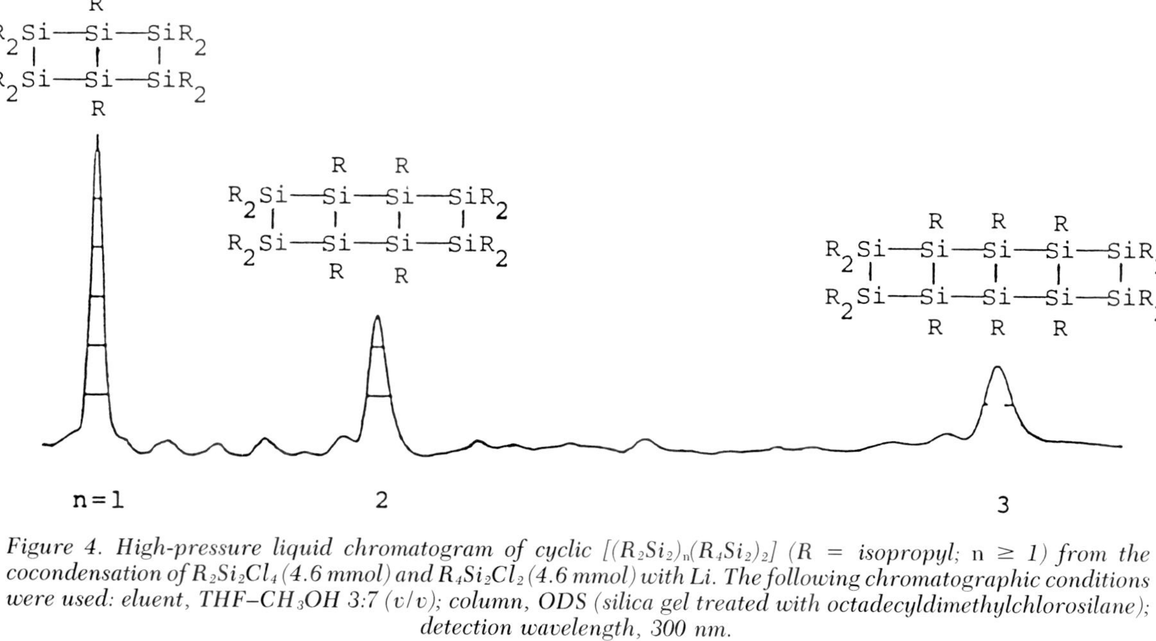

Figure 4. High-pressure liquid chromatogram of cyclic $[(R_2Si_2)_n(R_4Si_2)_2]$ (R = isopropyl; n ≥ 1) from the cocondensation of $R_2Si_2Cl_4$ (4.6 mmol) and $R_4Si_2Cl_2$ (4.6 mmol) with Li. The following chromatographic conditions were used: eluent, THF–CH_3OH 3:7 (v/v); column, ODS (silica gel treated with octadecyldimethylchlorosilane); detection wavelength, 300 nm.

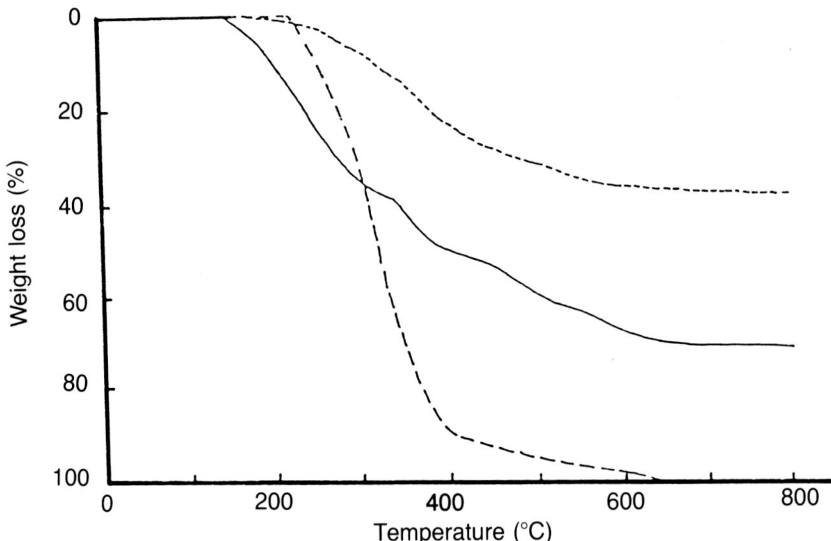

Figure 5. Thermogravimetric curves for polysilanes. Key: —, polysilastyrene; – – –, permethylpolysilane; and - - -, ladder polysilane. (Reproduced with permission from reference 17. Copyright 1983 American Ceramic Society.)

silicon compounds at 800 °C. The ladder silicon network in polymer **2** seems to be responsible for the observed highly efficient ceramic yield.

References

1. Trujillo, R. E. *J. Organomet. Chem.* **1980,** *198,* C27.
2. West, R.; David, L. D.; Djurovich, P. I.; Stearley, K. I.; Srinivasan, K. S. V.; Yu, H. *J. Am. Chem. Soc.* **1981,** *105,* 7352.
3. Trefonas, P., III; Djurovich, P. I.; Zhang, X. H.; West, R.; Miller, R. D.; Hofer, D. C. *J. Polym. Sci., Polym. Lett. Ed.* **1983,** *21,* 819.
4. Zhang, X. H.; West, R. *J. Polym. Sci., Polym. Chem. Ed.* **1984,** 22, 159.
5. Zhang, X. H.; West, R. *J. Polym. Sci., Polym. Chem. Ed.* **1984,** 22, 225.
6. Todesco, R. V.; Basher, R. *J. Polym. Sci., Part A* **1986,** *24,* 1943.
7. West, R. *J. Organomet. Chem.* **1986,** *300,* 327.
8. Miller, R. D.; Sooriyakumaran, R. *J. Polym. Sci., Polym. Lett. Ed.* **1987,** *25,* 321.
9. Harrah, L. A.; Zeigler, J. M. *Macromolecules* **1987,** *20,* 2037.
10. Matsumoto, H.; Yokoyama, N.; Sakamoto, A.; Aramaki, Y.; Endo, R.; Nagai, Y. *Chem. Lett.* **1986,** 1643.
11. Ishikawa, M.; Watanabe, M.; Iyoda, M.; Ikeda, H.; Kumada, M. *Organometallics* **1987,** *1,* 317.
12. Plinka, T. B.; West, R. *Organometallics* **1986,** *5,* 128.
13. Watanabe, H.; Akutsu, Y.; Shinohara, A.; Ohta, A.; Onozuka, M.; Nagai, Y. 52nd Annual Meeting of the Chemical Society of Japan, Kyoto, Japan; Chemical Society of Japan: Tokyo, 1986; Abstract 1K15.

14. Watanabe, H.; Akutsu, Y.; Shinohara, A.; Shinohara, S.; Yamaguchi, Y.; Ohta, A.; Onozuka, M.; Nagai, Y. *Chem. Lett.* **1988,** 1883.
15. Baney, R. H.; Gaul, J. H., Jr.; Hilty, T. K. *Organometallics* **1983,** *2*, 859.
16. Matsumoto, H.; Miyamoto, H.; Kojima, N.; Nagai, Y. *J. Chem. Soc. Chem., Chem. Commun.* **1987,** 1316.
17. West, R.; David, L. D.; Djurovich, P. I.; Yu, Y.; Sinclair, S. *Ceram. Bull.* **1983,** *62*, 899.
18. Yajima, S.; Hasegawa, Y.; Hayashi, J.; Iimura, M. *J. Mater. Sci.* **1978,** *13*, 2569.

RECEIVED for review May 27, 1988. ACCEPTED revised manuscript July 31, 1989.

28

Electronic Structures of Silicon-Based Polymers

Nobuo Matsumoto, Kyozaburo Takeda, Hiroyuki Teramae, and Masaie Fujino

Basic Research Laboratories, Nippon Telegraph and Telephone Corporation, 3-9-11, Midori-cho, Musashino-shi, Tokyo 180, Japan

Crystalline and amorphous silicons, which are currently investigated in the field of solid-state physics, are still considered as unrelated to polysilanes and related macromolecules, which are studied in the field of organosilicon chemistry. A new idea proposed in this chapter is that these materials are related and can be understood in terms of the dimensional hierarchy of silicon-backbone materials. The electronic structures of one-dimensional polymers (polysilanes) are discussed. The effects of side groups and conformations were calculated theoretically and are discussed in the light of such experimental data as UV absorption, photoluminescence, and UV photospectroscopy (UPS) measurements. Finally, future directions in the development of silicon-based polymers are indicated on the basis of some novel efforts to extend silicon-based polymers to high-dimensional polymers, one-dimensional superlattices, and metallic polymers with alternating double bonds.

SILICON-BACKBONE MATERIALS include silane oligomers, polysilanes, silicon clusters, and amorphous and crystalline silicons. These materials have been investigated independently in two different fields. Crystalline and amorphous silicon are studied in the field of solid-state physics (*1*), whereas polysilanes and related molecules are studied in the field of organosilicon chemistry (*2*). Crystalline silicon (c-Si) and amorphous hydrogenated silicon (a-Si:H) are well known as two of the most useful semiconductors for electronic and optical devices. Polysilanes have been investigated for application as SiC ceramic binders (*3*) and photoresists (*4*). The methods of synthesizing

0065–2393/90/0224–0515$07.75/0

these materials are very different. c-Si and a-Si:H films are deposited by molecular beam epitaxy (MBE) or chemical vapor deposition (CVD) in an ultrahigh-vacuum chamber, whereas polysilanes are synthesized by chemical reaction in a glass flask. Therefore, the materials have been considered unrelated.

In 1983, Wolford et al. (5) discovered a new type of hydrogenated amorphous silicon, called polysilane alloys, which include a large number of silicon hydride polymers. Their structures were studied by Matsumoto et al. (6), and it was clarified that polysilane alloys are materials with intermediate properties relative to those of amorphous silicon and organosilane polymers.

Four kinds of silicon-backbone materials are shown in Figure 1: a crystalline silicon (c-Si) substrate (Figure 1a); a hydrogenated amorphous silicon (a-Si:H) film (Figure 1b) and a polysilane alloy film (Figure 1c), both deposited on fused-quartz substrates; and an organosilane polymer film cast on a fused-quartz substrate (Figure 1d). Why do these materials exhibit different colors? This important question is the starting point of this chapter. The dimensional hierarchy model can explain the origin of the color differences shown in Figure 1.

The next question is how side-chain substitutions and skeleton conformations affect the band structures of polysilanes. Side chains provide two interesting effects (7): band gap reduction caused by substitution of larger alkyl side chains and skeleton–side chain interaction (i.e., σ–π mixing) in aryl polysilanes. This interaction was confirmed by UV photo spectroscopy (UPS) (8–9) and photoabsorption and luminescence measurements (*10, 11*). Skeleton conformations are related to thermochromism (*12–17*). The ab initio

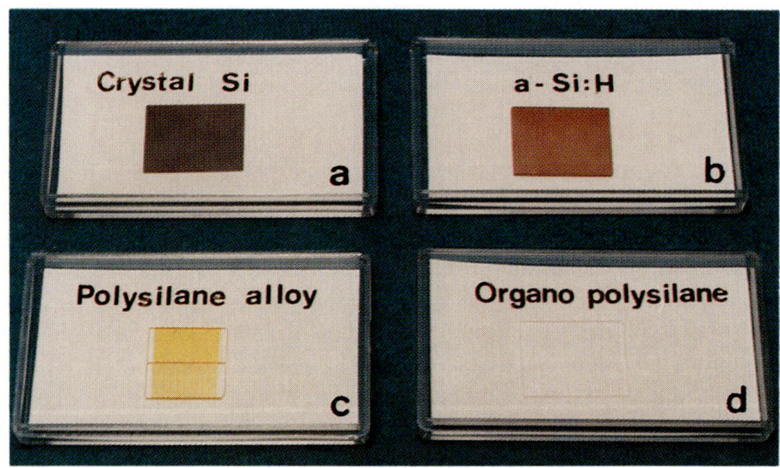

Figure 1. Si-backbone materials: (a) crystalline silicon, (b) amorphous hydrogenated silicon, (c) polysilane alloy, and (d) organosilane polymer.

study answers two questions: (1) Which conformation is more stable, *trans* or *gauche*? (2) What is the essential difference between the band structures of *trans* and *gauche* polysilanes?

Dimensional Hierarchy Model of Silicon-Based Polymers

The crystalline Si lattice structure, which is the same as that for diamond, is shown in Figure 2. The bonding between Si atoms is by σ bonds, with σ electrons delocalized over the whole lattice. The blue chain in Figure 2 corresponds to the backbone of polysilane. The red lattice plane is formed by bonding blue chains to each other. The total lattice is formed by piling up red planes. Figure 3 shows the schematic representation of the red plane lattice, which corresponds to the plane denoted as (111) by conventional crystal notations. This plane, called a *unit plane,* consists of unit chains shown in blue in Figure 2. A unit chain is formed by bonding unit dimers

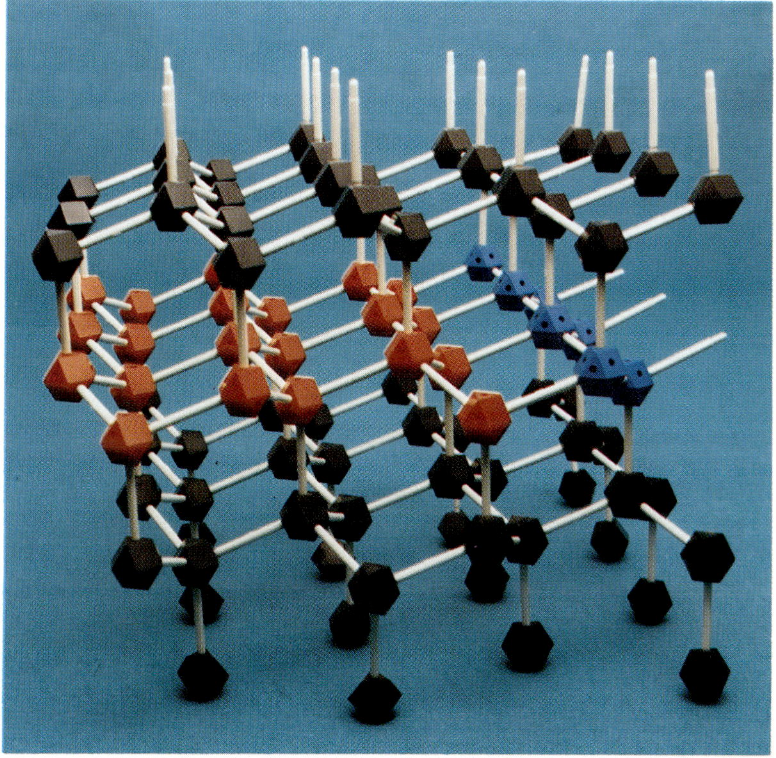

Figure 2. Crystal silicon lattice structure. The blue chain of silicon atoms corresponds to the backbone of polysilane. The red lattice plane is formed by bonding blue chains to each other.

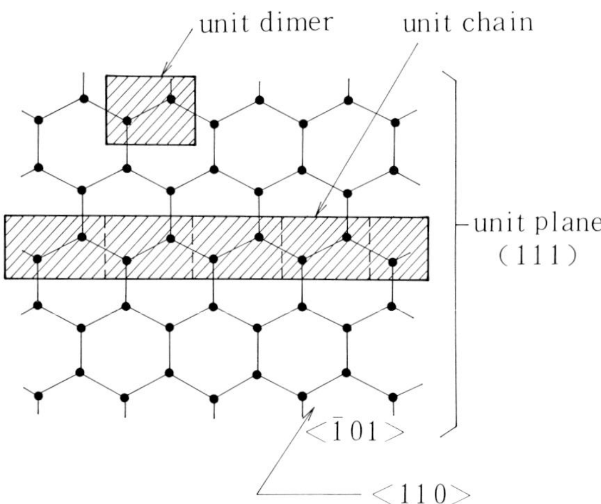

Figure 3. Schematic representation of the red plane lattice shown in Figure 2.

along <110> or <101> crystal axes. Sublattices cut from the crystalline silicon lattice can be described by using this unit structure.

The notation CL(u,v,w) represents the skeleton of the cluster, which is formed by (1) bonding u unit dimers into a finite unit chain, (2) connecting v unit chains into a rectangular unit plane, and (3) piling up w unit planes to form a three-dimensional cluster. Any tetrahedrally bonded cluster can be formed by polymerization of unit dimers, chains, and planes. Therefore, in a wide sense, Si-based polymers include oligomers, polymers, clusters, and even crystalline silicon.

Table I summarizes the hierarchy of the clusters CL(u,v,w). The network dimension changes from zero to three.

- CL(1,1,1) means a unit dimer itself, which corresponds to the skeleton of disilane. The dimension is zero, and the energy gap of disilane is 6.5 eV.

Table I. Typical Examples of CL (u,v,w)

Dimension	*CL* (u,v,w)	*Material*	E_g *(eV)*
0	CL (1,1,1)	Disilane	$E_0 = 6.5$
	CL (3,1,1)	Hexasilane	
1	CL (∞,1,1)	Polysilane	$E_1 = 4$
	CL (∞,2,1)	Ladder polymer	
2	CL (∞,∞,1)	Siloxene	$E_2 = 2.5$
	CL (∞,∞,n)	Superlattice	
3	CL (∞,∞,∞)	Crystal Si	$E_3 = 1.1$

- CL(3,1,1) means three unit dimers connected to form an oligomer, (i.e., hexasilane).
- CL(∞,1,1) means a unit chain, which corresponds to polysilane. The dimension is one, and the energy gap is about 4 eV.
- CL(∞,2,1) corresponds to ladder polymers, which are formed by connecting two unit chains.
- CL(∞,∞,1) is a unit plane, whose dimension is 2. The layered crystal siloxene, $[Si_6H_3(OH)_3]_n$ (*18, 19*), is a unique example of this two-dimensional lattice. The measured band gap is ~2.5 eV.
- CL(∞,∞,*n*) is realized by superlattice structures such as Si and Ge multiquantum wells, in which *n* values correspond to the silicon layer thickness.
- CL(∞,∞,∞) is three-dimensional crystal silicon, with an energy gap of 1.1 eV.

Table I shows that the band gap, the energy difference between HOMO (highest occupied molecular orbitals) and LUMO (lowest unoccupied molecular orbitals) levels, decreases monotonically with the increase in network dimension. This decrease is caused by the delocalization of skeleton σ electrons, which form both band edges. As is well known, eigenvalues of delocalized wave functions confined to a potential well are determined by the well size and potential-barrier heights. When delocalized wave functions are confined to a smaller area, the HOMO level moves downward and the LUMO level moves upwards, which results in the increase in band gap energy. This quantum size effect is given by

$$
\begin{aligned}
E_g(u,v,w) &= \frac{q^6}{A^2B^2C^2}(E_0 - 3E_1 + 3E_2 - E_3) \\
&+ \left[\frac{q^4}{A^2B^2} + \frac{q^4}{B^2C^2} + \frac{q^4}{C^2A^2}\right](E_1 - 2E_2 + E_3) \\
&+ \left[\frac{q^2}{A^2} + \frac{q^2}{B^2} + \frac{q^2}{C^2}\right](E_2 - E_3) + E_3 \qquad (1)
\end{aligned}
$$

in which $A = u + p$, $B = v + p$, $C = w + p$, $p = 2/(\pi - 2)$, and $q = \pi/(\pi - 2)$. $E_g(u,v,w)$ corresponds to the energy gap of CL(u,v,w). E_0–E_3 are the measured band gaps of the polymers in each network dimension (Table I).

An increase in u, v, and w results in a decrease in band gap energy, which reflects the delocalization of skeleton σ electron wave functions. The first term in equation 1 defines the effect of volume size, and the second and third terms define the effects of surface size and linear size, respectively. The calculated curve and experimental results for poly(dimethylsilane)s (*20*) are shown in Figure 4a. The abscissa is the number of silicon atoms, and the ordinate is band gap energy (i.e., absorption peak energy) in electron-volts. The calculated curve and experimental results for a-Si:H–a-Si:N:H superlattices (*21*) are shown in Figure 4b. The abscissa is the silicon layer thickness. In both figures, the calculated curve and the experimental data coincide closely.

Band gap dependence on network dimensions and unit numbers is summarized in Figure 5. Curve u shows the variation from disilane to polysilane. For curve v, $n = 1$ corresponds to polysilane, and $n = 2$ corresponds to a ladder polymer; the limiting case is siloxene. Curve w shows the energy gap changes for superlattices of different layer thickness, and it approaches the curve for crystalline silicon.

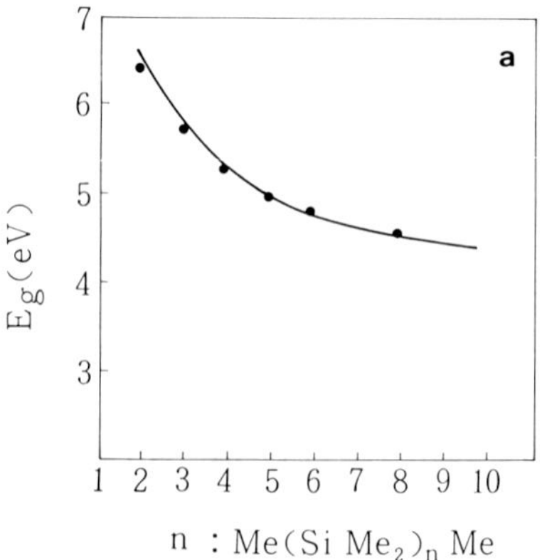

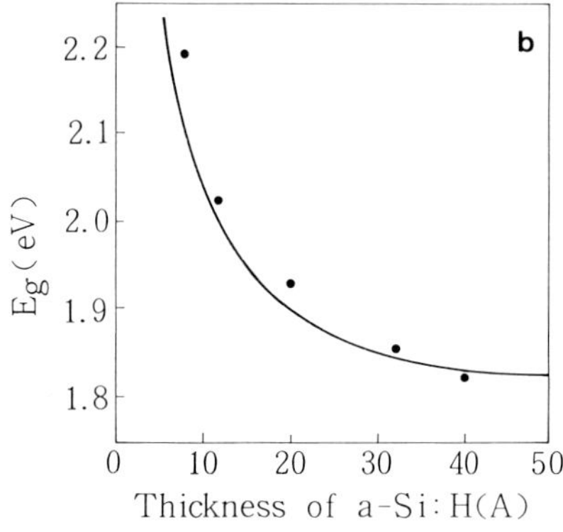

Figure 4. Band gap variations calculated by using equation 1 and experimental data for poly(dimethylsilane) (a) and a-Si:H–a-Si:N:H superlattices (b). (Data are from refs. 20 and 21.)

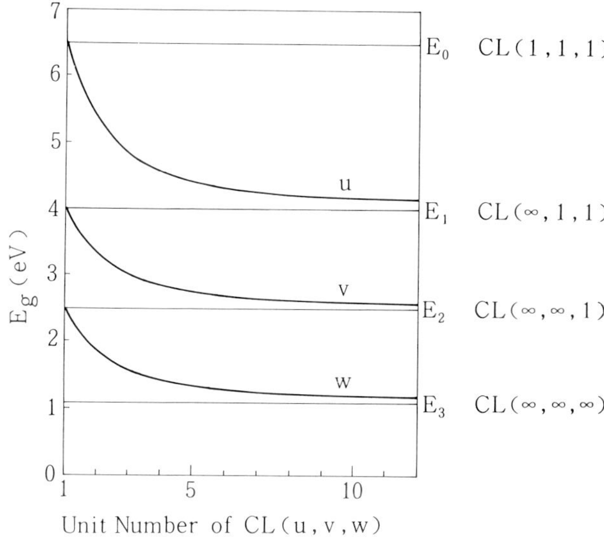

Figure 5. Band gap dependence of Si-based polymers on network dimensions and unit numbers.

The dimensional hierarchy of Si-based polymers is summarized in Figure 6. Disilane, polysilane, siloxene, and c-Si are arranged by network dimensions in the right circle. Oligomers have network dimensions between zero and one. Ladder polymers and superlattices have intermediate network dimensions. Many kinds of clusters also have network dimensions between zero and three. Although a-Si:H and polysilane alloy cannot be described by the unit dimer model, the effective network dimensions of these materials are between one and three. The network structure of polysilane alloy corresponds to that of cross-linked polymers, and so, the effective dimension is near unity. On the other hand, the network structure of a-Si:H, which includes many terminal hydrogen atoms and dangling bonds, topologically resembles that of c-Si. Therefore, the network dimension of a-Si:H is near three. The route from polysilanes to crystalline silicon through polysilane alloys and amorphous silicons shows that the decrease in the hydrogen content is accompanied by an increase in the Si network dimension and results in a decrease in the energy gap. In conclusion, the color differences of the four kinds of Si-backbone materials shown in Figure 1 correspond to this variation in network dimensions from one to three.

Electronic Structures of Polysilanes

The electronic structures of Si-based polymers are determined basically from the network dimensions discussed in the previous section. In addition, two

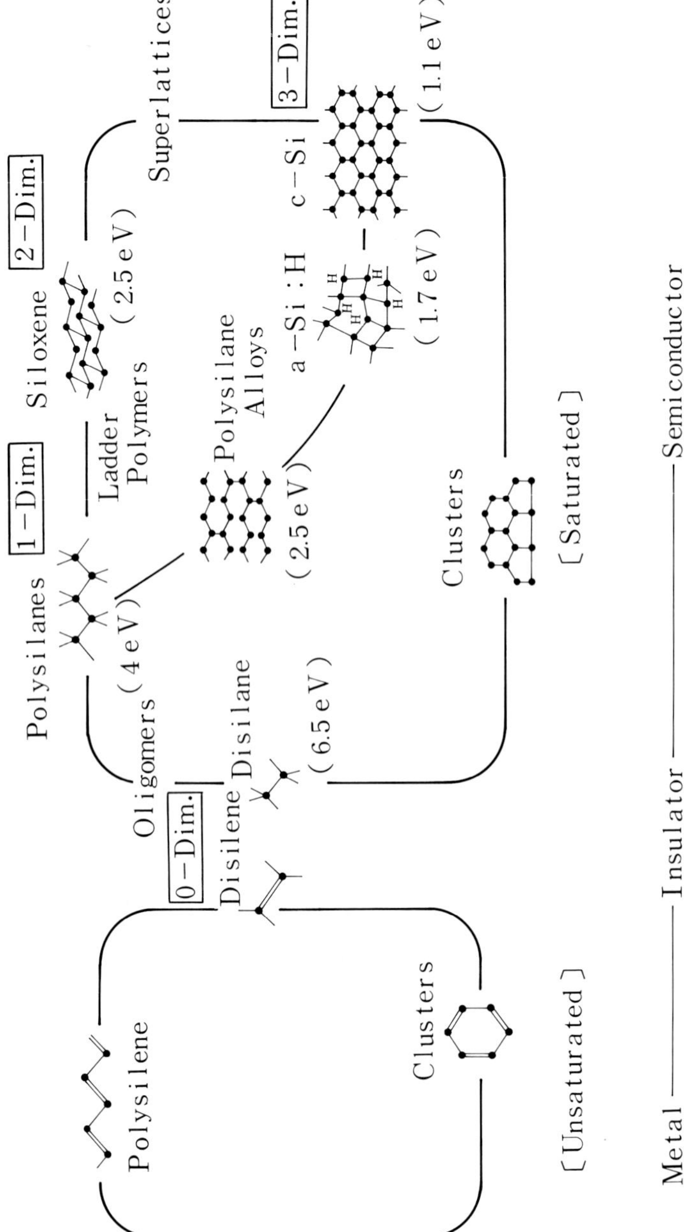

Figure 6. Dimensional hierarchy of Si-based polymers. The abbreviation Dim. denotes dimensional.

important factors must be considered in determining accurate and detailed electronic structures: side-chain and conformation effects. These effects are discussed in this section with respect to polysilanes, which correspond to one-dimensional polymers.

The Band Concept. The eigenfunctions for the delocalization of Si σ electrons along the skeleton are described by Bloch functions (22–23). A good quantum number is not a space coordinate but a wave vector *K*. An example of a band structure is shown in Figure 7. The band gap energy (E_g) is the difference between the edges of the conduction and valence bands.

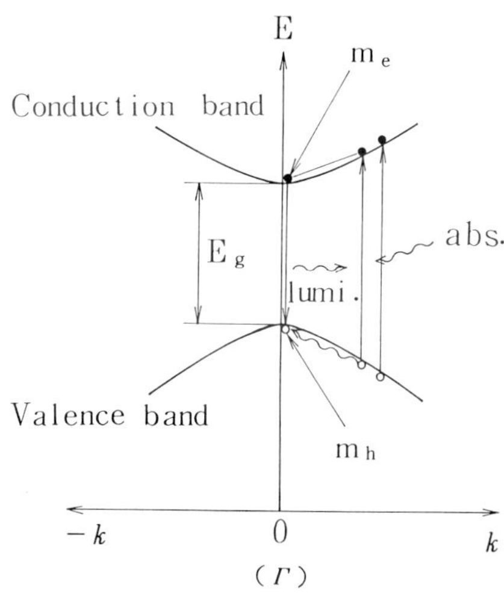

Figure 7. Schematic view of polysilane band structure. The abbreviations and symbols are defined as follows: m_e, *effective mass of electrons;* m_h, *effective mass of holes; lumi., luminescence; abs., absorption;* Γ, k = *0 point; and* k, *wave vector.*

The effective masses of electrons and holes are estimated by parabolic approximation; a large curvature corresponds to a small effective mass and a small curvature corresponds to a large mass. With this band concept, light absorption and luminescence are interpreted as follows: Light is absorbed by the transition from valence band to conduction band. Therefore, the broadening of the absorption spectrum originates basically from the one dimensionality of the joint density of states, which is described by $(E - E_g)^{-1/2}$. Excited electrons and holes relax to the bottom of the bands and then recombine radiatively. Therefore, the photoluminescence of the spectrum is very sharp. The energy difference between two peaks is called the Stokes shift.

Figure 8a shows a parent polysilane structure. The red particles are

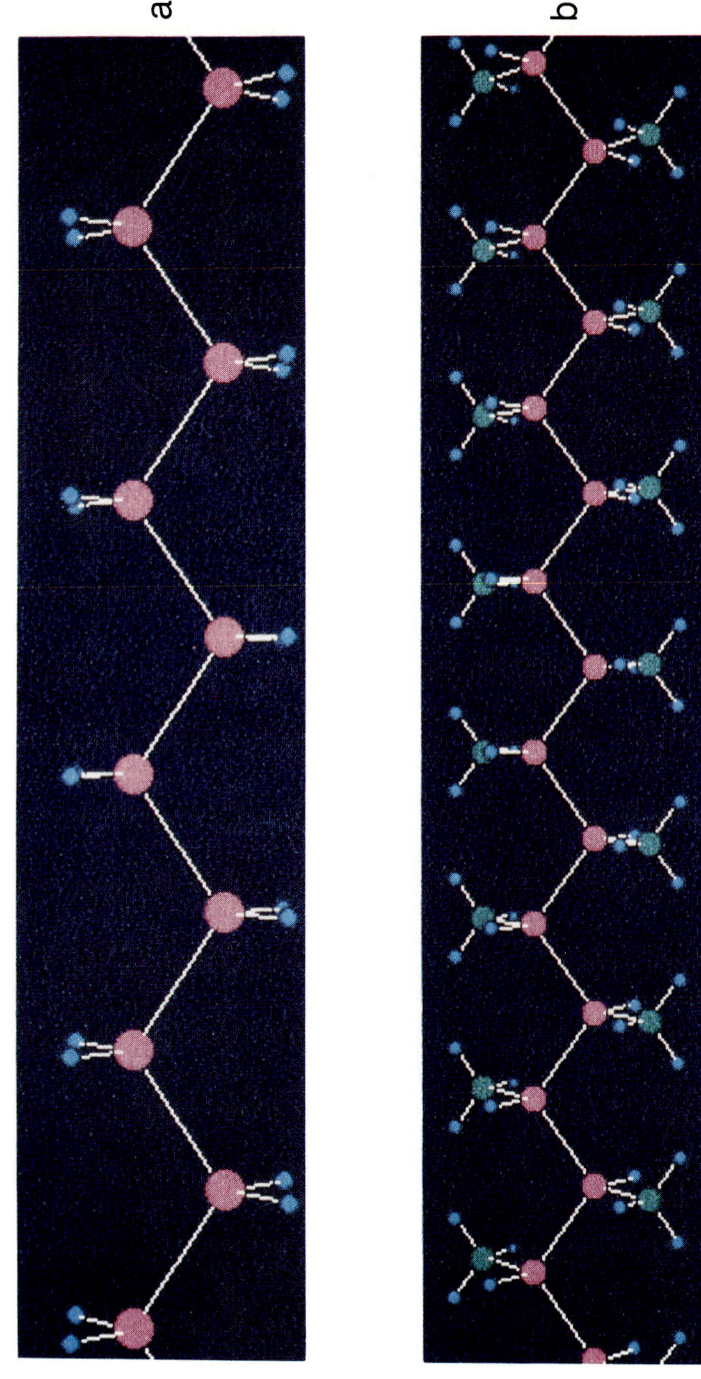
a
b

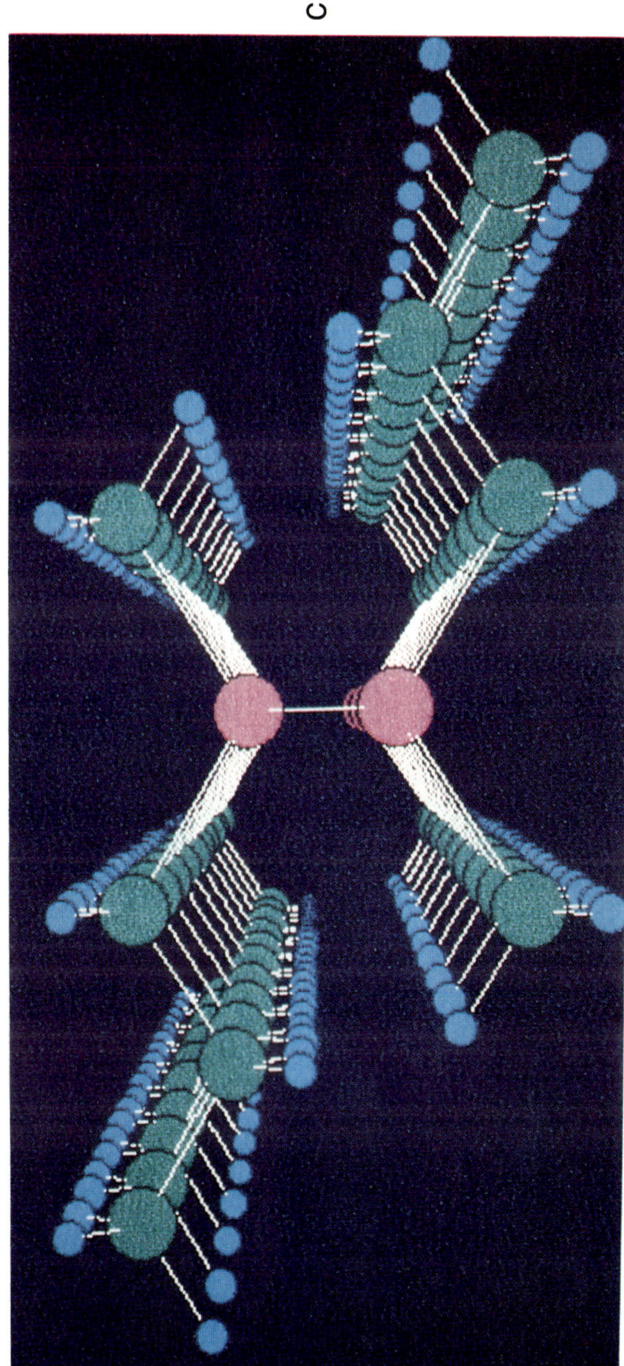

Figure 8. Side views of parent polysilane (a) and poly(monomethylsilane) (b) and a perspective end view of poly(methylpropylsilane) (c).

silicon atoms, and the blue particles are hydrogen atoms. Figure 8b is a schematic view of poly(monomethylsilane), in which methyl side chains substitute for the hydrogen atoms seen in Figure 8a. Figure 8c shows a perspective end view of poly(methylpropylsilane). The corresponding band structures (Figure 9) were calculated by a semiempirical LCAO (linear combination of atomic orbitals) method (7). The ab initio method cannot accurately estimate absolute energy levels of the excited states (i.e., the conduction band). The semiempirical method is more suitable for the estimation of band gap energy. Figure 9a shows the band structure of parent polysilane. Band edge states with different parities exist at the Γ point ($k = 0$). The figure shows that parent polysilane is a direct semiconductor with a 4.5-eV band gap. Both band edge states are composed mainly of skeleton Si atomic orbitals. The valence band edge is the σ bonding state of the skeleton Si $3p_x$ orbital, and the conduction band edge is the σ* antibonding state of the skeleton Si 3*s* atomic orbitals. This band structure shows that polysilane is a σ-conjugated polymer or a one-dimensional semiconductor with delocalized σ bands. Figure 9b shows the band structure of poly(monomethylsilane), and Figure 9c shows that of poly(methylpropylsilane).

Effect of Substitution on Band Gap. The substitution effects of alkyl side chains have two characteristic features: the unchanged state of the direct-type band structure and the systematic change in band gap value and the position of the pseudo π band, which are related to the electron-donating property of alkyl side chains. The effects of alkyl side chains on band gap energy are summarized in Figure 10. The band edge states are formed mainly of silicon atomic orbitals, which produce the skeleton band gap. This band gap tends to be compressed when larger alkyl groups are substituted. The amount of reduction, however, is not significant because of the weak electronic contribution from the side chains.

Substitution of aryl side chains results in a different band structure. A perspective end view of poly(diphenylsilane) is shown in Figure 11a. Electrons conduct along the red region under the influence of a potential barrier of phenyl groups. This electrical analogue of an optical fiber consists of an electrical core and an electrical clad. A perspective end view of poly(methylphenylsilane) is shown in Figure 11b. In these aryl polysilanes, two important points should be considered: the existence of states localized at phenyl side chains and the σ–π interaction between delocalized skeleton σ bands and localized π states.

The energy band structures of poly(methylphenylsilane) are shown in Figure 12. The original skeleton band gap is 3.6 eV, and the polymer also yields a direct-type band structure. The characteristic features of this elec-

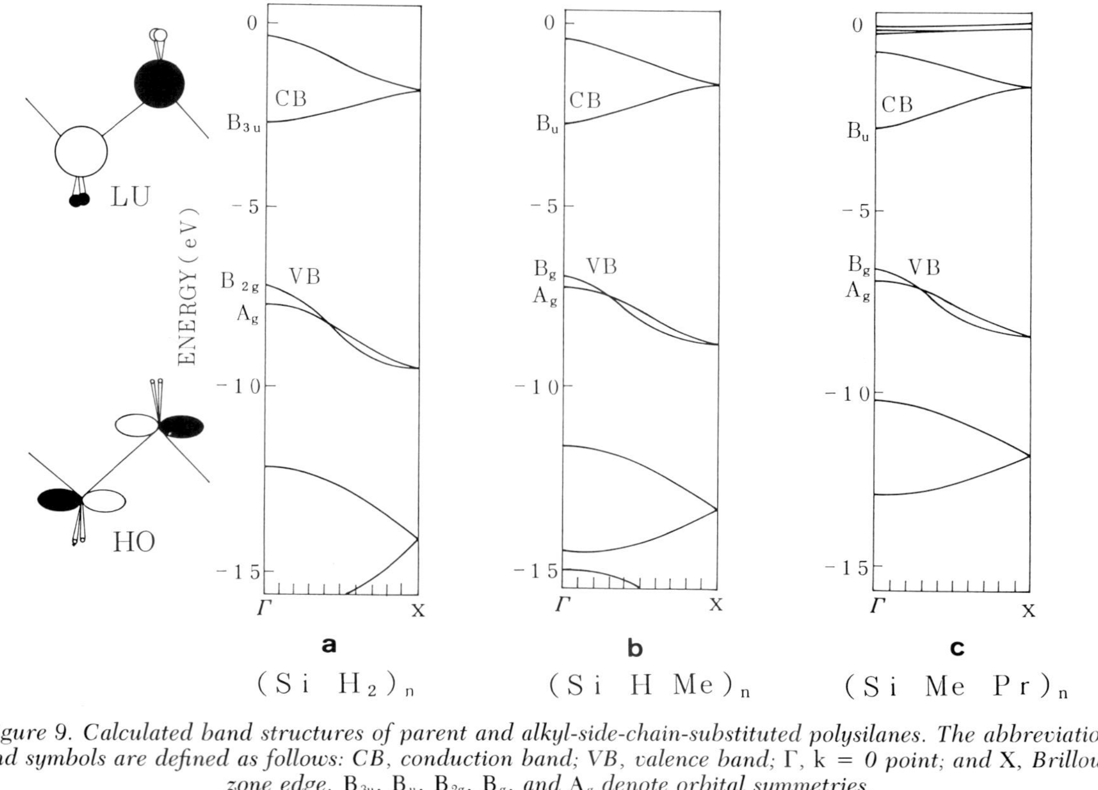

Figure 9. Calculated band structures of parent and alkyl-side-chain-substituted polysilanes. The abbreviations and symbols are defined as follows: CB, conduction band; VB, valence band; Γ, k = 0 *point; and* X, *Brillouin zone edge.* B_{3u}, B_u, B_{2g}, B_g, *and* A_g *denote orbital symmetries.*

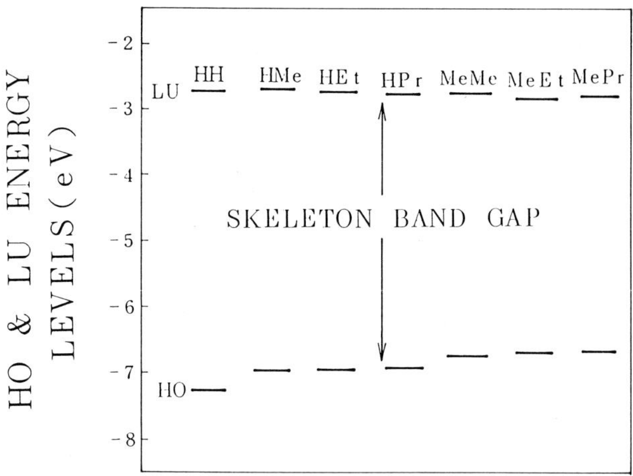

Figure 10. Band gap variation with alkyl-side-chain substitution. Abbreviations are defined as follows: HO, highest occupied; LU, lowest unoccupied; HH, polysilane; HMe, poly(monomethylsilane); HEt, poly(monoethylsilane); MeMe, poly(dimethylsilane); MeEt, poly(methylethylsilane); and MePr, poly(methylpropylsilane).

tronic structure are a σ–π band mixing at the valence band edge and an intrusion of unoccupied localized levels. The polysilane skeleton valence band edge is in the σ state, which is formed by bonding between skeleton Si $3p_x$ atomic orbitals, and it is well delocalized along the skeleton axis. The energy difference between delocalized σ HOMO and localized phenyl π states is much less than the corresponding values between alkanes and benzene.

Skeleton–side-chain interaction due to π-like coupling results in σ–π band mixing between the skeleton Si $3p_x$ and side-chain phenyl π HOMO states. The symmetric π HOMO can mix with the delocalized σ valence band. This mixing results in the formation of two delocalized σ bands. The asymmetry of the other π HOMO states cancels this σ–π mixing effect. The resulting state remains strongly localized in the individual phenyl side chains.

On the other hand, σ*–π* band mixing does not take place because of the orbital symmetry. π* states are localized at phenyl side chains and thus are not dispersed. The σ–π band mixing was confirmed by UPS measurement (Figure 13) (*8*, *9*). Feature A, which corresponds to the top of the valence band of poly(methylphenylsilane), is lifted by about 1 eV compared with the top of the valence band of poly(dimethylsilane) (A′). On the other hand, feature B, which is derived from electrons localized at phenyl sites, remains at the benzene HOMO level B′.

a

b

Figure 11. Perspective end views of poly(diphenylsilane) (a) and poly(methylphenylsilane) (b).

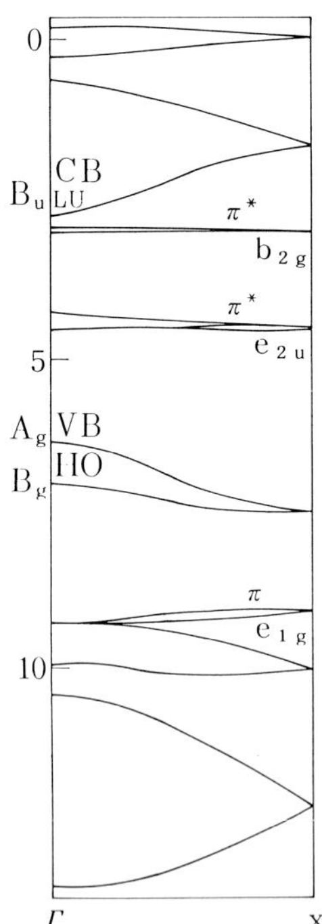

Figure 12. Calculated energy band structure of poly(methylphenylsilane). Abbreviations and symbols are defined as follows: CB, conduction band; LU, lowest unoccupied molecular orbital; VB, valence band; HO, highest occupied molecular orbital; Γ, k = *0 point; and* X, *Brillouin zone edge.* B_u, A_g, B_g, b_{2g}, e_{2u}, *and* e_{1g} *denote orbital symmetries.*

The skeleton–side-chain interaction is reflected in optical properties. The absorption and photoluminescence spectra of poly(methylpropylsilane) and poly(methylphenylsilane) are shown in Figure 14. For poly-(methylpropylsilane), the spectrum profiles are explained by the simple band model just discussed. However, for poly(methylphenylpolysilane), the spectrum profiles are very different. The absorption peak at 3.7 eV corresponds to a σ–σ* transition, and the second peak at 4.5 eV corresponds to a π–π* transition in phenyl side chains. The sharp photoluminescence peak originates from the σ*–σ transition.

The explanation of the origin of the broad visible luminescence, however, has been a very controversial issue (*10, 24*). Two kinds of luminescence should be distinguished: the luminescence measured after photodegradation and the luminescence from nondegraded polymers at very low temperatures (4–70 K). In nondegraded samples, the intensity of this broad luminescence decreases very rapidly with an increase in temperature from 4 to 70 K, and it is reversible in the heat cycle (*11*). On the other hand, the sharp peak

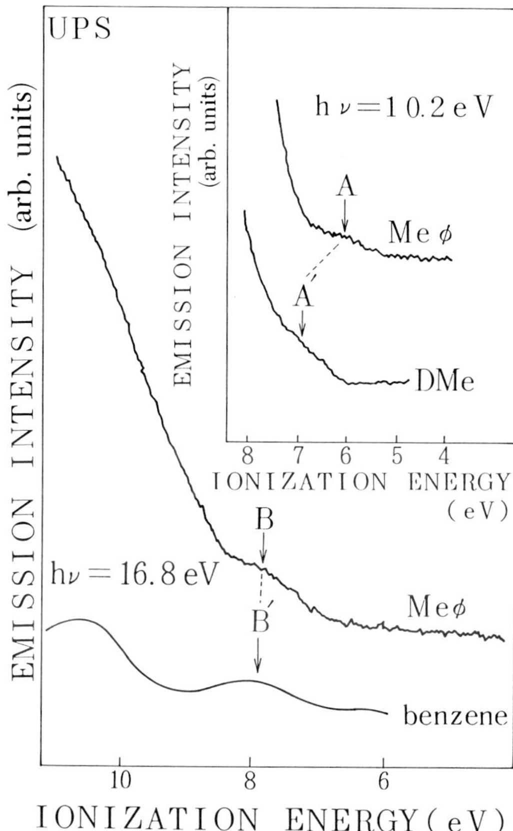

Figure 13. Confirmation of skeleton–pendant group interaction by UPS measurements. Abbreviations are defined as follows: Meϕ, poly(methylphenylsilane); and DMe, poly(dimethylsilane). (Reproduced with permission from reference 8. Copyright 1987 American Physical Society.)

does not change at all with variations in temperature. At temperatures above 70 K, degradation occurs; the sharp peak decreases rapidly, and another shoulder appears. The luminescence from degraded samples is irreversible and not reproducible and may result from impurities generated by photolysis of the polymer chains. However, the luminescence from nondegraded samples at low temperature is reproducible, a fact indicating that it originates from the intrinsic electronic structures of poly(methylphenylsilane).

The details of the mechanism have not been clarified sufficiently. We propose the following mechanism as one of the most plausible: The irradiated photons are absorbed by the π–π^* transition at phenyl side chains, the excited states (π, π^*) relax to charge-transfer states (σ, π^*), and then they relax radiatively to the ground state. The σ–π mixing between skeleton and pendant groups plays an essential role.

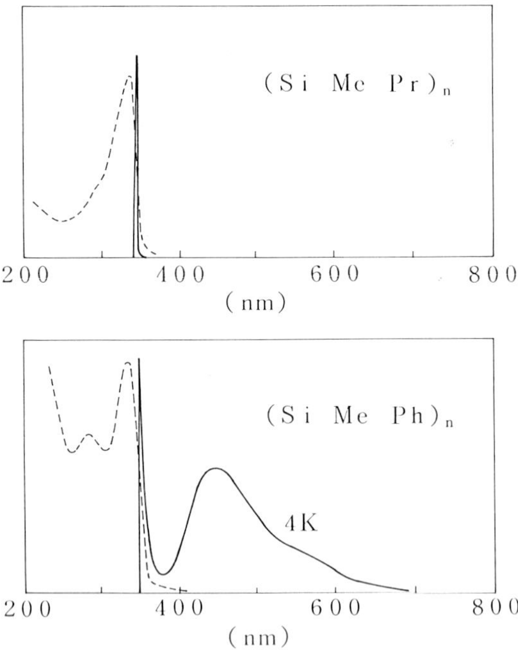

Figure 14. Absorption and luminescence spectra of poly(methylpropylsilane) (top) and poly(methylphenylsilane) (bottom).

Effect of Conformation on Band Structure. The *trans* and *gauche* parent polysilanes are shown in Figure 15. Thermochromism is one of the most controversial subjects in polysilanes. However, the following basic questions have not been answered yet: Which conformation is more stable, *trans* or *gauche*? What is the essential difference in their band structures?

An ab initio study of parent polysilanes was conducted recently by our group. Perspective end views of *trans* and *gauche* polysilanes are shown in Figure 16. Every bonding angle between Si atoms is tetrahedral in both types of polysilanes. In *gauche* polysilane, Si atoms projected on the plane are rotated 90° to the longitudinal axis. This rotation angle is called a helical angle and can be changed from 0 to 180°. The helical-angle dependence on total energy, calculated by using the minimal basis set (STO-3G) (Figure 17), shows that the *trans* type is more stable than the *gauche* type by 0.4 kcal/mol per unit cell. Calculated band structures for *trans* and *gauche* polysilanes show two characteristic features for band gap energy and effective mass. *gauche*-Type polysilanes have wider gap energies and flatter parabolic bands, which correspond to larger effective masses. This finding is interpreted physically as follows: HOMO levels at the top of the valence band are formed by Si 3*p* wave functions on the plane that includes the two nearest silicon atoms and itself. The 3*p* orbital lobe located at the nearest

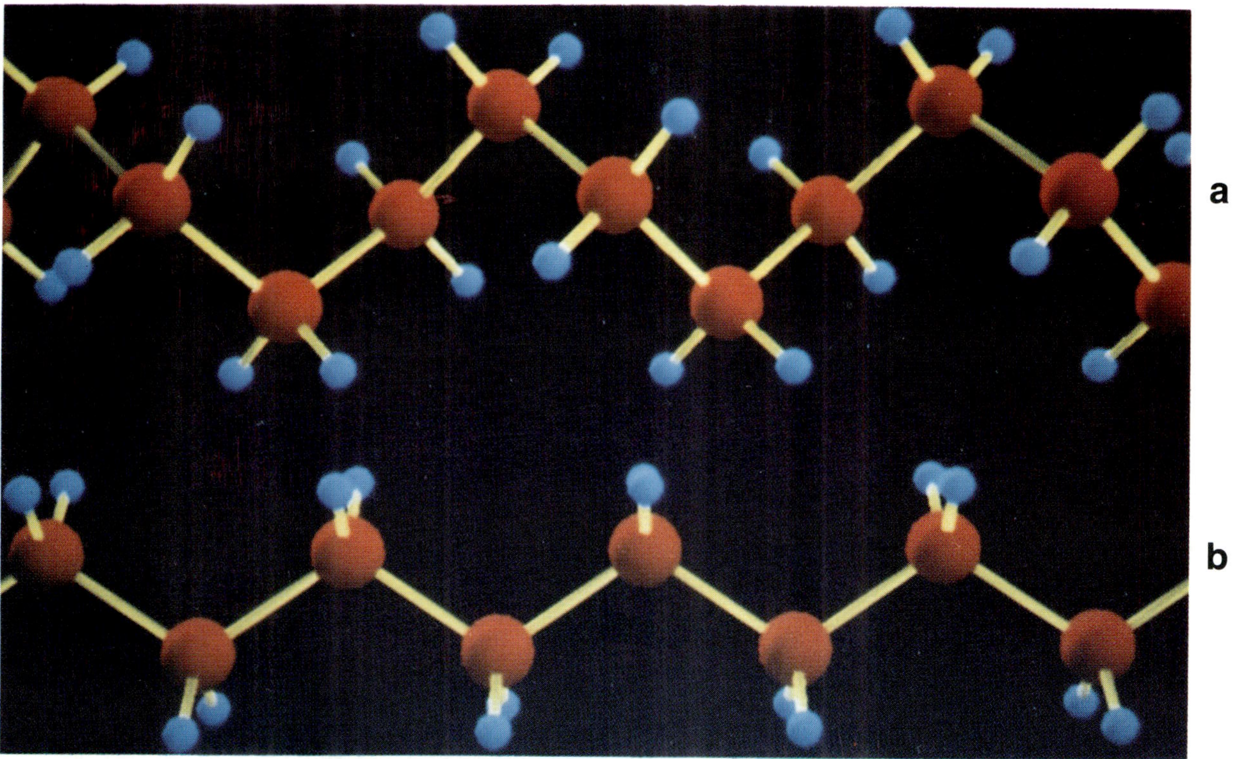

Figure 15. Side views of trans *(a) and* gauche *(b) polysilane.*

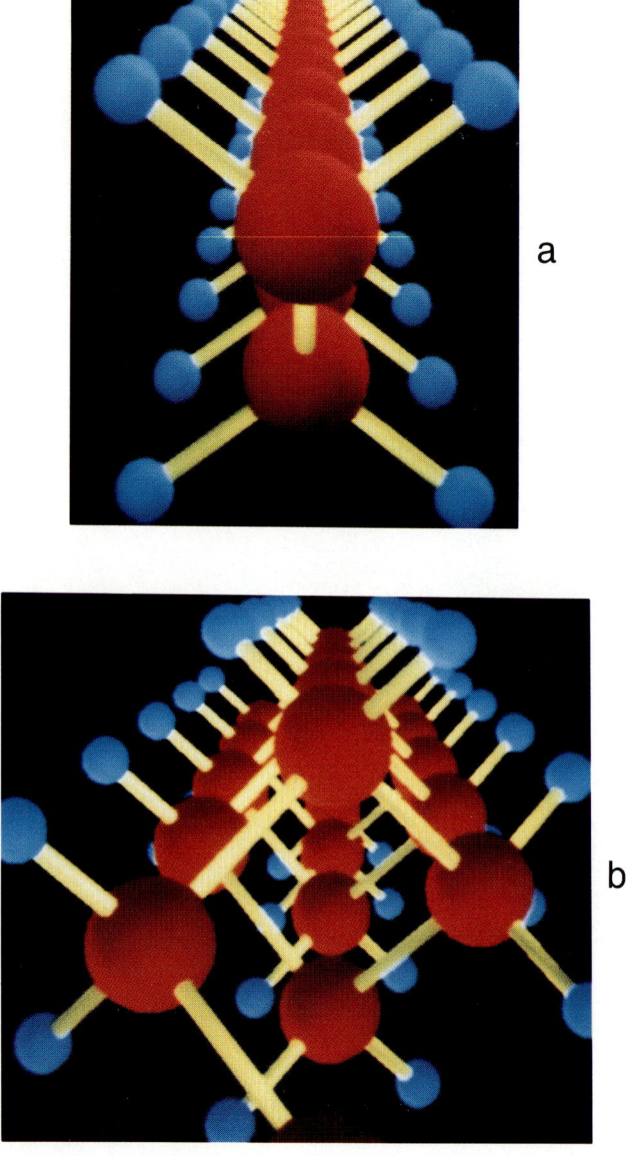

Figure 16. Perspective end views of trans *(a) and* gauche *(b) polysilane.*

silicon atom is also on the plane formed by the similar three silicon atoms. The two orbital lobes are not arranged in a straight line. The interaction of these orbitals, therefore, is less than that for the *trans*-type chain. Reduction in interaction between skeleton silicon atoms is accompanied by an increase in band gap energy and effective mass.

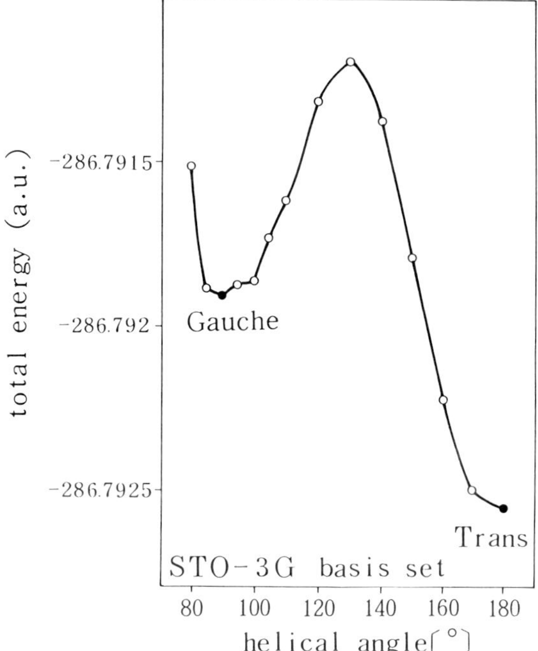

Figure 17. Calculated helical-angle dependence of parent polysilane on total energy. The abbreviation a.u. denotes arbitrary unit.

Future Directions for Silicon-Based Polymers

Polysilanes have two serious problems with respect to their use in semiconductor films: They are difficult to dope, and they are easily photodegraded. High-dimensional polymers such as ladder polymers may overcome these problems. Doping difficulty is caused by the wide band gap. Band gap energy decreases rapidly with an increase in lattice dimensions by forming ladder structures. Photosensitivity may be reduced by forming high-dimensional networks. The limiting case is the three-dimensional crystal silicon lattice, which is very stable.

Experimental approaches have been reported by two groups. The group at Gunma University (*25*) synthesized a prototype of ladder polymers, poly(bicyclosilane)s. Polymers with three, four, and five ladder steps were synthesized, and the bathochromic shifts were measured. Ikehata et al. (*26*) attempted substitutional doping by thermal neutron irradiation of $(SiH)_n$ ladder polymers synthesized through the reduction of trichlorosilane with lithium in THF (tetrahydrofuran). Thermal neutron capture by ^{30}Si will yield ^{31}P in this process for total flux amounts of up to $10^{18}/cm^2$. The concentration of phosphorus impurity is 0.1 ppm. No serious deterioration was caused by strong neutron irradiation. However, in this study, the detailed properties were not measured, and the structure of ladder polymers was not confirmed. Further experiments are currently being carried out.

Superlattice and low-dimensional physics are some of the most interesting subjects in solid-state physics. A challenging problem in this field is the formation of quantum wire and quantum box structures by using ultrahigh technology such as MBE, MOCVD (metallorganic chemical vapor deposition), and related frontier microprocessing. However, this problem has not yet been solved. Polysilane is probably a perfect quantum wire in itself. The absorption spectrum of polysilane clearly shows the characteristics of a one-dimensional quantum wire. Even a quantum box or a one-dimensional superlattice can be formed by chemical polymerization, which may be the simplest way.

An example of a one-dimensional superlattice structure is structure **1**, which is an ordered copolymer. The skeleton is formed by silicon and germanium atoms. A unit cell is three times larger than that of a homopolymer. The band structure of this ordered copolymer changes to the zone-folded profile, which may result in a characteristic absorption spectrum.

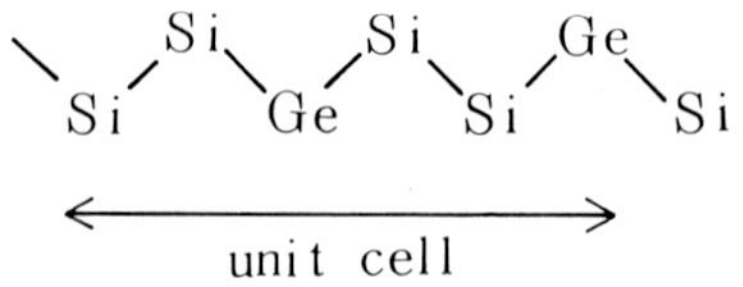

Si Si Ge
Si Ge Si Si

unit cell

1

One-dimensional superlattice structure

Recently, the alternating copolymer **2** was synthesized as a prototype (27). Copolymer **2** has an alternating structure of tetramethyldisilylene and dibutylsilylene units. This polymer was synthesized in three steps. First, dibutylhexamethyltrisilane was synthesized by coupling dibutyldichlorosilane and excess trimethychlorosilane in the presence of a sodium–potassium alloy. Second, dichlorodibutyltetramethyltrisilane was obtained by the chlorination of dibutylhexamethyltrisilane with anhydrous aluminum chloride as a catalyst. Third, the trisilane was polymerized by dechlorination using sodium metal. A trimodal molecular weight distribution was observed with peaks at about 3,000, 20,000 and 2,000,000. When the silicon atoms are substituted with germanium, **1** may be synthesized through a similar process.

Polysilene with alternating double bonds (**3**) has not been synthesized, and the synthesis may be very difficult. The calculated band structure of polysilene (28) is shown in Figure 18. The most striking feature is the overlapping of two bands, which is accompanied by double crossing at the Fermi

$$-\!\left[Si(CH_3)_2 - Si(CH_3)_2 - Si(C_4H_9)_2 \right]_n\!-$$

2

Alternating copolymer with ordered side chains (Me, CH_3; Bu, C_2H_5)

$$=Si-Si=Si-Si=Si-Si=$$

3

Schematic structure of polysilene with alternating double bonds

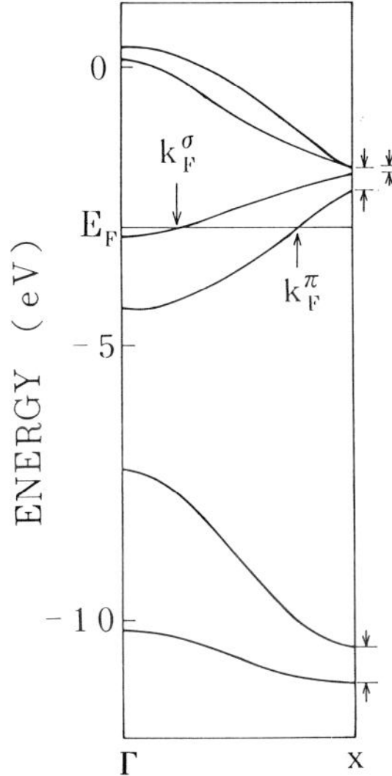

Figure 18. Calculated band structure of polysilene. Abbreviations and symbols are defined as follows: E_F, *Fermi energy;* k_F^σ, k *value at Fermi surface in* σ *band;* k_F^π, k *value at Fermi surface in* π *band;* Γ, k = *0 point; and* X, *Brillouin zone edge. (Reproduced with permission from reference 28. Copyright 1988 American Physical Society.)*

level. This feature indicates that polysilene may have metallic properties. On the other hand, in polyacetylene, the π band is in the region between the σ and σ^* bands, the gap of which is two times larger than that for polysilanes.

Of course, theoretical calculations often give wrong band structures. For example, one of the most reliable ab initio methods gives a very wide band gap, 10–15 eV, for polysilanes. Therefore, further studies are required to verify the metallic properties of polysilene. The calculation just mentioned was done by using the same semiempirical method for calculating polysilane band structures. Although parameters such as bond lengths and bond angles vary widely, overlapping-band feature does not. These results suggest strongly that polysilene has metallic properties.

The dimensional hierarchy of silicon-based polymers is summarized in Figure 6. The right circle corresponds to the saturated systems already discussed. The left circle corresponds to the unsaturated systems. The electronic properties of silicon-based polymers vary from conducting (metallic) and semiconducting to insulating. This figure shows that silicon atoms can form many kinds of materials with various properties. However, the study of silicon-based materials has been concentrated in a very small area of this figure.

Finally, the future development of silicon-based polymers can be based on the considerations just mentioned. Figure 19 identifies four types of silicon-based polymers with promising applications. Two problems exist for semiconducting polymers: long lifetime and narrow band gap. These problems should be solved by increasing the lattice dimension, for example by using ladder polymers. Functional polymers (polymers capable of various functions) could be developed by exploiting the effects of conformation variations and skeleton–pendant group interactions. Metallic polymers may be realized by synthesizing polysilenes, and ideal, low-dimensional materials will be useful in pure physics. If these kinds of polymers can be combined with silicon technology in the LSI (large-scale integrated circuit) field, new types of molecular devices can be developed.

Summary

Silicon-based polymers form a dimensional hierarchy from disilanes, to crystal silicon, and through polysilanes, ladder polymers, siloxenes, polysilane alloys, clusters, and amorphous silicons and include unsaturated systems, such as polysilenes, hexasilabenzenes, and so on. Their properties depend basically on the network dimensions and can vary from conducting (metallic) and semiconducting to insulating.

Side-chain substitutions and conformation variations perturb the detailed electronic structures of polysilanes. Substitution by larger alkyl side chains decreases the band gap slightly because of the electron-donating

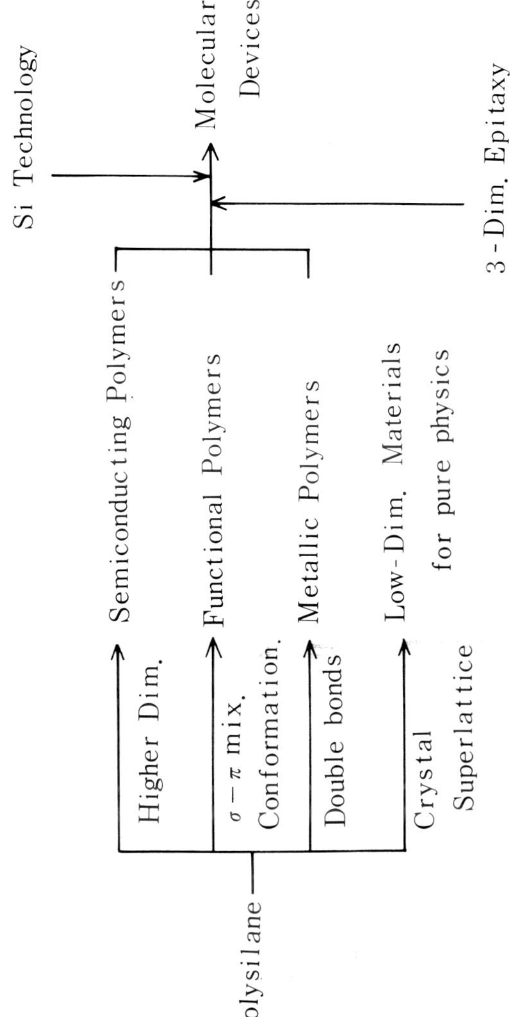

Figure 19. Future directions for silicon-based polymers.

effect. With aryl-side-chain substitution, skeleton–pendant group interaction by σ–π mixing results in a decrease in ionization potential and the appearance of highly efficient visible luminescence. The total energy difference between *trans* and *gauche* polysilanes is very small compared with that in the corresponding carbon system. *trans*-Type polysilanes are more stable and have narrower band gap energies and smaller effective masses compared with *gauche*-type polysilanes.

Future studies should concentrate on four types of silicon-based polymers: semiconducting polymers, functional polymers, metallic polymers, and ideal low-dimensional materials for pure physics. The range of Si-based polymers should be expanded rapidly to include high-dimensional polymers, one-dimensional superlattices, and unsaturated polymers.

References

1. Pankove, J. I. *Semiconductors and Semimetals;* Academic: New York, 1984; Vol. 21.
2. West, R. *J. Organomet. Chem.* **1986,** *300,* 327.
3. Yajima, S.; Hayashi, J.; Omori, M. *Chem. Lett.* **1975,** 1931.
4. Hofer, D. C.; Miller, R. D.; Willson, C. G. *Advances in Resist Technology, Proc. Soc. Photo-Opt. Instrum. Eng. (SPIECJ)* **1983,** *469,* 16.
5. Wolford, D. J.; Reimer, J. A.; Scott, B. A. *Appl. Phys. Lett.* **1983,** *42,* 369.
6. Matsumoto, N.; Furukawa, S.; Takeda, K. *Solid State Commun.* **1985,** *55,* 881.
7. Takeda, K.; Teramae, H.; Matsumoto, N. *J. Am. Chem. Soc.* **1986,** *108,* 8186.
8. Takeda, K.; Fujino, M.; Seki, K.; Inokiuchi, H. *Phys. Rev. B,* **1987,** *36,* 8129.
9. Loubriel, G.; Zeigler, J. M. *Phys. Rev. B* **1986,** *33,* 4203.
10. Kagawa, T.; Fujino, M.; Takeda, K.; Matsumoto, N. *Solid State Commun.* **1986,** *57,* 655.
11. Ito, O.; Terajima, M.; Matsumoto, N.; Takeda, K.; Fujino, M.; Azumi, T. to be published in *Macromolecules.*
12. Harrah, L. A.; Zeigler, J. M. *J. Polym. Sci., Polym. Lett. Ed.* **1985,** *23,* 209.
13. Miller R. D.; Hofer, D.; Rabolt, J. F.; Fickes, G. N. *J. Am. Chem. Soc.* **1985,** *107,* 2172.
14. Kuzmany, H.; Rabolt, J. F.; Farmar, B. L.; Miller, R. D. *J. Chem. Phys.* **1986,** *85,* 7413.
15. Bock, H.; Ensslin, W.; Feher, F.; Freund, R. *J. Am. Chem. Soc.* **1976,** *98,* 668.
16. Klingensmith, K. A.; Dowing, J. W.; Miller, R. D.; Michl, J. *J. Am. Chem. Soc.* **1986,** *108,* 7438.
17. Teramae, H.; Takeda. K. to be published in *J. Am. Chem. Soc.*
18. Ubara, H.; Imura, T.; Hiraki, A.; Hirabayashi, I.; Morigaki, K. *J. Non-Cryst. Solids* **1983,** *59 & 60,* 641.
19. Hirabayashi, I.; Morigaki, K. *J. Non-Cryst. Solids* **1983,** *59 & 60,* 645.
20. Drenth, W.; Noltes, J. G.; Bulten, E. J.; Creemers, H. M. J. C. *J. Organomet. Chem.* **1969,** *17,* 173.
21. Abeles, B.; Tiedje, T. *Phys. Rev. Lett.* **1983,** *51,* 2003.
22. Takeda, K.; Matsumoto, N.; Fukuchi, N. *Phys. Rev. B* **1984,** *30,* 5871.
23. Dewar, M. J. S. *J. Am. Chem. Soc.* **1984,** *106,* 669.
24. Harrah, L. A.; Zeigler, J. M. *J. Polym. Sci., Polym. Lett. Ed.* **1987,** *25,* 205.

25. Matsumoto, H.; Miyamoto, H.; Kojima, N.; Nagai, Y.; Goto, M. 8th International Symposium on Organosilicon Chemistry, June 1987, St. Louis.
26. Ikehata, S.; Iwata, T. *Synth. Met.* **1987,** *18,* 597.
27. Fujino, M.; Ban, H.; Sukegawa, K.; Matsumoto, N. *J. Polym. Sci.: Part C: Polym. Lett. Ed.* *26,* 109.
28. Takeda, K.; Kagoshima, S. *Phys. Rev. B* **1988,** *37,* 6406.

RECEIVED for review May 27, 1988. ACCEPTED revised manuscript March 20, 1989.

29

Band Structure and Optical Absorption Properties of Polysilane Chains

J. W. Mintmire[1] and J. V. Ortiz[2]

[1]Chemistry Division, Naval Research Laboratory, Washington, DC 20375
[2]Department of Chemistry, University of New Mexico, Albuquerque, NM 87131

*An LCAO (linear combination of atomic orbitals) local-density functional approach was used to calculate the band structures of a series of polymer chain conformations: unsubstituted polysilane in the all-*trans *conformation and in a 4/1 helical conformation, and all-*trans *poly(dimethylsilane). Calculated absorption spectra predict a highly anisotropic absorption for the all-*trans *conformation of polysilane, with the threshold absorption peak arising strictly from polarizations parallel to the chain axis. The absorption spectrum for the helical conformation is much more isotropic. Results for the dimethyl-substituted polysilane chain suggest that the states immediately surrounding the Fermi level retain their silicon-backbone σ character upon alkyl-group substitution, although the band gap decreases by ~1 eV because of contributions from alkyl substituent states both below the valence band and above the conduction band to the frontier states.*

EXPERIMENTAL AND THEORETICAL interest in organosilane polymer systems (*1*) has increased in recent years largely because of the potential technological applications of these materials as silicon carbide precursors, as photoresists in photolithography, as photoinitiators in radical-assisted polymerization, and as photoconductors in photocopying processes. Many of the technological applications for the organosilane polymers depend intimately on the electronic structure and resulting optical properties of these materials. The absorption threshold in the ultraviolet (UV) of a range of substituted polysilanes decreases in energy with both increasing molecular

0065–2393/90/0224–0543$06.00/0

weight and increasing substituent size (*2, 3*). The absorption and emission spectra suggest that photoexcitation may produce a localized triplet-defect state that leads to silicon–silicon bond scission and silylene production (*3, 4*).

Several dialkyl-substituted polysilanes, in particular poly(di-*n*-hexylsilane), exhibit a red shift of the absorption peak of up to 40 nm (corresponding to an energy shift of 0.4–0.5 eV) and a concomitant narrowing of the absorption line shape with decreasing temperature (*5–8*). These bathochromic shifts have been attributed to an order–disorder transition of the silicon backbone from a highly ordered system with nearly all-*trans* (or planar zigzag) conformations of the backbone to a disordered system with an entropic mix of *trans* and *gauche* conformers along the silicon backbone.

A variety of theoretical methods (*9–17*) have been used to investigate the electronic structure of both polymeric chain systems (*9–12*) and model molecular species (*13–17*). Theoretical studies of the polymeric chains have consisted of semiempirical calculations of the band structure and of the effects of substituents and conformation on the band structure (*10–12*) and ab initio calculations of the electronic structure and minimum energy conformations of the unsubstituted polysilane chain (*9*). As part of a broader theoretical study of the electronic structure and properties of organopolysilanes, we have begun a study of these systems by using our previously developed linear-combination-of-atomic-orbitals local-density functional (LCAO-LDF) method for chain polymers (*18, 19*). This method calculates both the band structure and total energy of a polymer chain possessing translational periodicity in one dimension; an equilibrium geometry can thus be obtained by a search for the minimum energy conformation. Absorption spectra can be estimated from interband transition cross-sections evaluated by using the one-electron wave functions and energy levels.

Experimental Procedures

In this study, we investigated a set of model polysilane chain systems that illustrate the basic physics and chemistry of some optical properties of these materials. In particular, we looked at the band structure for unsubstituted polysilane in an all-*trans* conformation, as well as in a 4/1 helical conformation with four silicon atoms contained in one translational repeat unit. In addition, we compared results for the dimethyl-substituted polysilane in an all-*trans* conformation with the results for the unsubstituted polysilane.

The bond lengths and band angles for these calculations were chosen from ab initio Hartree–Fock geometric optimizations for tetrasilane and other model molecular systems (*17*). The bond lengths used were 2.35 Å for the Si–Si bonds, 1.48 Å for the Si–H bonds, 1.89 Å for the Si–C bonds, and 1.08 Å for the C–H bonds. The Si–Si–Si bond angle was chosen to be 111.6°, the same as that obtained in our earlier geometric optimization of tetrasilane. For consistency with the tetrasilane results, the H–Si–H bond angle was chosen to be 108°, and the C–Si–C bond angle was also set to this value. All bond angles centered on the carbon nuclei were set to tetrahedral bond angles of 109.47°, with the hydrogens staggered relative to the bonds on the nearest silicon. The version of the LCAO-LDF method we used required translational symmetry; therefore the Si–Si–Si–Si dihedral angle for the helical conformation was set to 62.56° to generate a translational unit cell containing four silicons while being close to an all-*gauche* backbone conformation. These calculations were performed with a moderate-size orbital basis set: 10*s* 6*p* 1*d* for silicon, 7*s* 3*p* for carbon, and 3*s* for hydrogen.

Results and Discussion

Figures 1 and 2 depict our calculated band structures for the all-*trans* conformations of unsubstituted polysilane and poly(dimethylsilane). Because the reflection plane containing the silicon nuclei in the all-*trans* conformations commutes with the operations of the one-dimensional translation group, all one-electron wave functions will be either symmetric (σ-like) or antisymmetric (π-like) with respect to this reflection operation. Thus the bands in Figures 1 and 2 are labeled with solid lines or dashed lines, which indicate that the corresponding one-electron wave functions are σ-like or π-like, respectively.

As noted by other workers (*11*), the highest occupied band (valence

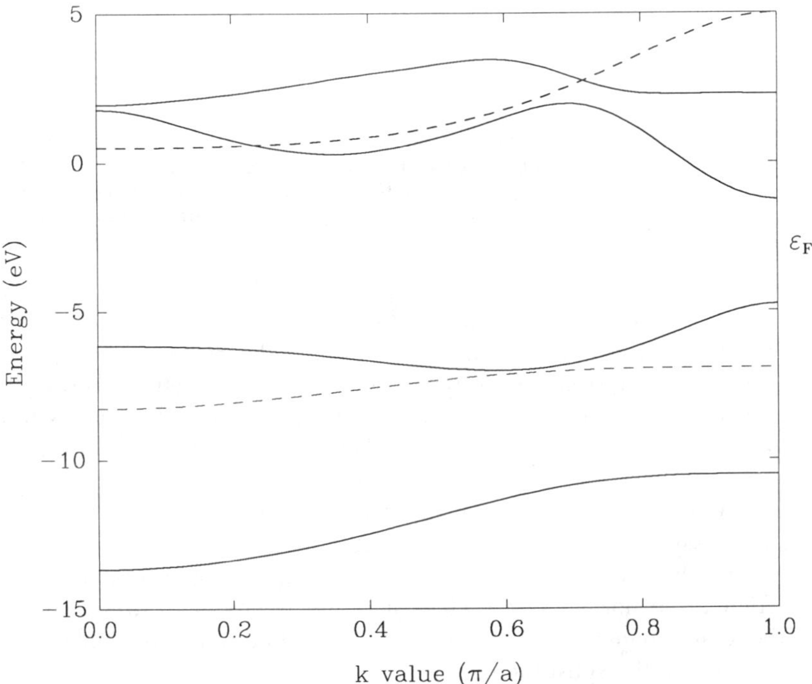

*Figure 1. Band structure for all-*trans *conformation of unsubstituted polysilane calculated by using the LCAO-LDF method. σ-like bands are denoted with solid lines. π-like bands are denoted with dashed lines. The Fermi level is denoted as* ϵ_F. k *represents the wave vector, and* a *is the length of the unit cell.*

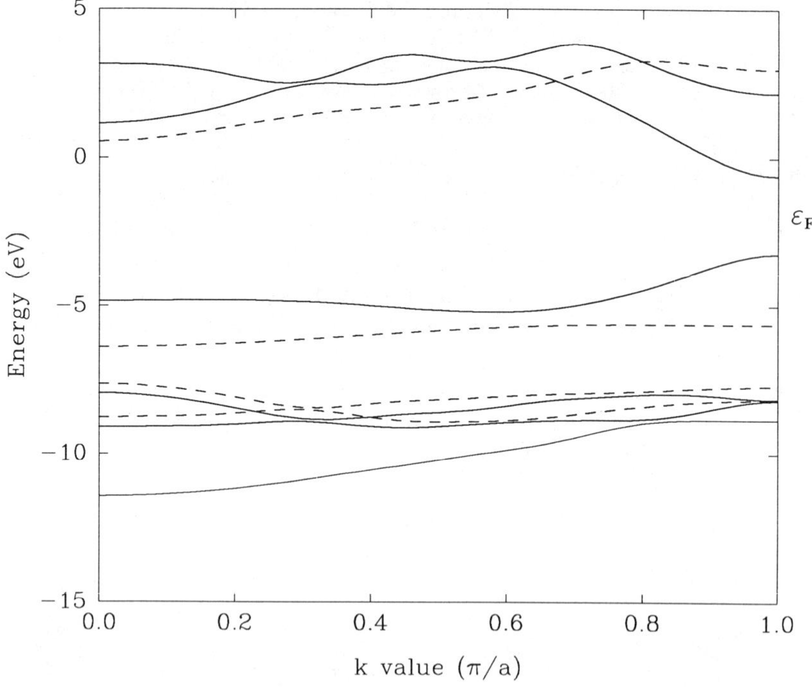

*Figure 2. Band structure for all-*trans *conformation of poly(dimethylsilane) calculated by using the LCAO-LDF method. σ-like bands are denoted with solid lines. π-like bands are denoted with dashed lines. The Fermi level is denoted as* ϵ_F. k *represents the wave vector, and* a *is the length of the unit cell.*

band) and the lowest unoccupied band (conduction band) are predominantly silicon-backbone bonding and antibonding states, respectively. Indeed, the highest state in the valence band at the edge of the Brillouin zone is a σ-like state that is antisymmetric with respect to the reflection planes perpendicular to the chain axis containing the silicon nuclei. This state is thus not only predominantly composed of a bonding combination of silicon p_z states (where the chain axis defines the z axis) but also has a small contribution from symmetric combinations of the carbon p_z orbitals in the dimethyl-substituted system. The lowest state in the conduction band, also at the edge of the Brillouin zone, is a σ-like state symmetric with respect to the reflection plane normal to the chain axis and is predominantly a silicon-backbone antibonding state that is slightly bonding to almost nonbonding with respect to the substituents.

The band gap decreases on going from the unsubstituted polysilane to the dimethyl-substituted polysilane from 3.52 to 2.64 eV, in line with previous semiempirical studies (*11, 12*). Local-density functional methods typically underestimate the band gap in insulators and semiconductors by 30–50% (*20*). For poly(dimethylsilane), with an experimentally indicated gap of 4.22 eV (*21*), that underestimation is 37%. Most of the decrease in gap on methyl substitution is attributable to an upward shift of the valence band edge, which is largely caused by mixing with lower lying bands, C–H bonding bands introduced from the methyl substituent. These states can be seen lying in the vicinity of −10 eV in Figure 2. Other than this shift in gap, no major changes occur in states immediately about the Fermi level in the valence and conduction bands of the polysilanes; the frontier states remain largely silicon-backbone bonding states immediately below the Fermi level and largely silicon-backbone antibonding states immediately above the Fermi level. Our calculations using a smaller basis set for the dimethyl-, diethyl-, and dipropyl-substituted polysilanes also indicate the same behavior.

This similarity of states immediately around the Fermi level manifests itself in the similarity of predicted extinction coefficients for polysilane and dimethyl-substituted polysilane (Figure 3). The optical conductivity $\sigma_i(\omega)$ is directly calculated for the three polarization directions by using oscillator strengths calculated for direct interband transitions and by neglecting local-field effects (*18*). This approach yields excellent results for π-conjugated polymers such as polyacetylene. The three curves in each diagram for the all-*trans* conformations correspond to the polarization parallel to the chain axis (solid lines), the polarization perpendicular to the plane of the silicon-backbone nuclei (dash-dotted lines), and the polarization in the plane of the silicon backbone (dashed lines); for the helical conformation, of course, the results for the two perpendicular polarizations are equivalent to within the numerical accuracy of our evaluation procedure. The extinction coefficient used in Figure 3 is thus a measure of the optical conductivity and is actually $\sigma_i\Omega/c$, in which σ_i is the calculated optical conductivity, Ω is the volume per silicon nucleus, and c is the speed of light in vacuum. Although the absolute magnitudes of the predicted extinction coefficients are not in particularly good agreement with experimental results, this approach allows us to analyze the expected line shape and to compare differences between similar polymeric systems.

The dominant low-energy feature of the absorption spectra for the all-*trans* conformations is the strong absorption peak for polarizations parallel to the chain (Figure 3). This peak arises from the excitation between the top of the valence band and the bottom of the conduction band. Because these two states differ in parity with respect to the reflection planes perpendicular to the chain axis and containing the silicon nuclei, large on-site matrix elements occur for the interband transition oscillator strengths for polarizations parallel to the chain axis and lead to a strong absorption peak. This strong threshold absorption for polarizations parallel to the chain axis is similar to that observed in π-conjugated polymers, but the large on-site matrix elements should lead to extinction coefficients larger than that observed in the π-conjugated systems, which have no contribution from on-site terms in the threshold absorption.

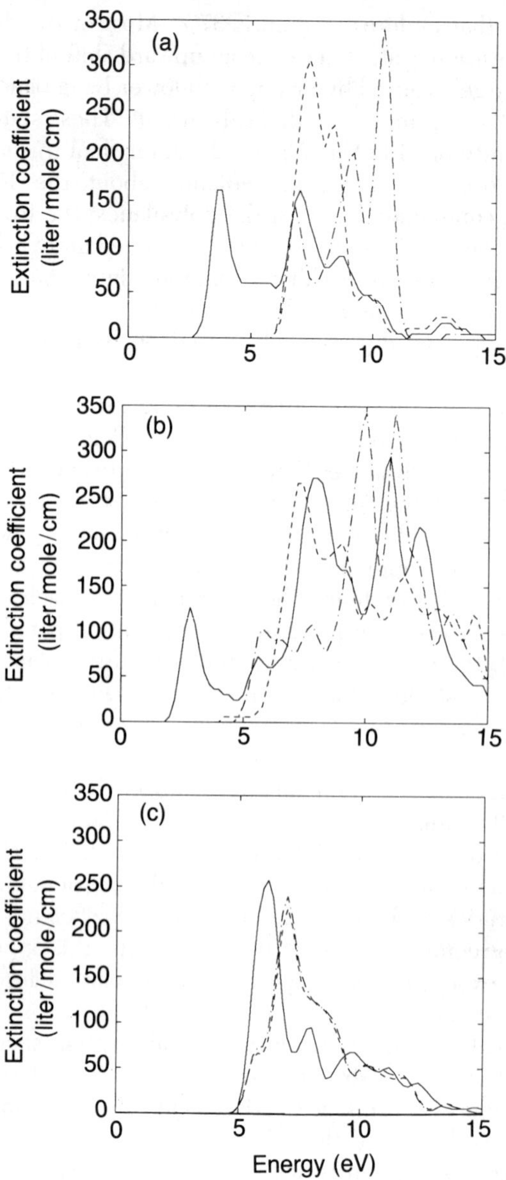

*Figure 3. Calculated absorption spectra for all-*trans *unsubstituted polysilane (a), all-*trans *poly(dimethylsilane) (b), and helical unsubstituted polysilane (c). Solid lines denote absorption for polarizations parallel to chain axis, dashed-and-dotted lines denote absorption for polarizations perpendicular to the plane of the silicon backbone (for the all-*trans *conformations), and dashed lines denote the remaining polarization. All curves have been broadened with a 0.7-eV full width at half-maximum Gaussian.*

This dominant feature is essentially the same for both the unsubstituted and dimethyl-substituted all-*trans* polysilane chains, and an equivalent feature is found when a smaller basis set is used for the dimethyl-, diethyl-, and dipropyl-substituted polysilanes. For the helical conformation, however, along with the larger band gap in this conformation (Figure 3c), a pronounced shift of the direct-gap absorption peak to higher energy is observed, with a trend toward a less anisotropic absorption.

These results support current interpretations of the bathochromic shifts observed in dialkyl-substituted polysilane. Experimental results for poly(di-*n*-hexylsilane) indicate that as the temperature is cooled below a transition temperature of roughly –35 °C, the major absorption peak shifts from a broad peak at about 310–320 nm (3.9–4.0 eV) to a narrower peak at about 350–370 nm (3.3–3.5 eV), with the red shift being attributed to a transition from a disordered system with a large population of *gauche* bond twists in the silicon backbone and in the alkyl substituent to a planar all-*trans* backbone conformation (*5–8, 15*). Results from polarized absorption spectra of stretch-oriented samples for the cooled samples exhibit absorbance only for polarizations parallel to the stretch (and presumably the chain axis) direction (*22*).

Summary

The local-density functional approach was used to compare the band structures of the all-*trans* conformation of unsubstituted polysilane with a 4/1 helical conformation and with an all-*trans* conformation of dimethyl-substituted polysilane. In line with previous theoretical studies, the electronic wave functions in the vicinity of the Fermi level are primarily silicon-backbone states, with the major effect of methyl substitution being a decrease in the gap. The predicted absorption spectra for the all-*trans* conformations of unsubstituted and dimethyl-substituted polysilane are similar for near-threshold absorption. Given this similarity, we believe that the shift in energy and strong anisotropy of threshold absorption that we predict for the two extremes of the all-*trans* conformation and the all-*gauche* model will also occur in alkyl-substituted systems, which are currently under investigation.

Acknowledgments

We thank J. Zeigler and G. Fearon for their part in organizing the workshop on which this volume is based. J. V. O. acknowledges partial support of this work, which was performed during the summer of 1987 under the auspices of the American Society for Engineering Education Summer Faculty Program at the Naval Research Laboratory.

References

1. For a review, see West, R. *J. Organomet. Chem.* **1986,** *300,* 327 and references therein.
2. Trefonas, P., III; West, R.; Miller, R. D.; Hofer, D. *J. Polym. Sci., Polym. Lett. Ed.* **1983,** *21,* 823.
3. Zeigler, J. M.; Harrah, L. A.; Johnson, A. W. *Proc. SPIE Conf., Adv. Resist Technol.* **1985,** *539,* 166.
4. Trefonas, P., III; West, R.; Miller, R. D. *J. Am. Chem. Soc.* **1985,** *107,* 2737.
5. Harrah, L. A.; Zeigler, J. M. *J. Polym. Sci., Polym. Lett. Ed.* **1985,** *23,* 209.
6. Miller, R. D.; Hofer, D.; Rabolt, J.; Fickes, G. N. *J. Am. Chem. Soc.* **1985,** *107,* 2172.
7. Rabolt, J. F.; Hofer, D.; Miller, R. D.; Fickes, G. N. *Macromolecules* **1986,** *19,* 611.
8. Trefonas, P., III; Damewood, J. R., Jr.; West, R.; Miller, R. D. *Organometallics* **1985,** *4,* 1318.
9. Teramae, H.; Yamabe, T.; Imamura, A.; *Theor. Chim. Acta* **1983,** *64,* 1.
10. Takeda, K.; Matsumoto, N.; Fukuchi, M. *Phys. Rev. B: Condens. Matter* **1984,** *30,* 5871.
11. Takeda, K.; Teramae, H.; Matsumoto, N. *J. Am. Chem. Soc.* **1986,** *108,* 8186.
12. Takeda, K.; Fujino, M.; Seki, K.; Inokuchi, H. *Phys. Rev. B: Condens. Matter* **1987,** *36,* 8129.
13. Bigelow, R. W.; McGrane, K. M. *J. Polym. Sci., Polym. Phys. Ed.* **1986,** *24,* 1233.
14. Bigelow, R. W. *Chem. Phys. Lett.* **1986,** *126,* 63.
15. Klingensmith, K. A.; Downing, J. W.; Miller, R. D.; Michl, J. *J. Am. Chem. Soc.* **1986,** *108,* 7438.
16. Mintmire, J. W.; Ortiz, J. V. *Macromolecules* **1988,** *21,* 1189.
17. Ortiz, J. V.; Mintmire, J. W. *J. Am. Chem. Soc.* **1988,** *110,* 4522.
18. Mintmire, J. W.; White, C. T. *Phys. Rev. B: Condens. Matter* **1983,** *28,* 3283.
19. Mintmire, J. W.; White, C. T. *Phys. Rev. B: Condens. Matter* **1983,** *27,* 1447.
20. Trickey, S. B.; Ray, A. K.; Worth, J. P. *Phys. Status Solidi B* **1981,** *106,* 631.
21. Pitt, C. G.; Jones, L. L.; Ramsey, B. G. *J. Am. Chem. Soc.* **1967,** *89,* 5471.
22. Harrah, L. A.; Zeigler, J. M. *Macromolecules* **1986,** *20,* 601.

RECEIVED for review May 27, 1988. ACCEPTED revised manuscript November 16, 1988.

30

Ground States and Ionization Energies of Polysilane Oligomers

J. V. Ortiz[1] and J. W. Mintmire[2]

[1]Department of Chemistry, University of New Mexico, Albuquerque, NM 87131
[2]Chemistry Division, Naval Research Laboratory, Washington, DC 20375

Ab initio calculations were performed on the structures and vertical ionization energies of small silane chains. Optimizations of Hartree–Fock double ζ plus polarization geometry were supplemented by correlated calculations on ground-state and vertical ionization energies. Energy differences between minima on the ground-state surface were insignificant, and rotation barriers were ~1 kcal/mol. Vertical ionization energies were more dependent on rotations about single bonds. For tetrasilane, the experimental photoelectron spectrum showed a mixture of rotational minima. Molecular orbital studies of pentasilane gave similar results. Parametrized calculations on longer chains confirmed the qualitative trends observed in the ab initio studies. Simple molecular orbital ideas, regardless of whether an atomic or bond function basis is used, suffice to explain the conformational dependence of vertical ionization energies.

INTEREST IN THE UNUSUAL spectral and chemical properties of polysilanes has been intensified by the emergence of several technological applications (*1–9*). Absorbance in the UV range by molecules with no π, *d*, or lone-pair electrons (*10–11*) is an especially provocative feature. Bond orbital models (*12*) suggest that there are similarities between these saturated chains and π-conjugated polymers such as polyacetylene. Delocalization of particle and hole states is suggested by the ESR (electron spin resonance) spectra of stable radical anions (*13–16*) and low-lying ionization energies (*17–23*). The conformational dependence of elementary excitation energies may be responsible for bathochromic shifts that accompany the phase transitions of substituted polysilanes in solution (*24–32*).

0065–2393/90/0224–0551$06.00/0

Much spectroscopic and theoretical work has concentrated on the conformational dependence of ground-, excited-, and ionic-state energies. Parametrized computational studies have treated the ground- and excited-state energies of oligomers (*33–45*) and the bands of infinite polymers (*46–47*). Until now, few ab initio studies have appeared on the electronic structure of any system larger than disilane (*48–57*).

Thorough ab initio calculations on small silanes are a valuable precursor to studies of longer chains (*58*, *59*). Two kinds of calculations are presented in this chapter: geometric optimization to obtain accurate structures and relative energies of rotational isomers and calculations of vertical ionization energies, which can be compared with photoelectron spectra and which disclose aspects of electronic structure that relate spectra to molecular geometry. The calculated electron-binding energies for tetrasilane were used to adjust a parametrized method, which was then applied to large polysilane chains. The conformational dependence of the energy of the highest occupied orbital is explained in terms of a simple bond orbital model.

Experimental Procedures

At the optimized HF/3-21G* geometries, single-point Hartree–Fock and MBPT(2) total energies were calculated with the 6-31G* basis (*60–69*). Electron propagator theory (EPT) (*70–75*) in the quasiparticle outer valence approximation (OVA) was used to calculate vertical ionization energies with relaxation and correlation corrections to Koopmans's theorem (*76*). Calculations on disilane and trisilane produced excellent results. HF/3-21G* Si–Si and Si–H distances of 2.345 and 1.48 Å, respectively, are in agreement with results of previous studies and experiment (*48–57*). Bond angles are close to 110°. Results of EPT–OVA with the 6-31G* basis on the first vertical ionization energy of each symmetry lie within 0.2 eV of experimental results (*17*, *18*). Sorting out the effects of dihedral-angle changes can begin with this reliable model.

Results and Discussion

***n*-Tetrasilane.** Optimizations of HF/3-21G* geometry were performed on the sequence *anti*, ecl-120 (eclipsed 120°), *gauche*, and ecl-0, which describes a steady decrease of the Si-backbone dihedral angle from 180° to 120° to 60° to 0°. For all four cases, the optimized Si–Si distances are about 2.35 Å, and the Si–H distances are about 1.48 Å. Si–Si–Si angles are close to 111°. Dihedral angles, even when not constrained by symmetry, are near 60°, –60°, and 180°.

All of the ground-state total-energy differences, regardless of basis set or correlation treatments, give essentially the same result. For example, MBPT(2) calculations with the 6-31G* basis favor the *gauche* conformation over the *anti* conformation by just 0.04 kcal/mol. The barrier between these two conformers is 0.6 kcal/mol, but the barrier between the two *gauche* forms is 1.2 kcal/mol. Steric factors will dominate considerations of the relative energies of substituted rotamers.

The three highest levels correspond to molecular orbitals (MOs) with nearest neighbor bonding interactions along the Si backbone. Two MOs are symmetric with respect to C_2 rotation (the symmetry operation that is conserved for all four structures), and the other is antisymmetric (Figure 1). Both of the MOs with symmetry label a (symmetric) are destabilized in the *anti* form, but the MOs with symmetry label b (antisymmetric) are stabilized. Simple MO notions suffice to explain these trends. For MOs with symmetry label b, the two terminal bonding regions are out of phase. These regions approach each other as one passes from right to left on the abscissa, and the one-electron level is destabilized. For MOs with symmetry label a, bonding relationships exist in each terminal Si–Si region, and symmetry requires that these two regions be in phase.

Because the total energies for *anti* and *gauche* conformers are virtually the same, the photoelectron spectrum will depend on the vertical ionization energies of both isomers. The results for both the *anti* (9.27, 10.46, and 10.73 eV) and *gauche* (9.49, 10.05, and 10.81 eV) conformers are in reasonable agreement with the maxima (*17, 18*) at 9.62, 10.3, and 10.85 eV. Quantitative errors in results of calculations using Koopmans's theorem are large, but a similar dependence of energy on dihedral angle is produced with the 3-21G* basis. The main flaws of results obtained by using Koopmans's theorem are the absolute energies and the relative displacements of the final states, but the dependence on dihedral angles is represented well.

***n*-Pentasilane.** Optimizations of HF/3-21G* geometry on the *anti–anti* and *gauche–gauche* conformers of the unbranched five-Si chain produce bond lengths, angles, and dihedral angles that do not vary significantly from those of the smaller chains. At this level, the *anti–anti* form is lower in energy by 0.16 kcal/mol, but the basis set and correlation improvements could reverse this order.

Because results obtained by using Koopmans's theorem give a reasonable qualitative description of how vertical ionization energies change with conformation in tetrasilane, the 3-21G* canonical orbital energies of pentasilane can perform a similar service in comparing the *anti–anti* and *gauche–gauche* conformers. The four highest occupied MOs are predominantly Si–Si nearest neighbor σ-bonding orbitals (Figure 2). Each symmetry label (a and b) has 1–4 bonding and antibonding cases. (In this notation, 1–2 refers to a single bond function, 1–3 refers to adjacent-bond functions, and 1–4 refers to next-to-adjacent-bond functions.) For symmetry label a, an avoided crossing exists between the two structures. The lower a level is 1–4 antibonding for the *anti–anti* conformer but becomes bonding for the *gauche–gauche* conformer. If the noncrossing rule is disregarded in following the transformation from the *anti–anti* conformer to the *gauche–gauche* conformer, the 1–4 bonding level for symmetry level a declines from −11.33 to −11.92 eV, and the other a level ascends from −11.54 to −10.74 eV. For b symmetry, the lower level is 1–4 antibonding and ascends from −11.74 to −11.34 eV. Finally, the

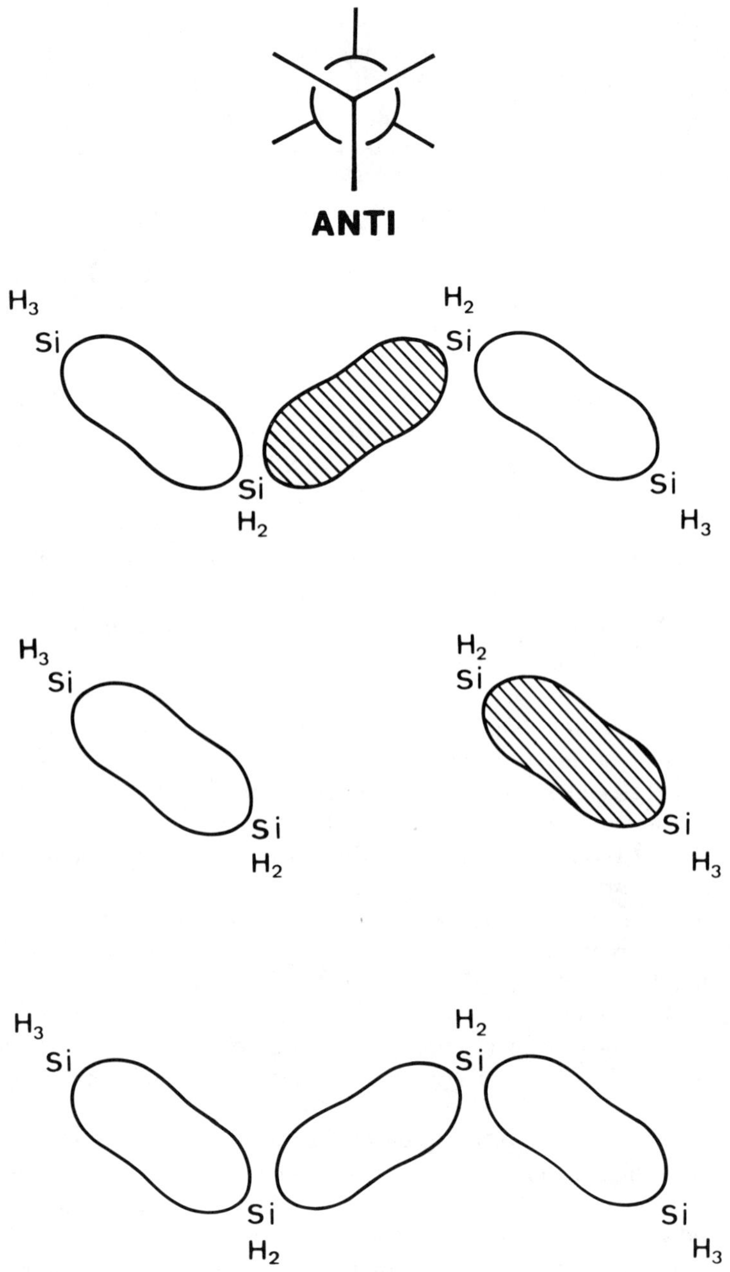

Figure 1. HOMOs of tetrasilane represented as linear combinations of Si–Si bond functions.

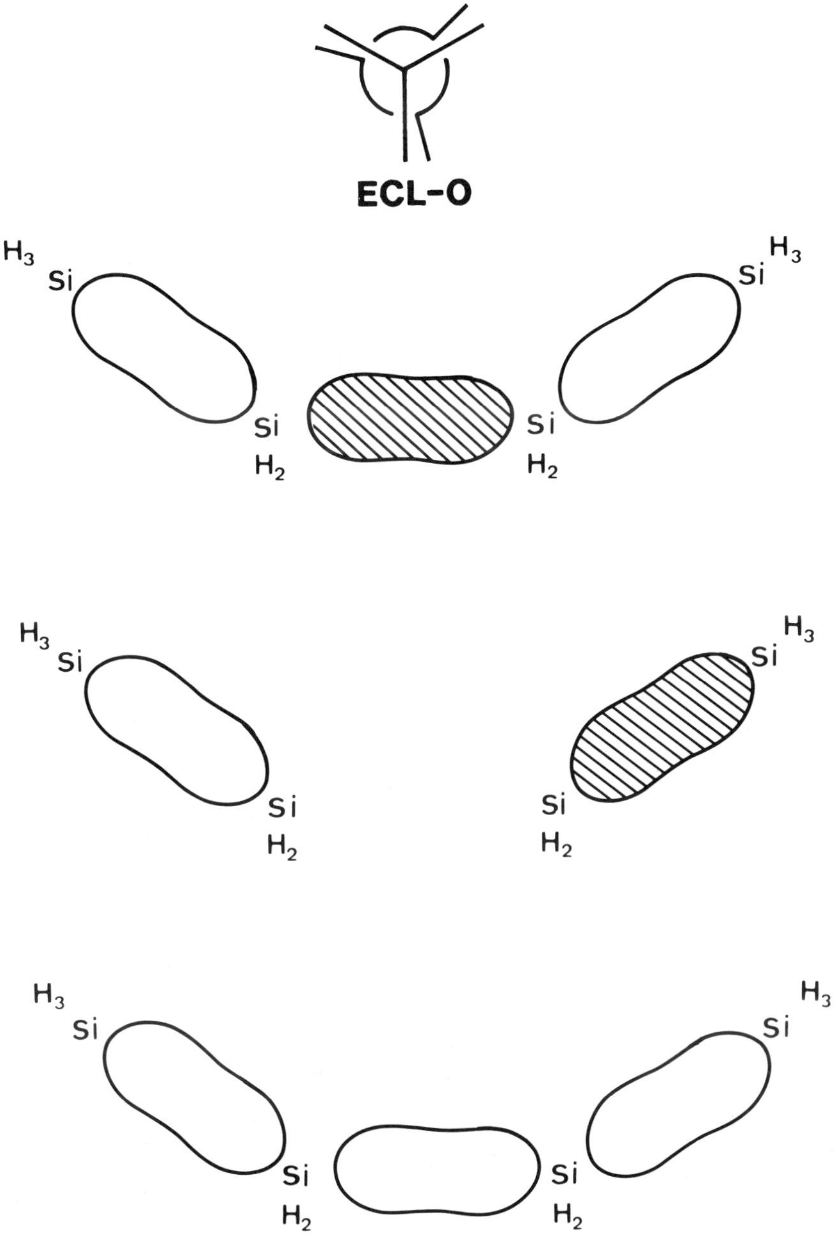

ECL-O

H_3 Si Si H_3
Si H_2 Si H_2

H_3 Si Si H_3
Si H_2 Si H_2

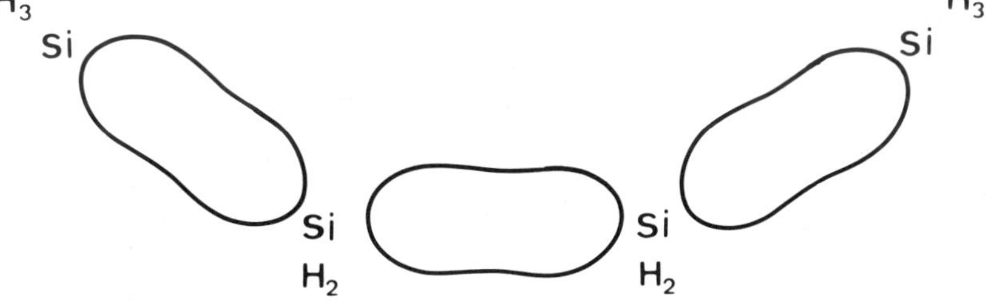

Figure 1.—Continued.

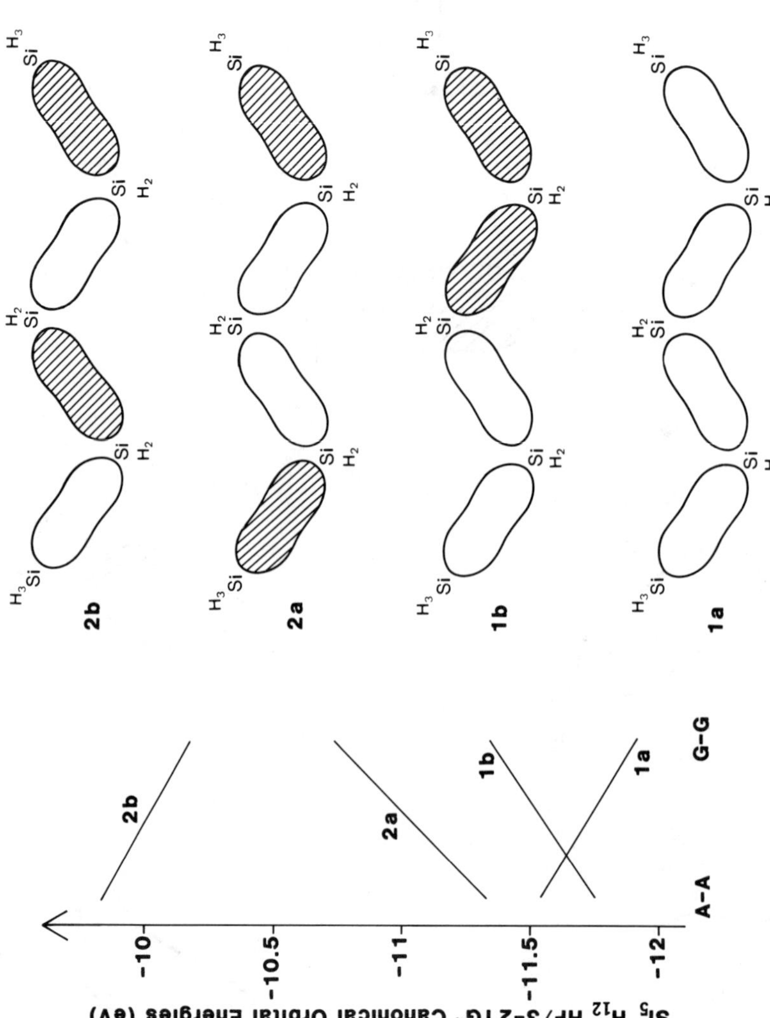

Figure 2. HOMOs of pentasilane represented as linear combinations of Si–Si bond functions. A–A and G–G denote anti–anti *and* gauche–gauche *conformations, respectively, and a and b refer to symmetry labels as described in the text.*

HOMO (highest occupied molecular orbital) with symmetry label b and 1–4 bonding character declines from −9.83 to −10.18 eV. Accentuation of the 1–4 interaction in the *gauche–gauche* conformer, in which nonadjacent Si–Si bonds come into closer proximity, controls the energies of these MOs.

Longer Chains. With the satisfactory ab initio results for disilane through pentasilane, a parametrized model of long-chain ionization energies and their conformational dependence can be built. The extended Hückel theory (*77–82*) makes interatomic interactions depend on overlap and is therefore a good choice for representing 1–4 interactions involving bonds along the Si backbone. H_{ii} parameters (*see* references 80 and 81 for a definition of these parameters) are adjusted (*80*) to produce tetrasilane orbital energies within 1 eV of the results of EPT–OVA calculations with the 6-31G* basis for the lowest final states of each symmetry. The proper dependence of the highest-occupied-orbital energies on the dihedral angle is obtained. For example, with the extended Hückel calculations, the HOMO falls by 0.30 eV during the transition from the *anti* conformer to the *gauche* conformer, compared with 0.22 eV in the EPT study. As a consequence of these trends, the total energy of the *gauche* form is lower by 0.1 eV. The extended Hückel theory is probably not suitable for total-energy calculations. A standard set of bond lengths and angles (*81*) obtained from the ab initio results was used in a study of longer chains.

Unbranched $Si_{10}H_{22}$, $Si_{20}H_{42}$, and $Si_{30}H_{62}$ were studied in all-*anti* and all-*gauche* forms. The HOMO energies are −7.883, −7.523, and −7.440 eV, respectively, for the *anti* conformation and −9.102, −9.052, and −9.041 eV, respectively, for the *gauche* conformation. Further changes as the chain length increases will be negligible. Overlap populations (OPs) pertaining to just this MO show the reasons for *gauche* stabilization. Two kinds of interatomic OPs are important: positive Si–Si neighboring OPs and negative Si–Si next-neighbor OPs. In the *gauche* form, the the Si–Si neighboring OP becomes slightly larger and the Si–Si next-neighbor OP becomes dramatically less negative. Average next-neighbor Si–Si OPs are −0.070, −0.037, and −0.025 for the three chains in the *anti* form but ascend to −0.021, −0.009, and −0.006 for the *gauche* form. Average neighboring Si–Si OPs are 0.080, 0.036, and 0.024 for the *anti* form and 0.089, 0.042, and 0.028 for the *gauche* form. The HOMOs for these three molecules are concentrated in the middle of the chain, so there is a range of OP changes. OPs are about twice the average in the center of the chain and fall almost to zero at the ends.

Two equivalent ways of interpreting these results are possible. A bond orbital picture (*12*) conceives of the HOMO of each chain as bonding lobes located between silicon atoms with alternating signs. The neighboring interatomic OP is the most positive for the straightforward reason that a bond function exists between the two atoms. The next-neighbor interatomic OP is smaller in absolute value and negative in sign, because the adjacent-bond functions are out of phase and give negative OP contributions, whereas the next-to-adjacent-bond functions are in phase and give small, positive OP contributions. Upon going from the *anti* conformation to the *gauche* conformation, the overlap between next-to-adjacent-bond functions is increased, because, for example, the 1–2 bond is closer to the 3–4 bond. This 1–4 bond orbital effect leads to a decrease in atomic next-neighbor antibonding character for the HOMO and, therefore, to energetic stabilization.

An equivalent explanation (though not as applicable to Si–Si bonding orbitals other than the HOMO or to shorter chains) emphasizes phase relationships between atomic *p* functions on Si atoms (*46*, *47*, *83*). The HOMO is constructed from Si *p* functions aligned with the chain axis. Another way of thinking of these *p* functions is as b_1 LUMOs (lowest unoccupied molecular orbitals) of SiH_2 fragments. In both conformations, alternating signs on each Si *p* function produce large, positive OPs between neighboring Si atoms and negative next-neighbor OPs with smaller absolute values. In the *anti* form, neighboring *p* functions have positive σ overlaps and small, negative π overlaps. In the *gauche* form, the positive σ overlap declines somewhat, but the π overlap becomes positive also. The total bonding interaction between the neighboring *p* functions is enhanced slightly in the *gauche* form because of the reorientation of the *p* functions. For next-neighbor interactions, the σ antibonding relationship that obtains in the *anti* form is diminished in the *gauche* form because the *p* functions are no longer pointing directly at each other. In other words, negative σ overlap is partially offset by positive π overlap.

Summary

After optimizations of ab initio geometry were carried out on small silane chains, calculations with improved basis set and correlation methods were used. Vertical ionization energies were in excellent agreement with experimental results. Total-energy results for tetrasilane suggest no significant difference between the energies of the *anti* and *gauche* conformers. Lower level calculations on pentasilane gave similar conclusions regarding the influence of dihedral angles on ground-state and vertical ionization energies. Parametrized calculations on longer chains confirmed the same trends. The influence of dihedral-angle changes on vertical ionization energies can be explained in terms of interactions of next-neighbor *p* functions or in terms of next-to-adjacent-bond functions.

Acknowledgments

The Naval Research Laboratory Research Advisory Committee provided a grant of computer time on the NRL Cray-XMP, which made this work possible. I (J.V.O.) thank John Zeigler and the University of New Mexico for facilitating my attendance at the symposium on which this volume is based.

References

1. West, R.; Carberry, E. *Science* **1975,** *189*, 179.
2. West, R. *J. Organomet. Chem.* **1976,** *300,* 327.
3. Zeigler, J. M.; Harrah, L. A.; Johnson, A. W. *Polym. Prepr.* **1985,** *26,* 345.
4. Zeigler, J. M.; Harrah, L. A.; Johnson, A. W. *SPIE Adv. Resist Technol. Proc. II* **1985,** *539,* 166.
5. Kepler, R. G.; Zeigler, J. M.; Harrah, L. A. *Bull. Am. Phys. Soc.* **1984,** *29,* 509.
6. Hofer, D. C.; Miller, R. D.; Willson, C. G. *SPIE Adv. Resist Technol. Proc. II* **1984,** *469,* 16.
7. Harrah, L. A.; Zeigler, J. M. 1984 International Congress of Pacific Basin Chemical Societies, Honolulu, Hawaii, Abstract 09F0016.
8. Hofer, D. C.; Miller, R. D.; Willson, C. G.; Neureuther, A. *SPIE Adv. Resist Technol. Proc. II* **1984,** *469,* 108.
9. Miller, R. D. *Polym. News* **1987,** *12,* 326.
10. Gilman, H.; Atwell, W. H.; Schwebke, G. L. *Chem. Ind.* (*London*) **1964,** 1063.
11. Gilman, H.; Atwell, W. H.; Schwebke, G. L. *J. Organomet. Chem.* **1964,** *2,* 369.
12. Pitt, C. G. In *Homoatomic Rings, Chains and Macromolecules of Main Group Elements;* Rheingold, A. L., Ed.; Elsevier: Amsterdam, 1977.
13. Carberry, E.; West, R. *J. Organomet. Chem.* **1966,** *6,* 582.
14. West, R.; Indriksons, A. *J. Am. Chem. Soc.* **1972,** *94,* 6110.
15. West, R.; Kean, E. S. *J. Organomet. Chem.* **1975,** *96,* 323.
16. Carberry, E.; West, R.; Glass, G. E. *J. Am. Chem. Soc.* **1969,** *91,* 5446.
17. Bock, H.; Ensslin, W.; Feher, F.; Freund, R. *J. Am. Chem. Soc.* **1976,** *98,* 668.
18. Ensslin, W.; Bergmann, H.; Elbel, S. *J. Chem. Soc., Faraday Trans. 2* **1975,** *71,* 913.
19. Pitt, C. G.; Carey, R. N.; Toren, E. C. *J. Am. Chem. Soc.* **1972,** *94,* 3806.
20. Pitt, C. G.; Bursey, M. M.; Rogerson, P. F. *J. Am. Chem. Soc.* **1970,** *92,* 519.
21. Bock, H.; Ensslin, W. *Angew. Chem. Int. Ed. Engl.* **1971,** *10,* 404.
22. Cowley, A. H.; Dewar, M. J. S.; Goodman, D. W.; Padolina, M. C. *J. Am. Chem. Soc.* **1974,** *96,* 2648.
23. Cowley, A. H. In *Homoatomic Rings, Chains and Macromolecules of Main Group Elements;* Rheingold, A. L., Ed.; Elsevier: Amsterdam, 1977.
24. Miller, R. D.; Hofer, D. C.; Rabolt, J. F.; Fickes, G. N. *J. Am. Chem. Soc.* **1985,** *107,* 2172.
25. Harrah, L. A.; Zeigler, J. M. *J. Polym. Sci., Polym. Lett. Ed.* **1985,** *23,* 209.
26. Trefonas, P.; Damewood, J. R.; West, R.; Miller, R. D. *Organometallics* **1985,** *4,* 1318.
27. Trefonas, P.; West, R.; Miller, R. D.; Hofer, D. *J. Polym. Sci., Polym. Lett. Ed.* **1983,** *21,* 823.
28. Schilling, F. C.; Bovey, F. A.; Lovinger, A. J.; Zeigler, J. M. *Macromolecules* **1986,** *19,* 2660.
29. Kuzmany, H.; Rabolt, J. F.; Farmer, B. L.; Miller, R. D. *J. Chem. Phys.* **1986,** *85,* 7413.

30. Lovinger, A. J.; Schilling, F. C.; Bovey, F. A.; Zeigler, J. M. *Macromolecules* **1986,** *19,* 2657.
31. Miller, R. D.; Farmer, B. L.; Fleming, W.; Sooriyakumaran, R.; Rabolt, J. F. *J. Am. Chem. Soc.* **1987,** *109,* 2509.
32. Rabolt, J. F.; Hofer, D.; Miller, R. D.; Fickes, G. N. *Macromolecules* **1986,** *19,* 611.
33. Herman, A.; Dreczewski, B.; Wojnowski, W. *Chem. Phys.* **1985,** *98,* 475.
34. Bigelow, R. W. *Chem. Phys. Lett.* **1986,** *126,* 63.
35. Bigelow, R. W.; McGrane, K. M. *J. Polym. Phys. B* **1986,** *24,* 1233.
36. Bigelow, R. W. *Organometallics* **1987,** *5,* 1502.
37. Dewar, M. J. S.; Lo, D. H.; Ramsden, C. A. *J. Am. Chem. Soc.* **1975,** *97,* 1311.
38. Gordon, M. S.; Neubauer, L. *J. Am. Chem. Soc.* **1974,** *96,* 5690.
39. Verwoerd, W. S. *J. Comput. Chem.* **1982,** *3,* 445.
40. Dewar, M. J. S.; Caoxian, J. *Organometallics* **1987,** *6,* 1486.
41. Klingensmith, K. K.; Downing, J. W.; Miller, R. D.; Michl, J. *J. Am. Chem. Soc.* **1986,** *108,* 7438.
42. Damewood, J. R.; West, R. *Macromolecules* **1985,** *18,* 158.
43. Damewood, J. R. *Macromolecules* **1985,** *18,* 1793.
44. Hummel, J. P.; Stackhouse, J.; Mislow, K. *Tetrahedron* **1977,** *33,* 1925.
45. Hoenig, H.; Hassler, K. *Monatsh. Chem.* **1982,** *113,* 129.
46. Takeda, K.; Matsumoto, N.; Fukuchi, M. *Phys. Rev. B* **1984,** *30,* 5871.
47. Takeda, K.; Teramae, H.; Matsumoto, N. *J. Am. Chem. Soc.* **1986,** *108,* 8186.
48. Nicholas, G.; Berthelat, J. C.; Durand, P. *J. Am. Chem. Soc.* **1976,** *98,* 1346.
49. Luke, B. T.; Pople, J. A.; Krogh-Jespersen, M. B.; Apeloig, Y.; Chandrasekhar, J.; von Ragué Schleyer, P. *J. Am. Chem. Soc.* **1986,** *108,* 260.
50. Halevi, E. A.; Winkelhofer, G.; Meisl, M.; Janoschek, R. *J. Organomet. Chem.* **1985,** *294,* 151.
51. Berkovitch-Yellin, Z.; Ellis, D. E.; Ratner, M. A. *Chem. Phys.* **1981,** *62,* 21.
52. Blustin, P. H. *J. Organomet. Chem.* **1976,** *105,* 161.
53. Topiol, S.; Ratner, M. A.; Moskowitz, J. W. *Chem. Phys.* **1977,** *20,* 1.
54. Collins, J. B.; von Ragué Schleyer, P.; Binkley, J. S.; Pople, J. A. *J. Chem. Phys.* **1976,** *64,* 5142.
55. Snyder, L. C.; Wasserman, Z. *Chem. Phys. Lett.* **1977,** *51,* 349.
56. Sax, A. F. *J. Comput. Chem.* **1985,** *6,* 469.
57. Teramae, H.; Yamabe, T. *Theor. Chim. Acta* **1983,** *64,* 1.
58. Mintmire, J. W.; Ortiz, J. V. *Macromolecules* **1988,** *21,* 1189.
59. Ortiz, J. V.; Mintmire, J. W. *J. Am. Chem. Soc.* **1988,** *110,* 4522.
60. Pietro, W. J.; Francl, M. M.; Hehre, W. J.; DeFrees, D. J.; Pople, J. A.; Binkley, J. S. *J. Am. Chem. Soc.* **1982,** *104,* 5039.
61. Gordon, M. S.; Binkley, J. S.; Pople, J. A.; Pietro, W. J.; Hehre, W. J. *J. Am. Chem. Soc.* **1982,** *104,* 2797.
62. Binkley, J. S.; Pople, J. A.; Hehre, W. J. *J. Am. Chem. Soc.* **1980,** *102,* 939.
63. Hehre, W. J.; Ditchfield, R.; Pople, J. A. *J. Chem. Phys.* **1972,** *56,* 2257.
64. Francl, M. M.; Pietro, W. J.; Hehre, W. J.; Binkley, J. S.; Gordon, M. S.; DeFrees, D. J.; Pople, J. A. *J. Chem. Phys.* **1982,** *77,* 3654.
65. Hariharan, P. C.; Pople, J. A. *Theor. Chim. Acta* **1973,** *28,* 213.
66. HF stands for Hartree–Fock.
67. MBPT(2) stands for second-order many-body perturbation theory, which is also known by the Hamiltonian partitioning scheme it employs, Moeller-Plesset (see references 68 and 69).
68. Bartlett, R. *Annu. Rev. Phys. Chem.* **1981,** *32,* 359.
69. Binkley, J. S.; Pople, J. A. *Int. J. Quantum Chem.* **1975,** *9,* 229.

70. von Niessen, W.; Schirmer, J.; Cederbaum, L. S. *Comput. Phys. Rep.* **1984,** *1,* 57.
71. Herman, M. F.; Freed, K. F.; Yeager, D. L. *Adv. Chem. Phys.* **1981,** *48,* 1.
72. Öhrn, Y.; Born, G. *Adv. Quantum Chem.* **1981,** *13,* 1.
73. Simons, J. *Theoretical Chemistry Advances and Perspectives;* Eyring, H., Ed.; Academic: New York, NY, 1978; Vol. 3.
74. Ortiz, J. V. Ph.D. Dissertation, University of Florida, Gainesville, 1981.
75. Cederbaum, L. S.; Domcke, N.; von Niessen, W. *J. Phys. B* **1977,** *10,* 2963.
76. Koopmans, T. *Physica* (*Utrecht*) **1934,** *1,* 104.
77. Hoffmann, R. *J. Chem. Phys.* **1963,** *39,* 1397.
78. Hoffmann, R.; Lipscomb, W. N. *J. Chem. Phys.* **1962,** *36,* 2179.
79. Hoffmann, R.; Lipscomb, W. N. *J. Chem. Phys.* **1962,** *37,* 2872.
80. Ammeter, J. H.; Burgi, H. B.; Thibeault, J. C.; Hoffmann, R. *J. Am. Chem. Soc.* **1978,** *100,* 3686.
81. Si parameters: $H_{ss} = -14.8$ eV, $H_{pp} = -7.6$ eV, $\zeta_{3s} = 1.592$, $\zeta_{3p} = 1.256$; H parameters: See references 77 and 78.
82. Standard bond lengths (in angstroms) and angles (in degrees): Si–Si, 2.345; Si–H, 1.478; Si–Si–Si, 111.5; Si–Si–H, 110.
83. Matsumoto, N., NTT Basic Research Laboratories, private communication, 1987.

RECEIVED for review May 27, 1988. ACCEPTED revised manuscript November 18, 1988.

PRECERAMIC POLYMERS

31

Synthesis of Some Organosilicon Polymers and Their Pyrolytic Conversion to Ceramics

Dietmar Seyferth

Department of Chemistry, Massachusetts Institute of Technology, Cambridge, MA 02139

This chapter gives an introduction to the preceramic polymer route to ceramic materials and focuses on the reasons why this new approach was needed and on the chemical considerations important in its implementation, with examples from research on organosilicon polymers. Novel polysilazanes have been prepared by the dehydrocyclodimerization reaction, a new method for polymerizing suitably substituted cyclooligosilazanes. The "living" polymer intermediate in this reaction has been used to convert Si–H-containing organosilicon polymers that are not suitable for pyrolytic conversion to ceramics into useful preceramic polymers.

SILICON-CONTAINING CERAMICS include the oxide materials, silica and silicates; the binary compounds of silicon with nonmetals, principally silicon carbide and silicon nitride; silicon oxynitride and the sialons; main group and transition metal silicides; and, finally, elemental silicon itself. Throughout the world, research activity on the preparation of all of these classes of solid silicon compounds by newer preparative techniques is vigorous.

Silicon carbide, SiC (*1*), and silicon nitride, Si_3N_4 (*2*), have been known for some time. Their properties, especially their high thermal and chemical stability, their hardness, and their high strength, have led to useful applications for both of these materials. The conventional methods for the preparation of SiC and Si_3N_4, the high-temperature reaction of fine-grade sand and coke (with additions of sawdust and NaCl) in an electric furnace (the

0065–2393/90/0224–0565$07.75/0

Acheson process) for SiC and usually the direct nitridation of elemental silicon or the reaction of silicon tetrachloride with ammonia (in the gas phase or in solution) for Si_3N_4, do not involve soluble or fusible intermediates. For many applications of these materials, this fact is not necessarily a disadvantage (e.g., for the application of SiC as an abrasive), but for some of the more recent desired applications, soluble or fusible (i.e., processable) intermediates are required.

Uses of Preceramic Polymers

The need for soluble or fusible precursors that will give the desired ceramic material on pyrolysis has led to a new area of macromolecular science—preceramic polymer chemistry (*3*). Such polymers are needed for a number of different applications. Ceramic powders by themselves are difficult to form into bulk bodies of complex shape. Although ceramists have addressed this problem by using the more conventional ceramics techniques with some success, preceramic polymers could, in principle, serve in such applications, either as the sole material from which the shaped body is fabricated, as a matrix in a shaped composite, or as a binder for the ceramic powder from which the shaped body is to be made. In each of these cases, pyrolysis of the green body would then convert the polymer to a ceramic material of the desired composition. The main shrinkage that occurs during pyrolytic conversion of a polymer to a ceramic is due to the change in density from polymer (~1 g/cm^3) to ceramic (~2–3 g/cm^3 before sintering). The mass loss that occurs during polymer pyrolysis also contributes to the shrinkage. If the preceramic polymer is used as a binder, shrinkage during pyrolysis should not be great.

Ceramic fibers of diverse chemical compositions are sought for application in the production of metal–, ceramic–, glass–, and polymer–matrix composites (*4*). The presence of such ceramic fibers in a matrix can result in very considerable increases in the strength (i.e., fracture toughness) of the resulting material. To prepare such ceramic fibers, a suitable polymeric precursor is needed, one which can be spun by melt-spinning, dry-spinning, or wet-spinning techniques into fibers, which then can be pyrolyzed (with or without a prior cure step).

Some materials with otherwise very useful properties, such as high thermal stability and great strength and toughness, are unstable with respect to oxidation at high temperatures. An example of such a class of materials is that of the carbon–carbon composites. If these materials could be protected against oxidation by infiltration of their pores and the effective coating of their surface by a polymer whose pyrolysis gives an oxidation-resistant ceramic material, then such carbon–carbon composite materials will have a wider range of applications.

Design of Preceramic Polymers

To have a useful preceramic polymer, considerations of structure and reactivity are of paramount importance. Not every inorganic or organometallic polymer will be a useful preceramic polymer. Although preceramic polymers are potentially high-value products if the desired properties result from their use, the more generally useful and practical systems will be those based on commercially available and relatively cheap starting monomers.

Preceramic Polymer Synthesis. Preferably, the polymer synthesis should involve simple, easily effected chemistry that proceeds in high yield. The preceramic polymer itself should be liquid, or if a solid, it should be fusible or soluble in at least some organic solvents, that is, it should be processable. Pyrolysis of the preceramic polymer should provide a high yield of ceramic residue, and the volatile products of pyrolysis preferably should be nonhazardous and nontoxic. In the requirement of high ceramic yield, economic considerations are only secondary. (Ceramic yield = [weight of pyrolysis residue] × 100/[weight of material pyrolyzed]. Therefore, ceramic yield represents the yield of the desired ceramic material plus that of any solid byproducts.) If the weight loss on pyrolysis is low, shrinkage will be less, as will be the destructive effects of the gases evolved during pyrolysis.

There are important considerations as far as the chemistry is concerned. First, the design of the preceramic polymer is of crucial importance. Many linear organometallic and inorganic polymers, even if they are of high molecular weight, decompose thermally by formation and evolution of volatile products of low molecular weight, usually unsaturated or cyclic species, and thus the ceramic yield is low. For example, high-molecular-weight, linear poly(dimethylsiloxane)s decompose thermally, principally by extruding small cyclic oligomers, $[(CH_3)_2SiO]_n$ (n = 3, 4, 5, . . .). When a polymer is characterized by this type of thermal decomposition, the ceramic yield will be low. Conversion of the linear polymer structure to a highly cross-linked structure by suitable chemical reactions will be necessary prior to pyrolysis.

Alternatively, a highly cross-linked but still soluble polymer can be prepared directly. In terms of the high-ceramic-yield requirement, the continuation of thermally induced cross-linking during the initial stages of the pyrolysis is desirable for the formation of nonvolatile networks, which result in maximum mass retention. Cross-linking requires the presence of appropriate reactive or potentially reactive substituent groups in the polymer. Thus, preceramic polymer design is an exercise in functional group chemistry.

In the design of preceramic polymers, achievement of the desired elemental composition in the resulting ceramics (SiC and Si_3N_4 in the present cases) is a major problem. For instance, in the case of polymers aimed at the production of SiC on pyrolysis, solid residues usually are obtained after pyrolysis that, in addition to SiC, contain an excess of either free carbon or free silicon, both of which are undesirable. To achieve the desired elemental composition, two approaches have been found useful in our research: (1) the use of two comonomers in the appropriate ratio in preparation of the polymer and (2) the use of chemical or physical combinations of two different polymers in the appropriate ratio.

Preceramic polymers intended for melt spinning require a compromise. If the thermal cross-linking process is too effective at relatively low temperatures (100–200 °C), then melt spinning will not be possible, because heating will induce cross-linking and will produce an infusible material prior to spinning. A less effective cross-linking process is required so that the polymer forms a stable melt that can be extruded through the holes of the spinneret. The resulting polymer fiber, however, must then be "cured", that is, cross-linked chemically or by irradiation to render it infusible so that the fiber form is retained on pyrolysis.

Preceramic Polymer Pyrolysis. Chemical options also are available for the pyrolysis step. Certainly, the rate of pyrolysis, that is, the time–temperature profile of pyrolysis, is extremely important. However, the gas stream used in pyrolysis also is of great importance. Inert- or reactive-gas pyrolysis can be carried out.

The gas stream used in the pyrolysis can determine the nature of the ceramic product. For example, our polymer of composition $[(CH_3SiHNH)_a(CH_3SiN)_b]_m$ gives a *black* solid (a mixture of SiC, Si_3N_4, and some free carbon) on pyrolysis to 1500 °C in an inert-gas stream (nitrogen or argon). However, when the pyrolysis is carried out in a stream of ammonia, a *white* solid remains, which usually contains less than 0.5% total carbon and is essentially pure silicon nitride. At higher temperatures (>400 °C), the methyl substituents on silicon are lost. Whether the substituents are lost via nucleophilic cleavage by NH_3 molecules or by reactions involving ammonia thermolysis products such as NH_2, NH, and H is not known. Such chemistry at higher temperatures can be an important and sometimes useful part of the pyrolysis process.

A different chemistry, ligand displacement, is involved in another example of reactive-gas pyrolysis from our boron ceramics project (5). Polymers of the type $[B_{10}H_{12}\cdot\text{diamine}]_x$ (in which the diamine is ethylene diamine or its *N*-methyl derivatives) give a mixture of B_4C, BN, and a small amount of carbon as a black residue on pyrolysis under argon. When pyrolysis is carried out in a stream of ammonia, a white residue, which is nearly pure boron nitride, is obtained. In this case, NH_3 displaces the diamine in a complicated process in which $(NH_4)_2(B_{10}H_{10})$ is formed as an intermediate. This intermediate then reacts with NH_3 pyrolysis products at higher temperature to give BN. In a further example from our research, pyrolysis of $(CH_3)_2N$-terminated polymers of the type $[Ti(NR)_2]_x$ gives golden-yellow TiN on pyrolysis in a stream of ammonia, but in a stream of argon, a black titanium carbonitride results (*6*).

In the pyrolysis of a preceramic polymer, the maximum temperature used is important. If the maximum temperature is too low, residual functionality (C–H, N–H, and Si–H bonds in the case of polysilazanes) will still be present. On the other hand, too high a pyrolysis temperature can be harmful because of solid-state reactions that can take place. For instance, if the polysilazane-derived silicon carbonitride contains a large amount of free carbon, a high-temperature reaction between carbon and silicon nitride (equation 1) (*7*) is a possibility.

$$Si_3N_4(s) + 3C(s) \leftrightarrows 3SiC(s) + 2N_2(g) \quad (1)$$

Characterization of Preceramic Polymers. The study of the pyrolysis products of preceramic polymers is not always straightforward. If crystalline species are produced (e.g., SiC and Si_3N_4 in the case of polysilazanes), then their identification by X-ray diffraction presents no problems.

However, in many cases the pyrolysis products are amorphous. For instance, the polysilazane-derived silicon carbonitride mentioned above crystallizes only when it is heated above 1450 °C, and its characterization is difficult. Elemental analysis poses problems, in part because the pyrolysis product is very porous. As a result of its high surface area, the material adsorbs moisture and volatiles very readily, and improper prior handling and preparation for analysis can result in misleading results.

Solid-state NMR spectroscopy is not routinely applicable to the study of such polymer-derived amorphous materials. In the case of our polysilazane pyrolysis product, the material was paramagnetic and gave a strong ESR (electron spin resonance) signal. A ^{29}Si NMR spectrum could not be obtained under experimental conditions that served well in obtaining a solid-state ^{29}Si NMR spectrum of crystalline β-SiC. Although it is possible to translate the elemental analysis of such amorphous silicon carbonitride materials into compositions comprising Si_3N_4, SiC, and free carbon, to do so is misleading. At this stage, carbon and nitrogen atoms (and oxygen atoms, if present) are bonded randomly to silicon in a three-dimensional covalent structure, according to Raman, ^{29}Si NMR, and electron microscopic data of Dow Corning workers for a hydridopolysilazane (HPZ)-derived pyrolysis product (*8*). This HPZ polysilazane is different from ours and is based on the reaction of $HSiCl_3$ with $[(CH_3)_3Si]_2NH$. In fact, a solid-state ^{29}Si NMR spectrum was obtained for the amorphous pyrolysis product of the HPZ polysilazane (*8*). Our amorphous silicon carbonitride was converted to crystalline material when it was heated to 1490 °C, and X-ray diffraction lines due to α-Si_3N_4 and β-SiC were observed.

The first useful organosilicon preceramic polymer, a silicon carbide fiber precursor, was developed by Yajima and his co-workers at Tohoku University in Japan (9). The fiber is manufactured by the Nippon Carbon Company and is known as Nicalon. As might be expected on the basis of the 2:1 C/Si ratio of the $(CH_3)_2SiCl_2$ starting material used in this process, the ceramic fibers contain free carbon as well as silicon carbide. A typical analysis (9) showed a composition of 1 SiC + 0.78 C + 0.22 SiO_2. The latter is introduced in the oxidative cure step of the polycarbosilane fiber.

Although it was one of the first, the Yajima polycarbosilane is not the only polymeric precursor to silicon carbide that has been developed. Another useful system that merits mention is the polysilane developed by Schilling and his co-workers at Union Carbide Corporation (*10*). As already mentioned, a useful polymeric precursor for silicon nitride has been developed more recently by workers at Dow Corning Corporation (*8*).

New Silicon-Based Preceramic Polymer Systems: Research at MIT

Preparation of Preceramic Polysilazanes. At MIT (Massachusetts Institute of Technology), our initial research on silicon-based preceramic polymers was aimed at developing a precursor for silicon nitride. To this end, we studied the ammonolysis of dichlorosilane, H_2SiCl_2 (*11*). This reaction had already been carried out on a millimolar scale in the gas phase and in benzene solution by Stock and Somieski in 1921 (*12*). We found that this reaction gave a much better yield of soluble ammonolysis product when it was carried out in more polar solvents such as dichloromethane or diethyl ether (*11*).

The ammonolysis product, which was isolated as a mobile oil, deviated from the ideal composition $(H_2SiNH)_n$, as determined by elemental analysis and ^1H NMR spectroscopy, being low in N and high in Si. Although this oil is stable indefinitely under nitrogen at –30 °C, at room temperature, it loses ammonia, with resulting cross-linking, to form a clear, insoluble glass. Pyrolysis of the oil to 1150 °C in a stream of nitrogen left a residue of α-Si_3N_4, as determined by X-ray diffraction, in about 70% ceramic yield. Although the H_2SiCl_2 ammonolysis product was not stable at room temperature (under nitrogen) as the neat liquid, it was stable in solution. Therefore, polysilazane fibers might be accessible via dry spinning in a suitable organic solvent. Also, such ammonolysis product solutions may be useful in the preparation of silicon nitride coatings and for the infiltration of porous bodies. Nevertheless, we did not pursue this research direction, mainly because H_2SiCl_2 is a potentially dangerous compound, as tests had shown (*13*). In independent studies in Japan (*14*), ammonolysis of the known H_2SiCl_2–pyridine adduct gave a viscous oil or a resinous solid that also underwent cross-linking at room temperature and gave α-Si_3N_4 on pyrolysis to 1300 °C.

In our continuing research, a compromise was made by replacing one of the hydrogen substituents of H_2SiCl_2 by a methyl group, that is, by using CH_3SiHCl_2 as the starting material. Using this commercially available, relatively cheap, and safe chlorosilane, we have developed a novel process for the preparation of useful polymeric precursors for silicon carbonitrides and, ultimately, Si_3N_4–SiC mixtures (*15*).

The ammonolysis of CH_3SiHCl_2 gives a mixture of cyclic and (possibly) linear oligomers, $(CH_3SiHNH)_n$ (*16*). After removal of the precipitated NH_4Cl, which is also produced, the ammonolysis product can be isolated as a clear, mobile liquid in high yield. Its C, H, and N analysis and its spectroscopic (^1H NMR and IR) data are in agreement with the $(CH_3SiHNH)_n$ formulation. Results of molecular weight determinations (by cryoscopy in benzene) of several preparations ranged from 280 to 320 g/mol (n = 4.7–5.4). The product is quite stable at room temperature, but it is sensitive to moisture and must be protected from the atmosphere. This mixture of $(CH_3SiHNH)_n$ oligomers is not suitable for ceramic preparation without further processing. On pyrolysis to 1000 °C in a stream of nitrogen, the ceramic yield was only 20%. Therefore, these $(CH_3SiHNH)_n$ oligomers had to be converted to a material of higher molecular weight.

The conversion of the cyclopolysilazanes obtained by ammonolysis of diorganodichlorosilanes to materials of higher molecular weight was investigated by Krüger and Rochow (*17*) some years ago, when there was interest in polysilazanes as polymers in their own right. Their procedure involved thermal ammonium-salt-induced polymerization, which in the case of hexamethylcyclotrisilazane gave polymers containing both linear and cyclic components. We applied this procedure to the CH_3SiHCl_2 ammonolysis product. A very viscous oil of higher molecular weight was produced, but the ceramic yield obtained on pyrolysis was a disappointing 36%.

The solution to the problem of converting the cyclo-$(CH_3SiHNH)_n$ oligomers to a useful preceramic polymer was provided by studies of earlier workers (*18*), who had described the conversion of silylamines of type **1** to cyclodisilazanes, **2**, in high yield by the action of potassium in di-*n*-butyl ether (equation 2).

$$2\ R_2\underset{H}{\underset{|}{Si}}\text{-}\underset{H}{\underset{|}{N}}R' \xrightarrow{K} H_2 + R_2Si\overset{NR'}{\underset{NR'}{\diamond}}SiR_2 \qquad (2)$$

1   **2**

In this dehydrocyclodimerization (DHCD) reaction, potassium metallates the NH functions to give **3**.

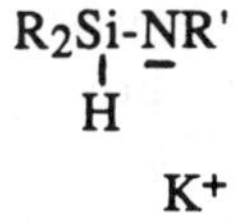

$$\underset{\displaystyle \mathrm{H}}{\mathrm{R_2Si}}\text{-}\overline{\mathrm{N}}\mathrm{R'} \quad \mathrm{K^+}$$

3

The amide ion **3** then either eliminates H^- from silicon to give an intermediate with a silicon–nitrogen double bond, $R_2Si{=}NR'$, which then undergoes head-to-tail dimerization to form **2** or, alternatively, reacts with a molecule of $R_2Si(H)NHR'$ to give an intermediate that undergoes cyclization to **2** with displacement of H^-.

The repeating unit in the cyclo-$(CH_3SiHNH)_n$ oligomers is **4**. For example, the cyclic tetramer is the eight-membered ring compound **5**.

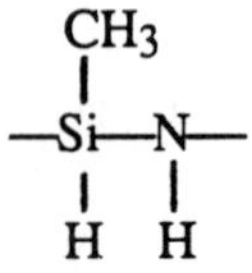

CH3
—Si—N—
H H

4

CH3 H H
Si—N
CH3
H—N Si
H H
Si N—H
CH3 N—Si
H H CH3

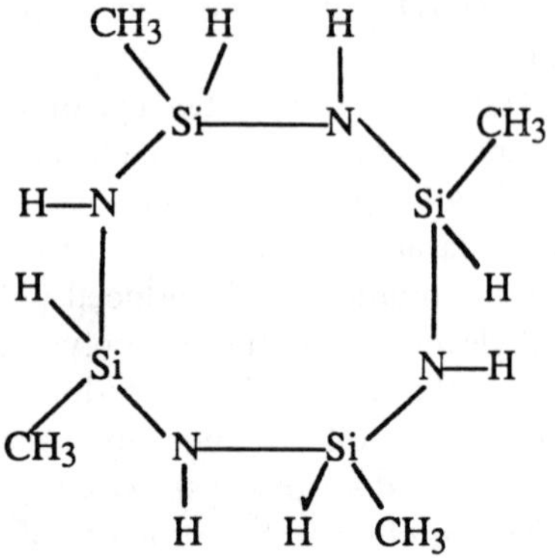

5

On the basis of equation 2, the adjacent NH and SiH groups provide the functionality that permits the increase in molecular weight of the cyclo-$(CH_3SiHNH)_n$ oligomers to be effected. Thus, two cyclo-$(CH_3SiHNH)_n$ oligomers may be linked together via a four-membered ring, as shown in **6**. However, six Si(H)–N(H) units remain in **6**, and so, in principle, extensive further linking of the species contained in the $(CH_3SiHNH)_n$ product of CH_3SiHCl_2 ammonolysis is possible.

Treatment of a solution of the CH_3SiHCl_2 ammonolysis product, cyclo-$(CH_3SiHNH)_n$, with a catalytic amount of a base (generally an alkali metal base) strong enough to deprotonate the N–H function in a suitable solvent results in immediate evolution of hydrogen. We believe that a sheet-like

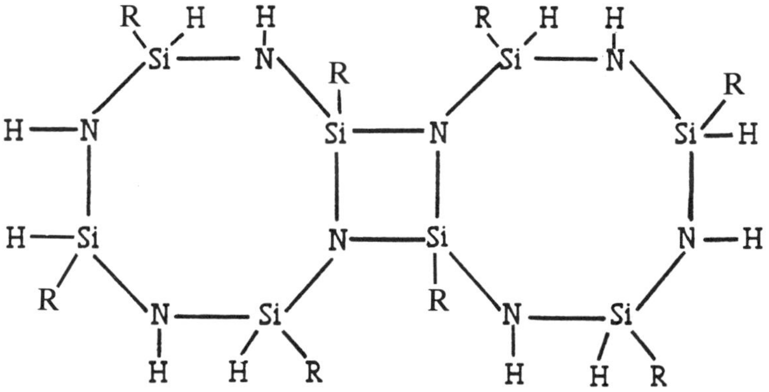

6

network polymer (which, however, is not flat) having the general functional unit composition $[(CH_3SiHNH)_a(CH_3SiN)_b(CH_3SiHNK)_c]_n$ is formed. When hydrogen evolution ceases, a "living" polymer with reactive silylamide functions is present. On reaction of the living polymer with an electrophile (CH_3I or a chlorosilane), a neutral polysilazane is formed and can be isolated in essentially quantitative yield. In our experiments, we generally have used catalytic amounts of potassium hydride (1–4 mol %, based on CH_3SiHNH) as a base. When THF (tetrahydrofuran) was used as reaction solvent, and after the addition of CH_3I, the product was isolated in the form of a white powder (with an average molecular weight of 1180), which was soluble in hexane, benzene, diethyl ether, THF, and other common organic solvents. Atmospheric moisture must be excluded because the cyclo-$(CH_3SiHNH)_n$ oligomers are readily hydrolyzed. The product polymer, on the other hand, is of greatly diminished sensitivity to hydrolysis.

The composition of the polysilazane product of an experiment of the kind just described, as ascertained by ^1H NMR spectroscopy, was $(CH_3SiHNH)_{0.39}(CH_3SiN)_{0.57}(CH_3SiHNCH_3)_{0.04}$. Its combustion analysis (C, H, N, and Si) agreed with this formulation. These results are compatible with a process in which $(CH_3SiHNH)_n$ rings are linked together via Si_2N_2 bridges, as shown in **6**. The experimental CH_3SiN/CH_3SiHNH ratio of ~1.3 indicates that further linking via Si_2N_2 rings must have taken place.

Pyrolysis of Preceramic Polysilazanes Prepared by the Dehydrocyclodimerization Reaction. Whatever the structure of the polysilazanes obtained by KH treatment of the cyclo-$(CH_3SiHNH)_n$ oligomers, these polymers are excellent ceramic precursors. Examination of the polymers from various preparations by TGA (thermogravimetric analysis) showed the weight loss on pyrolysis to be between only 15 and 20%. Pyrolysis appears to take place in three steps: a 3–5% weight loss (involving evolution of H_2) from 100 to 350 °C, a 2% weight loss from 350 to 550 °C, and a 9% weight loss from 550 to 900 °C. During the 550–900 °C stage a mixture of H_2 and methane was evolved. In addition to H_2, a trace of ammonia was lost between 350 and 550 °C.

In a typical bulk-pyrolysis experiment, pyrolysis was conducted under a slow stream of nitrogen. The sample was heated quickly to 500 °C and then slowly over 8 h to 1420 °C and was held at 1420 °C for 2 h. The ceramic product was a single body and black. Powder X-ray diffraction (Cu K_α with Ni filter) showed only very small, broad peaks for α-Si_3N_4. (At 1490 °C, lines due to β-SiC also appeared). Analysis by scanning electron microscopy showed little discernible microstructure, with only a few very fine grains appearing at high magnification. The bulk appearance of the ceramic suggested that pyrolysis took place after the polymer had melted. Many large holes and craters were evident in places where the liquid bubbles apparently had burst. In such experiments, ceramic yields usually were between 80 and 85%.

The polysilazane used in the experiment just described had been prepared in diethyl ether. It had a molecular weight of around 900 and went through a melt phase when it was heated, which was evident when it was heated in a sealed capillary. The material began to soften at ~65 °C and became more fluid with increasing temperature. The polysilazane prepared in THF is of higher molecular weight (1700–2000) generally, does not soften when heated to 350 °C, and gives an 83% yield of a black ceramic material on pyrolysis under nitrogen to 1000 °C.

Pyrolysis of the polysilazane may be represented by equation 3.

$$2(CH_3SiHNH)(CH_3SiN) \rightarrow Si_3N_4 + SiC + C + 2CH_4 + 4H_2 \qquad (3)$$

In this reaction, the ceramic yield (Si_3N_4 + SiC + C) would be 83 wt %. Analysis of the ceramic product gave 12.87% C, 26.07% N, and 59.52% Si. This analysis is compatible with equation 3 and leads to a *formal* ceramic constitution based on the four silicons of equation 3 of 0.88 Si_3N_4, 1.27 SiC, and 0.75 C or 67% Si_3N_4, 28% SiC, and 5% C on a weight percent basis.

Thus the chemistry leading to the desired ceramic product is quite satisfactory: most of the requirements mentioned earlier are met. Initial evaluation of the polysilazane shows that it has promise in three of the main potential applications of preceramic polymers: in the preparation of ceramic fibers and ceramic coatings and as a binder for ceramic powders.

On pyrolysis to 1100 °C, isostatically pressed (40,000 psi [275,800 kPa]) bars of the polymer gave a coherent, rectangular ceramic bar that had not cracked or bloated and could not be broken by hand.

In collaboration with ceramists at the Celanese Research Company, we found that our "meltable" polysilazane cannot be melt spun. Apparently, the thermal cross-linking process that is so effective in giving a high ceramic yield on pyrolysis takes place in the heated chamber of the spinning machine and quickly gives an infusible polymer. However, these polysilazanes can be dry spun. In this process, the solid polysilazane is dissolved in an appropriate solvent and then is extruded through a spinneret into a heated drying chamber in which the solvent is volatilized to leave the solid polymer fiber. These polymer fibers could be pyrolyzed to give ceramic fibers, and no cure step is required. Simpler experiments showed that fibers 30 to 61 cm long could be drawn from the sticky, waxy solid that remained when a toluene solution of the polysilazane was evaporated. Pyrolysis of these fibers under nitrogen produced long, flexible black fibers.

The polysilazane also serves well as a binder for ceramic powders. The preparation of such composites (by using commercial samples of fine α-SiC, β-SiC, and α-Si_3N_4 of 0.36–0.4-μm mean particle size) required appropriate dispersion studies. The ceramic powder was dispersed in a solution of toluene containing the appropriate weight of polysilazane. Then the toluene was evaporated with a rotary evaporator, and a waxy residue was left. Vacuum distillation removed the remaining solvent and left chunks of solid material, which was finely ground and pressed into a bar at 5000 psi (34,500 kPa). Isostatic pressing to 40,000 psi (275,800 kPa) followed, and then the bars were pyrolyzed in a tube furnace under nitrogen (10 °C/min up to 1100 °C). The maximum density (~2.4 g/cm^3) was achieved in these experiments with a polymer loading of 30%.

As already mentioned, our polysilazane undergoes thermal cross-linking too readily to permit melt spinning. If melt spinning is to be successful, a less reactive polysilazane with fewer reactive Si(H)–N(H) groups is required. To achieve this, we have studied polysilazanes derived by the dehydrocyclodimerization of the products of coammonolysis of CH_3SiHCl_2 and $CH_3(Un)SiCl_2$, in which Un is an unsaturated substituent such as vinyl, $CH_2{=}CH$, or allyl, $CH_2CH{=}CH_2$. These groups were used because a melt-spun fiber requires a subsequent cure step to render it infusible. If this is not done, the fiber would simply melt on pyrolysis, and no ceramic fiber would be obtained. The vinyl and allyl groups provide C=C functionality that can undergo catalyzed Si–H additions, reactions that, after melt-spinning, in principle should lead to extensive further cross-linking without heating.

To provide the required monomers, the following mixtures were ammonolyzed: (1) CH_3SiHCl_2 and $CH_3(CH_2{=}CH)SiCl_2$ and (2) CH_3SiHCl_2 and $CH_3(CH_2{=}CHCH_2)SiCl_2$. The $[(CH_3SiHNH)_x(CH_3(Un)SiNH)_y]_n$ oligomers were prepared using $CH_3SiHCl_2/CH_3(Un)SiCl_2$ ratios of 3, 4, and 6, although other ratios can be used. These ratios of reactants give materials that still contain many (CH_3)Si(H)–N(H)– units, and polymerization by means of equation 2 still will be possible. However, dilution of the ring systems of the cyclic oligomers with CH_3(Un)Si units will decrease the cross-linking that will occur on treatment with the basic catalyst.

The coammonolysis of CH_3SiHCl_2 and $CH_3(Un)SiCl_2$ in the indicated ratios was carried out by the procedure described for ammonolysis of CH_3SiHCl_2 (*15*). A mixture of mixed cyclic oligomeric silazanes, $[(CH_3(H)SiNH)_x(CH_3(Un)SiNH)_y]_n$, with more than one ring size, is to be expected in these reactions. In each preparation, the soluble, liquid products were isolated and used in the base-catalyzed polymerizations. When a $CH_3SiHCl_2/CH_3(Un)SiCl_2$ ratio of x:1 ($x > 1$) was used in the ammonolysis reaction, the $CH_3(H)Si/CH_3(Un)Si$ ratio in the soluble ammonolysis product usually was somewhat less than x:1.

The addition of the various ammonolysis products to suspensions of catalytic amounts of KH in dry THF resulted in hydrogen gas evolution with formation of a clear solution. Thermal treatment was followed by treatment with CH_3I to "kill" the living polymeric potassium silylamide present in solution. Generally, the resulting polysilazane was obtained in high yield (90%), and these products, which always were white powders, were soluble in organic solvents such as hexane, benzene, toluene, and THF. Their average molecular weights were in the 800–1200 range. The presence of unchanged vinyl and allyl groups was proven by their IR and ^{1}H NMR spectra. A typical elemental analysis led to the empirical formula $SiC_{1.5}NH_{4.4}$, which could be translated into the composition $[(CH_3SiHNH)(CH_3\{CH_2{=}CH\}SiNH)(CH_3SiN)_{2.3}]_x$.

Upon pyrolysis to 1000 °C under argon or nitrogen, these polysilazanes gave black ceramic materials in good yield (73–86% by weight). Analysis of the ceramic produced by such pyrolysis of one of these polysilazanes gave a formal composition of 67.5 wt % Si_3N_4, 22.5 wt % SiC, and 10.0 wt % unbound carbon. Curing by UV irradiation of fibers formed from these vinyl- or allyl-containing polysilazanes was possible. The following experiment shows this. Fibers were pulled from a concentrated syrup of a polysilazane derived by DHCD in THF from a 4:1 molar ratio of $CH_3SiHCl_2/CH_3(CH_2{=}CH)SiCl_2$ coammonolysis product. Some fibers were pyrolyzed without any further treatment. These fibers melted in large part and left very little in the way of ceramic fibers. Other fibers were subjected to UV irradiation for 2 h. On pyrolysis under argon, these fibers did not melt, and ceramic fibers were obtained. Therefore, this polysilazane is a good candidate for melt spinning.

To obtain a mixture of SiC and Si_3N_4 that is rich in Si_3N_4 by the preceramic polymer route, a polymer that is richer in nitrogen compared with the CH_3SiHCl_2 ammonolysis product, $(CH_3SiHNH)_n$, is required. In designing such a preceramic polymer, one would like to retain the facile chemical and thermal cross-linking system that the Si(H)–N(H) unit provides. The coammonolysis of CH_3SiHCl_2 and $HSiCl_3$ serves our purposes well. For $HSiCl_3$, ammonolysis introduces three Si–N bonds per silicon atom, so the ammonolysis product of mixtures of CH_3SiHCl_2 and $HSiCl_3$ will contain more nitrogen than the ammonolysis product of CH_3SiHCl_2 alone (*19*).

To define the optimum system, we have investigated the ammonolysis of mixtures of CH_3SiHCl_2 and $HSiCl_3$ in various ratios in two solvents: diethyl ether and THF. $CH_3SiHCl_2/HSiCl_3$ mole ratios of 6, 3, and 1 were examined. In both solvents, the 6:1 and 3:1 ratios produced polysilazane oils with molecular weights of 390–401 and 480 g/mol, respectively. When a 1:1 reactant ratio was used, waxes of somewhat higher molecular weight (764–778 g/mol) were obtained in both solvents. The reaction of a 1:1 mixture in diethyl ether gave a 40% yield of soluble product, but in THF, it was nearly quantitative.

The oils produced from the reactions of 6:1 and 3:1 mixtures in diethyl ether appeared to be stable on long-term storage at room temperature in the absence of moisture. However, the waxy product of the reaction of a 1:1 mixture in diethyl ether and all the coammonolysis products prepared in THF formed gels (i.e., became insoluble) after 3–4 weeks at room temperature, even when stored in a nitrogen-filled dry box.

The pyrolysis of the coammonolysis products was studied. The ammonolysis product of the 6:1 mixture would be the least cross-linked, because it contains the least amount of trifunctional component. As expected, low ceramic yields were obtained on pyrolysis of these products. Pyrolysis of the products of the 3:1 mixture gave increased ceramic yields, whereas pyrolysis of the most highly cross-linked ammonolysis products of the 1:1 mixture gave good ceramic yields of 72% for the product prepared in diethyl ether and 78% for that prepared in THF.

The ammonolysis products were submitted to the KH-catalyzed DHCD reaction in order to obtain more highly cross-linked products that would give higher ceramic yields on pyrolysis. In all cases, the standard procedure (*15*) was used (1% KH in THF and followed by quenching with CH_3I or a chlorosilane). In every reaction, the product was a white solid that was produced in virtually quantitative yield. As expected, the ^1H NMR spectra of these products showed an increase in the $SiCH_3/(SiH + NH)$ proton ratio, whereas the relative SiH/NH ratio was unchanged. In all cases, the molecular weights of the solid products were at least double that of the starting ammonolysis product; therefore, the desired polymerization had occurred. Although the increase in molecular weight in these DHCD reactions is not great, any increase is useful for further processing. The reactions bring the advantage that the oils are converted to more easily handled solids.

Pyrolysis of the white solids obtained in these KH-catalyzed DHCD reactions (under argon from 50 to 950 °C) produced black ceramic residues, with the exception of the solid derived from the ammonolysis of the 1:1 mixture in THF, which left a brown residue. The ceramic yields were excellent (all greater than or equal to 82%, with the highest being 88%).

Analysis of bulk samples of the ceramic materials produced in the pyrolysis of the various KH-catalyzed DHCD products showed that the goal of a higher formal Si_3N_4/SiC ratio had been achieved. For the polymers derived from the ammonolysis of a 1:1 mixture, the ceramic compositions were 86% Si_3N_4, 9% SiC, and 5% C (THF ammonolysis) and 83% Si_3N_4, 11% SiC, and 6% C (diethyl ether ammonolysis). For the polymers derived from the ammonolysis of 3:1 and 6:1 mixtures, the ceramic compositions were 77% Si_3N_4, 18–19% SiC, and 4–5% C (diethyl ether ammonolysis) and 74% Si_3N_4, 20% SiC, and 5–6% C (THF ammonolysis).

These polymers may be used in the preparation of high-purity silicon nitride if the pyrolysis is carried out in a steam of ammonia (a reactive gas) rather than under nitrogen or argon. Thus pyrolysis to 1000 °C of the DHCD product of the ammonolysis product (THF) of the 1:1 mixture in a stream of ammonia gave a white ceramic residue in high yield, which contained only 0.29% C, with the remainder being silicon nitride.

Other commercially available $RSiCl_3$ compounds are CH_3SiCl_3 (a cheap byproduct of the direct process) and $CH_2{=}CHSiCl_3$ (also an inexpensive starting material), and both were included in this study. In both cases, the ammonolysis products of 6:1, 3:1, and 1:1 mixtures of CH_3SiHCl_2 and $RSiCl_3$ were prepared and submitted to the DHCD procedure. The ceramic yields obtained on pyrolysis of the resulting polymers were high. For experiments with mixtures of CH_3SiHCl_2 and CH_3SiCl_3, yields were 78–86%. In all cases, a black ceramic residue resulted when pyrolysis to 1000 °C was carried out in a stream of argon. As expected, the carbon content (12–18% SiC and up to 9.5% free C) was higher than that of the ceramic derived from a mixture of CH_3SiHCl_2 and $HSiCl_3$. Nonetheless, higher Si_3N_4 contents (76–80%) compared with those obtained when CH_3SiHCl_2 is used alone (~67%) were obtained. To produce a ceramic material containing only Si_3N_4, the solid polysilazane derived from DHCD of the oil obtained by ammonolysis of a 6:1 mixture of CH_3SiHCl_2 and CH_3SiCl_3 was pyrolyzed to 1000 °C in a stream of ammonia. A white ceramic residue containing only 0.36% C by weight resulted.

Compared with the DHCD of the ammonolysis products of mixtures of CH_3SiHCl_2 and CH_3SiCl_3, DHCD of the ammonolysis products of 6:1, 3:1, and 1:1 mixtures of CH_3SiHCl_2 and $CH_2{=}CHSiCl_3$ gave white solids whose pyrolysis resulted in increased carbon content (9–13% SiC and 12–18% C by weight) and decreased Si_3N_4 content (69–73%).

Another approach to mixed systems of the type just discussed is the copolymerization of oligomeric cyclopolysilazanes and of alkylaminosilanes with the CH_3SiHCl_2 ammonolysis product, $(CH_3SiHNH)_n$, followed by DHCD. This approach is based on the ring-opening polymerization of cyclic oligosilazanes of Krüger and Rochow (*17*). In this process, the cyclic ammonolysis products (in which n is mainly 3 or 4) were heated with a catalytic amount of an ammonium halide, preferably NH_4Br or NH_4I, at temperatures of 160 °C or greater and at atmospheric pressure for 6–8 h. During the reaction, gaseous NH_3 was evolved, and an oily polymer that contained silicon–halogen bonds was formed. The Si–X functions were removed by treatment of the polymer with NH_3 in diethyl ether solution. The final products were colorless, waxy polysilazanes of molecular weight greater than 10,000. The structure of these polymers could not be determined with certainty, but linear and cyclic structural components were suggested (*17*). The polymers obtained were soluble in organic solvents such as benzene, petroleum ether, or carbon tetrachloride; they melted in the range 90–140 °C, and they were partly crystalline. The attempted NH_4Br-catalyzed copolymerization of a 4:1 mixture of $[(CH_3)_2SiNH]_3$ and $[(CH_3)(CH_2{=}CH)SiNH]_4$ as previously described (*17*) gave an insoluble, colorless rubber.

Primary alkylaminosilanes and -silazanes also have been reported to undergo a deamination (equation 4) and ring equilibration (equation 5) when heated with catalytic amounts of ammonium salts (*20*).

$$x\mathrm{RR'Si(NHR'')_2} \xrightarrow{\mathrm{NH_4X}} (\mathrm{RR'SiNR''})_x + x\mathrm{RNH_2} \quad (x = 3 \text{ or } 4;\ \mathrm{R''} \neq \mathrm{H}) \tag{4}$$

$$4(\mathrm{RR'NHR''})_3 \xrightarrow{\mathrm{NH_4X}} 3(\mathrm{RR'SiNR''})_4 \tag{5}$$

Such ammonium-salt-catalyzed processes very likely proceed via reactive silylammonium intermediates that may be useful in other silazane conversion processes. To obtain less highly functionalized polysilazanes suitable for melt spinning, such ammonium-salt-catalyzed processes were examined.

In the general procedure, the mixture of $(CH_3SiHNH)_n$ and the silylamino or silazane compound is heated under an inert atmosphere (dry nitrogen or argon) in the presence of ~2 wt % of an ammonium salt at temperatures of ~155–170 °C for 4.5–5 h. During the reaction, ammonia, and in those cases where a Si–NHR compound is the reaction partner, an amine (RNH_2), is evolved. The preferred ammonium salt is $(NH_4)_2SO_4$, but others, such as NH_4Cl or NH_4Br, may be used. $(NH_4)_2SO_4$ is preferred because it is easily removed by filtration when, upon completion of the deamination reaction, the reaction mixture is diluted with THF.

The coreactants used in this study were cyclo-$(CH_3SiHNCH_3)_n$ and cyclo-$[CH_3(CH_2{=}CH)SiNH]_n$, $[CH_2{=}CHSi(NH)_{1.5}]_n$, $CH_3(CH_2{=}CH)Si(NHCH_3)_2$, and cyclo-$[HSi(NHCH_3)NCH_3]_n$. The cross-linking products of ammonium-salt-catalyzed deamination were viscous liquids, in the case of the diorganosilane derivatives, and resinous waxes, in the case of the monoorganosilane derivatives.

In an alternative procedure, instead of the separate ammonolysis of CH_3SiHCl_2 and $CH_3(R)SiCl_2$, an appropriate mixture of the two was treated with ammonia. The coammonolysis product then was heated with $(NH_4)_2SO_4$ at 170–185 °C to give a soluble, cross-linked, partially deaminated product. The coammonolysis product was a slightly viscous, colorless oil; the deaminated product (T = 170–175 °C) was a light yellow, viscous liquid.

The products of the $(NH_4)_2SO_4$-catalyzed reactions of $(CH_3SiHNH)_n$ with other silazanes and aminosilanes were subjected to the KH-catalyzed DHCD reaction (*15*). This reaction converted the viscous liquids and waxes to soluble solids, as is generally the case when this procedure is applied to CH_3SiHCl_2-derived polysilazane systems.

Each step of this sequence of $(NH_4)_2SO_4$-catalyzed deaminative cross-linking and KH-catalyzed DHCD increases the ceramic yield obtained on pyrolysis of the reaction product. For instance, the ceramic yield obtained from a 4:1 (by weight) blend of $(CH_3SiHNH)_n$ and $[CH_3(CH_2{=}CH)SiNH]_n$ on pyrolysis to 1000 °C under argon was 49%. This mixture, after it had been heated with 1.5 wt % $(NH_4)_2SO_4$ at 165 °C for 4.5 h, was converted to a viscous liquid whose pyrolysis (under the same conditions) gave a ceramic yield of 69%. When this material in turn was treated with 1.3 wt % of KH in THF at room temperature (with a CH_3I quench), a solid product was obtained whose pyrolysis to 1000 °C gave a ceramic yield of 83%. Not only do the final solid DHCD products give a higher ceramic yield on pyrolysis, but they are less reactive toward atmospheric moisture and they are more readily handled than the ammonium salt products. The increase in ceramic yield on going from the ammonium salt product to the DHCD product was variable and depended on the other component used (in addition to $[CH_3SiHNH]_n$) and the relative amounts of each used. In general, the greater the percentage of the other product in the mixture, the less was this increase in ceramic yield.

The nature of the ammonium-salt-catalyzed deamination and cross-linking of silazanes and aminosilanes is not well understood. Experiments indicate clearly that NH_3 is evolved when $[CH_3(R)SiNH]_n$-type silazanes participate in the reaction and that CH_3NH_2 is evolved when methylaminosilanes or -silazanes are used. Such eliminations would lead to some cross-linking through conversion of Si_2N units to Si_3N, as shown in equation 6.

$$3\;\overset{|}{\underset{|}{-\mathrm{Si}}}\mathrm{NH}\overset{|}{\underset{|}{\mathrm{Si}}}- \rightarrow 2(-\overset{|}{\underset{|}{\mathrm{Si}}})_3\mathrm{N} + \mathrm{NH}_3 \qquad (6)$$

Thus the function of the $(NH_4)_2SO_4$-catalyzed reaction in the present case is to copolymerize the two reactants through such a deaminative cross-linking process. The product is not equivalent to the material obtained simply by coammonolysis. That is, the processes shown in Schemes I and II are not equivalent. The process in Scheme II gives a partially cross-linked product. In principle, a product with the same general constitution as that of Scheme II should be obtained by treating the product of Scheme I with a catalytic amount $(NH_4)_2SO_4$ at 170 °C. However, the detailed structure of the two products are not expected to be the same.

A major objective of this work was the development of melt-spinnable polysilazanes. Such materials were provided by the $CH_3SiH/CH_3(CH_2{=}$

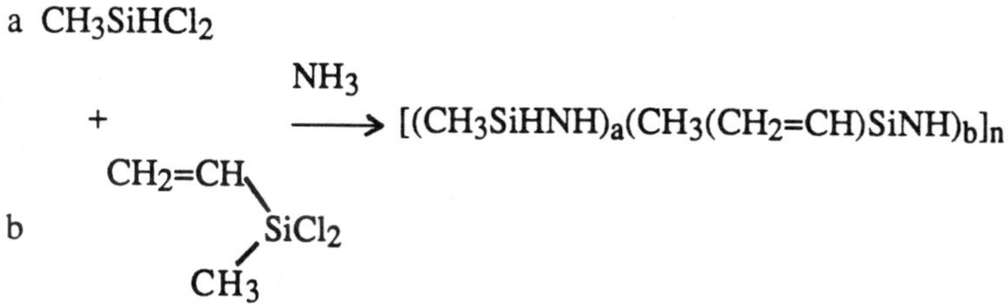

$$a\ CH_3SiHCl_2 + b\ CH_2{=}CH(CH_3)SiCl_2 \xrightarrow{NH_3} [(CH_3SiHNH)_a(CH_3(CH_2{=}CH)SiNH)_b]_n$$

Scheme I

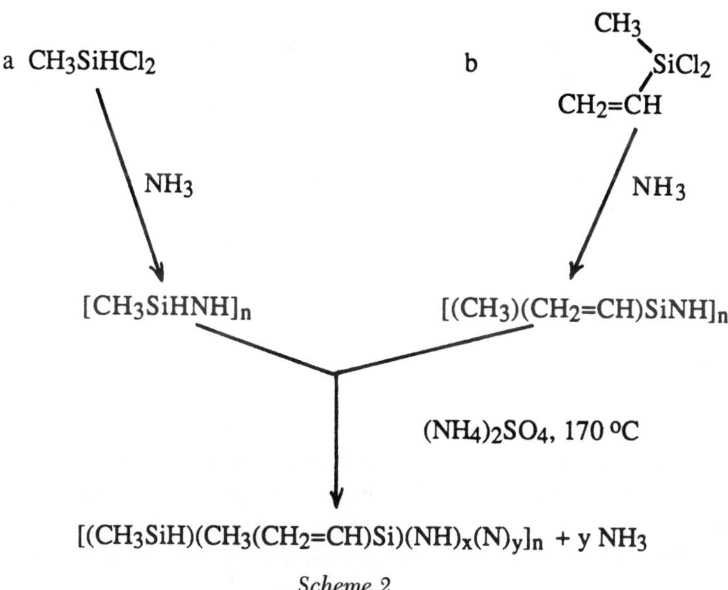

Scheme 2

CH)Si system, with the experimental approach either of Scheme I (followed by $[NH_4]_2SO_4$-catalyzed deamination and cross-linking and KH-catalyzed cross-linking) or of Scheme II (followed by KH-catalyzed cross-linking) being used. With Scheme II, 30 wt % of either $[CH_3(CH_2{=}CH)SiNH]_n$ or $CH_3(CH_2{=}CH)Si(NHCH_3)_2$ was used with the CH_3SiHCl_2 ammonolysis product, $(CH_3SiHNH)_n$. Treatment of the product with 1.6 wt % of KH in THF at room temperature gave a white solid that had a softening temperature (determined by TMA) of 79–96 °C and that gave a 73% ceramic yield on pyrolysis to 1000 °C under argon. With Scheme I, 1 mol of CH_3SiHCl_2 and 0.3 mol of $CH_3(CH_2{=}CH)SiCl_2$ gave a coammonolysis product that was treated first with 2 wt % of $(NH_4)_2SO_4$ at 170–185 °C for 4–5 h and then with 1.5 wt % of KH in THF at room temperature for 2.5 h to give as final product a white solid with a TMA softening temperature of 49–68 °C and ceramic yield after pyrolysis to 1000 °C under argon of 71%. On being heated, both of these solid products melted to a transparent liquid at 80–100 °C.

Decomposition or evolution of volatile products does not appear to occur at temperatures below 160 °C. Consistent with this observation is the TGA of these materials, which shows weight loss starting at about 150 °C. The final product of the approach using Scheme I was heated to 165 °C for 3 h under nitrogen. After the melt had cooled to room temperature, the resulting glassy solid was still soluble in hexane, a fact indicating that extensive thermal cross-linking had not taken place. These products are good candidates for melt spinning.

Fibers could be drawn from concentrated syrups of these products in toluene. The drawn fibers could not directly be converted to ceramic fibers, because they melted on being heated, and a cure step was required to render them infusible. The required cure could be accomplished by UV irradiation of the fibers. Following this treatment, pyrolysis of the fibers in a stream of argon gave black ceramic fibers, and pyrolysis in a stream of ammonia gave white ceramic fibers.

On pyrolysis under argon to 1000 °C, the polysilazanes prepared from the systems just discussed gave black ceramic products whose elemental analysis could be explained in terms of the following formal compositions (in weight percent): 62–65% Si_3N_4, 15–20% SiC, and 14–16% free carbon. In general, the white materials obtained on pyrolysis in a stream of ammonia contained less than 1% carbon.

Silicon Carbide Precursors. Our research on SiC precursors began with an examination of a potential starting material in which the C/Si ratio was 1, the ratio desired in the derived ceramic product. Available methylsilicon compounds with a 1:1 C/Si stoichiometry are CH_3SiCl_3 and CH_3SiHCl_2. In principle, CH_3SiHCl_2 could give $(CH_3SiH)_n$ cyclic oligomers and linear polymers on reaction with an alkali metal. In practice, the Si–H linkages are also reactive toward alkali metals. Thus, mixed organochlorosilane systems containing some CH_3SiHCl_2 have been treated with metallic potassium by Schilling and Williams (*21*). The CH_3SiHCl_2-based contribution to the final product was $(CH_3SiH)_{0.2}(CH_3Si)_{0.8}$, that is, about 80% of the available Si–H bonds had reacted. Such reactions of Si–H bonds lead to cross-linking in the product or to formation of polycyclic species if cyclic products are preferred. Nevertheless, we have used this known reaction of CH_3SiHCl_2 with an alkali metal as a route to new preceramic polymers (*22*).

When the reaction of CH_3SiHCl_2 with sodium pieces was carried out in THF, a white solid was isolated in 48% yield. This solid was poorly soluble in hexane, somewhat soluble in benzene, and quite soluble in THF. Its ^1H NMR spectrum indicated that extensive reaction of Si–H bonds had occurred. The $\delta(SiH)/\delta(SiCH_3)$ integration led to the constitution $[(CH_3SiH)_{0.4}(CH_3Si)_{0.6}]_n$. In this solid, the CH_3SiH units are ring and chain members that are not branching sites, and the CH_3Si units are ring and chain members that are branching sites. In our reactions, mixtures of polycyclic and linear (possibly cross-linked) polysilanes were expected to be formed. (Attempts to distill out pure compounds from our preparations were not successful. Less than 10% of the product was volatile at higher temperatures at 0.0001 mm of Hg [0.013 Pa].) The ceramic yield obtained when the $[(CH_3SiH)_{0.4}(CH_3Si)_{0.6}]_n$ polymer was pyrolyzed (TGA to 1000 °C) was 60%; a gray-black solid was obtained whose analysis indicated a composition of 1.0 SiC + 0.49 Si.

The reaction of methyldichlorosilane with sodium in a solvent system composed of six parts of hexane and one part of THF gave a colorless oil in 75 to over 80% yield, which was soluble in many organic solvents. In various experiments the average molecular weight (determined by cryoscopy in benzene) was 520–740 and the constitution (determined by ^1H NMR) ranged from $[(CH_3SiH)_{0.76}(CH_3Si)_{0.24}]_n$ to $[(CH_3SiH)_{0.9}(CH_3Si)_{0.1}]_n$. Compared with the product obtained in THF alone, the less cross-linked material gave much lower yields of ceramic product on pyrolysis to 1000 °C (TGA yields of 12–27% in various runs). Again, analysis showed that the product was a mixture of SiC and elemental silicon, with 1.0 SiC + 0.42 Si being a typical composition. These results are not especially promising, and further chemical modification of the $[(CH_3SiH)_x(CH_3Si)_y]_n$ products obtained in the CH_3SiHCl_2–Na reactions was required.

A number of approaches that we tried did not succeed, but during the course of our studies, we found that treatment of the $[(CH_3SiH)_x(CH_3Si)_y]_n$ products with alkali metal amides in catalytic quantities converts them to materials of higher molecular weight whose pyrolysis gives significantly higher ceramic yields. Thus, in one example, to 0.05 mol of liquid $[(CH_3SiH)_{0.85}(CH_3Si)_{0.15}]$ in THF under nitrogen was added a solution of about 1.25 mmol (2.5 mol %) of $[(CH_3)_3Si]_2NK$ in THF. The resulting red solution was treated with methyl iodide. Subsequent nonhydrolytic workup gave a soluble white powder with a molecular weight of 1000 whose pyrolysis to 1000 °C gave a ceramic yield of 63%.

The ^1H NMR spectrum of this product showed only broad resonances in the Si–H and Si–CH_3 regions. In the starting $[(CH_3SiH)_x(CH_3Si)_y]_n$ material, the observed ^1H-NMR integration ratios, $SiCH_3/SiH$, ranged from 3.27 to 3.74. This ratio for the product of the silylamide-catalyzed process ranged from 8.8 to 14. Both Si–H and Si–Si bonds are reactive toward nucleophilic reagents. In the case of the alkali metal silylamides, extensive structural reorganization involving both Si–H and Si–Si bonds of the $[(CH_3SiH)_x(CH_3Si)_y]_n$ polysilane that results in further cross-linking via Si–Si and SiH–Si redistribution or disproportionation must have taken place. The following initial processes can be envisioned:

$$(R_3Si)_2NK + -\overset{|}{\underset{|}{Si}}H \rightarrow -\overset{|}{\underset{|}{Si}}N(SiR_3)_2 + KH \qquad (7)$$

$$(R_3Si)_2NK + -\overset{|}{\underset{|}{Si}}H- \rightarrow (R_3Si)_2NH + -\overset{|}{\underset{|}{Si}}K \qquad (8)$$

$$(R_3Si)_2NK + -\overset{|}{\underset{|}{Si}}-\overset{|}{\underset{|}{Si}}- \rightarrow (R_3Si)_2N\overset{|}{\underset{|}{Si}}- + -\overset{|}{\underset{|}{Si}}K \qquad (9)$$

In each process, a new reactive nucleophile is generated: KH in equation 7 and a silyl alkali metal function in equations 8 and 9. These nucleophiles also could attack the $[(CH_3SiH)_x(CH_3Si)_y]_n$ system, and during these reactions, some of the oligomeric species that make up the starting material would be linked together to give products of higher molecular weight. Other processes are possible as well, for example, the silylene process as shown in equation 10.

$$(R_3Si)_2NK + H-\overset{|}{\underset{|}{Si}}-\overset{|}{\underset{|}{Si}}- \rightarrow (R_3Si)_2N\overset{|}{\underset{|}{Si}}- + H-\overset{|}{\underset{|}{Si}}-K \rightarrow \overset{\diagdown}{\underset{\diagup}{Si}} + KH \qquad (10)$$

Thus not only anionic species but neutral silylenes also could be involved as intermediates. In any case, extensive loss of Si–H takes place during the catalyzed process, a fact suggesting that the reaction is more than a simple redistribution. Further studies relating to the mechanism of the process must be carried out.

Although these potassium-silylamide-catalyzed reactions provided a good way to solve the problem of low ceramic yield in the pyrolysis of $[(CH_3SiH)_x(CH_3Si)_y]_n$, the problem of the elemental composition of the ceramic product remained (i.e., the problem of Si/C ratios > 1) because only catalytic quantities of the silylamide were used.

As noted previously, KH-catalyzed polymerization of the CH_3SiHCl_2 ammonolysis product gives a living polymeric potassium silylamide of the type $[(CH_3SiHNH)_a(CH_3SiN)_b(CH_3SiHNK)_c]_n$. In a typical example, $a = 0.39$, $b = 0.57$, and $c = 0.04$, and only a low concentration of potassium silylamide functions is present in the polymer. This polymeric silylamide reacts with electrophiles other than methyl iodide, for example, with diverse chlorosilanes, and it has been isolated and analyzed. Because the polymer is an alkali metal silylamide, it was expected to react also with $[(CH_3SiH)_x(CH_3Si)_y]_n$ polysilane-type materials. Not only would it be expected to convert the latter into material of higher molecular weight, but it also would be expected to improve the Si/C ratio (i.e., bring it closer to 1). As previously noted, pyrolysis of $[(CH_3SiHNH)_a(CH_3SiN)_b(CH_3SiHNCH_3)_c]_n$ gives a ceramic product in 80–85% yield containing (formally) Si_3N_4, SiC, and excess carbon. Thus, combinations of the two species, $[(CH_3SiH)_x(CH_3Si)_y]_n$ and $[(CH_3SiHNH)_a(CH_3SiN)_b(CH_3SiHNK)_c]_n$, and pyrolysis of the product (which we will call a graft polymer) after a CH_3I quench could, in principle, lead to a ceramic product in which the excess Si obtained in pyrolysis of the former and the excess C obtained in the pyrolysis of the latter combine to give SiC. Accordingly, experiments were carried out in which the two polymer systems, $[(CH_3SiH)_x(CH_3Si)_y]_n$ (that prepared in THF alone and that prepared in hexane–THF), and the living polymeric silylamide, $[(CH_3SiHNH)_a(CH_3SiN)_b(CH_3SiHNK)_c]_n$, were mixed in THF solution in various proportions (2.4:1 to 1:2 molar ratios) and allowed to react at room temperature for 1 h and at reflux for 1 h. After the reaction mixture was quenched with methyl iodide, nonhydrolytic workup gave a new polymer in nearly quantitative yield (based on the weight of material charged).

The molecular weight of these products was in the 1800–2500 range. Their pyrolysis under nitrogen gave ceramic products in 74–83% yield. Thus the reaction of the two polymer systems gives a new polymer that seems to be an excellent new preceramic polymer in terms of ceramic yield. The formal composition of the ceramic materials obtained is in the range 1 Si_3N_4 + 3.3–6.6 SiC + 0.74–0.85 C. Thus, as expected, the ceramic materials are rich in silicon carbide, and the excess Si that is obtained in the pyrolysis of the $[(CH_3SiH)_x(CH_3Si)_y]_n$ materials alone is not present. Therefore, the objective has been achieved. By proper adjustment of the ratios of starting material, the excess carbon content can be minimized (23).

In an alternative synthesis of $[(CH_3SiH)_x(CH_3Si)_y][(CH_3SiHNH)_a(CH_3SiN)_b]$ "combined" polymers, the poly(silylamide) was generated in situ in the presence of $[(CH_3SiH)_x(CH_3Si)_y]_n$. However, this method gave materials that were somewhat different. In one such experiment, a mixture of cyclo-$(CH_3SiHNH)_n$ oligomers (obtained by the ammonolysis of CH_3SiHCl_2 in THF) and the $[(CH_3SiH)_{0.76}(CH_3Si)_{0.26}]_n$ material in THF was treated with a catalytic amount of KH. After the reaction mixture had been treated with methyl iodide, the usual workup gave an 89% yield of hexane-soluble white powder with a molecular weight of ~2750. On pyrolysis, this material gave a 73% yield of a black ceramic.

The combined polymer prepared in this way ("in situ polymer") was in some ways different from the combined polymer prepared by the first method ("graft" polymer). Principal differences were observed in their ^1H NMR spectra and in the form of their TGA curves. These differences suggest that the two differently prepared polymers have different structures. Perhaps in the in situ preparation intermediates formed by the action of KH on the cyclo-$(CH_3SiHNH)_n$ oligomers were intercepted by reaction with the $[(CH_3SiH)_x(CH_3Si)_y]_n$ also present before the $[(CH_3SiHNH)_a(CH_3SiH)_b(CH_3SiHNK)_c]_n$ polymer (which is the starting reactant used in the graft procedure) has a chance to be formed to the extent of its usual molecular weight. Thus, fewer of the original CH_3SiHNH protons are lost or more of those of the $[(CH_3SiH)_x(CH_3Si)_y]_n$ system are reacted.

The TGA curves of the graft polymer are different as well. The graft polymer showed a small weight loss between 100 and 200 °C, which began at around 100 °C. This small initial weight loss occurred only at higher temperatures (beginning at ~175 °C) in the case of the in situ polymer. This difference in initial thermal stability could well have chemical consequences of importance with respect to ceramics, and both kinds of polymers may be useful as preceramic materials.

Physical blends of $[(CH_3SiH)_x(CH_3Si)_y]_n$ (solid polymer; THF preparation) and $[(CH_3SiHNH)_a(CH_3SiN)_b(CH_3SiHNCH_3)_c]_n$ also were examined. When about equimolar quantities of each polymer were mixed and finely ground together, pyrolysis to 1000 °C gave a 70% ceramic yield. A reaction between the two polymers probably already occurs at lower temperatures. When such mixtures were heated either in the absence of a solvent at 100 °C under nitrogen or in toluene solution at reflux, white powders were obtained, which were insoluble in hexane, benzene, and THF. The ceramic yields were 67% and 75%, respectively, for the polymers prepared in the absence of solvent and in toluene solution.

Further experiments showed that the combined polymers may be converted to black ceramic fibers. Pyrolysis of pressed bars of the combined polymer to 1000 °C gave a black product of irregular shape (with 75–76% ceramic yield). In other experiments, SiC powder was dispersed in toluene containing 20% by weight of the combined polymer. The solution was evaporated, and the residue, a fine powder of SiC with the combined polymer binder, was pressed into bars and pyrolyzed to 1000 °C. A slightly shrunken ceramic bar (6% weight loss) was obtained.

The living polymer intermediate in our polysilazane synthesis is useful in upgrading other Si–H-containing polymers for application as precursors for ceramic materials (*24*). An example of this application is the upgrading of poly(methyl hydrogen siloxane), $[CH_3Si(H)O]_n$, to give a useful precursor for silicon oxynitride, a ceramic that is more resistant to high-temperature oxidation than silicon nitride. On pyrolysis under argon to 1000 °C, such a linear polymer, with an average molecular weight of 2000–5000 (vendor data), left a black ceramic residue in only 13% yield. In this study, experiments were carried out with a $[CH_3Si(H)O]_m$ prepared using conditions under which the yield of the cyclic oligomers ($m = 4, 5, 6, \ldots$) is maximized (*25*), as well as with the commercial $[CH_3Si(H)O]_m$ polymer of higher molecular weight and with, presumably, high linear content.

In one approach, the polymeric potassium silylamide was prepared as described previously (from the product of CH_3SiHCl_2 ammonolysis in THF), and the $[CH_3Si(H)O]_m$ oligomers (high cyclic content) were added slowly to this living polymer solution. An immediate reaction with some gas evolution occurred. The resulting clear solution was treated with CH_3I to react with any remaining silylamide units. Silylamide/siloxane weight ratios of 1:1 and 1:5 were used. In both cases, the polymeric products were organic-soluble white solids of moderate average molecular weight (1700 and 2400, respectively), and pyrolysis under nitrogen or argon gave high char yields (78 and 76%, respectively). In this case, the cross-linking very likely occurs via the well-known base-catalyzed SiH–SiO redistribution. Pyrolysis to 1000 °C of a bulk sample of the 1:1 by weight polymer gave a black solid in 80% yield. Elemental analysis indicated a formal composition of 1 SiC + 0.84 Si_3N_4 + 2.17 SiO_2 + 2.0 C (this expression is not meant to reflect the composition in terms of chemical species present). On the other hand, pyrolysis under gaseous ammonia gave a white ceramic solid in 78.5% yield. Analysis confirmed that the ceramic product was a silicon oxynitride. The white solid contained less than 0.5% carbon.

In an alternative approach to this graft procedure, the in situ procedure was used. A mixture (~1:1 by weight) of the CH_3SiHCl_2 THF ammonolysis product, $[CH_3SiHNH]_m$, and the CH_3SiHCl_2 hydrolysis product, $[CH_3Si(H)O]_n$, in THF was added to a suspension of a catalytic amount of KH in THF. Hydrogen evolution was observed, and a clear solution resulted. After the reaction mixture was quenched with CH_3I, further workup gave the new polymer, a soluble white powder with an average molecular weight of 1670. Pyrolysis to 1000 °C (TGA) gave a black ceramic solid in 84% yield. Pyrolysis of a bulk sample under argon yielded a black ceramic (73% yield). Analysis indicated that the formal composition is 1 SiC + 1.03 Si_3N_4 + 1.8 SiO_2 + 2.63 C.

In place of the (mostly) cyclo-$(CH_3SiHO)_n$ oligomers, a commercial poly(methyl hydrogen siloxane) (Petrarch PS-122) of higher molecular weight and presumably consisting mostly of linear species may serve as the siloxane component. When a 1:1 by weight ratio of the preformed polymeric silylamide and the $[CH_3Si(H)O]_n$ polymer was used and the reaction mixture was quenched with CH_3I, the usual workup produced a soluble white solid with a molecular weight of 1540. Pyrolysis of a bulk sample yielded a black solid (73% yield) with a formal composition of 1 SiC + 1.5 Si_3N_4 + 3.15 SiO_2 + 3.6 C. Application of the in situ procedure to a 1:1 by weight ratio mixture of $(CH_3SiHNH)_n$ and $[CH_3Si(H)O]_n$ gave a soluble white powder with an average molecular weight of 1740; pyrolysis gave a black solid in 88% yield. The fact that the pyrolysis of these polymers under a stream of ammonia gives white solids, silicon oxynitrides that contain little if any, carbon, in high yield is of interest. The ceramic applications of the silicon oxynitride precursor systems have received further study by Yu and Ma (*26*).

The Nicalon polycarbosilane mentioned earlier, which was obtained by thermal rearrangement of poly(dimethylsilylene), is a polymeric silicon hydride, with $[(CH_3)(H)SiCH_2]$ as the main repeating unit (*9*). As such, it might also be expected to react with our $[(CH_3SiHNH)_a(CH_3SiN)_b(CH_3SiHNK)_c]_n$ living poly(silylamide). This was found to be the case (*27*). The commercially available polycarbosilane (sold in the United States by Dow Corning Corporation; our sample, a white solid, had a molecular weight of 1210 and gave a ceramic yield of 58% on pyrolysis) reacts with our poly(silylamide). A reaction carried out in THF solution initially at room temperature and then at reflux and followed by treatment of the reaction mixture with CH_3I gave after appropriate workup a nearly quantitative yield of a white solid that was very soluble in common organic solvents including hexane, benzene, and THF. When the polycarbosilane/poly(silylamide) ratio was ~1, pyrolysis of the product polymer gave a black ceramic solid in 84% yield, which analysis showed to have a formal composition of 1 SiC + 0.22 Si_3N_4 + 0.7 C. When the polycarbosilane/poly(silylamide) ratio was ~5, the ceramic yield was lower (67%).

In these experiments the cyclo-$(CH_3SiHNH)_n$ starting materials used to synthesize the poly(silylamide) had been prepared by CH_3SiHCl_2 ammonolysis in diethyl ether. When this preparation was carried out in THF, the final ceramic yields obtained by pyrolysis of the polycarbosilane–poly(silylamide) hybrid polymer were 88% (for a 1:1 reactant ratio) and 64% (for a 5:1 reactant ratio). The in situ procedure in which the CH_3SiHCl_2 ammonolysis product, cyclo-$(CH_3SiHNH)_n$, was treated with a catalytic quantity of KH in the presence of the polycarbosilane and followed by a CH_3I quench and the usual workup gave equally good results in terms of high final ceramic yields, whether the starting cyclo-$(CH_3SiHNH)_n$ was prepared in diethyl ether or THF.

A new polymer is clearly formed when the polycarbosilane and the polymeric silylamide are heated together in solution and then quenched with methyl iodide. ^1H NMR spectroscopy brought further evidence of such a reaction. A physical mixture of the polycarbosilane and the polysilazane, $[(CH_3SiHNH)_a(CH_3SiN)_b(CH_3SiHNCH_3)_c]_n$, also reacted when heated to 1000 °C and gave good yields of ceramic product. That appreciable reaction had occurred by 200 °C was shown in an experiment in which a 1:1 by weight mixture of the initially soluble polymers was converted by such thermal treatment to a white, foamy solid that was no longer soluble in organic solvents.

Finally, another procedure for upgrading the $[(CH_3SiH)_x(CH_3Si)_y]_n$ polysilanes is possible. With their many Si–H bonds, these polysilanes should be good candidates for cross-linking to a network polymer by means of hydrosilylation. Within that broad class of reactions, in terms of the polyolefinic compounds and the catalyst types that might be used (28), cyclo-$[(CH_3)(CH_2{=}CH)SiNH]_3$ was found to be an especially good polyolefin that reacted very effectively with these polysilanes when the addition reaction was initiated by azobis(isobutyronitrile), a radical catalyst (29). When the cross-linking reaction was carried out with an $SiH/SiCH{=}CH_2$ ratio of 6 or larger, the products were white solids that were all soluble in organic solvents such as hexane, benzene, and THF. The yields were quantitative. The products were of fairly low molecular weight, and in some cases, quite satisfactory ceramic yields (68 and 77%) were obtained when these materials were pyrolyzed to 1000 °C.

Pyrolysis of one of the products whose ^1H NMR spectrum and elemental analysis indicated a constitution of $[(CH_3SiH)_{0.73}(CH_3Si)_{0.1}[(CH_3)SiCH_2CH_2Si(CH_3)NH]_{1.17}]$ indicated a formal composition of 1 SiC + 0.033 Si_3N_4 + 0.04 C. Thus the stoichiometry and reaction conditions in this experiment gave a high yield of a ceramic product that was silicon carbide contaminated with only minor amounts of silicon nitride and free carbon. When such a pyrolysis was effected to 1500 °C, the ceramic product was at least partly crystalline. X-ray diffraction showed lines due to β-SiC only.

Epilogue

The design of useful (i.e., potentially commercializable) organosilicon preceramic polymers presents an interesting challenge to the synthetic chemist. Using examples from some of our research, I have attempted to convey some of the chemical considerations that are important if such a synthesis is to be successful. Our initial experiment in this area, the reaction of ammonia with dichlorosilane, was carried out by Christian Prud'homme in April 1981. This project then was taken over by a graduate student, Gary H. Wiseman, who, during the course of his Ph.D. research, developed the dehydrocyclodimerization route to useful preceramic polymers and widened our horizons into the fascinating area of ceramics. Another graduate student, Timothy G. Wood, studied the action of sodium on CH_3SiHCl_2 and carried out the first graft synthesis of the polysilazane–polysilane hybrid polymer as the last experiment of his thesis work. Further contributions to the development of this area were made by Joanne M. Schwark, a graduate student, and by several postdoctoral co-workers: Charles A. Poutasse, Yun Chi, Yoshio Inoguchi, Tom S. Targos, Gudrun Koppetsch, and especially Yuan-Fu Yu.

It was not the purpose of this chapter to give an exhaustive review of the preceramic polymer literature. The studies of some of the other workers in this area (LeGrow, Lipowitz, and others at Dow Corning; Yajima, Schilling, Arai, West, Sinclair and Brown-Wensley, and Harrod) have been cited in the discussion. Others currently active in the polysilazane area include King and co-workers at Union Carbide (*30*), Penn et al. (*31*) Arkles (*32*), Laine (*33*), Nakaido et al. (*34*), Lebrun and Porte (*35*), and Mazaev et al. (*36*). Finally, mention should be made of the pioneering work of Verbeek and Winter (*37*, *38*) on polysilazane $SiC–Si_3N_4$ precursors prepared by thermal condensation polymerization of the $CH_3SiCl_3–CH_3NH_2$ reaction product. These workers were also the first to report reactive gas pyrolysis, by using a stream of ammonia, of a polysilazane.

Research in the preceramic polymer area, if it is to come to fruition, must be an interdisciplinary endeavor. After the chemistry has been developed, there are important ceramics issues to be addressed. Good chemistry does not guarantee good ceramics! However, there is the possibility that the chemistry can still be modified to give good ceramics. Either the chemist must learn a good bit about ceramics characterization and processing or, better, he must work closely with the ceramist. I have been most fortunate to find very helpful colleagues in the ceramics area: Roy W. Rice, when he was still at the Naval Research Laboratory, and his co-workers; H. Kent Bowen and John S. Haggerty and their colleagues at the Ceramics Processing Research Laboratory at MIT; and John Semen of the Ethyl Corporation in Baton Rouge.

Acknowledgments

This work was carried out with the generous support of the Office of Naval Research, the Air Force Office of Scientific Research, the Rhoñe Poulenc Company, Union Carbide Corporation, Celanese Research Company, and AT&T Bell Laboratories. I acknowledge also with thanks and admiration the skilled and dedicated efforts of my co-workers mentioned in the Epilogue and the help and guidance we have received from our ceramist friends.

References

1. *Gmelin Handbook of Inorganic Chemistry,* 8th ed.; Springer-Verlag: Berlin, Federal Republic of Germany, Silicon, Supplement Volumes B2, 1984, and B3, 1986.
2. Messier, D. R.; Croft, W. J. In *Preparation and Properties of Solid-State Materials;* Wilcox, W. R., Ed.; Dekker: New York, 1982; Vol. 7, Chapter 2.
3. a. Wynne, K. J.; Rice, R. W. *Annu. Rev. Mater. Sci.* **1984,** *14,* 297.
   b. Rice, R. W. *Am. Ceram. Soc. Bull.* **1983,** *62,* 889.
4. Rice, R. W. *CHEMTECH* **1983,** 230.
5. Rees, W. S., Jr.; Seyferth, D. *J. Am. Ceram. Soc.* **1988,** *71,* C-194.
6. Seyferth, D.; Mignani, G. *J. Mater. Sci. Lett.* **1988,** *7,* 487.
7. Kato, A.; Mizumoto, H.; Fukushige, Y. *Ceram. Int.* **1984,** *10,* 37.
8. a. LeGrow, G. E.; Lim, T. F.; Lipowitz, J.; Reaoch, R. S. In *Better Ceramics Through Chemistry II;* Brinker, C. J.; Clark, D. E.; Ulrich, D. R., Eds.; Materials Research Society: Pittsburgh, PA, 1986; pp 553–558.
   b. LeGrow, G. E.; Lim, T. F.; Lipowitz, J.; Reaoch, R. S. *Am. Ceram. Soc. Bull.* **1987,** *66,* 363.
9. Yajima, S. *Am. Ceram. Soc. Bull.* **1983,** *62,* 893.
10. Schilling, C. L., Jr.; Wesson, J. P.; Williams, T. C. *Am. Ceram. Soc. Bull.* **1983,** *62,* 912.
11. Seyferth, D.; Wiseman, G. H.; Prud'homme, C. *J. Am. Ceram. Soc.* **1983,** *66,* C-13; U.S. Patent 4 397 828, 1983.
12. Stock, A.; Somieski, K. *Ber. Dtsch. Chem. Ges.* **1921,** *54,* 740.
13. Sharp, K. G.; Arvidson, A.; Elvey, T. C. *J. Electrochem. Soc.* **1982,** *129,* 2346.
14. a. Arai, M.; Sakurada, S.; Isoda, T.; Tomizawa, T. *Polym. Prepr. (Am. Chem. Soc., Div. Polym. Chem.)* **1987,** *28,* 407.
    b. Toa Nenryo Kogyo, K. K. *Jpn. Kokai Tokkyo* JP 59, 207, 812 (84, 207, 812), Nov. 26, 1984; *Chem. Abstr.* **1982,** *102,* 206067k.
    c. Arai, M.; Funayama, T.; Nishi, I.; Isoda, T. *Jpn. Kokai Tokkyo* JP 62, 125, 015 (87, 125, 015), June 6, 1987; *Chem. Abstr.* **1987,** *107,* 160180q.
15. a. Seyferth, D.; Wiseman, G. H. *J. Am. Ceram. Soc.* **1984,** *67,* C-132; U.S. Patent 4 482 669, 1984.
    b. Seyferth, D.; Wiseman, G. H. In *Ultrastructure Processing of Ceramics, Glasses and Composites 2;* Hench, L. L.; Ulrich, D. R., Eds.; Wiley: New York, 1986; Chapter 38.

16. Brewer, S. D.; Haber, C. P. *J. Am. Chem. Soc.* **1948,** *70,* 3888.
17. a. Krüger, C. R.; Rochow, E. G. *J. Polym. Sci. A* **1964,** *2,* 3179.
b. Krüger, C. R.; Rochow, E. G. *Angew. Chem. Int. Ed. Engl.* **1962,** *1,* 458.
c. Rochow, E. G. *Monatsh. Chem.* **1964,** *95,* 750.
18. Monsanto Company, Neth. Appl. 6 507 996, 1965; *Chem. Abstr.* **1966,** *64,* 19677d.
19. Seyferth, D.; Schwark, J. M. U.S. Patent 4 720 532, 1988.
20. Abel, E. W.; Bush, R. P. *J. Inorg. Nucl. Chem.* **1964,** *2,* 1685.
21. a. Schilling, C. L., Jr.; Williams, T. C., Report 1983, TR-83-1, Order No. ADA141546; *Chem. Abstr. 101,* 196820p.
b. U.S. Patent 4 472 591, 1984.
22. Wood, T. G. Ph.D. Dissertation, Massachusetts Institute of Technology, 1984. Very much the same study, with the same results, was carried out by Sinclair and Brown-Wensley at 3M prior to our investigation. This work came to our attention when the U.S. patent was issued (Brown-Wensley, K. A.; Sinclair, R. A. U.S. Patent 4 537 942, 1985). West (*39*) and Harrod (*40*) and their respective co-workers also have prepared polysilanes containing the Si–H functionality.
23. This "combined-polymer approach" has been patented. Seyferth, D.; Wood, T. G.; Yu, Y.-F. U.S. Patent 4 645 807, 1987.
24. Seyferth, D.; Yu, Y.-F.; Targos, T. S. U.S. Patent 4 705 837, 1987.
25. Seyferth, D.; Prud'homme, C.; Wiseman, G. H. *Inorg. Chem.* **1983,** *22,* 2163.
26. Yu, Y.-F.; Ma, T. I. In *Better Ceramics Through Chemistry II;* Brinker, C. J.; Clark, D, E.; Ulrich, D. R., Eds.; Materials Research Society: Pittsburgh, PA, 1986; pp 559–564.
27. Seyferth, D.; Yu, Y.-F. U.S. Patent 4 650 837, 1987.
28. a. Eaborn, C.; Bott, R. W. In *Organometallic Compounds of the Group IV Elements;* MacDiarmid, A. G., Ed.; Dekker: New York, 1968; Vol. 1, pp 213–278.
b. Lukevics, E.; Belyakova, Z. V.; Pomerantseva, M. G.; Voronkov, M. G. *J. Organomet. Chem. Libr.* **1977,** *5,* 1.
29. a. Seyferth, D.; Yu, Y.-F. In *Design of New Materials;* Cooke, D. L.; Clearfield, A., Eds.; Plenum: New York, 1987; pp 79–93.
b. Seyferth, D.; Yu, Y.-F. U.S. Patent 4 639 501, 1987.
c. Seyferth, D.; Yu, Y.-F.; Koppetsch, G. E. U.S. Patent 4 719 273, 1988.
30. King, R. E., III; Kanner, B.; Hopper, S. P.; Schilling, C. L. U.S. Patent 4 675 424, 1987.
31. Penn, B. G.; Daniels, J. G.; Ledbetter, F. E., III; Clemons, J. M. *Polym. Eng. Sci.* **1986,** *26,* 1191.
32. Arkles, B. *J. Electrochem. Soc.* **1986,** *133,* 233.
33. Laine, R. M.; Blum, Y. D.; Tse, D.; Glaser, R. In *Inorganic and Organometallic Polymers;* Zeldin, M.; Wynne, K. J.; Allcock, H. R., Eds.; ACS Symposium Series 360; American Chemical Society: Washington, DC, 1988; pp 124–142.
34. Nakaido, Y.; Otani, Y.; Kozakai, N.; Otani, S. *Chem. Lett.* **1987,** 705.
35. Lebrun, J.-J.; Porte, H. U.S. Patents 4 689 382, 1987; and 4 694 060, 1987.
36. Mazaev, V. A.; Kopylov, V. M.; Shkol'nik, M. I.; Miklin, L. S.; Domashenko, T. M.; Semenova, E. A.; Markova, N. V. *Izv. Akad. Nauk SSSR, Neorg. Mater.* **1987,** *23,* 581.
37. Verbeek, W. U.S. Patent 3 853 567, 1974.
38. Winter, G.; Verbeek, W. U.S. Patent 3 892 583, 1975.
39. West, R. *J. Organomet. Chem.* **1986,** *300,* 327.
40. Aitken, C.; Harrod, J. F.; Samuel, E. *J. Organomet. Chem.* **1985,** *279,* C11.

RECEIVED for review May 27, 1988. ACCEPTED revised manuscript October 24, 1988.

32

Polymeric Routes to Silicon Carbide and Silicon Nitride Fibers

William H. Atwell

Dow Corning Corporation, Midland, MI 48686–0995

The chapter gives a brief review of the preparation and use of silicon-containing preceramic polymers to prepare ceramic fibers. Several key issues are discussed that must be addressed if this area of technology is to continue to advance.

THE USE OF CHEMICAL APPROACHES to improve the processing, properties, and performance of advanced ceramic materials is a rapidly growing area of research and development. One approach involves the preparation of organometallic polymer precursors and their controlled pyrolysis to ceramic materials. This chapter will review the preparation and application of silicon-, carbon-, and nitrogen-containing polymer systems. However, the discussion is not exhaustive; the focus is on systems with historical significance or that demonstrate key technological advances.

Silicon-containing preceramic polymers are useful precursors for the preparation of ceramic powders and fibers and for ceramic binder applications (*1*). Ceramic fibers are increasingly important for the reinforcement of ceramic, plastic, and metal matrix composites (*2*, *3*). This chapter will emphasize those polymer systems that have been used to prepare ceramic fibers. An overview of polymer and fiber processing, as well as polymer and fiber characterization, will be described to illustrate the current status of this field. Finally, some key issues will be presented that must be addressed if this area is to continue to advance.

0065–2393/90/0224–0593$06.00/0

Routes to Silicon Carbide Fibers

The earliest work on silicon-carbide-related fibers was by Verbeek and Winter (*4*). Using the principles developed earlier by Fritz and co-workers (*5*), Verbeek and Winter (*4*) reported that the high-temperature pyrolysis of tetramethylsilane or methylchlorosilanes gives branched polycarbosilane (PCS) polymers containing a structure with alternating silicon and carbon atoms (equation 1).

$$(CH_3)_4Si \xrightarrow{\Delta} \text{polycarbosilane (PCS)} \xrightarrow[\text{pyrolysis}]{\text{spin aid}} \text{SiC fibers} \qquad (1)$$

Spinning of PCS, often with the use of an organic-polymer spinning aid such as poly(ethylene oxide), followed by curing and high-temperature pyrolysis gives black silicon-carbide-like fibers.

However, interest in this technology was ignited by the extensive and pioneering work of Yajima and co-workers (*6, 7*). In an early work, Yajima and Hayashi (*6*) applied the known Kumada rearrangement to Burkhard's poly(dimethylsilane) polymer and obtained melt-spinnable polycarbosilane polymers (equation 2).

$$[(CH_3)_2Si]_n \xrightarrow{\Delta} \text{PCS} \xrightarrow{\text{melt spin}} \xrightarrow{\text{cure and pyrolysis}} \text{SiC fibers} \qquad (2)$$

$$\downarrow Ti(OR)_4$$

$$\text{poly(titanocarbosilane)} \rightarrow \text{SiTiC fibers} \qquad (3)$$

The spun fibers were cross-linked (cured) by air oxidation and pyrolyzed to give silicon-carbide-type fibers. Yajima et al. (*7*) reported further that heteroatoms such as titanium could be incorporated into the polymers and ceramic fibers to enhance their stability (equation 3).

Following the pioneering works of Verbeek and Yajima, numerous investigators began to explore the scope of polymer systems that would provide useful ceramic compositions. West and co-workers (*8*) prepared polysilastyrene (PSS) polymers by the sodium coupling of phenylmethyldichlorosilane and dimethyldichlorosilane (equation 4).

$$(C_6H_5)(CH_3)SiCl_2 + (CH_3)_2SiCl_2 \xrightarrow{Na/K} \text{polysilastyrene (PSS)} \xrightarrow{\Delta} \text{SiC powders} \qquad (4)$$

One aspect of their elegant and extensive studies dealt with the pyrolysis of these polymers to ceramic powders, fibers, and film.

Schilling and co-workers (*9*) used the known disilylation reaction of chlorosilanes to produce polycarbosilane polymers with controlled levels of cross-linking (equation 5).

$$(CH_3)_2SiCl_2 + (CH_2{=}CH)(CH_3)SiCl_2 \xrightarrow{K} \text{polycarbosilane} \xrightarrow{\Delta} \text{SiC powders} \quad (5)$$

Baney and co-workers (*10*) used the known Si–Si and Si–Cl bond redistribution of methylchlorodisilanes to prepare poly(methylchlorosilane)s. Pyrolysis of these branched polysilane polymers gives nearly stoichiometric amounts of silicon carbide (equation 6).

$$\text{methylchlorodisilane fraction} \xrightarrow{R_4PCl \text{ catalyst}} \text{poly(methylchlorosilane)} \xrightarrow{\Delta} \text{SiC powders and fibers} \quad (6)$$

Routes to Silicon Nitride Fibers

The earliest report of fibers containing silicon and nitrogen is by Verbeek (*11*). Chlorosilanes and amines are used to prepare polycarbosilazane polymers, which are converted into amorphous SiCN-containing ceramic fibers (equation 7).

$$CH_3SiCl_3 + CH_3NH_2 \rightarrow (CH_3SiNCH_3)_x \text{ (polycarbosilazane)} \xrightarrow{\text{spin aid}} \xrightarrow{\Delta} \text{SiCN amorphous fiber} \quad (7)$$

Seyferth and co-workers (*12*) used the ammonolysis of dichlorosilane to prepare carbon-free polysilazanes that could be converted into silicon nitride (equation 8).

$$H_2SiCl_2 + NH_3 \rightarrow (H_2SiNH)_x \xrightarrow{\Delta} Si_3N_4 \text{ powder} \quad (8)$$

Gaul and co-workers (*13*, *14*) prepared a variety of polysilazanes by using chlorosilanes and hexamethyldisilazane to control polymer molecular weight, rheology, and spinnability (equation 9). Fibers were prepared from the poly(methyldisilylazane) (MPDZ) system (equation 10).

$$CH_3SiCl_3 + [(CH_3)_3Si]_2NH \rightarrow (CH_3SiNH)_x \xrightarrow{\Delta} \text{SiCN powder} \quad (9)$$

$$\text{methyldichlorosilane} + [(CH_3)_3Si]_2NH \rightarrow \underset{\text{(MPDZ)}}{(CH_3SiSiCH_3NH)_x} \rightarrow \text{SiCN amorphous powder} \quad (10)$$

Some key observations can be drawn from these early pioneering efforts.

- Pyrolysis of organosilicon polymers in nonoxidizing atmospheres provides a route to nonoxide ceramic compositions.
- The yield of inorganic char (residue) increases with increasing polymer cross-linking.
- Both amorphous and crystalline materials are obtained, depending on the polymer compositions and temperatures used.

Recent workers have extended this field in several key areas. Cannady (*15*) and LeGrow and co-workers (*16*) extended the use of hexamethyldisilazane and prepared poly(hydridosilazane) (HPZ) polymers (equation 11).

$$HSiCl_3 + [(CH_3)_3Si]_2NH \rightarrow (HSiNH)_{3/2}\ (HPZ) \xrightarrow{\text{cure},\ \Delta} Si_3N_4 \text{ fiber and powder} \qquad (11)$$

Gas-phase curing of polymer fibers with trichlorosilane followed by pyrolysis gave high-nitrogen, low-carbon, silicon-nitride-type fibers.

Burns and co-workers (*17*) prepared a series of alkyl-, aryl-, and arylalkyl-substituted polysilazane polymers (equation 12), and a mechanistic study of pyrolysis was carried out to determine the effect of substituents on char yield, char composition, and stability of the resulting ceramic powders.

$$RSiCl_3 + NH_3 \rightarrow (RSiNH)_{3/2} \xrightarrow{\Delta} \text{SiCN powders} \qquad (12)$$

Okamura and co-workers (*18*) have taken air-cured PCS polymer and, through pyrolysis in the presence of ammonia, prepared essentially carbon-free silicon oxynitride fibers (equation 13). However, if the PCS polymer fiber is cured by electron beam radiation (to prevent oxygen addition), the same ammonia pyrolysis conditions provide nearly stoichiometric quantities of silicon nitride fibers (equation 14).

$$\text{PCS} \xrightarrow{\text{oxygen}} \text{cured PCS} \xrightarrow{NH_3,\ \Delta} Si_2N_2O \text{ fiber} \qquad (13)$$

$$\text{PCS} \xrightarrow{\text{electron beam}} \text{cured PCS} \xrightarrow{NH_3,\ \Delta} Si_3N_4 \text{ fiber} \qquad (14)$$

Quite recently, Arai and co-workers (*19*) prepared higher molecular weight poly(dihydridosilazane) polymers through amine complexation (equation 15).

$$H_2SiCl_2 + NH_3 \rightarrow (H_2SiNH)_x \rightarrow Si_3N_4 \text{ powders and fibers} \qquad (15)$$

Dry spinning in the presence of spinning aids followed by pyrolysis gives low-carbon silicon-nitride-type fibers.

From this recent work, several additional key observations can be drawn:

- Char compositions can be controlled quite precisely through the proper choice of polymer composition, pyrolysis temperature, and atmosphere, as well as through an improved understanding of the mechanisms of polymer pyrolysis.
- Many of the inorganic chars are thermally unstable at elevated temperatures, generally as a result of poor control of stoichiometry, that is, high oxygen content and excessively high carbon/silicon ratios.
- Past workers have placed a high emphasis on polymer synthesis, although recent work is more balanced, with an increasing emphasis on detailed characterization of the char.

Fibers

The following section will provide an overview of fiber processing, as well as descriptions of the polymer-derived fibers that have been prepared. A general processing scheme is depicted in Figure 1.

Typically, spinning is carried out by melt-spinning extrusion through a spinneret to obtain polymer fibers with a diameter of 10–20 μm. However, solvent spinning has been reported (*19*). The fragile polymer fibers need to be cured (cross-linked) below their melting point to prevent coalescence

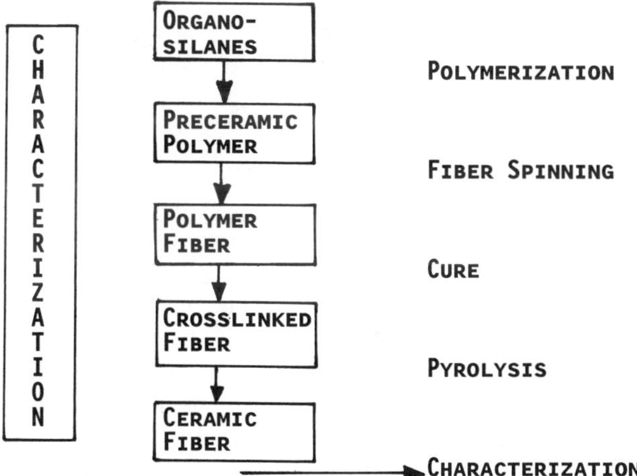

Figure 1. Process scheme for fiber production.

during pyrolysis and to increase char yields. Proper handling must be exercised to prevent damage to the fiber surface.

Pyrolysis must be carried out with careful control of atmospheric conditions. The gases formed during pyrolysis cause the fibers to undergo a weight loss of 20–40% (60–80% volume decrease), and the fibers achieve a final diameter of 10–15 μm. This diameter is desirable for many ceramic and plastic composite applications. A more complete discussion of this fiber processing has been presented recently (*20*).

A summary of some of the polymer-derived SiCN fibers is presented in Table I. This table includes the polymer precursor systems for these fibers, the compositional data, and the developing companies. Reliable elemental analyses of these fibers are often difficult to obtain because of incomplete sample combustion. In addition, many of these fiber compositions are being changed as their development proceeds.

Table I. Polymer-Derived SiCN Fibers

		Elemental Composition (wt %)					
Fiber	*Polymer*	*Si*	*N*	*C*	*O*	*Total*	*Company*
Nicalon	PCS	58.4	0.1	31.2	10.1	99.7	Nippon Carbon Company
Tyranno[a]	PCS–Ti	49.3	0.3	27.9	20.2	98.9	Ube Industries
Si_3N_4	$(H_2SiNH)_n$	62.5	34.3	0.5	3.1	100.4	Toa Nenryo Kogyo K.K.
SiCN	HP2	59.4	28.9	10.1	3.1	101.5	Dow Corning Corporation
SiC	PMS	69.1	—[b]	29.5	1.0	99.6	Dow Corning Corporation

[a]This fiber contains 1.2 wt % titanium.
[b]— means not determined.

Characterization. Characterization work at all stages of fiber processing, as illustrated in Figure 1, is important. The discussion in this chapter will be limited to certain aspects of polymer and ceramic fiber characterization, which will be illustrated with poly(methylsilane) (MPS) and HPZ polymers and the ceramic fibers derived from these polymers.

The characterization of polymer-derived ceramic fibers is an important but somewhat neglected area, although several publications have now dealt with the extensive characterization of preceramic polymers for ceramic fiber spinning and pyrolysis. A wide variety of characterization techniques have been used; in addition to elemental analyses and IR and NMR spectroscopy, gel permeation chromatography (GPC), thermal mechanical analysis (TMA), thermogravimetric analysis (TGA), and temperature–viscosity studies are valuable techniques. The TMA and temperature–viscosity characterization data are particularly useful for predicting polymer melt spinning conditions. As a result of these characterizations, a picture of the fiber structures is now emerging (*16*, *18*, *21*, *22*). For a more detailed discussion of these techniques, the reader is referred to a recent article (*16*) on HPZ polymer.

The results of X-ray diffraction (XRD), transmission electron microscopy

(TEM), high-resolution electron microscopy (HREM), and electron spectroscopy (XPS) have shown that the continuous phase in these fibers is amorphous. For some fibers (*21*), the evidence supports the presence of a microcrystalline β-silicon carbide, dispersed in the amorphous phase, with an average crystallite size of 2–4 nm (Figure 2). The presence of a microcrystalline carbon phase similar to graphitic carbon is supported (*21*) by elemental analyses and Raman, magic-angle-spinning NMR (^{13}C MAS NMR), and ESR (electron spin resonance) spectroscopy.

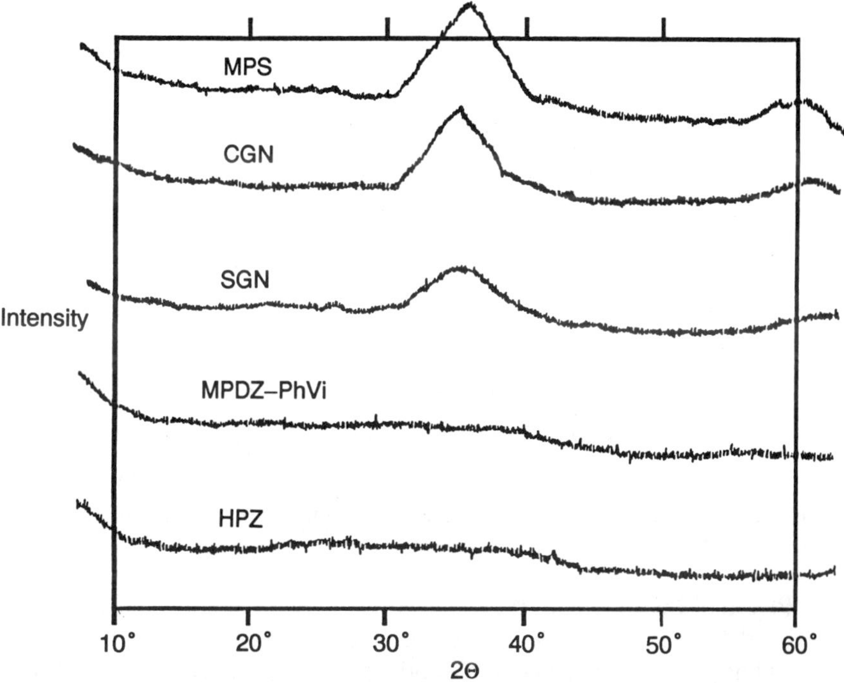

Figure 2. X-ray diffractograms of powdered ceramic fibers (Cu K_2). CGN is ceramic-grade Nicalon, and SGN is standard-grade Nicalon. MPDZ–PhVi is poly(methyldisilylazane)-derived fiber. See *reference 21 for complete details.*

The chemical bonding in these ceramic fibers appears to approach a structure in which individual silicon atoms are simultaneously bonded to C, N, and O in a random distribution. This structure is most dramatically demonstrated by ^{29}Si MAS NMR (*21, 23*) spectroscopy. In Si–C–O fibers, five tetrahedral silicon bond arrangements are possible: SiC_4, SiC_3O, SiC_2O_2, $SiCO_3$, and SiO_4. Figure 3 shows that all five arrangements are indicated in the spectra of standard-grade Nicalon (SGN) fiber. In contrast, the ^{29}Si MAS NMR spectrum of MPS-derived silicon carbide fibers (Figure

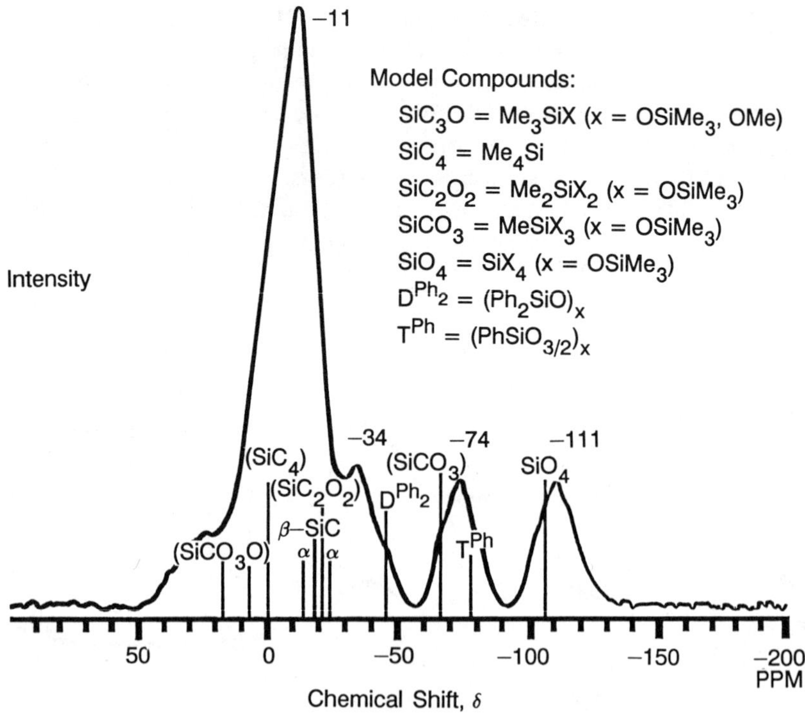

Figure 3. ^{29}Si MAS NMR spectrum of powdered SGN fibers.

4) is rich in SiC_4 structure, with smaller amounts of SiC_3O and SiC_2O_2. Additional data are available for other fibers (*16, 21, 23*).

Physical Properties. Polymer-derived fibers show low densities (determined by density gradient column technique) compared with that reported for crystalline silicon carbide and silicon nitride (~3.20 g/cm^3). Table II gives the reported values of tensile strength, elastic modulus, and density for several fibers. Numerous nitrogen and krypton BET (Brunauer–Emmett–Teller) surface area measurements have been carried out on these fibers. Essentially, the surface area is the geometric area. In addition, mercury porosimetry studies show the absence of significant surface-interconnected porosity (*24*). Present studies, including small-angle X-ray-scattering studies, suggest the existence of considerable closed porosity on the 2–4-nm level as an explanation for the low densities and low surface areas.

Finally, the surface and bulk compositions of these fibers can vary dramatically. Figures 5 and 6 show the scanning Auger microscopy (SAM) profiles of SGN and HPZ ceramic fibers, respectively. Because the fiber–matrix interfaces are critical in many applications (*2*), this type of characterization of fiber surface composition and chemistry must be carried out.

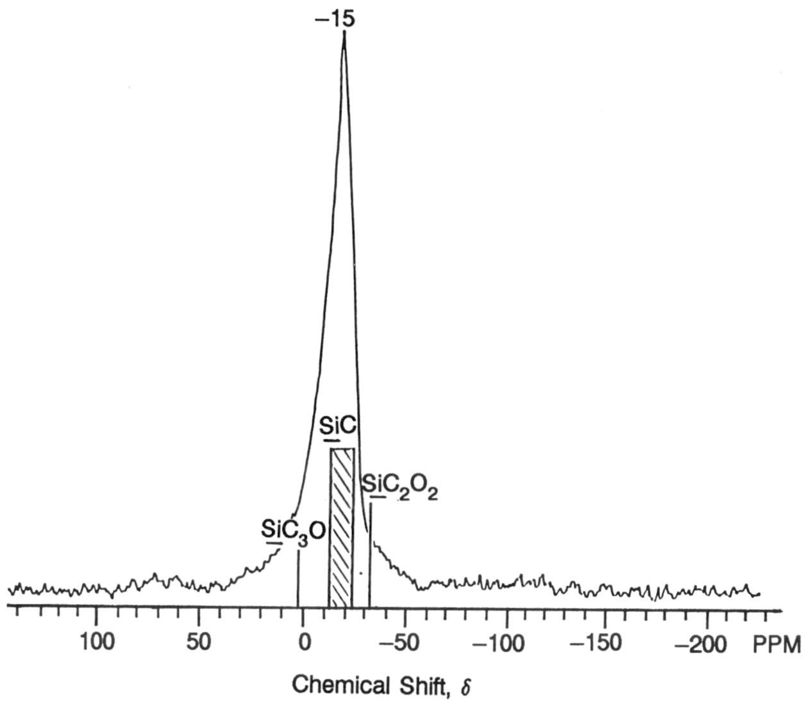

Figure 4. [29]*Si MAS NMR spectrum of powdered MPS-derived ceramic fiber.*

The preceding brief discussion, shows that a wide variety of characterization techniques have been used to characterize preceramic polymers and the derived ceramic fibers. Although some structural details remain elusive, the structural understanding of these fibers has advanced dramatically during the past 5 years.

Key Issues

The remaining discussion will focus on four key issues that must be addressed and understood to advance the technology of polymer-derived ceramic fibers.

Table II. Properties of SiCN Fibers

Fiber	*Tensile Strength (GPa)*	*Elastic Modulus (GPa)*	*Density (kg/m^3)*
Si_3N_4	2.45	196	2500
Si_3N_4	2.80	217	2620
Nicalon	2.98	196	2555
Tyranno	2.80	196	2400
SiC	1.75	210	2750

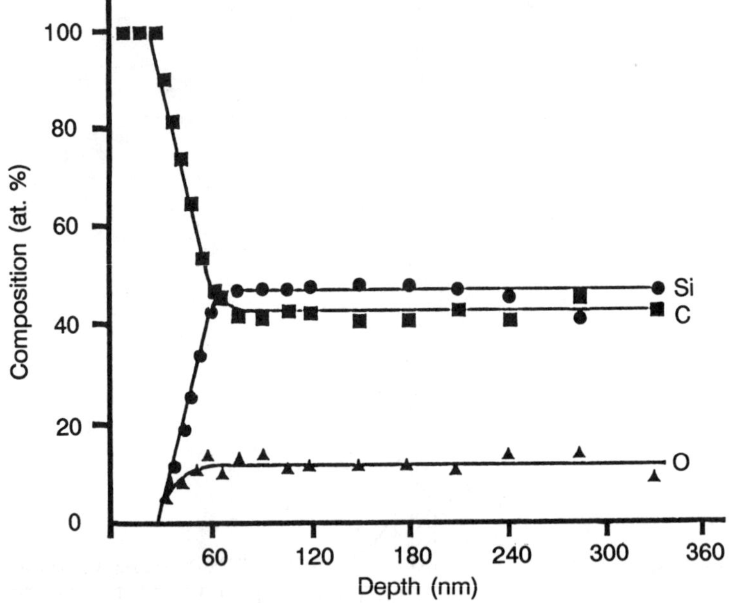

Figure 5. SAM profile of SGN fiber.

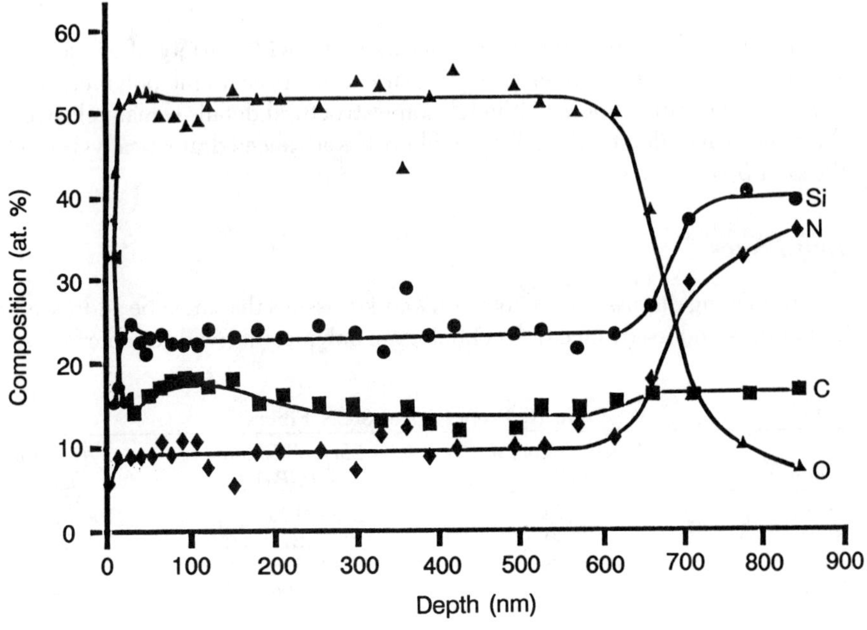

Figure 6. SAM profile of HPZ-derived ceramic fiber.

Polymer Properties. The first issue is that fiber processing is driven by polymer properties and chemistry. Some key polymer requirements related to the process scheme presented in Figure 1 are the following:

- Rheology—stable under melt-spinning conditions
- Chemical reactivity—uniform, rapid, and controllable cure in solid state
- Composition—uniform and highly pure
- Pyrolysis—high char yield of controllable chemical composition

Control of all these requirements is essential if suitable fiber processing and properties are to be developed.

Control of polymer purity and curability are the critical areas for future work. Some purity requirements will become quantified as this discussion proceeds. With regard to fiber cure, gas-phase chemical cure and energetic methods have been most effective to date (*16, 18*). Fiber processing involving the dry spinning of thermally unstable polymers (*4, 11, 19*) may be of future importance, because the need for a separate cure step is eliminated.

Fiber Strength. The second key issue is that fiber tensile strengths must be improved. This property in brittle materials is controlled by flaw size and distribution. The tensile strength of current fibers is limited by the purity of the polymer precursor (*20, 22*). An analysis of tensile strength versus flaw size suggests that flaws with diameters of ~1 μm limit fiber tensile strength to ~1.75 GPa. Reduction of flaw diameter to 0.1 μm gives a fiber strength of ~2.86 GPa (*20*).

Flaw frequency and flaw size can be limiting factors in determining the strength of these ceramic fibers. Tensile strength measurements are typically made with a single fiber having a gauge length of 2.54 cm. A frequency of two flaws per 2.54 cm would be expected if only 1 ppm of 1-μm-diameter spherical flaws is present in a 10-μm-diameter fiber. This flaw population would limit the attainable tensile strength to ~1.75 GPa. Fractographic analysis of ceramic fiber fracture surface by scanning electron microscopy (SEM) is an invaluable tool for the study of strength-limiting flaws (*20, 22*). Obviously, fiber processing also provides opportunities to introduce fiber-weakening flaws, and the attention required in each process step cannot be overemphasized.

Elastic Modulus. The third key issue relates to improvements required in the elastic modulus of these ceramic fibers. Studies indicate that modulus improvements will depend on careful control of composition and increases in fiber density (*24*). Earlier discussions in this chapter summarized the low densities of these ceramic fibers relative to their crystalline counterparts, as well as the existence of considerable pore volume.

In a recent study (*24*), significant increases in the density of HPZ-derived fiber was achieved without changes in weight or composition. The resultant modulus changes are shown in Figure 7. Extrapolation suggests that the predicted modulus for amorphous, fully dense HPZ-derived fiber corresponds to a value of ~360 GPa.

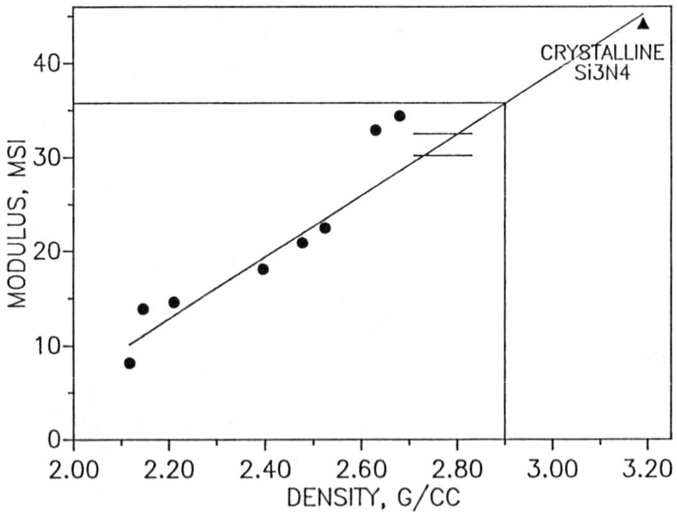

Figure 7. Density versus modulus for HPZ-derived ceramic fiber.

Thermal Stability. The last key issue relates to improvements required in the high-temperature stability of ceramic fibers. Thermal stability may refer to the percent retention of initial tensile strength either at temperature or subsequent to heat aging at temperature. The fibers known at present show a significant reduction in tensile strength after heat aging at 1200–1400 °C (*25–27*). The results can vary widely not only with fiber type but also with fiber batch and processing conditions, such as atmospheric conditions and sample geometry. Careful definition and control of protocols are required for reproducible results.

One detailed study (*28*) suggests that strength loss is a result of crystal growth and the formation of fine holes or pores around grain boundaries. Other studies (*25*) suggest that strength loss of Nicalon during nitrogen aging is due to growth of existing internal or surface flaws. However, when aged in argon, Nicalon fiber strength reduction appears to be due to grain growth and pores between grains. Although the mechanisms of strength reduction are not understood, this factor must be controlled if these fibers are to attain their full potential use in reinforced structures.

Summary

The polymer-derived approach is a viable method for the preparation of ceramic fibers for reinforcement applications. During the past decade, significant advances have been made in materials, processes, and characterization capabilities. This approach holds great promise for future advances; however, focus must be placed on addressing the key issues that have been identified.

Acknowledgments

I acknowledge the support of a Defense Advanced Research Projects Agency (DARPA)-funded Air Force contract (F33615–83–C–5006) administered by the Air Force Wright Aeronautical Laboratories/Materials Laboratory and the Dow Corning Corporation. I thank A. Zangvil, University of Illinois, for the SAM profiles.

References

1. For a recent review of this subject, see Baney, R. H.; Chandra, G. In *Encyclopedia of Polymer Science and Engineering;* Wiley: New York, 1988; Vol. 13, p 312.
2. Mah, T.; Mendiratta, M. G.; Katz, A. P.; Mazdiyasni, K. S. *Ceram. Bull.* **1987,** *66,* 304–317.
3. Petrisko, R. A.; Stark, G. L. *SAMPE Proc.* **1988,** *33,* 1015.
4. Verbeek, W.; Winter, G. Ger. Offen. 2 236 078, 1974.
5. Fritz, G.; Matern, E. *Carbosilanes Synthesis and Reactions;* Springer: New York, 1986.
6. Yajima, S.; Hayashi, J.; Omori, M. *Chem. Lett.* **1975,** 931.
7. Yajima, S.; Iwai, T.; Yamamura, T.; Okamura, K.; Hasegawa, Y. *J. Mater. Sci.* **1981,** *16,* 1349.
8. *Ultrastructure Processing of Ceramics, Glasses and Composites;* West, R. C.; Hench, L. L.; Ulrich, D. R., Eds.; Wiley: New York, 1984; p 235.
9. Schilling, C. L.; Wesson, J. P.; Williams, T. C. *Am. Ceram. Soc. Bull.* **1983,** *62,* 912.
10. Baney, R. H.; Gaul, J. H.; Hilty, T. K. *Organometallics* **1983,** *2,* 859.
11. Verbeek, W. U.S. Patent 3 853 567, 1974.
12. Seyferth, D.; Wiseman, G. H.; Prud'Homme, C. *J. Am. Ceram. Soc.* **1983,** *66,* C-13.
13. Gaul, J. H. U.S. Patent 4 312 970, 1982.
14. *Emergent Process Methods for High Technology Ceramics;* Davis, R. F.; Palmer, H., III; Porter, R. L., Eds.; Plenum: New York, 1984; Vol. 17, p 235.
15. Cannady, J. P. U.S. Patent 4 535 007, 1985.
16. LeGrow, G. E.; Lim, T. F.; Lipowitz, J.; Reaoch, R. S. *Am. Ceram. Soc. Bull.* **1987,** *66,* 363.
17. Burns, G. T.; Angelotti, T. P.; Hanneman, L. F.; Chandra, G.; Moore, J. A. *J. Mater. Sci.* **1987,** *22,* 2609.
18. Okamura, K.; Sato, M.; Hasegawa, Y. *Ceram. Int.* **1987,** *13,* 55.

19. Arai, M.; Hayato, N.; Osamu, F.; Takeski, I. Eur. Patent Appl. 86308752.4, 1986.
20. Salinger, R. M.; Barnard, T. D.; Li, C. T.; Mahone, L. G. *SAMPE Q.* **1988,** *19*, 27.
21. Lipowitz, J.; Freeman, H. A.; Chen, R. T.; Prack, E. R. *Adv. Ceram. Mater.* **1987,** 2, 121.
22. Sawyer, L. C.; Jamieson, M.; Brikowski, D.; Haider, I.; Chen, R. T. *J. Am. Ceram. Soc.* **1987,** *70*, 798.
23. Lipowitz, J.; Turner, G. L. *Polym. Prepr. (Am. Chem. Soc., Div. Polym. Chem.)* **1988,** *29*, 1.
24. Rabe, J. A.; Lipowitz, J.; Frevel, L.; Sander, W. *Proceedings of the 12th Annual Conference on Composites and Advanced Ceramics;* Cocoa Beach, FL, 1988, in press.
25. Lipowitz, J., Dow Corning Corporation, Midland, MI, unpublished results.
26. Clark, T. J.; Jaffe, M.; Rabe, J.; Langley, N. R. *Ceram. Eng. Sci. Proc.* **1986,** 7, 901.
27. Sawyer, L. C.; Chen, R. T.; Haimbuch, F., IV; Hagert, P. J.; Prack, E. R.; Jaffe, M. *Ceram. Eng. Sci. Proc.* **1986,** 7, 914.
28. Langley, N. R.; Filsinger, D. H.; Rabe, J. A.; Jaffe, M.; Clark, T. J. *Proceedings of a Joint NASA/Department of Defense Conference;* NASA Conf. Publ. 2445; Government Printing Office: Washington, DC, 1986; p 7.
29. *Proceedings of a Symposium on High Temperature Chemistry;* Ishikawa, H.; Ichikawa, H.; Teranishi, H.; Munir, Z. A., Eds.; Electrochemical Society: Pennington, NJ, 1988; Vol. 4, p 205.

RECEIVED for review December 20, 1988. ACCEPTED revised manuscript March 27, 1989.

33

New Polymer Precursors for Silicon Nitride

Bernard Kanner[1] **and Roswell E. King III**

Union Carbide Corporation, Tarrytown, NY 10591

A novel process for polysilazane preparation has been developed via transamination, which does not require the use of chlorosilane intermediates. Silicon nitride, Si_3N_4, especially as fibers or coatings, is obtained generally by the pyrolysis of polysilazane precursors. When chlorosilanes are used, cumbersome separation techniques must be used to remove the amine hydrochloride byproducts.

POLYSILAZANES HAVE NOW BEEN PREPARED by the catalyzed transamination of $HSi[N(CH_3)_2]_3$ [tris(dimethylamino)silane, Tris], or its derivatives, with ammonia, primary amines, hydrazines, or diamines. Chlorosilane intermediates are not required in this process, because Tris is readily prepared in excellent yield via a new direct reaction of silicon with dimethylamine.

Aminosilanes that contain Si–H groups are transaminated by a variety of catalysts, including CO_2, ammonium carbamates, carbamatosilanes, and organic and inorganic acids. Aminosilanes that do not contain an Si–H group are effectively transaminated only by strong organic or inorganic acid catalysts.

By the appropriate selection of reaction conditions and reactants, transamination processes can be directed to the selective formation of new silylamines, disilazanes, volatile oligomeric polysilazanes, soluble higher molecular weight polysilazanes, or highly cross-linked insoluble polymers.

Polysilazanes prepared by these processes are readily converted to silicon nitride or silicon-nitride-containing ceramic compositions upon pyrolysis at temperatures up to 1600 °C. High-purity α-Si_3N_4 (α-phase crystalline form of silicon nitride) has been efficiently prepared by these processes.

[1]Current address: 5 Robin Lane, West Nyack, NY 10994

0065–2393/90/0224–0607$06.00/0

Routes to Silicon Nitride

Within the past decade, research on the development of new Si_3N_4 compositions has greatly intensified. These efforts have resulted in a proliferation of new synthetic procedures. Preparative routes have been developed with increasing emphasis on the control of purity and physical properties. Silicon nitride powders, fibers, coatings, and composites each have their own characteristic requirements, which can dictate the choice of a particular route (*1*).

To produce silicon nitride powder, silicon can be nitridated at elevated temperatures in an atmosphere of nitrogen or ammonia:

$$3Si + 2N_2 \xrightarrow{1400\text{–}1700\ °C} Si_3N_4 \tag{1}$$

This route is useful for only silicon nitride powder and can not be used for silicon nitride fibers or coatings (*2–3*).

Silicon nitride powders have also been prepared from more readily available starting materials. The preparations involve the reduction of silica by carbon in the presence of nitrogen (*2*) or of silicic acid reacting with ammonia (*4–7*):

$$3SiO_2 + 6C + 2N_2 \rightarrow Si_3N_4 + 6CO \tag{2}$$

$$Si(OC_2H_5)_4 + 4H_2O \rightarrow Si(OH)_4 + 4C_2H_5OH \tag{3}$$

$$Si(OH)_4 + NH_3 \rightarrow Si_3N_4 + 6H_2O \tag{4}$$

These processes are carried out at 1000–1500 °C.

Other established routes to silicon nitride involve silicon sulfide intermediates (equations 5a and 5b) (*8, 9*), the reaction of silane with ammonia (equation 6) (*10*), or the reactions of various chlorosilanes such as silicon tetrachloride with ammonia (equations 7–8) (*11–17*):

$$Si + 2H_2S \xrightarrow{900\ °C} SiS_2 + 2H_2 \tag{5a}$$

$$3SiS_2 + 4NH_3 \xrightarrow{1200\text{–}1450\ °C} Si_3N_4 + 6H_2S \tag{5b}$$

$$3SiH_4 + 4NH_3 \xrightarrow{500\text{–}900\ °C} Si_3N_4 + 12H_2 \tag{6}$$

$$SiCl_4 + 6NH_3 \rightarrow \text{–}[Si(NH)_2]_x\text{–} + 4NH_4Cl \tag{7a}$$

$$3\text{–}[Si(NH)_2]_x\text{–} \xrightarrow{1200\text{–}1400\ °C} Si_3N_4 + 2NH_3 \tag{7b}$$

$$H_2SiCl_2 + 3NH_3 \rightarrow \text{–}[H_2SiNH]_x\text{–} + 2NH_4Cl \tag{8a}$$

$$3\text{–}[H_2SiNH]_x\text{–} \xrightarrow{1200\text{–}1400\ °C} Si_3N_4 + 3NH_3 \tag{8b}$$

The routes involving the reactions of chlorosilanes with ammonia generate large quantities of ammonium chloride, which must be removed carefully.

Polysilazane Routes to Silicon Nitride–Silicon Carbide

Interestingly, all of these routes provide intermediates for Si_3N_4 and SiC fiber formation. The procedures for fiber formation are analogous to those already described, except that methylsilicon groups are needed to yield silicon carbide. CH_3SiCl_3 (*18–20*), CH_3SiHCl_2 (*21–22*), or disilazanes (*23–27*) are used as starting materials.

$$CH_3SiCl_3 + 6CH_3NH_2 \rightarrow CH_3Si(NHCH_3)_3 + 3NH_4Cl \quad (9a)$$

$$CH_3Si(NHCH_3)_3 \rightarrow -[(CH_3NH)(CH_3)Si{-}N(CH_3)]_x- + CH_3NH_2 \quad (9b)$$

$$-[(CH_3NH)(CH_3)Si{-}N(CH_3)]_x- \rightarrow Si_3N_4 + SiC \quad (9c)$$

$$CH_3HSiCl_2 + NH_3 \xrightarrow{35\ ^\circ C,\ KH} -[(CH_3)(H)Si{-}N(H)]_x- + 2NH_4Cl \quad (10a)$$

$$-[(CH_3)(H)Si{-}N(H)]_x- \xrightarrow{1000\ ^\circ C} Si_3N_4 + SiC \quad (10b)$$

$$HSiCl_3 + 3(R_3Si)_2NH \rightarrow -[(R_3SiNH)(H)Si{-}N(SiR_3)]_x- + 3R_3SiCl \quad (11a)$$

$$-[(R_3SiNH)(H)Si{-}N(SiR_3)]_x- \xrightarrow{1600\ ^\circ C} Si_3N_4 + SiC \quad (11b)$$

Like the preparative procedures for polysilazane intermediates for silicon nitride, the reactions leading to silicon nitride–silicon carbide involve chlorosilanes and generate substantial quantities of amine hydrochlorides, which must be removed by extensive purification, because their presence in the later stages of the process can be deleterious (*28*).

The preparation of compounds with Si–N bonds, especially silazanes, without the intermediacy of chlorosilicon compounds has been difficult until quite recently. As part of a fundamental investigation of the reactions of silicon metal with simple small molecules, a new direct reaction of silicon was discovered (*29*). Silicon reacts with dimethylamine in a fluidized bed reactor to give Tris in excellent yield. The reaction must be catalyzed by copper.

$$(CH_3)_2NH + Si \xrightarrow{250\ ^\circ C,\ Cu} [(CH_3)_2N]_3SiH\ (>90\%) \quad (12)$$

The reaction proceeds in the same general fashion as do the reactions of methyl chloride and hydrogen chloride with silicon.

Because the chemical properties of Tris are virtually unknown, we (*30*) investigated this material and found that Tris does not undergo hydrosilation reactions readily. Chloroplatinic-acid-catalyzed reactions of Tris with simple olefins (equation 13) do not proceed to a significant extent, and the corresponding reactions with acetylenes (equation 14) are sluggish under ordinary conditions (*31*):

$$HSi[N(CH_3)_2]_3 + RCH{=}CH_2 \xrightarrow[75\text{–}200\ ^\circ C]{Pt} \text{no reaction} \tag{13}$$

$$HSi[N(CH_3)_2]_3 + RC{\equiv}CH \xrightarrow[75\ ^\circ C]{Pt} RCH{=}CHSi[N(CH_3)_2]_3\ (15\text{–}20\%) \tag{14}$$

However, at much higher temperatures, Tris reacts very cleanly with acetylene to give an excellent yield of vinyltris(dimethylamino)silane (vinyl-Tris):

$$HSi[N(CH_3)_2]_3 + HC{\equiv}CH \xrightarrow[200\ ^\circ C]{Pt} H_2C{=}CHSi[N(CH_3)_2]_3\ (>90\%) \tag{15}$$

The reaction is surprisingly clean under these vigorous hydrosilation conditions, with no evidence for the formation of 1,2-bis[tris(dimethylamino)]ethane (*32*). Thus, very satisfactory processes have been developed for both Tris and vinyl-Tris that do not involve chlorosilane intermediates. Both monomers are suitable intermediates for the preparation of polysilazane preceramic polymers that could be converted thermally to silicon nitride and mixtures of silicon nitride and silicon carbide.

Experimental Procedures

Ammonia, primary amines, and aliphatic and aromatic hydrocarbon solvents were obtained commercially and used without further purification. The preparation of Tris and vinyl-Tris has been described previously (*30*).

Transamination of Tris with Allylamine. A 10-L, three-neck, round-bottom flask fitted with a Friedrich condenser, heating mantle, magnetic stirrer, thermometer, and fittings for a dry-N_2 atmosphere was dried and flushed with N_2. Toluene (300 mL), $HSi[N(CH_3)_2]_3$ (199.1 g, 1.234 mol), and allylamine were charged into the flask. After the mixture was stirred for 30 min at room temperature, the carbamate salt $[(CH_3)_2NH_2]^+[CO_2NCH_3]^-$ (16.54 g, 0.12 mol) was added. Immediate gas evolution was observed. The reaction mixture was heated for 4 h at a reflux temperature of 110 °C. The reaction was monitored by GLC (gas–liquid chromatography) for the disappearance of the starting silane $HSi[N(CH_3)_2]_3$. After 4 h, the reaction was determined to be complete and was cooled to room temperature. The toluene solvent was removed by vacuum stripping (13 Pa) at 80 °C. The residual product was determined to be $HSi(NHCH_2CH{=}CH_2)_3$ by gas chromatography–mass spectrometry (GC–MS) and ^{29}Si and ^{13}C NMR spectroscopy.

Transamination of Tris with Aniline. A 1-L, three-neck, round-bottom flask fitted with a Friedrich condenser, heating mantle, magnetic stirrer, thermometer, and provision for a dry-N_2 atmosphere was charged with 400 mL of toluene, 381.9 g (4.10 mol) of aniline, and 220.5 g (1.37 mol) of Tris. The solution was stirred at 20 °C for 30 min, and 16.5 g (0.12 mol) of dimethylammonium dimethylcarbamate was added. As the reaction flask was heated to a reflux temperature of 117 °C, vigorous gas evolution was noted. Reflux was continued for 3 h until no further gas evolution was observed. Toluene and excess aniline were removed by vacuum distillation (13 Pa) at 120 °C. Upon cooling to room temperature, the resulting solid was washed liberally with pentane, filtered, and dried under vacuum to give a final yield of 328.3 g of white crystalline needles (mp = 80 °C) from toluene. The product was identified as compound A (*see* equation 20) by GC–MS and NMR and IR analyses.

Transamination of Tris with Methylamine by Inverse Addition. A 2-L, three-neck flask was fitted with a mechanical stirrer and two Claisen adapters. One Claisen adapter was fitted with a reflux condenser and a CH_3NH_2-sparging tube that could be immersed in the reaction solvent. The other Claisen adapter was fitted with a pressure-equalizing dropping funnel and a thermometer that could be immersed in the reaction solvent. The reflux condenser and addition funnel were also fitted with N_2 adapters. The flask was charged with heptane (500 mL) and *p*-toluenesulfonic acid (*p*-TSA) monohydrate (6.10 g, 0.032 mol) in a N_2 atmosphere. The reaction mixture was heated to reflux temperature, and CH_3NH_2 was introduced at a continuous rate.

The addition funnel was charged with $HSi[N(CH_3)_2]_3$ (300 mL, 1.58 mol), and $HSi[N(CH_3)_2]_3$ was added dropwise to the reaction mixture, with stirring and CH_3NH_2 sparging, over a period of 2 h. An immediate outgassing of $(CH_3)_2NH$ could be observed by temporarily shutting off the CH_3NH_2 sparge. Upon complete addition of Tris, the mixture was refluxed for an additional 2 h. A GC analysis of the reaction mixture showed no remaining $HSi[N(CH_3)_2]_3$ and the appearance of a series of peaks corresponding to compounds with higher boiling points. Heptane was removed under vacuum at ambient temperature. The various fittings were removed, and the crude product was filtered under N_2. Vacuum distillation of the filtrate yielded a colorless distillate (123.4 g, boiling range 90–120 °C at 13 Pa). Only 15.5 g of a white viscous higher molecular weight polysilazane remained as a residue in the distillation flask. The distillate was identified as a mixture of cyclic and linear oligomers, $[(CH_3NH)_2(H)Si{-}N(CH_3)]_nH$, in which n had values of 3 to 6, on the basis of GC–MS and ^{1}H, ^{13}C, and ^{29}Si NMR spectroscopic analyses.

Transamination of Tris with Methylamine via Standard Addition. Toluene (375 mL) and $HSi[N(CH_3)_2]_3$ (369.9 g, 2.29 mol) were combined under a N_2 atmosphere in an apparatus similar to that described in the previous section. The mixture was stirred and then charged with *p*-TSA monohydrate (9.0 g, 0.05 mol). The reaction mixture was heated to reflux temperature and then sparged with CH_3NH_2 continuously. The reaction mixture was maintained at reflux temperature for 4 h, at which time it was determined by GC that no $HSi[N(CH_3)_2]_3$ remained. The reaction mixture was cooled to room temperature, and the toluene solvent was removed under vacuum. The remaining white resinous product was washed with 200 mL of pentane, and the washing was discarded. The sample was dried further under vacuum at 13 Pa to yield 138.2 g of a white resin that hardened at <40 °C. The product was not distillable under vacuum. The structure of the product mixture was nearly identical to that obtained in the previous experiment (on the basis of ^{1}H, ^{13}C, and ^{29}Si NMR spectroscopic analyses) but was more highly condensed and cross-linked.

Discussion

Transamination Reactions. The transamination reactions of Tris were studied (33) to design processes for the preparation of preceramic polysilazanes. Because of the bulky dimethylamino groups, Tris is a sterically crowded molecule. This crowding very likely accounts for its sluggish hydrosilation reactivity.

The transamination reactions of Tris with typical secondary amines were extremely slow at best in the absence of catalyst. Even when catalyzed, these reactions remained sluggish and generally incomplete:

$$HSi[N(CH_3)_2]_3 + R_2NH \overset{\text{catalyst}}{\rightleftharpoons} HSi[N(CH_3)_2]_x(NR_2)_{3-x} \quad (16)$$

Transamination reactions are catalyzed by carbon dioxide or its salt with dimethylamine, presumably via an insertion process that involves silylcarbamate formation (*34*). The catalyzed transamination of Tris with diethylamine gives typical results:

$$\begin{aligned} HSi[N(CH_3)_2]_3 + 3HN(C_2H_5)_2 \overset{CO_2}{\rightleftharpoons} & \; HSi[N(CH_3)_2]_3 \text{ (22\%)} \\ & + HSi[N(CH_3)_2]_2[N(C_2H_5)_2] \text{ (51\%)} \\ & + HSi[N(CH_3)_2][N(C_2H_5)_2]_2 \text{ (24\%)} \\ & + HSi[N(C_2H_5)_2]_3 \text{ (2\%)} \end{aligned} \quad (17)$$

The reactions of ammonia and primary amines with Tris and vinyl-Tris are much more rapid and complete (*33*). Reduced steric crowding and the discovery of much more effective catalysts such as trifluoroacetic acid and *p*-TSA were important contributing factors.

Objectives of Preceramic Polymer Design. The conversion of Tris and vinyl-Tris to preceramic polysilazane polymers was investigated, with the objective of developing processes with the following properties:

- The chemistry is well understood and reproducible and can be readily monitored.
- The process provides cross-linking mechanisms that do not introduce oxygen (i.e., does not involve oxidation or hydrolysis).
- The conversion step to ceramic is efficient and produces ceramic products of high purity.
- The process is sufficiently flexible and can be directed to selective formation of ceramic powders, fibers, coatings, binders, composites, or coherent parts.

Process Variables. On the basis of the preceding objective, a wide range of transamination process variables were investigated. In addition to carbon dioxide, a wide variety of organic and inorganic acids effectively catalyzed transamination reactions (*35–37*). Transamination reactions involving ammonia or primary amines with either Tris or vinyl-Tris yield a very broad distribution of polysilazane polymers. Considerable control over this process can be realized through the proper selection of reactants and reaction conditions. Thus, although transaminations involving ammonia proceed very rapidly to form high-molecular-weight polysilazanes, some reactions with primary amines can be directed readily under mild conditions to yield the simple transaminated silylamines:

$$HSi[N(CH_3)_2]_3 + 3H_2C{=}CHCH_2NH_2 \xrightarrow[CO_2]{40\ ^\circ C} HSi(NHCH_2CH{=}CH_2)_3 + 3(CH_3)_2NH \quad (18)$$

$$H_2C{=}CHSi[N(CH_3)_2]_3 + 3C_4H_9NH_2 \xrightarrow[CO_2]{40\ ^\circ C} H_2C{=}CHSi(NHC_4H_9)_3 + 3(CH_3)_2NH \quad (19)$$

At low reaction temperatures and relatively short reaction times, the transamination process can be stopped at the first stage with some primary amines. To drive the reaction further, more-vigorous conditions are needed. When catalyzed by strong acids such as *p*-TSA, certain disilazanes can be isolated in excellent yield. These products include two solid derivatives that were obtained in high purity:

$$2HSi[N(CH_3)_2]_3 + 5C_6H_5NH_2 \xrightarrow{110\ ^\circ C,\ H^+} 6(CH_3)_2NH + [(C_6H_5NH)_2Si(H)]_2NC_6H_5\ (A; >90\%) \quad (20)$$

$$2H_2C{=}CHSi[N(CH_3)_2]_3 + \text{excess } CH_3NH_2 \xrightarrow{110\ ^\circ C,\ H^+} 6(CH_3)_2NH + [(CH_3NH)_2(CH{=}CH_2)Si]_2NCH_3\ (B; 90\%) \quad (21)$$

These solid disilazane derivatives are easily isolated by solvent removal and purified by recrystallization (disilazane A, m.p. 80 °C; disilazane B, m.p. 20 °C).

Both disilazanes A and B, which are readily prepared and isolated, may be cocondensed with other polymeric polysilazanes. By cocondensation, the controlled introduction of either (or both) vinyl and silyl hydride groups can be accomplished readily. In turn, these groups provide a convenient cross-linking mechanism for these polymers via platinum catalysis or thermal inducement of hydrosilation. This method eliminates the need for cross-linking via hydrolysis or oxidation, which introduces undesirable oxygen bonds. Other advantages will be treated in a later section.

Reaction Control for Product Selection. The catalyzed transamination of Tris with methylamine in solvent can be directed to give mixtures of polysilazanes of different molecular weights by altering one of the reaction conditions. When Tris is added slowly to an excess of CH_3NH_2 (inverse addition) with *p*-TSA as catalyst, an excellent yield of volatile polysilazanes is obtained:

$$\text{excess } CH_3NH_2 + HSi[N(CH_3)_2]_3 \xrightarrow[100\ ^\circ C,\ \text{heptane}]{p\text{-TSA}}$$

$$(CH_3NH)_2Si(H)\text{–}N(CH_3)\text{–}[(CH_3NH)(H)Si\text{–}N(CH_3)]_n\text{–}Si(NHCH_3)_2 + \text{–}[(CH_3NH)(H)Si\text{–}N(CH_3)]_n\text{–} + 3(CH_3)_2NH \qquad n = 3\text{–}6 \quad (22)$$

The polysilazane products are a mixture of cyclic and linear molecules, all of which are distillable. This property facilitates their isolation and characterization and provides more detailed information on the nature of the transamination process.

Volatility can be a problem during the thermal conversion of polysilazanes to silicon nitride. The efficiency of the process not only depends on losses due to stripping of excessive side groups but also decreases because of losses due to polymer volatility. Therefore, the polysilazanes must have sufficiently high molecular weights to minimize such losses.

By a simple process change involving the addition of CH_3NH_2 to Tris, the reaction can be directed to produce higher molecular weight, essentially nonvolatile polysilazanes:

$$HSi[N(CH_3)_2]_3 + \text{excess } CH_3NH_2 \xrightarrow[100\ ^\circ C,\ \text{toluene}]{p\text{-TSA}}$$

$$[\text{–}(CH_3NH)(H)Si\text{–}N(CH_3)\text{–}]_n\ (>90\%) + 3(CH_3)_2NH \quad (23)$$

The viscous, hexane-soluble, white resinous product is obtained by adding methylamine to Tris (regular mode of addition). The reaction is catalyzed by moderately strong acids, such as *p*-TSA, and the yield is virtually quantitative. The product is primarily linear with minor branching.

Highly branched and even cross-linked polysilazane polymers are also desirable, particularly for the formation of ceramic powders. The transamination of Tris and vinyl-Tris with ammonia is a convenient process for the preparation of these precursors. The reactions were effectively catalyzed by strong acids such as trifluoroacetic acid:

$$HSi[N(CH_3)_2]_3 + \text{excess } NH_3 \xrightarrow[\text{toluene}]{110\ ^\circ C}$$

$$\text{–}[(NH_2)(H)Si\text{–}N(H)]_x\text{–}\ (>90\%) + 3(CH_3)_2NH \quad (24)$$

The product is a white powder and very sensitive to moisture. The given structural formula is idealized, because most likely, extensive cross-linking occurs through reactions involving NH_2 and SiH groups. The cross-linked nature of the product is consistent with its essential insolubility in hydrocarbon solvents.

The transamination of vinyl-Tris with excess ammonia proceeds under similar conditions to give analogous polysilazanes:

$$H_2C{=}CHSi[N(CH_3)_2]_3 + \text{excess } NH_3 \xrightarrow[CF_3CO_2H]{110\ ^\circ C} -[(-NH)(H_2C{=}CH)Si{-}N(H)]-_n + 3(CH_3)_2NH \quad (25)$$

This reaction proceeds satisfactorily only when catalyzed by trifluoroacetic acid or similarly strong acids. The product is a granular white solid that is insoluble in hydrocarbon solvents and sensitive to hydrolysis.

Copolymeric polysilazanes can also be prepared. The transamination products of Tris and vinyl-Tris with methylamine, which have already been described, can be cocondensed to form a new copolymeric composition:

$$(CH_3NH)_2(CH{=}CH_2)Si{-}N(CH_3){-}Si(CH{=}CH_2)(NHCH_3)_2 + -[(CH_3NH)(H)Si{-}N(CH_3)]_x-$$

$$\downarrow 110\ ^\circ C,\ CF_3CO_2H$$

$$-[(CH_3NH)(H_2C{=}CH)Si{-}N(CH_3)]_x{-}[(CH_3NH)(H_2C{=}CH)Si{-}N(CH_3)]_y- \quad (26)$$

The product is a white resin that is soluble in hexane and can be cross-linked by further condensation or through reactions involving vinyl and silyl hydride groups.

The drawing of these precursor polysilazane polymers to form fibers and their subsequent pyrolysis to silicon nitride fibers is a complex process that will be reported separately.

Pyrolysis to Si_3N_4 and Si_3N_4–SiC. Precursor polymers are converted to ceramics by pyrolysis. The efficiency of conversion depends on several factors. Aside from simple stoichiometric efficiency, volatility losses, entrapment of impurities, and side reactions (such as oxidation and hydrolysis) can contribute to the apparent ceramic yield. Because of their high volatility, precursors such as the solid disilazanes previously described are unsuitable for conversion to ceramics, unless pyrolysis is carried out in a confined space. In unconfined pyrolysis, polymers of low molecular weight (<1000) are even less suitable as ceramic precursors.

Better pyrolysis results were obtained with the transamination products of Tris and vinyl-Tris with either methylamine or ammonia (33):

$$-[(NH_2)(H)Si{-}N(CH_3)]_x- \xrightarrow[N_2]{1000\ ^\circ C} Si_3N_4\ (82\%) \quad (27)$$

$$-[(CH_3NH)(H)Si{-}N(CH_3)]_x- \xrightarrow[N_2]{1000\ ^\circ C} Si_3N_4\ (69.2\%) + SiC\ (\text{minor product}) \quad (28)$$

$$-[(H_2N)(H_2C{=}CH)Si{-}N(H)]_x \xrightarrow[N_2]{1000\ ^\circ C} \underset{74\%}{Si_3N_4 + SiC} + C\ (\text{minor product}) \quad (29)$$

$$-[(CH_3NH)(H_2C{=}CH)Si{-}N(CH_3)]_x- \xrightarrow[N_2]{1500\ ^\circ C} \underset{70\%}{Si_3N_4 + SiC} + C\ (\text{minor product}) \quad (30)$$

The ceramic products of these pyrolyses were completely amorphous. Pyrolysis carried out at 1500 °C or lower yields only amorphous products. Amorphous materials can be converted to crystalline form by heating above the transition temperature. Pyrolysis must be carried out at 1550 °C or higher to obtain crystalline products. Thus, pyrolysis of the transamination product of Tris with ammonia at 1550 °C or higher gave high-purity, α-phase silicon nitride (*33*), as analyzed by powder X-ray diffraction (Figure 1 and Table I).

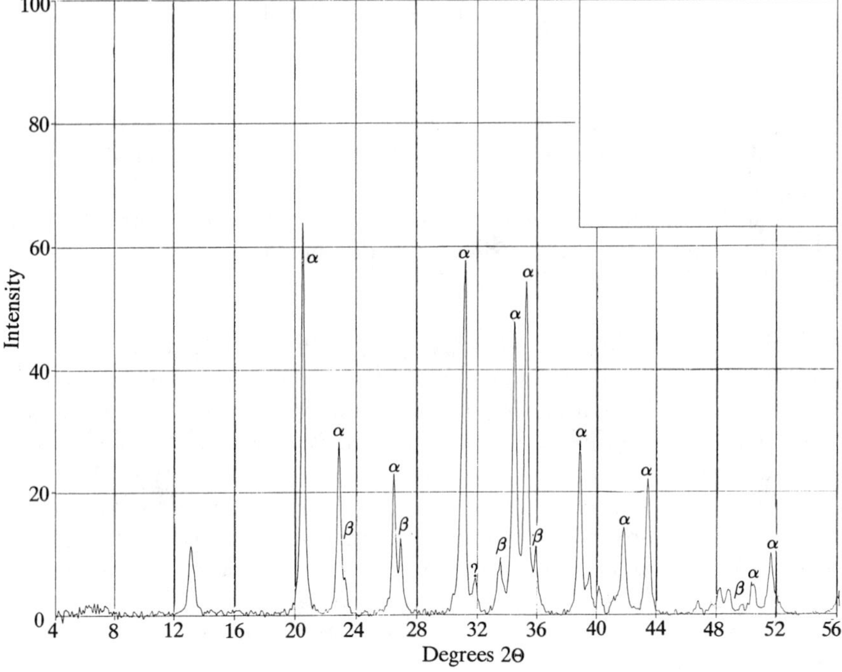

Figure 1. Representative X-ray powder diffraction spectrum of [Si(NH₂)₂], which is the product of pyrolysis at 1550 °C of the transamination product of Tris with ammonia. α and β refer to α- and β-phase silicon nitride, respectively.

Table I. X-Ray Diffraction Data

Peak Number	2θ (°)[a]	d *Spacing*[a]	2θ (°)[b]	d *Spacing*[b]	*Peak Area*	*Peak Height*	*Relative Intensity*
1	13.097	6.7599	13.100	6.7581	2,875	115	17.6
2	20.499	4.3325	20.467	4.3391	10,715	656	100.0
3	22.834	3.8945	22.800	3.9001	4,846	290	44.2
4	23.201	3.8337	23.234	3.8282	662	62	9.5
5	26.445	3.3703	26.426	3.3726	3,726	236	36.0
6	26.895	3.3149	26.899	3.3144	1,922	128	19.5
7	30.363	2.9437	30.314	2.9483	395	34	5.3
8	30.930	2.8910	30.890	2.8947	9,983	594	90.6
9	31.743	2.8188	31.748	2.8184	1,235	61	9.3
10	33.508	2.6743	33.466	2.6776	2,274	89	13.6
11	34.473	2.6016	34.437	2.6042	9,982	493	75.1
12	35.267	2.5448	35.240	2.5467	10,082	558	85.0
13	35.894	2.5018	35.899	2.5015	1,781	114	17.5
14	38.833	2.3190	38.794	2.3212	5,119	300	45.7
15	39.512	2.2807	39.448	2.2842	1,400	70	10.7
16	40.153	2.2457	40.132	2.2469	744	43	6.6
17	41.803	2.1608	41.792	2.1614	3,432	145	22.2
18	43.409	2.0845	43.383	2.0857	4,375	227	34.6
19	48.214	1.8874	48.169	1.8891	1,065	43	6.7
20	48.776	1.8670	48.758	1.8676	899	41	6.3
21	50.350	1.8122	50.320	1.8132	550	52	8.0
22	50.484	1.8077	50.541	1.8058	642	49	7.6
23	51.628	1.7703	51.626	1.7704	2,405	102	15.6

[a]Data were obtained from measurements of peak height.
[b]Data were obtained from measurements of peak area.

Other analyses, such as elemental analysis and IR spectroscopy, support these findings.

Conclusions

Tris transaminates readily with ammonia or primary amines when catalyzed by carbon dioxide or strong organic acids. The polysilazane products range from discrete solid disilazanes, to liquid distillable oligomers, and to highly cross-linked infusible polymers. Some of these polysilazanes can be pyrolyzed to amorphous silicon nitride or mixtures of silicon nitride and silicon carbide below 1550 °C or to crystalline ceramics above that temperature.

References

1. Kireev, V. V. *Advances in Polymer Chemistry;* Korshak, V. V., Ed.; MIR: Moscow, 1986; pp 238–245.
2. Segal, D. L. *Br. Ceram. Trans. J.* **1986,** *85*, 184.
3. Riley, F. L. *Progress in Nitrogen Ceramics;* Martinus Nijhoff: The Hague, Netherlands, 1983.
4. Crosbie, G. M. U.S. Patent 4 582 696, 1986.
5. Szweda, A.; Hendry, A.; Jack, K. H. *Proc. Br. Ceram. Soc.* **1981,** *31*, 107.
6. Johansson, T. U.S. Patent 4 530 825, 1985.
7. Woodhead, J. L.; Segal, D. L. *Chem. Br.* **1984,** *20*, 310.

8. Morgan, P. E. D. *Mater. Res. Soc. Symp. Proc.* **1984,** *32,* 213.
9. Morgan, P. E. D. U.S. Patent 4 552 740, 1985.
10. Prochazka, S.; Greskovitch, C. *Am. Ceram. Soc. Bull.* **1978,** *57,* 579.
11. Arkles, B. U.S. Patent 4 595 775, 1986.
12. Brown-Wensley, K. A.; Sinclair, R. A. U.S. Patent 4 611 035, 1986.
13. Moulson, A. J. *J. Mater. Sci.* **1979,** *14,* 1017.
14. Weiss, J. *Annu. Rev. Mater. Sci.* **1981,** *11,* 381.
15. Mazdiyasni, K. S.; Cooke, C. M. *J. Am. Ceram. Soc.* **1973,** *56,* 628.
16. Billy, M.; Brossard, M.; Desmaison, S.; Girand, D.; Goursat, P. *J. Am. Ceram. Soc.* **1975,** *58,* 254.
17. Rustad, D. S.; Jolly, W. L. *Inorg. Chem.* **1967,** *6,* 1986.
18. Penn, B. G.; Ledbetter, F. E., III; Clemons, J. M.; Daniels, J. G. *J. Appl. Polym. Sci.* **1982,** *27,* 3571.
19. Penn, B. G.; Ledbetter, F. E., III; Clemons, J. M. *Ind. Eng. Chem. Proc. Des. Dev.* **1984,** *23,* 217.
20. Penn, B. G.; Daniels, J. G.; Ledbetter, F. E., III; Clemons, J. M. *Polym. Eng. Sci.* **1986,** *26,* 1191.
21. Seyferth, D.; Wiseman, G. H. *J. Am. Ceram. Soc.* **1984,** *67,* C-132.
22. Seyferth, D.; Wiseman, G. H. U.S. Patent 4 482 669, 1984.
23. Cannady, J. P. U.S. Patent 4 540 803, 1985.
24. LeGrow, G. E.; Lim, T. F.; Lipowitz, J.; Reaoch, R. S. *Am. Ceram. Soc. Bull.* **1987,** *66*(2), 363.
25. Haluska, L. A. U.S. Patent 4 482 689, 1984.
26. Cannady, J. P. U.S. Patent 4 535 007, 1985.
27. Cannady, J. P. U.S. Patent 4 543 344, 1985.
28. Clarke, D. R. *J. Am. Ceram. Soc.* **1982,** *65,* C–21.
29. Kanner, B.; Herdle, W. H. U.S. Patent 4 255 348, 1981.
30. Kanner, B.; Herdle, W. H.; Quirk, J. M. In *Silicon Chemistry;* Corey, E. R.; Corey, J. V.; Gapar, P. P., Eds.; Ellis Horwood: West Sussex, England, 1988; Chapter 12.
31. Quirk, J. M.; Kanner, B. U.S. Patent 4 668 812, 1987.
32. Quirk, J. M.; Kanner, B.; DeMonte, A. P.; Mehta, K. R. U.S. Patent 4 558 146, 1985.
33. King, R. E., III; Kanner, B.; Hopper, S. P.; Schilling, C. L. U.S. Patent 4 675 424, 1987.
34. Kanner, B.; Prokai, B. U.S. Patent 3 957 842, 1976.
35. Tansjo, L. *Acta Chem. Scand.* **1959,** *13,* 29.
36. Tansjo, L. *Acta Chem. Scand.* **1960,** *14,* 2097.
37. Wiberg, N.; Uhlenbrock, W. *Chem. Ber.* **1971,** *104,* 2643.

RECEIVED for review May 27, 1988. ACCEPTED revised manuscript March 15, 1989.

34

Oxidation Reaction of Polycarbosilane

Hiroshi Ichikawa, Fumikazu Machino, Haruo Teranishi, and Toshikatsu Ishikawa

Research and Development Laboratory, Nippon Carbon Company, Ltd., Shin-urashima-cho, Kanagawa-ku, Yokohama 221, Japan

Polycarbosilane (PCS) is an organic polymer with a skeleton of Si–C bonds. SiC fiber (Nicalon) is produced from PCS in three steps: spinning, curing (rendering the material infusible), and heating. SiC fiber has high tensile strength, high elastic modulus, and excellent thermostability in high-temperature oxidizing atmospheres and is generally cured via oxidation of PCS by heating in air. The structure and mechanical characteristics of SiC fiber are largely affected by the conditions of PCS curing. The mechanism of PCS oxidation has not been clarified yet, and to clarify this point, the changes in PCS powder after oxidation in air were studied. The oxidation of PCS involves the following reactions: (1) oxidation of two Si–H groups followed by condensation to give Si–O–Si, (2) oxidation of Si–H to give Si–OH, (3) oxidation of Si–CH_3 to give Si–OH and formaldehyde, and (4) dehydration condensation of two Si–OH groups to give Si–O–Si. PCS becomes infusible by the cross-linkage of Si–O–Si groups. However, infusible PCS contained more than 7 wt % of oxygen, because it had more Si–OH than Si–O–Si. The condensation reaction of Si–OH groups also gives Si–O–Si after heat treatment in nitrogen. By applying this method, we obtained infusible PCS containing only 3–6 wt % of oxygen, which is less than that contained in fibers prepared by the conventional method.

STRUCTURAL MATERIALS made of fiber-reinforced composites with light weight and high strength have come into use in space rockets, aircrafts, automobiles, and sports articles in recent years. Silicon carbide (SiC) fiber, which was first developed by Yajima et al. (*1*, *2*), is drawing particular attention as a reinforcing fiber for composites. SiC fiber with β-SiC as the

0065–2393/90/0224–0619$06.00/0

main component may be obtained by spinning, curing, and heating polycarbosilane (PCS), an organosilicon compound (Figure 1). SiC fiber has been commercialized successfully and is presently on the market with the trade name of Nicalon (*3*, *4*).

SiC fibers are continuous fibers bundled in groups of 500 filaments, with each filament having a diameter of 15 μm. The fibers have high strength,

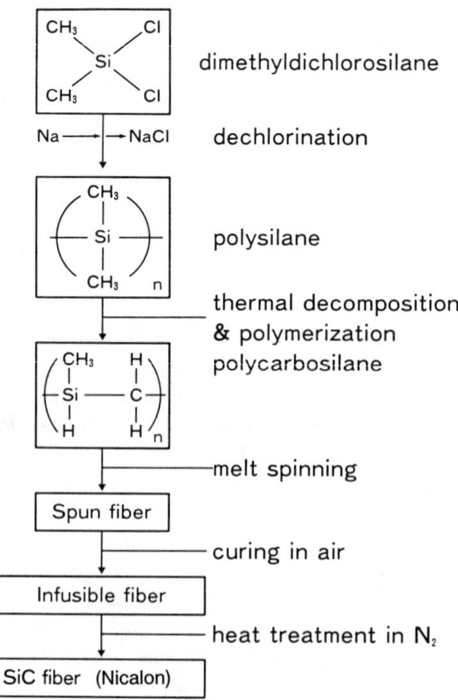

Figure 1. Production process for PCS and SiC fiber (Nicalon).

high modulus, high thermostability in a high-temperature oxidizing atmosphere (*5*), and good compatibility with metal or glass. These characteristics are not found in carbon fibers. Therefore, practical developmental research for applications of the fiber as a heat-resistant material for space rockets and as a reinforcing fiber for fiber-reinforced plastics (FRP), metals (FRM) (*6*, *7*), and ceramics (FRC)(*8*) is in progress.

In the production of SiC fiber, one of the most important steps is the curing process, by which melt-spun PCS fiber is heated without fusion or adhesion and converted to SiC fiber. The structure and mechanical characteristics of the SiC fibers are largely affected by the curing conditions (*9*).

The curing of PCS fiber generally involves oxidation by heating in air.

The reaction polymerizes PCS and and renders it infusible. However, the mechanism of the reaction has not been clarified yet. PCS structural changes during heat treatment in vacuum have been reported by Hasegawa et al. (*10*), and studies related to oxidation in air during SiC fiber production have been partially reported (*11*). This chapter will try to clarify the reaction mechanism of PCS oxidation during curing to produce SiC fiber.

Experimental Procedures

PCS Synthesis. Poly(dimethylsilane) (400 g) was synthesized by the dechlorination reaction of 1066 g of dichlorodimethylsilane with 390 g of fusing sodium in 2.5 L of xylene at 100–140 °C for 12 h (*12*). Poly(dimethylsilane) (300 g) in a 2-L autoclave was subjected to airtight pressure pyrolysis for 4 h at 470 °C. The pressure during the reaction was 0–62 kg/cm^2 (gauge pressure). A light-yellow solid PCS (160 g) was produced as follows: (1) A 200-g sample of reactant was dissolved in 1 L of hexane. (2) The insoluble substance was removed by filtration. (3) The low-molecular-weight fraction was removed by vacuum distillation at 133–400 Pa. The PCS product was analyzed by gel permeation chromatography (GPC), IR spectroscopy, thermogravimetric analysis, (TGA), and differential scanning calorimetry (DSC). The specific gravity and melting point of the product were determined.

Oxidation. PCS powder was used to gain basic knowledge about the curing process. PCS was crushed, pulverized in a mortar, and filtered through a sieve measuring less than 325 mesh. About 2 g of the sample was weighed into a glass measuring bottle (60-mm diameter by 40 mm), which was filled by vibration to make the powder settle evenly. The amount of powder at this time was ~70 mg/cm^2. The PCS-powder-filled measuring bottle was placed in a hot-air-circulating drier and subjected to oxidation for 1–8 h, with the temperature being raised by 10 °C increments (10 °C/h) from 150 °C to a fixed temperature (150, 160, 170, 180, or 190°C). After oxidation, the weight change of the sample was measured, and the IR spectrum and solubility in hexane and tetrahydrofuran (THF) were determined. Melting points were determined for and GPC analyses were performed on the measurable samples.

Heat Treatment under Nitrogen Atmosphere. Among the oxidized PCS powders, those with weight gains of 3–9.5% (six samples) were heated in a nitrogen atmosphere. The oxidized PCS powder (1 g) was placed in a heat-resistant glass container (28-mm diameter by 200 mm) and heated with a cylindrical band heater under a gaseous-nitrogen flow of 2 mL/min. The heating temperature was 260–480 °C. Samples with oxidation weight gains of <6.2% were heated at a rate of 20 °C/h, and samples with weight gains of >8.2% were heated at a rate of 100 °C/h. After heat treatment, the samples were examined by IR spectroscopy, solubility in THF, and visual observation to distinguish whether they had become infusible.

Measurement Techniques. PCS melting points were measured by light permeability with a Mettler automatic melting-point apparatus (model PF-61). The powdered sample, which was filtered through a sieve of less than 325 mesh, was introduced into a 1-mm-diameter glass capillary tube (filling density 0.54–0.57 g/cm^3; height 4.0–4.5 mm), and the temperature was raised at a rate of 10 °C/min. The temperature at which light began to be transmitted was named the melting-starting temperature (T_s), and the temperature at the inflection point at which the sample fuses and the light transmission curve rapidly rises was named the melting-finishing temperature (T_m).

For GPC analysis, a high-speed liquid chromatograph (Toyo Soda Company, model HLC-801 A) was used. The columns (G 2000 H and G 4000 H, two by two) were connected in series. The detector was a differential refractometer. The sample was dissolved in THF (65 mg of PCS per 3 mL of THF) and analyzed at a current speed of 1 mL/min and a pressure of 8 kg/cm^2. The average molecular weight was calculated on the basis of the elution of polystyrene molecular weight standards.

TGA and DSC were performed with a TGA–DSC equipment (Rigaku-denki Company) in air, with sample weights of 3 mg and temperature increments of 5 °C/min. IR spectra were obtained with a spectrophotometer (model TRA-1) manufactured by Nippon-bunko Company. KBr tablets with a density of 1 mg of PCS per 200 mg of KBr were used.

Solubility in hexane or THF was determined by adding ~100 mg of PCS powder to 20 mL of hexane or THF. Methanol was added to any sample that became a gel or was insoluble. Samples that became cloudy were determined as partly soluble, and samples that settled to give a transparent supernatant liquid were determined as insoluble.

Oxidation weight gain was calculated by precisely weighing both treated and untreated PCS samples after leaving them in a silica gel desiccator for more than 1 h and using equation 1

$$\Delta w/w = \frac{W_2 - W_1}{W_1} \times 100 \tag{1}$$

in which $\Delta w/w$ is the oxidation weight gain (in percent), W_1 is the weight before oxidation (in grams), and W_2 is the weight after oxidation (in grams).

Oxygen analysis was performed by using an oxygen–nitrogen analyzer (model EMGA-2200) manufactured by Horiba-Seisakushyo. A 3-mg sample was placed in a graphite crucible and heated to >2500 °C in He gas. Carbon dioxide emissions were measured by using an IR detector, and the amount of oxygen was calculated. Silicon was measured by a gravimetric method after alkali fusion. Carbon and hydrogen were measured by volumetric burning, which applies to JIS M 8813 of coal analysis (Liebig method).

Results and Discussions

PCS Characteristics. The characteristics of PCS powder are given in the list on page 623. In this list, D_4^{25} is density, and M_w and M_n are the weight-average and number-average molecular weights, respectively. M_ns were calculated from GPC data.

The decomposition of PCS was studied by TGA and DSC in air. Results are shown in Figure 2. The starting temperature in exotherm was 170–180 °C, and weight increase started at around 210 °C. A slight difference in the starting temperature was observed. This difference may be due to the rapid rise in temperature (5 °C/min). The weight increased with heating, but it reached a limit at >275 °C, which is the exothermic peak of DSC.

Characteristics of PCS	
D_4^{25}	1.116
Melting points (°C)	
T_s	176
T_m	194
Molecular weights	
M_n	1470
M_w	2980
IR absorbance ratios	
A_{2100}/A_{2950}	2.6
A_{1355}/A_{2950}	0.35
Elemental analysis (wt %)	
Si	45.9
O	0.63
C	41.4
H	8.36

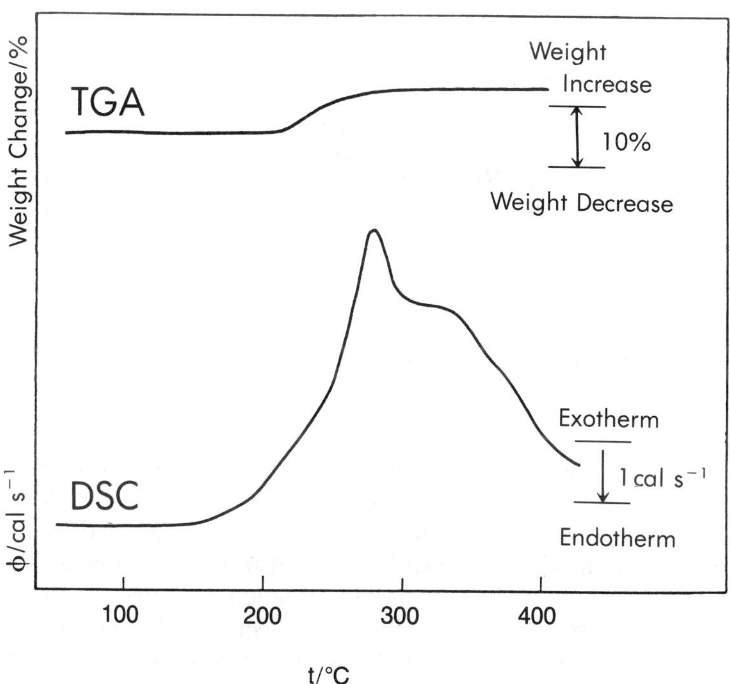

Figure 2. TGA–DSC curves for PSC oxidation in air. The heating rate was 5 °C/min.

PCS Weight Gain by Oxidation. PCS powder was oxidized for 1–8 h in air at 10 °C intervals from 150 to 190°C. The weight gain ($\Delta w/w$) was measured at each level (Figure 3). The weight gain $\Delta w/w$ directly increased with oxidation temperature and time. In particular, the rate of change of $\Delta w/w$ directly increased with oxidation temperature. The weight gain per hour from 0 to 3 h was calculated for each oxidation temperature (Figure 4).

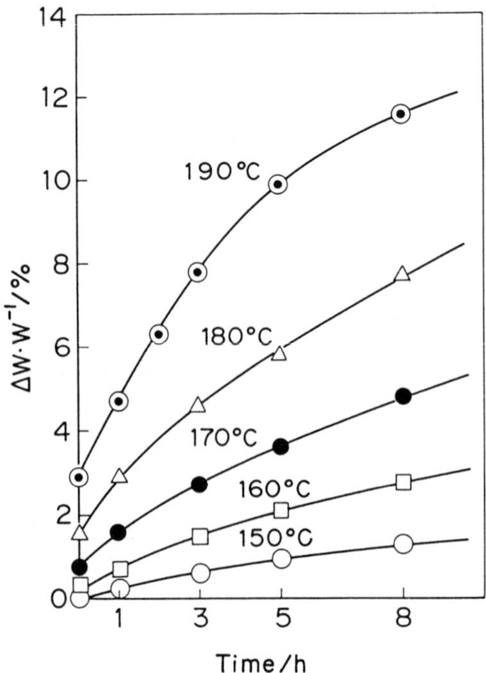

Figure 3. Dependence of Δw/w *on time of oxidation in air at various temperatures.*

The weight gain per hour is proportional to the oxidation reaction rate. However, the reaction rate of PCS reached an inflection between 170 and 180 °C and increased at temperatures of >180 °C. At temperatures of >180 °C, the reaction mechanism must be different from that at temperatures of <170 °C. The PCS weight gain during oxidation was attributed to the introduction of oxygen. Therefore, the oxygen content of some samples was measured by elemental analysis, and its relationship with $\Delta w/w$ was studied (Figure 5).

The curve of the plot of oxygen content versus $\Delta w/w$ makes a 45° angle with the x axis when $\Delta w/w < 7\%$. As is shown in the previous list (*see* page 623), the initial oxygen content of PCS before oxidation is ~0.6%. The

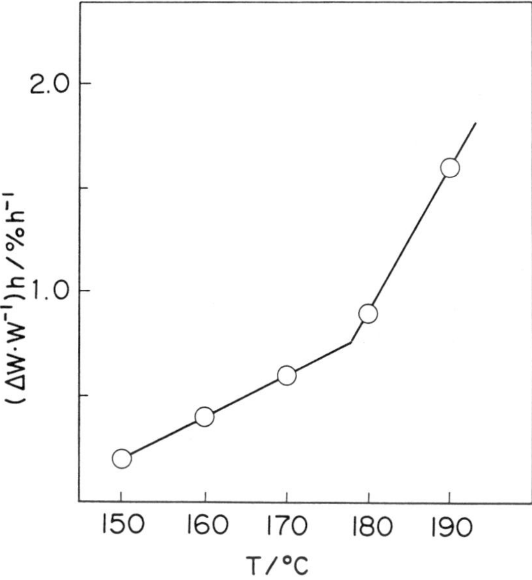

Figure 4. Relation between rate of increase of Δw/w *and temperature of oxidation.*

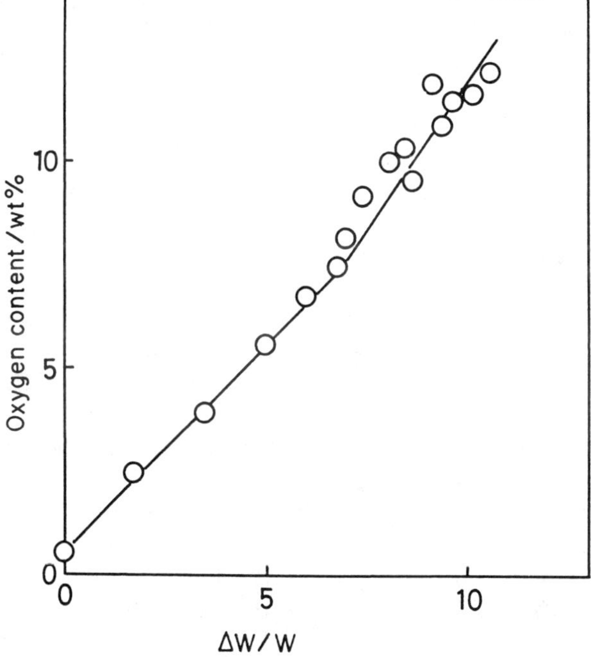

Figure 5. Relation between Δw/w *and oxygen content.*

scatter in data and the increase in oxygen content when $\Delta w/w = 7\%$ are presumed to be due to various reactions that become significant at high temperatures.

Melting Point, GPC, and Solubility of Oxidized PCS. PCS powder that was oxidized at 160–190 °C in air was studied with regard to oxidation weight gain ($\Delta w/w$), melting point, GPC, average molecular weight, and solubility in hexane and THF. Table I shows the effect of oxidation conditions on these properties.

Oxidized PCS melted only partially, with $\Delta w/w = 5.0\%$, and T_m disappeared (i.e., the sample was insoluble in hexane). Furthermore, at $\Delta w/w = 8.5\%$, T_s became undetectable, and in THF, the sample became a partially insoluble gel. Figure 6 shows the GPC chromatogram of the THF-soluble PCS. As oxidation continued, the peak at the high-molecular-weight side with an elution volume of 48 mL (M_n 9000) increased. The tailing expanded to the high-molecular-weight side, and M_w/M_n increased.

PCS is cured to keep its shape, as well as to prevent the surface of the fibers from adhering to each other when the spun fiber is heated subsequently at high temperature. Even a very small fusion causes defects on the fiber surface. Therefore, the disappearance of T_s indicates the end of infusion. This infusion-finishing point occurs when $\Delta w/w = 7.1\%$. Therefore, $\Delta w/w \simeq 7\%$ is the turning point for fusion and infusion.

IR Spectroscopic Analysis of Oxidized PCS. ***Characteristic IR Absorption Bands.*** The IR spectra of PCS before and after oxidation were obtained. The IR spectra of untreated PCS and PCS treated at 180 °C for 1–10 h are shown in Figure 7. The characteristic IR absorption bands of PCS are as follows (cm^{-1}): 2950, 2900 (C–H stretching vibration); 2100 (Si–H stretching vibration); 1355 (CH_2 deformation vibration of Si–CH_2–Si bond), 1260 (Si–CH_3 deformation vibration); 1020 (CH_2 wagging vibration of Si–CH_2–Si); 830 (rocking vibration of Si–CH_3).

In PCS synthesized from poly(dimethylsilane), some Si–Si bonds could remain, depending on the synthesizing conditions. Hasegawa et al. (*10*) studied the PCS structure on the basis of NMR and IR spectroscopic studies of various PCS types. They (*10*) reported that PCS PC-470 synthesized from poly(dimethylsilane) in an autoclave at 470 °C for 14 h is rearranged completely and contains no remaining Si–Si bonds. The IR absorbance ratios A_{2100}/A_{2950} and A_{1355}/A_{2950} of PCS synthesized in our experiment are 2.6 and 0.35, respectively (*see* list on page 623). These ratios are almost equal to the absorbance ratios (2.5 and 0.3, respectively) of PCS PC-470 prepared by Hasegawa et al. (*10*). Therefore, we conclude that the rearrangement of PCS is complete and that the possibility of any remaining Si–Si bonds is small.

The characteristic change in the IR spectrum of PCS after oxidation is

Table I. Changes in PCS Characteristics After Oxidation in Air

	Heat Treatment		$\Delta w/w$						*Solubility*	
Sample	*(° C)*	*Time (h)*	(%)	T_s (°C)	T_m (°C)	M_n	M_w	M_w/M_n	*Hexane*	*THF*
PCS				176	194	1,470	2,980	2.0	soluble	soluble
101	160	1	0.6	178	205	1,490	3,130	2.1	soluble	soluble
201	170	1	1.6	185	247	1,600	3,490	2.2	soluble	soluble
301	180	1	3.3	211	278	1,700	3,780	2.2	soluble	soluble
304	180	4	5.0	235	—[b]	1,920	4,930	2.6	insoluble	soluble
306	180	6	6.5	273	—[b]	2,660	16,600	6.3	insoluble	soluble
307	180	7	7.1	—[a]		—[c]	—[c]	—[c]	insoluble	gel
308	180	8	7.6	—[a]		—[c]	—[c]	—[c]	insoluble	gel
310	180	10	8.5	—[a]		—[c]	—[c]	—[c]	insoluble	gel
405	190	5	9.9	—[a]		—[c]	—[c]	—[c]	insoluble	insoluble
409	190	9	11.9	—[a]		—[c]	—[c]	—[c]	insoluble	soluble

NOTE: Mean molecular weights were measured by GPC.
[a]The sample did not melt.
[b]The sample did not melt completely.
[c]The sample could not be analyzed by GPC because it was not completely soluble in THF.

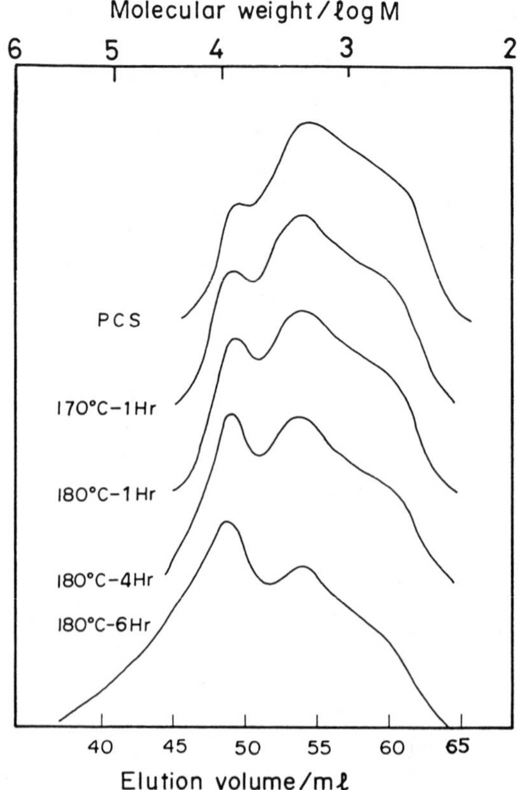

Figure 6. GPC elution curves for PCS at various conditions of oxidation in air.

that, as the Si–H peak at 2100 cm^{-1} decreases, new peaks appear. One new peak is a broad Si–OH stretching vibration at ~3400 cm^{-1}, and another is a C=O stretching vibration at ~1740 cm^{-1}. Still another new peak is due to Si–O–Si absorption at ~1000 cm^{-1}

Relation Between Absorbance Ratios and Oxidation Weight Gain. The change in the IR spectrum of PCS after oxidation was studied in terms of the relationship between the absorbance ratios at specific wave numbers ($A_\nu/A_{\nu'}$) and $\Delta w/w$. Characteristic absorptions at 2950 (C–H), 3400 (Si–OH), 2100 (Si–H), and 1260 ($Si–CH_3$) cm^{-1} were chosen as ν and ν'. The absorption peak at 2950 cm^{-1} originates from the silylenemethylene bond, which is the basic structure of PCS, and the C–H bond in $Si–CH_3$. This absorption band should be minimally affected by the oxidation reaction, and because of minimal observable change in this absorption, it was chosen as the reference absorption band (Figure 8).

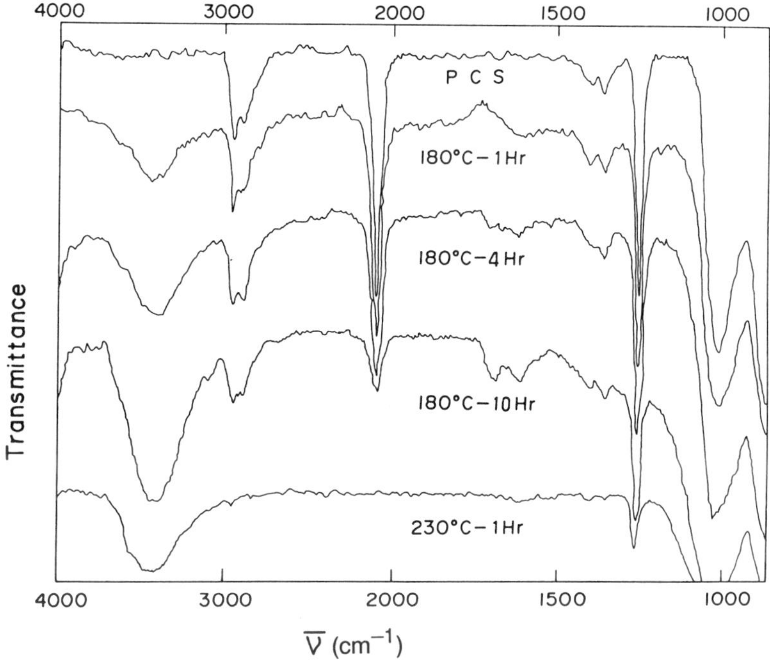

Figure 7. IR spectra of PCS after oxidation in air.

As the weight gain increased, A_{3400}/A_{2950} increased, and A_{2100}/A_{2950} decreased. These results mean that the Si–H groups decrease and the Si–OH groups increase. The ratio A_{1260}/A_{2950} was almost constant regardless of the increase in $\Delta w/w$. However, for the sample oxidized at <230 °C for 1 h (for which $\Delta w/w$ = 17%; Figure 7), the C–H absorption peak at 2950 cm^{-1} and the Si–CH_3 absorption peak at 1260 cm^{-1} clearly decreased. Therefore, to examine the change in Si–CH_3 groups in detail, the absorbance of the characteristic absorption was studied.

Before and after oxidation, the IR spectra of PCS powders in KBr tablets at a fixed density (1 mg of PCS per 200 mg of KBr) and thickness were obtained. The absorbance was calculated by using the Beer–Lambert formula ($A = \log_{10} I_o/I$). The siloxane bond (Si–O–Si) was expected to be formed as a result of the oxidation reaction. This bond shows a stretching vibration at 1000–1090 cm^{-1}. However, a strong absorption due to the wagging vibration of Si–CH_2–Si occurs at 1020 cm^{-1}, and because of this absorption, the siloxane peaks are not noticeable. Therefore, Si–O–Si bond formation was studied with A_{1020}/A_{830}, which was corrected by using the absorption peak of Si–CH_3 (stretching) at 830 cm^{-1} as the reference band. The changes in A_{1260}, A_{2100}, and A_{1020}/A_{830} are shown in Figure 9.

As the oxidation weight gain increased, A_{2100} (Si–H) decreased rapidly,

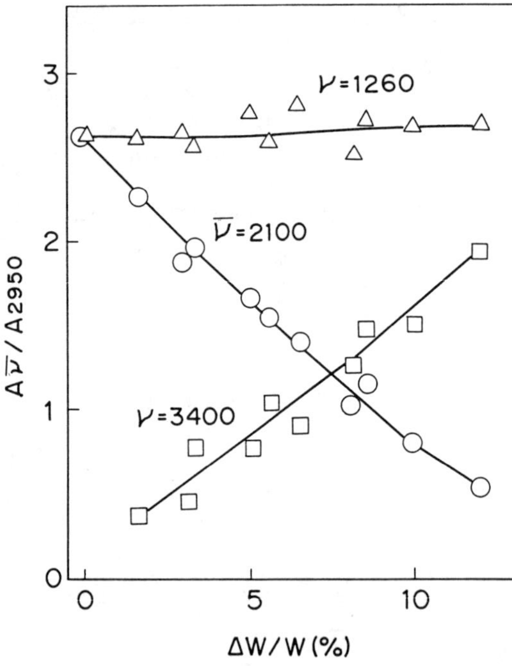

Figure 8. Relation between Δw/w *and* A_ν/A_{2950} *for PCS.*

but A_{1260} (Si–CH_3 deformation vibration) clearly decreased, in contrast to the results shown in Figure 8. A_{1020}/A_{830} also increased. These results indicate that the cross-linkage of the Si–O–Si bond increases with oxidation.

Proposed Mechanism of PCS Oxidation in Air. From the results of heat treatment in air, the oxidation of PCS probably proceeds as follows:

cross–linking reaction

$$\rangle SiH + HSi\langle + O_2 \rightarrow \rangle Si{-}O{-}Si\langle + H_2O \qquad (2)$$

$$\rangle SiOH + HOSi\langle \rightarrow \rangle Si{-}O{-}Si\langle + H_2O \qquad (3)$$

Si–OH bond formation

$$\rangle SiH + \tfrac{1}{2}O_2 \rightarrow \rangle SiOH \qquad (4)$$

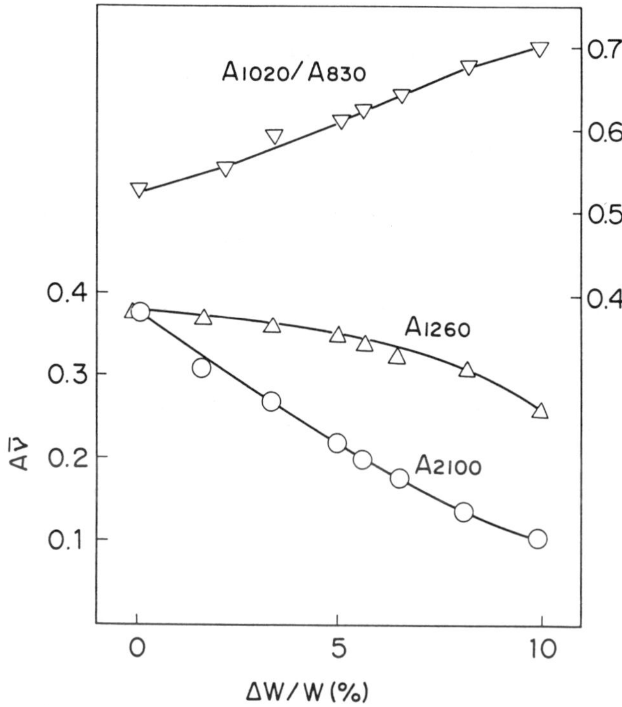

Figure 9. Relation between Δw/w *and absorbance in IR spectrum for PCS.*

$$\geq\!SiCH_3 + O_2 \rightarrow [\geq\!SiCH_2OOH \rightarrow\ \geq\!Si^{\cdot} + {}^{\cdot}OH + CH_2O] \rightarrow$$

$$\geq\!SiOH + CHO_2O \xrightarrow{O_2}\ \geq\!SiOH + CO_2 + H_2O$$

$$\downarrow \tfrac{1}{2}O_2$$

$$\geq\!SiOH + HCOOH \qquad (5)$$

Originally, the main reaction in PCS oxidation was believed to be the formation of Si–O–Si bonds by the oxidation of Si–H groups, which are the most active functional groups (equation 2). Figures 5 and 8 show that A_{2100} (Si–H) rapidly decreased with oxidation, but A_{3400}/A_{2950} (Si–OH bonds) increased notably. However, A_{1020}/A_{830} (Si–O–Si) increased slowly.

Because PCS oxidation involves a reaction between the gaseous and solid phases, the degree of freedom for the Si–H bond in the solid phase is small. Therefore, only the Si–H bonds that are positioned in three dimensions and are near each other can react with oxygen to form Si–O–Si bonds.

With large distances between three-dimensional Si–H groups, oxidation results in Si-OH bond formation (equation 4).

Figure 9 shows that A_{1260} decreased as $\Delta w/w$ increased. This result is explained by the disappearance of Si–CH_3 groups as they are oxidized (equation 5) to form Si–OH bonds and formaldehyde. This reaction was proven by the presence of formaldehyde, which was detected only at >180 °C in the exhaust gas of the furnace by the phenylhydrazine method during curing.

The weight increase accompanying the change from Si–CH_3 to Si–OH (equation 5) is 2 g/mol, which, compared with 14 and 16 g/mol for equations 2 and 4, respectively, is relatively small. The exothermic reaction without weight increase observed by TGA–DSC analysis (Figure 2) can be explained by this oxidation of the Si–CH_3 bonds. The inflection point in the oxidation rate between 170 and 180 °C (Figure 4) may be explained as follows: Only the reaction given by equation 2 occurs at <170 °C; at >180 °C, the reactions given by equations 3–5 also occur, and the reaction rate is accelerated. The variation in the relation between $\Delta w/w$ and oxygen content when $\Delta w/w$ = 7% (Figure 5) is explained similarly.

The studies of the oxidation reaction in air reveal that Si–OH groups are formed in large amounts by the oxidation of Si–H and Si–CH_3. Therefore, among the infusible PCS samples with a minimum oxygen content of 7 wt %, many Si–OH groups are present but do not contribute to the curing. This conclusion suggests that PCS curing is possible even with a lower oxygen content.

PCS Oxidation by Heating in Nitrogen. ***Results.*** The oxidized PCS samples were heated in a nitrogen atmosphere up to 500 °C. The experimental conditions and THF solubility of heat-treated PCS samples are shown in Table II. The solubility of each sample in THF changes from soluble, to partially soluble (becomes a gel), and to insoluble as the heating temperature rises. As previously stated, the increase in molecular weight of PCS is closely related to the increase in melting point. The curing was considered finished when PCS became gellike in THF. Therefore, these experimental results indicate that PCS is cured even by heat treatment in a nitrogen atmosphere.

After heat treatment, the appearance and flowing condition of the PCS powder were examined. A sample with $\Delta w/w$ = 2.9% clearly melted and congealed, but samples with $\Delta w/w$ > 3.2% remained in powdered form, with no trace of fusion. Even oxidized PCS with $\Delta w/w$ < 6.5% (Table I), which melted or partially melted at <300 °C, did not melt when heated slowly at the rate of 20 °C/h.

For comparison, a sample with $\Delta w/w$ = 6.2% was heated to 480 °C at a rate of 100 °C/h in nitrogen atmosphere. The PCS powder congealed and partially melted. Figure 10 shows the changes in the IR spectrum in relation to the increase in heating temperature of a PCS sample with $\Delta w/w$ = 8.2%,

Table II. Changes in PCS Characteristics After Oxidation in Nitrogen

Sample	$\Delta w/w$ *(%)*	*Temperature (°C)*	*Heating Rate (°C/h)*	*THF Solubility*	*Observation After Heat Treatment*
302	2.9			soluble	
A		260	20	soluble	solid (melted)
B		300	20	soluble	solid (melted)
401	3.2			soluble	
B		380	20	insoluble	powder
C		480	20	insoluble	powder
402	4.9			soluble	
A		290	20	gel	powder
C		480	20	insoluble	powder
307	6.2			soluble	
A		260	20	gel	powder
B		300	20	insoluble	powder
C		380	20	insoluble	powder
310	8.2			gel	
A		290	100	insoluble	powder
B		390	100	insoluble	powder
C		480	100	insoluble	powder
406	9.5			insoluble	
A		290	100	insoluble	powder
B		390	100	insoluble	powder
C		460	100	insoluble	powder

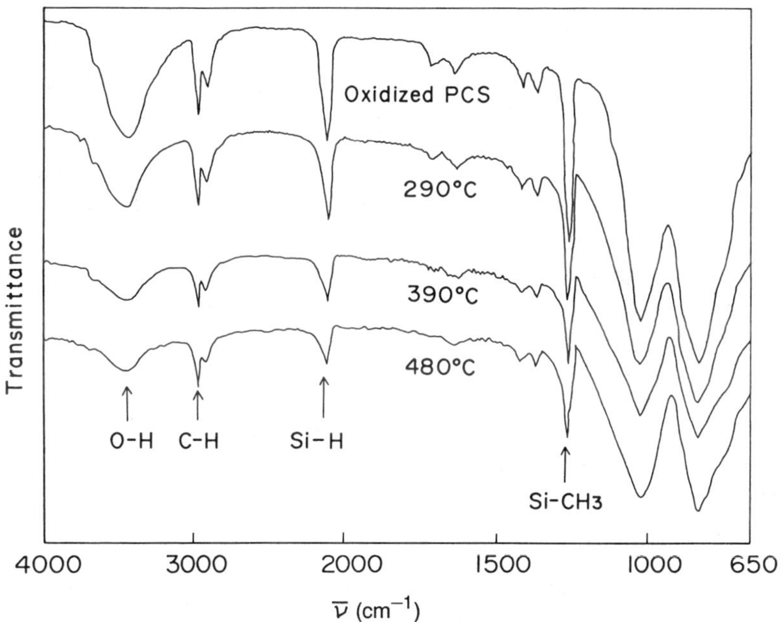

Figure 10. IR spectra of oxidized PCS after heating to 480 °C in nitrogen.

and Figures 11a–11c indicate the change in absorbance ratios for each sample. Peaks at 3400, 2950, 2100, 1260, 1020, and 830 cm^{-1} were chosen as characteristic absorptions, and among them, the absorptions at 2950 (C–H stretching) and 830 (Si–CH_3 rocking) cm^{-1} were chosen as the reference bands.

Figure 11a shows that A_{3400}/A_{2950} rapidly decreased as the heating temperature rose. The decrease of A_{3400}/A_{2950} indicates a decrease of Si–OH groups relative to C–H groups. Figure 10 shows that the absorbance at 2950 cm^{-1}, which was the reference band for the absorbance ratio, tended to decrease slightly with the decrease in the transmittance of light through all the samples. Therefore, the decrease in A_{3400}/A_{2950} can be regarded as a decrease in Si–OH groups. This result indicates that the cross-linkage of Si–O–Si caused by a dehydration condensation reaction among Si–OH groups still continues as the heating temperature rises.

Figure 11b shows that A_{2100}/A_{2950} decreased at >300 °C for the sample with $\Delta w/w < 6.2\%$ and for the sample with $\Delta w/w = 8.2\%$ until ~400 °C. This result means that during oxidation treatment PCS with a large amount of remaining Si–H starts to decrease at 300 °C and PCS with a few remaining Si–H starts to decrease at 400 °C.

Figure 11c shows that A_{1260}/A_{2950} also tended to decrease.

Figure 11d shows that A_{1020}/A_{830} increased as the heating temperature rose. As previously stated, the peak at 1020 cm^{-1} is caused by Si–O–Si stretching vibration and CH_2 deformation in Si–CH_2–Si. Also, the absorption at 830 cm^{-1} is caused by the Si–CH_3 stretching vibration. Therefore, this result indicates that Si–O–Si and Si–CH_2–Si groups are increasing relative to Si–CH_3 groups. The Si–CH_2–Si group also has an absorption at 1355 cm^{-1} for a different angle of vibration.

The Si–CH_2–Si absorption was hardly changed; the value of A_{1350}/A_{2950} for the sample was 0.3–0.4 and did not change during the experiment as the temperature rose. Because the Si–CH_3 absorption, which was the reference, was almost unchanged until 300 °C, the increase in A_{1020}/A_{830} until 300 °C is attributed to the increase in Si–O–Si groups.

Figure 11 also shows the THF solubility of heat-treated PCS samples. The solubility in THF is closely related to the degree of infusion. The value of A_{1020}/A_{830} had the highest correlation with THF solubility (Figure 11d). The samples for which $A_{1020}/A_{830} < 0.63$ were soluble in THF. When $A_{1020}/A_{830} = 0.66$–0.67, the samples became gels (partially soluble), and when $A_{1020}/A_{830} > 0.69$, the samples became insoluble. This trend proves that PCS becomes infusible with cross-linkage of Si–O–Si groups and polymerization.

Proposed Mechanism of PCS Oxidation in Nitrogen. From the results just discussed, the oxidation of PCS by heating to 500 °C in a nitrogen

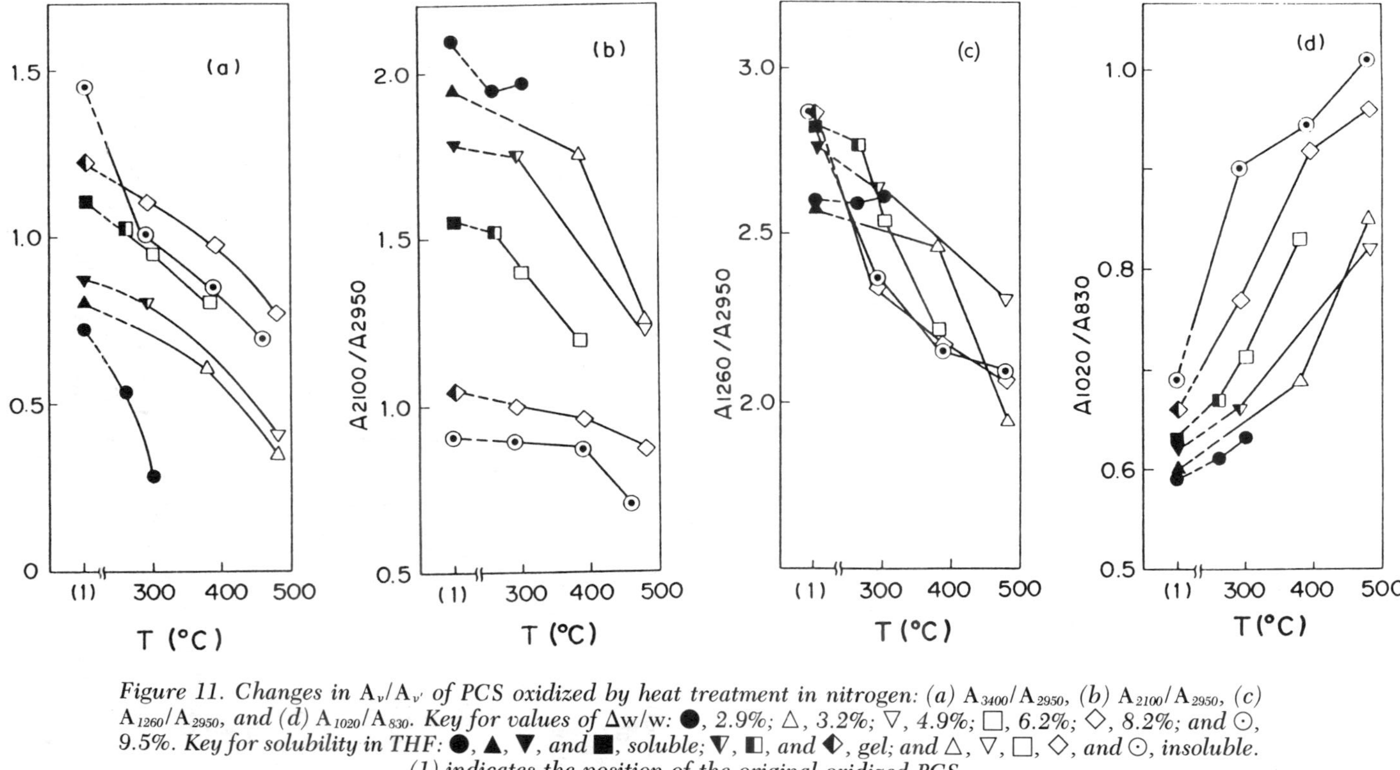

Figure 11. Changes in $A_{\nu}/A_{\nu'}$ *of PCS oxidized by heat treatment in nitrogen: (a)* A_{3400}/A_{2950}, *(b)* A_{2100}/A_{2950}, *(c)* A_{1260}/A_{2950}, *and (d)* A_{1020}/A_{830}. *Key for values of* $\Delta w/w$: ●, 2.9%; △, 3.2%; ▽, 4.9%; □, 6.2%; ◇, 8.2%; *and* ⊙, 9.5%. *Key for solubility in THF:* ●, ▲, ▼, *and* ■, *soluble;* ▼, ◧, *and* ◆, *gel; and* △, ▽, □, ◇, *and* ⊙, *insoluble.* (1) *indicates the position of the original oxidized PCS.*

atmosphere probably proceeds by the following mechanism:

$$\geq SiOH + HOSi\leq \rightarrow \geq Si{-}O{-}Si\leq + H_2O \quad (3)$$

$$\geq SiOH + HSi\leq \rightarrow \geq Si{-}O{-}Si\leq + H_2 \quad (6)$$

$$\geq SiOH + CH_3Si\leq \rightarrow \geq Si{-}O{-}Si\leq + CH_4 \quad (7)$$

$$\geq SiH + CH_3Si\leq \rightarrow \geq Si{-}CH_2{-}Si\leq + H_2 \quad (8)$$

The main reaction of PCS oxidation by heating in a nitrogen atmosphere involves the formation of Si–O–Si by a dehydration condensation reaction of Si–OH bonds (equation 3). This reaction can be deduced from the fact that the number of Si–OH bonds decreases notably compared with those of other groups. The decrease in Si–H ($\Delta w/w < 6.2\%$) and Si–CH_3 bonds at >300 °C may be explained by equations 6 and 7. At >400 °C, a decrease in Si–H and Si–CH_3 bonds was noted for every PCS sample. This finding suggests that Si–CH_2–Si bond formation from Si–H and Si–CH_3 proceeds by equation 8.

The previous results indicate that the Si–OH groups introduced by the oxidation reaction proceed to form Si–O–Si bonds through dehydration condensation even with heating in a nitrogen atmosphere.

As shown in Table II, even a sample with $\Delta w/w = 3.2\%$, which was fusible before treatment, became infusible without adhesion with gradual heating in nitrogen. This fact means that the Si–O–Si content required for infusion is equal to an oxygen content of 3 wt %.

Therefore, the actual oxygen content is less than one-half of the minimum oxygen content (7 wt %) in the infusible PCS. This conclusion indicates that SiC fibers with less oxygen content (3–6 wt %) compared with conventional SiC fiber can be produced.

Conclusion

The characteristics of PCS were studied after oxidation in air or nitrogen under various conditions. The results were used to elucidate the mechanisms of the oxidation and curing reactions. The following conclusions were reached:

- The melting point and molecular weight of PCS increases with an increase in weight during oxidation. PCS becomes infusible when the weight gain during oxidation is >7%. The weight gain caused by oxidation is almost the same as the amount of oxygen introduced.

- The oxidation of PCS proceeds according to the following steps:
  1. oxidation of two Si–H bonds followed by condensation to give Si–O–Si,
  2. oxidation of Si–H to give Si–OH,
  3. oxidation of the Si–CH_3 bond to give Si–OH and formaldehyde, and
  4. dehydration condensation of two Si–OH to give Si–O–Si.
- PCS becomes infusible by cross-linkage of Si–O–Si groups. However, the infusible PCS samples contain large amounts of Si–OH groups that do not contribute to curing.
- Si–O–Si bond formation by dehydration condensation of Si–OH groups continues during gradual heating in a nitrogen atmosphere. PCS becomes infusible even with oxidation weight gains of 3–6%.

On the basis of experimental findings, the amount Si–O–Si groups required for the curing of PCS is equivalent to a PCS oxygen content of 3 wt %. Therefore, SiC fibers with less oxygen content compared with conventional SiC fibers can be prepared.

References

1. Yajima, S.; Hayashi, J.; Omori, M. *Chem. Lett.* **1975,** 931.
2. Yajima, S.; Okamura, K.; Hayashi, J. *Chem. Lett.* **1975,** 1209.
3. Ishikawa, T.; Nagaoki, T. *Recent Carbon Technology;* JEC Press Inc.: Cleveland, OH, 1983; p 348.
4. Teranishi, H.; Ichikawa, H.; Ishikawa, T. *New Mater. New Proc.* **1983,** *2,* 379.
5. Ishikawa, T.; Ichikawa, H.; Teranishi, H. *Proc. Electrochem. Soc.* 88–5; *Proc. Symp. High Temp. Mater. Chem.* **1987,** *4,* 205–217.
6. Yajima, S.; Okamura, K.; Tanaka, J.; Hayase, T. *J. Mater. Sci.* **1980,** *15,* 2139.
7. Ishikawa, T. *CHEMTECH* **1989,** *19,* 38.
8. Prewo, K. M.; Brennan, J. J. *J. Mater. Sci.* **1982,** *17,* 2371.
9. Ichikawa, H.; Teranishi, H.; Ishikawa, T. *J. Mater. Sci. Lett.* **1987,** *6,* 420–422.
10. Hasegawa, Y.; Okamura, K. *J. Mater. Sci.* **1983,** *18,* 3633.
11. Taki, T.; Maeda, S.; Okamura, K.; Sato, M.; Matsuzawa, T. *J. Mater. Sci. Lett.* **1987,** *6,* 826–828.
12. Yajima, S.; Hasegawa, Y.; Hayashi, J.; Iimura, M. *J. Mater. Sci.* **1978,** *13,* 2569.

RECEIVED for review May 27, 1988. ACCEPTED revised manuscript May 11, 1989.

OTHER SILICON-BASED POLYMERS

35

Synthesis and Properties of Silicon-Containing Polyacetylenes

Toshio Masuda and Toshinobu Higashimura

Department of Polymer Chemistry, Kyoto University, Kyoto 606, Japan

The synthesis, structure, properties, and applications of Si-containing polyacetylenes are reviewed. Various Si-containing acetylenes were polymerized, namely, $HC{\equiv}CCH[Si(CH_3)_3](\text{n-}C_5H_{11})$ *(**2c**),* $HC{\equiv}C[[\text{o-}Si(CH_3)_3]C_6H_4]$ *(**3**), and* $CH_3C{\equiv}CSi(CH_3)_3$ *(**4a**). Compounds **2c** and **3** polymerize in high yields with Mo and W catalysts, whereas **4a** produces a polymer with Nb and Ta catalysts. The molecular weights of the polymers are in the range* 10^5*–*10^6*. The polymers possess the structure* $-(CR{=}CR')_n-$*. Poly(**2c**), poly(**3**), and poly(**4a**) are yellow, purple, and colorless solids, respectively. Unlike polyacetylene, these polymers are stable in air and soluble in many organic solvents. Poly(**4a**) shows extremely high oxygen permeability and ethanol permselectivity during ethanol–water pervaporation.*

THE METHODS FOR PREPARING POLYACETYLENE are well established, and the applications of polyacetylene are the subject of intensive research at present (*1*, *2*). In contrast, the study of substituted polyacetylenes has not advanced so much (*3–6*), probably because the preparation of high-molecular-weight polymers from substituted acetylenes has been difficult. In addition, relatively few studies have been made about Si-containing polyacetylenes, and no review article has been published.

Transition metal catalysts are useful for the polymerization of acetylenes. Ti catalysts are known to polymerize acetylene. Catalysts containing group V and VI transition metals (i.e., Nb, Ta, Mo, and W) polymerize substituted acetylenes (*3*, *4*). The group V and VI transition metal catalysts can be classified into three groups: (1) chlorides of Nb, Ta, Mo, and W; (2) 1:1 mixtures of the metal chlorides with organometallic cocatalysts (e.g.,

0065–2393/90/0224–0641$06.25/0

$(C_6H_5)_4Sn$); and (3) catalysts obtained by UV irradiation of the CCl_4 solution of $Mo(CO)_6$ and $W(CO)_6$.

The following relationships between the structure of acetylenic monomers and the activity of catalysts are observed usually: Ziegler catalysts can give rise to polymers from acetylene and monosubstituted acetylenes without bulky substituents. In contrast, Mo and W catalysts are very effective for monosubstituted acetylenes with bulky substituents and disubstituted acetylenes with less bulky substituents. Further, Nb and Ta catalysts are useful for various disubstituted acetylenes, including those with bulky substituents.

We have been interested in the synthesis of high-molecular-weight polymers from substituted acetylenes by use of group V and VI transition metal catalysts (*3, 4*). Such monomers include HC≡C(*tert*-C_4H_9), HC≡C[(*o*-CF_3)C_6H_4], CH_3C≡C(*n*-C_5H_{11}), $CH_3C{\equiv}CC_6H_5$, ClC≡C(*n*-C_6H_{13}), and $ClC{\equiv}CC_6H_5$. All these monomers are sterically fairly crowded. When suitable catalysts and solvents are chosen for the individual monomers, the polymer yields can become as high as 80–100%. The molecular weights of those polymers are in the range 5×10^5–2×10^6. In contrast, acetylenes that are sterically not so crowded, such as HC≡C(*n*-C_6H_{13}) and $HC{\equiv}CC_6H_5$, give polymers with lower molecular weight, probably because of side reactions like cyclization.

Many properties of polyacetylenes with bulky substituents are substantially different from those of polyacetylene. For example, the substituted polyacetylenes are soluble because of the interaction between the substituents and solvent. Furthermore, such polymers are usually only lightly colored and are stable in air at room temperature. These properties arise from twisted conformations assumed by the main chain because of the presence of substituents. The electrically insulating and nonparamagnetic properties of substituted polyacetylenes are attributable also to the same cause.

Polymerization of Si-Containing Monosubstituted Acetylenes

Among Si-containing acetylene monomers, only (trimethylsilyl)acetylene ($HC{\equiv}CSi(CH_3)_3$; **1a**) has appeared in the literature, if our studies are excepted.

$$HC{\equiv}C\overset{\displaystyle CH_3}{\underset{\displaystyle CH_3}{\overset{|}{\underset{|}{Si}}}}{-}R$$

1

R: **1a,** CH_3; **1b,** C_6H_5; **1c,** *tert*-butyl;

1d, *n*-C_6H_{13}; **1e,** $-CH_2CH_2C_6H_5$; **1f,** $-CH_2C_6H_5$

Russian researchers (*7*) have reported that $MoCl_5$ polymerizes this monomer to give a yellow polymer. Zeigler (*8*) has synthesized a totally soluble polymer by using a mixture of WCl_6 and methylmagnesium bromide. A French group (*9*) has used a metal carbene complex, $(CO)_5W{=}CR_1R_2$, as initiator. However, the polymers of **1a** [poly(**1a**)] so produced usually have low molecular weights, around 10^4, and are often partly insoluble in all known solvents.

The polymerization of **1a** and its homologues (**1b**–**1f**) was studied first (*10, 11*). Polymerizations were carried out with WCl_6 as catalyst in toluene at 30 or 80 °C for 24 h. All the monomers in Table I produce methanol-insoluble polymers in moderate to good yields.

Table I. Polymerization of $HC{\equiv}CSi(CH_3)_2R$ by WCl_6

R	*Yield (%)*	M_n[a]
CH_3[b]	47	partly insoluble
C_6H_5	44	partly insoluble
tert-C_4H_9	87	partly insoluble
n-C_6H_{13}	21	9800
$CH_2CH_2C_6H_5$	33	9000
$CH_2C_6H_5$	38	7100

NOTE: The reactions were carried out in toluene at 80 °C for 24 h ($[M]_0$ = 0.50 M; [catalyst] = 10 mM). Practically no or very little polymers were formed with Mo, Nb, or Ta catalysts.
[a]Values of M_n were measured by gel permeation chromatography.
[b]The reaction was carried out at 30 °C.

Poly(**1a**), poly(**1b**), and poly(**1c**), which possess inflexible R groups, are partly insoluble in all known solvents. In contrast, poly(**1d**), poly(**1e**), and poly(**1f**), whose R groups are flexible, dissolve completely in toluene. The molecular weights of these polymers, however, are not more than 10^4. All the poly(**1**)s are yellow. Whereas poly(**1d**) is rubbery, other poly(**1**)s are powdery.

Because the type **1** monomers do not give rise to high-molecular-weight polymers, we next considered 3-(trimethylsilyl)-1-alkynes (**2**) as monomers (*12*).

$$HC{\equiv}C\underset{\displaystyle Si(CH_3)_3}{\underset{|}{C}}HR$$

2

R: **2a,** CH_3; **2b,** *n*-C_3H_7; **2c,** *n*-C_5H_{11}; **2d,** *n*-C_7H_{15}

In these monomers, the silyl group is not directly bonded to the acetylenic carbon. Although linear 1-alkynes produce oligomeric products only, the type **2** monomers are expected to form high-molecular-weight polymers because of steric crowding. These monomers are prepared from the corresponding 1-alkynes by dilithiation followed by silylation at C-3.

When the type **2** monomers are polymerized with $MoCl_5$ and WCl_6 in toluene at 30 °C for 24 h, the yields of methanol-insoluble polymers become as high as 70–80% (Table II). For the polymerization of **2c**, use of $(C_2H_5)_3SiH$ as cocatalyst enhances both the yield and the molecular weight of the polymer. The $Mo(CO)_6$- and $W(CO)_6$-based catalysts are also effective.

Table II. Polymerization of HC≡CCH[Si(CH$_3$)$_3$]R by Mo and W Catalysts

R	*Catalyst*	*Yield (%)*	M_w *(× 10³)*[a]
CH_3	$MoCl_5$	80	partly insoluble
	WCl_6	70	partly insoluble
n-C_3H_7	$MoCl_5$	82	partly insoluble
	WCl_6	74	partly insoluble
n-C_5H_{11}	$MoCl_5$	77	98
	WCl_6	72	42
n-C_7H_{15}	$MoCl_5$	82	170
	WCl_6	80	55
n-C_5H_{11}	$MoCl_5$–$(C_2H_5)_3SiH$	90	320
	WCl_6–$(C_2H_5)_3SiH$	88	300
	$Mo(CO)_6$-$h\nu$[b]	75	100
	$W(CO)_6$-$h\nu$[b]	81	160

NOTE: The reactions were carried out in toluene at 30 °C for 24 h ($[M]_0$ = 0.50 M; [catalyst] = [cocatalyst] = 10 mM). Practically no or very little polymers were formed with Nb or Ta catalysts.
[a]Values of M_w were measured by gel permeation chromatography.
[b]The reaction was carried out in CCl_4.

Poly(**2a**) and poly(**2b**) are partly insoluble in toluene. In contrast, poly(**2c**) and poly(**2d**) are totally soluble in toluene, because they have a long, flexible alkyl chain. The molecular weights of these polymers are fairly high, 5×10^4–3×10^5. Free-standing films can be obtained from poly(**2c**) and poly(**2d**) by solution casting (usually, a molecular weight of over 10^5 is necessary for the formation of free-standing film from a substituted polyacetylene). All the poly(**2**)s are yellow solids.

The third type of monomer among Si-containing monosubstituted acetylenes is *o*-(trimethylsilyl)phenylacetylene (**3**).

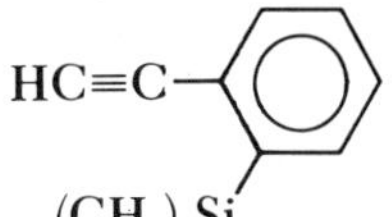

HC≡C–C$_6$H$_4$–Si(CH$_3$)$_3$

(CH$_3$)$_3$Si

3

Because phenylacetylene is not very crowded sterically, the molecular weight of the polymer is usually in the 10^4 range only (*3, 4*). *o*-Methylphenylacetylene, which has a somewhat larger steric crowding, produces a polymer with higher molecular weight. In this respect, we determined how high the molecular weight of poly(**3**) can be. The monomer is prepared from phenylacetylene by metallation at the terminal and *ortho* positions, silylation at both positions, and partial desilylation (*13*).

Table III shows results of the polymerization of **3** by various catalysts (*14*). Poly(**3**) is obtained almost quantitatively in the presence of W and Mo catalysts. Nb and Ta catalysts also give the polymer, although the yields are not high. The molecular weight of poly(**3**) exceeds 10^6 in many cases. Poly(**3**) is dark purple, in striking contrast with the whiteness or paleness of most disubstituted acetylene polymers. A tough film can be prepared by casting from the polymer solution.

Table III. Polymerization of $HC{\equiv}C[[o\text{-}Si(CH_3)_3]C_6H_4]$ by Various Catalysts

Catalyst	*Yield (%)*	M_w ($\times 10^3$)[a]
WCl_6	91	1500
$W(CO)_6$–$h\nu$[b]	94	1400
$MoCl_5$	86	1700
$Mo(CO)_6$–$h\nu$[b]	0	
WCl_6–$(C_2H_5)_3SiH$	100	1600
$MoCl_5$–$(C_2H_5)_3SiH$	100	1100
$NbCl_5$–$(C_2H_5)_3SiH$	35	1300
$TaCl_5$–$(C_2H_5)_3SiH$	15	180

NOTE: The reactions were carried out in toluene at 30 °C for 24 h ($[M]_0$ = 1.0 M; [catalyst] = [cocatalyst] = 10 mM).
[a]Values of M_w were determined by GPC.
[b]The reaction was carried out in CCl_4.

Living polymerization is a polymerization in which neither termination nor chain transfer occurs. Certain Mo catalysts effect the living polymerization of **3** (*15*). The molecular weight distribution (MWD) of the polymer formed with $MoCl_5$ alone is broad (Figure 1). The $MoCl_5$–$(n\text{-}C_4H_9)_4Sn$ catalyst narrows the MWD significantly, but it is yet imperfect. In contrast, when a three-component catalyst, $MoCl_5$–$(n\text{-}C_4H_9)_4Sn$–C_2H_5OH is used, the MWD becomes very narrow, with the polydispersity ratio (M_w/M_n) being close to unity. For $MoOCl_4$-based catalysts as well, a three-component catalyst, $MoOCl_4$–$(n\text{-}C_4H_9)_4Sn$–C_2H_5OH produces poly(**3**) with a narrow MWD.

Polymerization of Si-Containing Disubstituted Acetylenes

1-(Trimethylsilyl)-1-propyne ($CH_3C{\equiv}CSi(CH_3)_3$; **4a**) is a representative Si-containing disubstituted monomer (*16, 17*).

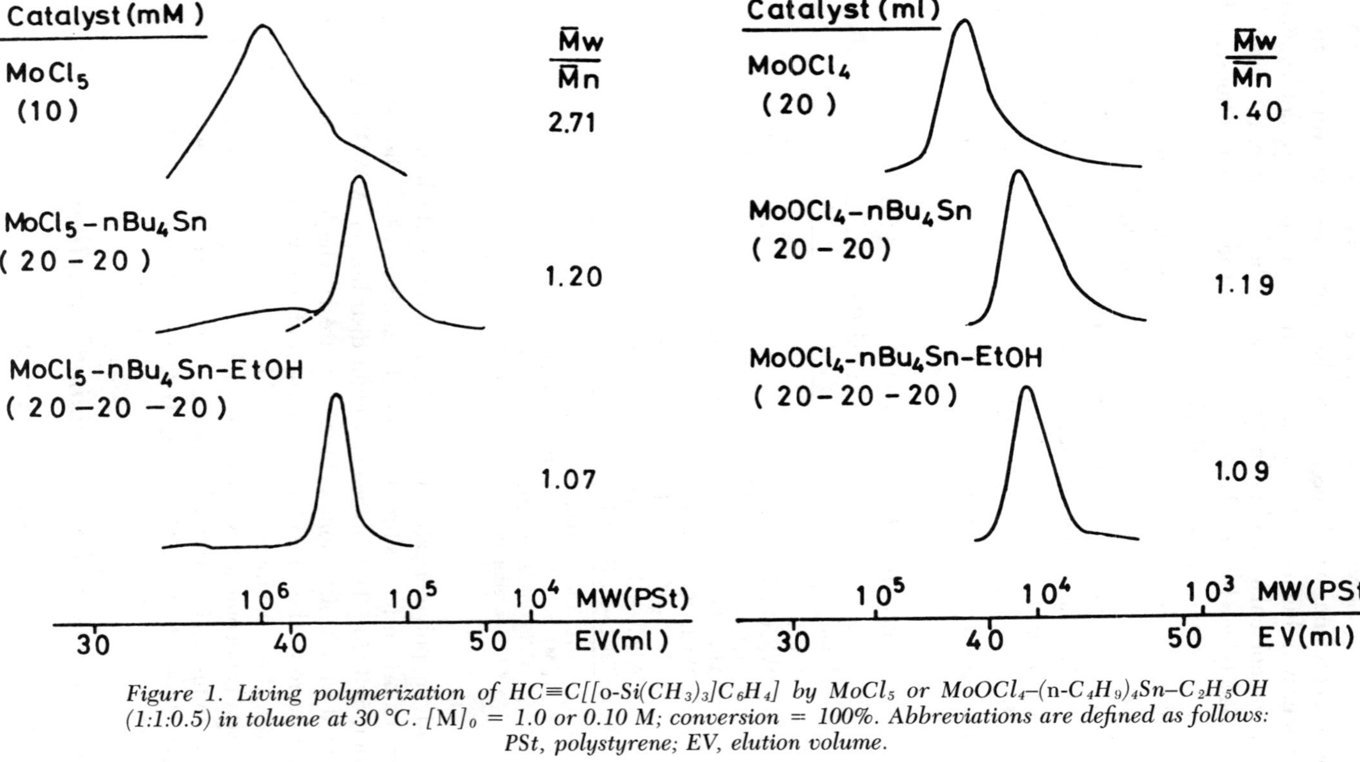

Figure 1. Living polymerization of $HC{\equiv}C[[o\text{-}Si(CH_3)_3]C_6H_4]$ *by* $MoCl_5$ *or* $MoOCl_4$–$(n\text{-}C_4H_9)_4Sn$–C_2H_5OH *(1:1:0.5) in toluene at 30 °C.* $[M]_0$ *= 1.0 or 0.10 M; conversion = 100%. Abbreviations are defined as follows: PSt, polystyrene; EV, elution volume.*

$$CH_3C{\equiv}C\overset{CH_3}{\underset{CH_3}{|}}Si{-}R$$

4

R: **4a,** CH_3; **4b,** n-C_6H_{13}; **4c,** C_6H_5;

4d, $-CH_2Si(CH_3)_3$; **4e,** $-CH_2-CH_2Si(CH_3)_3$

Because **4a** is remarkably hindered sterically, it does not polymerize at all with Mo or W catalysts. On the contrary, the chlorides and bromides of Nb and Ta give a soluble polymer virtually quantitatively (Table IV). The molecular weights of the polymer are in the 10^5–10^6 range.

Table IV. Polymerization of $CH_3C{\equiv}CSi(CH_3)_3$ by NbX_5 and TaX_5

Catalyst	*Yield (%)*	M_w (× 10^3)[a]
NbF_5	94	insoluble
$NbCl_5$	100	310
$NbBr_5$	100	280
NbI_5	0	
TaF_5	0	
$TaCl_5$	100	730
$TaBr_5$	95	410
TaI_5	0	

NOTE: The reactions were carried out in toluene at 80 °C for 24 h ($[M]_0$ = 1.0 M; [catalyst] = 20 mM). No polymer was formed with Mo or W catalysts.

[a]Values of M_w were determined by GPC.

Figure 2 illustrates the effects of cocatalysts on the polymerization of **4a** by $TaCl_5$ (*18*). The polymerization by $TaCl_5$ alone is virtually completed after 1 h under the conditions shown in Figure 2. The molecular weight of the polymer is about 10^6 regardless of polymer yield. When $(C_6H_5)_3Bi$ is added as cocatalyst, polymerization is much faster than that by $TaCl_5$ alone. The polymer formed has a superhigh molecular weight of up to 4 × 10^6, the highest molecular weight among those of all the substituted polyacetylenes.

The living polymerization of **4a** is possible by using $NbCl_5$ as catalyst and cyclohexane as solvent (*19*). In this polymerization, the molecular weight increases in direct proportion to conversion (Figure 3). The M_w/M_n ratio is as small as 1.2 regardless of conversion. Figure 3 (right) shows that the MWD is narrow and that the peak shifts toward the high-molecular-weight side with increasing conversion.

The polymerization of several homologues of **4a** (**4b**–**4e**) has also been examined (*20, 21*). All these monomers are polymerizable (Table V). As for

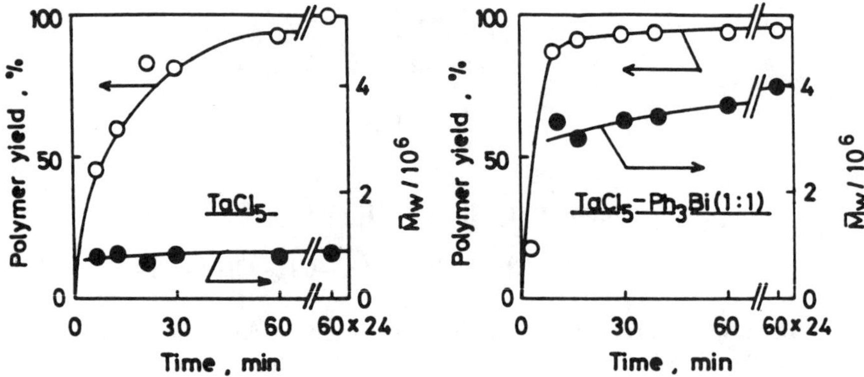

Figure 2. Effect of cocatalyst on the polymerization of $CH_3C{\equiv}CSi(CH_3)_3$ by $TaCl_5$ in toluene at 80 °C. $[M]_0$ = 1.0 M; [catalyst] = 20 mM. (Reproduced from reference 18. Copyright 1986 American Chemical Society.)

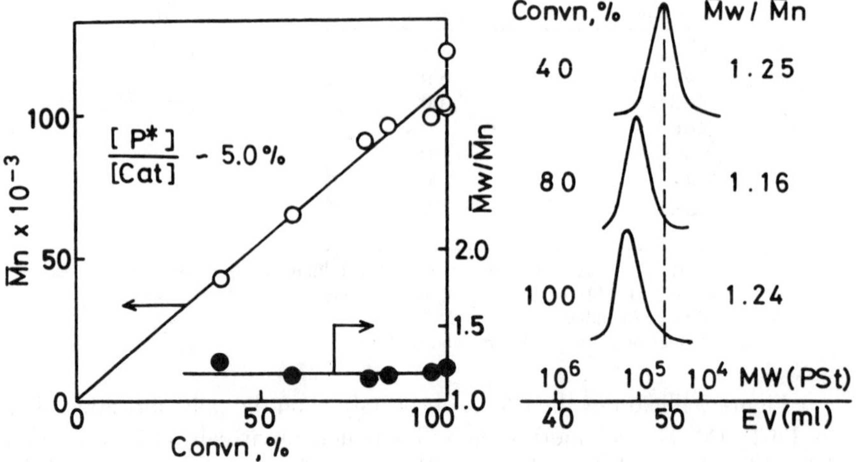

Figure 3. Living polymerization of $CH_3C{\equiv}CSi(CH_3)_3$ by $NbCl_5$ in cyclohexane at 25 °C for 24 h. $[M]_0$ = 1.0 M; [catalyst] = 20 mM. (Reproduced with permission from reference 19. Copyright 1988.)

Table V. Polymerization of $CH_3C{\equiv}CSi(CH_3)_2R$

R	*Catalyst*	*Yield (%)*	M_w ($\times 10^3$)[a]
n-C_6H_{13}	$TaCl_5$–$(C_6H_5)_3Bi$	75	1400
C_6H_5	$TaCl_5$–$(C_6H_5)_4Sn$	15	460
–$CH_2Si(CH_3)_3$	$TaCl_5$	100	1500
–$CH_2CH_2Si(CH_3)_3$	$TaCl_5$–$(C_6H_5)_4Sn$	58	400

NOTE: The reactions were carried out in toluene at 80 °C for 24 h ($[M]_0$ = 1.0 M; [catalyst] = [cocatalyst] = 20 mM).

[a]Values of M_w were measured by GPC.

the catalysts, either $TaCl_5$ alone or its mixtures with cocatalysts are effective. The polymer yields range from 15 to 100% depending on the kind of monomers, whereas the molecular weights of all the polymers are in the 10^5–10^6 range. The poly(**4**)s are all white solids that are soluble in low-polarity solvents. Free-standing films can be obtained by solution casting.

A relationship between the structure and polymerizability of sterically crowded acetylenes can be deduced as follows: $CH_3C{\equiv}CSi(CH_3)_3$, $CH_3C{\equiv}CSi(CH_3)_2(n\text{-}C_6H_{13})$, $CH_3C{\equiv}CSi(CH_3)_2(C_6H_5)$, and $HC{\equiv}C(tert\text{-}C_4H_9)$ are polymerizable, whereas $C_2H_5C{\equiv}CSi(CH_3)_3$, $CH_3C{\equiv}CSi(CH_3)_2(sec\text{-}C_3H_7)$, $CH_3C{\equiv}CSi(CH_3)_2(tert\text{-}C_4H_9)$, and $CH_3C{\equiv}C(tert\text{-}C_4H_9)$ are not polymerizable. The polymer molecular weight generally tends to increase with increasing steric crowding of the monomer. However, if the steric crowding exceeds a certain limit, the monomer will not polymerize. For instance, **4a** is polymerizable, whereas $C_2H_5C{\equiv}CSi(CH_3)_3$ is not polymerizable; **4b** and **4c** are polymerizable, whereas *sec*-C_3H_7- and *tert*-C_4H_9-substituted analogues are not polymerizable. $HC{\equiv}C(tert\text{-}C_4H_9)$ is polymerizable, whereas $CH_3C{\equiv}C(tert\text{-}C_4H_9)$ is not.

Structure and Properties of Polymers

Poly(**2c**), poly(**3**), and poly(**4a**) are interesting and representative Si-containing polyacetylenes in the sense that they are high-molecular-weight, totally soluble, film-forming polymers.

The IR, ^1H, and ^{13}C NMR spectra of poly(**2c**) are shown in Figure 4a. The C=C stretching, deformation of the CH adjacent to Si, and Si–C stretching are seen in the IR spectrum. The olefinic, alkyl, and $(CH_3)_3Si$ protons are all evident in the ^1H NMR spectrum. In the ^{13}C NMR spectrum, the olefinic, alkyl, and $(CH_3)_3Si$ carbons can be seen. Figure 4b is the corresponding set of spectra for poly(**3**). The IR spectrum confirms the presence of Si atoms. In the ^{13}C NMR spectrum, olefinic and phenyl carbons are detected in the region δ 150–120, whereas the $(CH_3)_3Si$ carbons are at δ 0. The IR spectrum of poly(**4a**) (Figure 4c) clearly shows the C=C stretching. The two broad peaks in the ^1H NMR spectrum of poly(**4a**) are assignable to two kinds of methyl protons. In the ^{13}C NMR spectrum, all of the two olefinic carbons and two kinds of methyl carbons appear at reasonable positions.

All the spectra in Figure 4 are consistent with the presence of the C=C bond in the polymers. No signals attributable to the C≡C bond are observed in any spectra. Data from elemental analysis of these polymers agree well with the theoretical values for the polymerization products. These results indicate that the polymers have the alternating double bond structure $-(CR{=}CR')_n-$.

Figure 5 presents the UV–visible spectra of the three Si-containing

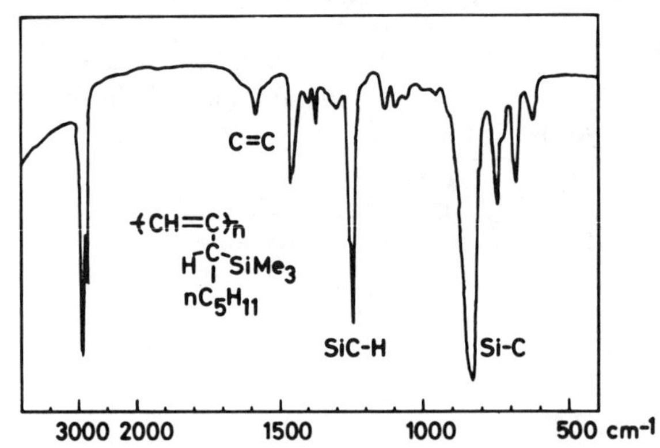

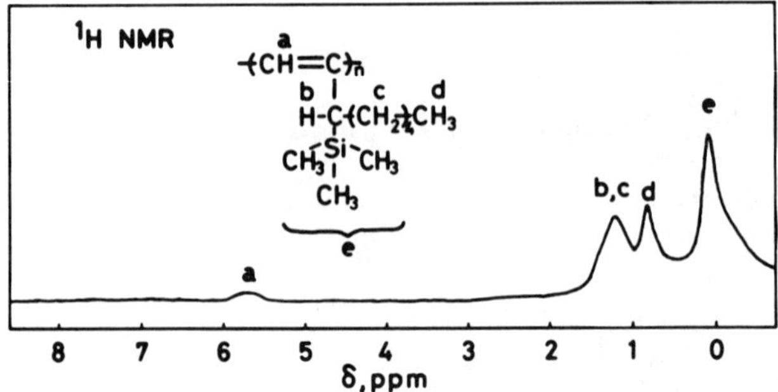

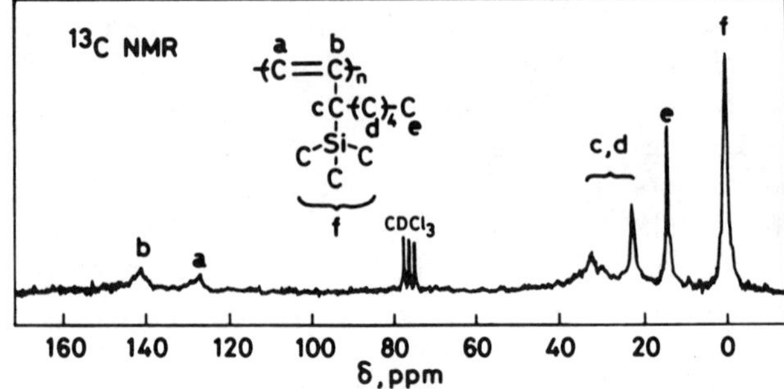

Figure 4a. IR, ^1H, and ^{13}C NMR spectra of $-[CH{=}CCH[Si(CH_3)_3](n\text{-}C_5H_{11})]_n-$. (Reproduced from reference 12. Copyright 1987 American Chemical Society.)

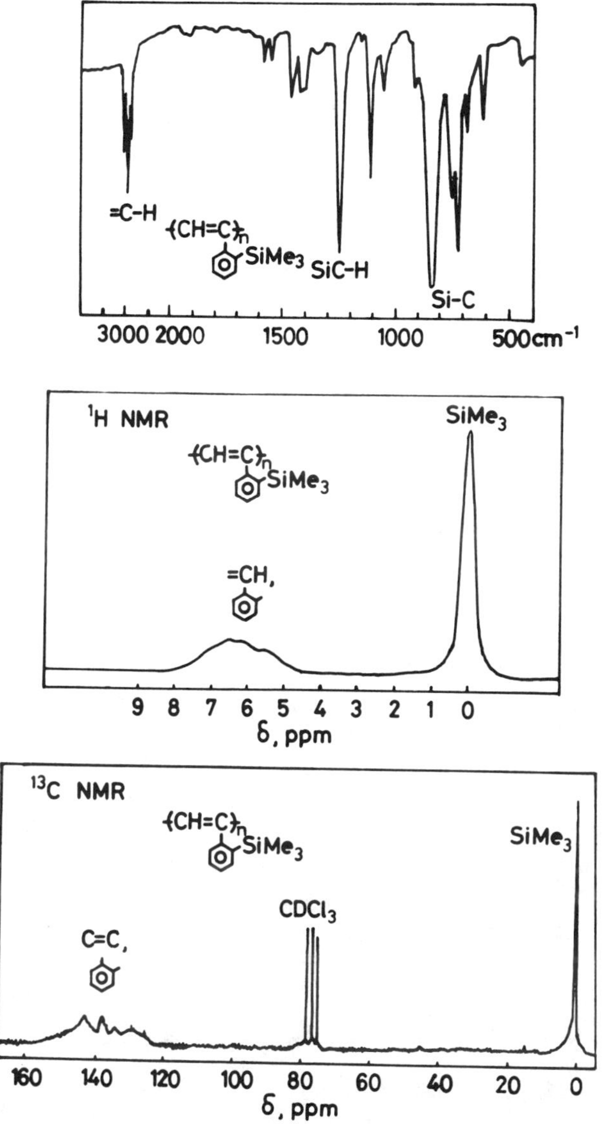

Figure 4b. IR, ¹H, and ¹³C spectra of $-[CH=C[[o\text{-}Si(CH_3)_3]C_6H_4]]_n-$. *(Reproduced from reference 14. Copyright 1989 American Chemical Society.)*

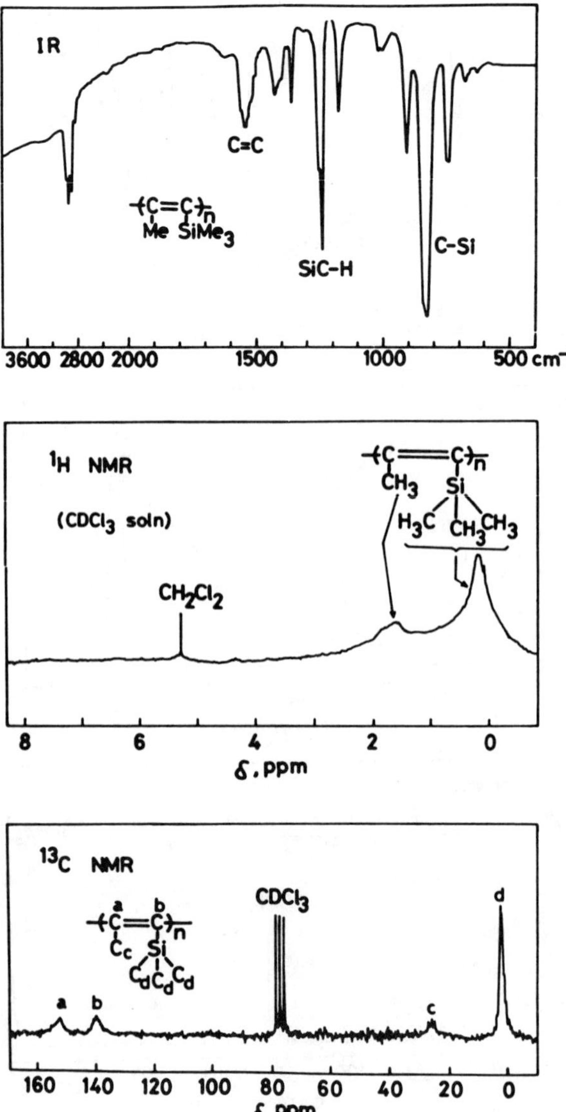

Figure 4c. IR, ¹H, and ¹³C NMR spectra of $-[CH_3C=CSi(CH_3)_3]_n-$. *(Reproduced from reference 17. Copyright 1985 American Chemical Society.)*

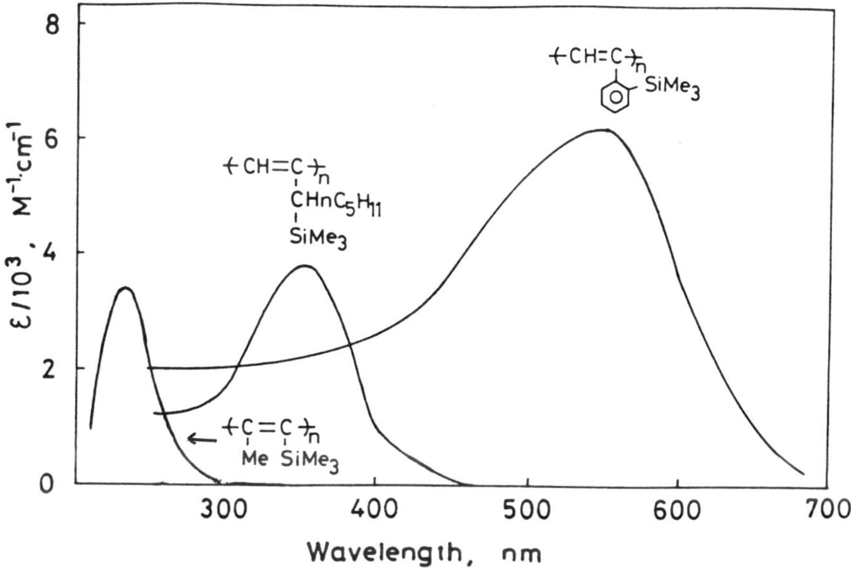

Figure 5. UV spectra of Si-containing polyacetylenes in tetrahydrofuran. ε *is molar absorptivity.*

polyacetylenes. Poly(**2c**) shows an absorption of medium strength in the near-UV region. Poly(**3**) exhibits a large absorption maximum around 550 nm, which extends to about 700 nm. In contrast, poly(**4a**) shows an absorption below 300 nm in the UV region. These spectra correspond well with the polymer colors shown in Table VI.

Table VI. Comparison of Properties of Si-Containing Polyacetylenes and Polyacetylene

Property	*Poly(2c)*	*Poly(3c)*	*Poly(4a)*	$-(CH{=}CH)_n-$
Maximum $M_w(\times 10^3)$	320	1600	4000	~10
Color	yellow	purple	white	black
Solubility				
in hexane	soluble	insoluble	soluble	insoluble
in toluene	soluble	soluble	soluble	insoluble
in anisole	insoluble	soluble	insoluble	insoluble
Crystallinity	amorphous	amorphous	amorphous	crystalline
Conductivity (S/cm)	3×10^{-18}	4×10^{-15}	1×10^{-17}	10^{-9}–10^{-5}
Spin density (g^{-1})	$<1 \times 10^{15}$	7×10^{17}	$<1 \times 10^{15}$	10^{18}–10^{19}

All of poly(**2c**), poly(**3**), and poly(**4a**) completely dissolve in many organic solvents such as aromatic hydrocarbons (e.g., toluene), halogenated hydrocarbons (e.g., $CHCl_3$), and tetrahydrofuran (Table VI). Furthermore, poly(**2c**) and poly(**4a**), which are aliphatic polymers, are also soluble in aliphatic hydrocarbons (e.g., hexane). On the other hand, poly(**3**), which is an aromatic polymer, does not dissolve in aliphatic hydrocarbons but does dissolve in slightly polar aromatic solvents like anisole (methoxybenzene). X-ray diffraction analysis indicates that all these polymers are amorphous. If the results for polyacetylene are taken into account, the excellent solubility and noncrystallinity of the polymers in this study evidently stem from the presence of bulky substituents.

All of the polymers in this study are much more thermally stable than polyacetylene (22). The results of thermogravimetric analysis in air (Figure 6) indicate that poly(**2c**) is somewhat less stable, probably because of the presence of the allylic proton. However, poly(**3**) and poly(**4a**) lose weight only above 300 °C and are fairly thermally stable.

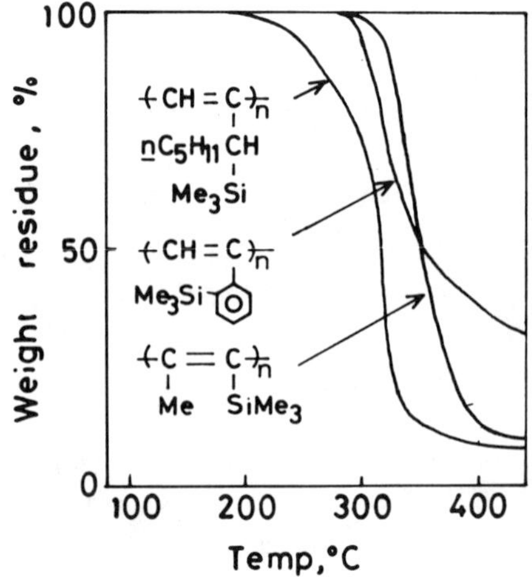

Figure 6. Thermogravimetric analysis (TGA) of Si-containing polyacetylenes in air at a heating rate of 10 °C/min.

Figure 7 shows the stress–strain curves for the three polymers (*23, 24*). As seen from the initial slopes of the curves, the Young's moduli of these polymers are similar (~500 MPa). In contrast, their elongations at break point are significantly different. Thus, poly(**3**) is rather hard and brittle, whereas poly(**4a**) is fairly ductile.

Figure 8 presents the temperature dependence of the dynamic viscoelasticity of the Si-containing polyacetylenes (*23, 24*). Poly(**2c**) shows a dispersion at low temperature because of the presence of the long *n*-pentyl group. From the sharp increase in tan δ, the glass transition temperature (T_g) of this polymer is about 150 °C. In contrast, poly(**3**) and poly(**4a**) hardly show dispersions at low temperature, and their T_gs are about 200 °C or higher. The high T_g values of these polyacetylenes compared with those of most vinyl polymers can be attributed to their stiff main chain.

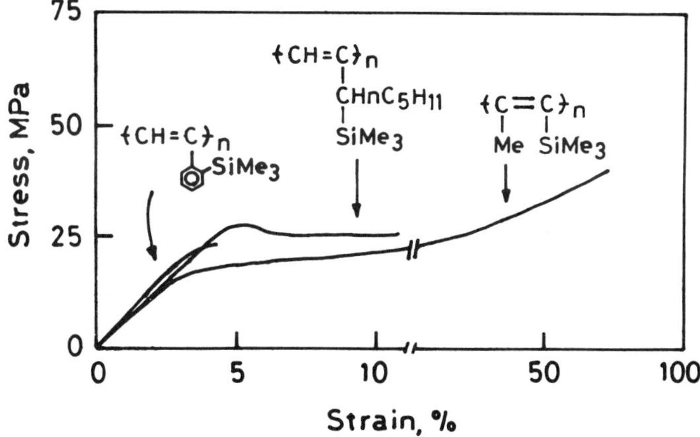

Figure 7. Stress–strain curves of Si-containing polyacetylenes at 25 °C and 86%/min. (Reproduced from reference 23. Copyright 1986 American Chemical Society.)

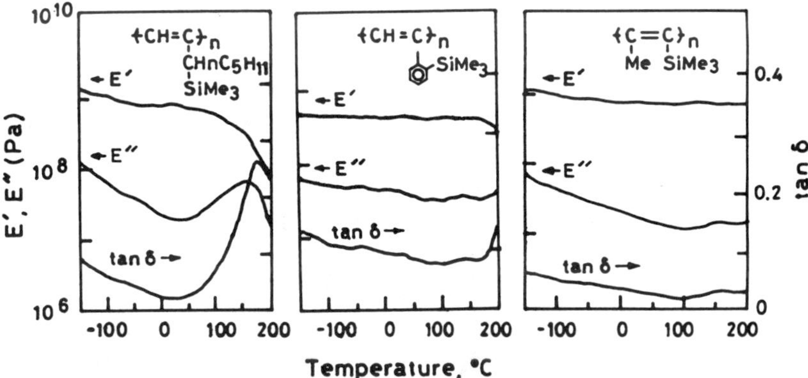

Figure 8. Dynamic viscoelasticity of Si-containing polyacetylenes. The measurements were made at temperatures from –150 to +200 °C at 110 Hz. (Reproduced from reference 23. Copyright 1986 American Chemical Society.)

Table VI includes the electrical and magnetic properties of the polymers studied. Polyacetylene is a semiconductor. In contrast, the conductivities of the Si-containing polyacetylenes are only in the order of 10^{-18}–10^{-15} S/cm, meaning that they are typical insulators. The unpaired-electron densities of the Si-containing polymers are also much lower than that of polyacetylene. Such low conductivities and unpaired-electron densities are probably attributable to the twisted polymer conformation due to the presence of substituents.

Applications of Polymers

Substituted polyacetylenes, especially poly(**4a**), have unique functions (*3*). One application is the use of the polymer membrane to separate substances.

The oxygen enrichment of air with membranes has been the subject of intensive research (*25*). In Figure 9, the abscissa stands for the oxygen-permeability coefficient (P_{O_2}), and the ordinate is the ratio of oxygen- and nitrogen-permeability coefficients (P_{O_2}/P_{N_2}). Poly(dimethylsiloxane) had the highest oxygen permeability among all the existing polymers. However, our studies show that the oxygen permeability of poly(**4a**) is about 10 times higher than that of poly(dimethylsiloxane) (*16, 26*). Other substituted polyacetylenes are more permeable compared with conventional polymers.

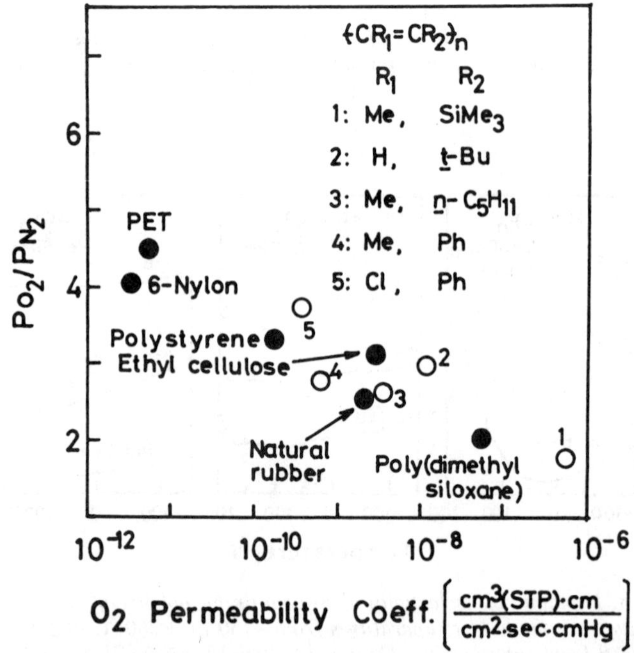

Figure 9. P_{O_2} *and* P_{O_2}/P_{N_2} *for various polymers at 25 °C. (Reproduced from reference 4. Copyright 1984 American Chemical Society.)*

Both poly(**4a**) and poly(dimethylsiloxane) have P_{O_2} values as high as 10^{-8}–10^{-7} barrer. However, their properties are quite different (*23, 27*) (Table VII). For instance, the T_g of poly(**4a**) is higher than 200 °C, and hence, it is glassy at room temperature.

Thus, poly(**4a**) can be easily cast into a thin, tough membrane. Furthermore, the gas permeation through this polyacetylene is connected with the dual-mode sorption mechanism involving Langmuir adsorption, whereas

Table VII. Comparison of Poly[1-(trimethylsilyl)-1-propyne] and Poly(dimethylsiloxane)

Property	*Poly[1-(trimethyl-silyl)-1-propyne*	*Poly(dimethyl-siloxane)*
Permeation of O_2[a]		
P	3.1×10^{-7}	3.6×10^{-8}
D	2.2×10^{-5}	1.9×10^{-5}
S	1.4×10^{-2}	1.9×10^{-3}
T_g	>200 °C	−127 °C
Thin-film formation	easy	difficult
Permeation mechanism	dual mode (Langmuir)	solution (Henry)

[a]P is expressed in barrer. *D* is expressed in cm^2/s. *S* is expressed in $cm^3(STP)/cm^3 \times$ cm Hg.

the permeation through poly(dimethylsiloxane) is explained simply in terms of Henry's law.

Poly(**4a**) would be a very promising membrane material for oxygen enrichment, but it has one problem: Its oxygen permeability (Figure 10) gradually decreases with time to become smaller than 1/10 of the original value after 100 days (*28*). Many researchers are investigating the mechanism of the permeability decrease, as well as the various methods for preventing it.

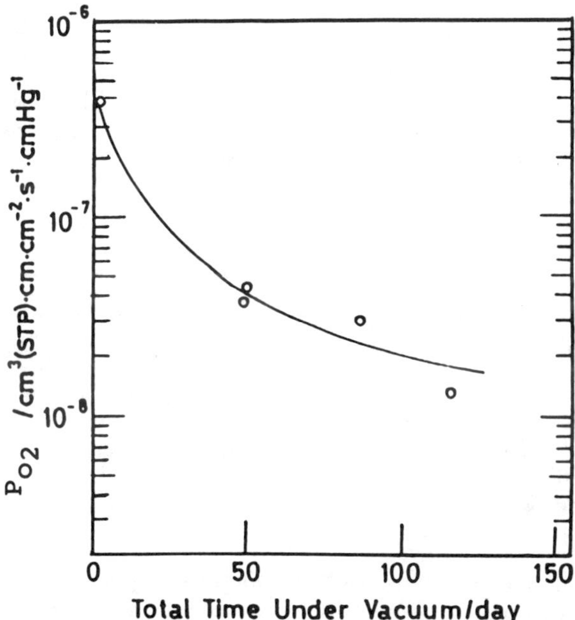

Figure 10. Time change of P_{O_2} *at 30 °C in* $-[CH_3C{=}CSi(CH_3)_3]_n-$ *(stored under vacuum at 30 °C). (Reproduced with permission from reference 28. Copyright 1986 Society of Polymer Science, Japan.)*

Another application of substituted polyacetylenes is in the pervaporation of the ethanol–water mixture. Pervaporation is a method for separating liquid mixtures by evacuating the downstream side of the separation membrane (*29*). The performance of a membrane is evaluated by the separation factor (α) and the specific permeation rate (R).

Figure 11 is a plot of α versus R for ethanol–water pervaporation through substituted polyacetylenes. If the α value is larger than unity, the membrame permeates ethanol preferentially. Previously, with the exception of poly(dimethylsiloxane), no polymer had exhibited ethanol permselectivity. Now, we know that poly(**4a**) permeates ethanol preferentially (*30*). The α and R values of poly(**4a**) are close to those of poly(dimethylsiloxane). In contrast, other polyacetylenes exhibit practically no permselectivity or allow water to permeate preferentially.

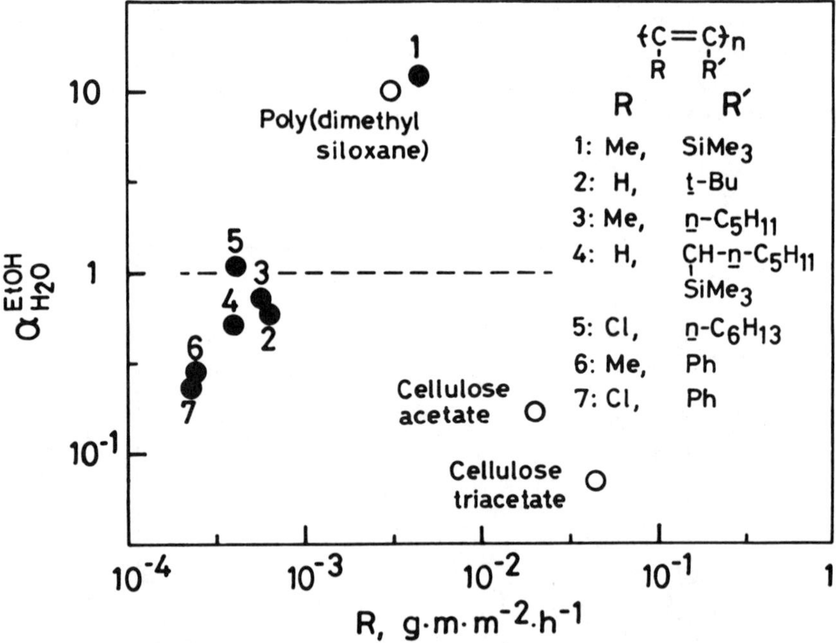

Figure 11. Plot of α *versus* R *in ethanol–water pervaporation through* $-(CR{=}CR')_n-$ *membranes (ethanol, 10 wt %; 133 Pa). (Reproduced with permission from reference 3. Copyright 1986 Springer-Verlag.)*

Figure 12 shows examples of the potential applications of substituted polyacetylenes studied, especially poly(**4a**). Oxygen enrichment (*26–28, 31–33*) is applicable to combustion furnaces, car engines, and respiration-aiding apparatus. The transport of oxygen dissolved in water (*33, 34*) can be applied to contact lenses and artificial lungs. Liquid-mixture separation (*30,*

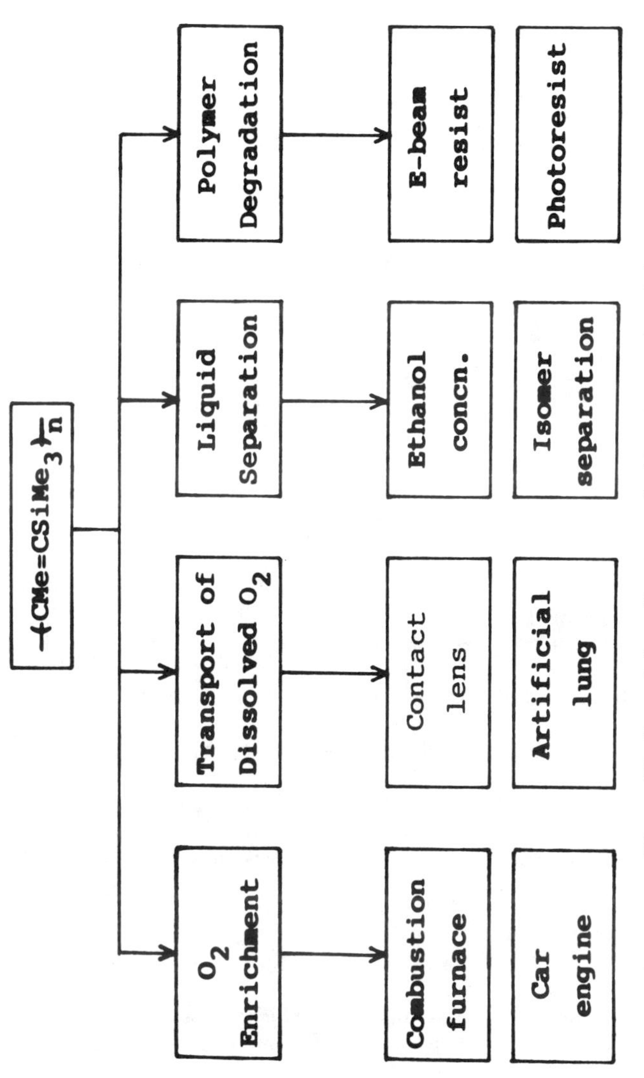

Figure 12. Potential applications of $-[CH_3C=CSi(CH_3)_3]_n-$.

35) can be applied to the concentration of ethanol from fermented biomass. Polymer degradation (*36–38*) has relevance to resist materials for microlithography. In relation to such applications of substituted polyacetylenes, some 25 papers in academic meetings, as well as about 75 Japanese patents, have appeared so far.

Acknowledgment

We thank E.-T. Kang of the National University of Singapore and K. Tamao of Kyoto University for many helpful suggestions.

References

1. Chien, J. C. W. *Polyacetylene;* Academic: New York, 1984.
2. Saxman, A. M.; Liepins, R.; Aldissi, M. *Prog. Polym. Sci.* **1985,** *11*, 57.
3. Masuda, T.; Higashimura, T. *Adv. Polym. Sci.* **1986,** *81*, 121.
4. Masuda, T.; Higashimura, T. *Acc. Chem. Res.* **1984,** *17*, 51.
5. Gibson, H. W.; Porchan, J. M. In *Encyclopedia of Polymer Science and Engineering,* 2nd ed.; Kvoschwitz, J. I., Ed.; Wiley: New York, 1984; Vol. 1, p 87.
6. Simionescu, C. I.; Percec, V. *Prog. Polym. Sci.* **1982,** *8*, 133.
7. Voronkov, M. G.; Pukhnarevich, V. B.; Sushchinskaya, S. P.; Annenkova, V. Z.; Annenkova, V. M.; Andreeva, N. J. *J. Polym. Sci., Polym. Chem. Ed.* **1980,** *18*, 53.
8. Zeigler, J. M. *Polym. Prepr. (Am. Chem. Soc., Div. Polym. Chem.)* **1984,** *25*(2), 223.
9. Liaw, D. J.; Soum, A.; Fontanille, M.; Parlier, A.; Rudler, H. *Makromol. Chem., Rapid Commun.* **1985,** *6*, 309.
10. Okano, Y.; Masuda, T.; Higashimura, T. *J. Polym. Sci., Polym. Chem. Ed.* **1984,** *22*, 1603.
11. Tajima, H.; Masuda, T.; Higashimura, T. *J. Polym. Sci., Polym. Chem. Ed.* **1987,** *25*, 2033.
12. Masuda, T.; Tajima, H.; Yoshimura, T.; Higashimura, T. *Macromolecules* **1987,** *20*, 1467.
13. Brandzma, L.; Verkruijsse, H. D. *Synthesis of Acetylenes, Allenes and Cumulenes;* Elsevier: Amsterdam, 1981; p 85.
14. Masuda, T.; Hamano, T.; Tsuchihara, R.; Higashimura, T. *Macromolecules* **1989,** *22*, 1036.
15. Masuda, T.; Yoshimura, T.; Fujimori, J.; Higashimura, T. *J. Chem. Soc., Chem. Commun.* **1987,** 1805.
16. Masuda, T.; Isobe, E.; Higashimura, T.; Takada, K. *J. Am. Chem. Soc.* **1983,** *105*, 7473.
17. Masuda, T.; Isobe, E.; Higashimura, T. *Macromolecules* **1985,** *18*, 841.
18. Masuda, T.; Isobe, E.; Hamano, T.; Higashimura, T. *Macromolecules* **1986,** *19*, 2448.
19. Fujimori, J.; Masuda, T.; Higashimura, T. *Polym. Bull.* **1988,** *20*, 1.
20. Masuda, T.; Isobe, E.; Hamano, T.; Higashimura, T. *J. Polym. Sci., Polym. Chem. Ed.* **1987,** *25*, 1353.
21. Isobe, E.; Masuda, T.; Higashimura, T.; Yamamoto, A. *J. Polym. Sci., Polym. Chem. Ed.* **1986,** *24*, 1839.
22. Masuda, T.; Tang, B.-Z.; Higashimura, T.; Yamaoka, H. *Macromolecules* **1985,** *18*, 2369.

23. Masuda, T.; Tang, B.-Z.; Tanaka, A.; Higashimura, T. *Macromolecules* **1986,** *19,* 1459.
24. Masuda, T.; Tang, B.-Z.; Matsumoto, T.; Higashimura, T., to be published.
25. *Industrial Gas Separations;* Whyte, T. E., Jr.; Yon, C. M.; Wagener, E. H., Eds.; ACS Symposium Series 223; American Chemical Society: Washington, DC, 1983.
26. Takada, K.; Matsuya, H.; Masuda, T.; Higashimura, T. *J. Appl. Polym. Sci.* **1985,** *30,* 1605.
27. Masuda, T.; Iguchi, Y.; Tang, B.-Z.; Higashimura, T. *Polymer* **1988,** *29,* 2041.
28. Shimomura, H.; Nakanishi, K.; Odani, H.; Kurata, M.; Masuda, T.; Higashimura, T. *Kobunshi Ronbunshu* **1986,** *43,* 747.
29. Frennesson, I.; Tragardh, G.; Hahn-Hagerdal, B. *Chem. Eng. Commun.* **1986,** *45,* 277.
30. Masuda, T.; Tang, B.-Z.; Higashimura, T. *Polym. J. (Tokyo)* **1986,** *18,* 565.
31. Ichiraku, Y.; Stern, S. A.; Nakagawa, T. *J. Membr. Sci.* **1987,** *34,* 5.
32. Langsam, M.; Robeson, L. M. *Polym. Prepr. (Am. Chem. Soc., Div. Polym. Chem.)* **1988,** *29*(1), 112.
33. Masuda, T. *Maku* **1988,** *13,* 195.
34. Tang, B.-Z.; Masuda, T.; Higashimura, T. *J. Polym. Sci., Polym. Phys. Ed.* **1989,** *27,* 1261.
35. Ishihara, K.; Nagase, Y.; Matsui, K. *Makromol. Chem., Rapid Commun.* **1986,** *7,* 43.
36. Higashimura, T.; Tang, B.-Z.; Masuda, T.; Yamaoka, H.; Matsuyama, T. *Polym. J. (Tokyo)* **1985,** *17,* 393.
37. Bower, T. N.; Baker, G. L. *Polym. Prepr. (Am. Chem. Soc., Div. Polym. Chem.)* **1986,** *27*(2), 218.
38. Gozdz, A. S.; Baker, G. L.; Klausner, C.; Bowden, M. J. *Adv. Resist Technol.* **1987,** *4,* 18.

RECEIVED for review May 27, 1988. ACCEPTED revised manuscript December 7, 1988.

36

Brominated Poly(1-trimethylsilyl-1-propyne)

Lithography and Photochemistry

Gregory L. Baker, Cynthia F. Klausner, Antoni S. Gozdz, John A. Shelburne III, and Trevor N. Bowmer

Bell Communications Research, 331 Newman Springs Road, Red Bank, NJ 07701–7040

Brominated derivatives of poly(1-trimethylsilyl-1-propyne), a silicon-containing polyacetylene, were examined as potential deep-UV resist materials. The polymers were prepared by free-radical bromination at allylic positions as verified by ^{13}C *and* ^{29}Si *NMR spectroscopy. Bromination greatly increased the sensitivity of the polymer toward photooxidation. The polymers formed high-quality films, retained the thermal stability of the unbrominated polymer, and were highly resistant to oxygen reactive-ion etching. Patterns developed in 0.3-μm-thick layers of the brominated polymer were transferred through 1.5 μm of hard-baked novolac and polyimide with good line width control. The increased sensitivity of brominated polymers toward photooxidation is probably due either to bromine-catalyzed singlet–triplet conversion in the polymers followed by trapping of the triplet excitation by oxygen or to trapping of allylic radicals by oxygen followed by a chain-degradative process.*

THE TREND TOWARD EVER-INCREASING FEATURE SIZES in current and future microelectronic devices poses challenges for optical lithography (*1*) and opportunities for silicon-containing polymers. Smaller feature sizes imply the use of shorter wavelength radiation for the patterning of those features, and thus visible photolithography has given way to mid-UV lithography (300–400 nm), which eventually will be supplanted by deep-UV lithography

0065–2393/90/0224–0663$06.00/0

($\lambda < 300$ nm). Each step in this evolution has required the preparation of new resist materials tailored to the wavelength of the exposing radiation.

Design Considerations for Deep-UV Resists

The switch to deep-UV lithography poses challenges beyond that of merely designing a sensitive lithographic material (2). As the wavelength decreases, the depth of focus for exposure tools becomes much smaller, and variables such as the planarity of the substrate become increasingly important. The rough surfaces characteristic of partially processed wafers become a major constraint. Imaging layers need to be thin for optimal resolution; however, the spinning of thin imaging layers over preexisting micrometer-sized features results in grossly nonplanar films. Thus, during exposure, portions of the film are in focus while other areas are out of focus. This situation yields distorted patterns.

Two- and three-layer lithographic schemes (Figure 1) have been devised to overcome the problems of topography on partially processed wafers. In either scheme, a thick planarizing layer is spun over the irregular surface to provide a level surface for the subsequent deposition of the imaging layer.

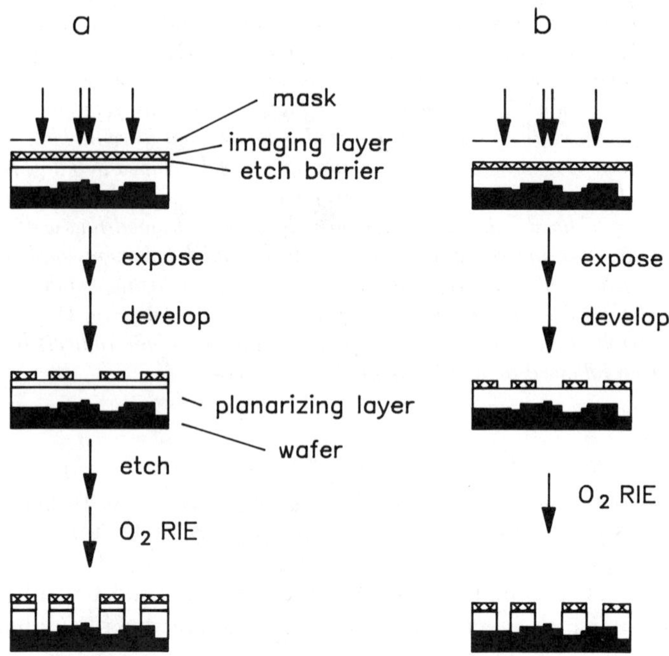

Figure 1. Three-layer (a) and two-layer (b) resist schemes for microlithography.

The image developed in the top layer is then transferred through the planarizing layer to the surface of the wafer by oxygen reactive-ion etching (O_2 RIE) with the organosilicon resist functioning as an etch mask. The anisotropic nature of the RIE process results in the transfer of a dimensionally correct pattern to the wafer surface (*3*). Two-layer processes have fewer processing steps than three-layer schemes, because the imaging and etch resistance functions are performed by a single layer.

In O_2-RIE environments, the surface of silicon-containing polymers is converted to SiO_x, in which x is slightly less than 2 (*4*). When the silicon content of the polymer is >10%, the oxide formed provides an excellent barrier to further degradation of the polymer in the RIE environment (*5*). For such polymers, the etch rate of the polymer may be 20–40 times less than that of the typical hydrocarbon polymers used as planarizing layers. This difference in etch rate allows images developed in a thin layer of silicon-containing polymer to be faithfully transferred from the imaging layer through the planarizing layer to the wafer surface.

Combining the lithographic and etch mask functions into a single polymer can be a major challenge, especially for deep-UV lithography. The latitude in resist design is limited, because at least 10 and preferably 15 wt % of the polymer structure must be reserved for silicon. A few materials, like silicon-substituted poly(methyl methacrylates) (*6*) and polysilanes (*7, 8*), have been used as positive two-layer resists for deep-UV lithography, but these materials suffer from either poor to moderate sensitivities to deep-UV radiation or an excessive absorption in the UV that limits exposure depth in the resist layer.

We became interested in poly(1-trimethylsilyl-1-propyne) [poly(TMSP)] after the initial report of its synthesis (*9*). Poly(TMSP) possesses many of the qualities needed in a practical deep-UV resist material: a high silicon content (~25%); thermal stability; excellent film-forming properties; and a chromophore, the conjugated backbone, that absorbs strongly in the deep-UV region. Our initial work with poly(TMSP) showed that it had one serious drawback; its sensitivity toward degradation by deep-UV radiation was poor, and doses of >1 J/cm^2 are required for the definition of a positive-tone pattern.

The sensitivity of resists has been improved occasionally by the introduction of halogens into the chemical structure. For example, the sensitivities of poly(α-methylstyrene), poly(*p*-methylstyrene), and poly(acrylate)s increase after chlorination, and a similar improvement in sensitivity might result from halogenation of poly(TMSP). Because halogenation could induce degradation by either cross-linking or chain scission, we were unable to predict whether the halogenated material might act in a positive or negative fashion. In any case, even if halogenation did not improve the sensitivity of poly(TMSP), a halogenated derivative would be an excellent intermediate (*10*) for the synthesis of useful derivatives of poly(TMSP).

Synthesis of Brominated Derivatives of Poly(TMSP)

As reported by Masuda et al. (*9*, *11*), 1-trimethylsilyl-1-propyne can be polymerized by $NbCl_5$, $TaCl_5$, and other catalysts to give a white air-stable polymer with molecular weights that can exceed 10^6. For lithographic investigations, we obtained poly(TMSP) by polymerization of the monomer with $TaCl_5$ in toluene at 80 °C. This method yielded a soluble product with a molecular weight of ~3 × 10^5. The simplest route to halogenated poly(TMSP) is free-radical bromination. The stability of poly(TMSP) in air, its lack of optical absorption at $\lambda > 280$ nm, and an examination of space-filling models of the polymer all implied that chemistry at the double bonds of the polymer would be limited and that allylic substitution should be favored. The steric crowding suggested by the models was so severe that the conjugated backbone should be chemically inert.

Under standard conditions (*N*-bromosuccinimide [NBS] and AIBN [azobis(isobutyronitrile)] in refluxing CCl_4), the desired brominated poly(TMSP) products were formed in high yield. Nearly quantitative yields were obtained up to bromination levels of ~50%, on the basis of one Br substitution per monomer unit, but unexpectedly, the maximum attainable bromination level was only ~60%, even when a large excess of NBS was used (Table I). We attributed this curious behavior to the extreme steric crowding at the allylic methyl groups in the polymer structure. The accessibility of the allylic sites apparently is so limited that monobromination is the exclusive reaction, and the presence of the bulky halogen decreases the likelihood of bromination at adjacent allylic sites. Similar results were obtained when benzoyl peroxide was used as the free-radical source.

Table I. Characteristics of Brominated Polymers

	Bromine Level		M_n	
Sample	*Theoretical*[a]	*Calculated*	$(\times 10^5)$[b]	M_w/M_n[c]
0-Br	0.00	0.00	1.7	1.7
10-Br	0.10	0.10	1.7	1.6
20-Br	0.20	0.20	1.4[c]	1.7
40-Br	0.40	0.38	1.9	1.5
100-Br	1.00	0.54	2.1	1.6

NOTE: M_n and M_w are number-average and weight-average molecular weights, respectively.
[a]A value of 1.00 corresponds to one Br per repeat unit.
[b]Data were obtained from membrane osmometry.
[c]Data were obtained from size exclusion chromatography.

Characterization of Brominated Poly(TMSP)

To understand the mechanism of the lithographic process, the selectivity of the bromination reaction and the structure of the brominated product had to be determined. The results of IR spectroscopy and elemental analysis were consistent with allylic bromination, but both techniques were unable to give detailed selectivity information. As expected, ^{13}C and ^{29}Si NMR spectroscopic experiments were more informative.

The ^{13}C NMR spectrum of poly(TMSP) consists of resonances for the vinyl carbons of the backbone, the trimethylsilyl group, and the allylic methyl group. Figure 2 shows the ^{13}C NMR spectrum of the aliphatic region for poly(TMSP) and several of its brominated derivatives. The resonance from the trimethylsilyl (TMS) group, a singlet, remained unchanged with bromination, consistent with no bromination at those sites. An examination of the alkyl region of the spectrum showed that a broad resonance centered at 39 ppm appeared and grew in intensity, whereas that at 25 ppm (allylic CH_3) decreased and shifted slightly downfield. The asymmetric nature of the resonance results from the presence of both *cis* and *trans* isomers in the polymer backbone. We assigned the broad resonance at 39 ppm to the allylic

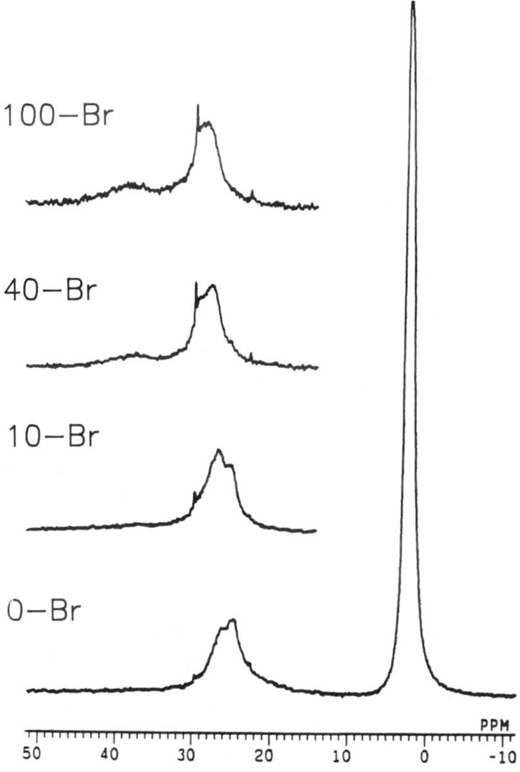

Figure 2. ^{13}C NMR (67.9 MHz) spectra of the aliphatic region for brominated samples of poly(TMSP). Spectra were obtained in $CDCl_3$ (30,000 scans; pulse width, 15 μs; pulse delay, 2 s). Sample designations are as follows: 100-Br, 100% Br; 40-Br, 40% Br; 10-Br, 10% Br; and 0-Br, 0% Br.

CH_2Br. The single, asymmetric peak observed in ^{29}Si NMR spectroscopic measurements (Figure 3) confirmed both the presence of isomers and the absence of bromination at that site. Thus selective allylic bromination was confirmed.

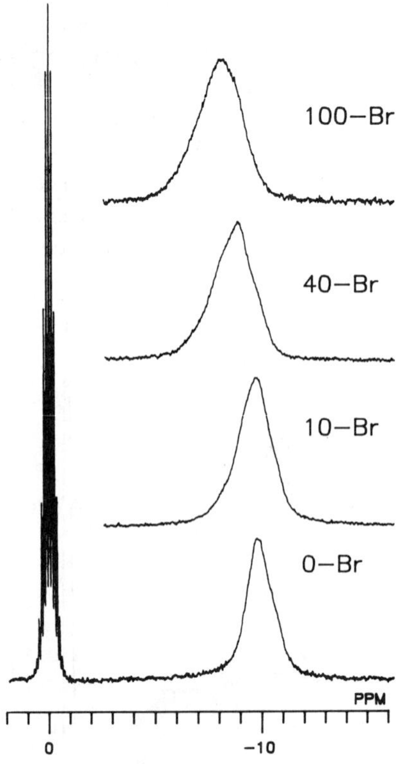

Figure 3. ^{29}Si NMR (53.7 MHz) spectra of brominated samples of poly-(TMSP). The spectra were obtained in $CDCl_3$ (20,000 scans; pulse width, 8 μs; pulse delay, 2 s). Sample designations are explained in the caption to Figure 2.

Bromination modified the physical properties of poly(TMSP), but those properties that initially attracted us to poly(TMSP) were largely retained in the brominated product. The UV–visible spectra of brominated polymers (Figure 4) exhibited a slight shift of oscillator strength to longer wavelengths, with the shift roughly correlated with the degree of bromination. Thus, brominated polymers were pale yellow compared with the colorless poly(TMSP). TGA (thermogravimetric analysis) scans showed that the thermal stability of the polymers was decreased by bromination (Figure 5), but even at the highest level of substitution, the polymers retained enough

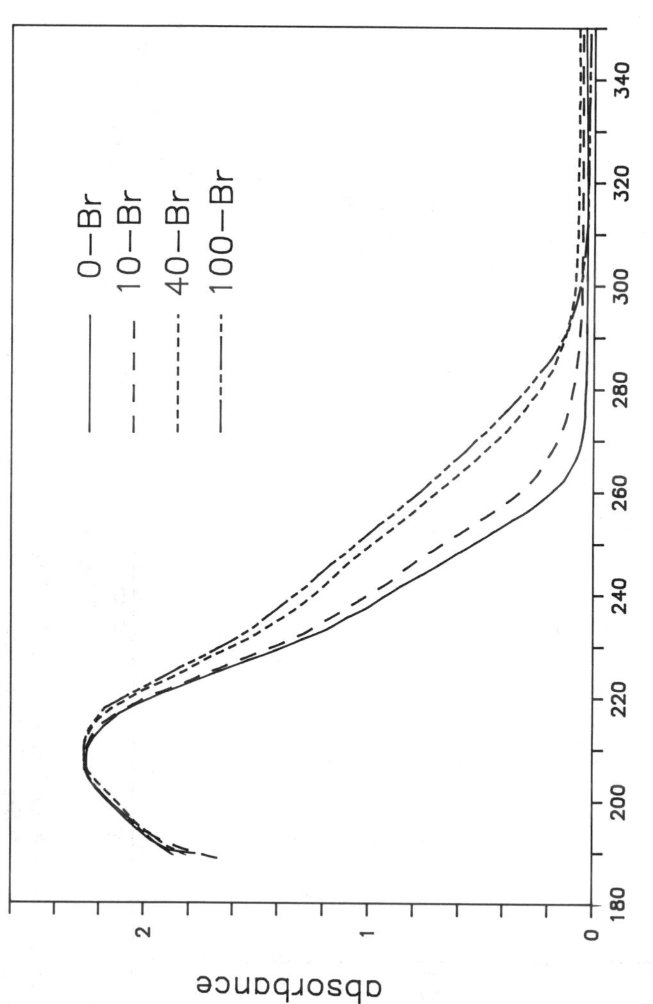

Figure 4. UV–visible spectra of brominated poly(TMSP) in cyclohexane (c = 0.03 g/L). *The absorbance scale is that for the unbrominated material; the absorption curves for the brominated polymers have been normalized to that of the unbrominated material for clarity. Sample designations are explained in the caption to Figure 2.*

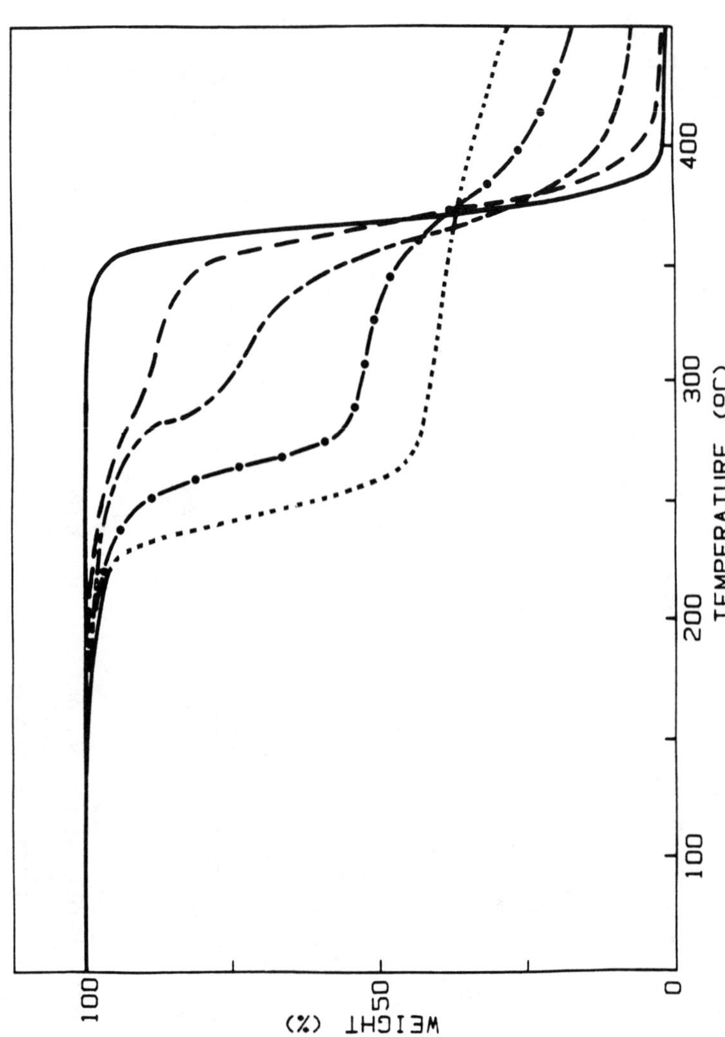

Figure 5. Results of TGA for brominated poly(TMSP) samples in N_2. The heating rate was 10 K/min. Key: —, 0-Br; – –, 10-Br; —·, 20-Br; —•, 50-Br; and - -, 100-Br. Sample designations are explained in the caption to Figure 2.

stability for most lithographic processing steps. Bromination might also be expected to decrease the polymer's stability under O_2-RIE conditions, but the etch rate was independent of the bromination level (Figure 6).

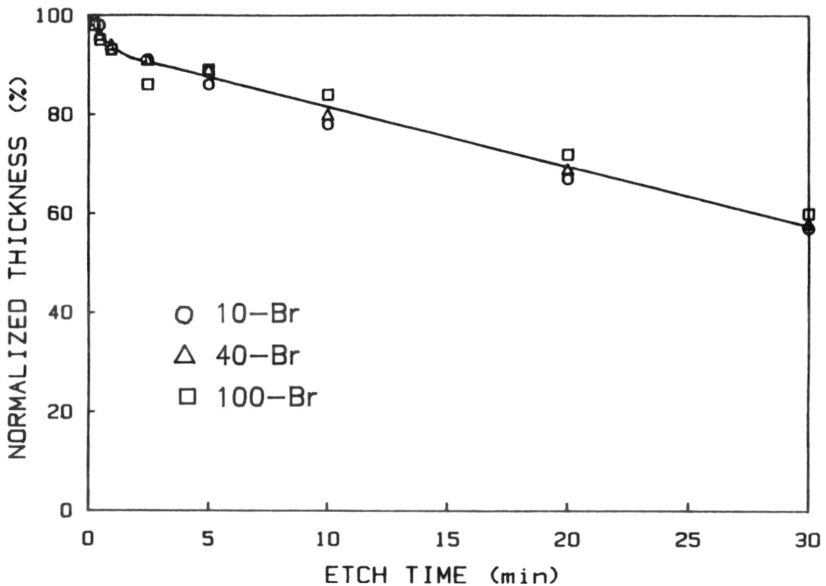

Figure 6. O_2 RIE of poly(TMSP) and its brominated derivatives. Etching was carried out in a Cooke Vacuum C71-3 apparatus at the following conditions: power, 0.14 W/cm^2; bias, –350 V; O_2 pressure, 2.7 Pa. Sample designations are explained in the caption to Figure 2.

Lithographic Evaluation of Poly(TMSP)

Improvements in the lithographic sensitivity of poly(TMSP) resulting from bromination were evaluated by a standard procedure. The *sensitivity*, a measure of the minimum dose necessary for the full development of a pattern in the resist layer, is evaluated typically by exposing regions of the sample to different doses of radiation, developing the pattern, and measuring the residual thickness of the polymer. In positive-acting resists, the exposed regions are removed, whereas in negative systems the unexposed regions are lost.

Plotted in Figure 7 are the sensitivity curves for brominated polymers exposed to deep-UV radiation. The samples were prepared by spin coating polished Si wafers with a 2% solution of the polymer in chlorobenzene,

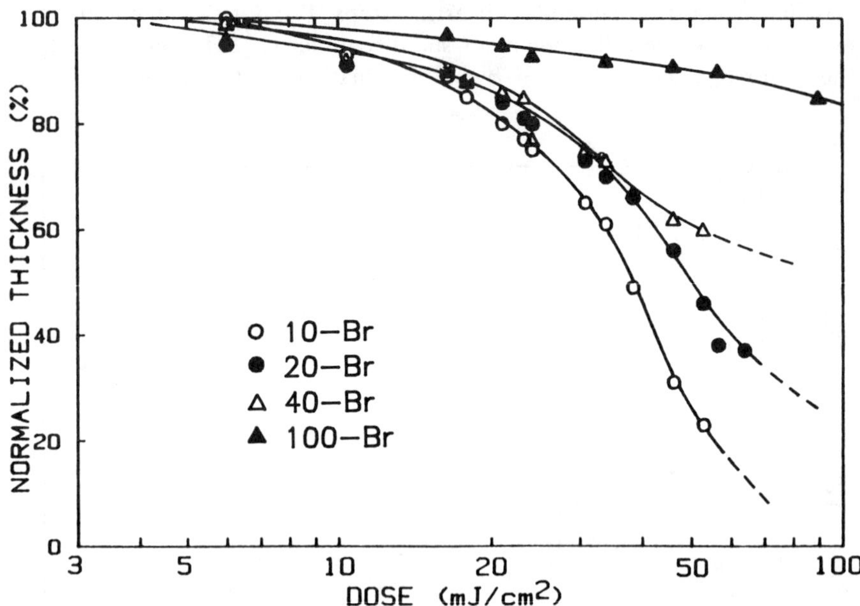

Figure 7. Sensitivity curves for brominated samples of poly(TMSP): film thickness, ≈300 nm; λ, 260 nm. Sample designations are explained in the caption to Figure 2.

followed by baking at 120 °C for 15 min to remove residual solvent. The samples were exposed in air to deep-UV radiation from a high-pressure Hg–Xe lamp (λ = 240–280 nm; Optical Associates, Inc., Santa Clara, CA) and were then developed in *n*-butyl alcohol. Bromination increased the sensitivity of poly(TMSP) more than 10 fold, with an *implied* sensitivity of 60 mJ/cm^2 for the sample with the least bromination (10% Br). Unexpectedly, an increase in bromine content correlated with a decrease in sensitivity. In all cases the sensitivities are described as "implied", because a thin residual film (apparently cross-linked) could not be removed during development. Later work showed that clean patterns could be obtained by baking the exposed wafers at 140 °C for 1 h before development. Not only was the residual film avoided by this additional step, but the sensitivity of the brominated polymers (Figure 8) increased markedly. Again, the lightly brominated samples had the best sensitivity (~10 mJ/cm^2).

The bilayer scheme was successfully demonstrated with lightly brominated poly(TMSP) (~10% Br). In these experiments, a 3000-Å layer of the brominated polymer was deposited on top of 1.5 μm of polyimide or novolac planarizing layers. The wafer was exposed to 25 mJ/cm^2 of deep-UV radiation and heated for 1 h at 140 °C. The pattern was developed in

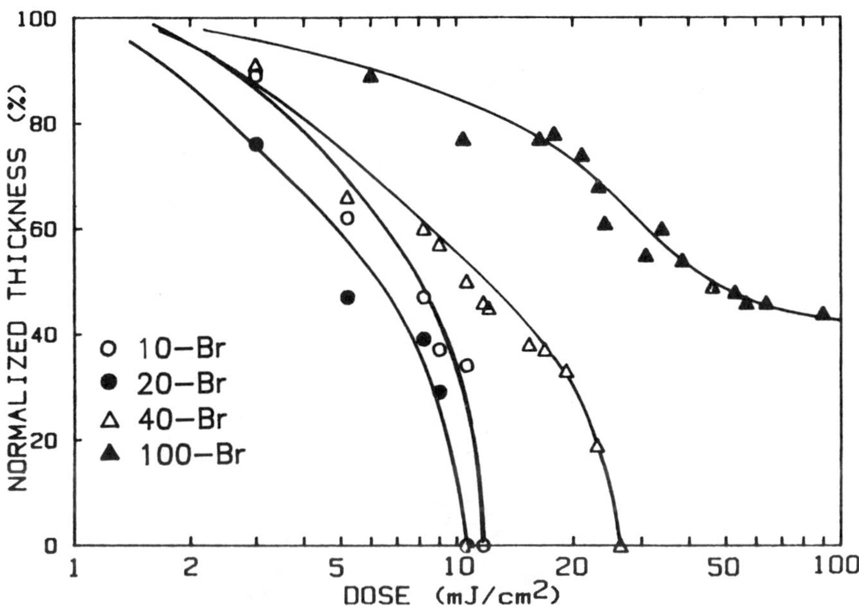

Figure 8. Sensitivity curves for brominated derivatives of poly(TMSP) after postexposure baking for 1 h at 140 °C. Sample designations are explained in the caption to Figure 2.

n-butyl alcohol. The pattern was transferred through the planarizing layer via O_2 RIE. A typical result is shown in Figure 9.

Mechanism of Lithographic Action

An understanding of the mechanism of the lithographic process is important, because it may suggest new resist materials. Several unusual aspects of the photolithographic behavior of brominated poly(TMSP) must be explained. Unlike most halogen-containing materials, the resist gave positive-tone images. The most sensitive formulations contained the least amount of bromine, and yet, the unbrominated polymer was insensitive toward photodegradation and required a postexposure baking step to prevent the formation of a cross-linked residual film.

Studies of polymer films before and after exposure to deep-UV radiation and heat treatment showed that photooxidation causes structural changes in the resist that lead to lithographic action. The IR spectra of a lightly brominated sample before and after exposure (Figure 10) showed a decrease in absorption at 1600 cm^{-1} and a large increase in bands near 1750 cm^{-1}. These changes indicate oxidation of the conjugated poly(TMSP) backbone and formation of carbonyl-containing photoproducts. Additional mechanistic infor-

Figure 9. SEM (scanning electron microscopy) micrographs of 1-μm line and space patterns transferred by O_2 RIE into 1.5 μm of planarizing material.

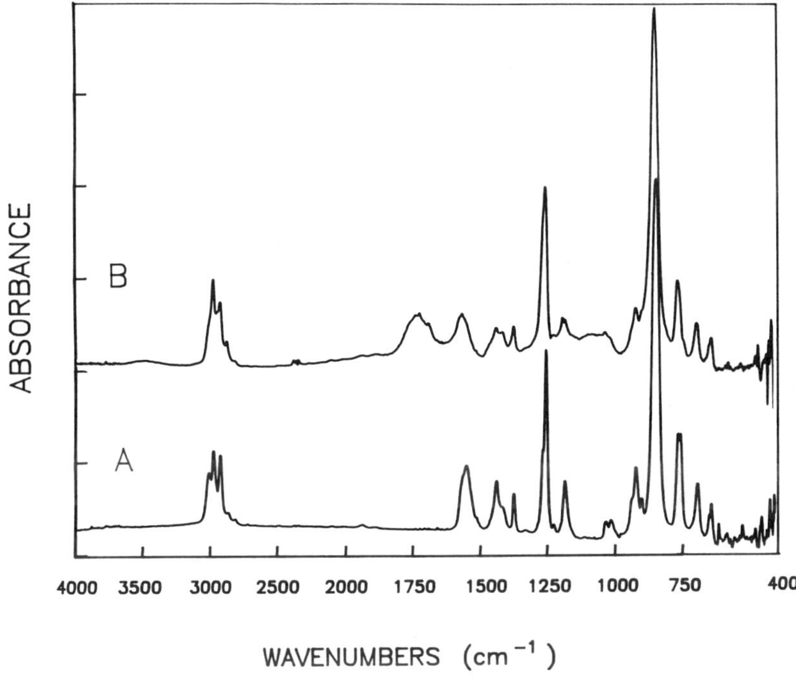

Figure 10. IR spectra of brominated poly(TMSP) (10% Br) before (A) and after (B) exposure to 50 mJ/cm² of deep-UV radiation and a 1-h heat treatment at 140 °C.

mation was obtained from studies of the molecular weight decrease during photooxidation (Table II). The number-average molecular weight of a sample (~40% Br) decreased 10 fold after exposure, but the polydispersity increased from 2.3 to 4.4, a result indicating that both cross-linking and chain scission occurred during the lithographic process. The last two entries in Table II show that the cross-linking is not permanent; heating the samples induced a further decrease in molecular weight and a decrease of polydispersity to 2, the value expected in the case of a random scission process.

Table II. Degradation of Brominated Poly(TMSP) (40-Br) Films

Sample Treatment	M_w (×10^4)	M_n (×10^4)	M_w/M_n
None	36	16	2.3
50 mJ/cm² at 260 nm	7.1	1.6	4.4
1 h at 110 °C	1.9	0.65	2.9
1 h at 140 °C	0.57	0.28	2.0

NOTE: M_n and M_w are number-average and weight-average molecular weights, respectively.

The data can be explained by a mechanism that combines elements of photochemistry and thermal decomposition. During exposure to UV radiation in air, photooxidation results in the formation of numerous thermally labile cross-links. The cross-linked portion of the sample corresponds to the residual film found in all samples that have not been subjected to a postexposure heat treatment. These cross-links (probably peroxides) decompose on heating and cause chain scission and the formation of products containing carbonyl species. The degraded products are highly polar and are soluble in *n*-butyl alcohol. These properties allow the development of the resist pattern without swelling unexposed regions. In samples with high degrees of bromination (>25%), irreversible cross-links (probably C–C bonds) also are formed. These cross-links lead to tailing of the sensitivity curves and to lower measured sensitivities.

The exact photochemical pathways in this system remain to be identified. On the basis of our current data, two different mechanisms can be envisioned, depending on whether the relevant chromophore is the allyl bromide or, instead, the polyene backbone of the polymer. Simple C–Br bond cleavage and trapping of the allylic radical by oxygen (Scheme I) leads to thermally labile cross-links and subsequent oxidation. In such a scheme, the excellent sensitivities of lightly brominated materials are difficult to explain unless a chain-oxidative process is invoked. Such a process could occur during the postexposure heat treatment. Because the most intense chromophore in the system is the polyene backbone ($\epsilon \sim 7600$), photoexcitation and trapping of excited states by molecular oxygen (Scheme I) provide a convenient route to peroxide formation and chain scission. In such a scheme, the high sensitivity of the brominated samples must correspond to a strong perturbation of the excited-state chemistry by the allylic bromides. One possibility is the conversion of an initial excited-state singlet to a triplet, a process that atoms such as bromine can accelerate through spin–orbit coupling. Preliminary experiments to probe for triplet formation have proven inconclusive, because neither poly(TMSP) nor its brominated derivatives show appreciable luminescence.

The actual photochemistry of the system may include contributions from both mechanisms. For example, absorption of photons by the conjugated backbone and subsequent oxidation would lead to the large decrease in molecular weight after exposure and heat treatment, but C–Br bond cleavage and coupling of the resulting allylic radicals provide an explanation for the irreversible cross-linking seen in highly brominated samples. With enough cross-linking, the polymer may act as a negative resist, similar to halomethylated polystyrenes. For a more definitive study, measurements of the formation of the excited states and their lifetimes are needed, as well as structural information that details the fate of allylic bromide during photooxidation.

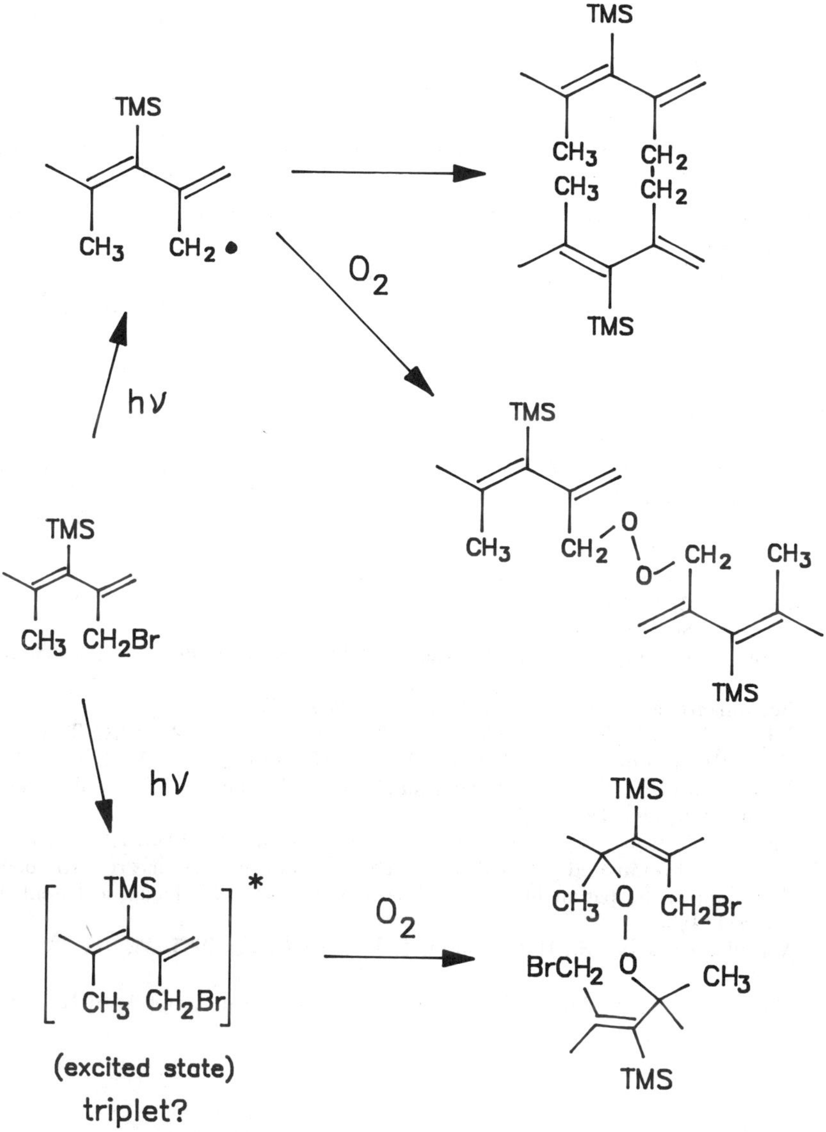

TMS

CH3 CH2•

hν

TMS

CH3 CH2Br

O_2

hν

TMS

CH3 CH2Br

(excited state)

triplet?

*

O_2

Scheme I. Proposed photochemical pathways for brominated poly(TMSP). TMS indicates the trimethylsilyl group.

Summary

Brominated poly(TMSP) is a promising photoresist for submicrometer deep-UV lithography. The brominated derivative is a positive-acting resist and is highly sensitive (<50 mJ/cm^2) at both 260 and 210 nm. At low bromination levels, the polymer retains the excellent resistance to O_2 RIE, thermal stability, and film-forming characteristics of the unbrominated polymer. The polymer has been used successfully in two-layer schemes to transfer images through thick planarizing layers and can be used to make features with dimensions as small as 0.5 μm. The dominant chemistry of the lithographic process appears to be photooxidation, with bromine acting as a sensitizer, but at higher bromination levels, C–Br bond cleavage results in cross-linking.

References

1. For a review of photolithography see *Introduction to Microlithography: Theory, Materials, Processes;* Thompson, L. F.; Willson, C. G.; Bowden, M. J., Eds.; ACS Symposium Series 219; American Chemical Society: Washington, DC, 1983.
2. For a review of deep-UV photolithography see Iwayanagi, T.; Ueno, T.; Nonogaki, S.; Ito, H.; Willson, C. G. In *Electronic and Photonic Applications of Polymers;* Bowden, M. J., Turner, S. R., Eds.; ACS Advances in Chemistry 218; American Chemical Society: Washington, DC, 1988; pp 109–224.
3. Chou, N. J.; Tang, C. H.; Paraszczak, J.; Babich, E. *Appl. Phys. Lett.* **1985,** *46,* 31.
4. Reichmanis, E.; Smolinsky, G. *Proc. SPIE* **1984,** *469,* 38.
5. Miller, R. D.; Hofer, D.; McKean, D. R.; Willson, C. G.; West, R.; Trefonas, P. T., III In *Materials for Microlithography;* Thompson, L. F.; Willson, C. G.; Fréchet, J. M. J., Eds.; ACS Symposium Series 266; American Chemical Society: Washington, DC, 1984; p 293.
6. Masuda, T.; Isobe, R.; Higashimura, T. *J. Am. Chem. Soc.* **1983,** *105,* 7473.
7. The use of brominated poly(TMSP) as a chemical intermediate has recently been described in the patent literature. Tanaka, H.; Morita, M. Japanese Patent 61 105 544, 1986.
8. Masuda, T.; Isobe, E.; Higashimura, T. *Macromolecules* **1985,** *18,* 841.

RECEIVED for review May 27, 1988. ACCEPTED revised manuscript April 12, 1989.

37

Stereoregular Anionic Ring-Opening Polymerization of Silacyclopent-3-enes

Xuehai Zhang,[1] Qingshan Zhou,[1] William P. Weber,[1,4] Raymond F. Horvath,[2] Tak-Hang Chan,[2] and Georges Manuel[3]

[1]Loker Hydrocarbon Research Institute and Department of Chemistry, University of Southern California, Los Angeles, CA 90089–1661
[2]Department of Chemistry, McGill University, Montreal, Quebec H3A 2K6, Canada
[3]Laboratoire des Organométalliques, UA 477 Université Paul-Sabatier, 31062 Toulouse Cedex, France

The ring-opening polymerization of 1,1-dimethyl-1-silacyclopent-3-ene, 1-methyl-1-phenyl-1-silacyclopent-3-ene, 1,1-diphenyl-1-silacyclopent-3-ene, and 1,1,3-trimethyl-1-silacyclopent-3-ene was catalyzed by alkyllithium reagents in the presence of hexamethylphosphoramide or N,N,N′,N′*-tetramethylethylenediamine and yielded poly(1,1-dimethyl-1-sila-*cis*-pent-3-ene), poly(1-methyl-1-phenyl-1-sila-*cis*-pent-3-ene), poly(1,1-diphenyl-1-sila-*cis*-pent-3-ene), and poly(1,1,3-trimethyl-1-sila-*cis*-pent-3-ene), respectively. The polymer products were characterized by ^{1}H, ^{13}C, and ^{29}Si NMR spectroscopy; gel permeation chromatography; and thermogravimetric analysis. The mechanism of this polymerization is discussed.*

THE ANIONIC RING-OPENING POLYMERIZATION of 1,1-dimethyl-1-silacyclopent-3-ene (**I**) has been reported recently, although the polymer has not been fully characterized (*1*). The properties of poly(1,1-dimethyl-1-sila-*cis*-pent-3-ene) (**II**), as well as several related systems, are reported in this chapter.

Anionic polymerization of **I** yields high-molecular-weight materials with weight-average (M_w) and number-average (M_n) molecular weights of 158,000

[4]Author to whom correspondence should be addressed.

0065–2393/90/0224–0679$06.00/0

and 69,000, respectively. By comparison, ring-opening metathesis polymerization of **I** (*2, 3*) gives low-molecular-weight materials.

Anionic polymerization requires a cocatalyst system composed of an alkyllithium reagent and either *N,N,N′,N′*-tetramethylethylenediamine (TMEDA) or hexamethylphosphoramide (HMPA). 1-Methyl-1-phenyl-1-silacyclopent-3-ene (**III**), 1,1-diphenyl-1-silacyclopent-3-ene (**IV**), and 1,1,3-trimethyl-1-silacyclopent-3-ene (**V**) were polymerized under similar conditions.

Experimental Procedures

^{1}H and ^{13}C NMR spectra were obtained with a JEOL FX-90FT (Fourier transform) spectrometer. ^{13}C NMR spectra were run with broad-band proton decoupling. ^{29}Si NMR spectra were obtained with a Brucker WP-270-SY FT spectrometer. Solutions of 10–15% in $CDCl_3$ or C_6D_6 were used to obtain ^{29}Si NMR spectra, whereas 5% solutions were used for ^{1}H and ^{13}C NMR spectra. A DEPT (distortionless enhancement by polarization transfer) pulse sequence was used to obtain ^{29}Si NMR spectra. This technique was effective, because all the silicon atoms are bonded to at least one methyl group (*4*). IR spectra were recorded on a Perkin–Elmer PE 281 spectrometer. The spectra were taken from neat oils between NaCl plates or from chloroform solutions in NaCl cells.

Gel permeation chromatographic (GPC) analysis of the molecular weight distribution of the polymers was performed with a Perkin–Elmer series 10 liquid chromatograph equipped with an LC-25 RI detector (25 °C), a 3600 data station, and a 660 printer. A Perkin–Elmer PL 10-μm particle mixed-pore-size cross-linked polystyrene gel column (32 cm by 7.7 mm) was used for the separation. The eluting solvent was reagent-grade tetrahydrofuran (THF) at a flow rate of 0.7 mL/min. The retention times were calibrated against known monodispersed polystyrene standards with M_ps of 194,000, 87,000, or 10,200 and for which the ratio M_w/M_n is less than 1.09.

Thermogravimetric analyses (TGA) were carried out with a Perkin–Elmer TGS-2 analyzer at a nitrogen flow rate of 50 cm^3/min. The temperature program for the analysis was 50 °C for 10 min followed by an increase of 5 °C/min to 600 °C.

Elemental analysis was performed by Galbraith Laboratories, Knoxville, TN. Satisfactory analytical results (±0.4%) were obtained for all new polymers.

Polymerization of I. **I** was polymerized in flame-dried equipment under N_2 at −40 °C as follows. A 25-mL round-bottom flask equipped with a poly(tetrafluoroethylene) (Teflon)-covered magnetic stirring bar and rubber septum was charged with **I** (1.2 g, 10.9 mmol) (*5, 6*), THF (10 mL), and either HMPA (5 drops) or TMEDA (5 drops). *n*-Butyllithium (0.8 mL, 1.2 M, 0.96 mmol) was added slowly to this mixture. The mixture quickly became thick. The mixture was stirred for 1 h at −40 °C and then warmed to −20 °C, and saturated aqueous ammonium chloride was added. The organic layer was separated, washed with brine and water, and dried over molecular sieves (4 Å). After filtration, the solvent was removed by evaporation under vacuum; 1.10 g (92% yield) of polymer was isolated. The yields of polymer (±2%) and their spectral properties were identical regardless of whether HMPA or TMEDA was used as cocatalyst. With *n*-butyllithium–TMEDA, a polymer with M_w and M_n of 158,000 and 69,000, respectively, was obtained, whereas with *n*-butyllithium–HMPA, a polymer with M_w and M_n of 120,000 and 30,400, respectively, was isolated.

The analysis of poly(1,1-dimethyl-1-sila-*cis*-pent-3-ene) (**II**) gave the following results. ^{1}H NMR spectrum (values are chemical shifts [δ] expressed in parts per million [ppm]): 0.125 (s [singlet], 6 H), 1.58 (d [doublet], 4 H, J [coupling constant] = 7 Hz), and 5.485 (t [triplet], 2 H, J = 7 Hz). ^{13}C NMR spectrum (δ, ppm): –3.26 (2 C), 16.72 (2 C), and 123.45 (2 C). ^{29}Si NMR spectrum (δ, ppm): 2.17. IR spectrum (values are frequencies [ν]): 1710 (br [broad]), 1670 (br), 1630 (C=C), 1345, 1248, and 840 cm^{-1}.

The polymerization of 1,1-dimethyl-1-silacyclopentane (*7*) was carried out as described for **I**, but only the starting material was recovered.

Poly(1,1,3-trimethyl-1-sila-*cis*-pent-3-ene) (VIII). **VIII** was prepared by polymerization of **V** (*5, 6*) as described for **I**; an 84% yield was obtained. With n-butyllithium–HMPA, a polymer with M_w and M_n of 23,500 and 11,400, respectively, was obtained, whereas with methyllithium–TMEDA, a polymer with M_w and M_n of 32,000 and 15,000, respectively, was found. The polymer products had identical spectral properties. Analysis gave the following results. ^{1}H NMR (δ, ppm): –0.056 (s), –0.018 (s), and 0.030 (s) at a peak intensity ratio of ~1:2:1 (6 H); 1.30 (d, 2 H, J = 8 Hz); 1.455 (d, 2 H, J = 1 Hz); 1.655 (d, 3 H, J = 1 Hz); and 5.01 (t, 1 H, J = 8 Hz). ^{13}C NMR (δ, ppm): –3.34, –2.36, –1.28, 17.34, 17.43, 18.26, 18.34, 21.41, 21.47, 22.33, 22.38, 117.09, 117.25, 117.40, 117.42, 130.58, 130.74, 130.96, and 131.08. ^{29}Si NMR (δ, ppm): 1.926. IR (ν): 1720–1610 (br), 1345, and 830 cm^{-1}.

Poly(1-methyl-1-phenyl-1-sila-*cis*-pent-3-ene) (IX). **IX** was prepared by polymerization of **III** (*8, 9*) as described for **I**; a 93% yield was obtained with n-butyllithium–HMPA. The polymer, which was purified by precipitation from methanol, was a viscous oil. Analysis gave the following results. M_w = 23,000. M_n = 10,000. ^{1}H NMR (δ, ppm): 0.226 (s, 3 H), 1.61 (d, 4 H, J = 6 Hz), 5.30 (t, 2 H, J = 6 Hz), and 7.5–7.27 (m [multiplet], 5 H). ^{13}C NMR (δ, ppm): –5.45, 15.30, 123.26, 127.65, 129.01, 133.88, and 137.84. ^{29}Si NMR (δ, ppm): –4.33. IR (ν): 1640, 1435, 1370, 1250, 1120, and 830 cm^{-1}.

Poly(1,1-diphenyl-1-sila-*cis*-pent-3-ene) (VI). **VI** was prepared as described previously by polymerization of **IV** (*5, 10*). Methyllithium with either HMPA or TMEDA and n-butyllithium with TMEDA were effective catalysts. The polymer was purified by addition of methanol to a THF solution of the crude reaction product. The polymer precipitated to give an 89% yield (mp = 130–136 °C). With n-butyllithium–TMEDA, a polymer with M_w and M_n of 13,400 and 7,900, respectively, was obtained. When methyllithium–TMEDA was used, M_w = 12,400 and M_n = 6,900 were found. Finally, with methyllithium–HMPA, M_w and M_n were 8000 and 4350, respectively. All the polymer products had identical spectral properties. ^{1}H NMR (δ, ppm): 1.76 (d, 4 H, J = 6 Hz), 5.31 (t, 2 H, J = 6 Hz), and 7.40–7.66 (m, 10 H). ^{13}C NMR (δ, ppm): 13.67, 123.48, 127.60, 129.22, 134.91, and 135.51. ^{29}Si NMR (δ, ppm): –10.54. IR (ν): 1420, 1360, and 1100 cm^{-1}.

Isomerization of VI. AIBN [azobis(isobutyronitrile)] (15 mg) and **VI** (100 mg) were dissolved in 4 mL of dry THF in a quartz tube. The solution was purged with argon and sealed with a rubber septum, and then, 10 mL of thiophenol was added. This solution was irradiated with a 450-W medium-pressure Hanovia Hg lamp for 24 h at 25 °C. After precipitation with methanol, a polymer with the following properties was obtained. mp = 62–70 °C. ^{1}H NMR (δ, ppm): 1.76 (br m, 4 H), 4.70 (br m), 5.3 (br m), and 7.3 (br m, 10 H) (the ratio of peak intensities for *cis*-vinyl (δ 5.3 ppm) and *trans*-vinyl (δ 4.7 ppm) hydrogens is 2.2). ^{13}C NMR (δ, ppm): 13.67, 17.9, 123.4, 124.7, 127.6, 129.2, 134.9, and 135.5. ^{29}Si NMR (δ, ppm): −10.54 and −11.2.

Discussion

The stereochemistry about the C–C double bonds of **II** was assigned by comparison of ^{13}C NMR chemical shifts with those of model compounds (*11*) (Chart I). Poly(1,1-diphenyl-1-sila-*cis*-pent-3-ene) (**VI**) (mp 130–136 °C) undergoes phenylthio-radical-catalyzed isomerization (*12, 13*) to a mixture of poly(1,1-diphenyl-1-sila-*cis*-pent-3-ene) and poly(1,1-diphenyl-1-sila-*trans*-pent-3-ene) (**VII**) (mp 62–70 °C). ^{13}C NMR chemical shifts support these assignments, as shown in Scheme I.

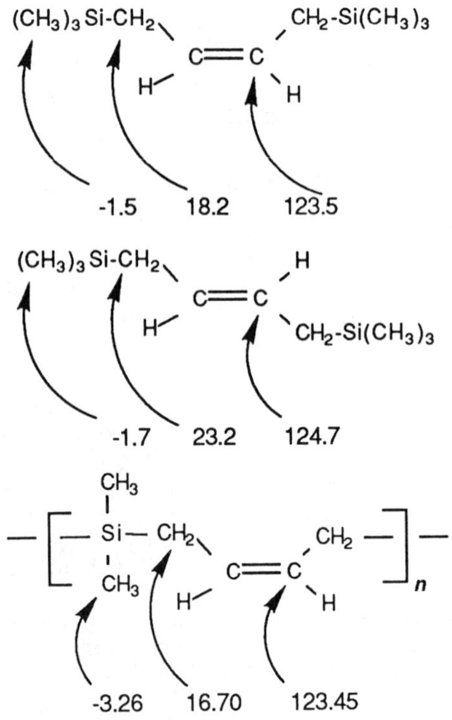

*Chart I. ^{13}C NMR chemical shifts of **II** compared with those of model compounds.*

The anionic ring-opening polymerization of **I** is unexpected because under similar conditions, allyltrimethylsilane undergoes metallation to yield an α-trimethylsilyl-substituted allyl anion (*14–16*). Relief of ring strain may be an important factor in facilitating this ring-opening polymerization. For example, the polymerization of silacyclobutanes is catalyzed by various nu-

δ : 13.67 123.5

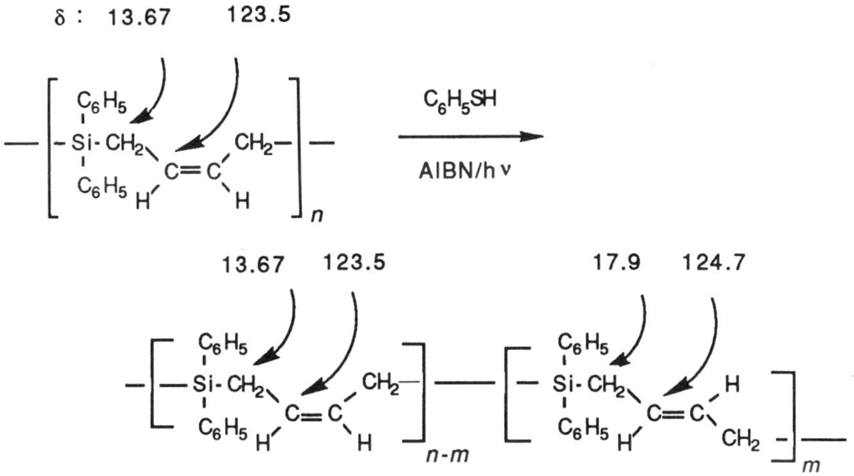

C_6H_5SH

AIBN/hν

13.67 123.5 17.9 124.7

n-m *m*

Scheme I. Isomerization of **VI** *as followed by* ^{13}C *NMR chemical shifts.*

cleophiles (*17–21*). However, we were unable to polymerize 1,1-dimethyl-1-silacyclopentane. The retention of stereochemistry of the double bonds in the polymer provides evidence against the intermediacy of a free allyl anion, which would be expected to undergo bond rotation to yield a polymer possessing both *cis* and *trans* double bonds. On the basis of these facts, chain propagation probably proceeds by reaction of a negatively charged pentacoordinated siliconate anion with a molecule of **I** to yield a new pentacoordinated siliconate anion (Scheme II). The molecular weights of these polymers are high with respect to the ratio of monomer to alkyllithium. This result suggests that the attack by alkyllithium reagents on the silyl center of the monomer may be slow or reversible.

The microstructure of polymer formed by anionic polymerization of **V** was analyzed by ^1H and ^{13}C NMR spectroscopy. Both ^1H and ^{13}C NMR spectra indicate that the methyl groups bonded to silicon may be in one of three distinct environments.

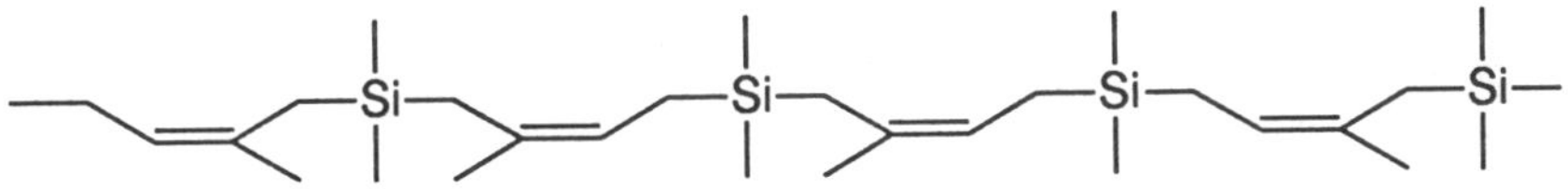

The thermal stability of these polymers was determined by TGA in a nitrogen atmosphere. Poly(1,1,3-trimethyl-1-sila-*cis*-pent-3-ene) (**VIII**) was stable up to 150 °C. **VIII** lost weight at a rate of 3–5%/50 °C increase in temperature between 150 and 350 °C. Rapid weight loss occurred between 350 and 425 °C. By 500 °C 97% weight loss had occurred. The TGA of **II** gave similar results. Poly(1-methyl-1-phenyl-1-sila-*cis*-pent-3-ene) (**IX**) was stable up to 200 °C, whereas **VI** was stable up to 250 °C. **IX** lost weight rapidly at >350 °C, whereas **VI** started to lose weight rapidly at 300 °C.

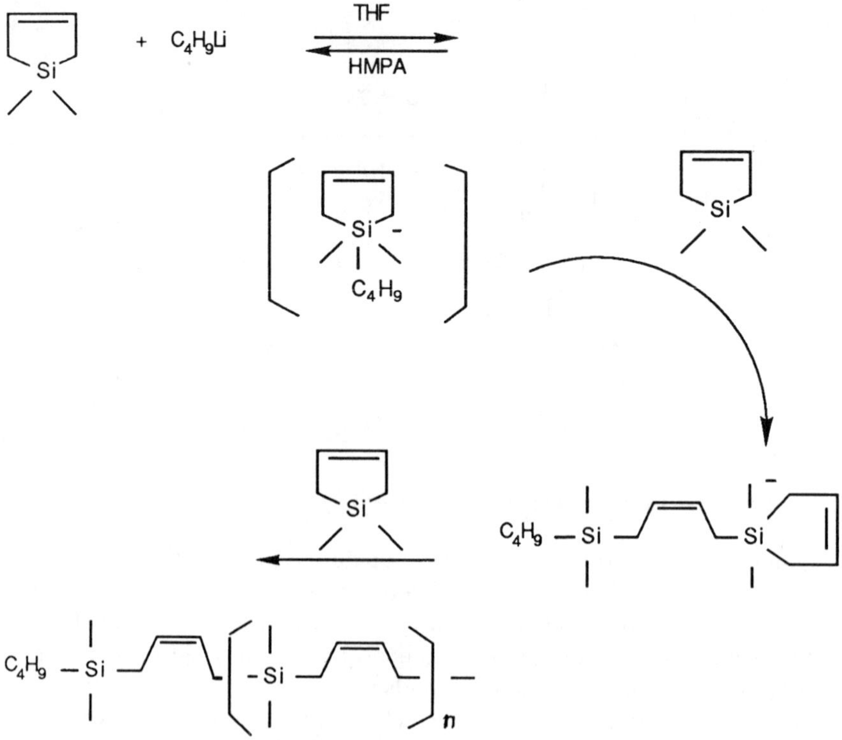

THF

HMPA

Scheme II. Polymerization of **I**.

Acknowledgments

This work was supported by the Air Force Office of Scientific Research (grant 89–0007). Manuel and Weber thank the North Atlantic Treaty Organization (NATO) for a travel grant. Chan and Horvath acknowledge the support of the Natural Science and Engineering Research Council of Canada and the Fond pour la Formation de Chercheurs et L'aide a la Recherche of Quebec.

References

1. Horvath, R. F.; Chan, T. H. *J. Org. Chem.* **1987,** *52*, 4489.
2. Lammens, H.; Sartori, G.; Siffert, J.; Sprecher, N. *J. Polym. Sci.* **1971,** *89*, 341.
3. Finkel'shtein, E. S.; Portnykh, E. P.; Ushakov, N. V.; Vdovin, V. M. *Izv. Akad. Nauk SSSR, Ser. Khim.* **1981,** *3*, 641.
4. Pegg, D. T.; Doddrell, D. M.; Bendall, M. R. *J. Chem. Phys.* **1982,** *77*, 2745.
5. Manuel, G.; Mazerolles, P.; Cauquy, G. *Synth. React. Inorg. Met.–Org. Chem.* **1974,** *4*, 133.
6. Weyenberg, D. R.; Toporcer, L. H.; Nelson, L. E. *J. Org. Chem.* **1968,** *33*, 1975.

7. Fessenden, R.; Coon, M. D. *J. Org. Chem.* **1961,** *26,* 2530.
8. Manuel, G.; Cauquy, G.; Mazerolles, P. *Synth. React. Inorg. Met.– Org. Chem.* **1974,** *4,* 143.
9. Manuel, G.; Mazerolles, P.; Darbon, J. M. *J. Organomet. Chem.* **1973,** *59,* C7.
10. Dunoques, J.; Calas, R.; Dedier, J.; Pisciotti, F. *J. Organomet. Chem.* **1970,** *25,* 51.
11. Marchand, A.; Gerval, P.; Joanny, M.; Mazerolles, P. *J. Organomet. Chem.* **1981,** *217,* 19.
12. Kobayashi, Y.; Okamoto, S.; Shimazaki, T.; Ochiai, Y.; Sato, F. *Tetrahedron Lett.* **1987,** 3959.
13. Golub, M. A. *J. Polym. Sci.* **1957,** *25,* 373.
14. Ayalon-Chass, D.; Ehlinger, E.; Magnus, P. *J. Chem. Soc., Chem. Commun.* **1977,** 772.
15. Lau, P. W. K.; Chan, T. H. *Tetrahedron Lett.* **1978,** 2383.
16. Ehlinger, E.; Magnus, P. *Tetrahedron Lett.* **1980,** 11.
17. Nametkin, N. S.; Vdovin, V. M.; Grinberg, P. L.; Babich, E. D. *Dokl. Akad. Nauk. SSSR* **1965,** *161,* 358.
18. Nametkin, N. S.; Vdovin, V. M.; Zav'yalov, V. I. *Izv. Akad. Nauk. SSSR, Ser. Khim.* **1964,** *191,* 2003.
19. Nametkin, N. S.; Vdovin, V. M.; Poletaev, Y. A.; Zavialov, V. I. *Dokl. Akad. Nauk. SSSR* **1967,** *175,* 1068.
20. Nametkin, N. S.; Vdovin, V. M.; Zav'yalov, V. I. *Dokl. Akad. Nauk. SSSR* **1965,** *162,* 824.
21. Topchiev, A. V. Ger. Patent 1,226,310, 1966; *Chem. Abstr.* **1967,** *66,* 76763j.

RECEIVED for review May 27, 1988. ACCEPTED revised manuscript March 13, 1989.

38

Stereoregular 1,4 Polymerization of 2-Triethylsilyl-1,3-butadiene

Yi-Xiang Ding and William P. Weber[1]

Loker Hydrocarbon Research Institute and Department of Chemistry, University of Southern California, Los Angeles, CA 90089–1661

Polymerization of 2-triethylsilyl-1,3-butadiene initiated by n*-butyllithium in hexane yielded (*E*)-1,4-poly(2-triethylsilyl-1,3-butadiene). The molecular weight distribution (as determined by gel permeation chromatography) of the polymers depended on the ratio of monomer to initiator concentrations. Polymers of low polydispersity (*M_w *[weight-average molecular weight]/*M_n *[number-average molecular weight] = 1.3–1.6) and with* M_w*s up to 110,000 were obtained. The 1,4 mode of polymerization and the geometry about the C–C double bonds of the polymer chain were established by* ^1H *and* ^{13}C *NMR spectroscopy, as well as by protodesilation of the polymer by treatment with HI. Polymerization in tetrahydrofuran yielded (*E*)- and (*Z*)-1,4-poly(2-triethylsilyl-1,3-butadiene).*

THE STEREOREGULAR POLYMERIZATION of 1,3-diene monomers is of considerable interest to polymer researchers. Anionic polymerization of isoprene initiated by alkyllithium reagents in hydrocarbon solvents yields polyisoprene of narrow molecular weight distribution and with a microstructure that has been shown by IR (*1*, *2*), ^1H NMR (*3*), and ^{13}C NMR (*4*) spectroscopy to consist of *cis*-1,4 (~80%), *trans*-1,4 (~15%), and 3,4 (~5%) units. Anionic polymerization of isoprene in donor solvents such as ether or tetrahydrofuran (THF) leads to a polymer with a microstructure consisting predominantly of

[1]Author to whom correspondence should be addressed.

0065–2393/90/0224–0687$06.00/0

the 3,4 unit (~60%) (*2, 3*). Thus the anionic polymerization of isoprene is regio- and stereoselective (*5*).

Experimental Procedures

^1H and ^{13}C NMR spectra were obtained with a JEOL FX-90Q spectrometer. ^{13}C NMR spectra were run with broad-band proton decoupling. ^{29}Si NMR spectra were obtained with a Brucker WP-270-SY spectrometer. Solutions of 10–15% in $CDCl_3$ were used to obtain ^{29}Si NMR spectra, whereas 5% solutions were used for ^1H and ^{13}C NMR spectra. Chloroform was used as an internal standard for ^1H and ^{13}C NMR spectra. All chemical shifts reported were externally referenced to tetramethylsilane (TMS). A DEPT (distortionless enhancement by polarization transfer) pulse sequence was used to obtain ^{29}Si NMR spectra (*6*).

IR spectra were recorded on a Perkin–Elmer PE 281 spectrometer. The spectra were taken from chloroform solutions in NaCl cells.

Gel permeation chromatographic (GPC) analysis of the molecular weight distribution of the polymers was performed with a Perkin–Elmer series 10 liquid chromatograph equipped with an LC-25 RI detector (25 °C), a 3600 data station, and a 660 printer. A Perkin–Elmer PL 10-µm particle mixed-pore-size cross-linked polystyrene gel column (32 cm by 77 mm) was used for the separation. The eluting solvent was reagent-grade THF at a flow rate of 0.7 mL/min. The retention times were calibrated against known monodispersed polystyrene standards with M_ps of 3,600,000, 194,000, 28,000, and 2,550 and for which M_w [weight-average molecular weight]/M_n [number-average molecular weight] < 1.09.

Thermogravimetric analysis (TGA) of the polymers was carried out with a Perkin–Elmer TGS-2 instrument at a nitrogen flow rate of 40 cm^3/min. The temperature program for the analysis was 100 °C for 10 min followed by an increase of 5 °C/min to 500 °C.

Elemental analysis was performed by Galbraith Laboratories, Knoxville, TN. Satisfactory analytical data (±0.5%) were obtained.

2-Triethylsilyl-1,3-butadiene (I). **I** was prepared from 1,4-dichloro-2-butyne and triethylsilane (*7*).

Polymerization of I. Polymerizations were carried out in flame-dried apparatus under a N_2 atmosphere. A 50-mL round-bottom flask equipped with a poly(tetrafluoroethylene) (Teflon)-covered magnetic stirring bar and a rubber septum was charged with **I** (6 g, 36 mmol) dissolved in 30 mL of hexane. *n*-Butyllithium (0.1 mL, 1.5 N) was added to this mixture at room temperature. The reaction was stirred for 60 h. Ether was then added, and the organic phase was washed twice with water, dried over anhydrous magnesium sulfate, and filtered. The solvent was removed by evaporation under reduced pressure. The residue was taken up in THF, and the polymer product, a viscous oil, was precipitated by addition of methanol. The THF–methanol supernatant liquid was decanted from the polymer; it contained 2.0 g of **I**. This precipitation procedure was repeated to obtain analytical samples. The polymer was dried at 57 °C under vacuum overnight, and 3.7 g (93% yield) was obtained.

Analysis of (*E*)-1,4-poly(2-triethylsilyl-1,3-butadiene) (*E*-**II**) gave the following results. ^1H NMR (values are chemical shifts [δ] expressed in parts per million [ppm]):

5.70 (br [broad] s [singlet], 1 H), 2.10 (br s, 4 H), 0.897 (t [triplet], 9 H, coupling constant [J] = 6.5 Hz), and 0.61 (q [quartet], 6 H, J = 6.5 Hz). ^{13}C NMR (δ, ppm): 141.79, 136.87, 30.14, 29.44, 7.50, and 3.22. ^{29}Si NMR (δ, ppm): 1.99. IR ($CHCl_3$) (values are frequencies [ν]): 1597, 1450, 1410, 1230, and 998 cm^{-1}.

Protodesilation of *E*-II. *E*-**II** (500 mg) was dissolved in 20 mL of CH_2Cl_2 in a 50-mL round-bottom flask equipped with a Teflon-covered magnetic stirring bar. Two milliliters of a 47% solution of aqueous HI was added. The mixture was stirred at room temperature for 96 h and then diluted with ether. The organic phase was washed with aqueous sodium bicarbonate and water, dried over anhydrous magnesium sulfate, and filtered. The solvent was removed by evaporation under reduced pressure to give (Z)-1,4-poly(butadiene) (**III**).

Analysis of **III** gave the following results. ^{1}H NMR (δ, ppm): 5.376 (br s, 2 H) and 2.07 (br s, 4 H). ^{13}C NMR (δ, ppm): 129.60 and 27.38. IR (film) (ν): 1650, 1440, 980, and 725 cm^{-1} (*7*).

Polymerization of I in THF. **I** was polymerized in THF at –25 °C, as described previously. A 75% yield of *E*- and Z-**II** was obtained. Analysis of the products gave the following results. M_w = 18,900; M_n = 9,170; *E*:*Z* = 1.3. ^{1}H NMR (δ, ppm): 5.99 (br s, Z-CH=), 5.71 (br s, *E*-CH=), 2.10 (br s, 4 H), 0.89 (t, 9 H, J = 6.5 Hz), and 0.60 (q, 6 H, J = 6.5 Hz). ^{13}C NMR (δ, ppm): 144.18, 143.69, 141.90, 141.79, 136.81, 136.65 135.94, 135.51, 38.48, 33.90, 33.51, 32.63, 30.51, 30.14, 29.43, 28.57, 7.66, 7.50, 4.30, and 3.27. ^{29}Si NMR (δ, ppm): 2.06, 1.96, 1.91, 1.89, 0.51, 0.38, and 0.30. IR ($CHCl_3$) (ν): 1600, 1450, 1412, 1233, and 997 cm^{-1}.

Results and Discussion

Anionic polymerization of 2-triethylsilyl-1,3-butadiene (**I**) in hexane at room temperature initiated by *n*-, *sec*-, or *tert*-butyllithium gave high yields of (*E*)-1,4-poly(2-triethylsilyl-1,3-butadiene) (*E*-**II**). Neither (Z)-1,4-poly(2-triethylsilyl-1,3-butadiene) (Z-**II**) nor 1,2 or 3,4 units were found. The reaction is both regio- and stereospecific.

The microstructure of the polymer was determined by ^{1}H, ^{13}C, and ^{29}Si NMR spectroscopy. In the olefinic region, only a single peak at 5.70 ppm (1 H) was observed. For comparison, the vinyl –CH of (*E*)-3-trimethylsilyl-3-octene is found at 5.60 ppm (*9*). Two nonequivalent vinyl carbons, as well as two distinct allylic carbons, were observed in the ^{13}C NMR spectrum. No peaks could be assigned to either 1,2 or 3,4 structures in either the ^{1}H or ^{13}C NMR spectra. Further, only a single silicon resonance was seen in the ^{29}Si NMR spectrum. Finally, protodesilation of the polymer by treatment with HI yielded (Z)-1,4-poly(butadiene) (**III**), whose structure was confirmed by IR spectroscopy (*8*). Protodesilation of vinyl silanes occurs stereospecifically with retention of configuration (*10, 11*). The data are consistent with *E*-**II** (equation 1).

$Si(C_2H_5)_3$ C_4H_9Li Hexane $Si(C_2H_5)_3$ n (1)

GPC analysis of *E*-**II** revealed a monomodal molecular weight distribution of low polydispersity (Table I). The molecular weight of the polymer depended on the ratio of the monomer to initiator. On the other hand, anionic polymerization of **I** in the presence of donor solvents such as THF gave moderate yields of *E*- and Z-**II**. Under these conditions, lower molecular weight polymers of higher dispersity were obtained. The ratio of *E* to Z units was determined by integration of the ^{1}H NMR signals; the Z vinyl –CH signal is at 5.99 ppm, whereas the *E* vinyl –CH signal is at 5.71 ppm. These assignments are consistent with chemical shifts reported for monomeric trimethylsilyl-substituted olefins (*9*, *12*). Similarly, anionic polymerization of 2-trimethoxysilyl-1,3-butadiene in THF yielded 1,4-poly(2-trimethoxysilyl-1,3-butadiene) with an *E*/*Z* ratio of 70:30 (*13*).

The ^{13}C NMR spectra of *E*- and Z-**II** are complicated. Eight vinyl and eight allylic carbon resonances were observed. This result is consistent with the ^{13}C NMR of (*E*)- and (Z)-1,4-polyisoprene (*14*) (equation 2).

$Si(C_2H_5)_3$ C_4H_9Li THF -25°C $Si(C_2H_5)_3$ n $Si(C_2H_5)_3$ m (2)

TGA indicates that these polymers are thermally stable up to 200 °C. Between 200 and 250 °C, about 3% weight loss occurred. Rapid and complete weight loss occurred between 250 and 350 °C.

Table I. Anionic Polymerization of I in Hexane at 25 °C

n-*Butyllithium Initiator (%)*	*Time (h)*	*Yield (%)*	M_w *(×10³)*	M_n *(×10³)*	M_w/M_n
0.5	60	92	108.8	67.3	1.61
1.5	30	93	18.3	14.0	1.28

Acknowledgment

This work was supported by the Air Force Office of Scientific Research grant 89–0007.

References

1. Tobolsky, A. V.; Rogers, C. E. *J. Polym. Sci.* **1959,** *40,* 73.
2. Stearns, R. S.; Forman, L. E. *J. Polym. Sci.* **1959,** *41,* 381.
3. Worsfold, D. J.; Bywater, S. *Can. J. Chem.* **1964,** *42,* 2884.
4. Morese-Seguela, B.; St-Jacques, M.; Renaud, J. B.; Prud'homme, J. *Macromolecules* **1977,** *10,* 431.
5. Morton, M. *Anionic Polymerization, Principles and Practice;* Academic: New York, 1983.
6. Pegg, D. T.; Doddrell, D. M.; Bendall, M. R. J. *Chem. Phys.* **1982,** *77,* 2745.
7. Batt, D. G.; Ganem, B. *Tetrahedron Lett.* **1978,** 3323.
8. Silas, R. S.; Yates, J.; Thorton, V. *Anal. Chem.* **1959,** *31,* 529.
9. Miller, R. B.; McGarvey, G. *J. Org. Chem.* **1979,** *44,* 4623.
10. Koenig, K. E.; Weber, W. P. *J. Am. Chem. Soc.* **1973,** 95, 3416.
11. Utimoto, K.; Kitai, M.; Nozaki, H. *Tetrahedron Lett.* **1975,** 2825.
12. Chan, T. H.; Mychajlowskij, W.; Amoroux, R. *Tetrahedron Lett.* **1977,** 1605.
13. Takenaka, K.; Hirao, A.; Hattori, T.; Nakahama, S. *Macromolecules* **1987,** *20,* 2034.
14. Tanaka, Y.; Sato, H. *Polymer* **1976,** *17,* 113.

RECEIVED for review May 27, 1988. ACCEPTED revised manuscript April 5, 1989.

39

Synthesis and Lithographic Evaluation of a Poly(acylsilane)

Antoni S. Gozdz,[1] Hans J. Reich,[2] and Michael D. Bowe[2]

[1]**Bell Communications Research, 331 Newman Springs Road, Red Bank, NJ 07701–7020**
[2]**S. M. McElvain Laboratories of Organic Chemistry, Department of Chemistry, University of Wisconsin, Madison, WI 53706**

The polymerization of two α,β-unsaturated acylsilanes, 1-trimethylsilyl-2-propen-1-one (vinyl trimethylsilyl ketone, VTMSK) and 1-trimethylsilyl-2-methyl-2-propen-1-one (isopropenyl trimethylsilyl ketone, IPTMSK) was studied. VTMSK polymerized readily under free-radical-initiation conditions to give a well-defined, highly soluble, high-molecular-weight polymer. Exposure of thin films of this polymer to mid- and deep-UV radiation or heating at <100 °C led to a rearrangement of the acylsilane moiety. The resulting polymer exhibited an increased solubility in 2-propanol and had imaging properties. The polymer exhibited good stability under O_2 reactive-ion-etching conditions. Attempts to obtain a high-molecular-weight polymer of IPTMSK by either free-radical or anionic initiation have not been successful.

PHOTODEGRADABLE POLYMERS with pendant carbonyl groups have found important technological applications in microlithography. Thus, poly(methyl methacrylate) and several of its derivatives (*1–3*), poly(isopropenyl methyl ketone) (*4, 5*) and poly(methyl glutarimide) (*6*) have been studied as high-resolution positive resists. Because these polymers exhibit only a weak optical absorption in the deep-UV region of the spectrum (200–310 nm), they cannot be exposed with the vast majority of exposure tools currently used in integrated-circuit manufacturing, and they require long exposure times with most commercial deep-UV sources. Also, these polymers are rapidly degraded in various dry-etching environments.

0065–2393/90/0224–0693$06.00/0

Low-molecular-weight acylsilanes (α-silyl ketones, **I** in Chart I) (*7*), on the other hand, exhibit molar absorptivities that are several times higher ($\epsilon \sim 130$) than those of their carbon analogues ($\epsilon \sim 30$). In acylsilanes, the silicon atom is directly bonded to the carbonyl carbon. This arrangement results in a large bathochromic shift in the electronic absorption spectrum of the carbonyl group (*8*). Depending on the electronic structure of the acylsilane, the $n \rightarrow \pi^*$ transition, which in the aliphatic ketones occurs at 280 nm, is shifted to the red region by 100–180 nm. These facts suggest that poly(α,β-unsaturated acylsilane)s (**II**) should exhibit optical absorption in the mid-UV range (310–400 nm) widely used in photolithography. They should also exhibit higher photosensitivities compared with their carbon analogues. Such polymers would also have a high silicon content, which would be advantageous in microlithographic applications (*9, 10*).

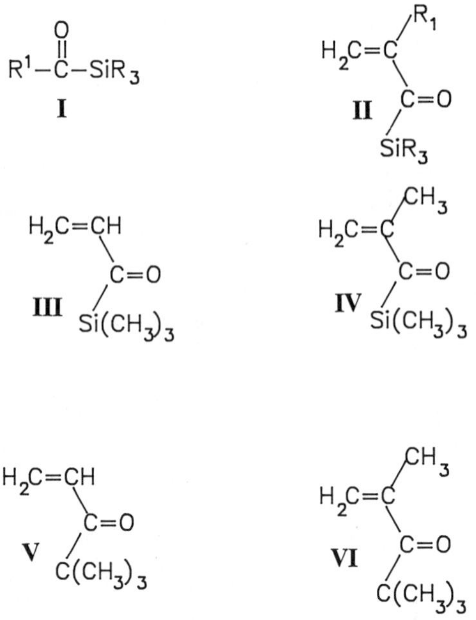

$R^1-C(=O)-SiR_3$ **I**

$H_2C=C(R_1)-C(=O)-SiR_3$ **II**

$H_2C=CH-C(=O)-Si(CH_3)_3$ **III**

$H_2C=C(CH_3)-C(=O)-Si(CH_3)_3$ **IV**

$H_2C=CH-C(=O)-C(CH_3)_3$ **V**

$H_2C=C(CH_3)-C(=O)-C(CH_3)_3$ **VI**

Chart I. Structures of acylsilanes and their carbon analogues.

Recently, we reported (*11*) the synthesis and preliminary characteristics of the first poly(acylsilane), poly(1-trimethylsilyl-2-propen-1-one) (PVTMSK), which was obtained by a free-radical polymerization of 1-trimethylsilyl-2-propen-1-one (*12*) (vinyl trimethylsilyl ketone [VTMSK], **III**). In this chapter, we report the lithographic evaluation of PVTMSK and our attempts to synthesize an α-methyl derivative of PVTMSK, poly(1-trimethylsilyl-2-methyl-2-propen-1-one) (poly[isopropenyl trimethylsilyl ketone] [PIPTMSK], **IV**).

Experimental Procedures

Materials. VTMSK (**III**) was synthesized in a four-step reaction sequence, as reported previously (*11*). The bright-yellow monomer was stored in a sealed glass ampoule over a free-radical-polymerization inhibitor at −15 °C. IPTMSK (**IV**) was obtained by a procedure reported by Danheiser et al. (*13*) and stored as just described. All the spectroscopic properties of VTMSK and IPTMSK were in agreement with the literature data (*12, 13*).

Polymerization. The monomer (or comonomers) and polymerization solvents were passed through a column packed with neutral-grade alumina to remove the inhibitor and other impurities. The monomer(s) and a solvent were degassed by several freeze–thaw cycles by using a vacuum line and were distilled at room temperature into an ampoule containing a known amount of AIBN (azobisisobutyronitrile), a free-radical-polymerization initiator. After additional degassing, the ampoule was sealed, wrapped in aluminum foil, and placed in a thermostatically controlled bath. Several examples of reaction conditions are given in Table I.

Table I. Free-Radical Polymerization of VTMSK

Solvent	*Initiator*	*Temp. (°C)*	*Time (h)*	*Yield (%)*	M_n ($\times 10^5$)	M_w/M_n
None	none	42	24	90	1.4	5.83
Hexane	AIBN (0.2%)	44	24	55.6	0.81	2.04
Toluene	AIBN (0.6%)	55	44	90	—[a]	—

[a]— indicates no data.

After polymerization, the solvent and the unreacted monomer were removed by room-temperature distillation under reduced pressure. The polymer was dissolved in chloroform, precipitated into methanol, and dried at 40 °C under vacuum. The monomers and polymers were protected from fluorescent lights during all operations.

Anionic polymerization of IPTMSK and styrene was carried out at −78 °C in a Schlenk flask attached to a vacuum line. The reaction was carried out in benzene or benzene–THF (tetrahydrofuran) with an *n*-butyllithium–18-crown-6 complex as the initiator. The reaction was carried out under argon atmosphere and was quenched by injecting methanol into the polymerization mixture.

Analyses. Polystyrene-equivalent molecular weights were determined by gel permeation chromatography (GPC) using THF solutions and Styragel [cross-linked poly(ethenylbenzene)] columns at 25 °C. IR spectra were recorded on a 983G Perkin–Elmer spectrophotometer using ~0.5-μm-thick films spin coated on NaCl disks from chlorobenzene solutions. UV spectra were taken for thin films spin coated on quartz disks with a Cary 2700 spectrophotometer. ^{1}H NMR (270 MHz, $CDCl_3$) spectra of PVTMSK were poorly resolved and very similar in appearance to those of poly(vinyl *tert*-butyl ketone) (chemical shifts [δ] at 0.23, 1.6, 1.85, and 2.73 ppm versus internal TMS [tetramethylsilane] standard). ^{13}C NMR spectra (67.8 MHz, $CDCl_3$) were more informative (δ at −2.54, 29.0, 50.2, and 248.8 ppm versus TMS).

Lithographic Evaluation. Films of PVTMSK were spin coated on polished silicon wafers by using 5 or 10% solutions in chlorobenzene. The lithographic sensitivity and resolution were determined by using both deep- and mid-UV radiation from OAI (Optical Associates, Inc., Santa Clara, CA) and Oriel light sources, respectively. The intensities of the deep-UV (200–300 nm) and mid-UV (>320 nm) sources at the wafer plane were 21 and 7.1 mW/cm^2, respectively. The patterns were exposed through chrome-on-quartz masks and developed by spraying with or dipping into 2-propanol, followed by drying. Film thickness was measured with an Alpha-Step 200 (Tencor Instruments) profilometer.

Oxygen reactive-ion etching (O_2 RIE) was carried out with a Cooke Vacuum Products (model C71-3) parallel-plate RIE reactor operating at 13.56 MHz. Oxygen pressure and flow rate were 2 Pa and 10 sccm (standard cubic centimeters per minute), respectively, and the RF (radio frequency) power density and self-bias were 0.15 W/cm^2 and –350 V, respectively.

Results and Discussion

Preparation. Although acylsilanes (**I**) and α,β-unsaturated acylsilanes (**II**) have been the subject of numerous synthetic and photochemical studies (*14, 15*), to the best of our knowledge, no acylsilane polymers have been characterized or reported prior to our recent work (*11*).

Two simple α,β-unsaturated acylsilanes, 1-trimethylsilyl-2-propen-1-one (**III**) and 1-trimethylsilyl-2-methyl-2-propen-1-one (**IV**) were chosen for polymerization studies. The polymerization of the carbon analogues of these α,β-unsaturated acylsilanes, that is, 4,4-dimethyl-2-propen-3-one (vinyl *tert*-butyl ketone, **V**) and 2,4,4-trimethyl-2-propen-3-one (isopropenyl *tert*-butyl ketone, **VI**) has been studied by Willson et al. (*16, 17*). These authors reported that whereas **V** readily polymerizes under free-radical-polymerization conditions, **VI** undergoes polymerization only under anionic-initiation conditions in the presence of a crown ether as a complexing reagent. On the basis of UV and NMR spectroscopic data, Willson et al. (*16, 17*) ascribed the difference in polymerization behavior to the nonplanar, unconjugated structure of ketone **VI** brought about by steric hindrance caused by the methyl group at C-2.

The α,β-unsaturated acylsilanes **III** and **IV** behaved similarly under free-radical-polymerization conditions. VTMSK (**III**) polymerized readily even without an added initiator (Table I), but IPTMSK (**IV**) did not.

VTMSK was polymerized in high yield both in bulk and in solution to give high-molecular-weight products. The polymerizations were always carried out at only slightly elevated temperatures (25–50 °C) to minimize the decomposition of the monomer and polymer. The spectral characteristics and other analytical data obtained for the products were in agreement with those expected for PVTMSK (Figure 1A). However, the separation and purification of PVTMSK were difficult, because the polymer exhibited an unusually high solubility in a range of polar and nonpolar organic solvents, including lower molecular weight alcohols and hexane. Although PVTMSK could be precipitated from chloroform solutions into an excess of methanol, such solutions decomposed rapidly even in the dark and at room tempera-

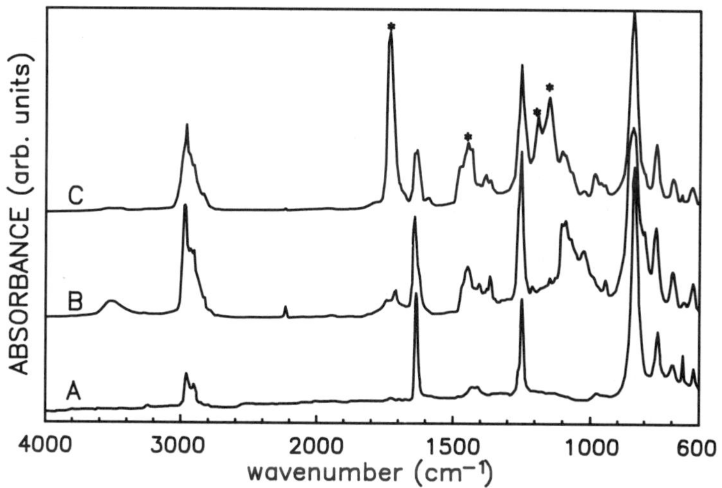

Figure 1. IR spectra of poly(acylsilane)s. (A) PVTMSK, (B) product of a free-radical polymerization of IPTMSK, and (C) a copolymer of IPTMSK with methyl methacrylate. The stars denote prominent PMMA bands.

ture. Solutions of PVTMSK in less reactive solvents such as hexane, acetone, and chlorobenzene were stable for several weeks when stored in the freezer.

Several attempts at the free-radical (benzene, 1% AIBN, 100 h, 55 °C) and anionic (1:4 THF–benzene, *n*-butyllithium–18-crown-6 complex [*16*, *17*], −78 °C) polymerization of IPTMSK afforded only traces (<2 %) of a viscous product. Free-radical copolymerization of IPTMSK with methyl methacrylate (MMA) (50 mol % MMA, benzene, 0.5% AIBN, 100 h, 45 °C) resulted in a low yield (<10 %) of a copolymer composed mainly of MMA units and, as indicated by the IR (Figure 1C) and NMR data, of largely decomposed IPTMSK units. The presence of IPTMSK during copolymerization with MMA significantly decreased the molecular weight of the resulting copolymer. These results suggest that IPTMSK is incapable of homopolymerization by either a free-radical or an anionic mechanism.

Because of the larger radius of the silicon atom, silicon-substituted derivatives of compounds **V** and **VI**, that is, acylsilanes **III** and **IV**, should exhibit less steric strain and deformation in their conjugated C=C–C=O system compared with their carbon counterparts. With less steric strain, polymerization is more facile. The data just presented indicate that this case is true only for VTMSK. The ground-state energies of compounds **III**, **IV**, **V**, and **VI** were calculated by using the semiempirical MNDO (modified neglect of differential overlap) method (*18*). Both α-hydrogen-substituted compounds **III** and **V** were found to have a planar geometry. However, the α-methyl derivatives **IV** and **VI** have π-electron systems that are strongly twisted in the lowest energy conformation (the calculated dihedral angles in the C=C–C=O system were 98.4 and 101.4° for **IV** and **VI**, respectively). The energy differences between the planar (forced) and twisted (minimum) conformations of the conjugated systems of **IV** and **VI** were 8.3 and 6.1 kcal/mol, respectively. These values indicate that both compounds strongly prefer a nonplanar conformation, and thus, they are expected to behave essentially as unconjugated compounds.

Thermal Stability. Results of thermogravimetric analysis (TGA) for PVTMSK are shown in Figure 2. Weight loss started at 100 °C in both O_2

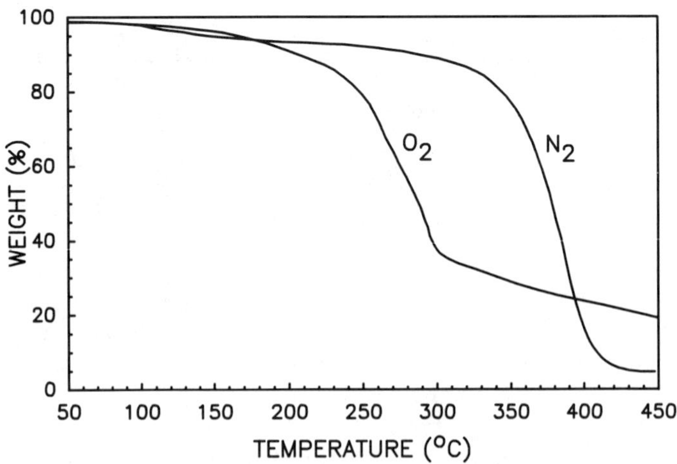

Figure 2. TGA curves for PVTMSK in oxygen and nitrogen atmospheres. Heating rate = 10 °C/min.

and N_2 atmospheres. Although the initial weight loss was faster in oxygen than in nitrogen (20% weight loss at 250 and 350 °C, respectively), the weight fraction of nonvolatile residue was higher in oxygen (20 versus 5% at 450 °C). This result is probably caused by the formation of nonvolatile SiO_x in an oxygen atmosphere. The IR and UV spectra of solid films of PVTMSK showed a decrease in the intensity of the absorption bands characteristic of the –C(=O)–SiR_3 group (at 1636 cm^{-1} and 365 nm in the IR and UV spectra, respectively) after 1 h of heating at temperatures as low as 80 °C (*11*). A reproducible glass transition was detected by differential scanning calorimetry (DSC) at 64 °C, and a strongly exothermic process was observed when PVTMSK was heated at >125 °C (Figure 3). This process occurred without a corresponding change in the weight of the sample and indicates that an exothermic rearrangement of the –C(=O)–SiR_3 group may have occurred.

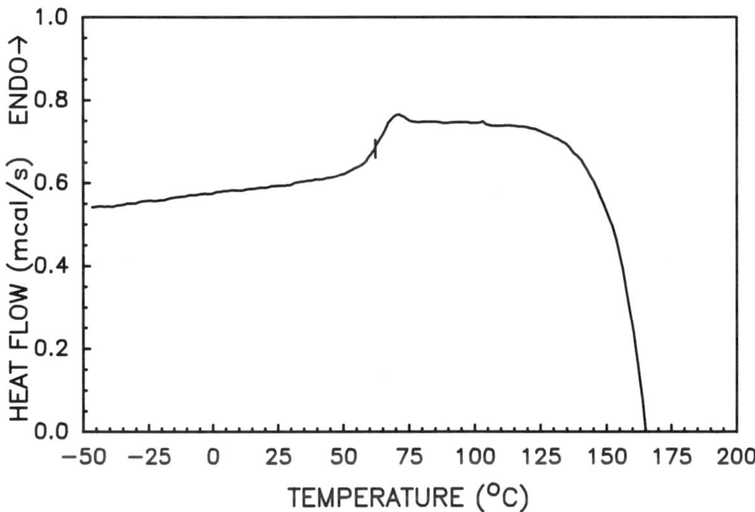

Figure 3. DSC curve for PVTMSK (previously heated to 130 °C). Heating rate = 10 °C/min.

Spectral Properties. The spectral properties of PVTMSK have an important bearing on the lithographic utility and behavior of this polymer. The IR and UV spectra of solid films of PVTMSK have already been discussed in some detail in our previous paper (*11*); thus only the salient points will be presented in this chapter. The most characteristic feature in the UV spectrum of PVTMSK is a long-wavelength electronic transition with a maximum at 365 nm (Figure 4). This $n \rightarrow \pi^*$ transition (*19*) provides the basis for the photosensitivity of PVTMSK to mid-UV radiation.

The intensity of the 365-nm band decreased significantly after exposure of PVTMSK to mid- or deep-UV radiation in air (Figures 4 and 5), with a parallel increase of absorption in the deep-UV region of the spectrum (200–325 nm). The spectra of UV-radiation-exposed films exhibit an isosbestic point at ~330 nm. Possibly, the $-C(=O)-SiR_3$ group is transformed into another carbonyl-bearing moiety not directly attached to a silicon atom.

Heat treatment of PVTMSK films resulted in spectral changes similar to those after mid- or deep-UV exposure. Thus, the intensity of the IR band at 1636 cm^{-1} ($\nu_{C=O}$) and of the UV band at 365 nm ($n \rightarrow \pi^*$) decreased after a PVTMSK film was kept for several hours at >80 °C. Both exposure to UV radiation and increased temperature gave rise to a new three-component band at 1718 cm^{-1} in the IR spectrum and caused an increased absorption at $\lambda < 325$ nm. Thus, these treatments lead to similar transformations of the acylsilane structure.

The dose resulting in a complete transformation of the acylsilane moiety

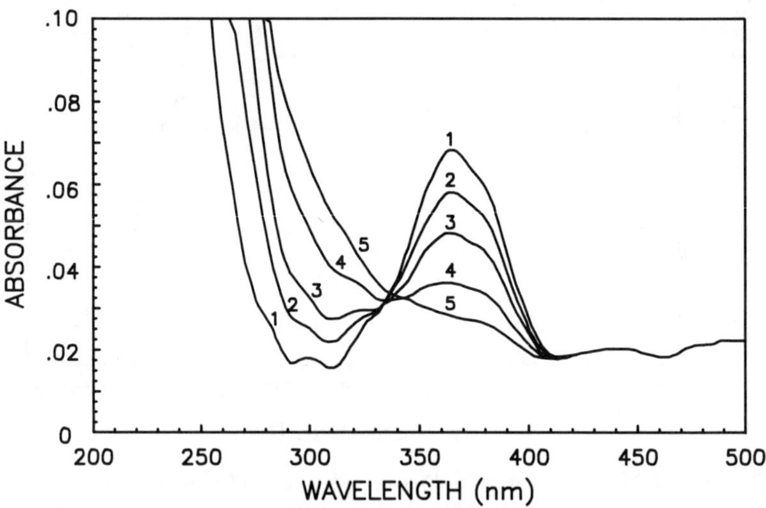

Figure 4. UV spectra of 1.2-μm-thick films of PVTMSK on quartz after exposure to increasing doses of deep-UV (200–300 nm) radiation. Samples 1–5 received deep-UV radiation doses of 0, 0.5, 1, 2, and 5 J/cm², respectively.

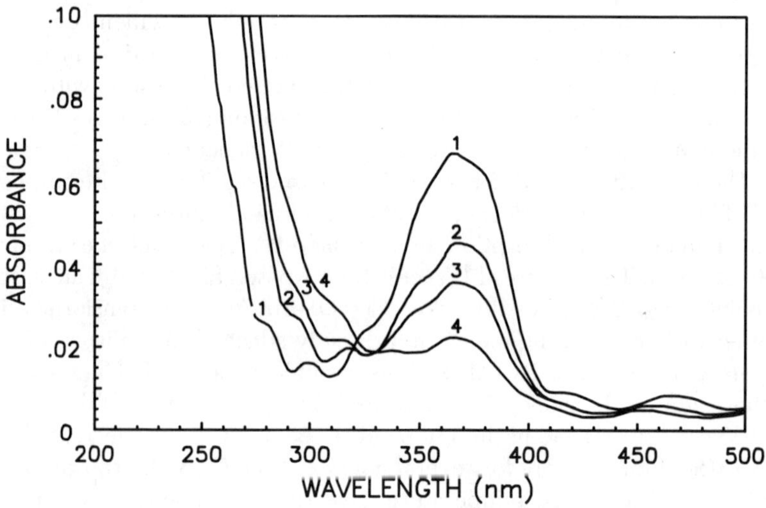

Figure 5. UV spectra of 1.2-μm-thick films of PVTMSK on quartz after exposure to increasing doses of mid-UV radiation (365 nm). Samples 1–4 received mid-UV radiation doses of 0, 1, 2, and 5 J/cm², respectively.

in PVTMSK in the solid state was >5 J/cm^2 for both deep-UV (200–300 nm) and mid-UV (365 nm) radiation (Figures 4 and 5). Although this dose is about 100 times higher than the optimum dose for microlithographic applications, the actual lithographic sensitivity for thin films of PVTMSK (<1 μm) was equal to a dose that is about 10 times lower, that is, 500 mJ/cm^2. This value is 8–10 times lower than that required to expose poly(methyl methacrylate) (PMMA).

The increased solubility of PVTMSK in 2-propanol after exposure to UV radiation in air may have been due to either a photochemically induced main-chain scission or a polarity change caused by the photochemical rearrangement of the acylsilane group. The GPC molecular weight of PVTMSK remained essentially unchanged after exposure of a thin resist film to doses of either 500 mJ/cm^2 at 365 nm or 5 J/cm^2 at 260 nm. In conjunction with the observed spectral changes in PVTMSK films after UV irradiation, these data strongly suggest that the rearrangement of the acylsilane group via a siloxycarbene into a group more polar or forming stronger hydrogen bonds, such as a silyl ether derivative, might be responsible for the imaging properties of PVTMSK.

Pattern Development. The development of patterns exposed in thin films of PVTMSK was difficult because of its high solubility in almost all organic solvents. Excessive thinning of the unexposed areas could be prevented by carrying out the development at low temperatures either by spray development with a 2-propanol mist while the wafer is spinning at 1000–2500 rpm or by dip development in 2-propanol at +5 °C. Spray development cooled the wafer to as low as 8–10 °C because of the fast evaporation of the solvent, but even at such low temperatures and short development times (5–15 s), a nonuniform thinning of the film occurred.

A lithographic sensitivity curve for PVTMSK is shown in Figure 6. The sensitivities for both deep-UV and mid-UV exposures were similar, although the deep-UV source emitted only ~5% of its radiation at 365 nm. The range marked by an arrow denotes the radiation dose for which the partially developed film had a very rough texture.

The imaging properties of PVTMSK were studied by spin coating 350-nm-thick films on silicon wafers or on silicon wafers precoated with a 1.5-μm-thick layer of hard-baked photoresist, exposing them to mid- or deep-UV radiation through a chromium-on-quartz lithographic mask, and developing the pattern as described earlier. This scheme was used to test the intended application of PVTMSK as an imaging material for two-layer resist applications. The densest patterns resolved were composed of 1-μm coded lines and spaces.

Aside from imaging properties, a resist for two-layer applications must exhibit excellent stability to an oxygen plasma under RIE conditions. Results of such tests are shown in Figure 7, in which the etching rate of PVTMSK

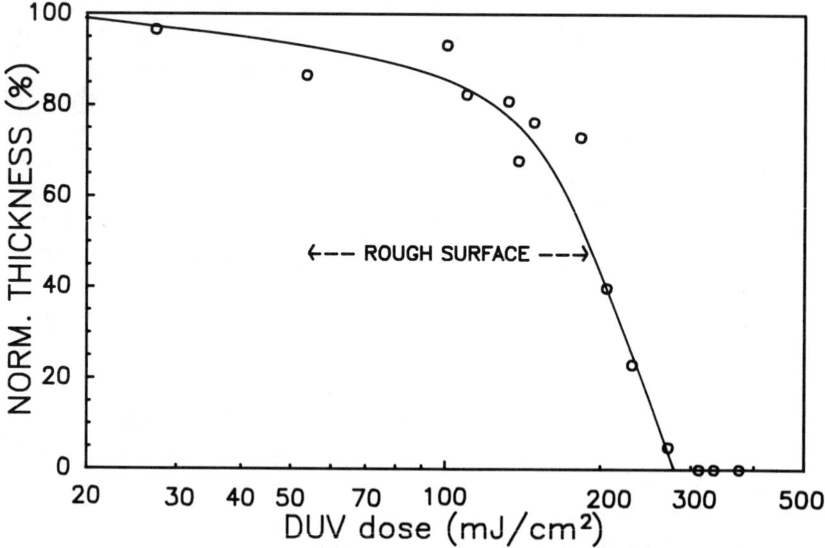

Figure 6. Lithographic sensitivity curve for PVTMSK exposed to deep-UV radiation. The pattern was spray developed for 10 s with 2-propanol at 2500 rpm.

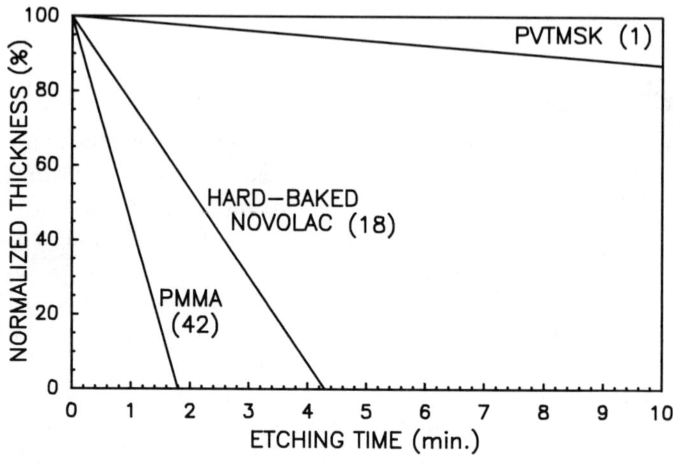

Figure 7. Comparison of etch rates of thin films of PVTMSK, hard-baked novolac, and PMMA by O_2 RIE. Pressure = 2 Pa; power = 20 W at 13.56 MHz; DC (direct-current) bias = –350 V; and O_2 flow rate = 10 sccm.

under standard O_2-RIE conditions is compared with that of PMMA and of a widely used hard-baked novolac-based photoresist. The etching rate ratio of PVTMSK to PMMA to photoresist was 1:42:18, which may not be sufficient in the demanding two-layer resist applications. Nonetheless, preliminary results shown in Figure 8 demonstrate that PVTMSK can serve both as an imaging layer and an etching barrier during the O_2-RIE pattern-transfer step.

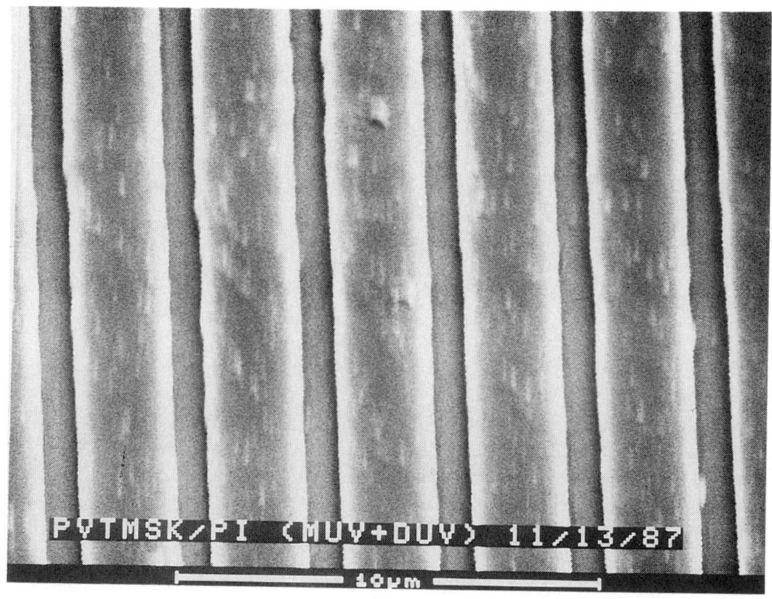

Figure 8. Microphotograph of 1-µm-wide line pattern obtained by exposing a 300-nm thick PVTMSK film to deep-UV (DUV) radiation on 500-nm-thick planarizing layer after O_2-RIE pattern-transfer step.

Conclusion

We have synthesized the first poly(acylsilane) by a free-radical polymerization of 1-trimethylsilyl-2-propen-1-one. However, our attempts at the free-radical and anionic polymerization of poly(1-trimethylsilyl-2-methyl-2-propen-1-one) failed.

The acylsilane moiety in poly(1-trimethylsilyl-2- propen-1-one) readily undergoes UV-radiation- and thermally induced rearrangement. The solid polymer is stable when stored at low temperature in the dark. PVTMSK exhibited an unusually high solubility in a wide range of solvents, a property that turned out to be a major obstacle in the practical utilization of this polymer as a mid- and deep-UV-sensitive resist. The lithographic sensitivity of this polymer to mid- and deep-UV radiation was low (300–500 mJ/cm^2). Thus, despite its low etching rate during O_2 RIE, the utility of PVTMSK as a practical resist material remains questionable.

Acknowledgments

We thank T. N. Bowmer for thermal analysis, J. Shelburne III for GPC data, G. L. Baker for critical reading of the manuscript, and J. Guillet (University of Toronto) for providing a sample of poly(vinyl *tert*-butyl ketone).

References

1. Willson, C. G. In *Introduction to Microlithography;* Thompson, L. F.; Willson, C. G.; Bowden, M. J., Eds.; ACS Symposium Series 219; American Chemical Society: Washington, DC, 1983; p 123.
2. Lin, B. J. *IBM J. Res. Dev.* **1976,** *20,*213.
3. Chandross, E. A.; Wilkins, C. W., Jr.; Hartless, R. L. *Can. J. Chem.* **1983,** *61,* 817.
4. Nakane, Y.; Tsumori, T.; Mifune, T. *Semiconductor Int.* **1979,** *2,* 45.
5. Watts, M. P. C. *Proc. SPIE, Int. Soc. Opt. Eng.* **1984,** *462,* 2.
6. de Grandpre, M. P.; Vidusek, D. A.; Legenza, M. W. *Proc. SPIE, Int. Soc. Opt. Eng.* **1985,** *539,* 103.
7. Brook, A. G. *J. Am. Chem. Soc.* **1957,** *79,* 4373.
8. Dexheimer, E. M.; Buell, G. R.; Le Croix, C. *Spectrosc. Lett.* **1978,** *11,* C41.
9. Lin, B. J. In *Introduction to Microlithography;* Thompson, L. F.; Willson, C. G.; Bowden, M. J., Eds.; ACS Symposium Series 219; American Chemical Society: Washington, DC, 1983; p 287.
10. Reichmanis, E.; Thompson, L. F. *Ann. Rev. Mat. Sci.* **1987,** *17,* 235.
11. Gozdz, A. S.; Reich, H. J.; Bowe, M. D. *Macromolecules* **1987,** *20,* 2031.
12. Reich, H. J.; Kelly, M. J.; Olson, R. E.; Holtan, R. C. *Tetrahedron* **1983,** *39,* 949.
13. Danheiser, R. L.; Fink, D. M.; Okano, K.; Tsai, Y.-M.; Szczepanski, S. W. *J. Org. Chem.* **1985,** *50,* 5393.
14. Brook, A. G. *Acc. Chem. Res.* **1974,** *7,* 77.
15. Cunico, R. F. In *Silicon Compounds: Register and Review;* Petrarch Systems: Bristol, PA, 1984; p 15.
16. Ito, H.; MacDonald, S. A.; Willson, C. G.; Moore, J. W.; Gharapetian, H. M.; Guillet, J. E. *Macromolecules* **1986,** *19,* 1839.
17. MacDonald, S. A.; Ito, H.; Willson, C. G.; Moore, J. W.; Gharapetian, H. M.; Guillet, J. E. In *Materials for Microlithography: Radiation-Sensitive Polymers;* Thompson, L. F.; Willson, C. G.; Fréchet, J. M. J., Eds.; ACS Symposium Series 266; American Chemical Society: Washington, DC, 1984; p 179.
18. The original MNDO program was written by W. Thiele. QCPE 353, an IBM-PC version with updated AAMPAC parameters (1986) was obtained from Serena Software, Bloomington, IN.
19. Brook, A. G.; Quigley, M. A.; Peddle, G. J. D.; Schwartz, N. V.; Warner, C. M. *J. Am. Chem. Soc.* **1960,** *82,* 5402.

RECEIVED for review May 27, 1988. ACCEPTED revised manuscript March 13, 1989.

40

Siloxane Surface Activity

Michael J. Owen

Dow Corning Corporation, Midland, MI 48686–0994

The structure, properties, and uses of poly(dimethylsiloxane) are surveyed in relation to its surface activity, with emphasis on the structural basis of this activity: the low intermolecular forces between methyl groups and the unique flexibility of the siloxane backbone. These basic attributes are used to explain the general interaction of poly(dimethylsiloxane) with high- and low-surface-energy substrates. The effects of variations of the pendant organic group and the backbone on surface activity are reviewed, and examples of current research and applications are given. The chief focus is on aliphatic fluorosilicones, because this class contains organosilicon polymers with the lowest surface energy. Information about organosilicon polymers with backbones other than siloxane is limited. The available data on polysilazanes, polysilalkylenes, and polysilanes are reviewed.

THE SURFACE PROPERTIES OF SILICONES, particularly the commercially available poly(dimethylsiloxane) (PDMS) materials, are best described by regarding polymer side chains or pendant groups as the primary surface-active entities, with the polymer backbone controlling the way in which these pendant groups are presented at a surface (*1*, *2*). The surface and interfacial properties and applications of PDMS are examined in this chapter in this context. PDMS has a very surface-active (low-surface-energy) pendant group, the methyl group, which is arranged along the most flexible backbone, the siloxane chain. The siloxane chain allows the methyl groups to be presented to their best effect. These structural features account for the many surface applications of PDMS.

To separate the roles of the pendant group and the backbone is a gross but useful simplification in explaining why PDMS has a critical surface tension of wetting similar to those of hydrocarbons with surfaces consisting of

0065–2393/90/0224–0705$09.75/0

closely packed methyl groups. Because of the extreme localization of the force fields in covalently bonded methyl groups, pendant methyl groups on PDMS also behave as an array of closely packed methyl groups with little direct effect from the siloxane backbone. This analysis based on pendant-group and backbone variations provides a convenient way of surveying other organosilicon polymers and is also a starting point for property–use predictions when such data are not available, especially for organosilicon polymers with backbones other than siloxane. The fluorosilicones are emphasized in this review, because this class contains organosilicon polymers with the lowest surface tension.

The surface activity of silicones is often exploited by using them as additives. For this reason, aspects of the two most important additive forms, copolymers and surfactants, are also included in this discussion. These two classes come together in the relatively low molecular weight PDMS–poly(alkylene oxide) block and graft copolymers that are commonly used as polyurethane foam stabilizers. Other short-chain silicone surfactants designed for aqueous systems and other silicone–organic copolymers are also available.

PDMS Surface Behavior

PDMS is an unusual macromolecular amphiphile composed of pendant organic methyl groups along the siloxane backbone. Its surface properties result in a wide variety of applications. Specific examples are listed in Table I (*3*), which demonstrates the diversity of interfaces at which PDMS-containing materials are active and the apparently contradictory nature of many of the applications: adhesion and release, foaming and antifoaming, etc.

Surface Tension. The most familiar surface characteristic of PDMS is its low liquid surface tension, ranging from 16 to 21 mN/m at room temperature, depending on molecular weight. This value is lower than the surface tensions of most other polymer systems; only aliphatic fluorocarbon species have lower surface tensions.

Figure 1 (*1*) shows the dependence of liquid surface tension at 20°C on the boiling point of a variety of CH_3- and CF_3-containing materials. As the temperature of a material approaches its boiling point, molecular motion increases, and surface tension decreases because of increased intermolecular separation. Thus, the surface tensions of materials at 20 °C increase with the boiling point. This point is important when simple liquids of different types are compared.

The low surface tension accounts for the tendency of PDMS-containing materials to accumulate at air–substrate surfaces. At condensed interfaces, such as that between organic oils and water, the same surface tension driving force does not exist for PDMS to accumulate at the interface. PDMS has a

Table I. Applications of PDMS-Based Surface-Active Additives

Interface	*Stabilizing Action*	*Destabilizing Action*
Air–water	Super-wetting agent, e.g., for polyethylene and polypropylene	Aqueous antifoams, e.g., in paper production
Air–oil	Solvent-based coating and leveling aid	Foam control in gas scrubbing at petrochemical plants
Air–polymer	Urethane foam formation	Defoaming during monomer stripping processes
Air–solid	Fertilizer anticaking aid	Masonry water repellant (for dewetting surface treatment)
Water–oil	Hydrocarbon oil-in-water emulsifiers[a]	Water-in-crude-oil deemulsifiers
Water–polymer	Silicone polymer fluid–water emulsifiers	Emulsion polymer coalescence aid[b]
Water–solid	Coal slurry stabilization[c]	Coal dewatering[c]
Polymer–solid	Pigment treatment (e.g., TiO_2) to enhance dispersion in paint resin	Nonspreading synthetic lubricant (e.g., for use as a clock oil)
Polymer–polymer	Organic pressure-sensitive adhesive release coating	Silicone pressure-sensitive adhesive

SOURCE: Reproduced with permission from reference 3. (Copyright 1986 Plenum.)
[a]This application has been claimed in patents.
[b]This application is still hypothetical.
[c]This application is being developed.

lower interfacial tension against pure water compared with nonpolar hydrocarbons, but more-polar oils can be significantly more surface active.

Copolymerization with other species is a major way of achieving the necessary interfacial activity to modify condensed interfaces and to vary compatibility with various phases. Incompatibility is another driving force toward surface or interfacial segregation. The balancing of these tendencies by controlling molecular weight, copolymer composition, etc., produces products with seemingly contradictory actions (Table I), which can be explained by the surface properties. For example, the formation and destruction of colloidal systems involve surface forces, and surface-active materials such as PDMS will affect both processes. Different attributes of PDMS are exploited in each application, with modifications made to enhance desirable attributes and to minimize undesirable attributes. These various attributes of PDMS relevant to these applications are the following:

- low surface tension and moderate interfacial tension against water;

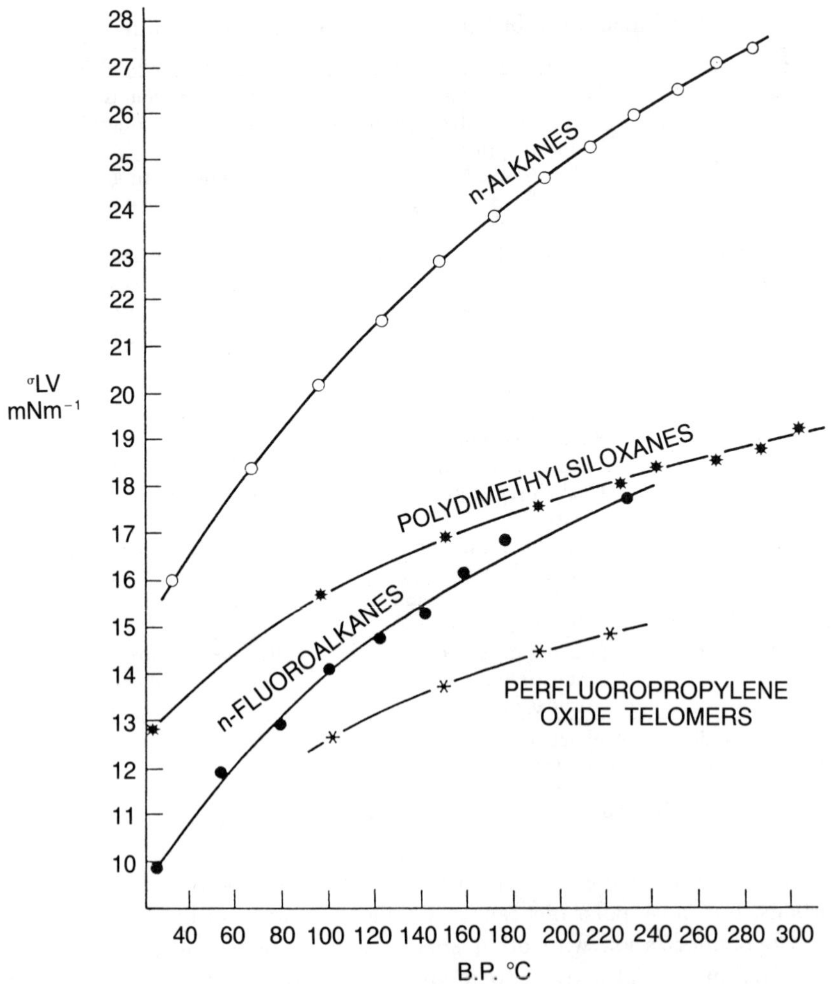

Figure 1. Dependence of surface tension at 20 °C on boiling point for a variety of CH_3- and CF_3-containing materials. (Reproduced from reference 1. Copyright 1980 American Chemical Society.)

- high water repellency;
- good wetting, spreading, and flow-out aid;
- large free volume;
- low apparent energy of activation for viscous flow;
- low glass transition temperature;
- free rotation about bonds;
- high compressibility and dampening action;

- liquid nature to high molecular weight;
- low boiling, freezing, and pour points;
- small temperature variation of physical constants;
- high dielectric strength, gas permeability, and resistance to shear;
- high thermal and oxidative resistance;
- low reactivity, toxicity, and combustibility;
- low environmental hazard; and
- insolubility in water.

This list is an extended version of Table II in reference 3. Some relevant numerical data are given in Table II (*4–16*). Not all the attributes in the list pertain to surface activity, but a product is rarely used for a surface-active application for surface characteristics alone. Usually a combination of appropriate surface properties and other useful aspects, such as high-temperature stability or long durability to weathering, determines which material is preferred for a specific application.

Molecular Basis of PDMS Surface Activity. Four structural characteristics of PDMS account for these attributes and are the link between the structure, properties, and uses of PDMS and most other silicone materials:

1. low intermolecular forces between methyl groups,
2. unique flexibility of siloxane backbone,
3. high energy of siloxane bond, and
4. partial ionic nature of siloxane bond.

The first two structural features explain much of the physical behavior of PDMS in various environments, whereas the last two explain the chemical consequences of the hybrid organic–inorganic nature of PDMS.

Low Intermolecular Forces Between Methyl Groups. The energy ranges of intermolecular forces across an interface are shown in Table III, which is based on one given by Good (*17*). The lowest of the London dispersion forces are the weakest intermolecular interactions between molecules and, for pendant groups in polymers, are those associated with aliphatic hydrocarbon and fluorocarbon groups. The weak intermolecular interactions between these groups are reflected in Zisman's critical surface tension of wetting (*18*). The surfaces with the lowest energy are those based on the constituent groups $-CF_3$, $-CF_2-$, $-CH_3$, and $-CH_2-$ and the various mixed

Table II. Selected Properties of PDMS

Property	*Value*	*Unit*	*Temp.*	*Comments*	*Ref.*
Surface properties					
Surface tension	20.3	mN/m	20 °C	70 mm^2/s (cs)[a] fluid	4
Interfacial tension (against water)	44.3	mN/m	20 °C	100 mm^2/s (cs) fluid	5
Critical surface tension of wetting	24	mN/m	20 °C	Elastomer	6
Polar component of surface tension	1.1	mN/m			
Dispersion force component of surface tension	21.7	mN/m	20 °C	Fluid baked on glass at 300 °C	7
Water contact angle	110	degree	room temperature	Fluid baked on glass at 300 °C	8
Surface viscosity	<10	μg/s	20 °C	High-molecular-weight fluid (10^7)[b]	9
Structural properties					
Si–O bond length	0.165	nm			
Si–C bond length	0.192	nm			

Si–O–Si bond angle	142.5	degree	Cyclic tetramer	10
O–Si–O bond angle	109	degree		
C–Si–C bond angle	106	degree		
Si–O bond energy	445	kJ/mol	Hexamethyldisiloxane	11
Si–C bond energy	306	kJ/mol	Trimethylchlorosilane	11
Surface area per monomer unit	0.23	nm^2	Spread configuration	4
	0.10	nm^2	Six-unit helix (both heptadecamer)	
T_g	150	K	Elastomer	12
Energy of rotation	0	kJ/mol	Elastomer	13
Activation energy of viscous flow	14.7	kJ/mol	100 mm^2/s (cs) fluid	14
Si–O bond polar character	41	%	Pauling calculation	15
Si–O bond dipole moment	9.3×10^{-30}	cm	Calculated from hexamethyldisiloxane molecular dipole moment	16
	(2.8)	debye		

SOURCE: The table is based in part on Table III in reference 3. (Used with permission. Copyright 1986 Plenum.)

[a]cs is centistokes, the usual unit for viscosity in the silicone industry.

[b]The value in parentheses is the molecular weight in daltons.

Table III. Types of Interaction at Interfaces

Interaction Type	*Bond Energy Range (kJ/mol)*	*Interaction Class*	*Adsorption Nature*
Ionic	600–1000	Primary bond	Chemisorption
Covalent	60–700	Primary bond	Chemisorption
Metallic	100–350	Primary bond	Chemisorption
Strong H bonds (involving F)	up to 40	Permanent dipole interactions	Physical adsorption
Other H bonds	10–25	Permanent dipole interactions	Physical adsorption
Other dipole–dipole interactions	4–20	Permanent dipole interactions van der Waals forces	Physical adsorption
Dipole–induced dipole interactions	1–2	van der Waals forces	Physical adsorption
London dispersion forces	0.1–40	van der Waals forces	Physical adsorption

SOURCE: The table is based on reference 17.

ones (–CFH–, etc.). The approach of Owens and Wendt (*7*) of separating the London dispersion forces and other (polar) components of surface tension of polymers also affirms the low surface energy of these groups. Examples in Table IV confirm that the order of decreasing London dispersion interaction is $-CH_2- > -CH_3 > -CF_2- > -CF_3$.

Table IV. Owens–Wendt Solid Surface Tension Data

Material	*Dispersion Component*	*Polar Component*
Poly(1,1-dihydroperfluorooctyl methacrylate)	9.1	0.3
Poly(tetrafluoroethylene)	12.5	1.5
PDMS	21.7	1.1
Paraffin wax	25.4	0
Polyethylene	32.0	1.1
Nylon 6,6	34.1	9.1
Poly(methyl methacrylate)	35.9	4.3
Polystyrene	41.4	0.6

NOTE: Data were taken from reference 7. Values are expressed in millinewtons per meter.

Thus among hydrocarbons, surfaces comprising closely packed methyl groups have the lowest surface tensions. For this reason, both PDMS and paraffin wax have similar surface tensions despite the difference in their molecular architectures. Essentially, both PDMS and paraffin wax have all-methyl surfaces.

Evidence for the low intermolecular forces between methyl groups in PDMS comes not only from low surface tensions but also from the lower boiling points of PDMS materials compared with those of organic materials of similar molecular weight. Noll (*19*) gives some useful data on this topic. The maintenance of liquid nature to unusually high molecular weights of linear PDMS polymers is a further consequence of this weak molecular interaction.

Like surface tension, viscosity is a manifestation of the physical interaction between molecules. The uniquely shallow slope of the viscosity–temperature curve of PDMS (*20*) is due in part to low intermolecular forces. The interfacial viscosity of PDMS ($\leq 10\ \mu g/s$) is the lowest known for a polymer film (*9*) and is also partly the result of low methyl–methyl interactions dominating polymer–polymer interactions in PDMS.

Unique Flexibility of Siloxane Backbone. The various properties of PDMS discussed in the preceding section are the result not only of low intermolecular forces but also of the extreme flexibility of the siloxane backbone. This flexibility is unique. Rotation about siloxane bonds in PDMS is virtually free, the energy being almost zero, compared with 14 kJ/mol for rotation about carbon–carbon bonds in polyethylene and >20 kJ/mol for poly(tetrafluoroethylene) (*13*). This free rotation is reflected in the glass transition temperature (T_g). However, the internal mobility of chains is not the only factor that determines T_g; polymer free volume, attractive forces between molecules, bulkiness of pendant groups, and chain length all contribute to T_g. Nevertheless, a low T_g indicates polymer flexibility.

Several classes of polymers have T_gs of <200 K: polydienes, polyacrylates, polymethacrylates, polyacrylamides, poly(vinyl ether)s, polyoxides, polyesters, polysulfides, polyphosphazenes, and silicones. The lowest reported value in the comprehensive compilation of Lee and Rutherford (*21*) is that of PDMS, whose T_g is only 2 K lower than that of polyethylene. But if the greater steric hindrance of methyl groups compared with hydrogen atoms is taken into account, the siloxane backbone is clearly more flexible than the carbon backbone. Silicones with T_gs lower than that of PDMS have been prepared. Table V (*21–23*) gives the T_gs of a variety of polyorganosiloxanes, together with all polymers with T_gs of <160 K in the list of Lee and Rutherford.

Table V. T_gs of Silicones and Low-T_g Polymers

Polymer	T_g *(K)*	*Ref.*
Poly(hydrogen methylsiloxane)	135	22
Poly(ethylmethylsiloxane)	138	23
PDMS	146	21
Polyethylene	148	21
Poly(propylmethylsiloxane)	153	23
Poly(thiodifluoromethylene)	155	21
Poly(oxytetramethyleneoxyadipoyl)	155	21
Poly(*cis*-1-pentenylene)	159	21
Poly(phenylmethylsiloxane)	187	21
Poly(trifluoropropylmethylsiloxane)	203	23
Poly(diethylsiloxane)	201	21

The prime role of the backbone is to present pendant organic groups such as methyl groups to their best advantage. The PDMS backbone performs this function better than hydrocarbon and other organic backbones by virtue of its superior flexibility. Compared with carbon–carbon and carbon–oxygen backbones (*1*), the siloxane backbone has the flattest bond angle and the longest bond. These structural features give siloxanes a more extended flexible chain system in which steric packing restrictions are less likely to impede the adoption by available methyl groups of a methyl-rich low-surface-energy configuration. The wide variability of the Si–O–Si bond angle compared with the tetrahedral angle of the carbon system may also contribute to backbone flexibility. This flexibility is due, in part, to the partially ionic nature of the Si–O bond, which will be discussed in the next section. Values from 105° to 180° have been quoted (*24*).

Siloxane Bond Energy and Partially Ionic Character. Low intermolecular forces and high chain flexibility explain most of the physical behavior of PDMS in various environments. The hybrid organic–inorganic nature of PDMS has chemical consequences that are primarily due to two other key properties of the siloxane backbone: its high bond energy and partially ionic character. For instance, these properties are clearly responsible for the substantial thermal stability of silicones.

Despite earlier uncertainties, the siloxane bond energy is now generally accepted as being significantly greater than those of carbon–carbon and carbon–oxygen bonds, as discussed in detail by Ebsworth (*25*). The value of 445 kJ/mol quoted in Table II for hexamethyldisiloxane is higher than the general values for carbon–carbon (346 kJ/mol) and carbon–oxygen (358 kJ/mol) bonds quoted by Ebsworth (*25*). The carbon–silicon bond energy of 306 kJ/mol (Table II) is less than the Si–O, C–C, and C–O bond energies. However, a methyl group on silicon has a higher thermal and oxidative stability compared with a methyl group that is part of a hydrocarbon chain. This stability is due to the partially ionic or polar nature of the siloxane backbone. The positively polarized silicon atom acts as an electron drain polarizing the methyl group and rendering it less susceptible to attack.

As pointed out by Noll (*26*), the Si–O interatomic distance is considerably smaller than the sum of the atomic radii. This discrepancy has been attributed to a resonance-type bond having both polar- and covalent-bond components. This bond type is expected because of the relatively large difference in the electronegativities of silicon and oxygen (1.7). According to Pauling (*15*), this difference gives a polar- or ionic-bond contribution of 41%. This electronegativity difference also suggests another way of explaining the flexibility of the siloxane backbone, because a purely covalent bond is completely directional in space, whereas a purely ionic bond has no direction.

The partially ionic siloxane backbone is susceptible to hydrolysis by water, particularly at extremes of pH. For example, the hydrolytic splitting of siloxane bonds in spread monolayers of high-molecular-weight PDMS on water becomes appreciable at $pH < 2.5$ and $pH > 11.0$ (*27*). This combination of higher thermal stability (compared with carbon-chain polymers) and susceptibility to nucleophilic or electrophilic attack is shared by virtually all inorganic polymers (*28*). Most feasible linkages resemble siloxanes not only in having higher bond energies (compared with carbon systems) but also in having considerable ionic nature. For these reasons, the thermal and hydrolytic degradation of such polymers is markedly dependent on the levels of impurities, residual catalysts, etc., in the polymers.

Interaction with Substrates. This broad subject cannot be covered comprehensively in this chapter. Scattered throughout this discussion are examinations of interactions with specific liquid or solid substrates. The purpose of this section is to relate, in a more general way, the fundamental properties discussed so far to substrate interaction.

PDMS can interact with substrates both through London dispersion forces from the induced dipoles in the methyl groups and through permanent dipoles in the partially polar siloxane backbone. At the polymer–air surface, configurations that maximize the packing of methyl groups at the surface are adopted, but at other interfaces, the backbone dipole assumes a more important role, because interaction energies involving permanent dipoles are stronger than dispersion forces associated with methyl groups (Table III). Thus, in general, with significantly polar substrates, PDMS will interact through its siloxane backbone with polar entities on the substrate surfaces, whereas with predominantly nonpolar substrates, interaction will be through methyl groups by London dispersion forces. Although ordinarily the London dispersion forces dominate (Table IV), the extreme flexibility of the siloxane backbone permits configurations facilitating the polar interaction. Thus, the interfacial tension of PDMS with water is lower than that of aliphatic hydrocarbons (*1*).

To a large degree, this distinction between significantly polar and predominately nonpolar substrates mirrors Zisman's classification of high- and low-energy surfaces (*18*). High-surface-energy materials, for example, liquid and solid metals; metal oxides, nitrides, and sulfides; silica; and glass, have surface tensions of >100 mN/m. Most organic materials, including organic polymers, oils, and waxes, are low-surface-energy materials, with surface tensions of <100 mN/m.

Current thinking on interaction with substrates ascribes the polar component primarily to acid–base interactions (*29*). This approach was applied recently to PDMS by Ross and Nguyen (*30*), who conclude that PDMS behaves as a hard Lewis acid. The silicon atom is the source of this acid character because of its small size, slight polarizability, and empty *d* orbitals, which can accept electrons from a base. In effect, this acidity explains the partially ionic character of the siloxane bond referred to earlier (*15*). The Lewis acid–Lewis base interaction between PDMS and basic solutes accounts for the solubility and surface activity of such systems.

One of the fullest studies of high-surface-energy systems is that of the adsorption from solution of PDMS on glass and silica (*31–35*). Because the strength of forces involved is similar to the strength of solvent–substrate interaction, solvent effects are very important in this system. Small amounts of more strongly polar groups on the polymer, such as silanol and amino functional groups, also have a large effect. In general, physical adsorption on isolated substrate silanols occurs, but with polymer silanol groups, condensation and chemisorption to these and other hydroxylated surfaces are possible, particularly at elevated temperatures and with acidic or basic surfaces with catalytic activity. Under such conditions, reequilibration of siloxane bonds can also occur.

The heat treatment of silicone-treated glass slides discussed in a later section (*see* Pendant-Group Variations) involves these effects (*8*). On certain metal oxides, notably copper oxide, changes have been noted at temperatures as low as 90–100 °C (*36*). This behavior is also the key to the action of silane coupling agents (*see* Pendant-Group Variations). Because of the way silicones are normally produced commercially, silanol groups on the polymer chain may produce strong interactions, particularly at elevated temperatures. Ross and Nguyen (*30*) point out that even when such polymer silanol groups are absent, the interaction of PDMS with silica can now be interpreted as the interaction of the PDMS Lewis acid with basic silanolate groups on the silica surface.

The main consequence of the low surface tension of PDMS resulting from dominant London dispersion force interaction of methyl groups is that this liquid polymer should spread over any high-surface-energy substrate, as shown by Zisman and co-workers (*18*) for materials such as metals, silica, and alumina. The only complication would be autophobic behavior. An autophobic liquid has a higher liquid surface tension compared with that of its adsorbed film and thus cannot spread when polymer adsorption occurs at the three-phase line of contact between liquid, vapor, and solid. Zisman et al. (*18*) have shown that PDMS is not autophobic (a fact that is also evident from Table II), but not all organosilicon polymers are nonautophobic (for example, poly(trifluoropropylmethylsiloxane) is autophobic). Both the dynamics of spreading of PDMS (*37–39*) and the means to prevent it have been investigated (*39, 40*).

PDMS will also spread over most low-surface-energy materials, except various fluorocarbon polymers, because of its lower surface tension. For the PDMS-nonspreadable materials, low-molecular-weight PDMS fluids provide a useful homologous series for determining the critical surface tension of wetting. For example, using PDMS fluids, Fox and Zisman (*41*) obtained a value of 17.4 mN/m for poly(tetrafluoroethylene), compared with 18.5 mN/m for *n*-alkanes. This ability to spread over most low-surface-energy liquids and solids is exploited in numerous applications, notably in antifoam formulations.

Pendant-Group Variations

Information about the surface activity of organosilicon materials comes from studies of various properties, including:

- surface tension of liquid (σ_{lv}),
- Zisman critical surface tension of wetting of solid (σ_c),
- Owens–Wendt (*7*) surface tension of solid (σ_{sv}) (and similar approaches [*42*]),

- surface tension of aqueous solutions (σ_{soln}),
- Langmuir trough studies of spread films on water,
- water contact angle,
- surface tension of solution and Langmuir trough studies with liquids other than water,
- other-liquid contact angle,
- interfacial tension of aqueous solutions and silicone fluids and of aqueous solutions and other oils, and
- adsorption on various solids.

These properties are listed in order of usefulness for comparative review purposes. Liquid surface tension is the most fundamental property, because it pertains only to the material in question (provided the material is adequately pure) and the technique used for measurement. All the other properties listed are dependent also on solvents, contact-angle test liquids, and liquid or solid substrates selected. For solids, approaches such as the Owens–Wendt analysis (*7*) have supplanted the Zisman method (*18*) in recent years, but data from the Zisman method for organosilicon polymers are more available compared with data from the Owens–Wendt approach. Some useful data on aqueous surface tensions and Langmuir troughs are also available. Data for other listed properties are of less fundamental use and rather scanty.

Much of the data concern copolymers of siloxanes functionalized with organic groups or higher alkyl groups with PDMS in materials that are predominately PDMS. The use of organofunctional siloxane frequently has no effect on surface tension, because PDMS is usually the more surface-active component, but it can have a pronounced effect on other properties, such as oil–water interfacial tension and adsorption on solids. The most useful data come from contact-angle studies of coupling-agent-type materials and Langmuir trough studies.

Studies of Coupling Agents. The studies of coupling agents concern polymerized alkoxysilanes derived from molecules of the structure $XSi(OR)_3$ in which R is usually either methyl or ethyl. With such highly cross-linked materials, concerns about the role of backbone flexibility are eliminated, because the materials form a tightly cross-linked, three-dimensional structure preventing reorientation once condensation has occurred. The silicon atoms carry the functional group X and no other complicating groups (such as the methyl group, as is usually the case with linear siloxanes with organic functional groups and higher alkyl substituents). A direct comparison of the functional groups can be made by assuming that functional groups occupy the outermost surface sensitive to contact-angle test liquids. With some of the polar organic functional groups that interact strongly, even chemically, with the substrate on which such films are cast, this assumption may not be reasonable.

Coupling agents are exciting materials with many uses. Organosilanes generally have two types of functional groups. The hydrolyzable alkoxy group (or halogen) will form the cross-linked siloxane network, and most common organosilanes carry three such groups. As well as forming a siloxane network, the hydroxyl or halogen groups can also react with many hydroxylated inorganic surfaces such as silica, alumina, and titania. The other X functionality is usually a nonhydrolyzable organic group that may be relatively inert, such as a hydrocarbon radical, or that may be very reactive to particular organic systems. This molecular architecture is designed for the most familiar use of these materials, as adhesion enhancers at polymer–glass fiber interfaces in reinforced composites. For this reason, these materials are often known as silane coupling agents, but with the very wide range of reactive and nonreactive organic substitutions available, many other uses can be expected.

The breadth of application for coupling-agent materials is evident in the review by Arkles (*43*), whose use categories include wetting, surface energy control, chromatography, liquid crystals, coupling agents, polymer applications (thermoset and thermoplastic composites), and solid-state catalysis. Plueddemann's book (*44*), which is devoted primarily to the use as adhesion promoters, also has a chapter on other applications that include uses as hydrophobic agents, immobilized catalysts, bound antimicrobial substances, soil treatments for water harvesting, charge-transfer chromatography, ion concentration and removal by bound chelate-functional silanes, and coated metal oxide electrodes. This book (*44*) surveys the chemistry of silane coupling agents and their chemistry at interfaces. Other compilations of the diverse chemistry and applications of silanes are two symposium proceedings edited by Leyden (*45*, *46*).

Two of the more-diverse studies of these materials are those of Lee (*47*, *48*) and Sacher (*49*), and the data in Table VI are taken from their reports and augmented by some results of Bascom (*50*) and Pittman (*51*). The data from different workers in this table must be compared with caution; glass substrate type and treatment, deposition procedure, use and type of catalyst, relative humidity, nature of reactive alkoxy functionality, contact-angle test liquids, etc., all vary. Also in some cases, aging, solvent rinsing, and use of substrates other than glass were studied. In Table VI, values for catalyzed films have been chosen when both catalyzed and uncatalyzed films were studied. The table contains only initial values on unrinsed glass. Despite the differences in conditions, the information provides a useful approximate ranking of the effect of functional groups on polysiloxanes that broadly parallels expectations from Zisman's studies (*18*) of the effect of various surface constituents on critical surface tension of wetting.

The biggest surprise in Table VI is the high surface tension associated

Table VI. Surface Tensions of Reactive Polysiloxanes on Glass Substrates at 20 °C

Pendant Functional Group	σ_c *(mN/m)*	σ_s *(mN/m)*	*Ref.*
$CF_3(CF_2)_6CH_2O(CH_2)_3$–	14	—[a]	50
$(CF_3)_2CFO(CH_2)_3$–	18	—	50
	15–16	—	51
CH_3–	22.5	24.0	47
CH_2=CH–	25–26	27.2–31.5	47
CH_2=$C(CH_3)COO(CH_2)_3$–	28.0	36.7	47
	38.9	—	49
$CH_3(CH_2)_2$–	28.5	31.7	47
$CF_3(CH_2)_2$–	33.5	34.7	47
$H_2N(CH_2)_2NH(CH_2)_3$–	33.5	30.8	47
	36.0	42.4	49
$CN(CH_2)_2$–	34.0	35.5	47
$CH_3C_6H_4$– (mixed)	34.0	—	47
$H_2N(CH_2)_3$–	35.0	34.2	47
	37.5	42.8	49
$CH_3COO(CH_2)_3$–	37.5	33.6	47
$H_2N(CH_2)_2NH(CH_2)_2NH(CH_2)_3$–	37.5	38.4	49
$(O)C_6H_3(CH_2)_2$–	39.5	36.5–52.8	49
C_6H_5–	40.0	33.2	47
$CH_2(O)CH(CH_2)_3$–	40.0	48.8	47
$(O)C_6H_9(CH_2)_2$–	40.0	53.7	47
$Cl(CH_2)_3$–	40.5	33.0	47
	43	—	50
	41.0	37.7–40.3	48
$HS(CH_2)_3$–	41.0	43.0	47
	33.6	40.2	49
$(CH_3)_2CHC_6H_4$– (mixed)	41.5	—	47
$BrCH_2C_6H_4$– (mixed)	42.0	—	47
$Br_2C_6H_3$– (mixed)	42.0	—	47
$CH_2(O)CHCH_2O(CH_2)_3$–	42.5	—	48
BrC_6H_4– (mixed)	43.5	—	47
$HOCH_2CH(OH)CH_2O(CH_2)_3$–	44.6	55.3	49
$ClC_6H_4(CH_2)_2$–	45–47	—	50

NOTE: σ_c is the critical surface tension of wetting, and σ_s is the solid surface tension obtained in a manner similar to that of Owens and Wendt (7) but with a series of hydroxy-containing liquids rather than just water and methylene iodide.

[a]— indicates that the measurement was not made.

with $CF_3(CH_2)_2$– substitution. The unusual case of poly(trifluoropropylmethylsiloxane) is dealt with in a later section. Lengthening the alkyl chain from methyl to propyl raises surface tension. A similar effect is noted with commercial alkylmethylsilicone fluids. Kovats and co-workers (52), using compounds of the type $X(CH_3)_2SiOR$, have shown that these surface tensions go through a maximum when X is hexyl and that when X is octadecyl, critical surface tensions similar to those when X is methyl are obtained. These results imply that both methyl- and octadecyl-substituted silicones have essentially methyl surfaces with similar degrees of packing. Octadecyl groups achieve this packing by alignment or crystallization of long hydrocarbon side-chains in a manner akin to that by solid hydrocarbons such as paraffin wax.

Contact-Angle Studies. Siloxanes with higher alkyl substituents have thermal properties that are also much more like those of paraffins than of PDMS, as shown by an early study by Hunter and co-workers (8). Figure 2 shows some of their data comparing the water contact angles of soda–lime glass surfaces treated with PDMS, poly(diethylsiloxane), and poly(octadecylmethylsiloxane). Compared with PDMS, poly(diethylsiloxane) and poly(octa-

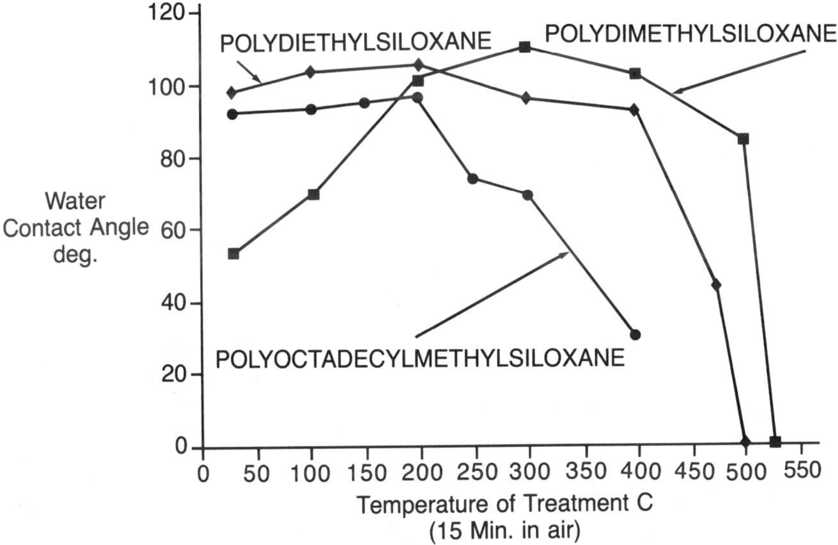

Figure 2. Water contact angles of soda–lime glass surfaces treated with 0.02% benzene solutions of polyorganosiloxanes. Data were taken from Table I of reference 8.

decylmethylsiloxane) have higher water contact angles on glass that has been treated at room temperature. This behavior is caused by bulkier groups that adversely affect chain flexibility and impede reorientation in the presence of water.

At higher treatment temperatures, differences between the water contact angles of the polymers are small, but at even higher temperatures, the hydrophobicity of poly(octadecylmethylsiloxane) declines more rapidly than that of poly(diethylsiloxane) and PDMS because of the poorer thermal stability of siloxanes with higher alkyl substituents. Despite this defect, sometimes the increased organic compatibility of alkylmethylsiloxane fluids outweighs their increased surface tension and diminished thermal stability, such as when they are used as release agents on substrates that are subsequently painted. This discussion highlights some of the difficulties in interpreting contact-angle data.

The ability of water vapor to penetrate a loosely cross-linked PDMS is exploited in gas-permeable water-repellent treatments. Water might also cause a surface reorientation of silicones containing hydrophilic groups. The surface reorientation of polymeric solids and its dependence on different environments is a familiar phenomenon (*53*) and particularly prevalent in silicones because of their high backbone flexibility. An example of this behavior is observed when silicone elastomers are exposed to an electrical discharge (corona) or to a plasma (*54–56*). These treatments increase water wettability, in part, because of surface hydroxyl formation, probably silanol. On exposure to air, these treated surfaces gradually lose wettability. The two most likely mechanisms are reorientation of surface hydrophilic groups away from the surface ("overturn") or migration of untreated polymer chains from the bulk. Both of these processes will be easier in silicones because of their high chain flexibility and low intermolecular forces.

Langmuir Trough Studies. Langmuir trough studies have featured in the investigation of silicone surface properties from the beginning (*4*). Although not all investigators agree on the traditional interpretation of these surface-pressure–surface-area isotherms (*57*), this approach, particularly the studies of Noll et al. (*58*, *59*), has contributed considerable information on the effect of pendant-group variations. Noll et al. (*58*, *59*) distinguished three classes of silicones on the basis of the characteristics of surface-pressure–surface-area isotherms in water (Table VII).

Class 1 is characterized by an isotherm with two transformation points, A_2 and B, between which the only appreciable surface pressure occurs. The polymer configuration at A_2 is a hydrated spread chain with enclosed water molecules, one molecule of water per atom of silicon. The point B is the limit of compression of spread chains. At this point, the polymer has a dehydrated configuration with all water molecules expelled from between the chains, extremely compressed, and meshed. At lower areas (higher pressures), the chains are lifted from the surface and coil up. Hydration of the polymer backbone is a direct consequence of its partially polar nature, and the point at which it ceases is very dependent on the size and nature of substituent groups. Class 1 has two subdivisions: 1a, for which the two transformation points are well separated, and 1b, for which the points are close together. The class comprises smaller, more-hydrophobic pendant groups that are affected by the flexible backbone during surface orientation.

Class 2 is characterized by isotherms with only one transformation point (A_2). At lower areas, there is a steady increase to high surface pressure with

Table VII. Classification of R_1R_2SiO Polysiloxanes by Langmuir Trough Behavior

R_1	R_2	Class
CH_3–	CH_3–	1a
	H–	1a
	$CH_2{=}CH$–	1a
	$CF_3(CH_2)_2$–	1a
	C_2H_5–	1b
	$ClCH_2$–	1b
	C_6H_5	1b
	$CN(CH_2)_2$–	2
	$HOCH_2$–	2
	H_2NCH_2–	2
	$X^- H_2N^+RCH_2$–	2
	$^+NHR_3{}^-SO_2NR'CH_2$–	2
	$CH_3(CH_2)_3$–	3
	$CH_3(CH_2)_5$–	3
	$C_6H_5(CH_2)_2$–	3
	$CH_3CH(C_6H_5)CH_2$–	3
	$BrCH_2$–	3
C_2H_5–	C_2H_5–	3
$CH_3(CH_2)_3$–	$CH_3(CH_2)_3$–	3

NOTE: The table is based on references 58 and 59. The classes are defined in the text.

no B point. This class comprises very hydrophilic pendant groups, and strong interaction with water prevents dehydration and coiling at these higher pressures.

Class 3 yields isotherms with no transformation points. The polymers spread slowly and do not produce appreciable surface pressures on short contact with the water substrate. This class includes larger aliphatic pendant groups, as well as aromatic groups. These polymers do not orient themselves on the water substrate. The molecular-interaction forces of this class with the water surface are lower than those of class 2. The bulkiness of the groups further reduces interaction and impedes reorientation.

Fluorosilicone Surface Activity

If polymer side-chains or pendant groups can be regarded as the primary surface-active entities, with the polymer backbone controlling the way in which these pendant groups are presented at a surface, then fluorosilicones should be the ultimate low-surface-energy polymers because they have groups with the lowest surface energy arranged on the most flexible backbone. For this reason, the surface activity of fluorosilicones is reviewed in more detail in this section.

Aliphatic fluorocarbon side chains directly pendant to the siloxane backbone, as in $[(CF_3)_2SiO]_n$, do not result in useful materials. Fluorinated substituents on C-1 and C-2 weaken the Si–C bond and cause the elimination of fluorine from these positions. The affinity of fluorine for silicon then forms SiF bonds, particularly with nucleophilic agents. Consequently, the hydrolytic stability of such materials is poor, especially under alkaline conditions. The thermal stability of 1- and 2-fluoroalkylsilicon compounds is also low, partly because of the high affinity of fluorine for silicon. Practical interest has thus focused on fluoroorganosiloxanes, in which the fluorocarbon substituent is sufficiently separated from the Si–C bond, usually by two carbon atoms. A selection of liquid surface tension (σ_{lv}) and aqueous-solution surface tension (σ_{soln}) data is given in Tables VIII and IX, respectively. The measured solid surface tension data have already been summarized in Table VI. Poly(trifluoropropylmethylsiloxane) is omitted from Table VIII and is separately discussed later (Table X [58]).

A surprising feature of Table VIII is how few low surface tensions are

Table VIII. Liquid Surface Tension (σ_{lv}) of Fluorosilicones at 20 °C

Compound	σ_{lv} *(mN/m)*	*Ref.*[a]
$CF_3(CH_2)_2Si(OCH_3)_3$	19.9	
Compound A[b]	19.2	60
Compound B[b]	20.3	60
Compound C[b]	21.7	60
$(CH_3)_3Si[OSiCH_3\{(CH_2)_3COOC_3F_7\}]_4OSi(CH_3)_3$	21.4	61
$(CH_3)_3Si[OSiCH_3\{(CH_2)_3COOC_3F_7\}]_8OSi(CH_3)_3$	20.6	61
$[H(CF_2)_4CH_2O]_4Si$	24.6	62
	24.5	63
$[H(CF_2)_4CHCH_3O]_4Si$	22.6	62
$(CH_3)_3SiOCH_2(CF_2)_5CF_2H$	21.0	64
$CH_3Si[OCH_2(CF_2)_5CF_2H]_3$	23.7	64
$C_7F_{15}(CH_2)_2Si(OCH_3)_3$	18.1	
$[C_7F_{15}COO(CH_2)_3(CH_3)_2Si]_2O$	19.9	
$[(CH_3)_3CO]_2[CF_3(CF_2)_6CH_2O]Si$	18.4	62
	18.4	63
$[(CH_3)_3CO]_2[H(CF_2)_6CH_2O]_2Si$	20.8	62
	21.0	63

[a]Unreferenced data are previously unpublished and were obtained by G. J. Quaal in 1964.

[b]

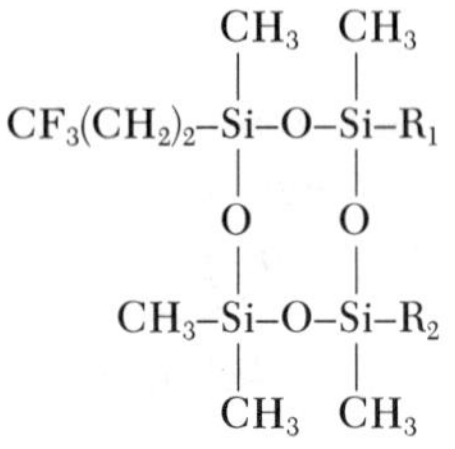

```
              CH3   CH3
               |     |
CF3(CH2)2–Si–O–Si–R1
               |     |
               O     O
               |     |
         CH3–Si–O–Si–R2
               |     |
              CH3   CH3
```

A: $R_1 = R_2 = CH_3–$
B: $R_1 = CF_3(CH_2)_2–$, $R_2 = CH_3–$
C: $R_1 = R_2 = CF_3(CH_2)_2–$

Table IX. Aqueous-Solution Surface Tension (σ_{soln}) of Fluorosilicone Surfactants

Surfactant[a]	σ_{soln} (*mN/m*)	*Temp.* (°*C*)	*Ref.*[b]
$[F(CF_2)_6CH_2OOCCH_2CH(^-SO_3{}^+Na)COO(CH_2)_3(CH_3)_2Si]_2O$	12.5	20	65
$F(CF_2)_6CH_2OOCCH_2CH(^-SO_3{}^+Na)COO(CH_2)_3(CH_3)_2SiO(CH_3)_2Si(CH_2)_3OH$	13.0	20	65
$[H(CF_2)_6CH_2OOCCH_2CH(^-SO_3{}^+Na)COO(CH_2)_3(CH_3)_2Si]_2O$	16.5	20	65
$H(CF_2)_6CH_2OOCCH_2CH(^-SO_3{}^+Na)COO(CH_2)_3(CH_3)_2SiO(CH_3)_2Si(CH_2)_3OH$	16.0	20	65
$[H(CF_2)_2CH_2OOCCH_2CH(^-SO_3{}^+Na)COO(CH_2)_3(CH_3)_2Si]_2O$	19.8	20	65
$H(CF_2)_2CH_2OOCCH_2CH(^-SO_3{}^+Na)COO(CH_2)_3(CH_3)_2SiO(CH_3)_2Si(CH_2)_3OH$	19.2	20	65
$CF_3(CH_2)_2(CH_3)_2SiO(CH_3)_2Si(CH_2)_3(OCH_2CH_2)_{11}OH$	23.9	25	
$CF_3(CH_2)_2(CH_3)_2SiO(CH_3)_2Si(CH_2)_3(OCH_2CH_2)_{11}OOCCH_3$	23.1	25	
$[CF_3(CH_2)_2(CH_3)_2SiO]_3Si(CH_2)_2SCH_2COO^{-}\,{}^{+}NH(C_2H_5)_3$	22.9	25	
$[CF_3(CH_2)_2(CH_3)_2SiO]_3Si(CH_2)_2SCH_2COO^{-}\,{}^{+}NH(CH_2CH_2OH)_3$	29.4	25	
$CH_3(CH_2)_2(CH_3)_2SiO(CH_3)_2Si(CH_2)_3(OCH_2CH_2)_{11}OH$	23.1	25	
$CH_3(CH_2)_2(CH_3)_2SiO(CH_3)_2Si(CH_2)_3(OCH_2CH_2)_{11}OOCCH_3$	24.7	25	

[a]Surfactants were used at a concentration of 1%.
[b]Unreferenced data are previously unpublished and were obtained by O. L. Flaningam in 1966.

included. The lowest value reported is 18.1 mN/m, and many values are greater than 20.4 mN/m, a reliable value for high-molecular-weight liquid PDMS (*66*). Zisman et al. (*18*) have conclusively demonstrated the superior surface activity of CF_3– and –CF_2– groups over CH_3–. However, other factors can explain the preponderance of relatively high values, including:

1. the low density of aliphatic fluorine-containing groups in many of these structures and the unlikelihood of a configuration in which such groups are packed sufficiently closely to the exclusion of most of the rest of the molecule with intrinsically higher surface energy;
2. the increased bulkiness of many fluorinated pendant groups, which will detract from siloxane chain flexibility and further limit the adoption of configurations with lowest surface energy;
3. the unresolved dipole at the –CF_2–CH_2– groups or other link, which is unavoidably present in any fluorosilicone of adequate stability; and
4. the increase in molecular interaction (and hence surface energy) by coordination of fluorine and silicon atoms from adjacent molecules.

The last factor was suggested by the Russian workers Lavygin et al. (*61*), whose work accounts for nearly half of the data in Table VIII.

Table VIII also reflects the well-established principle that –CF_2H-terminated fluorocarbon chains are significantly less surface active (higher surface tension) than the fully fluorinated equivalent. Not until $CF_3(CF_2)_6$– are surface tensions significantly lower than that of methylsiloxanes. Table VI shows that a branched group containing three fluorinated carbon atoms, $(CF_3)_2CF$–, is almost as effective as $CF_3(CF_2)_6$– in lowering the critical surface tension of wetting. The linear $CF_3(CF_2)_2$– group is not so effective in achieving low liquid surface tension (*61*); the polar ester linkage associated with the fluorocarbon group in this particular case must contribute considerably.

Generalizations from the aqueous-solution surface tensions in Table IX are risky, because values are as dependent on the hydrophilic–lipophilic balance (HLB) as on the intrinsic surface activity of the hydrophobe. The data in Table IX are consistent with earlier observations that longer perfluorinated groups are most effective in producing low surface tensions (in this case $CF_3(CF_2)_5$–) and that a terminal CF_2H– is detrimental.

The $CF_3(CF_2)_5$– group in the sodium sulfosuccinate derivatives of Gol'din and co-workers (*65*) is remarkably surface active. In these molecules, the silicon and fluorine atoms are far apart, a separation that may be preserved when adjacent molecules are oriented at the air–water surface and that could account for the observed surface activity. Whatever the explanation, the surface activity of sodium sulfosuccinate derivatives is as good as that of any fluorochemical surfactant. A comparison of Table IX with Table XIII (*see* Backbone Variations) shows that the $CF_3(CH_2)_2(CH_3)_2Si$– group produces higher surface tensions in aqueous solution than does $(CH_3)_3Si$–. This property may be explored further by comparing poly(trifluoropropylmethylsiloxane) and PDMS. The last two entries in Table IX are data for methylsiloxane-based surfactants equivalent to fluorinated surfactants, data for which are also in the table. A direct comparison of the effects of CH_3– and CF_3– groups suggests that these groups have similar effects on the aqueous-solution surface tension of propyl-containing siloxane surfactants.

Surface tension studies of the most common fluorosilicone, poly(3,3,3-trifluoropropylmethylsiloxane) (PTFPMS), give unexpected results. Compared with (PDMS), PTFPMS has a higher liquid surface tension, a similar critical surface tension of wetting, and a considerably lower solid surface tension, as determined by water and methylene iodide contact angles and the method of Owens and Wendt (*67*). These results are summarized in Table X (*7, 67, 72–74, 76, 77*), in which PTFPMS is compared with two other fluorocarbon polymers, poly(tetrafluoroethylene) (PTFE) and poly(chlorotrifluoroethylene) (PCTFE). PTFE behaves like PTFPMS, whereas PCTFE behaves like PDMS.

These differences could be artifacts of various measurements and derivations, but I believe that they are real and reflect orientational and packing-density differences in different states. Different degrees of freedom apply to the liquid and solid states. The interaction of fluorines with adjacent silicons (both in the next monomer unit and in neighboring macromolecules), as suggested by Lavygin and co-workers (*61*), will be easier in the liquid state than in the solid. The $CF_3(CH_2)_2$– group consists of two entities, one that is more surface active than CH_3– and one that is less surface active. –CH_2– may play a more significant role in the liquid state than in the solid, where the motion of the $CF_3(CH_2)_2$– group is reduced and an orientation stressing the external CF_3– may be induced. Similarly the contribution of the CF_3–CH_2 dipole will be different if the liquid and solid orientations are different.

Fluorosilicones have been used traditionally in applications that exploit the reduced organic compatibility and permeability conferred by the bulkier fluorocarbon group; for example, they have been used as lubricating oils in bearings subjected to washing by fuels and solvents and as sealants in a variety of applications that require resistance to fumes and splashes or even total immersion in fuels, oils, and solvents. Newer applications include their use as magnetic-medium lubricants, crude-oil antifoams, silicone pressure-sensitive adhesive release liners, and oily-dirt pick-up-resistant sealants and coatings.

Table X. Comparison of PTFPMS with Other Fluorocarbon Polymers

Polymer	Equilibrium Liquid Surface Tension	Ref.	Critical Surface Tension of Wetting: n-Alkanes	Ref.[a]	Other Test Liquids	Ref.[b]	Solid Surface Tension[c]	Ref.
PTFPMS	24 (300 mm^2/s)	1	21		21		14–15	67
PTFE	26 (mw $\rightarrow \infty$)[d]	68, 69	19	41	16–22	70	14	7
PDMS	21 (mw $\rightarrow \infty$)	66, 71	22		24	18	19–25	72–74
PCTFE	31	75	—		31	41	27	76, 77

SOURCE: Reproduced with permission from reference 67. Copyright 1988 Wiley.

NOTE: All surface tension values are given in millinewtons per meter and are rounded to the nearest whole number.

[a]Data for PTFPMS and PDMS were obtained by G. J. Quaal in 1964 and were first reported in reference 67.

[b]Datum for PTFPMS was obtained by G. J. Quaal in 1964 and was first reported in reference 67.

[c]Values were obtained by using the Owens–Wendt technique with water and methylene iodide as test liquids.

[d]The expression mw $\rightarrow \infty$ indicates that the data are values of liquid surface tension at infinite molecular weight of the polymer.

Backbone Variations

The surface activity of organosilicon polymers with backbones other than siloxane is not very well known. Interest in varying the backbone in organosilicon polymers does not normally stem from a desire to modify surface properties. Usually, the purpose of backbone variation is to increase thermal stability, as for example, with poly(silphenylenesiloxane) and poly(carboranesiloxane) copolymers. Because thermal stability is often achieved by increasing T_gs by using rigid backbones, most backbone variations will have a detrimental effect on polymer surface activity.

The plot of surface tension versus boiling point for simple liquids such as hexamethyldisilazane and hexamethyldisilmethylene suggests that these two simple liquids will have similar surface energies, which are 2–3 mN/m higher than that of PDMS (Table XI [*78–80*]). The T_gs agree with this inference. By keeping the pendant groups the same (methyl), the effects of intermolecular-force and steric-bulk variations will be minimized, and T_gs will reflect mainly the other key factor, backbone flexibility.

Table XI. Liquid Surface Tensions (σ_{lv}) of Organosilicon Dimers

Dimer	σ_{lv} *(mN/m)*	*Temp.*[a] *(°C)*	*Boiling Point (°C)*[b]	*Ref.*
$(CH_3)_3SiNHSi(CH_3)_3$	18.2	25	126	78
	17.6	24		—[c]
$(CH_3)_3SiCH_2Si(CH_3)_3$	18.9	20	132 (740)	79
	18.4	24		this work
$(CH_3)_3Si(CH_2)_2Si(CH_3)_3$	19.4	20	151 (760)	79
$(CH_3)_3SiOSi(CH_3)_3$	15.8	20	99	79
	15.7	20		4
	15.5	20		80

[a]Temperature of measurement.
[b]The numbers in parentheses indicate the pressure in millimeters of Hg. 1 mm of Hg = 133 Pa.
[c]The result is previously unpublished and was obtained by G. J. Quaal in 1964.

T_gs for polymers with architectures similar to PDMS are given in Table XII (*21, 81*). The T_gs of poly(dimethylsilazane) and poly(dimethylsilmethylene) lie between those of PDMS and poly(isobutylene) (critical surface tensions of wetting of 24 and 27 mN/m, respectively [*70*]). These values suggest that poly(dimethylsilazane) and poly(dimethylsilmethylene) will have critical surface tensions of wetting in the 25–26-mN/m range.

Surface Tension of Aqueous Surfactant Solutions. The 25–26-mN/m range is the best range that methylsilmethylene- and methylsilane-based surfactants can lower the surface tension of aqueous solutions, as demonstrated by the work of Maki and co-workers at the University of Osaka (*82–85*). Some selected data relevant to the comparison of methylsilmeth-

Table XII. T_gs of Polymers of the Structure $-[(CH_3)_2X-Y]_n-$

Polymer	*X*	*Y*	T_g *(K)*
PDMS	Si	O	146
Poly(dimethylsilazane)	Si	NH	191
Poly(dimethylsilmethylene)	Si	CH_2	173
Poly(dimethylsilpropylene)	Si	$(CH_2)_3$	203
Poly(isobutylene)	C	CH_2	200
Poly(isobutylene oxide)	C	CH_2O	264
Poly(acetone)	C	O	unknown
Poly(dimethylphosphazene)	P	N	227

NOTE: All T_gs were taken from reference 21 except that for poly(dimethylphosphazene), which comes from reference 81.

ylene- and methylsilane-based surfactants with methylsiloxane surfactants are given in Table XIII (*82–84*).

Surfactants and block polymers useful for lowering the surface tension of solutions have two components: the hydrophobe, which has a lower surface tension and is usually insoluble in aqueous solutions, and the hydrophile, which is the more compatible component. The lowering of surface tensions of solutions provides evidence of the degree of surface activity of the hydrophobe but is a less reliable way of inferring surface activity compared with direct surface tension measurement of the hydrophobic material, because surface tension lowering depends also on the concentration of surfactant used, the type and relative proportions of hydrophobe and hydrophile, the overall molecular weight of the surfactant, and the solvent used. Nevertheless, even the best surfactant of a given class cannot perform beyond certain limits, and these limits offer a useful measure of surface activity.

Surfactant surface activity is most completely presented in the form of the Gibbs adsorption isotherm, the plot of solution surface tension versus the logarithm of surfactant concentration. For many pure surfactants, the critical micelle concentration (CMC) defines the limit above which surface tension does not change with concentration, because at this stage, the surface is saturated with surfactant molecules. The CMC is a measure of surfactant efficiency, and the surface tension at or above the CMC (the low-surface-tension plateau) is an index of surfactant effectiveness (Table XIII). A surfactant concentration of 1% was chosen where possible from these various dissimilar studies to ensure a surface tension value above the CMC. Surfactants with hydrophobes based on methylsiloxanes can achieve a low surface tension plateau for aqueous solutions of ~21–22 mN/m. There is ample confirmation of this fact in the literature (*86, 87*).

Other organosilicon hydrophobes have not been studied much except by Maki and co-workers (*82–85*), who examined a much wider range of materials and situations than is considered in Table XIII. They studied cationic as well as nonionic surfactants and included such properties as foam

Table XIII. Aqueous-Solution Surface Tension (σ_{soln}) of Organosilicon Trimer Surfactants

Surfactant	σ_{soln} (*mN/m*)	*Temp.* (°C)	*Ref.*
$(CH_3)_3Si[O(CH_3)_2Si]_2(CH_2)_3OCH_2CH(CH_2OCH_3)(OCH_2CH_2)_{11}OH$	24.4	25	82
$(CH_3)_3Si[O(CH_3)_2Si]_2(CH_2)_3OCH_2CH(CH_2OCH_3)(OCH_2CH_2)_5OH$	23.4	25	82
$[(CH_3)_3SiO]_2(CH_3)Si(CH_2)_3(OCH_2CH_2)_{11}OH$	21.5	25	—[a]
$[(CH_3)_3SiO]_2(CH_3)Si(CH_2)_3(OCH_2CH_2)_7OOCCH_3$	20.4[b]	23	this work
$(CH_3)_3Si[CH_2(CH_3)_2Si]_2(CH_2)_3OCH_2CH(CH_2OCH_3)(OCH_2CH_2)_9OH$	27.6	25	83
$(CH_3)_3Si[CH_2(CH_3)_2Si]_2(CH_2)_3OCH_2CH(CH_2OCH_3)(OCH_2CH_2)_6OH$	26.7	25	83
$[(CH_3)_3SiCH_2]_2(CH_3)Si(CH_2)_3(OCH_2CH_2)_6OOCCH_3$	24.6[b]	23	this work
$(CH_3)_3Si[(CH_3)_2Si]_2(CH_2)_3(OCH_2CH_2)_8OH$	25	25	84

NOTE: Surfactants were used at a concentration of 1% except where indicated.
[a]The result is previously unpublished and was obtained by L. Flaningam in 1966.
[b]Surfactant concentration was 0.1%.

stability and bacteriostatic action for cationic surfactants and interfacial tension with silicone fluid, surface tension reduction of polyols, and wetting of polyethylene for nonionic surfactants. These properties reflect the uses of such materials, but for this examination of backbone variation, only the surface tension lowering of aqueous solutions is considered. Some peculiarities were noted with certain chain lengths, but generally, Maki et al. (*82–85*) found that the surface activities of poly(methylsilmethylene) and poly(methylsilane) are similar, and both are inferior to that of poly(methylsiloxane)s. The low-surface-tension plateau is 25–26 mN/m for poly(methylsilmethylene) and poly(methylsilane) and 21–22 mN/m for poly(methylsiloxane). The values in Table XIII from my work tend to be lower than those of Maki and co-workers (*82–85*). This difference probably reflects an aging effect, because more time for equilibration was allowed in my study.

Plasma-Polymerized Polymers. Another significant study of organosilicon backbone variations was reported by Wrobel (*88, 89*), who measured solid surface tensions of plasma-polymerized dimethylsiloxane, dimethylsilazane, and dimethylsilane monomers by using the Owens–Wendt approach (Table XIV [*7, 88–90*]). Changes on aging that were observed are possibly due to further reaction of radical species, but because these changes do not affect the conclusions presented in this chapter, only immediate postplasma data are given in the table. Wrobel's data agree with the trends presented earlier. Poly(dimethylsilazane) and poly(dimethylsilane) have surface tensions that are similar and higher than that of poly(dimethylsiloxane).

Table XIV. Surface Tensions of Plasma Polymer Films Deposited from Organosilicon Monomers

Starting Monomer	σ^d	σ^p	σ_{sv}
Hexamethyldisiloxane	27.4	1.7	29.1
Hexamethylcyclotrisiloxane	25.7	1.8	27.5
Octamethylcyclotetrasiloxane	25.7	1.8	27.5
Hexamethyldisilazane	28.3	4.6	32.9
Hexamethylcyclotrisilazane	30.1	5.3	35.4
Octamethylcyclotetrasilazane	38.0	7.0	45.0
Bis(dimethylsilyl)tetramethylcyclodisilazane	33.4	2.0	35.4
Hexamethyldisilane	30.7	1.9	32.6
Dodecamethylcyclohexasilane	34.3	7.1	41.4
Silane	32.9	8.0	40.9
Tetramethyldisiloxane[a]	24.9	0.8	25.7
PDMS[b]	20.5	1.6	22.1

NOTE: All data are from references 88 and 89 except where indicated. σ^d is the nonpolar component of surface tension, σ^p is the polar component of surface tension, and σ_{sv} is the Owens–Wendt solid surface tension. All values are in units of millinewtons per meter.

[a]Data are from reference 90.

[b]PDMS is a nonplasma polymer included for comparison (*7*).

Plasma-polymerized materials differ significantly from those polymerized by conventional methods in their surface properties, and surface tension values do not correspond. This difference may be due to the highly cross-linked nature of plasma polymers or to the incorporation of other entities from the carrier gas. These effects are more important than the intrinsic differences in backbone flexibility. Wrobel (*88*) presents ATR-IR (attenuated total reflection infrared) spectroscopic data indicating that silazanes and silanes cross-link more readily than do siloxanes under plasma conditions. Wrobel and his co-workers (*89*) have also used contact angles to study the thermal decomposition of plasma-polymerized organosilicon polymers.

These data are comprehensive, but they are not the only contact-angle data on plasma-polymerized organosilicon polymers. Hexamethyldisilazane, in particular, has been studied by several groups, and differences in results have not been reconciled yet. For instance, Eib and co-workers (*91*) obtained critical surface tensions in the 22–24-mN/m range, which is similar to that for conventionally polymerized PDMS, whereas Inagaki and co-workers (*92*) reported higher solid surface tensions closer to that reported by Wrobel (*88*). Varshney and Beatty (*93*) obtained critical surface tensions of wetting in the 24–28-mN/m range that could be reduced by polymerization in the presence of nitrogen and ammonia, but despite this reduction in critical surface tensions, the polymers had lower water contact angles.

Hirotsu (*94*) measured water contact angles only and got higher values (~100°) compared with that obtained by Varshney and Beatty (~95°). Wrobel (*88*) and Sachdev and Sachdev (*95*) obtained still lower values (84° and 78° respectively). Hirotsu's work did not suggest a minimum surface tension with thickness, whereas Eib and co-workers detected a minimum surface tension at 10–20-nm thickness. These papers also contain ESCA (electron spectroscopy for chemical analysis) and IR data that shed light on these differences. For instance Eib's (*91*) ESCA data suggests a structure akin to $[(CH_3)_2SiNH(CH_3)_2SiO]_n$, in keeping with the low value obtained for the surface tension. Some of the differences are due to different reaction conditions given in these papers, which provide a better appreciation of this debate.

Other Variations

I have focused on methyl derivatives of nonsiloxane organosilicon backbones to achieve a useful comparison of polymer backbones. There are studies on materials with pendant groups other than methyl and backbones other than siloxane. The most useful of these studies is the direct liquid-surface-tension measurement by Feher and co-workers (*96*) of silane oligomers from trisilane to heptasilane, including some branched species (Table XV) (*96–99*). The data are useful because they answer the question of the surface activity contribution of the Si–H group. The situation with SiH-containing siloxanes

Table XV. Liquid Surface Tension of Silane Oligomers at 20 °C

Oligomer	σ_{lv} *(mN/m)*	*Boiling Point (° C)*	*Ref.*
Disilane	21.7 at −33°C	−15	97
Trisilane	18.7	53	96
2-Silyltrisilane	19.1	102	96
n-Tetrasilane	20.9	90	96
2-Silyltetrasilane	21.1	146	96
n-Pentasilane	22.4	153	96
2-Silylpentasilane	22.8	185	96
n-Hexasilane	23.4	194	96
n-Heptasilane	24.2	227	96
Tetrachlorosilane	19.1	58	98
Hexachlorodisilane	24.0	149	99
Octachlorotrisilane	24.7	215	99

is confusing, because all data are for $H(CH_3)SiO$ systems, and effects such as the small size of the hydrogen atom allowing the methyl group to occupy better the outermost surface cannot be distinguished from the intrinsic effect of the hydrogen. On a diagram of surface tension versus boiling point (such as Figure 1), the surface tensions of silane oligomers are higher than that of PDMS and are much closer to the *n*-alkane line. This similarity is in line with the similar electronegativity differences of the Si–H and C–H bonds. Despite their similar surface tension effects, the bonds are of very different reactivity, and there is little hope of exploiting $-SiH_3$ as a new low-surface-energy pendant group akin to $-CH_3$. Some studies of perchlorosilanes (98, 99), also included in Table XV, show that the Si–Cl entity gives a surface tension that is only up to a few millinewtons per meter higher than that given by Si–H. This trend is in the same direction but less than that for the substituted-carbon case and is surprising because the electronegativity difference of Si–Cl is greater than that of C–Cl.

Summary

PDMS is the mainstay of the silicone industry, and the majority of its applications are related to its unusual surface properties. Most of these applications are not the result of surface behavior alone but come from desirable combinations of surface properties and other characteristics, such as resistance to weathering, high- and low-temperature serviceability, and high gas permeability. These applications are all a direct consequence of four fundamental structural properties of PDMS, namely: (1) the low intermolecular forces between the methyl groups, (2) the unique flexibility of the siloxane backbone, (3) the high energy of the siloxane bond, and (4) the partially ionic nature of the siloxane bond. These four structural features determine the surface behavior of PDMS and provide a framework for structure–property–use correlations.

PDMS was compared with other polymers, mostly other organosilicon polymers, although some organic and fluorocarbon polymers were included. Both the effects of pendant side groups and backbone variations were examined. PDMS emerged from this comparison as a polymer with particularly desirable surface properties brought about by having pendant groups with low surface energy arranged along the most flexible backbone. Only aliphatic fluorocarbon groups are intrinsically of lower surface energies compared with aliphatic hydrocarbon, and only polymers with extensive aliphatic fluorination, such as fluorosilicones, fluoroacrylates, fluoromethacrylates, fluoroethers, and fluorocarbons, have lower surface tensions. For this reason, fluorosilicones, including the unusual case of the most commercially available example, poly(trifluoropropylmethylsiloxane), were reviewed in greater detail.

Future Directions

The expansion of this unique set of surface properties is virtually limitless, and the next 40–50 years should give as varied results as the previous decades. A short list of our anticipations includes the following:

1. improvements in quantity and quality of surface property data,
2. calculation from first principles of polymer–substrate interactions,
3. considerable use of highly fluorinated fluorosilicones,
4. tailored organosilicon additives to other materials, particularly block polymers and surfactants,
5. more hybrids of organosilicon compounds and natural products,
6. more liquid–solid interfacial applications,
7. expansion of use in personal-care products and medical devices,
8. exploration of more-flexible backbones and pendant groups with even lower surface energies,
9. exploitation of polymers in which the silicon atom is in the pendant group and not in the backbone, and
10. development of applications of organosilicon plasma-polymerized materials.

Abbreviations and Symbols

ATR-IR	Attenuated total reflection infrared
CMC	critical micelle concentration
ESCA	electron spectroscopy for chemical analysis
HLB	hydrophilic–lipophilic balance
PCTFE	poly(chlorotrifluoroethylene)
PDMS	poly(dimethylsiloxane)
PTFE	poly(tetrafluoroethylene)
PTFPMS	poly(trifluoropropylmethylsiloxane)
T_g	glass transition temperature
σ_c	critical surface tension of wetting
σ_{sv}	Owens–Wendt solid surface tension
σ_{soln}	surface tension of aqueous solution
σ_{lv}	surface tension of liquid
σ^d	dispersion force (nonpolar) component of surface tension
σ^p	polar component of surface tension

References

1. Owen, M. J. *Ind. Eng. Chem. Prod. Res. Dev.* **1980,** *19,* 97.
2. Owen, M. J. *CHEMTECH* **1981,** *11,* 288.
3. Owen, M. J. In *Surfactants in Solution;* Mittal, K. L.; Bothorel, P., Eds.; Plenum: New York, 1986; Vol. 6, p 1557.
4. Fox, H. W.; Taylor, P. W.; Zisman, W. A. *Ind. Eng. Chem.* **1947,** *39,* 1401.
5. Siow, K. S.; Patterson, D. *J. Phys. Chem.* **1973,** *77,* 356.
6. Lee, L. H. *J. Adhes.* **1972,** *4,* 39.
7. Owens, D. K.; Wendt, R. C. *J. Appl. Polym. Sci.* **1969,** *13,* 1741.
8. Hunter, M. J.; Gordon, M. S.; Barry, A. J.; Hyde, J. F.; Heidenreich, R. D. *Ind. Eng. Chem.* **1947,** *39,* 1389.
9. Hard, S.; Neuman, R. D. *J. Colloid Interface Sci.* **1987,** *120,* 15.
10. Noll, W. *Chemistry and Technology of Silicones;* Academic: New York, 1968; p 458.
11. Beezer, A. E.; Mortimer, C. T. *J. Chem. Soc. A* **1966,** 514.
12. Polmanteer, K. E.; Hunter, M. J. *J. Appl. Polym. Sci.* **1959,** *1,* 3.
13. Tobolsky, A. V. *Properties and Structure of Polymers;* Wiley: New York, 1960; p 67.
14. Kataoka, T.; Ueda, S. *J. Polym. Sci., Part B* **1966,** *4,* 317.
15. Pauling, L. *J. Phys. Chem.* **1952,** *56,* 361.
16. Sauer, R. O.; Mead, D. J. *J. Am. Chem. Soc.* **1946,** *68,* 1794.
17. Good, R. J. In *Treatise on Adhesion and Adhesives;* Patrick, R. L., Ed.; Marcel Dekker: New York, 1967; Vol. 1, p 15.
18. Zisman, W. A. In *Contact Angle, Wettability, and Adhesion;* Advances in Chemistry 43; American Chemical Society: Washington, DC, 1964; p 1.
19. Noll, W. *Chemistry and Technology of Silicones;* Academic: New York, 1968; p 325.
20. Hardman, B. B.; Torkelson, A. In *Kirk–Othmer Encyclopedia of Chemical Technology,* 3rd ed.; Wiley: New York, 1982; Vol. 20, p 938.

21. Lee, W. A.; Rutherford, R. A. In *Polymer Handbook;* Brandrup, J.; Immergut, E. H., Eds.; Wiley: New York, 1975; p III–139.
22. Lee, C. L.; Haberland, G. G. *Polym. Lett.* **1965,** *3,* 883.
23. Stern, S. A.; Shah, V. M.; Hardy, B. J. *J. Polym. Sci., Part B* **1987,** *25,* 1263.
24. Voronkov, M. G.; Mileshkevich, V. P.; Yuzhelevskii, Yu. A. *The Siloxane Bond;* Livak, J., Transl.; Consultants Bureau: New York, 1978; p 12.
25. Ebsworth, E. A. V. In *The Bond to Carbon;* MacDiarmid, A. G., Ed.; Marcel Dekker: New York, 1968; Part 1, p 46.
26. Noll, W. *Chemistry and Technology of Silicones;* Academic: New York, 1968; p 306.
27. Rudoi, V. M.; Ogarev, V. A. *Kolloidn. Zh.* **1978,** *40,* 270.
28. Critchley, J. P.; Knight, G. J.; Wright, W. W. In *Heat Resistant Polymers;* Plenum: New York, 1983; p 441.
29. Fowkes, F. M. *J. Adhes. Sci. Tech.* **1987,** *1,* 7.
30. Ross, S.; Nguyen, N. *Langmuir* **1988,** *4,* 1188.
31. Perkel, R.; Ullman, R. *J. Polym. Sci.* **1961,** *54,* 127.
32. Kiselev, A. V.; Novikova, V. N.; Eltekov, Yu. A. *Dokl. Akad. Nauk. SSSR* **1963,** *149,* 131.
33. Tulbovich, B. I.; Priimak, E. I. *Zh. Fiz. Khim.* **1969,** *43,* 963.
34. Ashmead, B. V.; Owen, M. J. *J. Polym. Sci., Part A-2* **1971,** *9,* 331.
35. Brebner, K. I.; Chahal, R. S.; St-Pierre, L. E. *Polymer* **1980,** *21,* 533.
36. Willis, R.; Shaw, R. *J. Colloid Interface Sci.* **1969,** *31,* 397.
37. Bascom, W. D.; Cottington, R. L.; Singleterry, C. R. In *Contact Angle, Wettability, and Adhesion;* Advances in Chemistry 43; American Chemical Society: Washington, DC, 1964; p 355.
38. Sawicki, G. C. In *Wetting, Spreading, Adhesion, Comprising Papers,* Symposium at Loughborough University, Loughborough, England; Padday, J. R., Ed.; Academic: London, England, 1978; p 361.
39. Kitchen, N. M.; Russell, C. A. *Electr. Contacts* **1975,** *21,* 79.
40. Bernett, M. K.; Zisman, W. A. In *Contact Angle, Wettability, and Adhesion;* Advances in Chemistry 43; American Chemical Society: Washington, DC, 1964; p 332.
41. Fox, H. W.; Zisman, W. A. *J. Colloid Sci.* **1950,** *5,* 514.
42. Wu, S. *J. Polym. Sci., Part C* **1971,** *34,* 19.
43. Arkles, B. *CHEMTECH* **1977,** *7,* 766.
44. Plueddemann, E. P. *Silane Coupling Agents;* Plenum: New York, 1982.
45. Leyden, D. E.; Collins, W. T., Eds. *Silylated Surfaces;* Midland Macromolecular Monographs; Gordon and Breach: New York, 1980; Vol. 7.
46. Leyden, D. E. *Silanes, Surfaces and Interfaces;* Gordon and Breach: New York, 1986.
47. Lee, L. H. In *Adhesion Science and Technology;* Plenum: New York, 1975; Vol. 9B, p 647.
48. Lee, L. H. *J. Colloid Sci.* **1968,** *27,* 751.
49. Sacher, E. *Silylated Surfaces;* Midland Macromolecular Monographs; Gordon and Breach: New York, 1980; Vol. 7, p 347.
50. Bascom, W. D. *J. Colloid Sci.* **1968,** *27,* 789.
51. Pittman, A. G. In *High Polymers;* Wall, L. A., Ed.; Wiley-Interscience: New York, 1972; Vol. 25.
52. Riedo, F.; Czencz, M.; Liardon, O.; Kovats, E. S. *Helv. Chim. Acta* **1978,** *61,* 1912.
53. Andrade, J. D.; Gregonis, D. E.; Smith, L. M. In *Surface and Interfacial Aspects of Biomedical Polymers;* Andrade, J. D., Ed.; Plenum: New York, 1985; Vol. 1, p 101.

54. Lee, C. L.; Homan, G. R. *IEEE Electrical Insulation Society;* Annual Report (81CH1668-3), Conference on Electrical Insulation and Dielectric Phenomenon; IEEE Service Center: Piscataway, NJ, 1981; p 435.
55. Ikada, Y.; Matsunaga, T.; Suzuki, M. *Nippon Kagaku Kaishi* **1985,** *6,* 1079.
56. Owen, M. J.; Gentle, T. M.; Orbeck, T.; Williams, D. E. In *Dynamic Aspects of Polymer Surfaces;* Andrade, J. D., Ed.; Plenum: New York, 1988; p 101.
57. Granick, S.; Clarson, S. J.; Formoy, T. R.; Semlyen, A. J. *Polymer* **1985,** *26,* 925.
58. Noll, W.; Steinbach, H.; Sucker, C. *Prog. Colloid Polym. Sci.* **1971,** *55,* 131.
59. Noll, W.; Steinbach, H.; Sucker, C. *Colloid Polym. Sci.* **1970,** *236,* 1.
60. Lavygin, I. A.; Skorokhodov, I. I.; Kleinovskaya, M. A.; Potashova, G. A.; Yur'eva, A. M. *Zh. Fiz. Khim.* **1978,** *52,* 1542.
61. Lavygin, I. A.; Skorokhodov, I. I.; Sobolevskii, M. V.; Nazarova, D. V.; Lotarev, M. B.; Kudinova, O. M.; Vorapayeva, G. V. *Vysokomol. Soedin.* **1976,** *A18,* 90.
62. Jarvis, N. L.; Zisman, W. A. *J. Phys. Chem.* **1959,** *63,* 727.
63. Ellison, A.; Zisman, W. A. *J. Phys. Chem.* **1959,** *63,* 1121.
64. Lavygin, I. A.; Sobolevskii, M. V.; Skorokhodov, I. I.; Nazarova, D. V.; Voropayeva, G. V.; Lotarev, M. B.; Zaitseva, L. Y. *Plast. Massy* **1978,** *6,* 28.
65. Gol'din, G. S.; Averbakh, K. O.; Nekrasova, L. A. *Zh. Prikl. Khim. (Leningrad)* **1977,** *50,* 908.
66. Roe, R. J. *J. Phys. Chem.* **1968,** *72,* 2013.
67. Owen, M. J. *J. Appl. Polym. Sci.* **1988,** *35,* 895.
68. Wu, S. In *Polymer Blends;* Paul, D. R.; Newman, S., Eds.; Academic: New York, 1978; p 243.
69. Dettre, R. H.; Johnson, R. E. *J. Phys. Chem.* **1967,** *71,* 1529.
70. Shafrin, E. G. In *Polymer Handbook,* 2nd ed.; Brandrup J.; Immergut, E. H., Eds.; Wiley: New York, 1975; p III–221.
71. Johnson, R. E.; Dettre, R. H. *J. Colloid Sci.* **1966,** *21,* 367.
72. Duel, L. A.; Owen, M. J. *J. Adhes.* **1983,** *16,* 49.
73. Gauthier, L. A. M.S. Thesis, Central Michigan University, 1981.
74. Gauthier, L. A.; Falender, J. R.; Howell, B. A. *Polym. Prepr. (Am. Chem. Soc., Div. Polym. Chem.)* **1982,** *23,* 264.
75. Schonhorn, H.; Ryan, F. W.; Sharpe, L. H. *J. Polym. Sci., Part A-2* **1966,** *4,* 538.
76. Kaelble, D. H.; Dynes, P. J.; Cirlin, E. H. *J. Adhes.* **1974,** *6,* 23.
77. Kaelble, D. H. *J. Adhes.* **1970,** *2,* 66.
78. Myers, R.; Clever, H. *J. Chem. Eng. Data* **1969,** *14,* 161.
79. Aleksandrova, Z. A.; Gundyrev, A. A.; Nametkin, N. S.; Panchenkov, G. M.; Topchiev, A. V. *Issled. Obl. Kremniiorg. Soedin* **1962,** 219.
80. Mills, A. P.; McKenzie, C. A. *J. Am. Chem. Soc.* **1954,** *76,* 2672.
81. Neilson, R. H.; Hani, R.; Wisian-Neilson, P.; Meister, J. J.; Roy, A. K.; Hagnauer, G. L. *Macromolecules* **1987,** *20,* 910.
82. Maki, H.; Murakami, Y.; Ikeda, I.; Komori, S. *Kogyo Kagaku Zasshi* **1968,** *71,* 1675.
83. Maki, H.; Murakami, Y.; Ikeda, I.; Komori, S. *Kogyo Kagaku Zasshi* **1968,** *71,* 1679.
84. Maki, H.; Saeki, S.; Ikeda, I.; Komori, S. *J. Am. Oil Chem. Soc.* **1969,** *46,* 635.
85. Maki, H.; Horiguchi, Y.; Suga, T.; Komori, S. *Yukagaku* **1970,** *19,* 1029.
86. Bailey, D. L.; Peterson, I. H.; Reid, W. G. *Chem. Phys. Appl. Surf. Act. Subst., Proc. Int. Congr. 4th* **1967,** *1,* 173.
87. Kanner, B.; Reid, W. G.; Peterson, I. H. *Ind. Eng. Chem. Prod. Res. Dev.* **1967,** *6,* 88.

88. Wrobel, A. M. In *Physicochemical Aspects of Polymer Surfaces;* Mittal, K. L., Ed.; Plenum: New York, 1983; Vol. 1, p 197.
89. Wrobel, A. M.; Kowalski, J.; Grebowicz, J.; Kryszewski, M. *J. Macromol. Sci. Chem.* **1982,** *A17,* 433.
90. Yasuda, H.; Yamanashi, B. S.; Devito, D. P. *J. Biomed. Mater. Res.* **1978,** *12,* 701.
91. Eib, N. K.; Mittal, K. L.; Friedrichs, A. *J. Appl. Polym. Sci.* **1980,** *25,* 2435.
92. Inagaki, N.; Kishi, A.; Katsuura, K. *Int. J. Adhes. Adhes.* **1982,** *2,* 233.
93. Varshney, S. K.; Beatty, C. L. *Org. Coat. Appl. Polym. Sci. Proc.* **1984,** *47,* 151.
94. Hirotsu, T. *J. Appl. Polym. Sci.* **1979,** *24,* 1957.
95. Sachdev, K. G.; Sachdev, H. S. *Thin Solid Films* **1983,** *107,* 245.
96. Feher, F.; Haedicke, P.; Frings, H. *Inorg. Nucl. Chem. Lett.* **1973,** *9,* 931.
97. Lapidus, I. I. *Teplofiz. Svoistva Veshchestv. Mater.* **1972,** *5,* 119.
98. Lakomy, J.; Lehar, L. *Chem. Listy* **1965,** *59,* 985.
99. Niselson, L.; Sokolova, T.; Golubkov, Y. *Teplofiz. Svoistva Veshchestv. Mater.* **1972,** *5,* 128.

Received for review May 27, 1988. Accepted revised manuscript October 28, 1988.

41

Synthesis and Properties of Silphenylene–Siloxane Polymers

Rahim Hani and Robert W. Lenz

Polymer Science and Engineering Department, University of Massachusetts, Amherst, MA 01003

Several poly(silphenylene–siloxane)s were prepared by a low-temperature condensation polymerization of 1,4-bis(hydroxydimethylsilyl)benzene with bis(ureido)silane. The polymers were soluble in common organic solvents and were characterized by gel permeation chromatography (GPC), thermal analysis (differential scanning calorimetry [DSC] and thermogravimetric analysis [TGA]), and multinuclear NMR spectroscopy. The polymers have weight-average molecular weights (M_w) that range from 70,000 to 340,000. The DSC measurements show that the glass transition temperature, T_g, increases with increasing size of the side chain. The thermal stability of these polymers is quite high, as indicated by decomposition-onset temperatures (as determined by TGA) of 480–545 °C.

SILPHENYLENE–SILOXANE POLYMERS (**I**) are, in a sense, modified polysiloxanes in which every third siloxyl group in the main chain backbone is replaced by a suitable silphenylene group (*see* structure on page 742).

Polysiloxanes exhibit exceptional properties over an extremely wide range of temperatures because of their unique combination of high thermal stability and low-temperature flexibility. However, polysiloxanes cannot completely satisfy the needs for high-temperature elastomers that will perform in extreme thermooxidative environments for extended periods. This deficiency originates from the susceptibility of their backbone chains, which are composed completely of polarized siloxyl units, to degradation by ionic reactions when these materials are exposed to temperatures above ~200–250 °C. At and above such temperatures, polysiloxanes are degraded by

0065–2393/90/0224–0741$06.00/0

$$\left[-\underset{CH_3}{\overset{CH_3}{Si}}-C_6H_4-\underset{CH_3}{\overset{CH_3}{Si}}-O-\underset{R'}{\overset{R}{Si}}-O- \right]_n$$

I

	R	R′
Ia	CH_3	$CH_2CH_2CH_3$
Ib	CH_3	$CH_2CH{=}CH_2$
Ic	CH_3	CH_2CH_2CN
Id	CH_3	$CH_2CH_2CH_2CN$
Ie	CH_3	$CH_2CH_2CF_3$
If	C_6H_5	$CH{=}CH_2$
Ig	C_6H_5	$CH_2CH{=}CH_2$
Ih	$CH_2CH_2CH_2CN$	$CH_2CH_2CH_2CN$

either thermally induced rearrangement or depolymerization reactions, which yield thermodynamically more stable low-molecular-weight cyclic products and which result in the rapid disappearance of the large polymer molecules. This property reduces their long-term thermal stability. Thus, to improve their thermal stability while still retaining the desirable low-temperature elasticity, considerable efforts have been directed toward the preparation of polymers in which more thermally stable units are incorporated into siloxane-based backbones, and the silphenylene–siloxane polymers are very promising candidates (*1–4*).

Partial replacement of siloxane units with silphenylene groups also increases polymer crystallinity (*5, 6*), results in a higher viscosity (*7*), increases the mechanical properties, and improves thermal stability. On the other hand, incorporation of the rigid silphenylene units into the polysiloxane backbones could decrease the low-temperature elasticity of these polymers (*8*).

Previous reports from this laboratory described the preparation of silphenylene–siloxane polymers (**I**), in which both R and R′ are CH_3, R = CH_3 and R′ = $CH{=}CH_2$, or R = CH_3 and R′ = C_6H_5, by the low-temperature condensation polymerization of 1,4-bis(hydroxydimethylsilyl)benzene (**II**) and the appropriate bis(ureido)silane (**III**) (2, 9), as shown in equation 1.

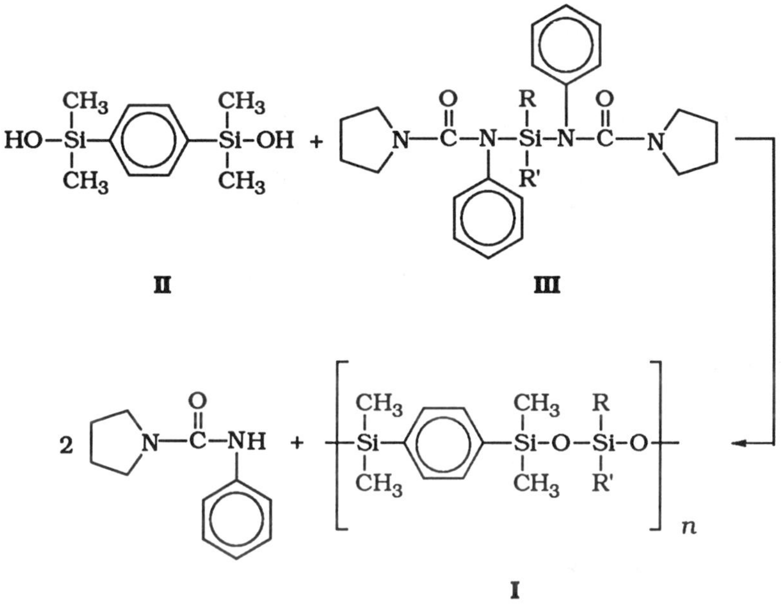

$$HO-Si(CH_3)_2-C_6H_4-Si(CH_3)_2-OH$$

II **III** **(1)**

$$[-Si(CH_3)_2-C_6H_4-Si(CH_3)_2-O-SiRR'-O-]_n$$

I

Copolymers of the monomer with two CH_3 groups and that with one CH_3 and one $CH{=}CH_2$ group could be cross-linked with peroxides to form elastomers with very good mechanical properties and stability (2). Indeed, all the polymers in this series showed excellent thermal stability both in air and in a nitrogen atmosphere (9). In a nitrogen atmosphere, for example, the polymer with CH_3 and $CH{=}CH_2$ groups lost <30% of its weight at 900 °C and retained most of its carbon content.

In this chapter, we report the results of a study of the synthesis of a more complete series of polymers, **Ia–Ih**, by the same low-temperature condensation polymerization reaction (equation 1). All these new polymers were characterized by NMR spectroscopy, gel permeation chromatography (GPC), thermal analysis (differential scanning calorimetry [DSC] and thermogravimetric analysis [TGA]), and elemental analysis.

Experimental Procedures

The dichlorosilane precursors and the 1,4-bis(hydroxydimethylsilyl)benzene monomer were purchased from Petrarch System; phenyl isocyanate was purified by vacuum distillation. Pyrrolidine was distilled from potassium hydroxide, and hexane and chlorobenzene were distilled from calcium hydride.

Polymer glass transition temperatures (T_gs) were determined with a Perkin–Elmer DSC-2 analyzer at a heating rate of 20 °C/min with liquid nitrogen as the cooling fluid. TGA was done with a Perkin–Elmer thermal analyzer at a heating rate of 15 °C/min under a nitrogen atmosphere.

^1H NMR spectra were obtained with a Varian CFT-20 spectrometer for polymer solutions in $CDCl_3$, and ^{13}C NMR spectra were obtained with a Varian XL300 for polymers in $CDCl_3$. GPC measurements were done with a Waters gel permeation chromatograph with a UV detector and five in-line columns with pore sizes of 500, 10^3, 10^4, and 10^5 Å. The samples were run in tetrahydrofuran at a flow rate of 1 mL/min. The columns were thermostatically set at 45 °C.

Monomer Synthesis. ***Bis(pyrrolidinyl)methylpropylsilane (IVa).*** A 1-L three-neck flask equipped with a mechanical stirrer, nitrogen inlet adaptor, and an addition funnel was charged with 500 mL of dry hexane and 142.2 g (2 mol) of pyrrolidine. The mixture was cooled to 0 °C, and 78.6 g (0.5 mol) of methylpropyldichlorosilane was added dropwise over a 1-h period. The reaction mixture was warmed to room temperature and stirred overnight. The solid amine hydrochloride precipitate was filtered under nitrogen and washed two times with 100 mL of dry hexane. The solvent was removed under reduced pressure, and the product was distilled under vacuum as a colorless liquid. Table I lists the boiling points and elemental analyses.

Bis(pyrrolidinyl)silanes (IVb–IVh). The same procedure used to prepare **IVa** was used. Table I lists the boiling points and elemental analyses of the compounds.

Table I. Boiling Points, Yields, and Elemental Analyses of Bis(pyrrolydinyl)silanes (IVa–IVh)

			C (%)		*H (%)*	
Compound[a]	*Boiling Point/Pressure*[b]	*Yield (%)*	*Calc.*	*Found*	*Calc.*	*Found*
IVa	85/0.9	84	63.65	63.68	11.57	12.14
IVb	79/0.3	86	64.22	64.09	10.78	11.10
IVc	116/0.01	83	60.71	60.44	9.76	9.85
IVd	105/0.02	71	62.10	62.01	10.02	10.35
IVe	61/0.05	84	51.40	51.13	8.27	8.52
IVf	117/0.3	65	70.53	70.47	8.88	9.03
IVg	125/0.02	83	71.27	71.14	9.15	9.24
IVh	225/0.2	55	63.11	62.99	9.27	9.42

[a]R and R′ for **IVa–IVh** are the same as for **Ia–Ih,** respectively (*see* structure **I**).
[b]Pressures are given in millimeters of Hg. 1 mm of Hg = 133 Pa.

Bis(1,1-tetramethylene-3-phenylureido)methylpropylsilane (Va). A 500-mL three-neck flask equipped with a magnetic stirring bar, nitrogen inlet adaptor, and an addition funnel was charged with 200 mL of dry hexane and 22.6 g (0.1 mol) of **IVa**. The mixture was cooled to 0 °C, and 23.8 g (0.2 mol) of phenyl isocyanate was added over a 40-min period. The mixture was warmed to room temperature and stirred overnight. The hexane was decanted, and the solid bis(ureido)silane was washed two times with 50-mL portions of dry hexane and dried under vacuum overnight. Table II lists the yield and elemental analysis.

Bis(ureido)silanes (Vb–Vh). The procedure for **Va** was used. In some cases, the compound did not precipitate from hexane, and so the solvent was removed, and the compound was dried under vacuum overnight. Table II lists the yields and elemental analyses of the bis(ureido)silanes, **Vb–Vh**.

Table II. Yields and Elemental Analyses of Bis(ureido)silanes (Va–Vh)

		C (%)		*H (%)*	
Compound[a]	*Yield (%)*	*Calc.*	*Found*	*Calc.*	*Found*
Va	74	67.20	66.91	7.81	7.51
Vb	60	68.69	68.48	7.54	7.62
Vc	92	65.65	65.18	6.99	6.86
Vd	90	66.22	65.96	7.20	7.08
Ve	95	60.21	60.06	6.41	5.96
Vf	68	70.55	70.49	6.71	6.85
Vg	87	70.96	70.81	6.92	7.14
Vh	60	66.39	65.62	7.08	7.06

[a]R and R′ for **Va–Vh** are the same as for **Ia–Ih**, respectively (*see* structure **I**).

Polymer Synthesis. ***Poly[1,4-bis(oxydimethylsilyl)phenylenemethylpropylsilane] (Ia).*** A 250-mL three-neck flask equipped with a magnetic stirring bar, a nitrogen inlet adaptor, and an addition funnel was charged with 11.3 g (0.05 mol) of 1,4-bis(hydroxydimethylsilyl)benzene (**II**) and 20 mL of dry chlorobenzene. Bis(ureido)silane **Va** (23.2 g, 0.05 mol) dissolved in 80 mL of chlorobenzene was placed in the addition funnel. The reaction mixture was cooled to –20 °C, and ~96% of bis(ureido)silane **Va** was added dropwise during a 3-h period. The mixture was stirred at –20 °C for 1.5 h, slowly warmed to 0 °C, stirred for an additional 3 h, warmed slowly to room temperature, and then stirred overnight. Samples of the reaction mixture were taken after each incremental addition of bis(ureido)silane (0.5 mmol) and injected into the GPC instrument. As the polymerization reaction proceeded, the polymer peak shifted toward shorter elution time (higher molecular weight).

This sequence of monomer addition and molecular weight determination by GPC was continued until the maximum molecular weight was observed. At this point, stirring was stopped, and the reaction mixture was filtered through a sintered glass funnel to separate the polymer solution from the urea byproduct. The solution was then added dropwise to 1 L of well-stirred methanol, from which a white, rubbery polymer precipitated. After decantation, the polymer was dried at 80 °C in a vacuum oven. Table III lists the yield and elemental analysis.

Polymers Ib–Ih. The procedure as described for **Ia** was used. Table III lists the yields and elemental analyses of **Ib–Ih**. Discrepancies exist between the calcu-

Table III. Yields and Elemental Analyses of Polymers Ia–Ih

		C (%)		*H (%)*	
Polymer	*Yield (%)*	*Calc.*	*Found*	*Calc.*	*Found*
Ia	82	54.14	52.83	8.44	8.57
Ib	70	54.49	51.86	7.84	7.69
Ic	70	52.29	51.96	7.21	7.17
Id	76	53.68	53.59	7.51	7.39
Ie	78	46.12	45.76	6.36	6.29
If	67	60.62	60.18	6.78	6.86
Ig	72	61.57	61.28	7.07	6.88
Ih	32	55.62	55.14	7.26	7.37

lated and experimentally determined analyses for C and H for some samples; the cause of these discrepancies is unknown.

Results

Preparation of Bis(pyrrolidinyl)silanes. The low-temperature polycondensation reaction of 1,4-bis(hydroxydimethylsilyl)benzene (**II**) with eight different bis(ureido)silanes (**Va–Vh**) in chlorobenzene was carried out to prepare the high-molecular-weight polymers **Ia–Ih**. The bis(pyrrolidinyl)silanes, **IVa–IVh**, were prepared by modification of published procedures (*10*) by the reaction of dichlorosilanes with pyrrolidine in hexane solution, as shown in equation 2.

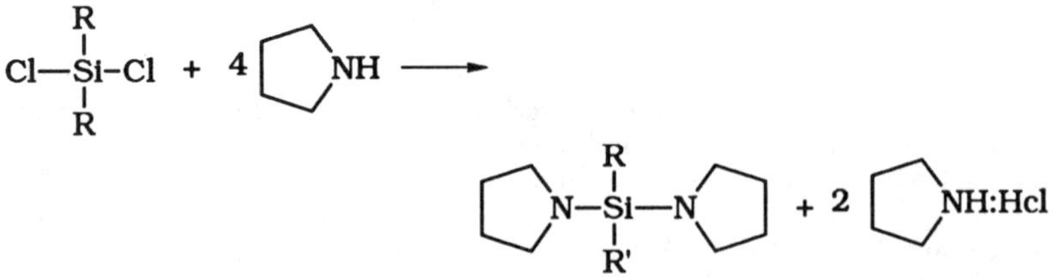

$$\mathrm{Cl{-}SiRR'{-}Cl} + 4\,\mathrm{C_4H_8NH} \longrightarrow \mathrm{(C_4H_8N){-}SiRR'{-}(NC_4H_8)} + 2\,\mathrm{C_4H_8NH{:}HCl} \quad (2)$$

The bis(pyrrolidinyl)silanes, **IVa–IVh**, which are colorless liquids and highly moisture sensitive, were purified by vacuum distillation. Compounds **IVa–IVh** were characterized by elemental analysis (Table I) and ^{1}H and ^{13}C NMR spectroscopy (Table IV). The NMR spectra confirmed the structure shown in the equation. Table IV lists the chemical shifts for the R and R′ groups in these compounds. The pyrrolidinyl group showed two resonances at about 1.7 and 2.9 ppm in the ^{1}H NMR spectra and at about 26 and 46 ppm in the ^{13}C NMR spectra for the two methylene groups, which, for simplicity, are not shown in Table IV.

Preparation of Bis(ureido)silanes. The eight new bis(ureido)silanes, **Va–Vh**, were prepared by modification of the published procedure (*10*) by the reaction of bis(pyrrolidinyl)silanes, **IVa–IVh**, with phenyl isocyanate in either hexane or ether solution, as shown in equation 3.

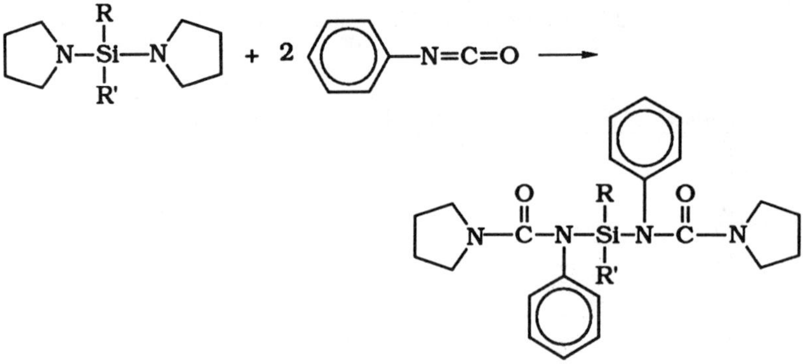

$$\mathrm{(C_4H_8N){-}SiRR'{-}(NC_4H_8)} + 2\,\mathrm{C_6H_5{-}N{=}C{=}O} \longrightarrow \mathrm{C_4H_8N{-}C(=O){-}N(C_6H_5){-}SiRR'{-}N(C_6H_5){-}C(=O){-}NC_4H_8} \quad (3)$$

The bis(ureido)silanes, **Va–Vh**, are white, solid or tacky compounds. These monomers are very reactive toward nucleophilic agents, which readily attack the Si–N bond; hence, these compounds are very reactive to traces of water to form the corresponding unsymmetrical urea, as shown in equation 4.

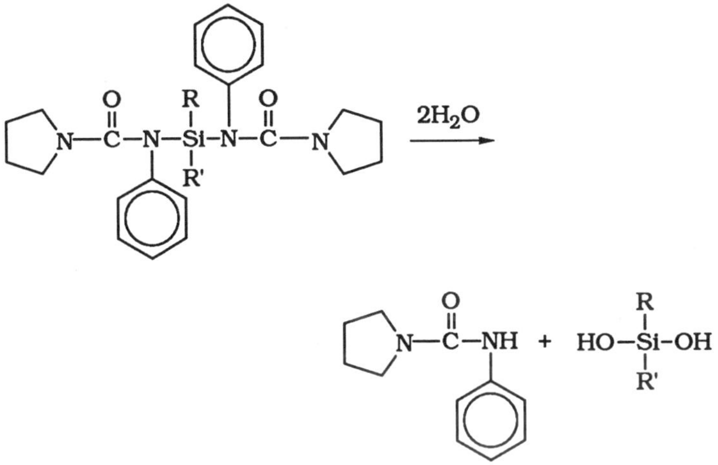

$$\xrightarrow{2H_2O} \qquad (4)$$

The presence of small amounts of urea impurity in the bis(ureido)silane monomer could be determined readily by ^{1}H NMR spectroscopy because the chemical shifts of the tetramethylene group in urea and bis(ureido)silane are considerably different (*9*).

All the monomers were characterized by elemental analysis (Table II) and ^{1}H and ^{13}C NMR spectroscopy (Table V). Table V lists the chemical shifts of the R and R′ groups in the monomers. The chemical shifts of the 1,1′-tetramethylene-3-phenylurea group, which is common for all of the compounds, are not reported in Table V for simplicity. This group has three resonances, which appear as multiplets at 1.6, 2.8, and 7.2 ppm for the two kinds of methylene groups and for the phenyl ring, respectively, in the ^{1}H NMR spectra and two singlets at 25 and 47 ppm for the two methylene groups, multiplets for the phenyl group at 124–143 ppm, and a singlet at 159 ppm for the carbonyl group in the ^{13}C NMR spectra.

Preparation of Silphenylene–Siloxane Polymers. The basic reaction for the preparation of exactly alternating silphenylene–siloxane polymers from bis(ureido)silanes, **Va–Vh**, and the disilanol monomer is step-growth polymerization. This reaction was carried out at −20 °C, at which temperature the possible side reactions described previously can be prevented or at least decreased in rate. To obtain a high-molecular-weight polymer, a solution of the bis(ureido)silane monomer was added slowly to a slurry of the disilanol monomer in chlorobenzene (*11*). After addition of 95%

Table IV. ¹H and ¹³C NMR Spectra of Bis(pyrrolydinyl)silanes IVa–IVh

Compound	*R and R′*	*¹H Chemical Shift*	*¹³C Chemical Shift*
IVa	CH_3Si	0.04	5.04
	$\mathbf{CH_2}CH_2CH_3$	0.53–0.72 (m)	17.17
	$CH_2\mathbf{CH_2}CH_3$	1.12–1.48 (m)	18.32
	$CH_2CH_2\mathbf{CH_3}$	0.95 (t)	17.05
IVb	CH_3Si	0.05	3.99
	$\mathbf{CH_2}CH{=}CH_2$	1.25 (OV)	22.87
	$CH_2\mathbf{CH}{=}CH_2$	4.32–5.19 (m)	128.91
	$CH_2CH{=}\mathbf{CH_2}$	3.78–3.93 (m)	113.09
IVc	CH_3Si	0.10	5.42
	$\mathbf{CH_2}CH_2CN$	1.12 (t)	10.54
	$CH_2\mathbf{CH_2}CN$	2.30 (t)	11.34
	$CH_2CH_2\mathbf{C}N$		121.39
IVd	CH_3Si	0.06	5.24
	$\mathbf{CH_2}CH_2CH_2CN$	0.76–0.88 (m)	13.86
	$CH_2\mathbf{CH_2}CH_2CN$	OV	20.33
	$CH_2CH_2\mathbf{CH_2}CN$	2.34 (t)	20.37
	$CH_2CH_2CH_2\mathbf{C}N$		119.73
IVe	CH_3Si	0.06	5.31
	$\mathbf{CH_2}CH_2CF_3$	0.82–0.93 (m)	6.12
	$CH_2\mathbf{CH_2}CF_3$	1.83 (m)	28.62 (q)[a]
	$CH_2CH_2\mathbf{C}F_3$		128.08 (q)[b]
IVf	C_6H_5	7.24–7.58 (m)	127.46–137.11 (m)
	$\mathbf{CH}{=}CH_2$	5.62–6.35 (m)	133.50
	$CH{=}\mathbf{CH_2}$	5.62–6.35 (m)	128.93
IVg	C_6H_5	7.24–7.59 (m)	127.46–137.11 (m)
	$\mathbf{CH_2}CH{=}CH_2$	1.93 (d)	22.11
	$CH_2\mathbf{CH}{=}CH_2$	5.59–6.12 (m)	128.91
	$CH_2CH{=}\mathbf{CH_2}$	4.76–4.96 (m)	113.44
IVh	$\mathbf{CH_2}CH_2CH_2CN$	0.69–0.90 (m)	12.47
	$CH_2\mathbf{CH_2}CH_2CN$	1.68 (OV)	19.97
	$CH_2CH_2\mathbf{CH_2}CN$	2.35 (t)	20.32
	$CH_2CH_2CH_2\mathbf{C}N$		119.53

NOTE: The chemical shifts in parts per million are relative to the signals for tetramethylsilane. Abbreviations are defined as follows: t, triplet; m, multiplet; dd, doublet of doublets; d, doublet; q, quartet; *J*, coupling constant; and OV, overlapping peaks.
[a] J_{CF} = 29.6 Hz.
[b] J_{CF} = 276.4 Hz.

of the bis(ureido)silane, the course of the polymerization was monitored by GPC. This analysis was particularly important at the end of the polymerization, when formation of a high-molecular-weight polymer is expected. Usually, small amounts of the bis(ureido)silane were added, and the molecular weight relative to polystyrene standards was determined by GPC until the molecular weight would not increase anymore after addition of more bis(ureido)silane.

The use of chlorobenzene as a reaction solvent had advantages because (1) the byproduct was insoluble in this solvent, and an easy way of separating the byproduct from the polymer was possible, and (2) the equilibrium was shifted to the formation of polymer.

Table V. ^1H and ^{13}C NMR Spectra of Bis(ureido)silanes Va–Vh

Compound	*R and R'*	*^1H Chemical Shift*	*^{13}C Chemical Shift*
Va	CH_3Si	0.27	−0.88
	$\mathbf{C}H_2CH_2CH_3$	0.93–1.34 (m)	18.04
	$CH_2\mathbf{C}H_2CH_3$	0.93–1.34 (m)	20.22
	$CH_2CH_2\mathbf{C}H_3$	0.86 (t)	16.96
Vb	CH_3Si	0.27	−0.95
	$\mathbf{C}H_2CH{=}CH_2$	2.13 (d)	22.19
	$CH_2\mathbf{C}H{=}CH_2$	5.51–6.18 (m)	OV
	$CH_2CH{=}\mathbf{C}H_2$	4.74–4.99 (m)	113.39
Vc	CH_3Si	0.27	−3.31
	$\mathbf{C}H_2CH_2CN$	1.28 (t)	11.23
	$CH_2\mathbf{C}H_2CN$	2.56 (OV)	11.79
	$CH_2CH_2\mathbf{C}N$		119.30
Vd	CH_3Si	0.28	−1.07
	$\mathbf{C}H_2CH_2CH_2CN$	1.34 (OV)	18.15
	$CH_2\mathbf{C}H_2CH_2CN$	OV	20.14
	$CH_2CH_2\mathbf{C}H_2CN$	2.23 (t)	20.47
	$CH_2CH_2CH_2\mathbf{C}N$		119.83
Ve	CH_3Si	0.16	−1.03
	$\mathbf{C}H_2CH_2CF_3$	1.26 (m)	10.63
	$CH_2\mathbf{C}H_2CF_3$	2.31 (m)	28.75 (q)[a]
	$CH_2CH_2\mathbf{C}F_3$		OV
Vf	C_6H_5	7.20–7.65 (m)	127.46–142.3 (m)
	$\mathbf{C}H{=}CH_2$	5.66–6.07 (m)	OV
	$CH{=}\mathbf{C}H_2$	5.62–6.07 (m)	124.03
Vg	C_6H_5	7.01–7.44 (m)	127.97–143.3 (m)
	$\mathbf{C}H_2CH{=}CH_2$	2.45 (d)	24.40
	$CH_2\mathbf{C}H{=}CH_2$	5.79–6.65 (m)	OV
	$CH_2CH{=}\mathbf{C}H_2$	4.81–5.14 (m)	114.18
Vh	$\mathbf{C}H_2CH_2CH_2CN$	1.22 (t)	16.25
	$CH_2\mathbf{C}H_2CH_2CN$	OV	20.15
	$CH_2CH_2\mathbf{C}H_2CN$	2.22 (t)	20.15
	$CH_2CH_2CH_2\mathbf{C}N$		119.62

NOTE: Abbreviations are defined at the bottom of Table IV. The chemical shifts in parts per million are relative to the signals for tetramethylsilane.
[a] J_{CF} = 29.34 Hz.

By the procedure described, eight new exactly alternating silphenylene–siloxane polymers, **Ia–Ih**, were prepared. Structural determinations were performed by elemental analysis (Table III) and by ^1H and ^{13}C NMR spectroscopy (Table VI). The results for thermal analyses by DSC and by TGA are given in Tables VII and VIII, respectively. GPC analysis showed that high-molecular-weight polymers were formed. The polymers were all viscous liquids, with consistencies ranging from that of easily flowing oils for low-molecular-weight products (usually <90,000) to that of relatively thick pastes for products with molecular weights of >90,000. All polymers were milky white and cloudy in appearance, and all were easily soluble in $CHCl_3$.

Table VI. ^{1}H and ^{13}C NMR Spectra of Polymers Ia–Ih

Polymer	*R and R'*	*^{1}H Chemical Shift*	*^{13}C Chemical Shift*
Ia	CH_3Si	0.01	−1.27
	$\mathbf{C}H_2CH_2CH_3$	0.52 (t)	17.99
	$CH_2\mathbf{C}H_2CH_3$	1.08–1.45 (m)	20.26
	$CH_2CH_2\mathbf{C}H_3$	0.83 (t)	16.59
Ib	CH_3Si	0.06	−0.78
	$\mathbf{C}H_2CH=CH_2$	1.46–1.82 (dd)	25.61
	$CH_2\mathbf{C}H=CH_2$	5.35–5.81 (m)	133.69
	$CH_2CH=\mathbf{C}H_2$	4.74–4.94 (bd)	113.98
Ic	CH_3Si	0.09	−0.33
	$\mathbf{C}H_2CH_2CN$	0.83 (bm)	11.00
	$CH_2\mathbf{C}H_2CN$	2.17 (t)	13.52
	$CH_2CH_2\mathbf{C}N$		121.02
Id	CH_3Si	0.05	−0.26
	$\mathbf{C}H_2CH_2CH_2CN$	0.48–0.73 (m)	16.94
	$CH_2\mathbf{C}H_2CH_2CN$	1.43–1.78 (m)	19.68
	$CH_2CH_2\mathbf{C}H_2CN$	2.18 (t)	20.17
	$CH_2CH_2CH_2\mathbf{C}N$		119.54
Ie	CH_3Si	0.05	−0.69
	$\mathbf{C}H_2CH_2CF_3$	0.74 (bm)	9.12
	$CH_2\mathbf{C}H_2CF_3$	1.84 (m)	27.65 (q), J_{CF} = 30 Hz
	$CH_2CH_2\mathbf{C}F_3$		127.48 (q), J_{CF} = 276 Hz
If	C_6H_5	7.21–7.60 (m)	OV
	$\mathbf{C}H=CH_2$	5.66–6.07 (m)	OV
	$CH=\mathbf{C}H_2$	5.66–6.82 (m)	OV
Ig	C_6H_5	7.10–7.61 (m)	127.46–137.21 (m)
	$\mathbf{C}H_2CH=CH_2$	1.73 (d)	24.16
	$CH_2\mathbf{C}H=CH_2$	5.34–5.95 (m)	129.56
	$CH_2CH=\mathbf{C}H_2$	4.69–4.84 (bm)	114.55
Ih	$\mathbf{C}H_2CH_2CH_2CN$	0.51–0.71 (m)	15.37
	$CH_2\mathbf{C}H_2CH_2CN$	1.36–1.76 (m)	19.32
	$CH_2CH_2\mathbf{C}H_2CN$	2.21 (t)	20.08
	$CH_2CH_2CH_2\mathbf{C}N$		119.24

NOTE: The chemical shifts in parts per million are relative to the signals for tetramethylsilane. Abbreviations are defined at the bottom of Table IV.

Table VII. GPC and DSC Characterization of Polymers Ia–Ih

Polymer	M_w	M_n	M_w/M_n	T_g (°C)
Ib	68,300	21,500	3.18	−66
Ia	142,000	85,400	1.66	−65
Ie	120,000	54,600	2.21	−51
Ig	120,000	66,600	1.80	−38
Ic	342,000	174,000	1.97	−37
Id	292,000	138,000	2.11	−37
If	86,000	19,800	4.35	−31
Ih	77,700	39,000	1.99	−30

NOTE: M_w and M_n are the weight-average and number-average molecular weights, respectively. Data were obtained by GPC and are relative to polystyrene standards.

Table VIII. TGA of Polymers Ia–Ih

Polymer	*Degradation Temperature (°C)*			*Polymer Remaining (wt %)*
	Onset	*50% Degradation*	*End*	
Ia	545	630	645	37
Ib	505	—[a]	592	59
Ic	495	—	635	60
Id	525	—	680	57
Ie	481	587	645	26
If	523	—	592	66
Ig	534	—	651	68
Ih	510	—	680	59

[a]— indicates that data could not be obtained.

In the ^{1}H NMR spectra (Table VI), two peaks at about 7.3 and 0.3 ppm for the phenyl protons and methylsilane groups, respectively, were always detected, whereas in the ^{13}C NMR spectra, peaks appeared at 132 and 140 ppm for the phenyl carbon atoms and at about –0.6 ppm for the methylsilane groups. These peaks were common for polymers **Ia–Ih** and are not reported in Table IV.

The results of TGA investigations are shown in Table VIII. Polymer degradation occurred by a single-step process, except for polymers **Ic**, **Id**, and **Ih**, which contained the cyanoalkyl side chain and showed more-complex TGA plots in the weight loss versus temperature behavior. In all cases, the thermal degradation process started at 480–545 °C and ended at 590–688 °C. The phenyl-substituted polymer exhibited an improved thermal stability and higher weight residue, as seen by a comparison of polymers **Ib** and **Ig** in Table VIII. The two polymers containing the cyanopropyl side chain were more stable than that containing the cyanoethyl group, as seen by a comparison of polymers **Ic** and **Id** in Table VIII. The onset of the degradation reaction started at a lower temperature for polymer **Ie** than it did for **Ia**, possibly because the presence of fluorine substituents may facilitate the decomposition of polymer **Ie**. The chemical reactions involved in these degradation processes are unknown.

The T_gs were measured by DSC at a heating rate of 20 °C/min, and the results are summarized in Table VII. The following conclusions were derived from the data in Table VII:

- Replacement of a methyl group by a phenyl group (polymers **Ib** and **Ig**) increases the T_g.
- Replacement of a propyl group by a 1,1,1-trifluoropropyl group increases the T_g (polymers **Ia** and **Ie**).
- Replacement of a methyl group by a cyanopropyl group increases the T_g (polymers **Id** and **Ih**).

In all the polymers, T_g was increased by the presence of large and bulky groups pendant to the polymer chain. This result agrees with the known trend that the presence of large and bulky groups pendant to a polymer chain generally increases the T_g of that polymer by decreasing either the backbone mobility or the interchain volume (*12–14*). On the other hand, the replacement of a vinyl group by an allyl group (polymers **Ig** and **If**) decreased the T_g, but in this case, the pendant allyl group in the silphenylene–siloxane polymers may plasticize the polymer internally rather than hinder free rotation.

Acknowledgments

We thank the Massachusetts Centers of Excellence Corporation and Millipore Corporation for financial support of this research at the University of Massachusetts. In addition, we also thank Anthony Allegrezza for his helpful discussion and Stephen Blazka for GPC measurements and the use of the facilities of the Materials Research Laboratory at the University of Massachusetts.

References

1. Dunnavant, W. R. *Inorg. Macromol. Rev.* **1971,** *1,* 165.
2. Koide, N.; Lenz, R. W. *J. Polym. Sci., Polym. Symp.* **1983,** *70,* 91.
3. Lictemalner, H. K.; Sprung, M. N. In Mark, H. F.; Gaylord, N. G.; Bikales, N. M. *Encyclopedia of Polymer Science and Technology;* Interscience: New York, NY, 1970; Vol. 12, p 465.
4. Barry, A. J.; Beck, H. N. In Stone, F. G. A.; Graham, W. A. G., Eds.; *Inorganic Polymers;* Academic: New York, 1962.
5. Price, F. P. *J. Polym. Sci.* **1959,** *37,* 71.
6. Sveda, M. U.S. Patent 2 562 000, 1951.
7. Gordon, A. F.; Clark, H. A. U.S. Patent 2 696 480, 1954.
8. Merker, R. L.; Scott, M. J.; Haberland, G. G. *J. Polym. Sci.* **1964,** A-2, 2, 34.
9. Dvornic, P. R.; Lenz, R. W. *J. Polym. Sci., Polym. Chem. Ed.* **1982,** *20,* 951.
10. Stewart, D. D.; Peters, E. N.; Beard, C. D.; Dunks, G. B.; Hedaya, G.; Kwiatkowski, G. T.; Mofitt, R. D.; Bohan, J. J. *Macromolecules* **1979,** *12,* 373.
11. Dvornic, P. R.; Lenz, R. W. *J. Appl. Polym. Sci.* **1980,** 25, 641.
12. Alkonis, J. J.; MacKnight, W. J.; Shen, M. *Introduction to Polymer Viscoelasticity;* Wiley–Interscience: New York, 1972; p 80.
13. Meares, P. *Polymers: Structure and Bulk Properties;* D. Van Nostrand: New York, 1967; p 262.
14. Dimarzio, E. A.; Gibbs, J. H. *J. Polym. Sci., Part A* **1963,** *1,* 1417.

RECEIVED for review May 27, 1988. ACCEPTED revised manuscript March 13, 1989.

42

Future Directions for Silicon-Based Polymers

Donald R. Weyenberg and Thomas H. Lane

Dow Corning Corporation, Midland, MI 48686–0994

Continued development of the field of polymer science requires an understanding of the structure, synthesis, and behavior of silicon-based polymers at four hierarchical levels of complexity: atomic, segmental, network, and domain. The inherent reactivity of monomers and their structure must be uncovered by physical and chemical studies of the behavior of polymer segments, networks, and domains and, finally, the complete system. The role of molecular modeling in bridging the gulf between polymer backbone design and physical behavior is discussed in this chapter. The commercial development and continued expansion of silicone applications are used to illustrate the dynamics of the cyclic process of science, invention, and innovation and to provide the basis for the further advancement of this exciting field of polymer chemistry.

SILICON-BASED POLYMERS have been of technological importance throughout recorded history. Glass and ceramic materials were an integral part of early civilizations, and pottery relics are still used to gain insight into the extent of the technological advancement of early societies. As human needs expanded beyond simple utensils and tools for survival, glass and ceramic technologies grew in response to the changing and demanding requirements of an evolving world. The new sciences of optics and electronics followed, with the most recent example of this synergy being optical fiber technology.

Significant Events That Shaped Silicon Chemistry

Two significant events during this century resulted in multiple-step changes in this technological area. The first event was the discovery and development

0065–2393/90/0224–0753$06.00/0

of silicones or poly(organosiloxanes). These exciting and novel materials were first described by Robinson and Kipping (*1*) during the first two decades of this century. Kipping's pioneering work in the use of the Grignard reagent for the formation of silicon–carbon bonds (*2*) allowed these materials to become commercially viable products. Corning Glass Works first realized the potential importance of silicones and developed the technology for these materials in the 1930s. Later, this technology was launched commercially, through the newly formed joint venture, the Dow Corning Corporation, in 1943.

The second key event was the semiconductor revolution. Silicon rapidly became the essential raw material for new electronic devices. Silicon chemistry grew as it became an integral part in the design and production of these electronic devices, which are housed on a thin chip of silicon.

The objective of this chapter is to describe some of the key trends, driving forces, and challenges that will set the future direction of the rapidly advancing field of silicon-based materials. In the process of identifying these key trends, two assumptions were made. The first assumption is that commercial applications will be a major impetus for the development of silicon-based polymers. The response of the late Philip Handler, former head of the National Academy of Science, to a question on science spending corroborates this assumption. Handler's reply (*3*) was

> *Science has great intrinsic value—because through new science, we better understand the universe and our place in it. As such, it should compete quite well with other disciplines for funding. However, that is not why we spend over 2.5% of our GNP on scientific research. We spend at that level because in addition to its intrinsic value, new science provides the basis of those material benefits which constitute our standard of living.*

The second assumption is that historical information can provide useful insight into the discovery cycle. Therefore, the history of silicone commercialization will be used to explain and illustrate some of the most important emerging trends.

Silicones and the Cycle of Science, Invention, and Innovation

The word *silicones*, first used by Kipping (*4*), has become a generic term for materials based on organosilicon chemistry and the poly(siloxane)s in general. These polymers with Si–O–Si backbones are the hydrolysis and polymerization products of organochlorosilanes, which are in turn prepared ultimately from silica.

Silica is reduced via a carbothermic process to silicon, which is converted to a variety of chlorosilanes. The major monomer, dimethyldichlorosilane, is produced in well over a billion pounds per year by several basic producers of silicones. Along with other organochlorosilanes, dimethyldichlorosilane is the basis of dozens of different siloxanes, which are formulated into hundreds of materials for thousands of applications in nearly every industry. The history of silicones has been summarized several times (5,6); only pertinent historical developments will be considered in this chapter.

Silicones were commercially introduced in the United States in 1943. Initial commercialization in any new field of science is a seminal event, because it sets in motion an autocatalytic cycle of science, invention, and innovation (Figure 1). Responsible product development and production demand more precise information, which in turn calls for a deeper and a more critical scientific understanding of materials and processes. This fundamental understanding builds a knowledge base, which allows more invention and greater commercial innovation. This cycle accelerates the demand for more and better science.

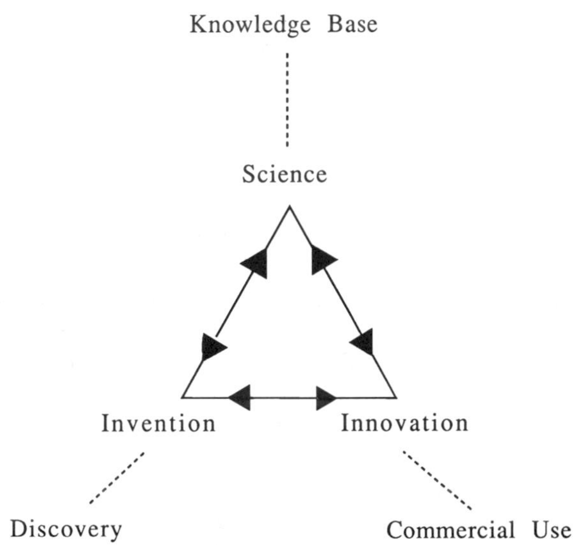

Figure 1. The cycle of science, invention, and innovation.

Science, invention, and innovation are separate but highly interdependent activities. Science is the knowledge base upon which all understanding is built. Invention is the discovery of new and useful things and their applications. Innovation involves the application of technology to a specific need. With innovation, new and useful things become available commercially.

This iterative, cyclic process accelerates the pace of all three activities. Good scientists, inventors, and innovators are equally ill at ease with defined boundaries, and so the triangles in Figure 1 tend to grow in scope and rapidly multiply. This dynamic cyclic process is central to the health, growth, and vitality of any new technology and is a major contributor to and stimulus for the development of the underlying scientific base.

Publications, patents, and sales are reasonable measures of the science, invention, and innovation supporting a given technology. Figure 2 summarizes the history of publication, excluding patent citations, in the area of organosilicon chemistry for each 5-year period over the last 50 years. The number of *Chemical Abstracts* citations per year increased from only a handful in the 1930s to over 4000 at present. The curve takes on an exponential character, which coincides with the first commercialization of silicones in the 1940s.

A similar curve has been generated for the number of U.S. patents dealing with organosilicon chemistry or materials over the same 50-year period. The results were much as anticipated: few patents dealing with silicones in the 1920s through the mid-1950s and a sudden growth in the number of issued patents to its current level of nearly 4000 (U.S. patents) per year. Sales are a reasonable measure of commercialization, but reliable global figures are not available. However, our best estimates show the same type of exponential growth, which has yielded the current multibillion-dollar industry.

The First Commercial Era of Silicones, 1943–1960. The 45-year commercial history of silicones is easily divided into several eras. The period from 1943 to 1960 was dominated by the classical silicones, poly(dimethylsiloxane) fluids, elastomers, and simple methyl and phenyl resins. Basic monomer processes were developed during this period.

Poly(dimethylsiloxane) was the backbone of this growing industry. The unprecedented inertness of this polymer under thermal, chemical, and biological environments, coupled with its unique physical behavior, led to a myriad of commercial applications. Its uses include mechanical applications, in which stability in hostile environments was the desired attribute; surface treatments, in which low surface energy was the important feature; and cosmetic and biomedical applications, in which the biological inertness of the material was exploited. During this period, this new and healthy industry was expanding in all directions.

Second-Generation Silicones, 1960–1980. The next era was dominated by second-generation silicones. The explosive growth during this period was fueled by several advances, including the development of fluorosilicones, silicone–polyether surfactants, and silanes with organic functional groups. However, our focus on this era will be limited to the tailoring of siloxane structures, because of the relevance of this development to this volume.

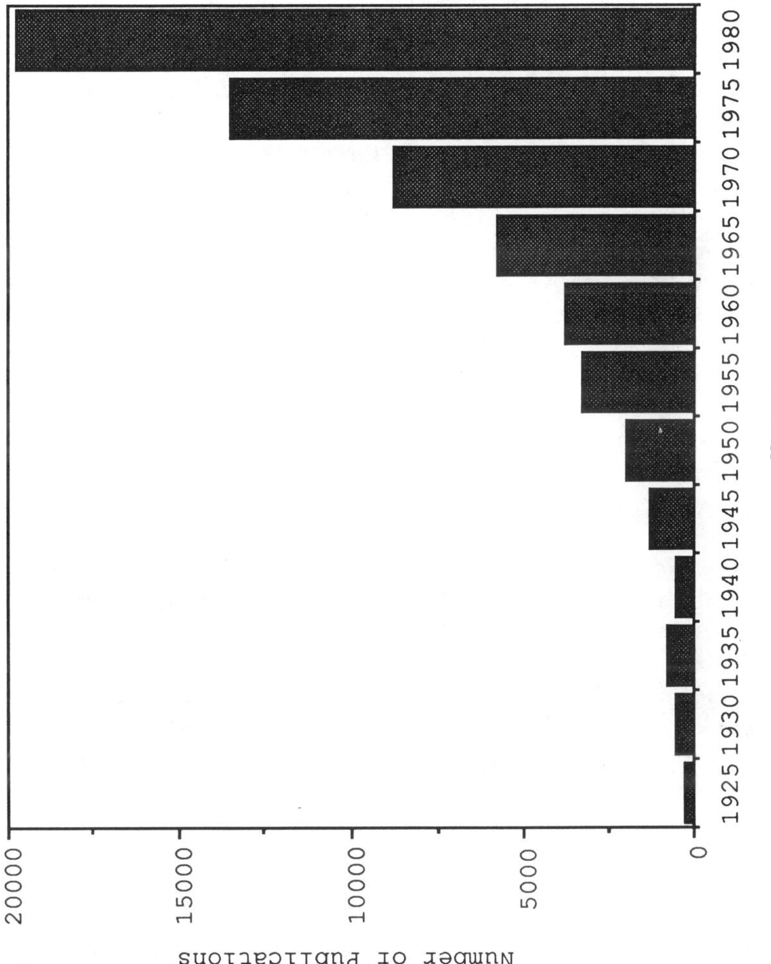

Figure 2. Trends in publications (excluding patents) in organosilicon chemistry.

Network Formation. The ability to tailor networks catalyzed the growth of this field. To make a strong net out of some long pieces of string, the challenge is to tie the strings together at some regular interval in a manner that would give the net strength. However, making this same net out of short strings presents a distinctly different problem. Now, all the ends of the short strings must be connected precisely in a reliable and repeatable fashion. The same challenge is presented by polymers.

Tying low-molecular-weight, and therefore flowable, molecules together into strong elastomers requires precise chemical manipulation of end groups. Siloxane polymerization techniques led to a variety of end groups, and the chemistry for tying these ends into networks was developed. The chemistry of network formation is generally one of three types: (1) a terminal silanol displacing an acyloxy, amide, or alkoxy group from a polyfunctional silane; (2) a polyfunctional acyloxy-, amide-, or alkoxy–siloxy-ended polymer reacting with water; or (3) a platinum-catalyzed coupling of hydride and vinyl-functionalized polysiloxanes. Liquid prepolymers could be converted in place to strong elastomers by any one of several techniques: by adding a catalyst, by heating, or by exposure to moisture.

The technology of network formation had a dramatic impact on the industry, because it offered the product designer an opportunity to design simultaneously the ultimate material property and the application. This technology marked the beginning of a trend in specialty materials that continues today.

Specialty Materials. The spectacular growth of silicone use in the construction industry is an excellent example of this new trend in specialty materials (5). Beginning with their initial use as weather-proofing sealants, all these silicone applications take advantage of their exceptional durability. Thus silicones are used in structures as sealants, adhesives, and coating materials. However, the ability of silicones to be cured in place and on site was the property that gave architects and construction engineers new degrees of freedom. The use of silicone adhesives for structural glazing allowed the all-glass exterior that is increasingly dominating our city skylines. After a few demonstrations in the 1960s, silicone sealant adhesives became the method of choice for securing exterior glass windows to high-rise structures.

Silicone coatings are also gaining acceptance as premium roof membranes. Applied as a moisture-curing one- or two-part system, elastomeric silicone coatings are often the final outer weather-proofing membrane in a roofing system that incorporates layers of polyurethane foam for insulation. Silicones are also playing an important role in the increasingly popular fabric roof systems.

Sealants have been used for difficult building joints with high movement, as well as for highways, where joint failure is a major problem and a tougher issue. High-elongation, low-modulus, high-adhesion, cure-in-place sealants are finding widespread application in both new highway constructions and in the repair of failed joints. The use of newer, sophisticated, and expensive silicones to replace less expensive materials, in this case asphalts, is now becoming standard practice, because the total cost over the lifetime of the system is dramatically reduced.

Liquid silicone rubber is a popular product with broad industrial use. These materials are pumpable liquids that can be injected into low-pressure molds for curing in a few seconds to several minutes at 100–200 °C. Liquid silicone rubbers are formed by hydrosilation. Network formation is by a platinum-catalyzed coupling of a siloxane hydride with a vinylsiloxane. The complete system contains reinforcing fillers and pigments to give an elastomer with properties similar to those of materials prepared from high-molecular-weight polymers. The network-forming chemistry comprises a subtle sequence of reactions that must take place in the same way and to the same extent every time the mold is filled. This step certainly involves one of the most demanding industrial uses of transition-metal catalysts. The network-forming chemistry has been the subject of some excellent studies (7–9), which both explain the system and demonstrate its use in preparing model networks.

The range of silicones for these form-in-place applications continues to grow (5). A variety of silicone foams is particularly useful in fire-resistant penetration seals. A promising new silicone is a latex that deposits a coherent and fully cured elastomer on loss of water. These new forms offer the convenience of silicones from water-based systems and are appearing as easy-to-use, water-based silicone caulks and as high-performance exterior-coating systems.

The protection of electronic devices has been a key application for specialty silicones, and this application continues to keep pace with the rate of device development (5). Silicones are used in various ways, ranging from resinous circuit board coatings to encapsulants, with the silicone gels representing a unique solution to a difficult problem, stress relief. These dielectric gels are prepared by hydrosilation and are lightly cross-linked poly(dimethylsiloxane)s. Their modulus is extremely low, but they are elastic in their behavior. They have the stress-relief characteristic of a liquid but the nonflow property of an elastomer. These jellylike materials maintain their physical profile over the broad temperature range of –80 to 200 °C.

These examples have been selected to emphasize a point: There has been and continues to be a very central, dominant, and underlying direction for this seemingly random explosive growth of second-generation silicones, and this direction is consistently toward more-specialized materials intended as integral parts of a system and optimized to enhance performance and reduce the cost of a specific system.

The Current Era of Silicon-Based Materials, 1980– . The exciting research frontiers presented in this book suggest several common themes and challenges. First, the commercial visions that will provide the sustained impetus for the advancing boundaries of this field are very much in place. In addition to the excitement that surrounds new science, a similar excitement surrounds a vision for new commercial applicatic ·. The various thrusts in this field share a pattern, which is characteristic of modern materials science and which is converging with the fundamental demands of the modern marketplace: the movement of materials science toward the understanding of increasingly higher levels of aggregation in asymmetric systems. Increasingly, structures beyond the molecular level are being described and understood, and the information from these activities is rapidly allowing design and synthesis at the supramolecular level.

In the marketplace, the inexorable trend toward the synthesis of systems builds upon and converges with the increasing sophistication of materials. This convergence is consistent with a thought-provoking article by Wrighton (*11*), which suggests that system synthesis is one of the key challenges in chemistry.

The Synthesis of Systems

Materials, components, and systems are often regarded as separate and sequential levels of integration. Materials are said to be synthesized, and systems are said to be assembled. The verb most commonly used with a component is *to form*. The old paradigm of "synthesize, form, and assemble" is no longer valid for the development of advanced materials; that distinction is disappearing. The old hierarchy is meaningless for microelectronic devices, and modern composite technology is challenging this distinction in larger structural systems.

The commercial opportunities and implications of this blurring distinction are immense for the materials technologist and for materials science. The materials technologist becomes an active partner in restructuring the design and manufacture of systems. The value of a specialty material can become very large if it eliminates the formation and assembly steps. In fact, system synthesis is the logical extension of that trend observed with specialty silicones. This exciting development will have an increasing effect in the field of silicon-based polymers.

Hierarchical Levels. The hierarchical structure in polymers is well recognized. The implications for new materials have been discussed recently by Bement (*12*) and Baer et al. (*13*), who emphasized four levels of structure based on dimension: molecular, nanometer, micrometer, and macromolecular levels. These authors (*12, 13*) illustrated these concepts with both synthetic and natural materials. Injection-molded liquid-crystal polymers were used as examples of a synthetic system, and a tendon was used as an example of a natural system. In each case, the unique physical properties are due to a hierarchy of highly anisotropic structures ranging from the smallest oriented molecular segment, which gives the structure its modulus and which extends over no more than 50 Å, to the microphase structures responsible for distribution of load, which are measured in hundreds of micrometers.

Atoms, Segments, Networks, and Domains. A related and simplified hierarchy seems applicable to the materials concepts described in this volume. Four levels make up the hierarchy of a system: atoms, segments, networks, and domains.

At the atomic level, the design of all materials begins with an understanding of the basic interatomic interactions. These bonding and nonbonding interactions set the framework for the material and define its physical and thermodynamic properties.

At the segment level, the cooperative motions of polymer segments that allow chain movement are the key determinant of the dynamic behavior of a polymer under mechanical stress. The concept of glass transition rests on these segmental motions, which reflect the steric inhibition of backbone conformational relaxation. The impact of backbone conformation on electronic and optical properties is becoming better understood.

At the network level, the ultimate mechanical properties of a homogeneous polymeric matrix are determined by the overall network of polymer chains that prevents the molecules from reacting independently to a stress. The network can be formed by the simple entanglement of glassy polymer segments, as in thermoplastics, or by the alignment of crystalline polymer segments. However, chemical network formation is being used increasingly either as an alternative to or as a complement of plastic formation. The mechanistic richness of substitution at silicon gives these polymer systems a clear advantage when chemical networking is desired.

At the domain level, controlled heterogeneity on the micrometer scale is, in fact, responsible for the unique properties of many, if not most, of our advanced structural materials. Fiber reinforcement, rubber toughening, crystallite reinforcement, and particulate reinforcement are familiar techniques. In each case, with proper adhesion at the interface, the resulting material provides the combination of strength, modulus, and toughness for the intended application.

Nature's materials, like wood or collagen, make beautiful use of these separate and very different domains. The final system, whether a composite airplane wing or a synthetic tendon, relies on all of these hierarchical levels for proper functioning. Its design requires a fundamental understanding at each level, as well as the ability to synthesize the desired molecular identities. Each emerging discipline discussed in this volume presents a unique set of scientific challenges, but among each set are usually some obstacles that limit the pace of advancement.

Atomic Structure. The control of atomic structure is fundamental to any system, and an incomplete understanding of atomic structure can limit advancement. For example, our understanding of preceramic polymers, up through the formation of networks, is improving; but the full exploitation of this chemistry is still limited by the lack of detailed knowledge of the structure of the resulting ceramic at the atomic level. Even with more familiar silicone polymer systems, synthetic barriers are encountered as polymers other than poly(dimethylsiloxane) are used. Stereochemical control is inadequate in the polymerization of unsymmetrical cyclic siloxanes to yield novel linear materials. Reliable synthetic routes to model ladder systems are insufficient.

Segmental Behavior. The understanding of segmental behavior is clearly a critical factor in many of the frontiers discussed, both because synthetic capability has broadened the potential array of new materials and because this understanding is a vital key in predicting the physical behavior and uses of polymeric materials. With computer modeling, significant improvements in this area can be achieved.

Molecular Modeling of Segmental Behavior. Ab initio calculations of both monomer and small siloxane chain segments have clarified the electrostatic interactions between the silicon and oxygen atoms of the siloxane chain. The flexibility of the siloxane bond is due to charge transfer from the lone pairs at oxygen to the covalent region between silicon and oxygen. This charge transfer destroys the sp^3 hybridization geometry at oxygen and shortens the Si–O bond, as shown by Grigoras and Lane (Chapter 7 of this volume). These data, in conjunction with skeletal, torsional, bending, and stretching information, give a more accurate description of polymer conformation.

In addition, the balance between the steric and electrostatic interactions, and not the van der Waals interactions, controls the conformation of linear siloxane. This finding means that the backbone conformation is neither helical nor hydrocarbonlike in nature (*trans–trans*). Rather, the backbone has a *cis–trans* conformation and forms randomly oriented planar segments (*14*).

Chain flexibility has been studied as a function of the substituent groups on silicon. These studies are a significant step towards a truly quantitative model for the prediction of siloxane chain flexibility.

These early attempts to predict segmental behavior from first principles are a milestone and suggest that the computational and synthetic chemists are becoming true partners in designing new materials. This partnership is beginning at a time when the availability and capability of supercomputers are growing tremendously. Clearly, the direction will be toward modeling to guide a synthetic program.

Networks and Domains. The synthesis and understanding of networks and discrete domains are well-illustrated in many chapters in this volume, especially for combinations of relatively well-studied homopolymer systems to provide new capabilities. The block copolymers of poly(dimethylsiloxane) with organic systems or the inorganic "ceramer" composites are the classic examples. However, most of the future structural systems, whether sol–gel materials, structural adhesives, or ceramic-in-ceramic composites, will require a detailed understanding at the network and domain levels. The overall challenge is to move our understanding and our ability to synthesize upward in this hierarchy, toward these higher aggregate levels.

The Challenge. The challenge is to continue the present movement toward the true merging of polymer chemistry and materials science. Historically, these disciplines have been moving together. Materials science, with its roots in engineering and its goal of relating structure and processing to property and use, has been moving down this hierarchical scale to the aggregation level in search of fundamentals. Polymer science, with its roots in chemistry and synthesis, has been moving up the hierarchical scale.

This movement is a key challenge for the entire field of advanced materials, but it is a particularly exciting challenge for silicon-based polymers. From the point of view of materials, silicon-based polymers span the three traditional domains: plastics, ceramics, and metals. Potential applications are equally diverse. Silicon-based polymers range from structural materials, to optoelectronic devices, and to speciality materials for biomedical applications. We are in a unique position to capture the benefits of this merger of materials and polymer science.

Summary

We have attempted to focus on the key forces that will shape our future challenges. We have tried to demonstrate the importance of commercialization as a driving force, not only for the technology, but for the underlying science. We have shown how this commercial push toward system synthesis is accelerating and encouraging the move toward higher levels of aggregation and toward understanding, modeling, and synthesizing more-complex aggregate levels of structure. Finally, although commercial innovation is a critical stimulus, particularly for individuals, science propels the system. The really important new inventions and innovations flow from science, not the other way around.

Acknowledgments

We thank John Zeigler and Gordon Fearon for organizing the symposium on which this book is based.

References

1. Robison, R.; Kipping, F. S. *J. Chem. Soc.* **1908,** *93,* 439.
2. Kipping, F. S. *J. Chem. Soc.* **1907,** *91,* 209.
3. *Science,* **1979,** *204*(4392), 474–479.
4. Kipping, F.; Lloyd, L. *J. Chem. Soc.* **1901,** *79,* 449.
5. Weyenberg, D. R. "Silicones: Past, Present, and Future" Presented at International Organosilicon Symposium, St. Louis, MO, June 1987; in press.
6. Liebhatsky, S. S. *Silicones Under the Monogram;* Wiley: New York, 1978.
7. Chandra, G.; Lo, P. Y.; Hitchcock, P. B.; Lappert, M. F. *Organometallics* **1987,** *6,* 191.
8. Macosko, C. W.; Benjamin, G. S. *Pure Appl. Chem.* **1981,** *53*(8), 1505–1518.
9. Mark, J. E.; Ning, Y. P. *Polym. Eng. Sci.* **1985,** *25*(13), 824–827.
10. Cush, R. *J. Plastic Rubber Int.* **1984,** 9(3), 14–17.
11. Wrighton, M. S. *Comments Inorg. Chem.* **1985,** *4*(5), 269.
12. Bement, A. L., Jr. *Metall. Trans., A* **1987,** *18A,* 363.
13. Baer, E.; Hiltner, A.; Keith, H. D. *Science* **1987,** *235,* 1015.
14. Grigoras, S.; Lane, T. H. *J. Comput. Chem.* **1988,** *25,* 9.

RECEIVED for review May 27, 1988. ACCEPTED revised manuscript October 28, 1988.

INDEXES

Author Index

Affiliation Index

Subject Index

A

C

D

G

H

M

Q

Copy editor and indexer: Ann Maureen Rouhi
Production editors: Donna Lucas, Ray Everngam, Jr., and Paula M. Bérard
Acquisitions editor: Robin Giroux
Cover design: Rebecca Lepkowski

Typesetting by Techna Type, Inc., York, PA
Printing and binding by Maple Press Company, York, PA

Paper meets minimum requirements of American National Standard for Information Sciences—Permanence of Paper for Printed Library Materials, ANSI Z39.48–1984 ♾

Related Titles

Polymers in Aqueous Media: Performance Through Association
Edited by J. Edward Glass
Advances in Chemistry Series 223; 575 pp; ISBN 0-8412-1548-0

Rubber-Toughened Plastics
Edited by C. Keith Riew
Advances in Chemistry Series 222; 440 pp; ISBN 0-8412-1488-3

Microelectronics Processing: Chemical Engineering Aspects
Edited by Dennis W. Hess and Klavs F. Jensen
Advances in Chemistry Series 221; 547 pp; ISBN 0-8412-1475-1

Electronic and Photonic Applications of Polymers
Edited by Murrae J. Bowden and S. Richard Turner
Advances in Chemistry Series 218; 372 pp; ISBN 0-8412-1400-X

Multiphase Polymers: Blends and Ionomers
Edited by L. A. Utracki and R. A. Weiss
ACS Symposium Series 395; 518 pp; ISBN 0-8412-1629-0

Polymer Association Structures: Microemulsions and Liquid Crystals
Edited by Magda A. El-Nokaly
ACS Symposium Series 384; 362 pp; ISBN 0-8412-1561-8

The Effects of Radiation on High-Technology Polymers
Edited by Elsa Reichmanis and James H. O'Donnell
ACS Symposium Series 381; 272 pp; ISBN 0-8412-1558-8